Risk Assessment in Setting National Priorities

ADVANCES IN RISK ANALYSIS

This series is edited by the Society for Risk Analysis.

A Continuation Order Plan is available for this series. A continuation order will bring delivery of each new volume immediately upon publication. Volumes are billed only upon actual shipment. For further information please contact the publisher.

Risk Assessment in Setting National Priorities

Edited by

James J. Bonin

and

Donald E. Stevenson

Shell Oil Company
Houston, Texas

PLENUM PRESS • NEW YORK AND LONDON

Library of Congress Cataloging in Publication Data

Risk assessment in setting national priorities / edited by James J. Bonin and Donald
E. Stevenson.
     p.   cm. – (Advances in risk analysis; v. 7)
    "Proceedings of the annual meeting of the Society for Risk Analysis, held
November 1–4, 1987, in Houston, Texas" – T.p. verso.
    Includes bibliographies and indexes.
    ISBN-13: 978-1-4684-5684-4   e-ISBN-13: 978-1-4684-5682-0
    DOI: 10.1007/978-1-4684-5682-0
    1. Risk assessment – Congresses. I. Bonin, James J. II. Stevenson, Donald E. III.
Society for Risk Analysis. IV. Series.
T174.5.R558 1989   89-8809
363.1 – dc20   CIP

Proceedings of the annual meeting of the Society for Risk Analysis,
held November 1-4, 1987, in Houston, Texas

© 1989 Plenum Press, New York
Softcover reprint of the hardcover 1st edition 1989
A Division of Plenum Publishing Corporation
233 Spring Street, New York, N.Y. 10013

Preface

The growing perception of the public and politicians that life is extremely risky has led to a dramatic and increasing interest in risk analysis. The risks may be very diverse — as demonstrated by the range of subjects covered at the annual meetings of the Society for Risk Analysis. There is a need to pause and see how well the present approaches are serving the nation. The theme, "Setting National Priorities," which was chosen for the 1987 SRA Annual Meeting, reflects the concern that in dealing with individual kinds of risks, society may be more concerned with the trees than the forest.

It is surprising how little attention is being given to the holistic aspects of risk. Who, for instance, is responsible for a national strategy to manage the reduction of health or other risks? Individual agencies have the responsibility for specific patterns of exposure, but these are not integrated and balanced to determine how the nation as a whole can obtain the greatest benefit for the very large investment which is made in risk-related research and analysis.

As a society we may be willing to spend billions of dollars to reduce a hypothetical risk to a small population but at the same time be unwilling to spend minimal amounts to provide primary health care to pregnant women, which could easily reduce infant mortality by perhaps 30-50% in some urban areas. The European Economic Community has produced an *Atlas of Avoidable Death in Europe* (published May 1988) which is an interesting illustration of a device which might be used to establish national priorities. What country, county, or town wants to head the list for infant mortality, smoking related deaths, motor vehicle accidents, or cirrhosis of the liver?

Although the United States of America has entered the "Age of Information," there still is no really comprehensive and reliable data base on the risks of living and working. (The present extension of disease registries will alleviate this problem to some extent.) There is also very little appreciation on how these risks have changed with time or compare with the risks of living in other countries. Thus, it is difficult to define with any certainty the norm and to determine which problems really impact the length and quality of life. While the emphasis on some present priorities has some beneficial side effects, it may in reality miss the real targets. This was demonstrated by the analysis discussed in the EPA report, "Unfinished Business: A Comparative Assessment of Environmental Problems," which was released in February 1987. Special interest groups from all sectors may distort reality to the extent that resources are diverted away from areas where real benefits could be achieved.

The 1983 National Academy of Sciences report, "Use of Risk Assessment in Government," marked a milestone in the understanding of risk analysis by defining an orderly process which separated the analysis of factual information from policy considerations. Unfortunately, it has become regarded as a rule book with procedural precedents that inhibit the advance of the state-of-the-art being applied by various authorities. The legitimacy of the process has been used to provide credibility to the end product. There has been no emphasis on feedback mechanisms during the process or on auditing or testing the results for "reasonableness." Since default or other assumptions play such a large part in the final answer, it is very important to lay out their consequences by the use of decision analysis or other techniques. Risk is often portrayed in two dimensions — hazard and exposure — whereas in reality all four dimensions need examination.

The NAS report emphasized the separation of risk assessment and risk management. It is increasingly clear that this is an uneasy distinction and that there is a need for continuing iteration and feedback between risk assessment, management, and communication. Risk assessment should provide a common meeting ground for all sectors of society since there is a common objective — the reduction of risk. We should be able to settle differences of opinion by obtaining and examining the facts rather than by rhetoric. The promise of risk assessment in this context still remains partly unfulfilled since there are deep differences in philosophy which remain unresolved.

There is an important challenge to the members of the Society for Risk Analysis to improve the approximation of assessed risk to the "true" risk. The future of the art is now in the balance. If it is not found to be a useful tool in forming and applying public policy, it will be displaced by other techniques. We trust that this challenge will be accepted by the Society for Risk Analysis in developing their programs for future annual meetings and thus truly be part of the process of "Setting National Priorities."

James J. Bonin
Program Chairman
1987 Annual SRA Meeting

Donald E. Stevenson
General Chairman
1987 Annual SRA Meeting

Contents

Science and Sociology: The Transition to a Post-Conservative Risk Assessment Era[a]

Terry F. Yosie[b]
U.S. Environmental Protection Agency
Washington, D.C.

I want to express my appreciation to the Society for Risk Analysis for the opportunity to present some thoughts on the issue of conservatism in risk assessment. I am using the term conservative as a synonym for the word "protective." The issue of conservatism is at the heart of what I perceive to be a central question before the Environmental Protection Agency, other regulatory agencies and members of the professional community concerned with the application of risk assessment to decision making. That question concerns our increased ability to substitute scientific data for assumptions, and the intellectual, institutional and societal difficulties encountered in making this substitution. The tensions along these factors comprise what I am calling the transition to a post-conservative era in risk assessment. Like all transitions, the outcome is still in doubt; however, through the citation of some current examples I believe it is possible to state some prominent characteristics of the transition.

In using the term risk assessment, I am referring to estimates of the probability that some effect will occur following exposure of individuals or populations to hazardous substances and situations.[1] My point of view is the result of personal experience, my institutional location as director of the EPA Science Advisory Board and as an interested student of institutional behavior. Some aspects of my argument are a logical extension of thoughts I expressed in a previous paper on EPA's risk assessment culture.[2]

I approach my topic with several preconceptions, or biases. One is the fallacy of believing that conservatism is the only thread running through EPA risk assessments, either now or in the past. EPA is not a monolithic organization, and neither do the assessments that it prepares conform to a single set of assumptions. Second, the factors within EPA that have sustained the use of conservative assumptions are changing and, therefore, much of what I say will characterize factors that both resist and promote change. Third, I will discuss how particular institutional or societal factors influence the rate, and sometimes the direction, of change in EPA's risk assessment practice.

---

a. Plenary address.

b. Remarks in this paper represent the personal views of the author and not necessarily the official policies of the U.S. Environmental Protection Agency.

CONSERVATIVE ASSUMPTIONS IN RISK ASSESSMENTS IN EPA

What are some of the conservative assumptions EPA and other bodies use or have used in risk assessment? They include some familiar friends or suspects, depending upon your point of view:[3]

- Assuming that plants operate at full capacity.

- Using emission rates that don't account for technology or controls that yield lower emissions.

- Using dispersion models for "model" sources than for specific plants.

- Assuming that people spend of all their time outdoors.

- Relying upon a Maximum Exposed Individual in lieu of population risks.

- Not accounting for concurrent exposures from multiple compounds and sources.

- Using a "one hit" model that assumes that a single molecule of a substance can cause cancer and predicts that risk is proportional to dose at lower levels of exposure.

- Assuming linearity at low doses in a multistage model.

- Using surface area over body weight.

- Counting both benign and malignant tumors.

- Using data from the most sensitive animal species.

There are many others.

The rationale for the development of such assumptions stemmed, in large part, from the need to bring intellectual order to the evolving field of health risk assessment in the 1970's. At that time, a number of scientists believed that risk assessment could provide a means to prevent the misuse of science by lawyers and regulators.[4] It was also useful for scientists to state the bases of logic that they would apply in evaluating chemicals. Given the decision making climate that existed at the time, it was not surprising that many of the assumptions that constituted the chain of logic were conservative, or protective, for they reflected the urgency that the nation felt toward cleaning up environmental problems.

Another rationale for conservative assumptions, also defensible today I believe, is that in the absence of valid information, there is a need to adopt a default position of "when in doubt, protect." However, in contrast to ten or fifteen years ago, there are more valid data to assess, and more understanding of some of the mechanisms by which disease resulting from environmental exposures can occur. Because of these developments, the pendulum is swinging away from the use of conservative default assumptions on an *a priori* basis. The use of such assumptions has a higher hurdle to jump than previously. The key issue at the present time is the following: when is it scientifically justified to use conservative default assumptions, and when do data exist to substitute for them?

This current debate over to use or not to use conservative assumptions also has its antecedents. Many groups, including industry, have argued that conservative assumptions lacked a strong scientific foundation. One of the first expressions of a scientific body that methodically challenged EPA's use of conservative assumptions, and that caught the attention of the agency's most senior managers, occurred on September 4-5, 1980. On those days, a subcommittee of the EPA Science Advisory Board, chaired by Dr. Sidney Weinhouse, met to review EPA risk assessments for acrylonitrile, methylene chloride, methyl chloroform, perchloroethylene, trichloroethylene and toluene. EPA continues to assess most of these compounds, and the SAB continues to review them. Therein lies the birth of a cottage industry.

The Weinhouse Subcommittee mounted a blow-by-blow scientific challenge to many of the aforementioned conservative assumptions although, in my mind it is not clear whether the subcommittee's predominant concern was the use of conservative assumptions, EPA's lack of explicitness about such assumptions, or reservations about quantitative risk assessments in general.[5] The Weinhouse Subcommittee, along with subsequent advice provided by the SAB's Environmental Health Committee, were contributing factors to at least two consequences within EPA: (1) the debate over conservatism in risk assessment became formalized over a broader array of compounds; and (2) EPA, over time, became more explicit in its statement of which assumptions it practiced, and the scientific limitations surrounding such assumptions.

Discussion of these and other issues overlapped in time with the deliberations of the National Research Council's Committee on the Institutional Means for Assessment of Risk to Public Health and, thus, it was not surprising that two of that committee's major recommendations were for regulatory agencies to develop more explicit statements of the scientific and other factors used in risk assessments, and to prepare guidelines that would codify risk assessment practice and reflect the latest scientific knowledge.[6]

I don't think it is necessary to recite the history of EPA's risk assessment guidelines before this body. I believe, however, that EPA's willingness to initiate and, in the case of cancer, to revise guidelines reflected several developments. For the cancer guidelines, there was the recognition that almost a decade had passed since their initial adoption and that new scientific concepts and data were available to write scientifically improved guidelines. Second was the desire to promote quality and consistency in EPA's assessments and to communicate the agency's approach to risk assessment to external scientists and the public. Underlying these goals was a recognition by agency scientists and managers (both career and political appointees) of the importance of science and peer review as a major influence in regulatory decision making.[7] In choosing risk assessment and risk management as the leitmotif of his second term at EPA, and in initiating the development of the guidelines, Bill Ruckelshaus invited the scientific community to sup at his table and, in the process, clothe himself and the agency in the protective garb of legitimacy that the scientific community could bestow if it reviewed and blessed the scientific basis of decision making. Third, preparation of the guidelines was an instrument for addressing both an internal administrative problem and an external political problem. Within the agency, they could serve as an instrument to control how various offices used scientific data and how they recommended policy choices to the administrator. Externally, they represented one of Washington's most venerable principles: be sure that your potential adversaries (be they industrial firms, environmentalists, Congress or the Office of Management and Budget) debate your ideas, and your agenda.

The guidelines, in my opinion, represented a major achievement. Aside from the difficult task of trying to reach consensus among a number of scientific disciplines and regulatory offices, a major attribute was their degree of explicitness concerning the assumptions relied upon in risk assessments. Admittedly, the major thrust of the guidelines was to reaffirm the conservative orientation of EPA's risk assessment practice, but to a significant degree the black box was opened. The guidelines encouraged a broader debate on the use of conservative assumptions, promised to be flexible in application, and solicited information to enable periodic updating. Much of the current debate over EPA risk assessments concern these same issues.

TRANSITION TO A POST-CONSERVATIVE RISK ASSESSMENT ERA

It is approximately two years since the risk assessment guidelines were initially utilized within EPA. Because of improved scientific understanding for some compounds, the agency is in the early stages of re-examining their use. EPA staff have recently initiated a formal review of issues and assumptions that are included in its guidelines. This represents EPA's first step in evaluating whether to re-open its guidelines.

There are, however, major difficulties that confront EPA in moving in this direction. First is the concern that some will express because of the fact that the guidelines are still relatively new. They will legitimately ask: have the guidelines been applied in enough risk assessments to judge both their utility and their limitations? Have enough new data emerged to provide EPA with a defensible scientific basis to reach different scientific conclusions?

A second source of EPA's difficulty in substituting data for conservative assumptions was highlighted by Secretary of State George Shultz during the Iran-Contra hearings. Expressing his frustration in failing to block the selling of arms to Iran he said, "Nothing is ever settled in Washington." A similar attitude exists in some quarters of EPA concerning the use of conservative assumptions. Didn't the guidelines settle this issue? Won't a re-opening of the guidelines unravel the scientific consensus that was achieved only with great difficulty?

It is hard for any non-EPAer to appreciate the pain associated with a formal re-opening of guideline positions. Enormous amounts of staff time and energy are consumed in preparing technical position papers to justify proposed changes, crafting compromise language, undergoing peer review, evaluating and responding to peer reviewer and public comments, and responding to comments from OMB staff, many of whom are not scientifically well informed.

A third source of EPA's difficulty in moving beyond the use of conservative assumptions will be in reconciling any proposed changes with one of the major reasons why the guidelines were initially developed—to achieve more predictability and consistency in risk assessment. When an agency such as EPA makes an official determination to re-examine and invite public comment upon its guidelines, it creates uncertainty for itself, the parties that it regulates, or those bodies having an interest in its regulatory decisions such as environmentalists or Congress. Added to this is the additional uncertainty that, within the next twelve to fifteen months, the agency will likely experience a significant turnover in senior leadership positions.

A final difficulty will be the perception of some EPA observers that substituting scientific data for conservative assumptions will lead to less rigorous environmental policy, or a belief that political appointees will interfere with the calculation of risk estimates. I see no evidence of either development.

Why then, do I believe that EPA is entering a post-conservative era in risk assessment? Perhaps now is the time to examine some current examples that I believe illustrate a broader trend. In the recent past, new data or hypotheses have emerged for each of the three substances I will review and have been forcefully presented to EPA by respected members of the scientific community. Having said in its guidelines that it would encourage the submittal of such information, EPA is now evaluating it in a cautious, one-step-at-a-time manner.

ARSENIC

In 1985 EPA's Risk Assessment Forum, a group of senior staff scientists, initiated a re-examination of scientific data related to the ingestion of inorganic arsenic. EPA had previously concluded, as recently as in a 1984 Health Assessment Document, that inorganic arsenic was associated with an elevated risk of lung cancer. That document also reported upon studies that stated an association between ingested inorganic arsenic and an increased incidence of non-melanoma skin cancer in a Taiwanese population. Various offices within EPA readily accepted the conclusion in the Health Assessment Document relative to the inhalational risks of arsenic, but they raised major issues about the strength of the evidence for ingestion related risks. These issues included: the validity of the Taiwan study reporting an increased incidence of non-melanoma skin cancer; its applicability to dose-response assessment in the U.S.; the interpretation and use of arsenic-related skin lesions; and the role of arsenic as an essential human nutrient. The Risk Assessment Forum examined each of these issues in a special review.

In conducting its work, the forum convenes more specialized panels of EPA scientists to review the relevant literature and may recruit outside experts to participate in the deliberations, either as consultants or as panel members. Draft reports are circulated to panel and forum members and revised. When a final draft is prepared, it is submitted to EPA's Risk Assessment Council, a group of senior scientists and managers, for final approval.

A rather unique feature in the development of this new document was the convening of a special workshop of extramural scientific experts on December 2-3, 1986. The objective of the workshop was to solicit expert comment on the issues undergoing special review. A second objective was a cross check on the scientific validity of alternative assumptions and conclusions under consideration within EPA.

Rather than review the succession of draft arsenic documents and how they changed, I would like to state the conclusions of the forum's special review:[8]

- There is a relationship between ingestion exposure to arsenic and an increased risk of skin cancer.

- The MLE of risk due to 1 μg/kg/day of arsenic ranges from 1×10^{-3} to 2×10^{-3}, or an order of magnitude lower than the estimates stated in the 1984 Health Assessment Document.

- The existing data on arsenic as an essential nutrient cannot resolve the question. If arsenic is essential, there is no scientific basis for choosing how to use this information relative to a dose-response assessment.

There is one more wrinkle to the story. For most of its cancer risk assessments, EPA does not make a distinction between cancer incidence and the predicted number of fatalities. Two exceptions to this practice are assessments of skin cancers associated with increased UV radiation resulting from stratospheric ozone depletion, and cancers at various sites of the body due to radionuclide exposures. In both instances, target organs can be identified.

Based upon EPA's re-analysis for inorganic arsenic, human evidence supports a conclusion that arsenic is associated with several forms of non-melanoma skin cancer, a form of cancer readily detectable, treatable and rarely fatal. A key risk management issue is whether EPA should, to some degree, discount these kind of cancer cases. At the present time, this issue is the subject of discussions among senior EPA officials. A key question they must resolve is whether there is a scientific basis to discount non-fatal cancer incidence.

Thus, two major issues at stake are whether the substitution of data for conservative assumptions should lead to a reduction in the risk estimate and, second, whether the data can subsequently support a policy decision to discount non-fatal skin cancers which would further reduce the risk estimate.

DIOXIN—2,3,7,8-TCDD

Dioxin, like poverty, pestilence, death, taxes, wars and rumors of wars and covert operations, will always be with us. I have lost count of the number of dioxin assessments and policy decisions EPA has made in the last decade. What is clear is EPA's and the scientific community's continuing challenge to stay afloat in the expanding sea of information and hypotheses that surround this compound.

During the past eleven months EPA has been preparing another risk assessment of dioxin to take into account more recent findings and hypotheses. This assessment would update the agency's 1985 Health Assessment Document for Dioxins that concluded that TCDD was a probable human carcinogen. Since the updated document is not yet available for public comment, I would like to outline some of the major scientific issues that have been debated within EPA, and then relate these issues to the broader question of conservatism in risk assessment.

The decision to prepare another dioxin update stemmed at least from several chief concerns: (1) The desire to reduce the Tower of Babel-like confusion resulting from the fact that so many different agencies and nations had arrived at different estimates of risk; (2) continued debate on whether definitive evidence existed to associate dioxin with significant human health risk; and (3) studies or evaluations by non-EPA scientists of the mechanism of action, low-dose extrapolation and bioavailability that supported alternative assumptions about the nature and degree of dioxin risk.

In July 1986 EPA's assistant administrator for Pesticides and Toxic Substances, John Moore, created a panel chaired by Dr. Henry Pitot to explore the status of scientific findings for dioxin in five major areas: health consequences, immunotoxicity, bioavailability, mechanism of action and appropriate risk assessment procedures. In general, the Pitot panel agreed with many of the conclusions in EPA's 1985 Health Assessment Document, but the panel differed in its approach to mechanism of action and low dose extrapolation. The Pitot panel concluded that 2,3,7,8-TCDD behaved largely as a cancer promoter, and that low dose extrapolation should be performed by a mathematical model that accounted for this mechanism.[9] It was unable, however, to suggest a model.

Because the Pitot panel was not a formal advisory committee, its conclusions had no formal standing in EPA's decision making process. Its views were influential, however, in motivating EPA to authorize on January 14, 1987 a formal review of these and other dioxin related issues.

The process for preparing such a review is now drawing to a close. If all goes well, EPA will release its draft document for public comment and formal Science Advisory Board review early next year.

There are a number of issues that EPA scientists reviewed at length. These issues can be grouped into three scenarios, each of which reflects a different degree of conservatism.

Scenario I presents reasons for *less* concern about the risks posed by TCDD. These include uncertainties about the applicability of the linear multistage assumptions and model; concerns about overestimates of risk; the possibility that the tumor counts include tumors attributable to irritation of nasal passages; several epidemiological studies reporting no excess cancer in persons exposed to TCDD; absence of reported cancers in chloracne victims accidentally exposed; and an apparent lack of adverse reproductive effects.

Scenario II includes a weight-of-the-evidence consideration for *more* concern about TCDD induced risks. Major elements of this scenario include: a data base showing adverse effects on many organ systems in different animal species: lack of information on the total consequences of TCDD promoter activity; uncertainty about the human impact of immune system effects observed in test animals; the suggestion of human carcinogenicity in some Swedish studies; and an inability to judge the impact of the long biological half-life of TCDD. In total, Scenario II would support EPA's existing potency estimate for TCDD.

Scenario III would combine the factors in Scenarios I and II and produce estimates in the middle of the range. If either the first or the last scenario is chosen, it would more closely align EPA's risk estimate with that of the Food and Drug Administration and the Centers for Disease Control and would reduce the current difference between EPA's upper limit incremental risk and that of a number of foreign countries; EPA's current estimate is 160-1600 times more conservative than estimates adopted by these nations.

These scenarios have been the subject of a lively, and healthy, intellectual debate within EPA as scientists and managers review the choices that have to be made on the basis of science and science policy considerations. At the present time, at least two observations can be made that are also related to the broader theme of post-conservatism: (1) because of the magnitude of the uncertainties, science alone will not determine the outcome. Rather, scientific data, in combination with science policy assumptions, must be used. (2) Because of the lack of a strong consensus among agency scientists and the symbolic importance of dioxin in the political process, the external scientific community will play a crucial role in deciding whether EPA will re-affirm or move away from its traditional conservative position on dioxin.

PERCHLOROETHYLENE

In contrast to dioxin, EPA's decision making on perchloroethylene was characterized by consensus at the staff scientist and policy levels. There was also a wide agreement both within and outside of EPA as to the major scientific issues of concern.

In 1986, EPA's Office of Research and Development submitted its Health Assessment Document for Perchloroethylene to the Science Advisory Board for review. In January, 1987 the Board issued its final report, which challenged a number of assumptions and conclusions used by EPA in its risk assessment. The SAB's major conclusions:[10]

- Questioned the significance of the rat kidney response observed in a National Toxicology Program study in male rats.

- Disagreed with EPA's interpretation of data that increases in mononuclear cell leukemias were associated with perchloroethylene exposures.

- Concurred that perchloroethylene inhalation is associated with a significant increase in the frequency of liver carcinoma in B6C3F1 mice, and that this information aids in extrapolation between routes of administration.

- Disagreed with EPA's overall weight-of-the-evidence judgment that perchloroethylene, using the risk assessment guidelines, is a probable human carcinogen.

7

Between EPA's receipt of the SAB's letter in January and Administrator Thomas' reply to the board in August, the agency received a large number of telegrams from dry cleaners across America and approximately 276 letters from individual congressmen and senators. The former urged EPA to adhere to the SAB's advice, while the latter urged a careful consideration of constituent concerns over classification of perchloroethylene as a probable human carcinogen.

EPA's dilemma was that other conclusions for perchloroethylene besides its own were also scientifically plausible, but that its position had mixed support within the scientific community and was unacceptable to a large segment of the regulated community and the Congress. If the agency was to flippantly reject the board's advice and proceed to reaffirm its earlier recommendation, the likelihood of adverse congressional reaction was high.

Administrator Thomas responded to the SAB's report on August 3 by asking for further scientific consultations on three issues:[11]

- Assuming that not all animal tumors are of equal significance to evaluating human hazard, what is the Science Advisory Board's current consensus position, based on scientific evidence or professional judgment, of the relative significance of male rat kidney or mouse hepatocellular tumors for human risk assessment?

- What is the board's view of the approach taken by EPA in using its guidelines to infer human carcinogenic potential from the total body of scientific evidence on perchloroethylene?

- Is there research underway or anticipated that will clarify these rodent tumor responses and their relationship to human health risk assessment? What additional research should be undertaken?

He indicated that he wanted the benefit of advice on these issues before accepting or rejecting the staff recommendation.

To develop its response to these questions, the SAB held a workshop on August 12 to review the state of the science on these and other issues with some of the leading researchers and research organizations in the country. The board is currently drafting a response to the Thomas letter and hopes to finalize it in the very near future.

I believe the perchloroethylene experience highlights several additional characteristics of post-conservatism. One is the extent to which an advisory process can, de facto, assume a policy making role. Scientific advisors lose control over the use of their own advice once it enters the political process. While not intending to become policy makers, their advice, in essence, can become a litmus test that administrators will be pressured to satisfy in order to sustain their decisions.

Second, because of the heightened importance of risk assessment, I believe that interest groups—be they industrial or environmental—will experience an increased temptation to intervene even earlier in the decision making process to insure that risk assessment assumptions or conclusions prevail which support particular policy decisions that they advocate. In the case of perchloroethylene, the dry cleaners' chief concern was not whether the compound was regulated. It was how the risk characterization was presented. The reason for their concern was that, rightly or wrongly, state and local environmental decisions, as well as public perceptions of risk, are greatly influenced by an EPA determination of a compound as a B_2, or probable human carcinogen. Such a conclusion may result in a series of domino-like decisions by state and local bodies, irrespective of site-specific exposures. A conclusion of C, or possible human carcinogen, may lead to no state or local action even if the compound warrants additional regulation.

SOME CHARACTERISTICS OF POST-CONSERVATIVE RISK ASSESSMENT

The above examples illustrate a number of scientific and institutional factors at work within today's EPA, and define many of the parameters within which a post-conservative era in risk assessment will emerge. What, then, are some of the characteristics of this era?

1. One is the increasingly fragile nature of the decision making process, particularly when contrasted to the period of a decade or more ago. At that time decision makers throughout government had more discretion in policy development and implementation. Today's decision makers correctly perceive that they lack the moral authority to act unilaterally or, in some cases to act, and instead must consult with an expanded array of constituencies, including scientists. Each of these constituencies has some ability to confer or deny legitimacy to a decision, or the technical foundation upon which a decision rests.

2. Second is the changing composition of coalitions within EPA to support or change decisions. Organizational theorists argue that institutions reach consensus only after elaborate bargaining and after coalitions have evolved to support a particular decision. These coalitions can shift, depending upon such factors as how a decision affects the interests of individual organizational units, the degree of participation and direction from senior management and perceptions of the acceptability of a decision to external parties.[12]

Within EPA, the coalition that existed to develop the guidelines may not be the same coalition that re-examines the use of conservatism in risk assessment. The development of the guidelines largely represented a technical effort among staff scientists seeking increased acceptance from the scientific community for their work, and a goal of policy makers to achieve consistency in assessing risk across many regulatory programs. In contrast, a primary motivation for re-evaluating conservative assumptions stems from concerns expressed by managers of the operating programs as they seek to apply risk assessment, and also to balance it with other factors involved in the implementation of standards. A recent story in the *Washington Post* citing EPA's arsenic review illustrates some of the choices between the scientific and policy perspectives. An agency scientist was quoted as saying that the arsenic decision would create a "bad precedent" that would lead to the review of "all other things that the agency regulates..." A senior career policy official, however, remarked that distinctions between fatal and non-fatal disease would be "appropriate" in deciding how best to allocate limited resources. As an example, he stated that a hazardous waste site creating a risk of non-fatal cancers would be "less of an urgency" than a site presenting a fatal cancer risk.[13] Both viewpoints are informed by risk assessment. The former examines risk within the scientific context of a single problem or pollutant and with the desire for consistency, whereas the latter views risk in a comparative framework and with a recognition that other factors besides science must be weighed.

3. I believe that risk assessment and risk management will increasingly be viewed as a continuum rather than as separate entities. The belief in such a separation has led to attempts to create boundaries between science and policy when, in fact, both are embedded within each other and cannot be separated to the degree that we sometimes believe. Identifying the uncertainty associated with selecting a margin of safety for a National Ambient Air Quality Standard or deciding whether a Superfund site receives a score of 28 or 28.5 to qualify for the National Priority List includes as many scientific issues as policy issues. Furthermore, the separation between risk assessment and risk management assumes that the flow of information is one way—from the risk assessor to the risk manager. The examples of arsenic, dioxin and perchloroethylene point out that allowing risk managers to pose questions to the risk assessors can also lead to improved scientific understanding.

4. Recognizing the fragile nature of its decision making, EPA has increased its reliance upon the scientific community. As I noted in my three examples, workshops, special panels and the SAB reviews have convened or will convene to evaluate new data and alternative assumptions for risk assessment. EPA is using these reviews to discern the permissible limits of scientific interpretation, to buffer itself against or prevent a storm cloud from breaking on Capitol Hill, or defend itself when an environmental group charges that the use of less conservative assumptions will lead to an increased rate of cancer or some other health effect. EPA's use of peer panels is no fluke. It reflects what is already a factor in decision making, and which will be a major characteristic in a post-conservative era of risk assessment: the use of science to help legitimize policy choices.

5. For all of the above reasons, post-conservative policy decisions will take longer. The revisions to the National Ambient Air Quality Standard for Particulate Matter is an example of this behavior. Begun in 1977, the revisions were completed only this year. When EPA finalized the particulate standard, the decision, with the exception of the steel industry and some environmental groups, was widely accepted. Such acceptance resulted, in part, from the thorough scientific review. Ironically, the length of the process confirmed for many people the value of either court or congressional deadlines to speed up EPA's decision making process.

6. There is increasingly a scientific basis to substitute data for conservative assumptions, and I have tried to cite some instances where EPA is attempting it. But questions will arise as to why EPA has considered changing its arsenic risk assessment but not its assessments of other compounds? Is the arsenic review a precedent for other compounds? Because the disengagement from conservative assumptions must be data driven, it must also proceed on a case-by-case basis. I can foresee revisions to risk estimates for a number of compounds leading to both higher and lower estimates of risk. In the short-term, however, EPA risk estimates may vary considerably, and it will be more difficult to achieve one of the major objectives of the risk assessment guidelines—greater consistency in risk assessment. In the longer run, this experience will provide the understanding necessary for a more coherent intellectual basis for decision making, involving not only EPA but the public at large. But we should not underestimate the short-term difficulties.

7. Finally, there is a tremendous challenge and need for EPA, the scientific community and the Society for Risk Analysis to convince the Congress and the public that substituting data for conservative risk assessment is a good idea. One of our major difficulties stems from different standards of value we use compared to Congress and the public. Scientists and regulators seek rational and consistent methodology and efficiency as criteria for good decision making. Congress and the public believe equity, and the distribution of equity, should be the desired outcome. In this context, the "right" to a clean environment emerges as an entitlement or a civil liberty, akin to rights claimed by other social groups for other worthy causes. If risk assessment becomes a contest between efficiency and equity, will Congress and the public conclude that, just as war is too important to be left to the generals, that risk assessment is too important to be left to the scientific community?[14]

I don't have the answer to this or many other questions I've posed. But with the increased use of risk-based decision making, and the transition to a post-conservative era in risk assessment, it is inevitable that science and politics will be seeing more of each other. Members of the scientific community will have an increased opportunity to reflect upon the thoughts of Abraham Lincoln who once pondered whether he controlled events or events controlled him.

Thank you.

REFERENCES

1. National Academy of Sciences/National Research Council, *Risk Assessment in the Federal Government: Managing the Process*, p. 3, Washington, D. C., 1983.
2. Terry F. Yosie, "EPA's Risk Assessment Culture," *Environmental Science and Technology* 21:526-531 (June 1987).
3. These and other examples are cited in Office of Science and Technology Policy, "Chemical Carcinogens: Review of the Science and Its Associated Principles," *Federal Register* 50:10372-10442 (1985); and Albert L. Nicholas and Richard J. Zeckhauser, "The Perils of Prudence: How Upper-Bound Assessments Distort Policies Toward Risk," John F. Kennedy School of Government, Harvard University, August 1986.
4. Roy E. Albert, "Approaches to Risk Assessment for Acid Aerosols," International Symposium on the Health Effects of Acid Aerosols, National Institute of Environmental Health Sciences, Research Triangle Park, N.C., October 19-21, 1987.
5. Transcript of discussions of the Subcommittee on Airborne Carcinogens, Science Advisory Board, U. S. Environmental Protection Agency, September 4-5, 1980.
6. National Academy of Sciences, pp. 129-131.
7. U. S. Environmental Protection Agency, "Guidelines for Carcinogen Risk Assessment," *Federal Register* 51:33993 (September 24, 1986).
8. U. S. Environmental Protection Agency, Risk Assessment Forum, *Special Report on Ingested Inorganic Arsenic: Skin Cancer and Nutritional Essentiality*, June 1987 Draft.
9. Office of Pesticides and Toxics Substances, *Report of the Dioxin Update Committee*, August 28, 1986.
10. Letter from Norton Nelson and Richard Griesemer to EPA Administrator Lee M. Thomas, January 27, 1987.
11. Letter from EPA Administrator Lee M. Thomas to Norton Nelson and Richard Griesemer, August 3, 1987.
12. For a discussion of these and other factors influencing organizational behavior, see Anthony Downs, *Inside Bureaucracy*, Boston: Little, Brown and Company, 1987.
13. Michael Weisskopf, "EPA Panel Shifts Course on Pollutant," *Washington Post*, October 5, 1987.
14. For an incisive discussion of these issues, see Arthur M. Okun, *Equality and Efficiency: The Big Tradeoff*, Washington: The Brookings Institution, 1975.

The Use of Focus Groups in Risk Communication

William H. Desvousges
Research Triangle Institute
Research Triangle Park, NC

V. Kerry Smith
North Carolina State University
Raleigh, NC

ABSTRACT

Over the last five years, risk communication has emerged as an important risk research issue. Recently, Covello, Slovic, and Von Winterfeldt [1987] have identified and summarized the most important risk communication issues. Chief among these was the need for empirical research on how to communicate risks effectively. An Environmental Protection Agency [1987] study, which noted the lack of such studies within the agency, reinforces this point. Because risk communication is becoming an important policy tool for government agencies to manage risks, the value of knowing more about communicating risks effectively is further enhanced.

This paper summarizes the findings of two studies that provide some insights on how to communicate risks effectively. The first study involves a technological hazard, the risks from exposure to hazardous wastes. The insights are drawn from both a series of focus groups and a random survey of 600 households in suburban Boston. The second study combines focus group findings with a longitudinal study of 2300 randomly selected homeowners in New York State. This second study involves the risk from exposure to naturally occurring radon gas.

The results from these two studies suggest two basic principles for communicating risk effectively: (1) the context of the risk affects communication and (2) both numerical and visual aids can enhance effectiveness. One of the most important factors affecting context is whether the risk is from a technological or natural hazard.

KEYWORDS: risk communication, hazardous waste and radon exposure, communication effectiveness, focus groups, environmental risks

INTRODUCTION

For many years, market researchers have used focus groups to learn about consumers' perceptions of various products or services. These small group discussions led by an experienced moderator have recently come to the attention of researchers interested in environmental risks. For example, Desvousges *et al.*[1] found focus groups to have an essential part in the process of developing a questionnaire on hazardous waste risks. Kunreuther *et al.*[2] and Mitchell and Carson[3] have endorsed focus groups or in-depth group interviews as valuable in developing survey questionnaires. Smith and Desvousges[4] conclude that focus groups also yield insights about how people answer questions that help to interpret survey responses related to risks.

Focus groups may also have a role in making risk communication more effective. The small size of the group and the group setting itself allow the consumers of risk communication to provide feedback to the risk communicators. Too often this direction in communication channels is absent because risk communicators are more concerned with educating the public rather than listening to them. Feedback can range from critiques of draft communication materials to illumination of perceptual/cognitive processing cues. Qualitative insights on the characteristics people perceive about risk also can be important to the design of communication materials or to their format.

This paper explores some qualitative insights gained from a series of focus groups involving the risks from hazardous waste and radon exposure. First, we briefly discuss the objectives of using focus groups in survey research and in the design of a risk communication program. Next, we review the advantages and disadvantages of focus groups and then turn to the specific role focus groups can have in designing risk communication. In particular, we will describe some of our findings on how people perceive hazardous wastes and radon risks, as well as their implications for visual aids that we have used that can communicate risks. We also offer some practical suggestions for others considering focus groups. Finally, we suggest some implications for risk communication.

FOCUS GROUPS: THEIR ROLES IN SURVEY DESIGN AND RISK COMMUNICATION

There has been intermittent interest in psychology, decision analysis, and related disciplines in developing methods for systematically translating lay persons' verbal reports into "data."[5,6] This research tries to answer the following question: Can analysts get people to tell them their thoughts or decision processes in a way that allows these responses to be systematically described and analyzed?

Originally developed in marketing research to provide a cost-effective way of evaluating new products or promotional campaigns, focus groups offer one method for answering this question. They provide a way to listen to people describe their thoughts or give their opinions. We have found that they provide useful information for designing and evaluating:

- Experimental designs for a survey.

- Questions and visual aids used for in-person, telephone, and mail questionnaires.

- Risk communication materials—e.g., brochures or public service announcements.

- Alternative channels for risk communication messages—e.g., various media or organizations.

- Alternative policies for mitigating risks—e.g., alternative siting processes or financing programs for mitigation.

This is just the beginning for finding new uses of focus groups in risk communication and related studies. For example, Desvousges and Frey describe how focus groups can be used in designing complex environmental risk surveys.[7] Basch argues that focus groups are underutilized as a technique for improving the theory and practice of health education.[8] The most important question for risk communication is how to use the technique effectively. Our experiences may begin to shed some light on this question.

FOCUS GROUPS: ADVANTAGES AND DISADVANTAGES

Focus groups allow participants to freely express their attitudes, experiences, and perceptions. They yield the opportunity to explore perceptual cues people use in answering questions or making decisions. This makes focus groups effective in pretesting the logic underlying the elements used in the structure of an experimental design, the description of survey questionnaires, or information materials such as brochures or the text of public service ads. They allow for people to evaluate the format, content, and overall general effectiveness of the materials. These groups also can help to ease the complexity of topics that will be covered in communication materials because they help to identify the language people commonly use.

Focus groups are flexible and easily implemented. For example, in evaluating reasons why people are not testing their homes for radon, focus groups allowed us to conduct a pair of back-to-back sessions with groups of testers and nontesters to evaluate similarities and differences in their perceptions of the risk and how these led them to their respective decisions. Within a week, we developed a topical agenda, recruited participants, conducted the sessions, and prepared both a written and a video record of the session.

While focus groups have advantages, they also have potential drawbacks. Problems may arise both in using the information and in their effects on the participants. These include:

- The moderator can bias people's responses.

- Without careful selection of locations, the groups represented can be specialized.

- There is no guarantee that a single meeting will be successful (e.g., there are quiet people).

- Specific individuals can dominate and distort the responses of others.

- The process of conducting the session can alter participants' perceptions of a problem (i.e., it is not necessarily a neutral method for obtaining information, leaving the respondents unaffected by the experiences).

Too often, the temptation arises to over-generalize from one or two groups to some larger population. Focus groups are intentionally nonrepresentative samples. Their findings cannot be generalized to the general population. We have found that they tend to yield responses which are more likely to be at the extremes in evaluating a questionnaire or visual aid. For example, people intentionally look for things that might be wrong and may be more sensitive than if you had simply asked them to use or read the materials.

ENVIRONMENTAL RISK FOCUS GROUPS: OVERVIEW

Over the past 4 years we have conducted more than 25 focus groups on environmental risk topics. Table 1 summarizes some of the features of these groups. The

Table 1. Focus Group Overview

Type of Risk	Location of groups	Number of groups	Number of participants
Hazardous wastes	North Carolina	15	172
	Massaachusetts	4	26
Radon	Pennsylvania	4	35
	New Jersey	2	18
	Maryland	2	22

hazardous waste focus groups, held in North Carolina and Massachusetts, consisted of 19 groups and almost 200 participants. These sessions explored various aspects of hazardous wastes, including people's awareness, knowledge, and risk perceptions. A primary objective of these groups was to explore various techniques for presenting risk information to make it easier for people to comprehend environmental risk concepts. These groups were part of a larger effort to use survey-based methods to measure the valuation of reductions in hazardous waste risks.[4,9]

The radon focus groups, conducted over an 18-month period, included eight sessions in Pennsylvania, New Jersey, and Maryland, and involved 75 participants. These sessions had multiple purposes, including the development of a systematic pretest of EPA's *Citizen's Guide to Radon*, the exploration of people's knowledge and awareness of radon issues, and the pretesting of experimental radon brochures that we have subsequently used in a study to evaluate communication effectiveness.[10]

RISK PERCEPTIONS

As noted earlier, one of the areas in which focus groups can be most effective is exploring people's perceptions of risk. From our two general areas of hazardous waste risks and the risk from radon exposure, we have obtained some insights into how people perceive each of these different kinds of risks.

Comments on hazardous waste risk perceptions can be grouped into four categories:

1. People think in concrete images/examples:

   - "Rusted barrels of chemicals."
   - "Illegal dumping."
   - "Times Beach, Love Canal, Woburn."

2. Hazardous Wastes are Pervasive:

   - "Enormous problem, I'll never see it end in my lifetime."
   - "Seems unavoidable, it's in the water, the ground."
   - "Tip of the iceberg."

3. People link risks to health endpoints:

- "Leukemia."
- "Cancer."
- "It gets into your system and then your brain."

4. Hazardous wastes create fear:

    - "That stuff is buried in the ground and you can't see it. That bothers me."
    - "There's so many unknowns. It's scary."
    - "Hopelessness, that's the first thing that comes to my mind."
    - "Cancer, leukemia. What horrible ways to die."
    - For man-made risks, people feel they are entitled to safety.

The comments on radon risks generally fall into two categories:

1. People consider several characteristics:

    - "It's natural. It comes from the ground."
    - "It's radioactive; it can cause cancer."
    - "Lung cancer and my children."
    - "It could be there 24 hours a day—every time we breathe."

2. People find radon hard to evaluate:

    - "I think about testing. . . but it hasn't hit close enough to home."
    - "I was kind of afraid. Do you really want to know?"
    - "The long-term aspect makes it easier to put off, but it bothers you more."
    - "It's easy to put off because you can't see or smell it. The health risk takes a very long time."

As these comments indicate, there are more similarities than differences between hazardous waste and radon risks. The similarities include the health endpoint that people associated with the risk (cancer), the latency period, the unknown character of the risk, and the newness of the risk. In addition, people consistently mentioned that they were even more concerned about their children's exposure than their own.

However, several important differences do emerge from these comparisons across the two sets of focus groups. For example, in talking about radon risk, people consistently mentioned that it was a natural risk, that it comes from the ground, and that it's associated with their house. They're more likely to link it to a flood or a tornado than they are to hazardous wastes. The most important implication from this difference concerns the question of entitlements to safety. That is, in evaluating policies associated with hazardous waste risks, individuals can exhibit reluctance to describe what they want because of their perceptions of the parties responsible for the risk. For example, when people discussed the responsibility for hazardous waste risks, they were much more likely to mention companies, the government, and, then only as a last resort, themselves. The level of government effectiveness, the importance of establishing corporate responsibility, and a perception of being entitled to situations without these risks consistently came up as areas of concern with hazardous waste risk. In contrast, when people thought about the risk from radon exposure, the situation was most aptly summarized by one focus group participant, "What are you going to do, sue God?"

RISK COMMUNICATION IMPLICATIONS

Our experiences with focus groups suggest that they can be quite valuable in evaluating different ways to present risk concepts. Figure 1 shows a preliminary risk ladder

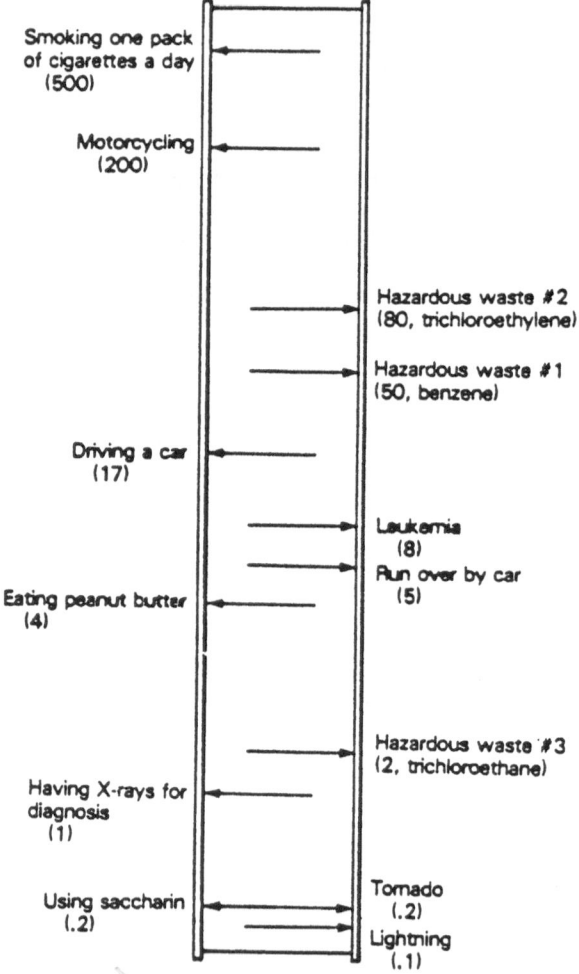

Fig. 1. Preliminary Version of Risk Ladder.

that we developed to try to communicate the risk from hazardous waste exposure. There are several features about this ladder that should be noted. It includes a variety of different sources and types of risks, in addition to the risks from hazardous waste exposures. Figure 2 shows the final ladder that we developed after a process of revisions using the focus groups.

There are rather striking differences between our first draft and our final draft of the risk ladder. For example, the focus group participants pointed out that in the first draft there was not enough diverse risk information—they wanted more lower risks to be included and they wanted risks that were more likely to be suited to their kinds of occupations.

Other significant changes include:

- The visual sense of the ladder is strikingly different. The arrows are no longer pointing in opposite directions. There is one consistent visual focus on the center of the ladder.

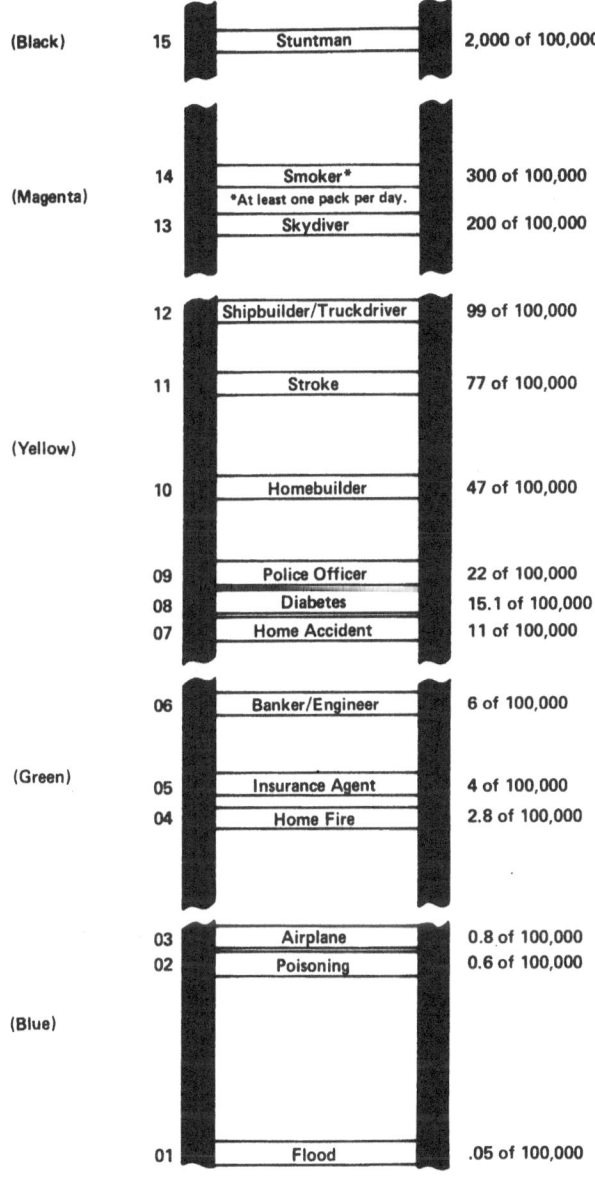

Fig. 2. Final Version of Risk Ladder.

- The breaks in the ladder are structured so a wide range of values for risk could be presented and yet the scale and transition between levels could be understandable. People found it easier to deal with the ladder with separations. We also found the use of colors (not shown here) along the ladder helped reinforce the differences.

Finally, the risk from consumption of peanut butter is not on the final ladder. People found the character of the risk from eating peanut butter to be so different than that from the risk of hazardous waste exposure that it completely disoriented them. After the first two focus groups, in which we spent almost an hour of each 2-hour session listening to people so confused by the signal that was coming from peanut butter on the ladder, we removed it.

19

The ladder has some potential disadvantages, such as including risks with different characteristics (smoking vs. floods). Whether the range of risks is wide enough is another potential shortcoming. At this stage of risk communication research, we do not have the answers for how the ladder could be made more effective. This implies the need for a systematic evaluation of the performance of visual aids that are used to help people form risk perceptions. Such an evaluation would systematically alter the inclusion of various risks and evaluate which version performed better. (How such performance would be measured is another important question.) At this point, all we have is our intuitive qualitative finding of risk with certain characteristics that seemed to work reasonably well in eliciting people's perceptions of the risk of dying from hazardous waste exposure.

The second general finding that we have for communicating risk from the focus groups is that different people process risk information differently. Figure 3, from our hazardous waste study, helps to illustrate this point. The figure includes the numerical representation (both fractions and percents), the visual representation of the circle, and the verbal representation with the words over each circle. The process leading to our decision to include both the fractions and the percents illustrates the value of the focus groups. After conducting several sessions using these circles without the percentages (given in Fig. 4), we noticed when we collected the visual aids that many people had calculated the percentages. In the next focus group, we provided the percentages and asked people whether they were helpful. We found that some people found the percentages and fractions helpful, others found the visual sense given by the darkened piece of the "pie" helpful, while others found the words and verbal descriptions that went along with the description most helpful. To communicate risks effectively, it may be necessary to use multiple techniques of visual information such as the circles or the ladders, quantitative information, the numerical equivalence of the risk, and verbal examples of the risk as well. Once again, this is an empirical question for future research.

SOME LESSONS FOR IMPLEMENTING FOCUS GROUPS

To help others avoid some of our mistakes, we offer the following reflections on implementing focus groups. Our experiences are summarized below:

- Work with civic groups, church organizations, and social organizations to reach target segments. By making a modest contribution to the group, people feel a greater sense of responsibility for attending and contributing to the session.

- Keep the groups relatively small. We have found eight to ten to be the most effective size group.

- Send people a confirming letter and a brochure about our organization; this reduces anxiety about intentions. People invariably brought the materials with them and mentioned after the session that they were less concerned about being targeted for a sales pitch.

- Make sure the moderator is represented as a nonexpert in the risk area. Having people ask the moderator questions severely reduces the effectiveness of the session. We have also found it most helpful for the moderator to be a member of the research team. This allows for more flexibility in following up unanticipated areas of discussion that are germane to the research objectives.

- Don't try to hold focus groups with respondents who might have difficulty with a topic. Generally, we found these to be the least informative sessions because the participants were unable to verbalize why they were having difficulty or simply felt

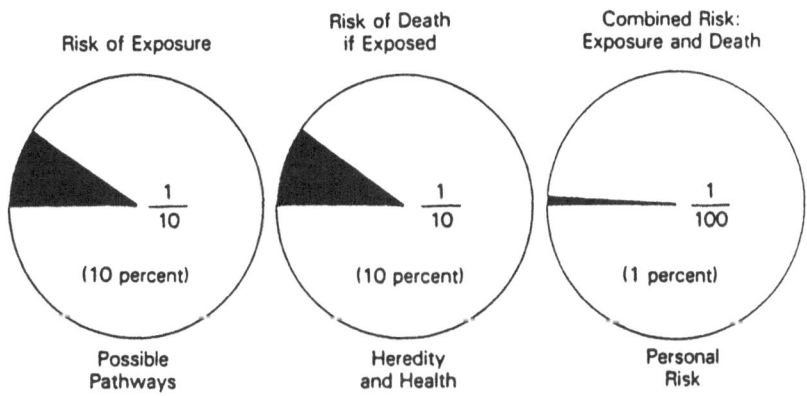

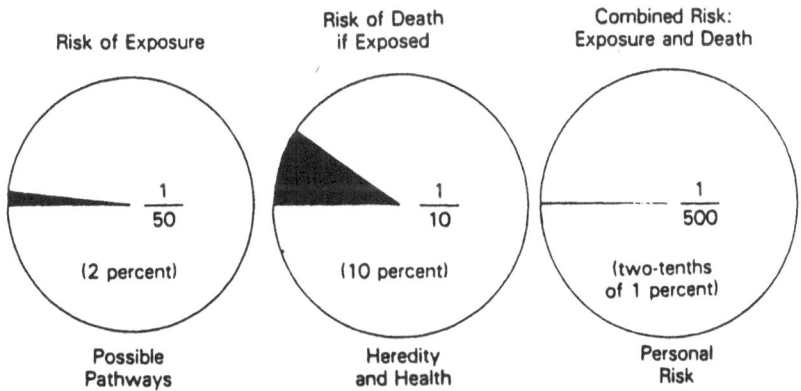

Fig. 3. Preliminary Version of Risk Circles.

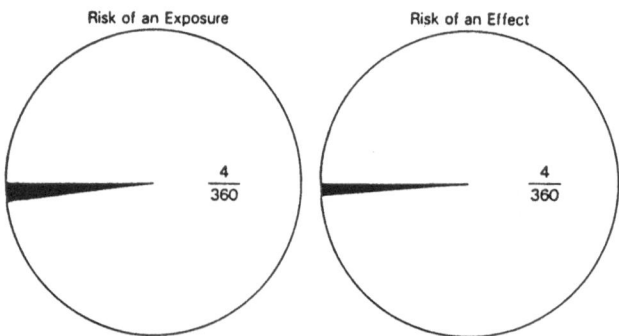

Card A
Hazardous Waste Risks

Risk of an Exposure

Risk of an Effect

$\frac{4}{360}$

$\frac{4}{360}$

Payment required: $50 per year in higher prices and taxes

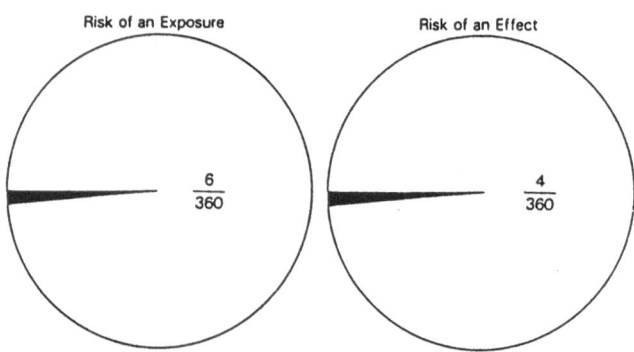

Card B
Hazardous Waste Risks

Risk of an Exposure

Risk of an Effect

$\frac{6}{360}$

$\frac{4}{360}$

Payment required: $100 per year in higher prices and taxes

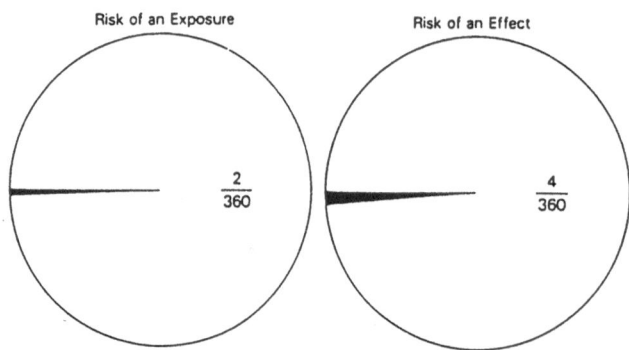

Card C
Hazardous Waste Risks

Risk of an Exposure

Risk of an Effect

$\frac{2}{360}$

$\frac{4}{360}$

Payment required: $175 per year in higher prices and taxes

Fig. 4. Final Version of Risk Circles.

- uncomfortable in a group setting. One-on-one in-depth interviews may be a better alternative for targeting these individuals.

- Make sure the organizational structure of a group knows about the session and its objective. No one showed up for a session involving high school teachers because the teacher helping with the arrangements did not clear the session with the school principal. After learning of the session, he threatened to censure them if they attended.

- Make multiple records for a session. Videotaping, audiotaping, or having analysts directly observe the sessions had no effect on the quality of the session. When possible, videotape the sessions, as this provides an effective way for reviewing the sessions later.

- Have clear objectives and a written agenda to keep the sessions on track and ensure that all important topics are covered.

- Select a relaxed setting with as informal a format as possible. Community halls, church halls, or local meeting places all work fine. Refreshments help to break the ice.

- Keep the session to 2 hours. We've found that a break is generally unnecessary, although a short one can sometimes help reorient the discussion if people are tending to pursue extraneous matters.

- Stay afterwards; this can alter impacts and ease anxieties.

These ideas are based solely on our experiences and not the result of a systematic, formal evaluation. However, they are generally consistent with principles found in marketing applications.[11]

OVERALL IMPLICATIONS

Our experience suggests that focus groups can be valuable tools in evaluating effectiveness of risk communication. We found them to provide the opportunity to explore perceptual cues to listen to the everyday language people use to discuss environmental risks, as well as to observe people using probability information.

Our focus group experiences suggest that much of the disagreement between a technician's risk assessment and the lay public's perception can be traced to disagreements on the relative weights that people place to different characteristics of the risk. Technicians argue that if people would only understand that the probabilities of exposure to hazardous waste are so much smaller than the probabilities of dying from an automobile accident, then we could communicate risk more effectively. As a technician, he's placing his entire weight on the importance of the magnitude of the probability estimate itself. The lay public, on the other hand, is much more likely to rate the catastrophe of the risk, its unknown quality and its long-term potential consequences (cancer or birth defects) as being much more important than the likelihood. Until the technical community begins to recognize this difference in characteristics and the weights that people have attached to those characteristics, then communication between the two groups will be difficult. Our focus group experiences suggest that too much emphasis is on educating the public about the size of probabilities and not enough attention is being given to listening to the public about their feelings concerning the severity of the risks to them.

Focus groups are not a substitute for more systematic quantitative research in evaluation. Nonetheless, they can be effective complements to more systematic research tools such as survey methods.

REFERENCES

1. W. H. Desvousges, V. K. Smith, D. H. Brown, and D. K. Pate, "The Role of Focus Groups in Designing a Contingent Valuation Survey to Measure the Benefits of Hazardous Waste Management Regulations," draft technical report prepared for U.S. Environmental Protection Agency, Research Triangle Institute, Research Triangle Park, NC (1984).

2. H. Kunreuther, R. Ginsberg, L. Miller, P. Sagi, P. Slovic, B. Borkan, and N. Katz, *Disaster Insurance Protection: Public Policy Lessons*, John Wiley and Sons, New York (1978).

3. R. C. Mitchell and R. T. Carson, "Valuing Drinking Water Risk Reduction Using the Contingent Valuation Method: A Methodological Study of Risks from THM and Giardia," final report prepared for U.S. Environmental Protection Agency, Resources for the Future (1986).

4. V. K. Smith and W. H. Desvousges, "An Empirical Analysis of the Economic Value of Risk Changes," *Journal of Political Economy* 95(1):89-114 (1987).

5. J. W. Payne, "Information Processing Theory: Some Concepts and Methods Applied to Decision Research," in *Cognitive Processes in Choice and Decision Behavior*, Thomas S. Wallsten, ed., Lawrence Erlbaum Assoc., Hillsdale, NJ (1980).

6. K. A. Ericsson and H. A. Simon, "Verbal Reports As Data," *Psychological Review* 87:215-51 (1980).

7. W. H. Desvousges and J. H. Frey, "Integrating Focus Groups and Environmental Risk Surveys," unpublished manuscript, Research Triangle Park, NC (1988).

8. C. E. Basch, "Focus Group Interview: An Underutilized Research Technique for Improving Theory and Practice in Health Education," *Health Education Quarterly* 14:411-48 (1987).

9. V. K. Smith, W. H. Desvousges, and A. M. Freeman, III, "Valuing Changes in Hazardous Waste Risks: A Contingent Valuation Analysis," draft report prepared for Office of Policy Analysis, U.S. Environmental Protection Agency, under cooperative agreement No. CR-811075, prepared by Vanderbilt University, Nashville, TN, and Research Triangle Institute, Research Triangle Park, NC (1985).

10. V. K. Smith, W. H. Desvousges, A. Fisher, and F. R. Johnson, "Communicating Radon Risk Effectively: A Mid-Course Evaluation," prepared for Office of Policy Analysis, U.S. Environmental Protection Agency, under cooperative agreement No. CR-811075, prepared by Vanderbilt University, Nashville, TN, and Research Triangle Institute, Research Triangle Park, NC (1987).

11. J. B. Higgenbotham and K. K. Cox, "Focus Group Interviews: A Reader," American Marketing Association, Chicago (1979).

Development of Cleanup Criteria for the Port Hope Remedial Program

D. B. Chambers, L. M. Lowe, and G. G. Case
SENES Consultants Limited
Richmond Hill, Ontario, Canada

R. W. Pollock
Atomic Energy of Canada Limited
Ottawa, Ontario, Canada

ABSTRACT

In 1975 radioactive contamination was identified in the Town of Port Hope, Ontario. A Federal-Provincial Task Force subsequently carried out cleanup work on about 400 properties. However, ten large on-land sites and the Port Hope Harbour were identified as requiring remedial work at some later date. This paper describes the approach used to develop cleanup criteria for these residual sites. It also outlines the basis for the proposed primary criteria and through pathways analysis shows how the primary criteria relate to soil contamination levels. Finally, various cost effectiveness analyses are described.

KEYWORDS: Soil contamination, radioactive contamination, cleanup criteria, exposure pathways, cost effectiveness

INTRODUCTION

The Town of Port Hope, Ontario lies on the north shore of Lake Ontario, approximately 100 km to the east of Toronto. From 1933 to the mid 1950s, the town was the site of a radium refinery operation. In addition, starting in 1942, the refinery plant site has been home to various generations of uranium refining and conversion facilities.

During the early years of operation, the regulations and practices for storage and disposal of radioactive materials were not as stringent as they now are, and processing residues and other contaminated wastes were placed in various areas of the town. In some instances, these wastes were used as fill material around homes and in low-lying areas. Additionally, modifications and expansions of the processing operations produced contaminated scrap building materials which were salvaged and reused in local structures.

The presence of this radioactive contamination was discovered in 1975 and the Federal Government decided that remedial action would be undertaken. A Federal-Provincial Task Force on Radioactivity headed by the Atomic Energy Control Board (AECB) was established to develop cleanup criteria, and to carry out remedial work at properties exceeding these criteria in Port Hope and in the uranium mining communities of

Elliot Lake, Ontario and Uranium City, Saskatchewan (Eaton, 1982). The Task Force established cleanup criteria (AECB, 1977) for both radon daughter levels inside buildings and gamma radiation.

Subsequently, over 4,000 properties in Port Hope were surveyed for compliance with the criteria and remedial work was carried out on about 400 properties. However, the radioactive waste management facility at the Chalk River Nuclear Laboratory of Atomic Energy of Canada Limited, where the Port Hope waste was trucked, had a limited capacity. Remedial work was thus concentrated on residential and commercial properties. Large volumes of contaminated soil in vacant areas and the contaminated sediments in the Port Hope Harbour were left for cleanup at a later date.

The Low-Level Radioactive Waste Management Office (LLRWMO) was formed by the Federal Government in 1982. Part of its mandate is to complete the cleanup of historical wastes in Port Hope. Table 1 provides a historical overview of the Port Hope situation.

Table 1. Historical Overview

1932-1953	Radium refining carried out.
1933-1939	On-site disposal of radium residues.
1939-1948	Residue disposal in-town.
1942	Uranium recovery begins.
1948-1954	Residue disposal at Welcome Site.
1955-Present	Residue disposal at Port Granby Site.
1938-Present	Several building demolitions, residue and waste transfer activities.
1975	Radioactive contamination discovered.
1976-1981	Remedial actions carried out by Task Force.
1980	Eldorado directed to decommission Port Granby Site.
1982	LLRWMO formed.
1986	(Spring) Port Hope area waste disposal project announced. Historic wastes in Port Hope to be cleaned up by LLRWMO. Disposal facility to be developed by Eldorado Resources Limited (ERL) for wastes from Welcome and Port Granby Sites and from Town of Port Hope.
1986 (Fall)	ERL disposal facility cancelled by Federal Government. Task Force appointed to recommend siting process.

Over the years since the original cleanup in 1975, the cleanup of radioactive contamination and the implications of natural radioactivity have received much attention from regulatory authorities and the scientific community worldwide. Consequently, to develop plans for completing the cleanup of Port Hope, the LLRWMO initiated a review of the previously used Port Hope cleanup criteria and their application. SENES Consultants Limited were retained by the LLRWMO to perform this review. The results of these investigations, detailed in SENES (1987), are summarized in this paper.

PRIMARY CLEANUP CRITERIA

Existing Radiological Cleanup Criteria

Table 2 is a summary of criteria reported in the literature which have been either applied or proposed for application to limit the radiation exposure of members of the public in a variety of settings. The reader is referred to the reference materials for further discussion of the various criteria and the context for which their application was intended. However, a few comments of a general nature are appropriate. First, most international agencies are moving, or have already moved, to adopt the concept of effective dose equivalent (commitment) recommended by the International Commission for Radiological Protection (ICRP). For example, this is clearly the intent of the Atomic Energy Control Board (AECB, 1986) and the U.S. Nuclear Regulatory Commission (USNRC, 1986). Second, with this approach, dose and risk are easily related through the use of ICRP (or other) risk factors.

Ideally, the cleanup criteria selected for Port Hope would require no restrictions on any credible future land use. The primary cleanup criteria proposed below were developed with that objective in mind. However, this objective may not be fully achievable, particularly at the landfill site. Land use restrictions, or amendments to local building construction codes, might thus be required to reasonably assure compliance with the primary criteria.

Despite the different values selected by various agencies (Table 2), some underlying concepts emerge. First, for conditions of chronic exposure, annual effective dose equivalent exposures due to contamination in excess of about 1 mSv y^{-1} (excluding radon) above background would generally be considered undesirable. Second, suggested design levels for future exposure situations (e.g., the building of houses on currently unoccupied but contaminated lands) are in the order of 0.02 WL (including background) for indoor radon daughters. These are the numeric values that were suggested in this study for primary remedial action criteria in Port Hope. The risks associated with exposure at these levels can be estimated as follows.

Risk Associated with the Primary Cleanup Criteria

The ICRP risk factor of 1.65×10^{-2} Sv^{-1} is intended to apply for full expressions of harm for average individuals irrespective of sex or age and includes the possibility both of cancers in the exposed persons and of hereditary detriment to the subsequent two generations (ICRP, 1977). Based on this risk factor, chronic annual exposure to 1 mSv y^{-1} carries a risk of

$$1 \times 10^{-3} \text{ Sv y}^{-1} \times 1.65 \times 10^{-2} \text{ Sv}^{-1} = 1.65 \times 10^{-5} \text{ y}^{-1} \ .$$

A remedial action criterion of 1 mSv y^{-1} would thus ensure that no one was subject to a lifetime risk of greater than about 20 per million per year. This would be the risk to the most exposed individual; the risk to the average individual following remedial action would be much smaller. While not zero, such annual risks are relatively small, comparable to the risks from annual exposure to normal background radiation exclusive of radon (risk of cancer), smoking one package of cigarettes per year (risk of cancer) or traveling 2000 km per year by automobile (risk of fatal accident) (risk factors adapted from Upton, 1982).

Continuous exposure (8760 h y^{-1}) to a radon daughter concentration of 0.02 WL results in an exposure of about 1 WLM y^{-1}. For exposures in nonoccupational settings, the (lifetime) risk of death from lung cancer associated with exposure to radon daughters is in

Table 2. Summary of Criteria to Control Public Radiation Exposure

Dose Criteria

AECB
Present (and proposed) regulations
- 5 mSv y^{-1} maximum due to operation of a nuclear facility

U.S. NRC
Proposed reference level for nuclear facilities
- 1 mSv y^{-1} (above bkd.) as action level for licensee

U.S. DOE
Annual limit due to contam. lands
- 1 mSv y^{-1} (above bkd.) average over a lifetime

U.S. NCRP
Remedial action level for external radiation
- 5 mSv y^{-1} (incl. bkd.) from members of the uranium series

ICRP
Limit for committed effective dose equivalent
- 1 mSv y^{-1} (above bkd.) for repeated exposures over prolonged periods

Indoor Radon Daughter Criteria

Federal/Provincial Task Force
- 0.02 WL annual average (incl. bkd.) as a primary criterion

AECB
Present regulations
- 0.02 WL annual average due to operation of a nuclear facility

U.S. EPA
Naturally occuring in-house radon daughter
- 0.02 WL (due to bkd.) remedial action guideline

U.S. DOE
For contaminated soils
- 0.03 WL maximum (incl. bkd.)
- 0.02 WL target (incl. bkd.) using "reasonable efforts" to reach

U.S. NCRP
Remedial action level
- 2 WLM y^{-1} (incl. bkd.) (equivalent to 0.04 WL for continuous exposure)

ICRP
Equilibrium equivalent radon (EER) concentration
- 200 Bq m^{-3} (0.054 WL) due to bkd. for existing exp. situations
- 100 Bq m^{-3} (0.027 WL) due to bkd. for future exp. situations

U.K. NRPB
Dose equivalent from background radon daughters
- 20 mSv y^{-1} (400 Bq m^{-2} EER or 0.11 WL) action level for existing dwellings
- 5 mSv y^{-1} (100 Bq m^{-3} EER or 0.027 WL) design level for new buildings

Sweden
Limits for natural, background sources
- 400 Bq m^{-3} EER (0.11 WL) for existing homes
- 100 Bq m^{-3} EER (0.027 WL) design level for new buildings

WHO
Limits for natural, background sources
- 400 Bq m^{-3} EER (0.11 WL) for prompt remedial action
- 100 Bq m^{-3} EER (0.027 WL) for existing homes if simple remedial measures are possible
- 100 Bq m^{-3} EER (0.027 WL) design level for new buildings

Table 2. (Cont.)

Contamination Criteria

Federal/Provincial Task Force Gamma outside bldgs.	–	100 uR h^{-1} (incl. bkd.) above bare ground
U.S. EPA Clean-up of radium-contaminated lands	–	5 pCi g^{-1} (0.19 Bq g^{-1}) above bkd. for any 100 m^2 area averaged over any 15 cm layer thick surface layer
	–	15 pCi g^{-1} (0.56 Bq g^{-1}) above bkd. for any 100 m^2 area averaged over 15 cm layer below the 15 cm thick surface layer
Gamma inside habitable bldgs.	–	20 uR h^{-1} above bkd.
U.S. DOE Clean-up for Th-232, Th-230, Ra-228, Ra-226	–	5 pCi g^{-1} (0.19 Bq g^{-1}) above bkd. for any 100 m^2 area avg. over 15 cm surface layer
	–	15 pCi g^{-1} (0.56 Bq g^{-1}) above bkd. for any 100 m^2 area avg. over any 15 cm layer below the 15 cm surface layer
Gamma inside habitable bldgs.	–	20 uR h^{-1} above bkd.

Sources Used

AECR (1978) (and Amendments)	USNCRP (1984)
AECB (1986)	NEA (1983, 1985)
ICRP 26 (1977)	APCA (1986)
ICRP 39 (1984)	UKNRPB (1983,1987)
USNRC (1986)	PEARCE (1987)
USEPA (1983)	Swedjemark (1985)
USDOE (1986)	WHO (1985)
USGAO (1986)	
Gilbert et al., (1985)	

the order of 1×10^{-4} WLM^{-1} (Evans *et al.*, 1981). This risk approaches that associated with the annual risk of fatality from motor vehicle traffic accidents experienced by an average member of the public in Ontario (based on 1984 statistics of 1203 deaths in a population of 8.9×10^6, or an individual annual risk of 1.35×10^{-4}) (Province of Ontario, 1984).

Nonradioactive Contamination

It was beyond the scope of this study to carry out a review of the potential toxicity and exposure-effect relations of the various nonradioactive species present in contaminated soils in Port Hope. However, a procedure to permit screening for nonradioactive contaminants of potential interest to cleanup activities was needed.

The approach taken was first to review the available soil and sediment data from Port Hope and by comparison to typical or normal concentrations determine which contaminants were potentially present at elevated concentrations. Reference annual intakes for those contaminants were then derived from guidelines for Canadian drinking water quality, set by Health and Welfare Canada (1980). Only contaminants for which the water quality guidelines were established because of health-related reasons were considered. It was assumed that the intake of contaminants at or below the maximum levels permitted in the drinking water guidelines would safeguard human health. On this basis, uranium and six nonradioactive contaminants, namely arsenic, boron, cadmium, chromium, cobalt and lead, were identified for further consideration.

PATHWAYS MODEL DEVELOPMENT

The primary criteria have to be related to secondary criteria, such as soil contamination levels, in order to direct the cleanup operation in the field. This was done by means of pathways analysis.

There are numerous scenarios and variations in scenarios that can be postulated for potential exposures to people who may in the future live on contaminated soil. This is because the contamination in Port Hope is present in large amounts, such as the landfill site and ravines, in smaller amounts and various distributions on individual properties, and in the sediments in the Port Hope Harbour. In addition, varied lifestyles increase the range of potential exposure scenarios. It was thus considered necessary to develop a base-case scenario, with well-defined parameters, to estimate potential exposure from various cleanup criteria and exposure pathways. The base case for on-land sites was chosen to be a model residence, situated on homogeneously contaminated soil containing 1 Bq g^{-1} of uranium-238 and of each of its daughters. The issue of potential exposures resulting from contaminated sediments in the Port Hope Harbour was addressed separately.

The pathways model considered the major sources and pathways of exposure to people in Port Hope. Some pathways, such as external exposure from immersion in contaminated airborne dust or from swimming, are known to be insignificant and were not carried through the pathways modelling exercise. Others, such as drinking water from a well and consumption of contaminated milk or meat, are not applicable to an urban setting.

The Approach to Pathways Modelling

Pathways by which residents could be exposed as a result of the contamination of on-land sites include:

- external gamma radiation
- indoor radon daughters
- ingestion of garden produce grown in contaminated soil
- direct ingestion of dirt (via hand-mouth transfer)
- inhalation of airborne dust

For ingestion pathways, both chemical and radiological soil contaminants were considered. The pathways model parameters selected for this analysis are shown in Table 3. The reader interested in the details of the pathways model calculations is referred to SENES (1987). However, the following brief comments are provided.

Receptor Characteristics

The exposures to two types of residents (also referred to as "receptors") were considered: an adult and a young child. Characteristics, summarized in Table 3, were assigned to these receptors such that they could be considered to be representative members of the critical group; that is, they are subject to conditions (e.g. location relative to contamination source, types of food consumed) under which they would receive the maximum *average* exposure of many comparable groups. The adult receptor was assigned the characteristics of a 20 to 39 year old male; the young child receptor was assigned the characteristics of a 2 to 3 year old.

Table 3. Input Parameters for Radiological Pathways Analysis

ADULT RECEPTOR CHARACTERISTICS   CALCULATED VALUES
(20-39 y age bracket)

Air Intake Rate $(m^3\ d^{-1})$	23	
Produce Intake Rate $(kg\ y^{-1})$	215	
Fraction of Time Spent at Site	0.67	
Fraction of Site Time Indoors	0.94	Fraction of Site Time
Hours per Day Working in Garden	8	Outdoors 0.06
Days per Week Working in Garden	2	
Weeks per Year Working in Garden	22	
Dirt Ingested when Outdoors	480	
(mg/ 8 h day)		

CHILD RECEPTOR CHARACTERISTICS
(2-3 y age bracket)

Air Intake Rate $(m^3\ d^{-1})$	5	
Produce Intake Rate $(kg\ y^{-1})$	110	
Fraction of Time Spent at Site	1.00	
Fraction of Site Time Indoors	0.88	Fraction of Site Time
Hours per Day Playing in Yard	8	Outdoors 0.12
Days per Week Playing in Yard	5	
Weeks per Year Playing in Yard	26	
Dirt Ingested when Outdoors	250	
(mg/ 8 h day)		

HOUSE DATA

House Width (m)	10	
House Length (m)	10	Emanating Area (m^2) 180
Basement Height (m)	2	Basement Volume (m^3) 200
Air Exchange Rate (h^{-1})	1.0	
Radon Decay Constant (h^{-1})	0.0076	
Radon Wall Attenuation (-)	0.30	
Radon Emanation Rate	0.50	
$(Bq\ m^{-2}\ s^{-1}\ per\ Bq\ g^{-1})$		
Radon/Radon Daughter Equilibrium	0.50	
Fraction		

CONTAMINATION DATA

Concentration in Soil $(Bq\ g^{-1})$	1.0
(U-238 + dtrs)	
Outdoor Dust Concentration	50.0
$(ug\ m^{-3})$	
Contaminated Fraction of Airborne	0.50
Dust	
Fraction of Outdoor Dust Indoors	0.75
Respirable Fraction of Airborne	1
Dust	

EXTERNAL GAMMA EXPOSURE

External DCF $(uSv\ h^{-1}\ per\ Bq\ g^{-1})$	0.30
(U-238 + dtrs)	
Geometric Correction Factor	1
(outdoors)	
Fraction of Outdoor Time Over	1
Contamination	
Reduction Factor When Indoors	0.1

FOOD DATA

Fraction of Initial Contamination	
Remaining After Food Preparation	0.50
Fraction of Produce from Site	0.25

Table 3. (Cont.)

TRANSFER COEFFICIENTS		U-nat	Th-230	Ra-226	Pb-210	Po-210
Soil-to-Plant Transfer Factor	(-)	4.6E-04	4.0E-05	2.3E-03	1.1E-03	1.5E-04
Adult DCF for Ingestion (sol.)	(uSv Bq^{-1})	7.8E-02	3.5E+00	3.2E-01	1.6E+00	5.3E-01
Adult DCF for Ingestion (insol.)	(uSv Bq^{-1})	6.8E-03	1.4E-01	3.2E-01	1.6E+00	5.3E-01
Adult DCF for Inhalation (insol.)	(uSv Bq^{-1})	4.0E+01	8.1E+01	2.6E+00	3.5E+00	2.5E+00
Child DCF for Ingestion (sol.)	(uSv Bq^{-1})	1.3E-01	9.2E+00	1.2E+00	6.9E+00	8.0E-01
Child DCF for Ingestion (insol.)	(uSv Bq^{-1})	1.0E-02	3.7E-01	1.2E+00	6.9E+00	8.0E-01
Child DCF for Inhalation (insol.)	(uSv Bq^{-1})	1.9E+02	3.1E+02	8.2E+00	1.5E+01	6.9E+00

<u>Note</u>:

1. Exponential notation: 4.6E-04 = 4.6×10^{-4}.
2. U-nat = natural uranium.

External Gamma Radiation

For the base-case exposure scenario, no reduction in the outdoor exposure rate or duration of exposure due to the finite size of the outdoor contamination was taken into account. However, such factors were considered for smaller sources. When the receptor is indoors, the protection afforded to him by buildings from outdoor gamma radiation varies considerably (USNCRP, 1984; UNSCEAR, 1982; USDOE, 1983). For this analysis, a reduction factor of 0.1 was selected for outdoor gamma exposure to receptors indoors.

Indoor Radon Daughter Exposure

For the case of unrestricted land use, it was considered possible for some future resident to build a home on contaminated soil. It was assumed the receptor lived in a single story house 10 m wide and 10 m deep with a full concrete block basement with walls 2 m high (Table 3). These dimensions are not unlike the dimensions of many houses.

The main factors that control indoor radon levels are the concentration of radium (the radon parent) in the soil around a building (and indeed in the building material itself) and the air exchange rate of the building. A simple single compartment steady-state model which balanced radon influx with fresh air changes was used to estimate indoor radon and radon daughter levels. Once the air exchange rate was specified, the radon and radon daughter concentrations could be calculated with the parameters shown in Table 3. For one air change per hour, which is typical of family residences, Krisiuk (1980) calculates an equilibrium factor of 0.54. This value is close to the annual average value of $F = 0.5$ actually observed in Port Hope (SENES, 1987).

Inhalation of Airborne Dust

For the present analysis, it was assumed (based on local measurements) that the outdoor dust concentration was 50 μg m^{-3} (Table 3). It was also assumed that 50% of the suspended dust was from contaminated local soils and that the indoor dust concentration was 75% of the outdoor level (Hawley, 1985). Finally, it was assumed that all the airborne dust was respirable.

Ingestion of Garden Produce

A garden that provided fruits and vegetables consumed by both the adult and the child, was assumed. Given that most people include in their diets fruits and vegetables not suited to growing in southern Ontario and that most prefer to eat some fresh produce throughout the year, it was assumed that 25% (Table 3) of all of the fruits and vegetables consumed by the receptors come from the home garden. This value would likely overestimate the consumption rate of local produce of most residents in a suburban neighborhood.

The selection of soil-to-plant transfer factors B_i is often a contentious issue since they are highly dependent on site-specific conditions. Two studies which addressed the specific issue of radionuclide uptakes in garden produce from contaminated soils in Port Hope were used to derive the B_i values for this analysis: a 1986 vegetation and soil sampling program carried out as part of this study and a study by Tracy et al. (1983). These studies were supplemented by a comprehensive literature survey done by Baes et al. (1984).

Radionuclide Dose Conversion Factors

The dose conversion factors (DCF) used to convert annual radionuclide intakes by inhalation and ingestion to committed effective dose equivalents were taken from Johnson and Dunford (1983) except for those for the ingestion of soluble thorium-230, which were taken from Johnson (1986) (see Table 3). The latter study uses more extensive information on thorium uptake in the body to estimate a revised DCF for thorium ingestion through environmental pathways. In selecting the appropriate DCFs, it was assumed that radionuclides ingested via garden produce were soluble. Those ingested through direct ingestion of contaminated dirt or inhaled as airborne dust were assumed to be insoluble since only relatively insoluble contaminants would persist in the surface soil layer.

Total Radiation Doses and Exposures

The contributions to dose and exposure through the pathways considered in the base-case analysis are shown in Table 4. Relative to the primary criteria of 1 mSv y^{-1} and 0.02 WL, exposure to radon daughters is the most significant pathway. Exposure to external gamma radiation is the most significant of the non-radon pathways, representing 65% and 53% of the dose for the adult and child receptors, respectively. Of the remaining pathways, produce and dirt ingestion represent 18% and 13% of the dose for the adult and 15% and 30% for the child. Inhalation is a minor pathway (<5%) for both the adult and the child.

Intake of Nonradiological Contaminants

The intakes of nonradiological contaminants from all pathways were estimated with the same procedures used for the radiological contaminants. A base-case soil concentration of 100 ppm was assumed for each of the selected nonradiological contaminants. Using the pathways model and measured contamination levels in Port Hope soils showed that the estimated intakes of nonradioactive contaminants from all pathways were well below reference levels derived from Canadian drinking water guidelines (SENES, 1987).

PROPOSED SOIL CLEANUP CRITERIA

Major On-Land Sites

Excluding the landfill site, there are nine major on-land areas in Port Hope which may require cleanup. In the absence of some form of institutional control, it is not difficult to imagine that a dwelling could be constructed at some of these contaminated sites. The basic scenario then for potential exposure to the waste is living in a house constructed on contaminated soil, ingesting contaminated dirt, consuming contaminated garden produce and being exposed to gamma radiation from contaminated soil.

Table 4. Estimated Doses and Exposures for the Base-Case Scenario [1,2]

ADULT RECEPTOR

Pathway	U-nat	Th-230	Ra-226	Pb-210	Po-210	Total	Rn-222 ($Bq\ m^{-3}$)	Rn-Dtr (WL)	Rn-Dtr ($WLM\ y^{-1}$)
external: outdoors	-	-	-	-	-	105.6	-	-	-
indoors	-	-	-	-	-	165.5	-	-	-
inhalation: outdoors	0.7	0.7	0.0	0.0	0.0	1.4	-	-	-
indoors	7.9	8.0	0.3	0.3	0.2	16.8	482.3	0.065	2.12
produce ingestion	1.9	3.8	19.8	47.3	2.1	74.9	-	-	-
dirt ingestion	0.3	3.0	6.8	33.8	11.2	55.0	-	-	-
Total[2] ($uSv\ y^{-1}$)	10.8	15.4	26.8	81.5	13.6	419.3	482.3	0.065	2.12

CHILD RECEPTOR

Pathway	U-nat	Th-230	Ra-226	Pb-210	Po-210	Total	Rn-222 ($Bq\ m^{-3}$)	Rn-Dtr (WL)	Rn-Dtr ($WLM\ y^{-1}$)
external: outdoors	-	-	-	-	-	312.0	-	-	-
indoors	-	-	-	-	-	231.3	-	-	-
inhalation: outdoors	2.1	1.7	0.0	0.1	0.0	3.9	-	-	-
indoors	11.4	9.3	0.2	0.5	0.2	21.7	482.3	0.065	2.96
produce ingestion	1.6	5.1	37.9	104.4	1.7	150.7	-	-	-
dirt ingestion	0.7	12.0	39.0	224.3	26.0	301.9	-	-	-
Total ($uSv\ y^{-1}$)	15.8	28.1	77.2	329.1	27.9	1021.5	482.3	0.065	2.96

Note:
1. Based on soil concentration of 1 Bq g^{-1} each of uranium-238 and daughters.
2. All values in uSv y^{-1} unless otherwise indicated.

The base-case scenario, which assumes that the receptors live in a house constructed in soil containing 1 Bq g^{-1} of uranium-238 and of each of its decay products, results in an indoor radon level of about 480 Bq m^{-3} and a corresponding radon daughter level of 0.065 WL. Thus, to meet the 0.02 WL criterion, the radium level in soil would need to be 0.02/0.065, or approximately 0.3 Bq g^{-1} (8 pCi g^{-1}). Non-radon doses total about 0.41 mSv y^{-1} for the adult and 1.02 mSv y^{-1} for the child. Achieving the 1 mSv y^{-1} criterion (for which the child is the critical receptor), would require a radium level of 1/1.02 or approximately 1 Bq g^{-1}. Thus, the 0.02 WL criterion is the most restrictive for the basic scenario analyzed. In effect, unrestricted site usage could be achieved if the cleanup were carried out to an average radium-in-soil concentration of about 0.3 Bq g^{-1} (8 pCi g^{-1}).

It is important to acknowledge that there is a degree of uncertainty associated with the derivation of the secondary criteria or soil contaminant levels. The suggested 0.3 Bq g^{-1} (8 pCi g^{-1}) criterion was based on a pathways model that required the selection of values for a large number of parameters used in the base-case exposure scenario. These parameter values were chosen in a realistic but conservative manner. Thus, even if some soil with radium levels slightly higher than 0.3 Bq g^{-1} remained after cleanup, it would still be unlikely that the primary criteria would be exceeded. Moreover, because the derived contaminant levels on which cleanup activities would be based are considered as upper-limit values, the average levels of the residual contamination remaining after cleanup would be lower than the cleanup criteria.

Small-Scale Sites

It was noted previously that the reference dwelling for the pathways model was assumed to have dimensions of 10 m × 10 m. A garden of about 10 m × 15 m would be required to provide all of the reference produce intake. Thus, for areas of contamination smaller than say 250 m^2, not all of the pathways relevant to larger areas of contamination would apply. For example, for a property with an area of contamination of about 100 m^2, the reference receptors could either live in a house built on the contaminated area or consume produce from a garden situated on the contaminated area but not both. As the area of contamination decreases still further, only part of the basement could be built in contamination and hence the radon influx would decrease in proportion to the area of contamination adjacent to the basement. This would result in a corresponding increase in the reference soil cleanup criteria. Since structures could be built in the future on currently vacant lots, the same cleanup criteria would also apply to these sites.

Since existing homes in which radon daughter levels exceeded the 0.02 WL criterion have already been cleaned up in previous remedial programs, the most likely exposure scenario at small-scale sites is, in the short term, exposure via external gamma radiation and uptake of contaminants by the garden pathway or the dirt ingestion pathway.

The effect of source-receptor geometry for the external gamma pathway should also be considered when evaluating small areas of contamination. For example, the exposure rates at the edge of a 10 m × 10 m source and 5 m away from the same source (along the center line) are about 44% and 1% respectively of those for an infinite plane source. A smaller area of contamination would also lower the likely duration of exposure to the contamination.

In summary, it was concluded that:

1. If the area of contamination is greater than about 100 m^2 and the contamination extends to the depth of a normal basement (approximately 2 m), then it is possible to conceive of a structure being built on the area of contamination. Hence, the soil cleanup criterion of 8 pCi g^{-1} (0.3 Bq g^{-1}) of radium should apply.

2. Situations are also encountered where there is a thin layer of radium-contaminated wastes spread over a large (i.e. > 100 m^2) area. In this situation the 1 mSv y^{-1} criteria may be limiting and a soil level of 1 Bq g^{-1} would apply.

3. For small areas of contamination, the option of building a house on the contamination, or growing a garden, are no longer of major concern. The two dominant pathways become external gamma radiation and ingestion of contamination via the hand-mouth pathway, if the small area happens to be extensively used. For example, one could visualize the placing of a child's sandbox in the center of a contaminated area. In this case, the critical pathways are external radiation outdoors and dirt ingestion which contribute nearly 80% of the dose when the indoor external exposure pathway is deleted. (Gamma exposure rates decrease rapidly with the area of the source and with distance from the source.) Small sources are also not likely to produce much windborne dust.

    The dose from the outdoor gamma and dirt ingestion pathways alone can be calculated as follows. A soil level of 1 Bq g^{-1} was estimated to result in about 0.3 mSv y^{-1} from outdoor gamma radiation and 0.3 mSv y^{-1} from dirt ingestion for a total of 0.6 mSv y^{-1}. Hence, the soil level could be as high as 1/0.6 or 1.7 Bq g^{-1} (46 pCi g^{-1}) and still meet the 1 mSv y^{-1} criterion. This soil criterion would apply to areas perhaps as small as 2 m $\times$ 2 m, at which point the area might be considered as a "spot" [see (4) below].

4. With regard to spots, there are two possibilities. First, the spot is truly a spot of identified contamination located essentially on the ground surface. It is anticipated that all such known spots would be cleaned up. The second possibility is that the spot is only a surface expression of some buried but as yet unidentified contamination. Whether or not such an area would be cleaned up and to what extent the cleanup should be carried out is uncertain. In view of this possibility, properties with "spots" should be re-surveyed to characterize the spots and the extent of contamination before remedial work is initiated.

Figure 1 is a plot of the number of structures versus measured gamma exposure rates. It should be recognized that there is some uncertainty in this figure especially at low gamma levels. The figure does, however, show that the number of sites that would require cleanup increases rapidly with decreasing gamma levels.

Figure 1 also shows the number of sites with contaminated areas less than 100 m^2 and greater than 100 m^2 for each of eight small-scale site cleanup options. For example, of the 198 sites exceeding the 25 µR h^{-1} level, only 52 have contaminated areas greater than 100 m^2. It is clear that few properties would require cleanup if areas of contamination less than 100 m^2 were excluded, unless they exceeded 1 Bq g^{-1} contamination concentration.

Preliminary cost estimates as a function of cleanup option have been prepared for the cleanup of small sites (MacLaren, 1987). The cleanup costs assume removal of contaminated areas down to the given gamma level and transport to the Port Hope boundary. The remedial actions are not complete when the waste materials arrive at the Port Hope boundary. It is still necessary to transport the wastes to a disposal site and to dispose of them within that site. Based on analyses of the design and cost of alternative disposal scenarios (SENES *et al.*, 1985; SENES, 1986), it appears most unlikely that the combined costs of transporting the Port Hope wastes from the town boundary to a disposal site and then to dispose of the wastes would be less than $100 m^{-3} of waste. Depending on the location and design of the disposal facility, the costs could range up to $500 m^{-3} and higher.

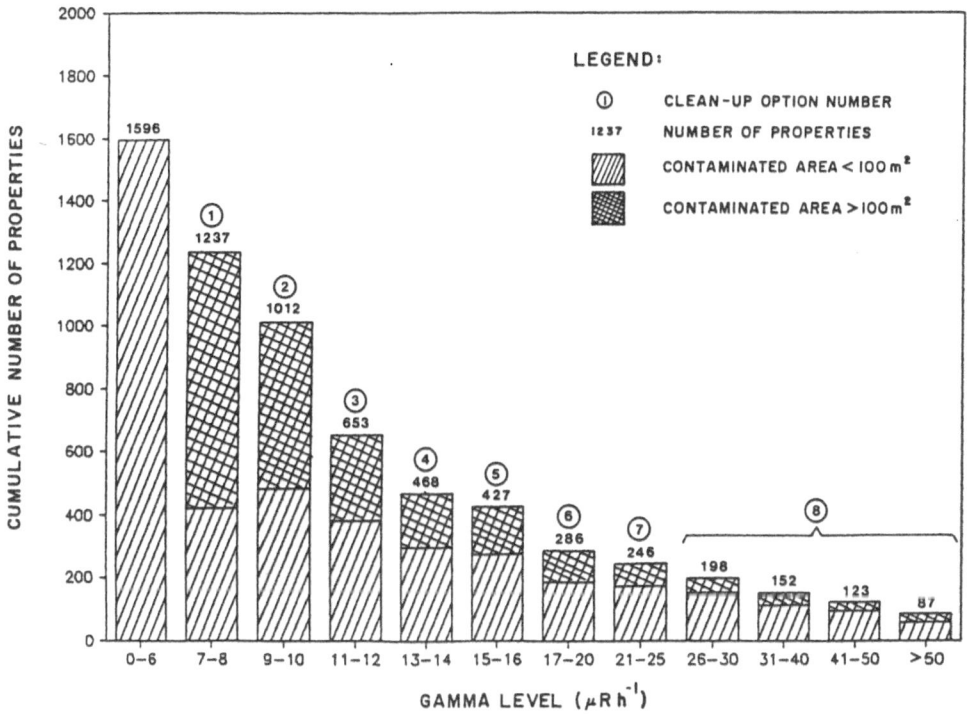

Fig. 1. Number of Small-Scale Sites Requiring Cleanup vs Gamma Criterion.

The foregoing data can be considered from a crude cost-effectiveness point of view. First, based on the calculations of the pathways model, the total dose and radon-daughter exposure resulting from uniform contamination of 1 Bq g^{-1} of uranium-238 and of each of its daughters is 720 μSv y^{-1} and 2.54 WLM, respectively, averaged over the adult and child receptors (Table 4). Noting that 1 WLM is equivalent to a dose of 5.5 mSv for environmental exposures (NEA, 1983), the total annual effective dose from all pathways is therefore 14.7 mSv y^{-1} per Bq g^{-1} of contamination. However, 1 Bq g^{-1} of uranium-238 and of each of its daughters produces an external gamma exposure rate of 43 μR h^{-1}. Therefore, the total effective dose is 14.7 mSv y^{-1} per 43 μR h^{-1} of external exposure, or 340 μSv y^{-1} per μR h^{-1}. This assumes complete linearity of all exposure pathways.

Second, Fig. 1 shows sites on which at least one gamma measurement exceeds the given gamma level. The pathways model is, however, based on an average gamma and soil contamination level. From the 1986 LLRWMO database, the average gamma level at, for example, the 198 locations in cleanup option number 8 (those above a criterion of 25 μR h^{-1}) is 12.95 μR h^{-1}. If these 198 locations were cleaned up using a criterion of 25 μR h^{-1}, the average exposure would become 9.63 μR h^{-1} and the average reduction in exposure at each location would be 12.95 − 9.63 = 3.32 μR h^{-1}. If it is assumed that each location houses an average of 4 people (2 adults and 2 children), and remembering that 1 μR h^{-1} of gamma radiation corresponds to 340 μSv y^{-1} of total dose, the total dose saved would be

$$3.32 \ \mu R \ h^{-1} \times 198 \times 4 \times 340 \ \mu Sv \ y^{-1} \ per \ \mu R \ h^{-1} = 0.89 \ person\text{-}Sv \ y^{-1}.$$

Finally, if the total cost of remedial measures were $2,105,000 (from MacLaren, 1987) plus transportation and disposal costs of $100 m^{-3}) for a criterion of 25 μR h^{-1}, the

Table 5. Cost-Effectiveness of the Cleanup of Small-Scale Sites

Exposure Rate Criterion (μR h^{-1})	Average Exposure Rate of Locations Above Criterion (μR h^{-1})		Dose Saved[1] (person-Sv per year)	Annualized[2] Cost ($ y^{-1})		Cost Per Dose Saved ($ Sv^{-1})	
	Before Clean-up	After Clean-up		A	B	A	B
6	7.07	5.59	2.49	2,935,440	10,781,840	1.2x10^6	4.3x10^6
8	7.54	6.00	2.12	1,723,360	6,212,960	8.1x10^5	2.9x10^6
10	8.57	6.53	1.81	873,920	3,085,120	4.8x10^5	1.7x10^6
12	9.61	7.13	1.58	601,680	2,112,080	3.8x10^5	1.3x10^6
14	9.77	7.36	1.40	502,160	1,750,160	3.6x10^5	1.3x10^6
16	11.44	8.31	1.22	324,880	1,128,080	2.7x10^5	9.3x10^5
20	11.90	8.81	1.03	258,720	892,320	2.5x10^5	8.6x10^5
25	12.95	9.63	0.89	168,400	565,200	1.9x10^5	6.3x10^5

Notes:

1. See text.

2. Based on an 8% annual interest rate and total clean-up costs (MacLaren, 1987) plus transportation and disposal costs of: A) $100 m^{-3} and B) $500 m^{-3}. See text.

corresponding annualized cost would be $168,400 for an assumed investment return of 8% per year. The cost per unit dose reduction is then $168,400 per 0.89 person-Sv or $1.9 $\times$ 10^5 per person-Sv. The results of this and similar calculations for other cleanup levels are shown in Table 5. In all cases, the cost per person-Sv saved is at least a factor of 10 larger than commonly suggested values of $10^3 to $10^4 per person-Sv (e.g. ICRP, 1983; Voilleque and Pavlick, 1982; Jammet and Lombard, 1987).

Based on this crude analysis, while cleanup to the highest criterion of 25 μR h^{-1} is the most cost-effective (Table 5), all of the cleanup levels would result in cost-benefit ratios that are higher than the requirements of the ALARA (as low as reasonably achievable) principle if $10^3 to $10^4 per person-Sv are used to define benefits. This analysis does not, however, suggest that cost-effectiveness alone should be used to select cleanup criteria. Other factors, such as the environmental effects, operational hazards of various cleanup options, and social factors, should also be considered.

Landfill Site

In terms of radioactivity, the landfill site is not different from other major on-land site areas in terms of what soil cleanup criteria should apply. Hence, for unrestricted land use, "accessible" areas of contamination should not contain more than 8 pCi g^{-1} (0.3 Bq g^{-1}) of radium-226. Analyses similar to that described for the small-scale sites were performed and similar results obtained. It should be noted that at the landfill site, the contaminated soil is generally found at depth and in pockets and deposits interspersed with clean layers of soil. Thus, an average value of 8 pCi g^{-1} can be achieved without removing all material down to this concentration.

Port Hope Harbour

Based on various analyses (Environment Canada, 1986; Health and Welfare, 1984; SENES, 1987), it was concluded that leaving the contaminated sediments in place would represent minimal environmental hazards. The issue of the possible future misuse of the contaminated sediments is then the driving force for potential remedial actions. Space does not permit a discussion of this aspect and the reader is referred to SENES (1987) for details.

SUMMARY

Three different classes of potential remedial action sites were investigated: large-scale on-land sites, such as ravines and large open areas; small-scale sites, such as individual residences; and the Port Hope Harbour. The Port Hope landfill was also considered separately because of the magnitude and special nature of the wastes (co-mingled with municipal refuse).

Criteria recommended by various national and international agencies to limit the exposure of members of the public to naturally occurring radioactivity were reviewed. Primary criteria of 1 mSv y^{-1} effective dose equivalent (excluding radon daughters) and an average annual indoor radon daughter concentration of 0.02 WL were suggested for application to the Port Hope remedial works program.

Pathways analysis was used to relate these primary criteria to secondary criteria such as soil contamination levels. The analysis attempted to use realistic parameter values consistent with available data from Port Hope. For large areas of contamination, the pathways model indicated that a radium level in soil of 0.3 Bq g^{-1} (8 pCi g^{-1}) would be required to meet the 0.02 WL criterion. For large areas, but with contamination only to a shallow depth which would eliminate the indoor radon daughter pathway, a radium level in soil of 1.0 Bq g^{-1} (27 pCi g^{-1}) would be consistent with the primary criterion of 1 mSv y^{-1}. For small areas of contamination, where the garden produce pathway would also be eliminated, a soil radium level of up to 1.7 Bq g^{-1} (46 pCi g^{-1}) would still achieve the primary criteria.

The analyses also indicated that remedial actions undertaken to control exposure to radioactive contaminants would also reduce exposure to the nonradioactive contaminants to levels that would safeguard human health. Based on engineering and environmental considerations, a number of cleanup options were identified for each of the potential remedial works categories. An analysis of the cost-effectiveness of the various cleanup options was performed to assist the decision-makers in assessing the viability of the options. The analysis indicated that once the primary criteria for exposure to the critical individuals were met, further remedial actions to lower potential exposures would not be cost-effective. Commonly proposed monetary equivalents for dose saved under the various cleanup options was used as a basis for comparison in the analyses.

While an examination of the occupational and environmental effects associated with the performance of the various cleanup options was beyond the scope of this study, these studies are included in the overall scope of work being done by the LLRWMO. Studies to define waste volumes and to develop conceptual designs and cost estimates for cleanup have also been done. The information in all of these studies will be consolidated by the LLRWMO for submission to the Federal Government, where it will be taken into account, together with social factors, in arriving at a decision on the future work to be undertaken.

REFERENCES

Air Pollution Control Association (APCA), 1986, "Indoor Radon," Proceedings of an APCA International Specialty Conference, SP-54, Philadelphia, Pennsylvania, February.

Atomic Energy Control Board (AECB), 1986, "Proposed General Amendments to the Atomic Energy Control Regulations," AECB Consultative Document C-83, April 28.

Atomic Energy Control Board (AECB), 1977, "Criteria for Radioactive CleanUp in Canada," AECB Information Bulletin 77-2, April 7.

Atomic Energy Control Regulations (AECR), 1978, "Atomic Energy Control Act; Atomic Energy Control Regulations, Amendment," *Canada Gazette*, Part II, **112(2)**:406-411.

Baes, C. F., III, Sharp, R. D., Sjoreen, A. L. and Shor, R. W., 1984, "A Review and Analysis of Parameters for Assessing Transport of Environmentally Released Radionuclides Through Agriculture," Prepared for the U.S. Department of Energy by the Oak Ridge National Laboratory, ORNL-5786.

Eaton, R. S., 1982, "Radon and Radon Daughters in Public, Private and Commercial Buildings, in Communities Associated with Uranium Mining and Processing in Canada," Atomic Energy Control Board, Published in book *Natural Radiation Environment*, 1982 Bhabha Atomic Research Centre, pp. 489-496.

Environment Canada, 1986, "A Discussion of the Environment Canada Investigation Into Radionuclide Levels in Fish Collected from Port Hope Harbour," Environmental Protection Service, Ontario Region, December.

Evans, R. D., Harley, J. H., Jacobi, W., McLean, M. S., Mills, W. A., and Stewart, C. G., 1981, "Estimate of Risk from Environmental Exposure to Radon-222 and Its Decay Products," *Nature* **290**:98-100.

Gilbert, T. L., Eckerman, K. F., Hansen, W. R., Healy, J. W., Kennedy, W. E., Jr., Napier, B. A., and Soldat, J. K., 1985, "A Manual for Implementing Residual Radioactivity Guidelines," (Review Draft), Prepared for the U.S. Department of Energy, 30 September.

Hawley, J. K., 1985, "Assessment of Health Risk from Exposure to Contaminated Soil," *Risk Analysis* **5(4)**:289-302.

Health and Welfare Canada, 1984, "Environmental Radioactivity in Canada—1982," Radiological Monitoring Annual Report, 84-EHD-105.

Health and Welfare Canada, 1980, "Guidelines for Canadian Drinking Water Quality—1978, Supporting Documentation," Cat. No. H48-10/1978-1E.

International Commission on Radiological Protection (ICRP), 1984, "Principles for Limiting Exposure of the Public to Natural Sources of Radiation," ICRP Publication 39, Annals of the ICRP, Vol. 14, No. 1.

International Commission on Radiological Protection (ICRP), 1983, "Cost-Benefit Analysis in the Optimization of Radiation Protection," ICRP Publication 37, Annals of the ICRP, Vol. 10, No. 2/3.

International Commission on Radiological Protection (ICRP), 1977, "Recommendations of the International Commission on Radiological Protection," ICRP Publication 26, Annals of the ICRP, Vol. 1, No. 3.

Jammet, H., and Lombard, J., 1987, "Towards a General Model of Health Detriment Cost Evaluation," *Health Physics* **52**:91-101.

Johnson, J. R., 1986, "A Literature Review of Thorium Uptake, Retention and Excretion," Draft Report for Eldorado Resources Limited, November.

Johnson, J. R., and Dunford, D. W., 1983, "Dose Conversion Factors for Intakes of Selected Radionuclides by Infants and Adults," Atomic Energy of Canada, AECL-7919, January.

Krisiuk, E. M., 1980, "Airborne Radioactivity in Buildings," *Health Physics* **38**:199-202.

MacLaren Engineers, 1987, "Port Hope Remedial Program —Summary of Waste Volumes and Cost Estimates for Proposed Remedial Works," Report No. 5 to the Low-Level Radioactive Waste Management Office, Atomic Energy of Canada Limited, July.

Nuclear Energy Agency (NEA), 1985, "Metrology and Monitoring of Radon, Thoron and Their Daughter Products," Report by a Group of Experts of the OECD-NEA, Paris.

Nuclear Energy Agency (NEA), 1983, "Dosimetry Aspects of Exposure to Radon and Thoron Daughter Products," Report by a Group of Experts of the OECD-NEA, Paris.

Pearce, F., 1987, "A Deadly Gas Under the Floorboards," *New Scientist* **5**:33-35, February.

Province of Ontario, 1984, "Vital Statistics for 1984."

SENES Consultants Limited, 1987, "Development of Cleanup Criteria for the Port Hope Remedial Program," Report to the Low-Level Radioactive Waste Management Office, Atomic Energy of Canada Limited, June.

SENES Consultants Limited, 1986, "Evaluation of Low-Level Radioactive Waste Disposal Alternatives," Report to the Low-Level Radioactive Waste Management Office, Atomic Energy of Canada Limited, January.

SENES Consultants Limited, Golder Associates and Chem-Nuclear Systems Inc., 1985, "Conceptual Design Study of a Low-Level Radioactive Waste Disposal Facility," Report to the Low-Level Radioactive Waste Management Office, Atomic Energy of Canada Limited, January.

Swedjemark, G. A., 1985, "Radon and Its Decay Products in Housing," Doctoral Dissertation for the Department of Radiation Physics, University of Stockholm, Sweden, March.

Tracy, B. L., Prantl, F. A., and Quinn, J. M., 1983, "Transfer of Ra-226, Pb-210, and Uranium from Soil to Garden Produce: Assessment of Risk," *Health Physics* **44**:469-477.

United Kingdom National Radiological Protection Board (NRPB), 1987, "Exposure to Radon Daughters in Dwellings," ASP No. 10.

United Kingdom National Radiological Protection Board (NRPB), 1983, "Human Exposure to Radon Decay Products Inside Dwellings in the United Kingdom," NRPB-R152, February.

United Nations Scientific Committee on the Effects of Atomic Radiation (UNSCEAR), 1982," Ionizing Radiation Sources and Biological Effects," United Nations Publication.

Upton, A. C., 1982, "The Biological Effects of Low-Level Ionizing Radiation," *Scientific American* **246**(2):41-49, February.

U.S. Department of Energy (DOE), 1986, "Technical Approach Document — Uranium Mill Tailings Remedial Action Project," Report 050425.0000.

U.S. Department of Energy (DOE), 1983, "Pathways Analysis and Radiation Dose Estimates for Radioactive Residues at Formerly Utilized MED/AEC Sites," Argonne National Laboratory, U.S. DOE Report ORO-832.

U.S. Environmental Protection Agency (EPA), 1983, "Standards for Cleanup of and Buildings Contaminated with Residual Radioactive Materials from Inactive Uranium Processing Sites," Federal Register **48**(3), 5 January. As referenced in U.S. DOE (1986).

U.S. General Accounting Office (GAO), 1986, "Air Pollution: Hazards of Indoor Radon Could Pose a National Health Problem," Report to the Pennsylvania Congressional Delegation House of Representatives, GAO/RCED-86-170, June.

U.S. National Council on Radiation Protection and Measurements (NCRP), 1984, "Exposures from the Uranium Series with Emphasis on Radon and Its Daughters," NCRP Report No. 77, 15 March.

U.S. Nuclear Regulatory Commission (NRC), 1986, "10 CFR *et al.*, Standards for Protection Against Radiation; Proposed Rule; Extension of Comment Period and Republication," Federal Register **51**(6):1082-1216, 9 January.

Voilleque, P. G., and Pavlick, R. A., 1982, "Societal Cost of Radiation Exposure," *Health Physics* **43**(3):405-409.

World Health Organization (WHO), 1985, "Working Group On Indoor Air Quality — Summary Report," ICP/CEH 002m/70(S), 86491, 20 September, Dubrovnik.

Combining Physiology, Carcinogenic Mechanism, and Interindividual Variation in Cancer Dose-Response Extrapolations

Robert L. Sielken, Jr.
Sielken, Inc.
Bryan, TX

ABSTRACT

The Individualized Response Model (IRM) can incorporate much of the available scientific information into the cancer dose-response modeling portion of quantitative risk assessment. IRM allows for (1) age-dependent administered dose levels, background doses, and susceptibilities, (2) the physiological and pharmacokinetic conversion of administered and background doses to delivered doses, (3) transitions from delivered doses to biologically effective dose, (4) interindividual variation in background doses and susceptibilities, and (5) use of extensions of the two-stage growth model introduced by Moolgavkar and Knudson (1981). The Individualized Response Modeling can utilize quantal dose-response models (probit, logit, Weibull, multihit, and multistage models) or time-to-response models (Hartley-Sielken, multistage-Weibull, and Weibull-Weibull models) or the two-stage growth model. If the more biologically based two-stage growth model is used, the chemical's suspected carcinogenic mechanism(s) determines the portion(s) of the two-stage growth model—stem cell proliferation, transition from a normal stem cell to an intermediate/ initiated cell, intermediate cell proliferation, and transition from an intermediate cell to a malignant cell—affected by the biologically effective dose and the form of the relationship between biologically effective dose and the probability of a specified carcinogenic response.

The quantitative implications of Individualized Response Modeling are illustrated. Particular attention is given to time-varying dose levels, physiologically-based delivered dose scales, impacts of chemicals on cell proliferation rates, and population risks based on a distribution of individual susceptibilities as opposed to the most susceptible individual.

KEYWORDS: Quantitative cancer risk assessment, dose-response modeling, pharma-cokinetics, susceptibility, age-dependence

INTRODUCTION

A chemical's carcinogenic effects on humans is to be evaluated. Let *EX* represent the example chemical.

A human carcinogenic effect depends on three factors: (i) the biologically effective dose, (ii) the probability of a specified carcinogenic response for a given biologically effective dose, and (iii) the proportions of the exposed population receiving particular biologically effective doses.

Human physiological studies can provide information on the biologically effective dose (i). Scientific studies of the human population at risk can help quantify the proportions of the exposed population receiving particular biologically effective doses (iii). On-the-other-hand, there is seldom enough human epidemiological information to establish the probability of a specified carcinogenic response for a given biologically effective dose (ii) directly in humans. Usually animal studies must be used to determine (ii).

Because (ii) can not usually be determined on the basis of human evidence, animal studies must be conducted. The determination of the probability of a specified carcinogenic response for given biologically effective doses in experimental animals requires (a) animal physiology to determine the relationship between administered doses and biologically effective doses, (b) an animal study (usually a chronic bioassay) to provide experimental evidence on the cancer incidence for the experimental biologically effective dose levels, and (c) a dose-response model founded on carcinogenic mechanism to quantify the relationship between cancer probability and biologically effective dose.

The evaluation of a chemical's carcinogenic effects in humans usually requires both human and animal studies. The animal studies provide the quantitative model for the relationship between cancer incidence and biologically effective dose. Figure 1 depicts the roles of animal physiology, carcinogenic mechanism, chronic animal bioassays, human physiology, and interindividual variation.

ANIMAL STUDIES

Chemical EX will have its probability of a specified carcinogenic response for a given biologically effective dose estimated on the basis of a chronic animal bioassay.

Determining the Biologically Effective Dose in Experimental Animals

Figure 1 indicates the role of the determination of the biologically effective dose in experimental animals.

In the chronic animal bioassay each animal has a specified exposure protocol. Let $d(t)$ represent the administered dose at time t. In a 24-month bioassay with "continuous" exposure, $d(t)$ would be a constant with respect to t—the experimental dose levels of chemical EX were constant and equaled 0, 10, 50, and 100 ppm in the diet.

For chemical EX there is no exposure from any sources other than the administered dose. Thus, the background dose $d_0(t)$ is zero for all times t.

The physiologically-based pharmacokinetic modeling of the amount of chemical EX reaching the cancer target site indicates that the relationship between the delivered dose and the administered dose follows approximately a Michaelis-Menten relationship (Fig. 2a). In addition, the delivered dose for a particular administered dose is dependent on the animal's age. The proportional multiplier for the delivered dose is shown in Fig. 2b. The resulting delivered dose for a particular administered dose is the corresponding value in Fig. 2a times the age-dependent multiplier in Fig. 2b.

In the animal bioassay for chemical EX, there are no individual differences in the delivery process or the intracellular treatment of the delivered dose. Hence, the individual

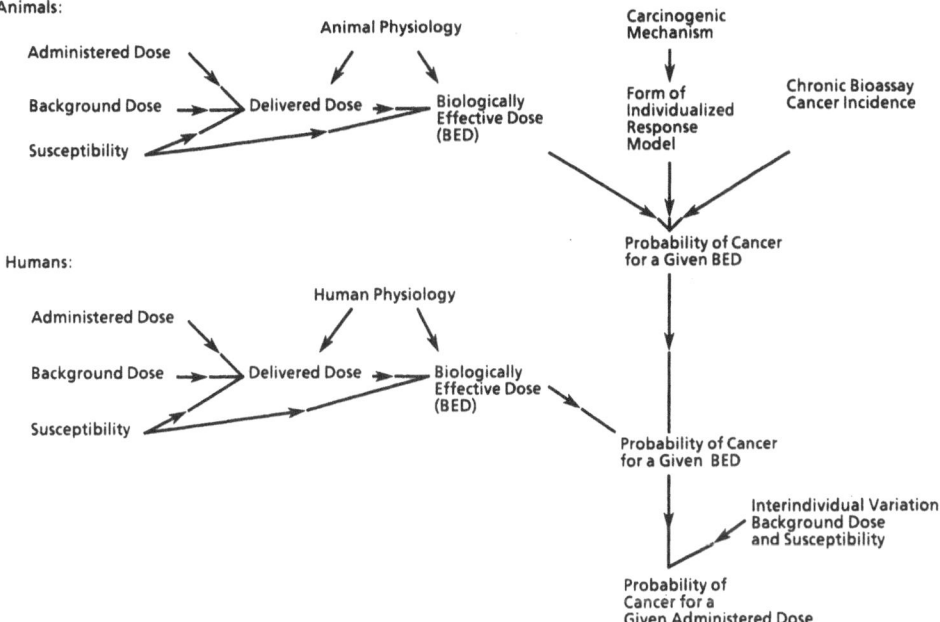

Fig. 1. Combining Physiology, Carcinogenic Mechanism, and Interindividual Variation in Cancer Dose Response Extrapolations.

susceptibility is the same for every individual. Furthermore, there is no intracellular interference with the delivered dose or the cancer-related activity it generates, therefore, the individual susceptibility frontier is zero (see Sielken (1987) for additional discussion on the role of susceptibility). With no impact due to individual susceptibility, the biologically effective dose equals the delivered dose for chemical *EX*.

The Carcinogenic Mechanism and the Form of the Individualized Response Model

The Individualized Response Model can describe the relationship between cancer occurrence and the biologically effective dose in the form of a quantal dose response model (probit, logit, Weibull, multihit, or multistage models) or a time-to-response model (Hartley-Sielken, multistage-Weibull, or Weibull-Weibull model) or an extension of the two-stage growth model introduced by Moolgavkar *et al.* (1979 and 1981).

The two-stage growth model is the most biologically reflective form of the Individualized Response Model. As illustrated in Fig. 3, the carcinogenic process is represented in the two-stage growth model in terms of normal cells, intermediate cells, and malignant cells. The proliferation of these cells and the transition rates from normal to intermediate to malignant cells may be age-dependent and/or dependent on the biologically effective dose. The model components and their potential dependencies are depicted in Fig. 3 and indicated in mathematical terms as follows:

$$P[t; \text{BED}(d, d_0, s)] = 1 - \exp\{- \int_0^{t-\tau} \lambda_2[u, \text{BED}(u)] \int_0^u \lambda_1[v, \text{BED}(v)]$$

$$\times N_0[v, \text{BED}(v)] \times \exp\{ \int_v^u \{\text{BR}[w, \text{BED}(w)] - \text{DR}[w, \text{BED}(w)]\} dw \} dv du\}$$

45

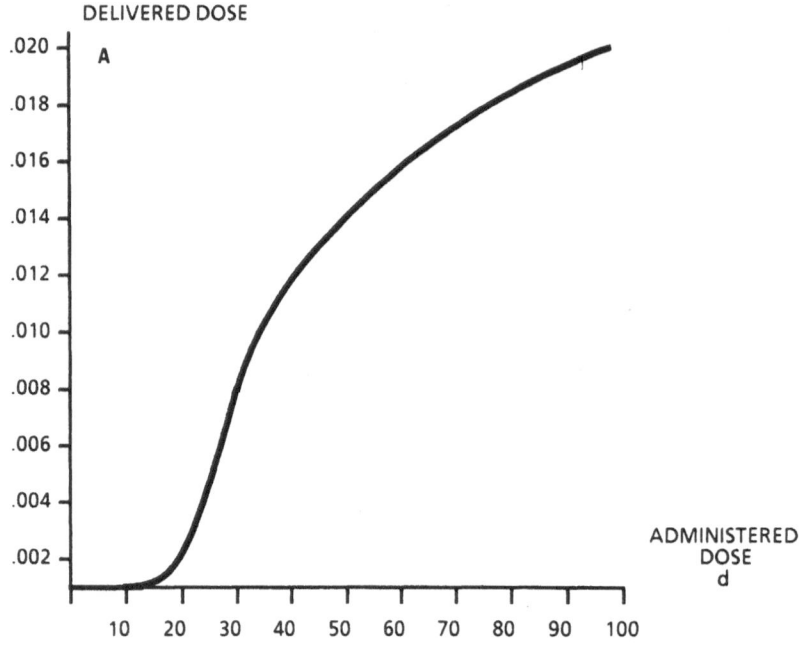

Fig. 2a. The Physiological Relationship Between Delivered Dose and Administered Dose.

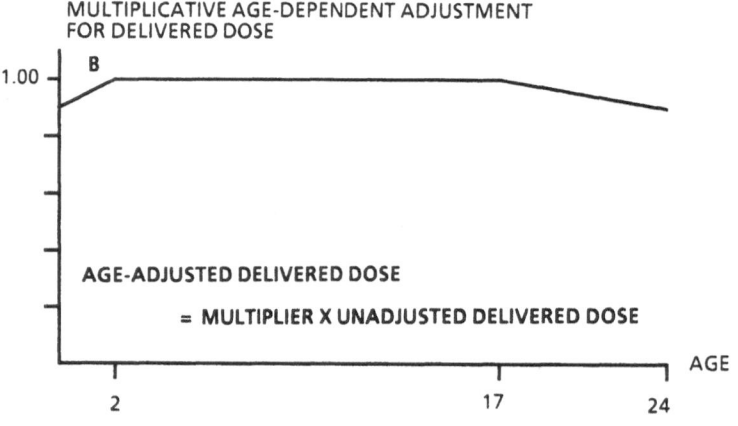

Fig. 2b. The Age-Dependent Multiplier of the Delivered Dose.

BIOLOGICALLY BASED MODELING THROUGH TIME

NUMBER of NORMAL STEM CELLS

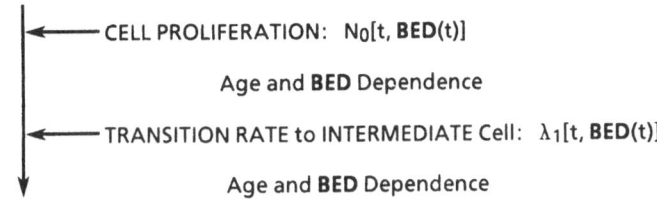

NUMBER of INTERMEDIATE CELLS

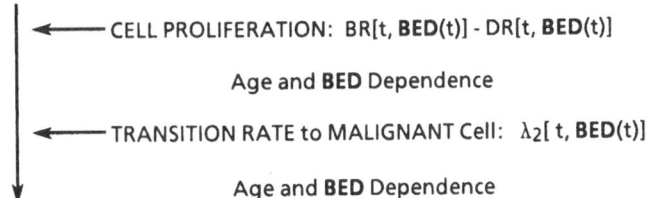

NUMBER OF MALIGNANT CELLS ARISING PER UNIT TIME

$P[t; \mathbf{BED}(d, d_0, s)]$

$\mathbf{BED}$ = BIOLOGICALLY EFFECTIVE DOSE

$\mathbf{d}$ = ADMINISTERED Dose
$\mathbf{d_0}$ = BACKGROUND Dose
$\mathbf{s}$ = SUSCEPTIBILITY

ALL Possibly TIME DEPENDENT

Fig. 3. Overview and the Components (N_0, λ_1, BR, DR, and λ_2) of the Two-Stage Growth Model.

where $P[t; \text{BED}(d, d_0, s)]$ is the probability of the specified carcinogenic response occurring by time t when the biologically effective dose (BED) corresponds to the possibly age-dependent values of the administered dose d, the background dose d_0, and the susceptibility s and where:

$N_0[v, \text{BED}(v)]$ = number of normal stem cells at time v,

$\lambda_1[v, \text{BED}(v)]$ = rate per cell per unit time that stem cells are transformed at time v into intermediate cells,

$\text{BR}[w, \text{BED}(w)]$ = birth rate (replication rate) of intermediate cells at time w,

$\text{DR}[w, \text{BED}(w)]$ = death rate (including terminal differentiation) of intermediate cells at time w,

$\lambda_2[u, \text{BED}(u)]$ = rate per cell per unit time that intermediate cells are transformed at time u into malignant cells,

τ = lag time between formation of a malignant cell and carcinogenic response.

The suspected carcinogenic mechanism implies the particular model component that is dependent on the biologically effective dose. Thorslund *et al.* (1987) suggest the following correspondence:

   i.   COCARCINOGENS, which induce regenerative hyperplasia as the normal consequence of a tissue's attempt to repair toxic damage, affect N_0.

   ii.   INITIATORS, which induce mutation, oncogene activation, or chromosomal translocation, affect λ_1.

   iii.   PROMOTERS, which increase the number of transformed preneoplastic cells through clonal expansion, affect BR–DR.

   iv.   COMPLETERS, which increase the rate of transformation from the penultimate stage to a malignant cell, affect λ_2.

   v.   INHIBITORS, which reduce the rate of transformation to malignancy by different mechanisms, cause decreases in N_0, λ_1, BR–DR, or λ_2.

Chemical *EX* impacts the multistage carcinogenic process by increasing the number of transformed preneoplastic cells (intermediate cells) through clonal expansion. The rate of this expansion is dependent on the biologically effective dose. Hence, the two-stage growth model component—BR–DR, i.e., the birth rate of intermediate cells minus their death rate—is dependent on the biologically effective dose.

For chemical *EX* the other two-stage growth model components are independent of biologically effective dose and only age-dependent. The age-dependence of the number (N_0) of normal cells, the transition rate (λ_1) from a normal cell to an intermediate cell, and the transition rate (λ_2) from an intermediate cell to a malignant cell are shown in Fig. 4. The magnitudes of N_0, λ_1, and λ_2 apart from their age-dependence are unknown constants, say N_0^*, λ_1^*, λ_2^*. These constants appear in the two-stage growth model as a single unknown constant $\lambda_2^* \times \lambda_1^* \times N_0^*$.

For chemical *EX* the intermediate cell proliferation rate is a linear function of age plus a linear function of the biologically effective dose which diminishes linearly with age

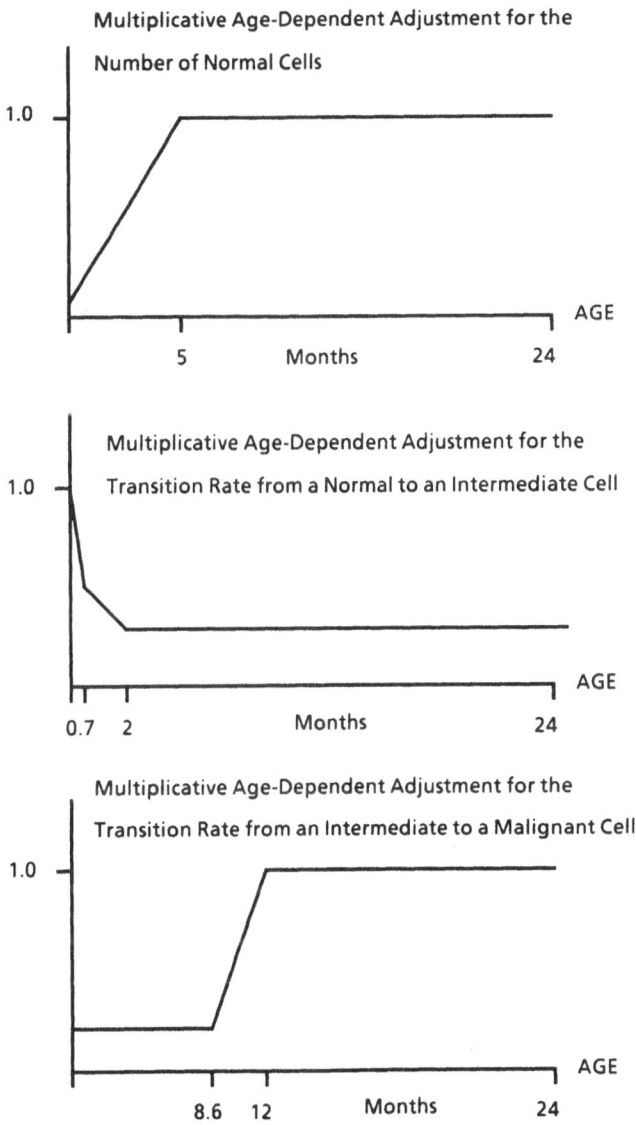

Fig. 4. The Age Dependency of the Non-Dose-Dependent Components of the Two-Stage Growth Model for Chemical EX.

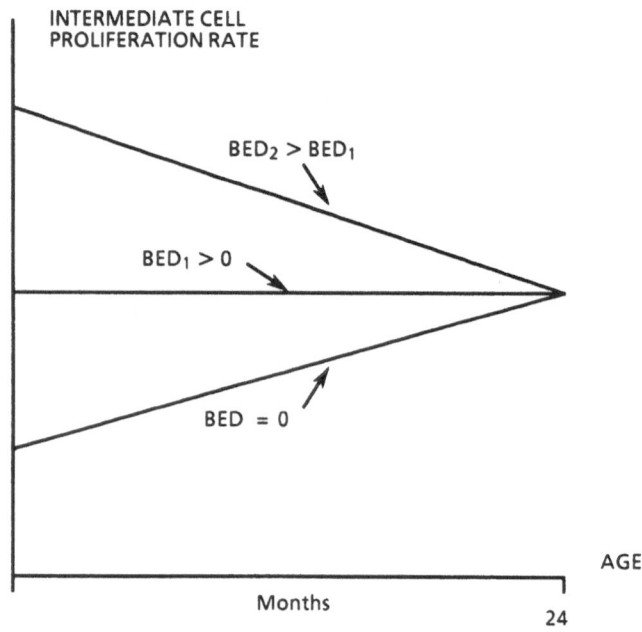

Fig. 5. Chemical EX's Promotion of Intermediate Cell Proliferation.

and disappears by the end of the animal's lifespan (say, 24 months). Mathematically, chemical *EX*'s promotion of the intermediate cell proliferation at age t is

$$\beta_0 + \beta_1 \times t + \beta_2 \times \text{BED} \times (24-t)$$

where β_0, β_1, and β_2 are unknown non-negative constants. Figure 5 illustrates the impact of the biologically effective dose on intermediate cell proliferation rate. The intermediate cell proliferation is an exponential function of this rate.

Estimating the Unknowns in the Individualized Response Model from the Cancer Incidence in the Animal Chronic Bioassay

The cancer incidence data in the animal chronic bioassay is used to estimate the four unknowns—β_0, β_1, β_2 and $[\lambda_2^* \times \lambda_1^* \times N_0^*]$—in the two-stage growth model for chemical *EX*, namely

$$P(t; \text{BED}) = 1 - \exp\left\{-[\lambda_2^* \times \lambda_1^* \times N_0^*] \times \int_0^t f_{\lambda_2}(u) \int_o^u f_{N_0}(v) \times f_{\lambda_1}(v)\right.$$

$$\left. \times \exp\left\{\int_v^u [B_0 + B_1 \times w + B_2 \times \text{BED}(w) \times (24 - w)]dw\right\}dvdu\right\}$$

where

$N_0^* \times f_{N_0}(v)$ = number of normal stem cells at time v,

$\lambda_1^* \times f_{\lambda_1}(v)$ = rate per cell per unit time that stem cells are transformed at time v into intermediate cells,

$[B_0+B_1 \times w+B_2 \times \text{BED}(w) \times (24-w)]$ = birth rate minus death rate for intermediate cells at time w,

$\lambda_2^* \times f\lambda_2(u)$ = rate per cell per unit time that intermediate cells are transformed at at time u into malignant cells.

The maximum likelihood estimates of these four unknown constants are identified when the specified form of the Individualized Response Model is fit to the experimental data on the frequency and time of occurrence of the specified carcinogenic response.

Figure 1 includes the merger of the biologically effective dose, carcinogenic mechanism, and chronic bioassay cancer incidence data to form an estimated Individualized Response Model for the probability of cancer for a given biologically effective dose.

The fitted model reflects the carcinogenic mechanism and estimates the probability of the specified carcinogenic response as a function of the biologically effective dose. It is this functional relationship that is combined with human biologically effective dose values to yield estimates of human cancer risks.

The determination of human BED values is discussed next.

HUMAN STUDIES

Combining the relationship between cancer probability and biologically effective dose with human values for the biologically effective dose provides a risk characterization for humans.

Determining the Biologically Effective Dose for Humans

Figure 1 indicates the role of the determination of the biologically effective dose for humans.

The biologically effective dose in humans exposed to an administered dose may be different than the biologically effective dose in experimental animals exposed to the same administered dose. The difference may be due to differences in the physiologically-based delivery processes, the background doses, the susceptibilities, and the way in which an individual's susceptibility acts on the delivered dose.

Human administered doses are not usually constant throughout the lifespan. For example, the chemical *EX*'s administered dose is the total amount of daily intake due to *EX* being in surface soil. Because soil ingestion is expected to be very much greater in early childhood and because this exposure route is the major exposure route, the administered dose is age-dependent. Age makes a proportional change in the administered dose. The multiplicative age-dependent adjustment to administered dose is shown in Fig. 6a. For example, the adjustments are approximately 1.0 at age 3 and 0.03 at age 30. Thus, whatever the administered dose is as a result of the surface soil concentration of *EX*, the administered dose at age 30 is only 3% of the administered dose at age 3.

The human background dose is also seldom constant over time. Changes in background dose are expected because of changing environmental levels, dietary changes, recreational activity changes, etc. The background dose for chemical *EX* decreases proportionally with approximately a ten-year half-life as shown in Fig. 6b.

For chemical *EX* the delivery process acts on the sum of the administered dose and the background dose because the delivery processes are the same for this chemical. The

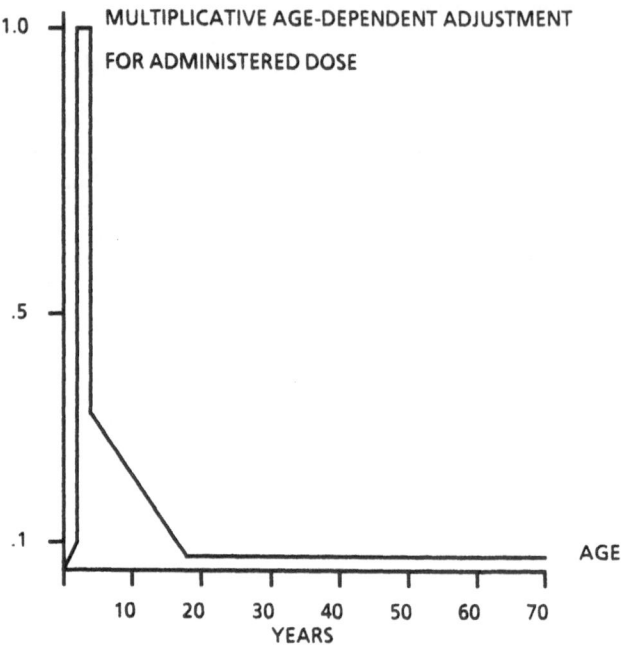

Fig. 6a. The Age Dependency of the Human Administered Dose.

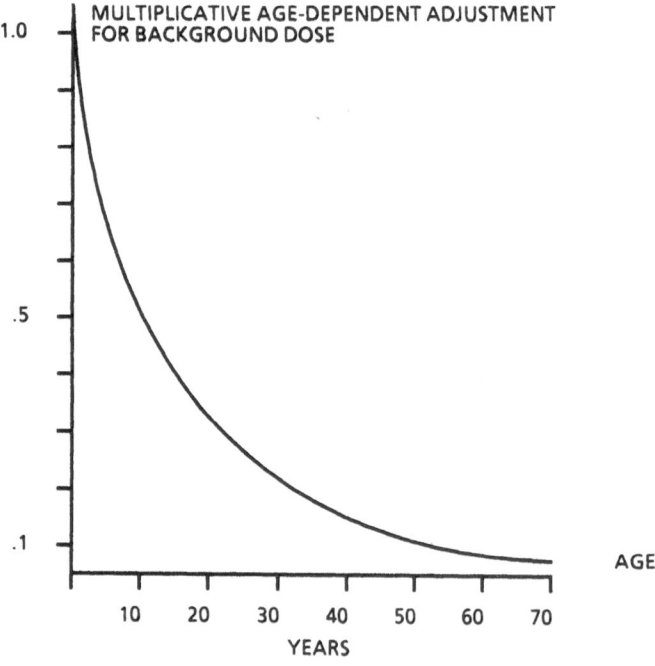

Fig. 6b. The Age Dependency of the Human Background Dose.

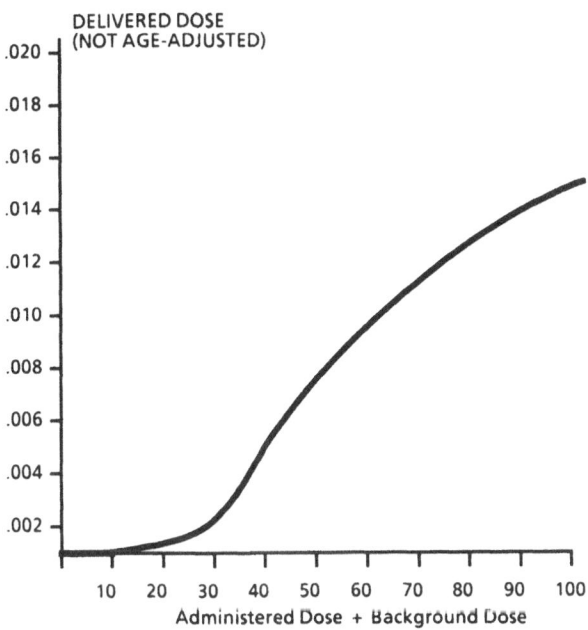

Fig. 7a. The Physiological Relationship Between Delivered Dose to the Target Site and the Administered and Background Dose.

non-age-adjusted delivered dose has approximately a Michaelis-Menten pharmacokinetic relationship with the sum of the administered and background doses (Fig. 7a). Age changes the properties of the delivery process, making it proportionally less effective before age 10 and after age 50. The specific multiplicative age-dependent adjustment to the non-age-adjusted delivered dose is shown in Fig. 7b. The age-adjusted delivered dose is the product of the corresponding numerical values in Figs. 7a and 7b. An example of the age-adjusted delivered dose for a non-age-adjusted administered dose of 100 and a non-age-adjusted background dose of 2 is shown in Fig. 7c.

The individual's susceptibility for chemical *EX* is a susceptibility frontier. The individual's delivered dose corresponding to the transition from the dose region of lower carcinogenic effectiveness to the dose region where the mechanisms resisting or suppressing carcinogenesis are overwhelmed is the individual's susceptibility frontier. The biologically effective dose increases slowly and approximately linearly for very small delivered doses, increases somewhat faster as the delivered dose increases to the susceptibility frontier, and increases still faster and linearly for delivered doses exceeding the susceptibility frontier. The relationship for chemical *EX* between biologically effective dose and the individual's susceptibility and delivered dose is indicated in Fig. 8a for an individual with a non-age-adjusted susceptibility frontier equal to 0.008.

The individual susceptibility is age-dependent. For chemical *EX* the susceptibility frontier is proportionally adjusted according to age as shown in Fig. 8b. For example, an individual's susceptibility frontier has decreased by nearly 50% at age 35 and continues to decrease thereafter, making an older individual more susceptible to a delivered dose than a younger individual experiencing the same delivered dose.

Individual susceptibilities are not assumed to be the same for every individual in the population at risk. For chemical *EX* the individual susceptibility frontier is normally distributed in the population with a mean of 0.001 and a standard deviation of 0.0002. Figure 9a shows this distribution of individual susceptibilities among the population.

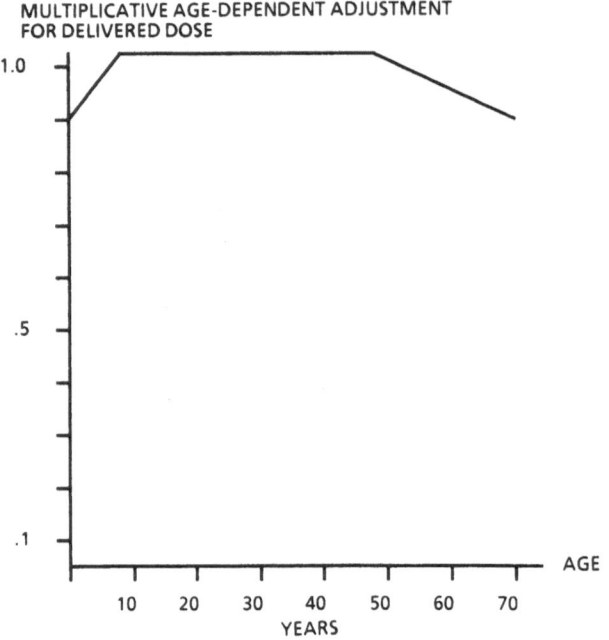

Fig. 7b. The Age Dependency of the Human Delivered Dose.

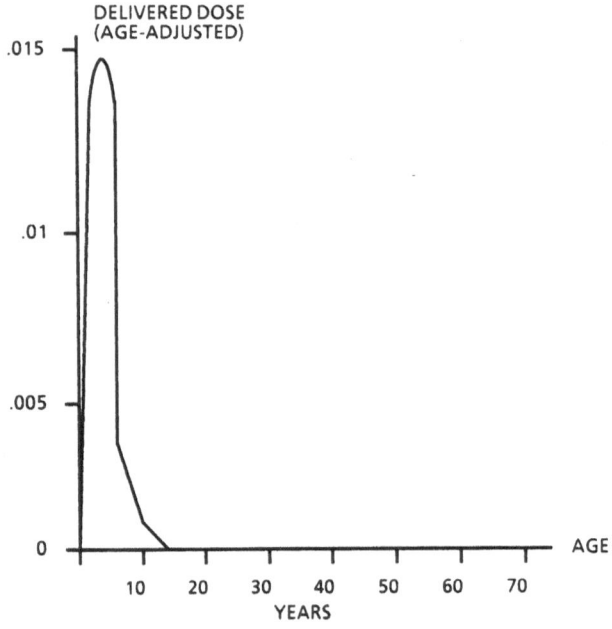

Fig. 7c. The Human Age-Dependent Delivered Dose for a Non-Age-Adjusted Administered Dose of 100 and a Non-Age-Adjusted Background Dose of 2.

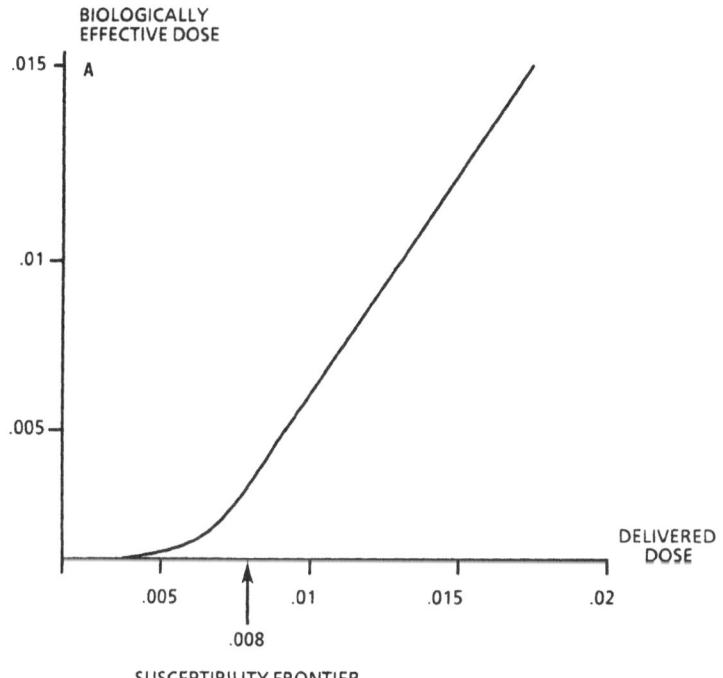

Fig. 8a. The Relationship Between Biologically Effective Dose and the Individual's Susceptibility and Delivered Dose.

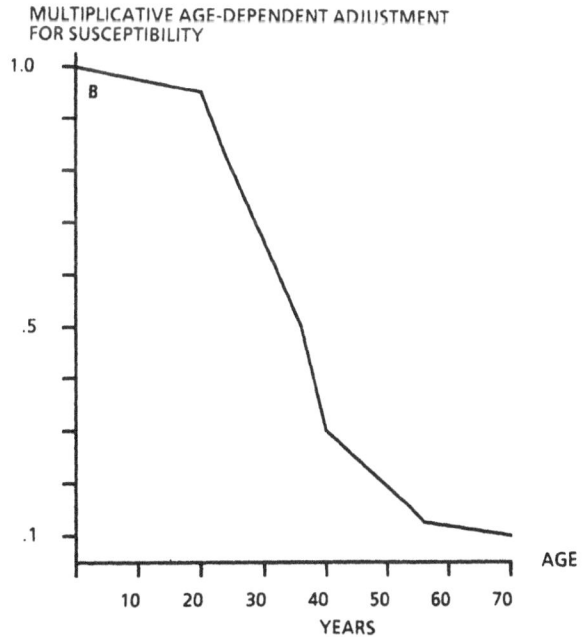

Fig. 8b. The Age-Dependent Adjustment to the Human Individual Susceptibility.

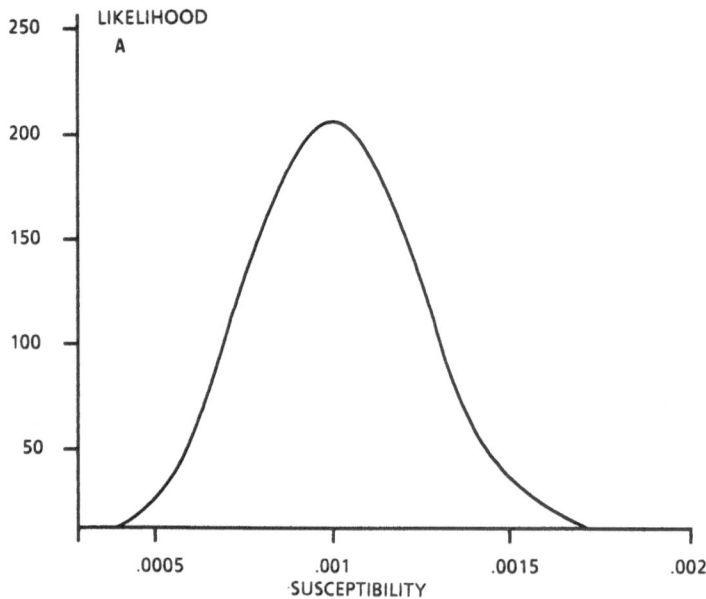

Fig. 9a. The Distribution of Individual Susceptibilities in the Human Population at Risk: A Normal Distribution with Mean 0.001 and Standard Deviation 0.0002.

An individual's biologically effective dose at a particular age is a function of not only administered dose and susceptibility but also background dose. The background dose varies within the population according to the lognormal distribution shown in Fig. 9b. In the population, 50% of the individuals have background doses less than 1.0, and 95% have background doses less than 10.0.

Combining the Human Biologically Effective Dose with the Estimated Individualized Response Model to Estimate Human Risks

The relationship between the probability of a specified carcinogenic response and an age-dependent biologically effective dose is the relationship estimated from the experimental data. For chemical *EX* a proportional change in the age dependence of the probability was made to reflect the approximate 70-year human lifespan versus the 24-month lifespan of the experimental animals. The resulting lifespan-adjusted probability was combined with human biologically effective doses to estimate human risks, as indicated in Fig. 1.

There are several possible risk characterizations—for example, added probabilities as functions of administered dose and time, mean response free periods, virtually safe doses, and mean free doses. These risk characteristics not only can be estimated, as well as bounded for individual values of susceptibility and background dose, but also can be averaged over the values of the individual susceptibilities and background doses in the population, as indicated in Fig. 10. The latter averages estimate the expected proportion of the population that will develop the specified carcinogenic response.

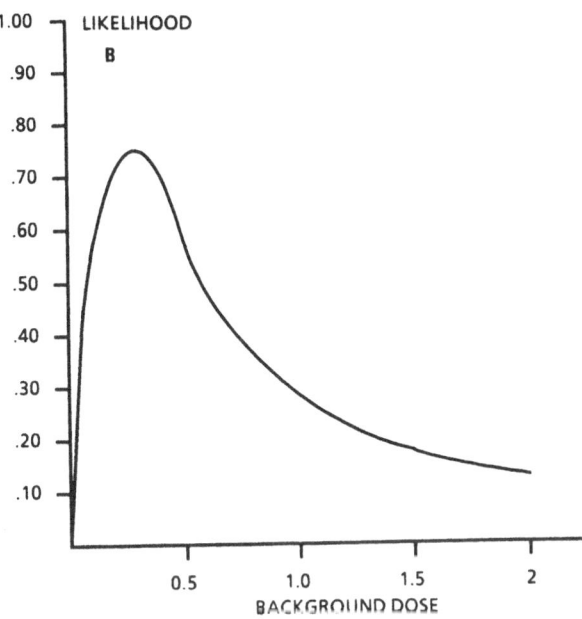

Fig. 9b. The Distribution of Individual Background Doses in the Human Population at Risk: A Lognormal Distribution with Median 1.0 and 95th Percentile 10.0.

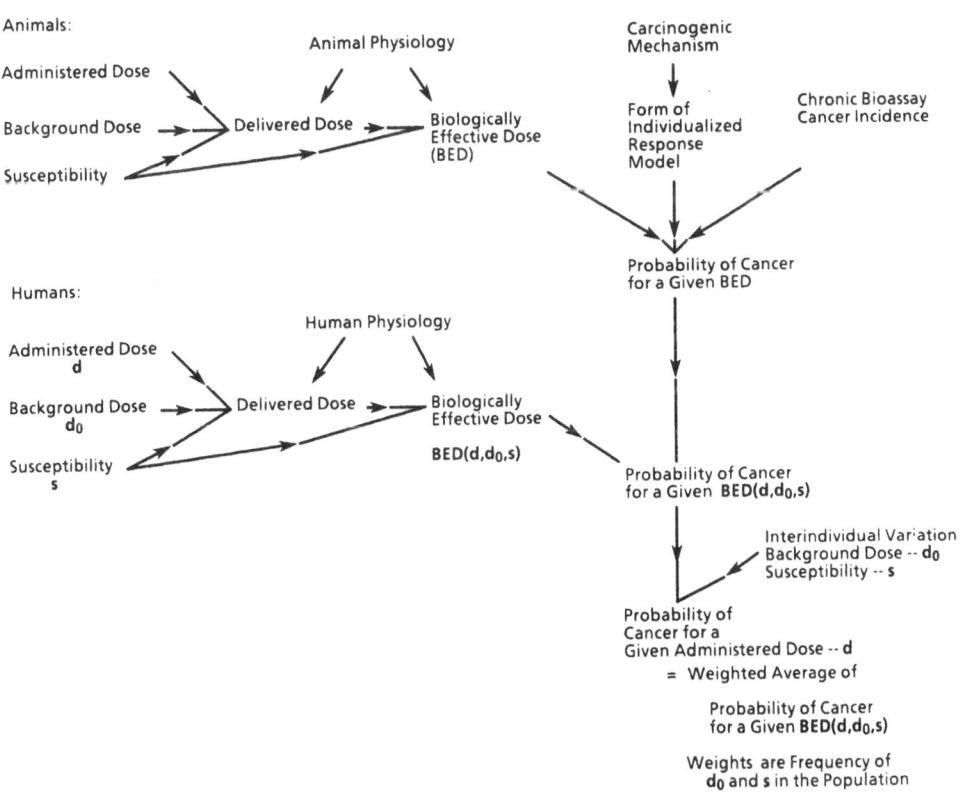

Fig. 10. Reflecting Interindividual Variation in Cancer Dose-Response Extrapolations.

REMARKS

The primary objective of the preceding discussion is to illustrate how research results on physiological processes, carcinogenic mechanisms, and interindividual (intraspecies) variation in background doses and susceptibilities in both animals and humans can be combined in quantitative cancer risk assessment. The example for chemical *EX* illustrates how the different scientific findings can be merged.

The Individualized Response Model and the associated GEN.T software can accommodate many additional functional forms for the components of the quantitative risk assessment. The example for chemical *EX* indicated only one form for each of its components.

The numbers for chemical *EX* are not important. The capability to incorporate the multitude of scientific findings into a risk assessment is important. With the machinery and the opportunity to incorporate more science into quantitative risk assessment, more scientific research will be done and more informed risk management decisions made.

REFERENCES

1. S. H. Moolgavkar and D. J. Venzon, *Math Biosci.* **47**:55-77 (1979).
2. S. H. Moolgavkar and A. G. Knudson, *J. Nat. Cancer Inst.* **66**:1037-1052 (1981).
3. R. L. Sielken, Jr., *Environ. Sci. Technol.* **21**:1033-1039 (1987).
4. T. W. Thorslund, C. C. Brown, and G. Charnley, *Risk Analysis* **7**:109-119 (1987).

On Basing Extrapolation of Risk for a Chemical Carcinogen on Total Dose (Constant Dose Rate × Continuous Exposure Duration): Implications and an Extension[a]

Kenneth G. Brown
Chapel Hill, NC

Robert P. Beliles
U.S. Environmental Protection Agency
Washington, DC

ABSTRACT

It is often assumed that the probability of observing a specific cancer-related response by time t under constant dose rate d depends on d and t only through the product dt, the "total dose." Loosely attributable to Haber from another context, it is shown here that for several classes of time/dose response curves, Haber's rule implies severe model restrictions for mathematical consistency. A generalized-Haber's rule (G-H rule) is suggested instead that is calculated from the specific time/dose-response model used to describe the data. This approach requires only a change in dose and time metameters, from d to $D(d)$ and from t to $T(t)$, respectively. Determination of the functions D and T is demonstrated for several models, for some corresponding expressions of extra risk, and for assumptions of additive and independent background mechanisms.

The G-H rule is of the form $D(d)T(t) = C$, for C a constant ($C > 0$), and defines the "tradeoff" between changes in d and t that give the same change in risk (the probability the event will occur by time t), or in extra risk, which is often the measure of interest.

KEYWORDS: Haber's rule, statistical methods, carcinogen risk assessment, background mechanisms, time/dose-response

INTRODUCTION

The U.S. Environmental Protection Agency's Guidelines for Carcinogen Risk Assessment appeared in 1986 in the *Federal Register* (FR).[1] As noted in that document (p. 33993), the format of the Guidelines is similar to that proposed by the National

---

a. This work was initiated under U.S. EPA Contract 68-01-6826. The views expressed are those of the authors. Agency policy should not be inferred. The authors gratefully acknowledge the support provided by Environmental Monitoring and Services, Inc., Chapel Hill, NC, toward completion of this work.

Resource Council (NRC),[2] and the recent report by the Office of Science and Technology Policy (OSTP)[3] forms an important basis for the content. The OSTP report is the product of the U.S. Interagency Staff Group on Carcinogens, the membership of which is given in the OSTP report, and in the group's report published elsewhere.[4] The fact that the report contains 752 references lends some perspective to the magnitude of the literature considered relevant to the topic.

The Guidelines state, and the NRC and OSTP documents reaffirm, that given current scientific knowledge, there are many questions that are currently unanswerable but that inferences must be made to ensure that progress continues in the assessment process. Aside from continued research to improve understanding of the mechanisms of cancer, uncertainty in risk assessment can also be reduced by identification and understanding of the implications of assumptions in the methodology currently in use. Sources of uncertainty in the quantitative aspects of risk assessment may arise because of failure to adequately represent important biological characteristics, or because of the assumptions intrinsic to the statistical methods used. More attention seems to have been focused in the literature on the former source of uncertainty than the latter, although both are important.

The Guidelines recommend use of the cumulative dose received over a lifetime, expressed as average daily exposure prorated over a lifetime, as a measure of exposure to a carcinogen (unless there is evidence to the contrary in a particular case [Ref. 1, p. 33998]). This time-weighted-average rule has been critically reviewed by Atherley[5] from the standpoint of historical evidence. Its basis is the presumption that toxicity, or probability of a toxic response, depends on total dose. Known as Haber's rule,[6] this assumption is the topic of this report. Dose rate as used in this report, denoted by d, will refer to administered dose, or target dose when the latter is available.

In the sections to follow we assume the constant-dose protocol described in Ref. 1 above.

MODEL-DEPENDENT GENERALIZATIONS OF HABER'S RULE

Let T^* be the time to occurrence of a specific response, such as occurrence of a clinically detectable bladder neoplasm, and define $P(t; d) = Pr\{T^* \leq t; d\}$ to be the absolutely continuous cumulative distribution function of T^* at dose d. If $\lambda(t; d)$ is the hazard function at time t, then $\Lambda(t; d) = \int_0^t \lambda(u; d)du$ gives the cumulative hazard function and

$$P(t; d) = 1 - \exp(-\Lambda(t; d)) \ . \tag{1}$$

Haber's rule,[6] $dt = C$, where $C > 0$ is a constant, is consistent with Eq. (1) *if and only if* $\Lambda(t; d)$ *depends on t and d through their product dt, for $d, t > 0$.* In that case, d and t are interchangeable without altering the value of Λ or P. In other words, a given proportional increase or decrease in either d or t has the same effect on the probability of observing a tumor, $P(t; d)$. Unfortunately, this property is consistent with only very limited time/dose-response models used for $P(t; d)$.

To illustrate, consider the multistage-Weibull distribution described by

$$\Lambda(t; d) = \sum_{i=0}^{k} c_i d^i t^k, c_i \geq 0, i = 0, ..., k \ . \tag{2}$$

Haber's rule implies that we need $c_i = 0$, $i = 0, ..., k - 1$, leaving $\Lambda(t; d) = c_k d^k t^k$, $d, t > 0$. It is usually assumed in applications (e.g., the linearized multistage model) that $c_i > 0$, which would imply further that $k = 1$ and give $\Lambda(t; d) = c_1 dt$. The resultant model is simply the "one-hit" or "one-stage multistage" model with zero background. If background is nonzero ($c_0 > 0$), then $\Lambda(t; d) = (c_0 + c_1 d)t$. A simple extension of Haber's rule suffices in this case, given by $D \cdot T = C$, where $D = c_0 + c_1 d$ and $T = t$.

In general, let D be a transformation of d but not t, and let T be a transformation of t but not d. Either D or T may depend on parameters as well. To generalize Haber's rule to the full multistage-Weibull, define $D(d) = \sum_{i=0}^{k} c_i d^i$, $T(t) = t^k$, so that Eq. (2) becomes $\Lambda(t; d) = D(d)T(t)$. A generalized Haber's rule is then of the form

$$DT = C > 0, \qquad \text{for } D > 0, T > 0 ,$$

and a proportionate increase or decrease in either D or T changes the cumulative hazard and hence the probability of response by the same quantity. We have simply changed the dose and time metameters to D and T respectively. Since the dose-rate d and time t are generally of interest, the numerical calculation of solving $D(d)$ or $D^{-1}(d)$, or $T(t)$ or $T^{-1}(t)$ (when defined) has been added. Parameters other than d, such as c_i in $D(d)$ in the example, need to be estimated. Since most cancers have a non-negligible induction period, time t should ideally be measured from the start of exposure plus the induction period. This makes the common assumption that induction period does not depend on dose rate.

To illustrate further, consider a hypothetical example of lifetime exposure of rats in several dose groups for $t = 24$ months. Suppose we use the multistage-Weibull distribution and estimate $c_0 = 0.02$, $c_1 = 0.1$, $c_2 = 0.06$ and $k=4$. Then $\Lambda(t; d) = (0.02 + 0.1d + 0.06d^2)(24)^4$. What is the value of the TD_{50}, the dose at which 50% of the animals are expected to exhibit a specific cancer-related response? We want $0.5 = 1 - \exp(-DT)$ or $DT = \ln 2$. $D(d) = (0.02 + 0.1d + 0.06d^2)$ and $T(t) = T(24) = 24^4$. The value of TD_{50} is the solution d_0 to $D(d_0)(24)^4 = \ln 2$. In this case d_0 can be solved by the binomial formula. In more complex problems, if an algorithm for the roots of a polynomial is not available, it may be quicker to iteratively search for a solution.

If the dose rate is double the TD_{50}, i.e., $d_1 = 2d_0 = 2TD_{50}$, by what time would we expect half the animals to exhibit the response of interest? We need to solve for t_1 in $D(2d_0)T(t_1) = D(d_0)T(24)$. Since $T(t_1) = t_1^4$, we have

$$t_1 = \left(\frac{D(d_0)T(24)}{D(2d_0)} \right)^{\frac{1}{4}} = 24 \left(\frac{0.02 + 0.1d_0 + 0.06d_0^2}{0.02 + 0.2d_0 + 0.24d_0^2} \right)^{\frac{1}{4}} ,$$

where d_0 is the value of the TD_{50} from above.

As a last example, let $P(24;5)$ be the probability of response by 24 months with dose rate 5. What would be the reduction in the response probability for only 18 months of exposure? Let $P(24;5) = 1 - \exp(-D(5)T(24))$, where D and T are defined by the model and the parameters of $D(5)$ and $T(24)$ are estimated from the data. We want the value of $P(24;5) - P(18;5)$. Let C_0 be the true (but unknown) value of $D(5)T(24)$. Using $k = 4$ as an example gives $T(t) = t^4$ and $T(18) = 18^4$. Then $D(5)T(18) = C_0 T(18)/T(24) = C_0(18/24)^4 = C_0(.75)^4$. The reduction is

$$1 - \exp(-C_0) - [1 - \exp(-(.75)^4 C_0)] = \exp(-(.75)^4 C_0) - \exp(-C_0) .$$

Since C_0 is unknown, we need to replace it by the value of $D(5)T(24)$ estimated from the data.

GENERALIZED-HABER'S RULE FOR SOME DIFFERENT CLASSES OF RESPONSE MODELS AND BACKGROUND MECHANISMS

In the previous section the multistage-Weibull distribution, assuming that treatment-induced response is independent of the mechanism of spontaneous occurrences (independent background), was used for illustration. A generalized-Haber (G-H) rule may differ by both the class of model and type of background mechanism assumed. In some cases the G-H rule corresponding to extra risk, defined by $(P(t; d) - P(t; 0))/(1 - P(t; 0))$, is a simpler expression, or extra risk is otherwise of more interest than $P(t; d)$.

Type of Background Mechanism

It is usually assumed that treatment-induced tumors occur either by a process that is independent of spontaneous occurrences, or that the dose d behaves mechanistically as if added to a background dose δ (fixed) already present. Let $P^*(t; d)$ be the probability that dose rate $d(d > 0)$ induces a response by time t, and define $P^*(t; 0) = 0$. If $P(t; d)$ is the probability of response at dose d by time t, due to either a treatment effect or a spontaneous occurrence, then for independent background

$$P(t; d) = P(t; 0) + (1 - P(t; 0))P^*(t; d) \quad d \geq 0 .$$

The risk of a spontaneous tumor by time t is $P(t; 0)$. The term $1 - P(t; 0)$ is usually referred to as Abbott's correction. The corresponding expression for additive background is

$$P(t; d) = P^*(t; d + \delta) \quad d \geq 0 .$$

The spontaneous tumor rate is $P(t; 0) = P^*(t; \delta)$.

To incorporate additive or independent background, first write the model for response *induced* by dose d, $P^*(t; d)$, $d > 0$. If the G-H rule corresponding to this model is $D(d)T(t) = C$, for some functions D and T as described previously, then the G-H rule for additive background is simply $D(d + \delta)T(t) = C$. The model $P(t; d)$, $d \geq 0$, is formed by replacing d in $P^*(t; d)$ by $d + \delta$.

Independent background may require a complicated expression for the relationship between d and t that does not completely reduce to the form $D(d)T(t) = C$. The derived expression will still be referred to as a G-H rule for convenience. Define $P^*(t; d)$ as above with cumulative hazard $\Lambda^*(t; d)$. To incorporate independent background, the cumulative hazard becomes $\Lambda^*(t; d) + \Lambda_I(t)$, where $\Lambda_I(t)$ is the cumulative hazard at $d = 0$.[7] With independent background, extra risk equals $P^*(t; d)$, $d > 0$, so the G-H rule for the model $P^*(t; d)$ and the extra risk for $P(t; d)$ are equal. The reader is referred to Ref. 7 for a more complete description of the following models to be discussed.

Log-Linear Model

Let T have a log-linear distribution of the form

$$\log T = \alpha + \beta \log d + \sigma W \quad \text{for } d > 0 , \tag{3.1}$$

where W is any error distribution whose support is the real line. For example, if $W \sim N(0,1)$, then $\Pr\{W \leq w\} = \Phi(w)$ and T has a log-normal distribution with mean $\alpha + \beta \log d$ and geometric standard deviation σ. If W has an extreme value distribution, then $\Pr\{W \leq w\} = 1 - \exp(-e^W)$ and T has a Weibull distribution.

The cumulative hazard rate function for the log-linear model is

$$\Lambda(t;d) = \Lambda_0(t \exp[-(\alpha + \beta \log d)])$$
$$= \Lambda_0(T(t)D(d)) \ , \tag{3.2}$$

where Λ_0 is the cumulative hazard of $\exp(\sigma W)$, $T(t) = t$, and $D(d) = \exp[-(\alpha + \beta \log d)]$, $d > 0$.

The G-H rule is $D(d)T(t) = C$, or equivalently, $\log D(d) + \log T(t) = C^*$ (C^* is used here because C is an arbitrary constant > 0, but $C^* = \log C < 0$ if $C < 1$). This gives $-\alpha - \beta \log d + \log t = C^*$, so that constant contours of the time/dose-response curve are over the line in the $(\log t, \log d)$ plane where $\log t$ is linear in $\log d$, with slope β. The simple Haber's rule, $dt = C$, implies $T(t) = t$ and $D(d) = e^{-\alpha}d$ from which it follows that $\beta = -1$. Hence the effect of the simple Haber's rule is to replace the parameter β in Eq. (3.1) by -1. For additive background, the G-H rule for $P(t;d)$ is $D(d+\delta)T(t) = C$, for D and T as given above. For extra risk under additive background, the more complicated expression $\Lambda_0[D(d+\delta)T(t)] - \Lambda_0[D(\delta)T(t)] = C$ is required, where Λ_0 is defined in Eq. (3.2).

For independent background, the background rate is defined by $P(T;0)$. Let Λ_I be the integrated hazard rate function for background response at time t, and

$$P(t;0) = 1 - \exp(-\Lambda_I(t)).$$

The G-H rule with independent background for $P(t;d)$ is $\Lambda_0(D(d)T(t)) + \Lambda_I(t) = C$, $d > 0$. The G-H rule for extra risk is just $D(d)T(t) = C$, $d > 0$.

Cox's Proportional Hazards Regression Model

In 1972, Cox[8] proposed a model for survival analysis of censored data that related failure time to regression variables or covariates. Prentice, Peterson and Marek[9] give specific applications to the analysis of dose-response experiments. The model specifies a regression relationship for the hazard rate at time t of the form

$$\lambda(t;d) = \lambda_0(t) \exp[\underline{z}(t;d)'\underline{\beta}] \ ,$$

where λ_0 is an unspecified baseline hazard function, $\underline{\beta}$ is a p-vector of regression parameters, and $\underline{z}(t;d) = (z_1(t;d), .., z_p(t;d))'$ is a vector of regression variables or covariates whose components may depend on both time t and dose d. As is often done, we will define the covariates so that $\underline{z}(t;d) = \underline{0}$ at $d = 0$, making $\lambda_0(t)$ correspond to a background rate of response. The hazards at different doses are proportional if $\underline{z}(t;d)$ depends on dose, but not time, which will be denoted by writing $\underline{z}$ as $\underline{z}(d)$. In that case

$$\Lambda(t;d) = \Lambda_0(t) \exp[\underline{z}'(d)\underline{\beta}] \ .$$

The G-H rule for $P(t;d)$ can be written for this case as $D(d)T(t) = C$, where $D(d) = \exp[\underline{z}'(d)\underline{\beta}]$ and $T(t) = \Lambda_0(t)$. For extra risk, the corresponding expression is $[D(d) - 1]T(t) = C$. The Cox model is very versatile in a more general form.

Multievent Models

A multievent model presumes that the response of interest results from the occurrence of a number of biological events. If there are k events with independent occurrence times, $T_1, .., T_k$, then the response occurs at $T = \max(T_1, .., T_k)$ if the events occur independently. Let $\Lambda_i(t; d)$ denote the cumulative hazard for T_i, $i = 1, .., k$ at dose d. For independent events, with occurrence times $T_1, .., T_k$ exponentially distributed, then

$$P(t; d) = \prod_{i=1}^{k} [1 - \exp(-\Lambda_i(t; d))] ,$$

which is approximately $\prod_{i=1}^{k} \Lambda_i(t; d)$ for all $\Lambda_i(t; d)$ small. If we assume further that T_i is exponentially distributed with cumulative hazard $\Lambda_i(t; d) = \psi_i(d)t$, then

$$P(t; d) \simeq C(d)t^k ,$$

where $C(d) = \prod_{i=1}^{k} \psi_i(d)$.[7]

The Armitage-Doll multistage model of carcinogenesis[10] assumes that a cancerous lesion results from the occurrence of k mutations at certain gene loci in a single somatic cell. The rate of occurrence for the ith stage is often assumed to be linear in dose, in the manner suggested by Neyman and Scott.[11,12] For a target tissue of n cells acting independently, with exponential mutation times as described above, this gives

$$\Lambda(t; d) = nC(d)t^k = \prod_{i=1}^{k} (\alpha_i + \beta_i d)t^k ,$$

where $\alpha_i > 0$, $\beta_i \geq 0$, $i = 1, .., k$. The G-H rule for $P(t; d)$ small is $D(d)T(t) = C$, where $D(d) = \prod_{i=1}^{k} (\alpha_i - \beta_i d)$, $\alpha_i > 0$, $\beta_i \geq 0$, and $T(t) = t^k$. This model is only defined for a positive background rate, corresponding to cumulative hazard rate function $D(0)T(t) = \left(\prod_{i=1}^{k} \alpha_i \right) t^k$. Additive background is written into the model and hence for some region near $d = 0$ the response model is approximately linear for t fixed.[12] Unfortunately, the "low dose region" where linearity is a good approximation depends on the model parameter values and hence may vary between very small to the entire dose range.[12] To see that the model assumes additive background, note that $D(d + \delta)$ for $\delta > 0$ is of the same functional form as $D(d)$. The G-H rule for extra risk when $P(t; d)$ is small is $[D(d) - D(0)]T(t) = C$, with $D(d)$ and $T(t)$ as given above.

The multistage model of Armitage and Doll described above can be written with $\prod_{i=1}^{k} (\alpha_i + \beta_i d)$ replaced by a polynomial $\sum_{i=0}^{k} c_i d^i$, subject to certain nonlinear constraints on the coefficients c_i corresponding to $\alpha_i > 0$ and $\beta_i \geq 0$.[7,13] The simpler constraints $c_i \geq 0$ on the polynomial $\sum_{i=0}^{k} c_i d^i$ were proposed by Guess and Crump[14,15] to make estimation tractable by the method of maximum likelihood. Their approach is usually presented as the "linearized multistage model," which is found by fitting the model with the value of c_1, the linear coefficient of d, so large that any larger value would be rejected for a 0.05 size test; i.e., $p < 0.05$ would result. This procedure produces an upper bound on the multistage model that has maximum slope at the origin and is linear near the origin. The linearized multistage model is recommended as the default procedure in the U.S. EPA Guidelines (Ref. 1, p. 33997).

The Guess-Crump model formulation using $\sum_{i=0}^{k} c_i d^i$ is frequently presented as being slightly more general than the Armitage-Doll model because it describes a more general family of curves,[16,17] i.e., for any parameter values of the Armitage-Doll model, the model can be written for the same curve by appropriately defined values of c_i and k. The converse is not true. Any case violating the nonlinear constraints in Ref. 13 is a counterexample. More simply, any case where $c_j > 0$ and $c_i = 0$ for some $i < j$ is a simple counterexample. Consequently, some precaution is in order in appealing to the biological underpinnings attached to the Armitage-Doll model when the Guess-Crump model is used. A second precaution is that while the general argument for low dose linearity in Ref. 12 applies to the Armitage-Doll model with mild conditions, because the model is written for additive background, that general result does not apply to the Guess-Crump model because it is written for independent background. This distinction of the models is shown in Ref. 13. The independent background presumption of the Guess-Crump model follows directly from the definition of independent background above with $P(t; 0) = 1 - \exp(c_0 t^k)$. The Crump and Guess model with time as a component is the multistage-Weibull model used for illustration above (see Eq. 2). The G-H rule for $P(d; t)$ is $D(d)T(t) = C$, where $D(d) = \sum_{i=0}^{k} c_i d^i$ and $T(t) = t^k$. For extra risk, the rule is $[D(d) - D(0)]T(t) = C$.

The models treated in this section should serve to illustrate how the G-H rule is determined. Some other time/dose-response models that may be of interest are given in Refs. 7, 18, 19, and 20.

DISCUSSION

We have proposed a generalization to Haber's rule for constant dose over lifetime as an alternative that is consistent with a dose/time-response for the data. The generalization is simply to replace $dt = C$ by $D(d)T(t) = C$ for suitably defined, model-specific transformations of d and t. In the example models where expressions for D and T are given for probability of response and for extra risk, it is shown that the type of background assumed will generally affect the expression, and that in at least one case, the log-linear model with independent background, a somewhat more complicated expression involving D and T is required than $DT = C$. Since the principle is the same, however, we include such cases under the notion of a generalized Haber's rule for simplicity.

One might ask if there is not a generalization to Haber's rule that would also accommodate non-constant exposure patterns. In the scenario where there is a pre- or post-exposure period, however, additional factors become relevant, such as which stage(s) is affected by dose, and whether age at time of exposure alters susceptibility. If the agent of interest is a reactive metabolite and the metabolic rate depends on age, then age may be a significant factor. In a post-exposure period, the clearance rate of the delivered dose from the target site may be significant. That is, aside from whether dose primarily affects an early or late stage(s), the half-life of the delivered dose, which varies widely across chemicals, may make a difference if tumor rate increases with duration of delivered dose. For an arbitrary exposure pattern, especially of short exposure duration, the pharmacokinetics of absorption, distribution, metabolism, and elimination could be expected to play a greater role, as well as the detoxification and repair mechanisms. The factors affecting response require increasing knowledge of the mechanisms of toxicity and pharmacokinetics specific to the test compound, and possibly to the species and strain treated, as exposure patterns depart from the constant dose rate protocol. Of course, human exposure is most often not constant, and not for a full lifetime. Hence it is unlikely that any general formula will apply to extrapolation of risk from constant-lifetime exposure to arbitrary, probably even simple, time-dependent exposure patterns. This conclusion underscores the need to treat carcinogenic compounds on an individual basis, and the dependence of the time-dose relationship on an agent's pharmacokinetics and pharmacodynamics, and on the typical time-dependent pattern(s) of human exposure.

REFERENCES

1. U.S. Environmental Protection Agency, "Guidelines for Carcinogen Risk Assessment," *Federal Register* **51**:33991-34003 (1986).
2. National Research Council, *Risk Assessment in the Federal Government: Managing the Process*, National Academy Press, Washington, DC (1983).
3. U.S. Office of Science and Technology Policy, "Chemical Carcinogens: A Review of the Science and Its Associated Principles," *Federal Register* **50**:10372-10442 (1985).
4. U.S. Interagency Staff Group on Carcinogens, "Chemical Carcinogens: A Review of the Science and Its Associated Principles," *Environmental Health Perspectives* **67**:201-208 (1986).
5. G. Atherley, "A Critical Review of Time-Weighted Average As an Index of Exposure and Dose, and of Its Key Elements," *J. of Amer. Indust. Hygiene Assoc.* **46**:481-487 (1985).
6. F. Haber, "Zur Geschichte des Gaskrieges (On the History of Gas Warfare)," in *Fünt Vorträ aus den Jahren 1920-1923 (Five Lectures from the Years 1920-1923)*, 76-92, Springer-Verlag, Berlin (1924).
7. J. Kalbfleisch, D. Krewski, and J. Van Ryzin, "Dose-Response Models for Time-to-Response Toxicity Data," *The Canadian J. of Stat.* **11(1)**:25-49 (1983).
8. D. R. Cox, "Regression Models and Life-Tables," (with discussion), *J. of the Royal Stat. Soc.* **34**:187-220, Series B (1972).
9. R. L. Prentice, A. V. Peterson, and P. Marek, "Dose Mortality Relationships in RFM Mice Following ^{137}Cs Gamma-Ray Irradiation," *Radiation Res.* **90**:56 (1982).
10. P. Armitage and R. Doll, "Stochastic Models for Carcinogenesis," in *Proc. of the 4th Berkeley Symp. on Math. Stat. and Prob.* **4**:19-38, University of California Press, Berkeley and Los Angeles (1961).
11. J. Neyman and E. G. Scott, "Statistical Aspect of the Problem of Carcinogenesis," in *Proc. of the 5th Berkeley Symp. on Math. Stat. and Prob.* **5**:745-776, University of California Press, Berkeley and Los Angeles (1965).
12. K. S. Crump, D. G. Hoel, C. H. Langley, and R. Peto, "Fundamental Carcinogenic Processes and Their implications for Low Dose Risk Assessment," *Cancer Research* **36**:1973-2979 (1976).
13. D. Krewski and J. Van Ryzin, "Dose Response Models for Quantal Response Toxicity Data," in *Curr. Topics in Prob. and Stat.*, M. Csörgő, D. Dawson, J. N. K. Rao, and E. Salek, Eds., North-Holland, Amsterdam (1981).
14. H. A. Guess and K. S. Crump, "Low Dose-Rate Extrapolation of Data from Animal Carcinogenicity Experiments—Analysis of a New Statistical Technique," *Math. Biosciences* **30**:15-36 (1976).
15. H. A. Guess and K. S. Crump, "Maximum Likelihood Estimation of Dose Response Functions Subject to Absolutely Monotonic Constraints," *Annals of Stat.* **5**:101-111 (1978).
16. K. S. Crump and R. B. Howe, "The Multistage Model with a Time-Dependent Dose Pattern: Application to Carcinogenic Risk Assessment," *Risk Anal.* **4**:163-176 (1984).
17. K. S. Crump, A. Guess, and K. L. Deal, "Confidence Intervals and Tests of Hypothesis Concerning Dose-Response Relations Inferred from Animal Carcinogenicity Data," *Biometrics* **33**:437-451 (1977).
18. P. Z. Daffer, K. S. Crump, and M. J. Masterman, "Asymptotic Theory for Analyzing Dose-Response Survival Data with Application to the Low-Dose Extrapolation Problem," *Math. Biosciences* **50**:207-230 (1980).
19. R. W. Whitmore and J. H. Matis, "Compartmental Analysis of Carcinogenic Experiments: Formulation of a Stochastic Model," *J. of Math. Biol.* **12**:31-43 (1981).
20. H. O. Hartley and R. L. Sielken, Jr., "Estimation of 'Safe Doses' in Carcinogenic Experiments," *Biometrics* **33**:1-30 (1977).

The Approach to Risk Analysis in Three Industries: Nuclear Power, Space Systems, and Chemical Process

B. John Garrick
Pickard, Lowe and Garrick, Inc.
Newport Beach, CA

ABSTRACT

The aerospace, nuclear power, and chemical processing industries are providing much of the incentive for the development and application of advanced risk analysis techniques to engineered systems. Risk analysis must answer three basic questions: (1) What can go wrong? (2) How likely is it? and (3) What are the consequences? The result of such analyses is not only a quantitative answer to the question "What is the risk?" but, more importantly, a framework for intelligent and visible risk management. Because of the societal importance of the subject industries and the amount of risk analysis activity involved in each, it is interesting to look for commonalities, differences, and, hopefully, a basis for some standardization. Each industry has its strengths: the solid experience base of the chemical industry, the extensive qualification and testing procedures of the space industry, and the integrative and quantitative risk and reliability methodologies developed for the nuclear power industry. In particular, most advances in data handling, systems interaction modeling, and uncertainty analysis have come from the probabilistic risk assessment (PRA) work in the nuclear safety field. In the final analysis, all three industries would greatly benefit from a more deliberate technology exchange program in the rapidly evolving discipline of quantitative risk analysis.

KEYWORDS: Nuclear power, risk analysis, chemical industry, space industry, probabilistic risk assessment (PRA)

INTRODUCTION

It is the purpose of this paper to review how risk and safety analysis is performed in the three major industries of nuclear power, space flight, and chemical and petroleum processes. The underlying reason for such a review is the belief that efficiencies and safety enhancements may result from a greater exchange of risk assessment technology between these industries. The thrust of this discussion relates to the engineered systems involved in the three industries.

The industries are very different. The chemical industry epitomizes the highly competitive private sector and its bottom-line emphasis; the nuclear power industry is unique by the degree to which it is regulated; and the space industry is essentially a government business just beginning to have commercial implications. Institutional differences are extreme; however, from a societal standpoint, these industries have an enormous number of things in common. Each has a very heavy dependence on technology, fulfills fundamental societal needs, and has safety implications with a far-reaching impact on public opinion and support.

In reviewing the risk and safety analysis activities of these industries, particular attention is given to the use of such quantitative approaches as probabilistic risk assessment (PRA) as it has evolved in the nuclear power industry.

PRACTICES

In general, the practices in the three industries are very different although there is beginning to be some convergence. The prime mover for the convergence is the increasing use of numerical and quantitative methods of analysis. The nuclear industry safety activities have always involved an extensive amount of detailed analysis, drawing from most of the basic disciplines of science and engineering. Driven partly by its "fear anything nuclear" image and by the regulatory environment in which it resides, nuclear safety has been in center stage of the development of nuclear technology. As a result, advanced risk assessment tools such as PRA have developed and have been solidly embedded in the risk management of that industry.[1] Currently, the United States Nuclear Regulatory Commission (NRC) is expected to require individual plant examinations based either on plant-specific PRAs or on extrapolations from referenced PRAs. As a result, the nuclear industry is moving rapidly to risk and safety analysis and risk management process heavily dependent on quantitative methods of analysis. The target is a quantitative answer to three basic questions: What can go wrong? How likely is it? and What are the consequences?

The situation is quite different in the space and chemical fields. Here, practices are rooted in a much stronger experience base, especially for the chemical industry. Unlike the nuclear industry with a beginning as a devastating destructor of human life and property, these industries had the positive beginning of serving mankind immediately. Largely due to the positive public relations accompanied with new technology in these industries, safety to the public was not an immediate issue, and there was not the same kind of pressure to make safety a separate science. To be sure, flight safety has always been a major factor in the evolution of aircraft, but it has never loomed as a gross public risk as in the nuclear industry. The major advantage of the chemical process industry is the extensive amount of experience that exists. Protection of the public and the environment has always been a concern; yet, not until major accidents at Flixborough, England; Seveso, Italy; and, more recently, at Bhopal, India, did the public seem to give much thought to the chemical industry as a major public risk issue.[2] The Bhopal and the Challenger accidents have been instrumental in bringing the chemical and space industries more into the "risk" limelight. One result is greater activity in the risk and safety disciplines for both industries.

Although these industries are now adopting more risk-based methods of safety analysis, the basic approach has been more qualitative than quantitative. The work that is done by the National Aeronautics and Space Administration (NASA) in the name of safety, reliability, and quality assurance is thorough; however, it lacks an integrative and systems engineering thought process for tying things together. Fortunately, NASA is now considering the augmentation of its traditional methods with more quantitative and integrative approaches such as PRA. Pilot applications of PRA are being performed on selected space shuttle subsystems. The first two pilot studies are expected to be completed in December of this year (1987) and will provide a basis for considering greater use of PRA-based risk analysis methods.[3]

The chemical industry also relies mostly on qualitative methods but is making strides to upgrade their whole hazards and risk analysis capability. The Center for Chemical Process Safety has published a qualitative hazards analysis procedure manual and is now preparing a guide and data base for quantitative risk assessment.[4] While these quantitative methods are being developed, the industry continues to use classical safety engineering methods and good practices guides as preventive measures to control risk and identify areas for safety improvement.

METHODS AND TECHNIQUES

The methods employed by the nuclear industry in risk and safety work can be broadly categorized as engineering analysis, reference plant analysis, and PRA. Engineering analysis is the backbone of compliance work for licensing. This type of analysis supports claims by the applicant that "there is reasonable assurance that there is no undue risk to the health and safety of the public." Such analysis has as its objective that there is not an accident that will exceed the so-called "design basis accident." This deterministic analysis primarily involves detailed computer models of plant thermal hydraulic conditions following transients and loss-of-coolant accidents.

Since 1977, PRA has been used increasingly as an important risk management tool in the nuclear power industry. However, it will be some time before PRA equals design basis analyses as a basis for licensing. PRA has made enormous strides, and it is clear to the nuclear industry that it is the optimal method available to answer the public question of "What is the risk?" PRA provides a logical framework with which an analyst can systematically identify the likelihood and consequence of specific scenarios. Theoretically, all scenarios of interest to risk and safety will be considered and quantitatively ranked by importance. From this ranked list of scenarios, decision makers are provided with a basis for the allocation of resources toward the improvement of safety and reduction of risk. PRA's strengths are that it is integrative and quantitative — integrative in that it has the ability to consider the whole system and quantitative in the sense of addressing in a scientific way consequence, likelihood, and the uncertainty associated with the quantification.[5]

In comparison to the nuclear industry, the backbone of the risk and safety methods employed in the space industry is qualitative failure mode and effects analysis (FMEA) and hazard analysis. The FMEA work is hardware oriented and consists of assuming a failure mode for individual components and assessing worst case effects.[6] In contrast to FMEA, hazard analysis consists of identifying undesired events, hazardous conditions, or accident scenarios and systematically identifying hazard causes, effects, and recommended corrective actions. The hazard analysis goes beyond the hardware and addresses software requirements, coding errors, environmental impacts on operations, crew errors, and procedural anomalies.[7]

The purpose of using both the FMEA and hazard analysis tools is to provide a check and balance of the completeness of the overall risk assessment. The products of an FMEA and a hazard analysis will provide information for each system component, its function, associated failure modes, criticality, cause of failure, effect on other components or systems, and any corrective actions if applicable. Since hazard reports are generated for each hazardous condition and FMEAs are produced for each component, thousands of pages of analyses are generated for large systems. To interpret the voluminous information, critical items lists rank the criticality of component failures. This hierarchical system groups failures according to qualitative measures of likelihood and consequence.[8]

For over 20 years, experts in the chemical process industry have performed safety and reliability studies; however, because of the extensive experience base, safety decisions in the chemical process industry have been based on expert judgment. Except for toxic effects of chemicals, analytical modeling for safety or risk in the chemical industry has focused on on-site consequences, e.g., plant damage and production loss. Only occasionally has the industry quantified the likelihood and consequences of releases. Today, chemical industry analysts are selectively performing analytical qualitative risk assessments similar to those performed in the space community. The primary tools employed are FMEAs and hazard and operability (HAZOP) studies. The FMEA used is very similar to that used by the space industry. However, the HAZOP goes beyond identification of hazards: it establishes the likelihood and consequence of the event using conservative methods. It also identifies the operability problems that, though not hazardous, could compromise the chemical plant's ability to achieve design productivity.

Like the space industry, the chemical industry has been doing some quantitative risk analysis on a selective basis. The full extent to which the PRA thought process will be embraced by the chemical industry is not yet known, but the trend toward more comprehensive and quantitative risk models is clear.[9]

INDUSTRY--SPECIFIC FACTORS AFFECTING RISK ANALYSIS

Each of the three industries has unique characteristics with respect to hazards, safety systems, and procedures that affect risk analysis. Both the space and nuclear industries are characterized by extensive use of redundancy and diversity for safety-related systems. Typically, even when subjected to multiple failures, system performance may not be severely degraded. In contrast, chemical safety systems are not complex (e.g., leakage of a storage tank or failure of a shutoff valve can lead to an atmospheric release), and few equipment redundancies exist; hence, single failures can have great consequences from economic, environmental, and public health perspectives. With respect to safety procedures, unlike the space and nuclear industries with many years of formal quality assurance practices, the facilities and systems in the chemical process industry typically do not follow a standardized approach to ensure that proper safety equipment or procedures are in place to either prevent a hazardous release or to mitigate release consequences.

An overview comparison of the hazard characteristics for the three industries is shown in the following table:

Comparison of Hazard Characteristics

Hazard Characteristics	Chemical	Nuclear	Space
Single-Concentrated Hazard Locations	Sometimes	Always	Always
Distributed Sources of Hazard	Almost Always	Reactor Only	Rarely
Chemical Toxicity	Often	Rarely, Radiation Effect Dominates	Always but Secondary to Fire and Explosions
Fires	Often	Only as They Result from Core Melt Effects	Major Hazard
Explosions	Often	Only as They Result from Core Melt Effects	Greatest Source of Hazard
Radioactivity	Rarely	Always	Rarely
Changing Configuration or Operating Mode	Not Important except in Transportation	Not Important except in Transportation and Spent Fuel Pool	Very Important
Human Error	Important	Important	Important

In general, the chemical process industry is characterized by large, complex, and varied inventories of hazardous chemicals with hazard sources located throughout the plant. Each component that either transports, stores, or processes a hazardous substance can be a source of hazard. Release of chemical inventories can result in a plethora of hazardous reactions. The properties of these materials under extreme conditions are often not known. Consequences of such reactions are often unpredictable since human and environmental responses to chemical dose levels vary widely, especially at low dose levels; this condition also exists for low dose levels in the nuclear industry.

Typically, the space industry is concerned with a single hazard location, as a rocket, that may have several hazard sources as propellants or explosives. The primary anomalies of concern are fires and explosions; unlike the chemical or nuclear industry that are ground based, it is generally not possible to mitigate the effects of fires or explosions in space applications. Also, the industry is unique since the engineered systems are subjected to a variety of environments during operation; changing configurations and operating modes often impose exotic design requirements on the vehicle's components and systems.

In contrast, the nuclear industry is characterized by a primary hazard location—the nuclear reactor core and its attendant inventory of radioactive material. The nuclear industry has adopted a unique approach to minimizing the likelihood of a release of its radioactive material. The principal safeguard against releases in the event of a severe accident is an elaborate containment system.

IMPLEMENTATION DIFFICULTIES

Most of the difficulties in implementing comprehensive and quantitative approaches to risk management are believed to be more institutional or cultural than they are technical. Although it was in the space and defense industries that quantitative systems safety analysis received its initial impetus, NASA's bad experience with some probability estimates during the Apollo program resulted in that agency turning away from "numerical and probabilistic methods" in safety and reliability work. The decision is believed unfortunate; the result is that progress is much slower than it should be. Furthermore, safety and reliability staffs do not seem to command the same respect in the space program, for example, as they do in the nuclear industry. As a result, there is a lack of clout in the decision-making process at the program level. Fortunately, increased budgets for risk analysis, key changes in management positions involving risk and safety, and the showing of interest by NASA in risk technology as applied to other industries are all very positive signs.[10]

Probably, the single biggest problem in nuclear safety continues to be the public "fear of anything nuclear." There continues to be a considerable fraction of the public that does not trust the nuclear power industry and its claims about safety in spite of a good record. The very complex and cumbersome licensing process probably adds to the public suspicion. As in the space industry, political conflicts delay the decision-making progress. Faced with these obstacles, the nuclear industry has made enormous progress from a technical standpoint. The industry has done a better job in the last few years of answering the basic risk question with the adoption and practicing of a PRA-based approach to risk assessment. As the technique moves more into the use of the risk models on a continuous basis for risk management, even more progress is expected. It is believed that the results are key to increased public confidence and simplification of the licensing process.

Similar to other industries, difficulties for risk analysis progress in the chemical industry are institutional. Some of the obstacles are that the industry is highly competitive and bottom-line oriented, and products and processes are often proprietary. Hence, there is a tendency to oversimplify or restrict dissemination of the analyses that are performed with the possible exception of the area of chemical toxicity.

Even so, there are many positive signs of change in the chemical and petroleum industry. The associated technical societies and trade associations are very active in training, conducting studies, and holding national meetings to upgrade the level of safety and risk analysis in their industry. There is an increasing number of chemical plant-specific studies aimed at getting a better handle on the types and frequencies of releases that can occur. Many chemical companies are installing sophisticated dispersion assessment tools at their plants to better anticipate and respond to emergencies should they arise. Finally, the new national Society for Risk Analysis is receiving a significant amount of its membership from the chemical and petroleum industry.

SOME REQUIREMENTS FOR POSITIVE RISK MANAGEMENT

Risk assessment and management of complex technical systems is a thinking business that cannot be fully prescribed or "cook booked." One requirement for success is to avoid getting trapped into thinking that such can be accomplished. In the right amount, it is important to have guidelines and handbook resources available to the risk analysts for their work. It is also important to be very precise on the questions that are to be answered—perhaps, the single weakest element of many risk and safety analyses.

All three industries need to upgrade their technical staffs who perform risk and safety analysis and implement risk management. Although, in general, the nuclear industry is very strong in the risk assessment phase, it is less so when it comes to implementation. In particular, it needs to push for greater competence within the plant staffs to be a champion of the usually competently developed risk models. Also, there needs to be a breakaway from the compliance state of mind.

Examples of areas for improvement in the space industry include more emphasis on training in contemporary PRA methods and less dependency on a mechanical approach to risk assessment. There should be additional application of subsystem and limited scope PRAs to build confidence in the methods and efforts to specialize and coordinate the data base to better support the overall effort.

In the chemical industry, there needs to be a continued effort to recognize safety as a bonafide technical discipline. There is progress here, but management needs to be more visible in the process. On the technical side, there needs to be more emphasis on providing the link between the likelihood of consequences and the magnitude of consequences.

CONCLUSION

The era of quantitative risk assessment and risk management is here. Technologists are being held accountable to the public for answering the basic question of "What is the risk?" of their engineered systems. The PRA thought process is the most effective means for answering this question in a comprehensive and quantitative manner.

A thorough understanding of the components and processes constituting an engineered system, along with a quantitative statement about the failures and scenarios important to risk, enable decision makers to properly allocate resources. Although the nuclear industry has taken great steps in this direction, combined strengths of the space, chemical, and nuclear industries will be required if an effective synergism of risk assessment technology is to occur. The space industry has set the standards throughout all industries for the rigorous testing of components. Exhaustive test procedures effectively screen out faulty designs and inferior components prior to flight testing; flight testing prompts engineers to make final design changes. The success of the aircraft, guided missile, satellite, and manned space programs has been largely due to the extensive qualification and acceptance test procedures. Until the Challenger accident, NASA had been confident in using qualitative safety and reliability methodologies. Currently, NASA recognizes that qualitative approaches do not yield information most crucial to questions such as "Is it safe to launch?" It is hopeful that NASA's pilot PRA studies will indicate to NASA how these questions can best be answered.

Similarly, the chemical process and petroleum industry also recognizes the need to quantify risk. Unlike the nuclear and space industries, the chemical industry is intensely competitive. This condition forces chemical executives to answer the key question "How much can be spent on safety?" and still stay in business. In some cases, the decision has been to drop a product line because the risk was too great. Urgency in answering this economic and health impact question has motivated chemical industry analysts to introduce simplified methodologies that enable completion of limited scope and inexpensive quantitative risk assessments.

The technology exchange opportunities between the three industries are numerous. Data bases can be greatly expanded, modeling methods can be shared, the form of the results can be standardized (at least in some areas), closer coordination can hasten the stabilizing of the methodology, and problem scoping can be standardized, leading to increased efficiency. Also, the more difficult analytical tasks such as uncertainty analysis and dependent failure analysis may have more obvious solutions if there were greater interaction among the industries. Such interaction should hasten increased confidence in the safety of all three industries.

ACKNOWLEDGEMENTS

The author wishes to acknowledge the invaluable assistance of Brian A. Fagan, Willard C. Gekler, Mardy Kazarians, and Dennis C. Bley of Pickard, Lowe and Garrick, Inc., in preparation of this paper.

REFERENCES

1. Garrick, B. J., "Lessons Learned from 21 Nuclear Plant PRAs," International Topical Conference on Probabilistic Safety Assessment and Risk Management, Zurich, Switzerland, August 30-September 4, 1987.
2. Lees, F. P., *Loss Prevention in the Process Industries*, Volumes 1 and 2, Butterworth & Co., Ltd., London, 1980.
3. Pickard, Lowe and Garrick, Inc., and McDonnell Douglas Astronautics Company, "Engineering Services, Numerical Risk Assessment of the Auxiliary Power Unit (APU) for the United States Space Shuttle," prepared for the National Aeronautics and Space Administration, December 1987.
4. Battelle Columbus Division, "Guidelines for Hazard Evaluation Procedures," prepared for The Center for Chemical Process Safety of the American Institute of Chemical Engineers, 1985.
5. Pickard, Lowe and Garrick, Inc., "Methodology for Probabilistic Risk Assessment of Nuclear Power Plants," PLG-0209, June 1981.
6. National Aeronautics and Space Administration, "Instructions for Preparation of Failure Modes and Effects Analysis (FMEA) and Critical Items List (CIL)," NSTS 22206, October 10, 1986.
7. National Aeronautics and Space Administration, "Instructions for Preparation of Hazard Analysis for the Space Transportation System," Preliminary, JSC 22254, November 1986.
8. U.S. Department of Defense, "Military Standard System Safety Program Requirements," MIL-STD-882B, March 30, 1984.

9. U.S. Environmental Protection Agency, "Superfund Amendments and Reauthorization Act, III (SARA III)," 1986.
10. Committee on Science and Technology, U.S. House of Representatives, "Investigation of the Challenger Accident," Ninety-Ninth Congress, October 29, 1986.

An Epidemiologic Approach to Dealing With Multimedia Exposure to Multiple Related Chemicals

Stan C. Freni[a]
U.S. Department of Health and Human Services
Centers for Disease Control
Atlanta, GA

ABSTRACT

In investigating the health risk from environmental exposure, one is usually confronted with the fact that people are exposed to many chemicals besides the one of interest. Therefore, if the statistical analysis of an association between exposure and disease is to be meaningful, and if the results are to include a quantitative risk estimate, some measure of composite exposure must be developed. The development of such a measure is described in the context of a retrospective cohort study on the effects of exposure to drinking water contaminated with multiple volatile organic compounds. The main problem in developing a composite measure appears to be the lack of standardization in the assessment and presentation of toxicologic parameters. Three of these parameters are discussed: the LD_{50}, the NOEL, and the pulmonary retention rate.

KEYWORDS: Multiple exposure, multimedia exposure, exposure assessment, epidemiology, dose-equivalent

INTRODUCTION

There is probably not a single environmental situation in which people are exposed to just one chemical, regardless of the route of exposure. People are also often exposed to multiple chemicals with similar toxicologic profiles. For instance, exposure to chloroform in chlorinated drinking water goes together with exposure to other halogenated methanes; 2,3,7,8-dioxin exposure usually implies concurrent exposure to other dioxins. There are many categories of chemically and biologically related chemicals, such as polyaromatic hydrocarbons, polychlorinated biphenyls (PCBs), steroids, organophosphorus pesticides, and chlorinated volatile organic compounds (VOCs), the subject of this paper. There are no established ways in epidemiology for dealing with exposure to multiple chemicals, an issue that has hardly been the subject of interest. Conventional methods for dealing with multiple exposures can be categorized as follows:

---

a. Current address: Center for Toxicological Research, Jefferson, AR.

1. Ignore the exposure assessment for individual chemicals, and assume the existence of a single imaginary substance, e.g., polluted water. The studies on cancer and Mississippi River water and the Woburn study[10,15] are examples of this approach. The results of such studies relate, at best, to the association of cancer and polluted water, and not to the chemicals in that water. Inferences from such a study regarding a specific chemical or group of chemicals are purely speculative.

2. Ignore concomitant exposure and rely on matching procedures for cases and noncases, or exposed and unexposed cohorts, as was done in a matched case-control study on 1,1,1-trichloroethane.[14] This requires the assumption of, on average, the same concomitant exposure among groups or individuals to be compared. It is extremely unlikely that this is a valid assumption, even when people were matched for occupation. Further, this method may, at best, only lead to conclusions about the effect of the chemical of primary interest, not of the chemical mixture.

3. Deal with concomitant exposure in the analysis by handling variables expressing exposure to chemicals other than the one of primary interest as covariates in a multivariable model. The study of herbicide use and risk of lymphoma and sarcoma by Hoar et al.[11] is an example of this approach. Its basic problem is that in a multivariable analysis, using either logistic regression or proportional hazard models, the effect of covariates is measured in a multiplicative fashion. But rarely, if ever, is the combined effect of chemicals multiplicative. Theoretically, interaction terms could be used to solve this problem, although such a solution would be extremely difficult to realize. The number of interaction terms grows rapidly with the number of chemicals in the mixture. A mixture of four chemicals would already require 11 interaction variables in addition to the four exposure variables. Because the number of variables in a model is determined by the study size, the number of participants in the study will very soon be the limiting factor for the number of chemicals that can be studied.

There is, thus, a clear need for an alternative approach specifically tailored to environmental scenarios. The U.S. Environmental Protection Agency (EPA) announced its use of the following equation for estimating risks from chemical mixtures:[4]

$$\text{Composite dose} = \text{dose}1/E1 + \text{dose}2/E2 + \text{dose}3/E3 + \text{dose}(i)/E(i) \tag{1}$$

In this equation, $E(i)$ represents some parameter of equivalence, for instance, the Allowable Daily Intake (ADI), an occupational standard, or (if cancer is the outcome of interest) a cancer risk estimate. This approach seems sound, but these parameters are compromised to various degrees by factors unrelated to toxicity, such as safety factors, selection of an extrapolation model, and technological or economical considerations.

ALTERNATIVE PARAMETERS OF DOSE-EQUIVALENCE

The LD_{50} and TD_{50} (lethal or toxic dose to 50% of the treated animals), as well as the NOEL and LOEL (no or lowest observed effect level), are factual measures of toxicity. We have evaluated the suitability of these parameters in the context of our study on the health effects of exposure to multiple VOCs in drinking water. These VOCs are chlorinated C2-aliphatics occurring in any combination of at most seven chemicals at concentrations of 1 to 2000 parts per billion (ppb). These chemicals have quite similar toxicologic profiles, and probably also have the same mechanism of action. Altogether, it would be logical and reasonable to assume additivity of the composite effect of these chemicals.

Of the parameters mentioned, the LD_{50} is the one most readily available. That dose-equivalents based on a LD_{50} may be related to dose-equivalents for chronic disorders has been demonstrated for cancer by Zeise *et al.*, who found a remarkably close correlation between the LD_{50} and carcinogenic risk.[25] A search for data on the LD_{50} of the seven VOCs yielded widely varying values. Although the listing of LD_{50} values in computerized data banks, government reports, and handbooks suggests mutual comparability, a search of the original data sources revealed a striking lack of standardization, especially with regard to the duration of the observation period and the age and sex of the test animals. Only one laboratory (Mellon Institute, University of Pittsburgh) standardized its testing protocol for the specific purpose of comparing the toxicity of chemicals.[20-22] Unfortunately, only two of the VOCs in our study were included in that laboratory's program, and no new reports on serial LD_{50} testing have been published since 1969. The following is an overview of LD_{50} data found for oral rat studies, expressed in g/kg and (if known) by gender.

TCA (trichloroethane):12.3 and 14.3 male rats; 10.3 and 11 female rats
TCE (trichloroethylene): 7.2 male rats (Mellon Institute)
PCE (perchloroethylene): 13 and 8.85
11-DCA (1,1 dichloroethane): 1.12 and 0.725
12-DCA (1,2 dichloroethane): 0.68 and (Mellon Institute, male rats) 0.77
CIS (cis-1,2 dichloroethylene): 0.7 for an undefined cis/trans mixture
DCE (1,1 dichloroethylene): various values ranging from 0.2 to 1.8

The effect of lack of standardization can be substantial, as reflected in a range as wide as 200 to 1800 mg/kg for DCE. Considering the lack of standardization, these values show dose-equivalents of approximately 10 for higher chlorinated compounds to 1 for the other VOCs. In evaluating the LD_{50} as a parameter of equivalence in studying low-dose effects, it should be recognized that LD_{50} levels were sufficiently high to have exceeded the saturation limit of several or all VOCs, whereas the very much lower doses used in NOEL studies are below the saturation limit for humans. This is certainly true for the still lower VOC levels in drinking water. For this reason, the ED_{50} (50% effective dose), the LOEL and the NOEL are preferred over the LD_{50}. They regard a specific health event, not death, and are lower on the dose scale and thus closer to the dose levels of concern in human studies. However, the search for studies yielding an ED_{50} for effects other than cancer was unfruitful, and reported LOELs pertained to different health events, and lower dose levels were usually not studied.

A search for NOEL data was more successful, but the lack of standardization was even more apparent than with the LD_{50}, especially with regard to the sensitivity of the toxicity indicators. Virtually all NOELs were inhalation NOELs, which necessitated the conversion to oral NOELs. This was achieved by using the following equation for the amount of VOC absorbed through the lung, using an average lifetime body weight of 70 kg and a respiration volume of 1 m^3/hr, and assuming equitoxicity per weight unit of VOCs absorbed in the lung and in the gut:

$$\text{VOC absorbed (lungs)} = (\text{NOEL} \times \text{CF} \times \text{Hr} \times \text{D} \times \text{R}) / (7 \times 70) \text{ mg/kg/d} \qquad (2)$$

where
CF = chemical-specific conversion factor from parts per million (ppm) to mg/m3;
Hr = hours exposure per day;
D = days exposure per week;
R = proportion of inhaled chemical retained in the body.

Equation (2) assumes that discontinuous exposure is related to continuous exposure by direct proportionality. This assumption, used throughout regulatory risk assessment, is probably invalid because recovery processes in dose-free periods (of which the extent is unknown) are ignored. A review of pertinent literature revealed that the pulmonary retention in humans ranged from 36 to 75% for TCE, 60 to 80% for PCE, 38 to 60% for DCE, and 25 to 30% for TCA.[3,5-8] No useful data were found for the other chemicals.

These wide ranges raise the question of what the real pulmonary uptake is. Further, although it is commonly assumed that the percent uptake is constant,[5-7] it has been shown that by increasing the concentration of DCE from 25 to 300 ppm, the percent retention after 3 hours of exposure decreased from 77 to 50%.[3] Thus, the uptake at the many orders of magnitude lower exposure levels prevailing in bathrooms appears yet to be established.

Below is a summary of oral NOELs, estimated from literature data from chronic exposure studies (some animal studies were semichronic), assuming a pulmonary absorption of 50% for CIS, 11-DCA, and 12-DCA (a rounded average of the rates known for the other VOCs). As it is still unknown what the quantitative effect is of interspecies differences in metabolism, respiration rate, lung surface/body weight ratio, and sensitivity of health effect indicators, NOELs for both humans and animals have been listed.

TCA — *humans*: the higher of two inhalation studies yielded a NOEL of 250 ppm (8 hr/weekday), equal to an oral NOEL of 33.1 mg/kg/d. The effect parameters included ECG, clinical chemistry, blood pressure, and a health questionnaire emphasizing cardiovascular and CNS function.[14]

TCE — *humans*: several occupational inhalation studies yielded a NOEL; the highest was 25 ppm (8 hr/weekday), equal to an oral NOEL of 6.1 mg/kg/d. The effect parameters included a health questionnaire and clinical examination with emphasis on urinalysis, hematology, CNS symptoms and behavioral performance, and limited blood chemistry.[23]

— *rats*: an inhalation NOEL of 55 ppm (8 hr/weekday) was found, equal to a human oral NOEL of 13.5 mg/kg/d. The effect parameters included hematology and clinical chemistry.[13]

PCE — *humans*: in several occupational inhalation studies, the highest tested exposure yielding a NOEL varied from 18 ppm to 32 ppm (8 hr/weekday). The 18-ppm level equals an oral NOEL of 7 mg/kg/d. No higher exposure was tested. The effect parameters were neurological tests, blood glucose, hematocrit, and behavioral performance.[24]

— *rats*: an inhalation NOEL of 70 ppm (8 hr/weekday) was found, equal to a human oral NOEL of 27.1 mg/kg/d. The parameters were histology, WBC count, and serum values of bilirubin, glucose, and calcium.[2]

11-DCA — *rat, guinea pig, and rabbit*: an inhalation NOEL of 1000 ppm (6 hr/weekday) was found, equal to a human oral NOEL of 125.8 mg/kg/d. The NOEL in cats was 500 ppm, equal to a human NOEL of 62.9 mg/kg/d. The parameters were hematology, clinical chemistry, and histology.[12]

12-DCA — *rat, rabbit, cat, and guinea pig*: an inhalation NOEL of 100 ppm (6 hr/weekday) was found, equal to a human oral NOEL of 12.4 mg/kg/d. The parameters were hematology, clinical chemistry, and histology.[12]

DCE — *rat and dog*: an oral NOEL of 10 mg/kg/d was found. (No conversion to human oral NOEL is needed, these were already oral studies.) The effect parameters were histology and clinical chemistry.[19]

Given the lack of standardization and the need to convert inhalation data to an oral NOEL, using either averaged or assumed pulmonary retention rates, the above data suggest little quantitative difference in the chronic toxicity of the various VOCs. In particular, the tenfold difference in the LD_{50} between higher and lower chlorinated VOCs is not confirmed in the NOEL. No NOEL was found for CIS, which necessitated an assumption of equitoxicity with the other dichloro compounds, based on its LD_{50}. In summary, considering the irrelevance of LD_{50} data for low-dose chronic effects, and the approximately 1:1 dose-equivalence found for the NOEL, it appears appropriate to adopt equitoxicity when calculating a composite exposure.

VOC CONCENTRATION AS A FUNCTION OF TIME

The next step is how to estimate chemical-specific exposures. The options are to measure current exposure or estimate the exposure accumulated over time. The first option, used in most risk assessments and epidemiologic studies, is also the least defensible, especially if the exposure level has changed over time. In this study of a scenario with increasing levels of groundwater contamination, the second option was selected. For this purpose, a model was developed to describe a chemical's concentration in groundwater as a function of time. The development of this model is described in detail elsewhere.[9] In short, a concentration-time (CT) curve was constructed from data of nearby city wells, which were monitored for some years. With test results of only one recent water sample, this curve was used to predict, for individual residential wells, when contamination of the well started and how concentrations changed over time. The area under the CT-curve represents the accumulated exposure. Individual exposure estimates were made by integrating the area under the curve between the points in time that a person was actually exposed to contaminated water. With a curve for each of the chemicals, the seven chemical-specific exposures can be calculated in units of ppb months for each individual. These values can easily be converted into a dose by taking into account the volume of water consumed per day, the body weight, the duration of the exposure, and appropriate conversion factors to arrive at a unit dose of 1 mg/kg. On the basis of a dose-equivalence ratio of 1, as discussed earlier, the total composite exposure is then the simple sum of the chemical-specific exposures.

SECONDARY ROUTES OF EXPOSURE

A complete exposure assessment includes the assessment of secondary routes of exposure. Measurement of exposure through ambient air and food was omitted because of the prohibitively high cost of testing an individual's weekly food basket and the ambient air. These sources were estimated to contribute, at most, a few percent of the total VOCs from contaminated water. Only a few food items contain measurable amounts of VOCs, and most of the VOCs will evaporate in the process of cooking. Ambient air VOC levels are known to be negligibly low. VOCs in soil at the individual residences are irrelevant, as direct soil ingestion is known to be minimal, and oral exposure and skin absorption from dust deposited on the skin is zero because of the volatility of the VOCs. The study design did include gathering data on occupational exposure to VOCs and to household commodities containing VOCs. The discussion of this issue is beyond the scope of this presentation. Exposure to VOCs through skin absorption and inhalation by individuals bathing or showering has been claimed to be substantial.[1,17]

The extent to which VOCs are retained in the body through absorption in the lungs depends on several factors: the concentration in water, the temperature of water and air, the water/air and air/blood partition coefficient, the duration of exposure, the preexisting concentration in the air, the extent of ventilation, the volume of water used, and the kind of activity (showering or bathing). The compound effect is the pulmonary retention rate. As discussed earlier, a search for retention rates yielded a wide range of values, clearly the result of differences in the experimental conditions. Most of the data required for calculating the absorbed amount of VOCs under low-dose condition are thus unknown, and they can be expected to differ not only between persons but also for the same individual from day to day. Assuming fixed rates,[1,11] regardless of the very low VOC levels in bathrooms and a duration of exposure too short to reach equilibrium compatible with the partition coefficients, is not compatible with a scientifically sound risk assessment.

The estimation of oral dose-equivalents for skin absorption is complicated by several problems for which there seems to be no adequate solution yet. Brown et al.[1] claimed that the skin absorption rate is proportional to the VOC concentration in water, that is, the ratio absorption rate/concentration (the permeation constant) is constant and calculated to be 0.001 for each of the four compounds reviewed. This is not logical, as it disregards the combined effect of the water/skin lipids and lipids/blood partition coefficients that differ between chemicals. Actually, it appears that the claim of constancy resulted from improper rounding of permeation constants. Adding just one decimal would have revealed a distinct and consistent *decrease* of the "constant" with *decreasing* concentration. For instance, the "constant" for ethylbenzene dropped as much as 28% when the concentration decreased by 29%. Considering that in environmental scenarios VOC levels are four orders of magnitude lower than in the experiments reviewed by Brown et al., this dependence of the permeation rate on the concentration may be such that there would be no observable absorption at these low levels.

Because of this lack of reliable physical parameters, it was not possible to calculate the absorbed dose and dose-equivalents of VOCs from secondary exposure. The problem was remedied by creating a new exposure variable equal to the frequency times the duration of showers or baths, a semiquantitative variable used as a covariate in a multivariable model. This solution is admittedly less satisfactory than using a composite dose for the three routes of entry combined based on dose-equivalents, because of previously discussed problems related to covariates in multivariable analysis. This approach is a pragmatic one, however, necessitated by the lack of experimental data, not by deficiencies of epidemiologic study designs.

A special problem arises when the toxicologic profiles of chemicals are not similar. Additivity may still be the case for a specific health effect, e.g., liver damage, but not for others. It is known, for example, that the toxicity of a mixture of PCBs and VOCs is more than additive, though still far from multiplicative.[18] This problem has no solution other than waiting until more is known from the toxicology of mixtures. Meanwhile, for the purpose of this study of the health effect of VOCs, the effect of concurrent exposure to chemicals other than the VOCs of interest is analyzed by entering the exposure of these chemicals as covariates in a multivariable model, as described in the beginning of this paper.

ACKNOWLEDGEMENT

This report was, in part, supported by funds from the Comprehensive Environmental Response, Compensation, and Liability Act trust fund, and from the Michigan Department of Public Health.

REFERENCES

1. Brown, H. S., Bishop, D. R., and Rowan, C. A., "The Role of Skin Absorption As a Route of Exposure for Volatile Organic Compounds (VOCs) in Drinking Water," *Am. J. Public Health* **75**:479-484 (1984).
2. Carpenter, C. P., "The Chronic Toxicity of Tetrachloroethylene," *J. Ind. Hyg. Toxicol.* **19**:323-326 (1937).
3. Dallas, C. E., Weir, F. W., Feldman, S., Putcha, L., and Bruckner, J. V., "The Uptake and Disposition of 1,1-Dichloroethylene in Rats During Inhalation Exposure," *Toxicol. Appl. Pharmacol.* **68**:140-151 (1983).

4. Environmental Protection Agency, "Guidelines for the Health Risk Assessment of Chemical Mixtures," *Fed. Reg.* **51**:34014-34025 (1986).

5. Environmental Protection Agency, "Health Assessment Document for 1,1,1-Trichloroethane," Final Report, Office of Research and Development, 1984 (NTIS PB84-183565, EPA 600/8-82-003F).

6. Environmental Protection Agency, "Health Assessment Document for Tetrachloroethylene (Perchloroethylene)," Final Report, Office of Research and Development, 1985 (NTIS PB85-249704, EPA 600/8-82-005F).

7. Environmental Protection Agency, "Health Assessment Document for Trichloroethylene," Final Report, Office of Research and Development, 1985 (NTIS PB85-249696, EPA 600/8-82-006F).

8. Environmental Protection Agency, Draft Criteria Document for Dichloroethylenes, Office of Drinking Water, 1984 (NTIS PB 84-199546).

9. Freni, S. C., and Phillips, D. L., "Estimation of the Time Component in the Movement of Chemicals in Contaminated Groundwater," *Environ. Health Perspect.* **74**:211-221 (1987).

10. Gottlieb, M. S., Carr, J. K., and Clarkson, J. R., "Drinking Water and Cancer in Louisiana," *Am. J. Epidemiol.* **116**:652-667 (1982).

11. Hoar, S. K., Blair, A., Holmes, F. F., *et al.*, "Agricultural Herbicide Use and Risk of Lymphoma and Soft-Tissue Sarcoma," *J. Am. Med. Assoc.* **256**:1141-1147 (1986).

12. Hofmann, H. Th., Birnstiel, H., and Jobst, P.. "Zur Inhalationtoxicität von 1,1- und 1,2-Dichloroäthan," *Arch. Toxikol.* **27**:248-265 (1971).

13. Kimmerle, G., and Eben, A., "Metabolism, Excretion, and Toxicology of Trichloroethylene After Inhalation, I. Experimental Exposure on Rats," *Arch. Toxicol.* **30**:115-126 (1973).

14. Kramer, C. G., Ott, M. C., Fulkerson, J. E., and Hicks, N., "Health of Workers Exposed to 1,1,1-Trichloroethane: A Matched-Pair Study," *Arch. Environ. Health* **33**:331-342 (1978).

15. Lagakos, S. W., Wessen, B. J. and Zelen, M., "An Analysis of Contaminated Well Water and Health Effects in Woburn, Massachusetts," *J. Am. Stat. Assoc.* **81**:583-596 (1986).

16. Lawrence, C. E., and Taylor, P. R., "Empirical Estimation of Exposure in Retrospective Epidemiologic Studies," in *Environmental Epidemiology*, F. C. Kopfler and G. F. Craun, Eds., Lewis Publishers, Chelsea, Michigan, pp. 239-246 (1986).

17. McKone, T. E., "Quantifying Personal Indoor Air Exposure to Radon and Volatile Organic Compounds Derived from Tap Water," Proceedings of the Annual Meeting of the Society for Risk Analysis, Houston, Texas, November 3, 1987.

18. Moslen, M. T., Reynolds, E. S., and Szabo, S., "Enhancement of the Metabolism and Hepatotoxicity of Trichloroethylene and Perchloroethylene," *Biochem. Pharmacol.* **26**:369-375 (1977).

19. Quast, J. F., Humiston, C. G., Wade, C. E., *et al.*, "A Chronic Toxicity and Oncogenicity Study in Rats and Subchronic Toxicity Study in Dogs on Ingested Vinylidene Chloride," *Fund. Appl. Toxicol.* **3**:55-62 (1983).

20. Smyth, H. F., and Carpenter, C. P., "The Place of the Range Finding Test in the Industrial Laboratory," *J. Ind. Hyg. Toxicol.* **26**:269-273 (1944).

21. Smyth, H. F., and Carpenter, C. P., "Further Experience with the Range Finding Test in the Industrial Laboratory," *J. Ind. Hyg. Toxicol.* **30**:63-68 (1948).

22. Smyth, H. F., Carpenter, C. P., Weil, C. S., *et al.*, "Range-Finding Toxicity Data: List VII," *Am. Ind. Hyg. Assoc. J.* **30**:470-476 (1969).

23. Takamatsu, M., "Health Hazards in Workers Exposed to Trichloroethylene Vapor, II," *Kumamoto Med. J.* **15**:43-54 (1962).

24. Tuttle, T. C., Wood, G. D., Grether, C. B., *et al.*, "A Behavioral and Neurological Evaluation of Dry Cleaners Exposed to Perchloroethylene," DHEW (NIOSH) Publication 77-214, 1977 (NTIS PB 274794).

25. Zeise, L., Wilson, R., and Crouch, E., "Use of Acute Toxicity to Estimate Carcinogenic Risk," *Risk Analysis* **4**:187-270 (1984).

Incorporating Risk Into the Development of Soil Cleanup Guidelines for Trace Organic Compounds

B. G. Ibbotson and D. M. Gorber
SENES Consultants Limited
Richmond Hill, Ontario, Canada

D. W. Reades
Golder Associates
Mississauga, Ontario, Canada

ABSTRACT

Two petroleum refinery facilities located west of Metropolitan Toronto recently have been closed and there is a general desire to convert parts of both sites to residential use. Soil sampling has revealed elevated concentrations of metals and organic compounds on site. Soil concentrations necessary for sites to be acceptable for other uses have been established for several metals and some general parameters; however, similar guidelines do not yet exist for specific organic compounds. To expedite the process of identifying appropriate concentrations of organic compounds, the petroleum companies engaged a team of consultants to develop an approach that could determine soil guidelines for the two sites.

Initially, more than 40 compounds were reviewed in terms of physical, chemical, and toxicological properties. For some compounds, information also was available concerning their occurrence in refinery wastes and phytotoxic effects. Eventually, ten compounds were selected to demonstrate the approach being developed. For the ten selected compounds, toxicological profiles were prepared that included the information needed to determine "acceptable" exposure levels. Acceptable daily intake values were calculated for compounds with suitable data following a method recommended by the World Health Organization. For carcinogens, virtually safe dose values were also calculated using a mathematical model from the U.S. EPA.

A pathways model was developed to assess potential exposures of future site residents from various routes of exposure. While designed to estimate the exposures of typical, suburban Ontario residents, the model also included several conservative factors that tend to overestimate total exposures. The results produced by the model indicated that the relative importance of individual pathways is largely influenced by compound properties. For example, the inhalation of vapours was the major pathway for volatile compounds such as benzene, while the direct ingestion of soil and dust accounted for the major portions of the exposure estimates for relatively immobile compounds such as benzo(a)pyrene.

Based on the exposure estimates of the pathways model and the acceptable exposure levels, ranges of acceptable on-site concentrations were derived. The concentrations were defined as those that would result in residents receiving the acceptable exposure levels. Since the initial determination of the acceptable exposure levels, it has become evident that other factors need to be considered before guidelines can be finalized. Those factors include ambient conditions, limitations of analytical procedures, and possibilities of adverse effects other than those pertaining to human health risks. In addition, there remain obstacles related to understanding an approach based on risk that could prove to be as challenging to overcome as those pertaining to the quantification of exposure and risk.

KEYWORDS: Soil cleanup guidelines, routes of exposure, industrial site redevelopment, acceptable exposure levels, background concentrations, organic compounds

INTRODUCTION

The decline in demand for petroleum products has resulted in the closure of several oil refineries in Canada. Two such sites are located west of Metropolitan Toronto. Originally built in rural areas, both sites now are located in the midst of communities and there is a general desire to redevelop parts of both properties for residential housing.

At designated sections of each site, oily wastes were applied to the surface soil and tilled to provide mixing and encourage the degradation of organic compounds through chemical and biological processes. This method, commonly referred to as landfarming, has been used in Canada and other countries for many years and is generally considered to be an effective, economic method of treatment and disposal for many oily wastes. Some constituents of organic (oily) wastes undergo relatively rapid and complete degradation; others degrade slowly, and the low levels of metals often present do not degrade.

The Ontario Ministry of the Environment (MOE) has drafted guidelines for the decommissioning of major industrial sites including oil refineries. The guidelines include recommended soil cleanup levels for some general indicator parameters (such as oil and grease) and some heavy metals, but not for any specific organic compounds.

The refineries thus were faced with the situation in which various organic compounds were known or suspected to be present in site soils, but without the criteria to determine if conditions were acceptable for redevelopment. When the MOE indicated that criteria for specific compounds could take several years to produce, the refineries offered to sponsor the development of an approach for establishing site-specific soil cleanup criteria for organic compounds. The refineries chose to develop an approach based primarily on the risks of adverse human health effects and "acceptable" levels of exposure. It was presumed that such an approach would generate guidelines as stringent as those generated by other approaches or types of concerns.

SELECTION OF COMPOUNDS TO BE STUDIED

Previous to trying to identify criteria, the refineries had been provided with a list of 43 organic compounds by the MOE and had initiated sampling programs to determine concentrations of those compounds in site soils. Preliminary site data suggested that not all of the 43 compounds were present at the sites. Neither did it seem necessary (or efficient) to investigate all 43 compounds to demonstrate whatever approach might be developed. As a result, an early objective was to produce a "short list" of compounds that would provide sufficiently broad ranges of environmental and toxicological parameters to demonstrate the utility of the approach to be developed and to illustrate that the approach could be applied to other compounds if required.

Table 1. Original List of 43 Compounds and Those Selected for the Short List

Monocyclic Aromatics

   ** benzene **
     toluene
     xylenes
     ethylbenzene
     styrene

Polycyclic Aromatics

   ** naphthalene **
     1-methyl naphthalene
     anthracene
   ** phenanthrene **
     benzo(a)anthracene
     7,12-dimethylbenzo(a)anthracene
   ** chrysene **
     methyl chrysenes
     pyrene
     fluoranthene
     dibenz(a,h)anthracene
   ** benzo(a)pyrene **
     benzo(b)fluoranthene
     benzo(k)fluoranthene
     dibenz(a,h)acridine
     indene

Phenolics

     phenol
   ** cresols **
     2,4-dimethylphenol
     4-nitrophenol
     2,4-dinitrophenol

Phthalate Esters

     dimethyl phthalate
     diethyl phthalate
     di(n)butyl phthalate
     di(n)octyl phthalate
   ** bis(2-ethylhexyl) phthalate **
     butyl benzyl phthalate

Halogenated Aliphatics

     chloroform
     1,2-dichloroethane
   ** ethylene dibromide **

Halogenated Cyclics

     chlorobenzene
     dichlorobenzenes

Miscellaneous

     benzenethiol
     carbon disulfide
   ** 1,4-dioxane **
   ** methyl ethyl ketone **
     pyridine
     quinolines

Note: Asterisks indicate compounds selected for the short list.

Several types of information were sought in the literature for all 43 compounds to determine which were best-suited for the short list. Information collected concerning the physical and chemical nature of each compound included aqueous solubility, octanol-water partition coefficient (K_{ow}), and Henry's Law constant. Information also was gathered concerning the health effects that can result from either short- or long-term exposures to each compound. All the collected information was used in the development of the short list of 10 compounds highlighted in Table 1.

DETERMINING ACCEPTABLE EXPOSURE LEVELS

Before an "acceptable" concentration in soil of a compound can be determined, it is first necessary to ascertain the acceptable level of exposure to that compound. The assessment of "acceptable" exposures and risks to human health is an inexact exercise largely based upon theoretical assumptions concerning the extrapolation of animal data to humans. It is usually necessary to extrapolate down several orders of magnitude to evaluate the implications of relatively low exposures such as those of interest in this study.

One way to achieve the necessary extrapolation is to identify a no-observed-adverse-effects-level (NOAEL) and divide that level by an appropriate safety factor (often a factor of 100, 1000, or more) to calculate an acceptable daily intake (ADI). The safety factor approach assumes that there exists an exposure threshold, below which no adverse effects occur. Many North American regulatory agencies prefer to apply this approach to non-carcinogens only. On the other hand, the World Health Organization (WHO) has endorsed this type of approach for assessing carcinogens and the WHO procedure is used by authorities in many countries.[1]

Extrapolation of low-dose effects can also be accomplished using mathematical models. All of the models attempt to predict the number of test animals that will respond at low exposure levels based on the number of responses observed at high dose levels. The models reveal little about predicted human response at any exposure level and require judgments based on broad biological assumptions. Many of the models do not consider the possibility of a threshold response to a chemical because of the statistical problems encountered in the determination of NOAEL values. The assumed absence of a threshold means that absolute safety (or zero risk) cannot be achieved unless exposure is zero. Inherent in the use of such mathematical models, therefore, is the concept of a virtually safe dose (VSD) associated with some suitably low or acceptable level of risk.

A VSD is defined as the exposure at which the level of risk is so small as to be considered "virtually safe," of no consequence, or acceptable. To determine VSD values requires that a model be selected and an acceptable level of risk be established. Many models have been endorsed by various health authorities to calculate VSD values for certain applications. For this study, the U.S. EPA Global-82 program was selected. The program uses a linearized, multistage model for low dose assessment and is generally perceived by toxicologists to produce extremely conservative (low) VSD estimates.

Just as many models are available for extrapolation, so too have several levels of risk been used to establish VSD values. While there is no consensus as to what constitutes an acceptable level of risk, most estimates place the value in the range of 1×10^{-4} to 1×10^{-7} for the probability of death in a lifetime. For this evaluation, the VSD was defined as the lower 95% confidence level of the dose corresponding to the lifetime risk of 1×10^{-6} (one-in-a-million) as calculated by the Global-82 program.

Calculation of ADI and VSD Values

Determining whether to use the ADI approach or one of the mathematical models to calculate a VSD is influenced by many factors, including the types(s) of toxicological information available and the type(s) of health effects that have been reported. Accordingly, the available literature was reviewed and toxicological profiles were prepared for each of the 10 compounds on the short list. Particular attention was paid to long-term studies designed to demonstrate the relationships between the dose applied and the responses observed in the subjects and having adequate numbers of experimental subjects to definitively characterize the toxicity of the compound in question. All of the published studies were evaluated based on the assumption that the administered dose (amount of substance given to the subject) is totally absorbed by the subject irrespective of the route of exposure (method of application). This may considerably overestimate the risks for volatile compounds applied dermally (to the skin) or given by inhalation or for those compounds that are poorly absorbed.

Table 2. Acceptance Daily Exposures for Compounds on the Short List

Compound	VSD* (in μg/kg)	ADI (in μg/kg)
benzene	0.034 and 0.32*	0.28 (adult) 0.1 (child)
naphthalene	not applicable	140
phenanthrene	not applicable	40
chrysene	not applicable	10
benzo(a)pyrene	8.7×10^{-5}	2.8×10^{-4} (adult) 1.0×10^{-4} (child)
cresols	not applicable	51
bis(2-ethylhexyl) phthalate	1.5	106
ethylene dibromide	2.4×10^{-5}	no suitable data
1,4-dioxane	4.6	9.6
methyl ethyl ketone	not applicable	46

*Acceptable exposure is defined as the 95% lower confidence limit of the VSD at a lifetime risk of 10^{-6} as calculated using the U.S. EPA Global-82 program.

**The VSD of 0.32 μg/kg for benzene is based upon a U.S. National Toxicology Program bioassay. The VSD of 0.034 μg/kg for benzene is based on human data.

Adequate information about health effects was found to permit ADI values to be calculated for nine of the compounds on the short list (see Table 2). There are no suitable data for deriving an ADI for ethylene dibromide, while there are sufficient data to calculate separate ADI values for adults and children for benzene and B(a)P.

VSD values could be calculated for the five carcinogens on the short list. Sufficient data are available to calculate separate VSD values for benzene based on animal and human data.

REFINERY SITE PATHWAYS MODEL

Either refinery site (or portions thereof) eventually may be converted to residential land use. In such a setting, future residents could be exposed to compounds present in site soils through various pathways (routes of exposure) such as the inhalation of dust or vapours, direct ingestion of dirt, and the ingestion of plants grown in local soil. Each pathway is influenced by many factors related to the environment, the compound being evaluated, and the characteristics of the person being exposed. Often these factors are combined into a "pathways model." To assess potential exposures of future residents for these two sites, the EXPOSE (*EX*Posure to *O*rganic *S*ubstances in the *E*nvironment) model was developed.

The factors and equations used in the pathways model are based, where possible, on the results of other investigations of exposures under similar circumstances, and on assumptions about the lifestyles of future site residents. When estimating possible exposures, it often is necessary to select specific values for some factors from ranges of values that have been reported. It is general practice in pathways analysis to make such selections so that exposure estimates are likely to be higher than actual exposures. In turn, this should tend to lead to actions that over-protect individuals rather than provide inadequate protection. Taken to extremes, such a "conservative" approach can lead to gross overestimations of exposures or portray people with unrealistic lifestyles. For this study, a conservative approach was followed but tempered by estimating exposures to residents whose lifestyles are likely and appropriate in the context of suburban communities in southern Ontario.

The Receptors

The pathways model is directed towards estimating the exposures to two types of future residents (also referred to as "receptors"): an adult and a young child. The adult and the child are assigned behaviors and living conditions such that they represent the greatest cumulative intakes for individuals living at either site. For example, the receptors are assumed to live in a house on-site, never leave the site, and eat produce from their garden.

The adult receptor has been assigned the characteristics of a man in the 20- to 39-year age bracket. These include a weight of 70 kg, a daily breathing rate of 23 m^3, and the consumption of 0.6 kg of produce daily. The adult is assumed to spend all of his time either in the house or in the yard. During the summer, he spends some time working in the garden. The garden is assumed to provide between 1 and 20% of the produce consumed by the receptors. During the winter, the adult is assumed to spend all of his time in the house.

The young child receptor has been assigned the characteristics of a child about 2 to 3 years old. These include a weight of 10 kg, a daily breathing rate of 5 m^3, and the consumption of 0.3 kg of produce daily. Like the adult, the child is assumed to spend all of his time in the house or in the yard. During the summer, the child spends five days a week playing in the yard. These are referred to as "active" summer days. The other two days a week are spent totally in the house and are referred to as "passive" summer days. During the winter, all of his time is spent in the house.

The pathways by which the receptors could realize exposures in this setting include direct ingestion of local soil and indoor dust, inhalation of vapours and particulate matter when indoors and outdoors, dermal absorption from soil and dust on exposed skin, and the ingestion of plants grown in local soil. All the pathways considered in the model are indicated in Fig. 1.

The Environment

The outdoor environment is based upon average meteorological and air quality conditions experienced in the Metropolitan Toronto area, although various simplifying assumptions are made where greater detail would not meaningfully improve exposure estimates. For example, the year has been divided into two seasons only, summer and winter, each lasting six months.

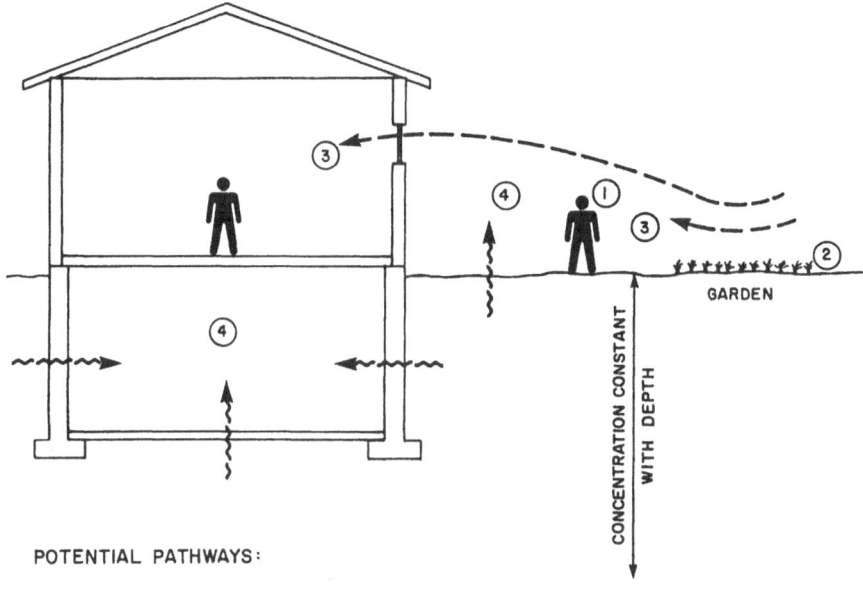

POTENTIAL PATHWAYS:

(1) DIRECT INGESTION OF SOIL
DERMAL EXPOSURE TO SOIL
INHALATION OF PARTICULATE MATTER

(2) INGESTION OF GARDEN PRODUCE

(3) DIRECT INGESTION OF DUST
DERMAL EXPOSURE TO DUST
INHALATION OF PARTICULATE MATTER

(4) INHALATION OF VAPOURS (BOTH OUTDOORS AND INDOORS)

Fig. 1. Pathways in the EXPOSE Pathways Model.

The receptors are assumed to live in a single-story house with a full, concrete block basement. Relevant aspects of indoor air quality include the indoor concentration of total suspended particulate (TSP) matter and vapour, dustfall rates, average dust coverings on indoor surfaces, and the origins of indoor dust.

For the initial approach to calculating acceptable concentrations, concentrations of organic compounds in site soil were assumed to be constant both vertically and horizontally and not to change with time. These assumptions helped minimize model complexity and were known to be conservative; that is, they would lead to overestimations of exposure.

Other Factors That Affect Exposures

Direct ingestion exposures stem from the inadvertent or deliberate ingestion of dirt and dust. This pathway is particularly relevant to young children who have a predilection for eating foreign material such as dirt and frequently put their fingers in their mouths. The adult may transfer dirt and dust from the hands to the mouth through activities such as eating or smoking.

Direct ingestion of dirt and dust by the child is assumed to be 250 mg of dirt while outdoors plus 50 mg of dust while indoors on active days. On passive summer days and throughout the winter, the child is indoors all day and is assumed to ingest 100 mg of dust.

The adult is assumed to ingest dirt from the inside surfaces of his hands on those days when he works in the garden and dust from his hands on days spent indoors.

For both the adult and the child, it is assumed that 50 to 100% of any organic compound in the ingested soil or dust is absorbed. Actual absorption rates are compound specific and probably include values less than 50%.

Dermal exposures result from the presence of soil or dust on the skin and the subsequent dermal absorption. This pathway is most relevant to outdoor activity during the summer. While absorption rates vary according to compound, an average of 0.5%/hour has been recommended as a simplifying assumption for adults.[2] For children, a rate of twice that for adults, or 1%/hour, has been used elsewhere and is used in this analysis.

Inhalation exposures occur continuously as a result of the presence of TSP matter and vapours in the air. Exposures are estimated from the amount of air breathed and the concentrations of TSP matter and vapours either outside or indoors. For both the adult and the child, it is assumed that 75% of inhaled TSP matter is retained in the lungs and that 100% of a compound associated with the retained inhaled soil or dust particles is absorbed. It is also assumed that 50 to 100% of inhaled vapours are absorbed.

Total Dose Estimates

Based on the data and considerations noted above, equations to estimate exposures via each pathway were developed and assembled into the EXPOSE model. For each pathway, an average daily dose is calculated and expressed in units of mg of soil per kg of receptor body weight per day. For example, exposures from the direct ingestion of dirt or dust are calculated as:

| amount ingested daily | × | absorbed fraction | × | percentage of days annually this occurs | + | receptor's weight |

Exposures via all pathways are converted to average daily dose estimates and summed to produce total dose estimates. These estimates are in the form of ranges to reflect the assumptions that absorption of compounds from directly ingested soil and dust ranges from 50 to 100%, that absorption of inhaled vapours ranges from 50 to 100%, and that the backyard garden can provide from 1 to 20% of the fruits and vegetables consumed by the receptors. The estimates are expressed in units of mg of local soil per kg of receptor body weight per day. These values must be multiplied by the concentration of a compound in the soil to produce exposure estimates for that compound.

INTERPRETATION OF PATHWAYS MODEL RESULTS

Initial Acceptable On-site Soil Concentrations

Acceptable concentrations of compounds in soils initially were calculated from exposure levels described in Table 2 divided by the total dose estimates. For example, if the acceptable exposure of a compound is 5 μg/kg/d and the dose estimate is 2 mg (of soil)/kg/d, the acceptable soil concentration is 2500 μg/g (i.e. 5 + 2 × 1000 to convert to units of μg/g).

Because both the acceptable exposure levels and the total dose estimates were presented as ranges, acceptable concentrations also were expressed as the ranges shown in Table 3. Values have been rounded off slightly to avoid giving the impression of undue accuracy.

Table 3. Acceptable On-Site Soil Concentration Ranges

Compound	Original Range (μg/g)	Modified Range (μg/g)	Reasons*
benzene	0.04 to 0.13	0.04 to 0.16	ambient
naphthalene	5400 to 13960	5400 to 13960	
phenanthrene	1870 to 4600	1870 to 4600	
chrysene	470 to 1150	470 to 1150	
benzo(a)pyrene	0.004 to 0.01	0.016 to 0.18	ambient and other
cresols	700 to 4390	700 to 4390	
bis(2-ethylhexyl)phthalate	70 to 12200	70 to 200	phytotox
ethylene dibromide	0.00006 to 0.0002	0.00006 to 0.0002	
1,4-dioxane	5.8 to 36.6	5.8 to 36.6	
methyl ethyl ketone	52 to 270	52 to 270	

*"Ambient" indicates ambient concentrations cause an increase to the upper end of the range; "other" indicates that the VSD calculation was refined and caused an increase of the entire range; "phytotox" indicates that phytotoxic concerns cause a decrease of the upper end of the range. Modifications not yet made concerning degradation and aesthetics.

Major Pathways and Key Parameters

The results of the pathways model indicated that estimates of exposure are highly compound-specific. One common feature of all the total dose estimates is that the child would experience the highest doses. Contributing factors include the child's relatively low weight, the relatively large amounts of soil and dust assumed to be ingested, and the amount of the time spent outdoors in summer which promotes several exposure mechanisms. Because the child is the critical receptor in this analysis and the acceptable exposures for the child are as low as or lower than those for the adult, the adult was not considered further in deriving acceptable soil concentrations.

A formal sensitivity analysis was not used to characterize the influence of specific parameters, but several key aspects can be derived from inspection of the model results. Although the contribution of each pathway varies from compound to compound, the results of the pathways model indicate that the total exposure estimates are dominated by one or two pathways. For the highly volatile compound such as benzene, the inhalation of vapours is the major pathway. For highly soluble compounds such as 1,4-dioxane and MEK, ingestion of garden produce dominates. The direct ingestion of soil and indoor dust account for the largest portions of the exposure estimates for relatively long-lived and immobile compounds such as B(a)P. The inhalation of TSP matter and dermal absorption contribute very little to total exposure estimates for all types of compounds.

The domination of the inhaled vapours pathway for volatile compounds largely results from the estimated indoor exposures. Vapour inhalation is estimated to be greatest during the six months of winter during which time the receptors are assumed not to leave the house.

The relative importance of the ingestion of garden produce to the estimates of total dose is strongly influenced by the calculated plant uptake factors and the amount of produce derived from the garden. Two plant uptake mechanisms are estimated in the model. Uptake via the roots is estimated to increase when the parameter K_{ow} decreases. Thus compounds with relatively low K_{ow} values, such as MEK and 1,4-dioxane, are estimated to have high uptake factor values with the resulting dominance of the garden produce pathway. For compounds with large K_{ow} values, such as B(a)P, chrysene, and DEHP, root uptake is estimated to be small. The second uptake mechanism, foliar deposition, predominates for these latter types of compounds but does not equal the root uptake estimated for the more mobile, soluble compounds with low K_{ow} values.

MODIFICATIONS TO THE INITIAL APPROACH

While the approach based on human health concerns appeared to identify guideline ranges that were in line with expectations, further reflection made it clear that several other factors would need to be considered before the ranges could be finalized or used as cleanup guidelines. Factors to be considered included ambient or "background" concentrations, the concentrations at which other types of adverse effects occur, the limitations of analytical techniques, the implications of ignoring degradation, and aesthetic concerns.

As a result, a series of supplementary investigations have been undertaken since the original acceptable concentration ranges were published in July 1986. Some of those studies have recently been completed and the ways to modify the acceptable ranges have tentatively been developed.

Ambient Concentrations in the Environment

Background or natural concentrations in soil have not been reported widely for many of the compounds on the short list. The refineries have recently sponsored a study of background levels for all 43 organic compounds in the communities where the sites are located. For most compounds, the background concentrations were found to be below (in some cases far below) the original ranges but there are some notable exceptions. For example, ambient levels of B(a)P typically range from 0.005 to 0.1 µg/g while the acceptable range noted in Table 3 is lower at 0.004 to 0.01 µg/g. Setting acceptable concentrations or cleanup guidelines at or below background levels could prove unworkable and impractical. As a result, it will be necessary to increase the upper values of ranges to reflect ambient concentrations. The ambient data also appear to indicate that an upward adjustment may be appropriate for the acceptable range for benzene.

Other Types of Adverse Effects

The health-based approach originally developed in this study was felt to be capable of generating "acceptable" concentrations that would be lower than those that cause phytotoxic effects. For many organic compounds, this probably is true, but there are some compounds which are well-known causes of adverse effects on plant health. For example, phthalate esters, which are found in plastics, have been associated with difficulties in raising plants in green houses. A review of the literature on phytotoxic effects of phthalate esters indicated that effects begin to occur at concentrations of approximately 500 µg/g of DEHP in soil. In this instance, the lower end of the original range (70 to 12,200 µg/g) appears to be appropriate but the upper end of the range will need to be adjusted downward to be less than the concentrations at which phytotoxic effects occur.

It is also possible that aesthetic considerations may lead to modifications of "acceptable" concentrations. The MOE has established nonquantitative requirements such as no odor being evident when a soil sample is held up to the nose and the absence of a sheen when a soil sample is immersed in water. To comply with such tests may require concentrations of organic compounds in soil to be lower than the concentrations based on health risk concerns. To date, the aesthetic requirements have not been translated into concentrations.

Limitations of Analytical Techniques

Analytical techniques now allow many compounds to be measured in the parts per billion or parts per trillion range. The experience of this study indicates that a risk-based approach is capable of identifying acceptable soil concentrations that challenge or exceed current analytical techniques. For example, the acceptable soil concentrations for EDB of 0.00006 and 0.0002 $\mu g/g$ (60 to 200 parts per trillion) can be measured but the resulting data are difficult to reproduce and merit a relatively low level of confidence. For some compounds, analytical techniques may be unable to differentiate between isomers, although the isomers may be known to present distinctly different health concerns. In such cases, it may be necessary to analyze samples using special techniques, re-evaluate the information used to establish the "acceptable" concentrations, or utilize the modifications discussed in this paper.

Implications of Degradation

The concentrations of organic compounds in site soil were assumed in the original version of the model to be constant with time. This simplified the modelling and the interpretation of health risks associated with site-related exposures but ignored the reality of degradation and the corresponding decrease in exposure likely to occur. For example, a compound with a half-life in soil of 50 days will be reduced by 99% in 350 days. Six of the compounds (benzene, naphthalene, phenanthrene, cresols, DEHP, and MEK) have half-lives of less than 50 days. B(a)P, the compound on the short list with the longest reported half-life of approximately 450 days, will be reduced by 99% in approximately ten years. It also must be remembered that reported half-lives may be based on tests in laboratories under conditions such as constant temperature that do not reflect actual site environments and therefore the results may not represent events that occur on site. By deliberately ignoring degradation in the modelling of receptor exposures, the risks associated with the original ranges overstate the actual risks to site residents by considerable margins for most compounds.

Soil concentration data are now becoming available for the refinery sites that span the past two years. The data will allow greater confidence to be placed in degradation rates that are occurring in site soils. Consideration is being given to refining the model such that degradation is taken into account. This will involve estimating exposures over a period equal to a certain number of half-lives (for example, ten half-lives represents 99.92% disappearance), rather than assuming constant soil concentrations over the lifetime of a receptor. This modification will increase "acceptable" concentrations of all compounds by a factor of approximately 2 to 10.

DEMONSTRATING APPROACH VIABILITY

Before the overall approach and the results that it generates can be used to identify cleanup guidelines, its suitability will have to be demonstrated to several groups. To date the approach has been presented only to provincial agency representatives, and they have expressed general satisfaction with the underlying basis of the approach and agreed that the EXPOSE model is sufficiently conservative and protective of human health. But there remains the major task of presenting the approach to elected municipal and provincial representatives, as well as to members of the public.

While the overall approach was been developed based on sound scientific and engineering principles, it is anticipated that some reviewers will still greet it with an intuitive feeling of suspicion or mistrust. The fact that an approach based on health risk concerns is capable of identifying "acceptable" concentrations that far exceed ambient levels may be seen as indicating that the approach is misguided. Provincial agencies may feel it necessary to adjust guidelines downward for no reason other than it may be difficult to convince decision makers or members of the public that such high levels are indeed "acceptable." It is recognized that such intangible influences have the potential to lead to modifications of "acceptable" concentrations far greater than those generated by scientific considerations.

One way to put the "acceptable" concentrations produced by the approach into perspective may be to describe relative risks. The pathways included in the EXPOSE model are not the only ways in which future residents would be exposed to organic compounds. Regardless of where a person lives, many of the compounds found in site soil also are present in the air, drinking water, and food as a result of natural processes, man's activities, or both. While there is insufficient information to estimate exposure for many compounds, there is sufficient information for two of the carcinogenic compounds on the short list that data can be used to put the estimated exposures of the receptors into perspective.

Typical Ontario residents are estimated to be exposed to 4.9 µg/kg/day of benzene and 0.012 of B(a)P. These values indicate that a resident of either site with these two compounds present at the original "acceptable" concentration ranges would have exposures from the soil equal to about 0.7 to 2% greater than a resident living elsewhere. As such, the exposures associated with the compounds in the soil would represent only minimal incremental increases in overall risk from these two compounds.

CURRENT STATUS AND FUTURE DEVELOPMENTS

The original approach has evolved from one based on human health considerations to one in which the risks of adverse health effects provide initial estimates of "acceptable" concentrations that then are subject to changes stemming from various nonhealth related considerations. These factors include the recognition that it is not be feasible to establish guidelines at or below ambient (background) levels; it may not be acceptable to establish guidelines that are far above background levels; that other types of adverse effects may occur at concentrations less than those associated with human health effects; and that it would be impractical to establish criteria at concentrations that cannot be measured accurately and confidently. Several of these concerns are being incorporated into the overall approach in the form of rules for modifying the original health-based results. Modifications made to date are summarized in Table 3. It also must be emphasized that the approach utilizes information and assumptions that are site-specific. The resulting "acceptable" concentrations should not be used at other sites without proper analysis and review of the appropriateness of the assumptions.

Once all of the modifications have been investigated and the form of the overall approach has been finalized, the process of presenting the approach to decision makers and the public will begin. It is anticipated that the guidelines will face a wide variety of challenges before receiving the acceptance of regulators, decision makers, and the public.

REFERENCES

1. World Health Organization (WHO), 1974, "Assessment of the Carcinogenicity and Mutagenicity of Chemicals," Report of a WHO Scientific Group Tech. Rep. Series No. 546.
2. Hawley, J. K., "Assessment of Health Risk from Exposure to Contaminated Soil," *Risk Analysis* **5**(**4**):289-302 (1985).

Quantifying and Comparing the Benefits of Risk Reduction Programs to Prioritize Expenditures

Ronald J. Marnicio
Envirosphere Company
Dublin, OH

Research Methodologies Study Group[a]
Carnegie-Mellon University
Pittsburgh, PA

ABSTRACT

Many techniques are being used to quantify the costs and benefits of risk reduction programs to allow competing programs to be compared and ranked for purposes of allocating limited available resources. Often, insufficient attention is paid to selecting appropriate measures for scaling program benefits, identifying implicit value-laden assumptions associated with the methodologies, and exploring the implications of inherent uncertainties. This paper presents the results of a case study of these issues which critically evaluates an approach developed at General Motors that uses cost and actuarial data to compare the cost effectiveness of various risk reduction programs. An earlier "best estimate" application of the approach to a set of risk reduction programs addressing diverse technological risks is reexamined to explicitly characterize and factor in the associated uncertainties. The strengths and limitations of this benefits quantification and prioritization approach are discussed. Considerations which are generally applicable to comparing and ranking the costs and benefits of programs addressing risks of highly diverse origin are described, and their impact on the selection of an appropriate cost/benefit quantification and comparison methodology are presented.

KEYWORDS: Cost effectiveness, life expectancy, longevity, mortality, prioritization, uncertainty analysis

INTRODUCTION

Assessments of risk in the public and private sectors are comprised of both risk analysis and risk management activities. Here, risk analysis is taken to encompass all of the activities undertaken to identify the nature, extent, and magnitude of the physical, chemical, or biological impact or assault associated with a given hazard. Risk management, then, involves evaluating the risk and making a judgment as to how best to reduce, eliminate, or

---

a. Participants in the Study Group were Deborah Amaral, Felix Dayo, Theresa Donahoe-Nestor, Jennings Ellis, Conrad Eustis, Eden Fisher, Jeffrey Funk, Carol Giron, Steve Goldstein, Ronald Marnicio, David Meeker, Gil Miller, Granger Morgan, William Mura, and Giri Tayi.

tolerate it. Institutions and individuals charged with risk management must routinely make decisions about which of the possibly many diverse risks within their purview should be addressed first, given the limited resources for risk reduction programs. A need to prioritize and rank risks quickly becomes clear. An ability to rank the various risks according to their "importance" would allow the most "important" risks to be allocated resources ahead of risks of lesser "importance."

Many approaches to prioritization have been developed and are being used in the management of risks. These approaches most often center on the use of cost/benefit or cost effectiveness analyses. Cost/benefit approaches require that the cost of the proposed risk reduction or elimination program be calculated and that the benefits of reducing or eliminating the risk be evaluated and quantified. The difficulties associated with quantifying the benefits of risk reduction programs in monetary terms are obvious. Having identified no generally superior approach for comparing and ranking risk reduction program benefits and allocating resources among them, many techniques have been proposed and refined by risk managers over the years. These techniques are generally appropriate for application under prescribed circumstances in certain risk management situations. However, given their availability, it is easy for one of these techniques to be adopted and used in an assessment outside the scope in which it was developed and proposed. Without a complete appreciation for the implications of applying the technique or methodology in a new context, the risk manager cannot be certain that the assessment will enhance the risk management decision and actually reflect what the risk manager considers to be "important."

One such methodology that was proposed to refine the basic cost effectiveness approach focused on "increased longevity" as an alternative measure of the benefits derived from large-scale risk reduction programs.[1] This methodology, developed at General Motors (GM), is based on the results of an examination of actuarial data. It provides a means for estimating the societal benefit of risk reduction programs that can be used in conjunction with program costs to rank and prioritize expenditures.

This paper briefly reviews this methodology and an application of its use in the context of risk expenditure prioritization. The caveats to the use of the methodology are examined. The results of a reanalysis of the development and application of this approach to identify its strengths and weaknesses for various types of risks and risk reduction programs are presented.

RISK REDUCTION BENEFITS

The Role of Life Tables in Benefits Quantification

The technique typically employed to estimate the increase in longevity due to averting or eliminating a particular risk of a premature death involves the construction of "life tables." Life tables make use of age-specific mortality rates (ASMRs), which are the number of individuals of a particular age who die each year due to a particular cause, normalized to the total population of people of that age who are at risk. These ASMRs are used to quantitatively describe the relative risk of dying at each age due to various causes. These tables[2,3] also can be used to compare the nature of various risks and to provide a source of data for estimating the potential benefits to be gained through programs to reduce them.

Life tables may contain a calculated quantity specified as "added years of life,"[2] which is related to eliminating the deaths attributable to a particular risk. This statistic was used in the development of the GM methodology. The "added years of life" statistic is calculated by simulating the duration of a statistically average person's life by applying the appropriate ASMRs for all causes of death, and then recalculating his life expectancy with all the ASMRs for the particular risk set equal to zero. The calculated life expectancy assuming the elimination of the one risk will always exceed the life expectancy calculated assuming all risks are in force. This difference in calculated life expectancy is the "added years of life" statistic for the risk in question. This calculation is based on the assumption that the ASMRs for the set of risks *not* nullified are independent of the ASMRs for the risk being examined. This calculational methodology has been discussed by a number of investigators in different contexts.[4,5,6]

If multi-decrement life tables were sufficiently comprehensive to address all the risks of premature death of possible interest, the ranking of potential program benefits would be a relatively simple exercise in cases where increased longevity provides a suitable metric. However, the need to compare and evaluate new risks and evolving risk reduction programs would require highly disaggregated, up-to-date life tables. Because life tables are quite costly to construct and highly impractical to disaggregate to any great degree, simpler, less data-intensive methods for estimating this longevity gain statistic were sought.

THE GM LONGEVITY GAIN CORRELATION

The GM approach[1] to estimating the longevity gain statistic in a simpler, less data-intensive manner began with the 1964 multi-decrement life tables.[2] It was anticipated that the increase in longevity due to eliminating a particular risk would be a function of the risk's mortality rate and the age distribution of its impact relative to the exposed population. This hypothesis was tested by plotting the tabulated longevity gain statistics against the corresponding CMRs for the various risks described in the life tables, as shown in Fig. 1. A reasonably linear relationship was exhibited over three orders of magnitude of the CMR. Some scatter about the line was seen to arise, which was attributed to differences in the way certain risks impact various age groups in the population. The expression for this linear relationship was found to be:

$$\text{LONGEVITY GAIN} = 0.02 \times \text{CMR} \ ,$$

where the CMR represents the number of annual deaths resulting from the risk in a population of 100,000 exposed individuals. The generality of the correlation was tested by plotting the same quantities for populations in the U.S., Canada, and Australia for four different risks (neoplasms, cardiovascular fatalities, violence, and motor vehicle accidents), for both men and women and for several different years. Although the data exhibited somewhat more scatter about the line, the original correlation was seen to be robust. The linearity of the correlation was postulated to reflect the fact that small changes in longevity could be approximated by the first term in a Taylor series if the change in mortality rate were sufficiently small.[1]

Given that this correlation appeared to be reasonably general, it offered a means for estimating the average population longevity change associated with the elimination of particular risks which are not explicitly addressed in multi-decrement life tables. This correlation could then also be applied to estimate the potential longevity change derivable from programs designed to reduce or eliminate various categories of risk.

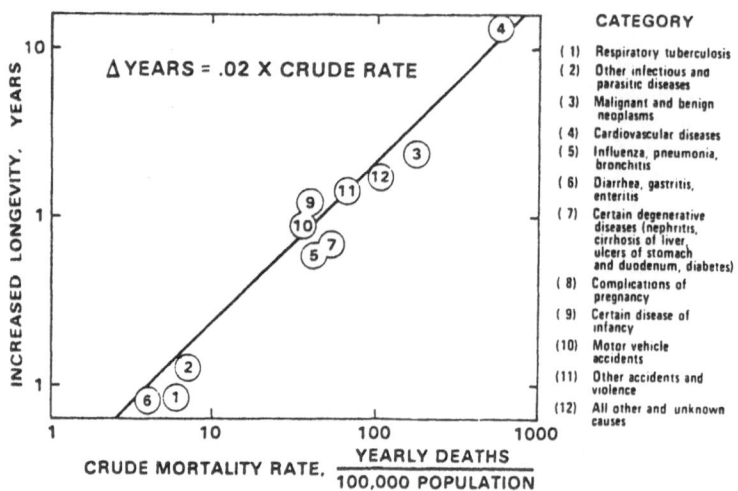

Fig. 1. Relationship Between Increased Longevity Due to Eliminating Each Specific Cause of Death and Its Crude Mortality Rate (Data for U.S. Males — Adapted from Ref. 1).

SAMPLE APPLICATION OF THE COST EFFECTIVENESS APPROACH

Estimates of Cost to Increase Longevity

In order to gain some perspective on the use of this correlation in comparing risk reduction programs, the costs and longevity gain benefits of ten medical, environmental, and safety programs were estimated and compared in an advertisement for the General Motors Research Laboratories, reproduced in Fig. 2. The cost effectiveness measures for the ten risk reduction programs were plotted against their impact on average U.S. longevity, as calculated using the GM correlation. Data for this analysis were necessarily drawn from a variety of studies in several sectors. Typically, the information was based on several different methods of cost estimation. The calculated estimates for the existing or proposed programs examined ranged in cost effectiveness over five orders of magnitude. The developer of the approach warned that the accuracy of some of the estimates may be questioned, but that the results would still represent a useful aid to policy decisions.[1] This advertisement also presented the longevity gain correlation, noting that it served two purposes: (1) it provided a perspective of the days and years gained from various diverse risk reduction programs; and (2) when combined with cost estimates, it would allow the cost effectiveness of those programs to be ranked. The advertisement concluded that "through such unbiased comparisons, policy makers can obtain a clearer picture of which programs offer the greatest potential gain for a fixed budget and, thereby, have a better basis for decision."[7] This approach was later drawn on in relation to spending for nuclear safety,[8] and the establishment of quantitative safety goals in regulations.[9]

CAVEATS

In presenting the derivation and application of this approach to benefits estimation and cost effectiveness prioritization, the developer highlighted a number of caveats:[1]

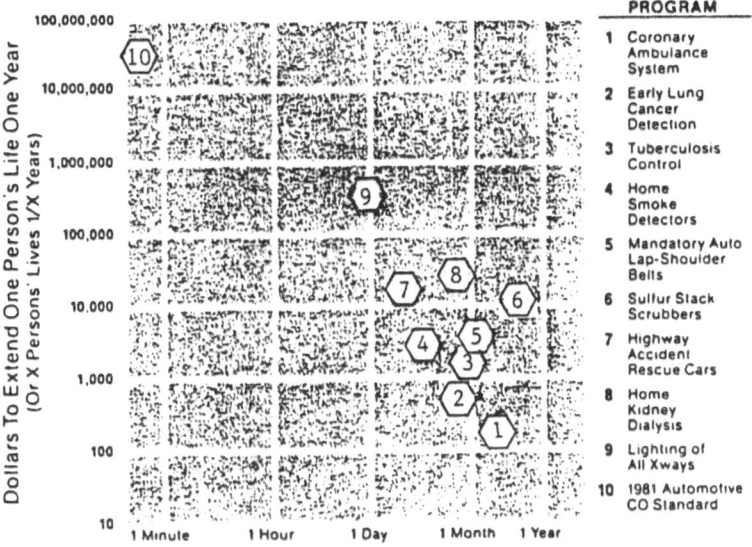

Fig. 2. Risk Reduction Program Cost Effectiveness Comparison (Adapted from Ref. 7).

- The calculation of the longevity gain statistic assumes that the effects due to the risks not eliminated are independent of the risk removed in each case.

- Subgroups in the population may exhibit mortality rates and risk susceptibilities which differ significantly from those characteristic of the general population.

- Individuals' risk perceptions do not always compare favorably with the magnitude of the actuarial risk.

- Only one measure or attribute of the benefits [longevity] has been addressed; other factors may enter into decisions regarding policy.

- Cost estimates can be very difficult to compare when they are extracted from a variety of sources.

- Uncertainties in both the effectiveness measures and elasticities should be emphasized; inputs to decision making about these particular programs could not be developed due to the large remaining uncertainties.

EXPLORING THE LINEARITY OF THE GM CORRELATION

A Further Analysis of the Original Data

Before more extensive use was made of the GM correlation for longevity gain, an attempt was made to explain or justify its apparent linearity. The first exercise involved deriving an algebraic expression for the longevity gain statistic. This was done by parametrically reconstructing the life table calculations for a simplified hypothetical scenario in which there were only three age categories and three possible causes of death, one of which was assumed to be eliminated. The fate of an hypothetical cohort was then "simulated" alternately assuming the presence and absence of the risk being singled out, and the resulting longevity gain was expressed as a function of the fraction of people in each age category that die of each particular cause of death. The longevity gain statistic was seen to be represented by a power series in the CMR for the risk being eliminated. In addition, the coefficients of the additive CMR terms were influenced by the risk profiles of the risks *not* eliminated. Order of magnitude calculations indicated that the linear CMR term dominated the higher order terms. Assuming a linear relationship, the coefficient was shown to be a function of the nature of the risk being examined and the risks that coexist with it. The study also showed that risks to the younger segments of the population displayed larger coefficients than those affecting primarily older individuals.

A second exercise was also undertaken which was far less fundamental, but possibly more intuitive. By combining a set of common variables relating to risks and mortality to create a meaningful parameter with the units of the longevity gain statistic, a relationship similar in form and character to that derived by GM through the risk-specific data analysis was obtained. This exercise led to the same conclusions.[b]

ASSESSING UNCERTAINTIES IN PROGRAM COSTS AND BENEFITS

Reanalysis of Selected Risk Reduction Programs

In addition to exploring the basis of the GM correlation as a means of quantifying the benefits of risk reduction programs, the study group reexamined the risk reduction programs which were compared and ranked in the IEEE Spectrum advertisement[7] (see Fig. 2). The explicit goal of the group was to consider and quantify the uncertainties associated with both the benefits and the costs of the programs. As a common target, 80% confidence intervals for both the cost effectiveness and the estimated longevity gain were sought. These resulting confidence "zones" have been overlaid on the original plot and presented in Fig. 3.

A few aspects of this follow-up analysis must be emphasized. The results presented on this plot reflect an assessment of the information that was *originally available* to the group. The program costs were expressed in terms of 1979 dollars to be compatible with the original estimates. New or updated information which has become available since that time would quite likely influence the shape or size of the confidence "zones" shown in Fig. 3. However, since the purpose of the exercise was to explore the effect of uncertainties on the quantification and prioritization process, these refinements were not deemed to be critical to the current discussion. In addition, only seven of the original ten programs represented in Fig. 2 were reexamined due to the relatively small size of the study group.

Before discussing the general lessons learned from the effort, a brief synopsis of the seven reexaminations is presented below to provide the necessary background. Each synopsis contains a brief description of the risk reduction program and a summary of the results of the uncertainty evaluation (see footnote cited above).

Coronary Ambulances and Mobile Coronary Care Units (CA/MCCU)

Description — CA/MCCUs are emergency vehicles that are especially equipped and staffed to aid heart attack victims in the crucial period before they reach a hospital.

---

b. More detailed descriptions of these exploratory exercises and the approaches used to calculate the uncertainties in the benefits and costs of individual programs are presented in a working paper of the author.

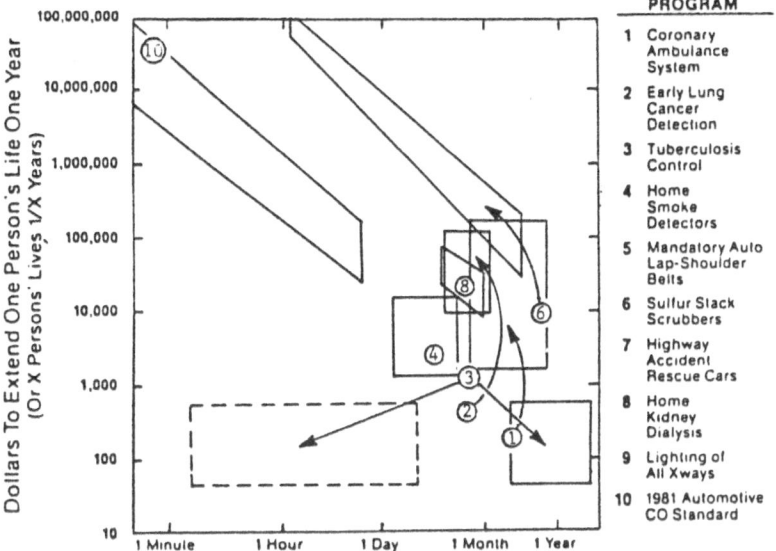

Fig. 3. Results of the Uncertainty Analysis Showing the 80% Confidence Zones.

The plot contains the following:

Y-axis: Dollars To Extend One Person's Life One Year (Or X Persons' Lives, 1/X Years) — from 10 to 100,000,000

X-axis: Average Longevity Gain For Total U.S. Population — 1 Minute, 1 Hour, 1 Day, 1 Month, 1 Year

PROGRAM

1 Coronary Ambulance System
2 Early Lung Cancer Detection
3 Tuberculosis Control
4 Home Smoke Detectors
5 Mandatory Auto Lap-Shoulder Belts
6 Sulfur Stack Scrubbers
7 Highway Accident Rescue Cars
8 Home Kidney Dialysis
9 Lighting of All Xways
10 1981 Automotive CO Standard

	ORIGINAL ESTIMATE	SUMMARY BENEFITS	COSTS	UNCERTAINTIES
# 1	Not Covered	OK	Higher	~ 1 Ord. Mag.
# 2	Not Covered	OK	Higher	~ 1 Ord. Mag.
# 3	Not Covered	Inadequate	Lower	1-2 Ord. Mag.
# 4	Covered	OK	OK	~ 1 Ord. Mag.
# 6	Not Covered	Lower	Higher	3+\ Ord. Mag.
# 8	Covered	OK	OK	<1\ Ord. Mag.
#10	Covered	OK	OK	~3\ Ord. Mag.

Results — An 80% confidence interval on program benefits spanned over an order of magnitude in increased longevity and the 10% and 90% percentiles in normalized program cost differed by a factor of 15. The plotted confidence "zone" did not include the corresponding point originally shown in Fig. 2.

Early Lung Cancer Detection

Description — This program involved screening a large number of individuals using chest X-rays or sputum cytology in an effort to detect and localize cancerous growths in the lung at an earlier stage so that they could be more effectively treated.

Results — The original estimate for longevity gain was within the 80% confidence interval developed. However, the cost effectiveness plotted in the advertisement was below the range identified in this analysis. The 80% confidence interval for both benefits and costs spanned approximately one order of magnitude.

Tuberculosis Control

Description — Tuberculosis (TB) is generally controlled through the use of powerful antibiotics and, in certain cases, isolating the affected individual. Identifying and testing the persons in contact with known carriers of the disease also plays an important role in its control.

Results — The GM correlation did not give satisfying results in the case of this contagious disease, even after extensive reinterpretation of the "CMR" parameters to reflect morbidity. The benefits uncertainty calculated using the *mortality* based correlation is shown in Fig. 3 as dashed. The solid zone (morbidity based) represents the study group's best judgment. Roughly one order of magnitude in both benefits and costs was estimated, and neither the solid nor the dashed confidence zone contained the original point.

Home Smoke Detectors

Description — This program involved the installation and maintenance of smoke detectors in homes.

Results — The confidence zone for this point was seen to span roughly one order of magnitude along both dimensions and to include the originally plotted point.

Sulfur Dioxide Stack Gas Scrubbers

Description — This risk reduction activity involved the application of an SO_2 removal system to the stack of a coal-fired power plant.

Results — The resulting confidence zone was seen to be skewed in shape due to the use of the uncertain CMR in the estimation process for both the benefits and the costs of this risk reduction program. Due to limitations in the data available at the time of the original study, especially for costs, the zone is probably not as large as it would be if it were to be recalculated today. The 80% confidence interval for the longevity gain spanned over three orders of magnitude. The originally plotted point from the ad was not within the calculated confidence zone.

Kidney Dialysis

Description — This program consisted of providing kidney dialysis treatments to victims of certain kidney failures.

Results — Based on the literature surveyed, relatively little uncertainty was associated with this program compared to the others. The 80% confidence intervals for both cost effectiveness and longevity gain spanned only about a factor of five. The diagonally oriented confidence zone (denoted with an '\' in the summary table of Fig. 3) was due to the joint influence of the average life extension parameter for individual kidney dialysis patients on both the X and Y axis quantities. The 80% confidence zone covered the original point.

Tighter Automotive CO Standard

Description — This program involved the contemplated tightening of the CO emission standard for automobiles in 1981, from 15 to 3.4 grams/mile. Additional controls, affecting the maintenance and fuel economy of the vehicles, were assumed to be needed on new cars to bring about these reductions.

Results — The confidence interval for program cost effectiveness spanned nearly an order of magnitude. The 80% confidence band for the longevity gain exhibited a distinct upper bound, but extended far to the left due to a 10% chance that the change in mortality resulting from the program would be zero. The diagonally oriented confidence which contained the originally plotted point reflected the joint quantitative dependence of the two quantities on the longevity gain.

IMPLICATIONS ON THE PRIORITIZATION OF EXPENDITURES

The Prioritization Criteria

In reworking the sample analysis, a number of factors were highlighted which indicated limitations to the use of this type of approach for prioritizing risk reduction expenditures. At first, it seems to be reasonable that the risk reduction programs that are most cost effective and do the most to increase overall life expectancy should be given the highest priority for implementation. It is only in attempting to apply these duel criteria to a set of risks of diverse nature, impact, and methods of control that the potential shortcomings and unexpected results of the approach are brought out. The highlighted factors are now discussed to flag instances in which the longevity gain benefits correlation and the overall cost effectiveness prioritization approach may yield misleading or unsatisfying results.

The Longevity Gain Correlation

The GM longevity gain correlation is elegant in its simplicity. It provides an actuarially based measure of the significance of a broad spectrum of mortal risks on the life expectancy of an average individual. Only a minimum of data on the affliction, its crude mortality rate, is needed to apply the correlation in its simplest form. Subsequent analysis of the correlation provided a technical basis for its linearity and showed that the multiplicative coefficient could be "fine tuned" when the age distribution of the risk is known. However, this simplicity could cause one to lose sight of the implicit assumptions associated with its use. A few of these caveats were presented earlier. Reiterating, this benefits estimation approach assumes: that longevity is the *only* attribute of the risk to be considered in the prioritization decision; that the cause of death under examination acts independently of other risks; and that programs benefiting subgroups of the population must be justified on the basis of their effect on the overall population.

Estimating risk reduction program benefits solely on the basis of their impact on mortality can make certain types of programs appear to be very effective while damning others in comparison. One obvious distinction arises in a comparative analysis of potentially fatal accidents and diseases. In many cases, a program to reduce the risk of a fatal accident will result in an averted death when it is successful, such that the life expectancy of the would-be victim is not subsequently affected. When the program fails, a death would result and a mortality change would be a direct indicator of the program's performance. Home smoke detectors serve as a pertinent example of this case. Conversely, aversion of a potentially fatal disease, such as through kidney dialysis treatments, would prolong life, but may not allow the patient to live an otherwise "normal" life expectancy due to the continued presence of the disease and the threat of complications. This example serves to support a benefits quantification technique which somehow accounts for the "quality" of life extension, as well as the amount of increased longevity.[10] Many such approaches have been discussed and applied. The GM correlation would generally lead to more "satisfying" results in comparisons of accidents and diseases if the programs resulted in "clean escapes" in the case of accidents and "cures" in the case of diseases, rather than treatments of chronic afflictions.

Given a mortality based approach for estimating risk reduction benefits, programs to reduce CO or sulfur emissions will not rank so well in comparison to other programs, as was seen in Fig. 3. The impact of these environmental risks of greatest concern is generally not mortality, but *morbidity* and the added stress placed on persons susceptible to other potentially fatal risks. The CMR for CO in this study, for example, had to be estimated by considering the distribution of ambient concentrations and the fraction of the population that was particularly susceptible to respiratory, acute cardiovascular, or central nervous system conditions. The implicit assumption of independent risks would not appear to hold up in this instance. Some deaths attributed to the more outwardly recognizable risks, such as heart attack, should possibly be apportioned between the many contributory risk factors which could increase overall stress to the point where the cataclysmic event occurs. This would increase the estimated benefit of risk reduction programs like those for CO and SO_2 emissions in which the risk cannot always be directly linked to changes in mortality.

Another limitation to estimating benefits based on mortality was clearly demonstrated by the reanalysis of tuberculosis treatment programs. TB is a contagious disease which has been effectively controlled over the years to the point where relatively few deaths now result from it. Fig. 3 displays two different confidence zones developed using the GM correlation. The solid line confidence zone reflects the use of adjusted *morbidity* statistics for TB in place of its CMR. The dashed line zone reflects the use of current *mortality* estimates. The solid zone represents the accumulated benefit of TB control programs that have been implemented over time. This approach recognizes that significantly higher mortality would result from this contagious disease if the control programs did not currently and previously exist. The dashed line results reflect the marginal benefit of further TB control expenditures beyond the levels that would maintain the current level of risk. An analysis for prioritizing expenditures must clearly differentiate between rankings based on overall or marginal benefits (assuming current levels of effectiveness).

The TB case study also demonstrated that externalities associated with risk reduction programs can further limit the utility of the longevity gain correlation. Controlling the risks from a contagious disease benefits more than those who actually contract the disease, because the likelihood that others in the population may ultimately contract it is also reduced. It was clear in this instance that using mortality rates in conjunction with the correlation did not provide an accurate measure of societal benefits.

The GM correlation scales benefits based on expected longevity gains of an "average" individual. In some cases, a large fraction of the population may possess "average" characteristics. In the case, for instance, of home smoke detectors, such a benefits measure may be quite meaningful. In other cases, the diversity of the population (in terms of subgroup susceptibilities or host factors) results in a very small fraction of the population being "average." Prioritization based on "average" benefits may not then reflect the criteria desired by the decision maker. For example, the results for coronary ambulance service assumed maintaining response times typical of urban and suburban settings. A parallel analysis for rural areas would certainly not appear as attractive due to the greater average distances involved and the effect of distance (and time) on both the cost and medical effectiveness of the program. In this example, the program clearly has an implicit demographic feature that should not be ignored in making the decisions. Such an oversight could occur as the result of using the overall longevity gain estimation framework. Similarly, lung cancer screening for smokers, which is most accurately reflected in Fig. 3, rates higher than programs to screen the whole population. This role of host factors in risk has been reflected in the setting of X-ray screening policies in recent years to target particularly susceptible subpopulations. Here again, the general population based benefits measure may not be responsive to the decision maker's needs.

Subpopulation susceptibilities also confound recent efforts to quantify the value of various risk reductions for purposes of making *personal* risk comparisons and judging risk acceptability,[11] or providing "anxiety benchmarks" to be used as the basis for adjusting one's lifestyle to offset society-wide risks.[12] These considerations necessarily involve individuals' risk perceptions and host factors. The GM correlation does not directly address this aspect of personal risk decision making, and assumes a hypothetical individual with "statistically average" host factors.

The lung cancer screening program reanalysis highlighted another aspect of longevity gain estimation that does not appear to be adequately captured by the GM correlation. Detecting a case of lung cancer through screening, as opposed to discovering it in a more advanced state when clinical complications begin to appear, can lead to a greater probability of a cure (a "cure" in this context typically means that the patient survives at least 3 or 5 years following treatment). Developing an appropriate CMR from the "cured" rate and interpreting any subsequent results rarely will be straightforward.

Estimates of Cost Effectiveness

One factor in prioritizing expenditures to reduce risks is equity: who benefits from the risk reduction program and who bears the cost. The programs examined were quite diverse, particularly in regard to "public" versus "private" cost and benefits considerations. In the case of home smoke detectors, for example, an individual or family typically bears the cost and receives the benefits (some benefits are passed on to the community from this private action, in terms of less demand on and risk to the fire company, and a reduced probability that a neighbor's house will catch on fire). Other programs are more public sector in nature, such as sulfur emission controls from a power plant. The general public receives the benefits of cleaner air and absorbs the costs ultimately in their electric rates. Kidney dialysis treatment is largely subsidized by the public, through federal or insurance monies, while benefiting particular individuals. The cost of emission control equipment for automobiles is paid by the individuals who buy new cars, while essentially the entire urban population accrues the benefits. Equity considerations often play an important and sometimes primary role in decisions to prioritize expenditures.

Another difficult aspect of cost estimation is where to draw the line on included costs. It was very difficult in the reanalysis to maintain a consistent scope of included costs across the entire set of risk reduction programs due to their diverse nature, the manner in which data on the cost of each was reported, and different judgments about which procedures were required and which were optional. For example, lung cancer or TB screening and treatment costs seemed to clearly include the cost of disposable diagnostic materials and the fees and wages of the medical personnel who administer them. Inclusion of other expenses was less clear, such as an amortized fraction of the costs of X-ray machines and related equipment and any follow-up visits to a physician. In the case of TB, some states required hospitalization during treatment while others did not. Whether or not to include this latter type of cost confounds the analyses and leads to difficulties in comparing risk reduction programs involving different levels of screening, diagnosis, and treatment. Published program costs often aggregate expenditures in a number of different ways, making it difficult to attribute costs separately to these various components.

The quality of the life gained as the result of the programs must be determined to be or not to be appropriate for consideration in any particular cost effectiveness comparison. Difficulties arise in comparisons when some programs result in no deterioration in the quality of the victim's remaining life while others do. For instance, if a seat belt prevents a death, generally the quality of the person's remaining life would not (in the long run) be different than it would have been had the accident not occurred (however, some counter examples can clearly be identified). But with kidney dialysis, some patients can no longer hold certain types of jobs or participate in particular activities. Evaluating the individual and societal cost of these changes in available lifestyle in a consistent manner is extremely difficult to do when the risk manager needs to take these costs into explicit account.

Another subjective component of the cost analysis is how to consider and evaluate the inconvenience and discomfort associated with the risk reduction programs. It is generally quite difficult to quantify these features of the diverse programs, even in nonmonetary terms. For example, how does one objectively gauge the worth of not having to buckle a seat belt upon entering a car, or estimate the value of the discomfort associated with a lung cancer screening procedure? The value of the lung cancer screening, with its inherent cost, discomfort, and inconvenience, would certainly not seem so great to the individual who is discovered to have lung cancer at a curable stage.

The reanalysis of the TB program highlighted the case of the contagion. The plotted results compared the marginal costs of this program to the complete implementation costs of the other programs. This cost for TB most closely approximates the expenditures that would have to be made at *this point* in time. If the screening program were to be eliminated, based on this cost effectiveness, a TB outbreak could result with significantly greater economic and health costs. In selecting a methodology for cost effectiveness comparisons, flexibility is required to enable cases such as those involving a contagion to be fully considered.

The kidney dialysis program presented a case in which institutional preferences for a particular form of program hindered the ability of first order cost effectiveness comparisons to adequately reflect a patient's economic choices. Federal policy under Medicare required kidney dialysis treatment to be administered in hospitals in order for the patient to be reimbursed. In many cases, however, home dialysis treatment was just as effective but less costly. Thus, the cost effectiveness methodology must be cognizant of, and be compatible with, individuals' economic constraints and anticipate their behavior.

Several cost and benefits externalities were seen to be associated with the programs. Screening for one disease (lung cancer, for example) can detect symptoms of other diseases (such as TB or emphysema). In addition, better or more frequent medical care as part of treatment for one disease is likely to lower the probability of contracting others or decrease their impact. On the other hand, false diagnoses from screening programs may lead to unnecessary follow-up procedures that cause mental anxiety and physical pain, as well as greater expense to the individual or society. Such externalities must be treated in a uniform manner in systematic cost and benefit evaluations.

Another cost consideration in public expenditure decision making is that the public has historically been willing to incur substantially more cost to avert the death of an identified or nameable victim than for a "statistical death." Instances such as the recent three-day effort of numerous individuals to rescue an 18-month-old girl who had fallen into a dry well exemplify this behavior. In this study, programs directed at named individuals (such as kidney dialysis or coronary ambulance care) may prove to be more publicly acceptable or attractive than programs to reduce air pollutants, regardless of the results of a cost effectiveness analysis.

Cost effectiveness comparisons of risk reduction programs generally do not make sufficient allowance for near-term technological advancement or "learning curve" effects in relation to the screening or treatments involved. In many fields of medicine, technological advancements occur at a tremendous rate, and it would not be prudent to base the decision on future expenditures on the cost and performance of techniques that will be soon obsolete. This aspect is generally difficult to explicitly factor into an analysis due to the unknowns associated with the unproven techniques.

Uncertainties in Benefits and Costs

The uncertainties shown in Fig. 3 stem from unknowns about the values of the various parameters used in the benefits or cost models or, alternatively, from uncertainty about the appropriateness of the models themselves. For most programs, uncertainties about the CMRs, the unit costs, or the spared victims' projected life expectancy were assessed and reflected in the calculations. Generally, the 80% confidence intervals for either costs or benefits spanned at *least* an order of magnitude. The use of the CMR in the longevity gain correlation for TB was questioned and a new model formulation (based on morbidity) was proposed. Quite different results were evident for the alternative models using mortality versus adjusted morbidity data. A lack of independent data on certain programs (such as CO and SO_2 emission controls) required that the costs and benefits be estimated in a correlated fashion, yielding the diagonally oriented confidence zones shown.

The ultimate question raised by the reanalysis revolves around the prioritization decision. If one had to decide on the relative appropriations for the seven reevaluated programs, would they differ from what would have been identified from the original "best estimate" plot? Clearly, incorporation of the uncertainties tended to reduce the apparent differences in costs and impacts among the programs. Based on the information plotted in Fig. 3, it would be difficult to argue that there were significant differences between the programs for kidney dialysis, lung cancer screening, coronary ambulance care, and home smoke detectors at this level of confidence. The TB program was very difficult to compare to the others due to its relation to a contagion and the somewhat "subsidized" nature of its costs over the past decades. The paucity of data and the difficulty in mortality attribution with regard to environmental emissions led to considerable uncertainty.

Delving further into the nature of the specific programs tended to accentuate their differences rather than the points they had in common, and made the comparison process somewhat more difficult. This effort then highlighted the importance of the various nuances relating to benefits estimation and cost evaluation which, although not captured in the "best estimate" analysis originally presented, would appear to play an important role in decisions to prioritize expenditures. An unavoidable problem in making comparisons of this nature is ensuring that apples are being compared to apples, and that costs and benefits are being attributed in a consistent manner. While this was the clear intent of the study group, a vigilant effort probably only succeeded in reducing the problem to a comparison of Red Delicious apples to Granny Smith or MacIntosh apples. In the end, the prioritization decision would depend on whether you were packing a lunch box or baking a pie. So it is with risks.

REFERENCES

1. Schwing, R. C. "Longevity Benefits and Costs of Reducing Various Risks," *Technology Forecasting and Social Change* 13:333-345, 1979.
2. Preston, S. H., N. Keyfitz, and R. Schoen, *Causes of Death: Life Tables for National Populations*, Seminar Press, Inc.: New York, 1972.
3. Namboodiri, K. and C. M. Suchindran, *Life Table Techniques and Their Applications*, Academic Press, Inc.: Orlando, FL, 1987.
4. Tsai, S. P., E. S. Lee, and R. J. Hardy. "The Effect of a Reduction in Leading Causes of Death: Potential Gains in Life Expectancy," *American Journal of Public Health* 68(10):966-971, October, 1978.
5. Cohen, B. L. and I. S. Lee. "A Catalog of Risks," *Health Physics* 36:707-722, June, 1979.
6. Birnbaum, Z. W. "On the Mathematics of Competing Risks," *Vital Health Statistics* 2(77):1-77, U.S. National Center for Health Statistics, 1979.
7. General Motors Research Laboratories. "How to Figure the Cost of Living ... A Longer Life," *IEEE Spectrum*, p. 8A, August, 1979.
8. Siddall, E. "Control of Spending on Nuclear Safety," *Nuclear Safety* 21(4):451-460, July-August, 1980.
9. Okrent, D., G. Apostolakis, and N. D. Okrent. "On the Usefulness of Quantitative Safety Goals for State Regulation of Energy Systems," *Journal of Hazardous Materials* 10(2-3):279-316, 1985.
10. Zeckhauser, R. and D. Shepard. "Where Now For saving Lives?" *Law and Contemporary Problems* 40(4), Issue on "Valuing Lives," Autumn, 1976.
11. Wilson, R. "Commentary: Risks and Their Acceptability," *Science, Technology, and Human Values* 9(2):11-22, Spring, 1984.
12. Spangler, M.B. "Policy Issues Related to Worst Case Risk Analyses and the Establishment of Acceptable Standards of de minimus Risk," in *Uncertainty in Risk Assessment, Risk Management, and Decision Making*, Plenum Publishing Corporation: New York, 1987.

Assessment of Risk from Short-Term Exposures

James J. Chen, Ralph L. Kodell, and David W. Gaylor
National Center for Toxicological Research
Jefferson, AR

ABSTRACT

One of the major problems encountered in cancer risk assessment is the estimation of risk for short-term exposures based upon results from chronic bioassays in which animals are exposed for a lifetime (generally two years for rodents). In the absence of data on cancer incidence for short-term exposures, the assumption is generally made that the cancer risk is proportional to the total accumulated dose (or equivalently, to the average dose rate) of a carcinogen, regardless of the ages at the times of exposures. This paper investigates the impact of using the total accumulated lifetime dose assumption to estimate cancer risk for short-term exposure for two carcinogenesis models, the Armitage-Doll multistage model and the Moolgavkar-Venzon-Knudson (M-V-K) birth-death mutation model. Two assumptions are discussed: (A) *the fraction of lifetime dose rate* assumes that the risk from a fractional lifetime exposure at a given dose rate is equal to the risk from full lifetime exposure at that same fraction of the given dose rate; and (B) *the fraction of lifetime risk* assumes that the risk from a fractional lifetime exposure at a given dose rate is equal to that same fraction of the risk from full lifetime exposure at the same dose rate. These two assumptions are equivalent when risk is a linear function of dose. Thus both can be thought of as generalizations of the assumption that cancer risk is proportional to the total accumulated lifetime dose.

KEYWORDS: Multistage model, birth-death-mutation model, fraction of lifetime dose rate, fraction of lifetime risk, risk ratio

INTRODUCTION

The multistage model (Armitage and Doll[1]) has been widely used to describe the carcinogenic process. Whittemore and Keller[2] provided a mathematical development of the multistage model from which risk from intermittent exposure can be estimated from continuous exposure. Following their work, Day and Brown[3] and Crump and Howe[4] applied the multistage model to investigate the effect of using the data from chronic bioassays to estimate cancer risk for short-term exposures. The risk estimates depend on the number of stages in the carcinogenic process, the particular stages which are affected by the chemical, and the age at exposure.

Recently, Moolgavkar, Venzon, and Knudson[5,6] proposed a two-stage birth-death-mutation model (M-V-K model) as an alternative to the multistage model to assess cancer risks. Like the multistage model, this model is based on an underlying molecular biological theory; it contains biologically meaningful parameters that can be estimated from experiments outside the chronic bioassay (Thorslund *et al.*[7]).

The excess cancer risk from exposure to a carcinogen depends on the dose rates, length of exposure, and age at exposure. A simplifying assumption for predicting the risk for short-term exposure from a long-term experiment is that the risk is proportional to the total accumulated lifetime dose (the average daily dose) regardless of the ages at the times of exposures. This paper studies the impact of using the total accumulated lifetime dose assumption to assess cancer risk based on two carcinogenesis models, the multistage model and the M-V-K model.

THE RISK RATIO FOR SHORT-TERM EXPOSURE VERSUS CONTINUOUS EXPOSURE

The probability that a tumor will develop by time t for a constant exposure at dose rate x from time t_0 to time t_0+m can be expressed by

$$P(x,t,t_0,m) = 1 - \exp\{-H'(x,t,t_0,m)\}$$

where $H'(x,t,0,t)$ is the cumulative hazard function and m is the length of exposure. The risk from a continuous dose rate x can be expressed as $P(x,t,0,t)$. The probability $P(0,t,0,0)$ represents the background cancer risk. Following Crump and Howe,[4] the excess risk due to the short-term exposure, $P(x,t,t_0,m) - P(0,t,0,0)$, may be approximated by the cumulative excess hazard rate, $H(x,t,t_0,m)$, for small values of x.

One common practice in risk assessment for short-term exposures is to assume that the probability of cancer is proportional to the total accumulated lifetime dose (or average daily dose). If the dose rate is constant, then total dose = dose rate × length of exposure. Two assumptions can be identified which are more general than this assumption, but which give rise to this assumption in the special case of a linear dose-response relationship. The assumption of fraction of lifetime dose rate (Assumption A) assumes that the risk from a fractional lifetime exposure at a given dose rate is equal to the risk from full lifetime exposure at the same fraction of the same dose rate. For example, assume that the average human lifespan is 70 years. The risk for 1/7th (10 years), of a lifetime exposure (regardless of the ages at the times of exposures) to 7 ppm of a carcinogen is equal to the risk from a lifetime exposure (70 years) at a dose rate of 1 ppm. The assumption of fraction of lifetime risk (Assumption B) assumes that the risk from a fractional lifetime exposure is equal to the same fraction of risk from full lifetime exposure at the same dose rate. For example, the risk for 1/7th of a lifetime exposure at 7 ppm is equal to 1/7th of the risk from a lifetime exposure at dose rate of 7 ppm.

The total accumulated lifetime dose exposed at a dose rate x from time t_0 to time t_0+m is xm. Let $f = m/t$ denote the fraction of the exposure period to the total lifetime. Then the equivalent dose rate for lifetime exposure is fx. The assumption of the fraction of lifetime dose rate can be examined by evaluating the risk ratio $R = H(x,t,t_0,m)/H(fx,t,0,t)$. The assumption of the fraction of lifetime risk can be examined by evaluating the risk ratio $R' = H(x,t,t_0,m)/f\,H(x,t,0,t)$. Assumption A (B) is appropriate only when $R=1$ ($R'=1$). $R>1$ ($R'>1$) represents an underestimate of risk, and $R<1$ ($R'<1$) represents an overestimate of risk. The two ratios, R and R', are equal when the cumulative excess hazard function is a linear function of dose. It can been seen that $R>R'$ if $H(x,t,0,t)$ is sublinear and $R<R'$ if $H(x,t,0,t)$ is supralinear.

THE MULTISTAGE MODEL

The Armitage-Doll multistage model has been widely used in cancer risk assessment. This model assumes that the formation of a tumor (cancer) cell is a result of several random biological events (stages) that occur in order and that each event is linearly related to dose. The probability of a tumor for the multistage model is

$$P(x,t,0,t) = 1 - \exp\{- H(x,t,0,t)\}$$

where $H(x,t,0,t) = b_0 + b_1 x + ... + b_k x^k$, and k is the number of the stages required to form a tumor.

Consider the multistage model in which only one stage, say r, is affected by exposure. Following Day and Brown[3] and Crump and Howe,[4] Kodell et al.[8] showed that

$$H(x, t, t_0, m) = a_1...a_k(b_r/a_r) \times \sum_{j=0}^{r-1} \left[\frac{(t_0)^j (t - t_0)^{k-j}}{j!(k - j)!} - \frac{(t_0 + m)^j (t - t_0 - m)^{k-j}}{j!(k - j)!} \right],$$

and

$$H(fx, t, 0, t) = a_1..a_k(b_r/a_r) \, fx \sum_{j=0}^{r-1} \frac{t^{k-j}}{j!(k - j)!} .$$

If a carcinogen is first-stage dose dependent ($r = 1$), then

$$R = \frac{(t - t_0)^k - (t - t_0 - m)^k}{(m/t) \, t^k} .$$

Note that $R = R'$ for the multistage model; therefore, only R will be given. It can be shown that the risk ratio, R, decreases as t_0 increases. Therefore, the use of the fractional lifetime dose rate assumption will underestimate the true risk for early-life exposure and overestimate the true risk for late-life exposure.

If a carcinogen is last-stage dose dependent ($r = k$), then

$$R = \frac{(t_0 + m)^k - (t_0)^k}{(m/t) \, t^k} .$$

In this case, the risk ratio R increases as t_0 increases; therefore, the use of the fractional lifetime dose rate assumption will overestimate the true risk for early-life exposure and underestimate the true risk for the late-life exposure.

Figure 1 is a plot of the ratio of the risk from a 10-year exposure to the risk from 70 years of continuous exposure with 1/7th the dose rate using the six-stage model with the first and the sixth stages dose dependent. In general, when an early stage is dose dependent, early exposure will have more effect than later exposure, and vice versa. The use of the fractional lifetime dose rate for early-life exposure to an early-stage dose-dependent carcinogen or late-life exposure to a late-stage dose-dependent carcinogen will

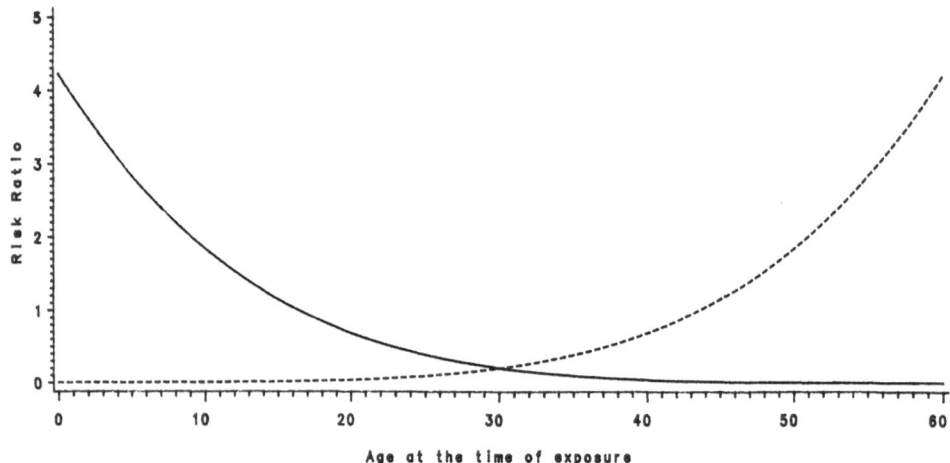

Fig. 1. The Ratio of Risk from a Short-Term (10-Year) Exposure to the Risk from 70 Years Continuous Exposure with 1/7th the Dose Rate Using the Armitage-Doll Six-Stage Model. (Solid line, first stage is dose dependent; dashed line, sixth stage is dose dependent.)

underestimate the true risk. Moreover, Kodell *et al.*[8] showed that the true risk is less than k times the risk based on total accumulated lifetime dose; i.e,

$$H(x,t,t_0,m) \leq k\, H(xm/t,t,0,t).$$

This inequality also holds when more than one stage is dose dependent.

MOOLGAVKAR-VENNZON-KNUDSON (M-V-K) MODEL

The M-V-K model assumes that the formation of a tumor (cancer) cell is a result of two mutations, each occurring at the time of cell division; the growth rate of the first mutated cell may be increased by subsequent exposure to a promoter. The probability of a tumor for the M-V-K model is

$$P(x,t,0,t) = 1 - \exp\{- H(x,t,0,t)\}\ ,$$

where $H(x,t,0,t) = \int_0^t \int_0^v \lambda_1 \lambda_2 \exp\{\int_0^v \Lambda(x)dw\}dudv$, and $\lambda_i(x) = a_i + b_i x$, $i = 1,2;\ \Lambda(x) = g + hx$.

The function $\lambda_i(x)$ represents the ith mutation rate, $i = 1,2$; and $\Lambda(x)$ represents the cell growth rate. For small background cell growth rate, g, the M-V-K model becomes the Armitage-Doll two-stage model.

Consider the M-V-K model in which only the first mutation stage (initiation) is affected by exposure to a carcinogen. A chemical that affects only the first mutation stage is called an initiator in the context of the M-V-K model. Chen *et al.*[9] showed that

$$H(x,t,t_0,m) = a_2 b_1/g^2\, x\, [\exp\{g(m+n)\} - \exp\{gn\} - gm],$$

where n is the elapsed time from the termination of exposure t_0+m to t. It can be seen that $H(x,t,t_0,m)$ increases with n and m, but is independent of t_0. For fixed m and $t=t_0+m+n$,

$H(x,t,t_0,m)$ decreases as t_0 increases. That is, when the first stage is dose dependent, early exposure will have more effect than later exposure. The risk ratio is

$$R = \frac{\exp\{g(t - t_0)\} - \exp\{g(t - t_0 - m)\} - gm}{f[\exp\{gt\} - gt - 1]} \cdot$$

Note that $R = R'$ when only one of the mutation stages is dose dependent. It can be shown that R decreases as t_0 increases. Therefore, the use of the fractional lifetime dose rate assumption will underestimate the true risk for early-life exposure and overestimate the true risk for the late-life exposure.

A chemical that affects only the second mutation stage is called a completer. If the second mutation stage is dose dependent, then

$$H(x,t,t_0,m) = a_1 b_2\, x\, [\exp\{g(t_0+m)\} - \exp\{gt_0\} - gm]$$

Here it can be seen that $H(x,t,t_0,m)$ increases with t_0 and m, but is independent of n. For fixed m, $H(x,t,t_0,m)$ increases as t_0 increases. When the last stage is dose dependent, late-life exposure will have more effect than early exposure. The risk ratio is

$$R = \frac{\exp\{g(t_0 + m)\} - \exp\{gt_0\} - gm}{f[\exp\{gt\} - gt - 1]} \cdot$$

The risk ratio R increases as t_0 increases; therefore, the use of the fractional lifetime dose rate assumption will overestimate the true risk for early-life exposure and underestimate the true risk for the late-life exposure.

Figure 2 is a plot of the ratio of the risk from a 10-year exposure to the risk from 70 years of continuous exposure with 1/7th the dose rate for both an initiator (first stage) and a completer (second stage) with the constant cell growth rate $g = 0.05$. When only one stage is dose dependent, the use of the fractional lifetime dose rate assumption for the early-life exposure to an initiator or the late-life exposure to a completer can lead to an underestimate of the risk. Moreover, the deviation from the true risk increases as the background cell growth rate, g, increases (Fig. 3). When both stages are dose dependent, the risk ratio is bounded by the larger of the two risk ratios of one-stage dose dependence.

For the dose-dependent promotion, Chen et al.[9] showed that

$$
\begin{aligned}
H(x,t,t_0,m) = {} & a_1 a_2/g^2\, [\exp\{(g + hx)m)\}\, \{\exp\{g(n + m)\} - 1\}\, \{\exp\{gt_0\} - 1\} \\
& - \exp\{g(t_0 + m + n)\} + \exp\{gn\} + \exp\{gt_0\} + g(n + m) - 1] \\
& + a_1 a_2/[g(g + hx)]\, [\exp\{(g + hx)m)\} - 1]\, [\exp\{gt_0\} - \exp\{gn\} - 2] \\
& + a_1 a_2/(g + hx)^2\, [\exp\{g + hx)m\} - (g + hx)m - 1]\ .
\end{aligned}
$$

It can be seen that for fixed t_0, $H(x,t,t_0,m)$ increases with m and n.

Assume that the average human lifespan is 70 years. The risk ratios for a promoter are given in Table 1 for $g = 0.01, 0.05, 0.1$; $t_0 = 0, 20, 60$; $m = 5, 10, 35$; $hx = 0.01, 0.05, 0.1$. Note that hx represents the effect of the promoter at dose x; the ratio hx/g represents the relative cell growth rate per (combined) unit of exposure. Table 1 indicates that the use of either the fraction of lifetime dose rate assumption or the fraction of lifetime risk assumption to estimate the risk for a short-term exposure to a promoter can both

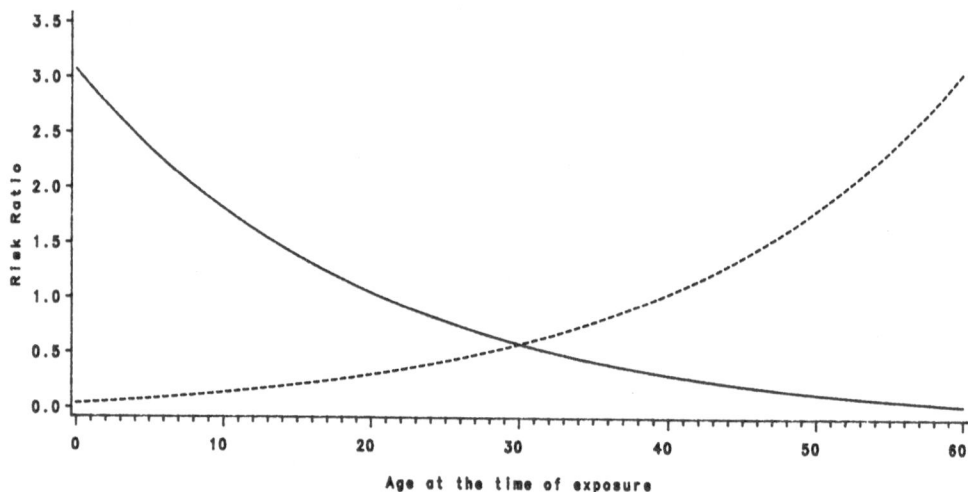

Fig. 2. The Ratio of Risk from a Short-Term (10-Year) Exposure to the Risk from 70 Years Continuous Exposure with 1/7th Dose Rate Using the M-V-K Model with Constant Cell Growth Rate $g = 0.05$. (Solid line, first stage; dashed line, second stage.)

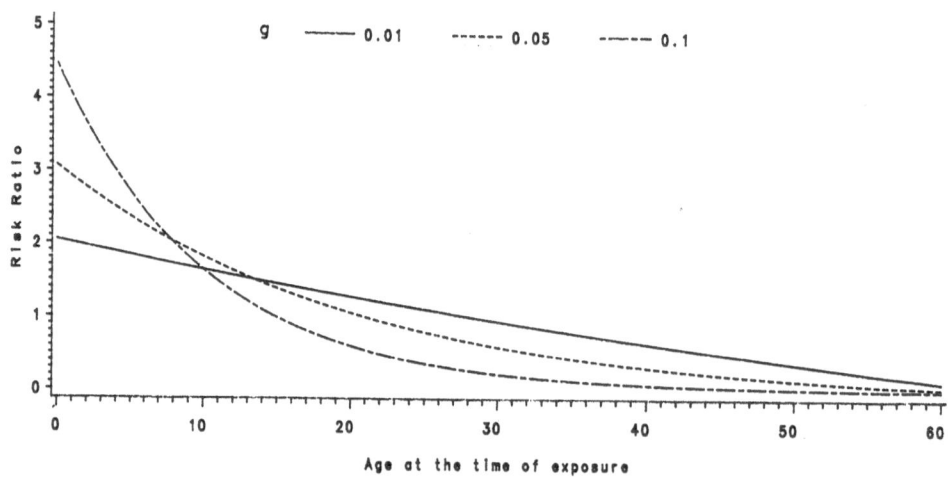

Fig. 3. The Ratio of Risk from a Short-Term (10-Year) Exposure to the Risk from 70 Years Continuous Exposure with 1/7th the Dose Rate Using the M-V-K Model with the First Stage Dose Dependent and $g = 0.01, 0.05, 0.1$.

underestimate and overestimate the true risk. R is greater than R' in the range of g and hx considered in this paper (low dose region, to be discussed later); thus, the use of the fraction of lifetime risk assumption gives more conservative estimates of risk than does the use of the fraction of lifetime dose rate assumption.

Table 1. Ratios R and R' of the Excess Risk from a Short-Term Exposure to a Promoter Over the Age Interval $[t_0, t_0+m]$ to the Risk from a Lifetime (70-Year) Exposure*

g	t_0	m	$hx = 0.01$ R	$hx = 0.01$ R'	$hx = 0.05$ R	$hx = 0.05$ R'	$hx = 0.1$ R	$hx = 0.1$ R'
0.01	0	5	109.54	91.13	20.91	6.59	9.86	0.52
		10	50.18	42.31	9.31	3.15	4.25	0.26
		35	8.76	7.92	2.13	1.07	1.58	0.25
	20	5	77.01	64.07	15.71	4.95	8.13	0.43
		10	34.58	29.16	7.45	2.52	4.24	0.26
		35	4.74	4.28	2.16	1.09	2.51	0.39
	60	5	8.99	7.48	2.19	0.69	1.37	0.07
		10	0.39	0.33	0.40	0.14	0.42	0.03
0.05	0	5	4.29	3.42	0.99	0.24	0.58	0.02
		10	2.27	1.84	0.75	0.20	0.57	0.02
		35	1.30	1.14	1.12	0.49	1.22	0.15
	20	5	4.10	3.27	1.86	0.46	1.64	0.05
		10	2.58	2.09	1.65	0.44	1.66	0.07
		35	1.50	1.33	1.68	0.73	2.21	0.27
	60	5	0.94	0.75	0.70	0.17	0.69	0.02
		10	0.43	0.35	0.43	0.12	0.44	0.02
0.1	0	5	0.46	0.35	0.32	0.07	0.31	0.01
		10	0.59	0.46	0.52	0.12	0.51	0.01
		35	1.02	0.88	1.05	0.41	1.12	0.11
	20	5	1.35	1.04	1.29	0.26	1.32	0.03
		10	1.33	1.04	1.34	0.30	1.43	0.04
		25	1.31	1.13	1.52	0.60	1.93	0.19
	60	5	0.74	0.57	0.75	0.15	0.76	0.02
		10	0.51	0.40	0.51	0.11	0.50	0.01

*R based on fraction of lifetime dose rate assumption; R' based on fraction of lifetime assumption.

The values of R are greater than 1 for $t_0 = 20$; i.e., the use of the fraction of lifetime dose rate assumption will underestimate the risk when the exposure occurs in midlife. For $g = 0.01$ and $t_0 = 0$, R is greater than 1. Hence, for a smaller background cell growth rate, the risk will be underestimated if the exposure occurs in early life. Conversely, for $g = 0.1$ and $t_0 = 0$, R is usually less than 1. For a larger background cell growth rate, the risk can be overestimated if the exposure occurs in late life. For $g = 0.05$ and 0.1, the minimum of R is 0.31 (risk overestimated by about a factor of 3), and the maximum is 4.29 (risk underestimated by about a factor of 4.)

The values of R' generally decrease with g and hx. Using the fraction of lifetime risk assumption, the true risks are underestimated when g and hx are small (the upper left of Table 1), and they are overestimated when g and hx are large (the lower right of Table 1). For $hx = 0.1$, since $R < 1$, the true risks are overestimated; the overestimation can be up to a factor of 100 when $g = 0.1$.

SUMMARY

The cancer risk is assumed to be proportional to the total accumulated dose (or equivalently, to the average dose rate) of a carcinogen, regardless of the ages at the times of exposure. This assumption is often used to estimate risk for short-term exposure. The question is how much the risk from a short-term exposure deviates from the risk estimated from a lifetime tumor rate using dose rate averaged over a lifetime. This paper investigates the impact of using this approach for both the Armitage-Doll multistage model and the Moolgavkar-Venzon-Knudson birth-death mutation model. Based on the multistage model, the use of the average lifetime dose rate in the case of an early-life exposure to a late acting carcinogen (or vice versa) can cause excess risk to be overestimated by several orders of magnitude. However, this still results in a conservative procedure from the standpoint of protecting public health. In the case of early-life exposure to an early-stage carcinogen, midlife exposure to a mid-stage carcinogen, or late-life exposure to a late-stage carcinogen, the excess cancer risk can exceed that which would be predicted from average lifetime exposure rates but the underestimation will be less than a factor k, where k is the number of stages in the carcinogenesis process. Based on the M-V-K model, the cancer risk estimate is generally within a factor of 10, but it can differ up to a factor of 100.

REFERENCES

1. P. Armitage and R. Doll, "Stochastic Models for Carcinogenesis," in *Proceedings of the Fourth Berkeley Symposium on Mathematical Statistics and Probability* 4:19-38, University of California Press (1961).
2. A. Whittemore and J. B. Keller, "Quantitative Theories of Carcinogenesis," *SIAM Review* 20:1-30 (1978).
3. N. E. Day and C. C. Brown, C. C., "Multistage Models and Primary Prevention of Cancer," *Journal of the National Cancer Institute* 64:977-989 (1980).
4. K. S. Crump and R. B. Howe, R. B., "The Multistage Model with a Time-Dependent Dose Pattern: Applications to Carcinogenic Risk Assessment," *Risk Analysis* 4:163-176 (1984).
5. S. H. Moolgavkar and D. J. Venzon, "Two-Event Models for Carcinogenesis: Incidence Curves for Childhood and Adult Tumors," *Mathematical Biosciences* 47:55-77 (1979).
6. S. H. Moolgavkar and A. G. Knudson, "Mutation and Cancer: A Model for Human Carcinogenesis," *J. Natl. Cancer Inst.* 66:1037-1052 (1981).
7. T. W. Thorslund, C. C. Brown, and G. Charnley, "The Use of Biologically Motivated Mathematical Models to Predict the Actual Cancer Risk Associated with Environmental Exposure to a Carcinogen," *Risk Analysis* 7:109-119 (1987).
8. R. L. Kodell, D. W. Gaylor, and J. J. Chen, "Consequences of Using Average Lifetime Dose Rate to Predict Risk from Intermittent Exposures to Carcinogens," *Risk Analysis* 7:339-345 (1987).
9. J. J. Chen, R. L. Kodell, and D. W. Gaylor, "Using the Biological Two-Stage Model to Assess Risk from Short-Term Exposures," *Risk Analysis* (accepted, 1987).

Societal Risk from a Thermodynamic Perspective

Elmer L. Offenbacher
Temple University
Philadelphia, PA

Paul Slovic
Decision Research
Eugene, OR

ABSTRACT

The problem of understanding risk and improving risk communication is approached by analogy to thermodynamics. We first emphasize the distinction between risk as probability, analogous to an intensive variable such as temperature, and risk as the flow of adverse consequences, analogous to the flow of thermal energy, or heat. Five parallels between thermodynamics and risk systems are described: global relevance (matter and lives), dual perspectives (macro and micro), new "dimensions" (temperature and probability), unifying variables (energy and resources), and derived variables (entropy and severity). Caveats for applying the paradigm are also discussed.

KEY WORDS: societal risk, thermodynamic paradigm, definitions of risk

INTRODUCTION: WHY ARE WE SEARCHING FOR A THERMODYNAMIC PARADIGM?

Concerns about societal risk are increasing at an accelerating pace, fueled by the nuclear age, the electronic revolution, biotechnology and the information explosion. The distribution of societal risks and the allocation of responsibilities for reducing them are lively political issues. Participants in the resolution of these issues have expertise in many disciplines, including engineering, statistics, and the natural and social sciences (particularly health-related fields, psychology, economics, political science, etc.). To facilitate effective communication among such a diverse professional group, it might be helpful to use concepts drawn from the classical and multidisciplinary subject of thermodynamics.

Because the discipline of thermodynamics spans virtually all of the natural sciences, it might serve as a template for the subject of risk. The need for such a template becomes ever more acute as our decision-making circuits become overloaded by the continuing amplification of new risks. This amplification is driven by such advances in science and technology as picogram detection capabilities for measuring toxins and by the rapidity of societal communication networks and the growth of strong special-interest groups. These developments have generated valiant attempts to generalize risk situations. This has led to the development of methodologies with broad applicability, to the establishment of numerous governmental agencies for controlling and managing risks, and to a heightened risk consciousness by society as a whole. In this web of risk-related discussions, misunderstandings often arise because of a lack of precision in terminology or the adoption of narrow views of risk situations. Perhaps we can gain a broader perspective by comparing the history and structure of the classical subject of thermodynamics with the development of the interdisciplinary subject of risk. Such a comparison may also constrain us to think more precisely on what we are trying to communicate when we explicate various kinds of risk assessments. It may also suggest novel approaches for characterizing society's risk-management responses.

In this paper we confine our considerations of the thermodynamic analogy to the presentation of some features that societal risk and thermodynamics have in common. These features provide some supportive arguments for tne relevance of a thermodynamic paradigm. In a companion paper we will propose the construction of several distinct societal risk probabilities for the same specific risk. We will introduce the idea of societal response coefficients by the use of a unifying variable, societal resources, akin to the thermodynamic concept of internal energy.

There is a fundamental similarity between applied thermodynamics and societal risk management. In thermodynamics one deals with changes in the physical state of a substance in response to changes in its physical surroundings. Similarly, in risk management the focus is on changes in society brought about by changes in environmental and social conditions. In both subjects the utilitarian purpose of theory is to predict the responses (of the substance or of the society) from a knowledge of some of its (thermodynamic or social) characteristics.

In thermodynamics the predictions can be very powerful because the important variables specifying the physical surroundings (i.e., the temperature and pressure) are well understood and can be measured rather precisely. This is obviously not the case with society's social and physical environmental conditions. However, in doing thermodynamics one focuses on the collective behavior of a system of particles independent of their individual behaviors. In this sense it is comparable to a description of the societal risk state that does not detail the risk of specific individuals.

In thermodynamics there are very general relationships from which one can predict responses of a substance to forces that change its state. For example, the isothermal compressibility of a substance can be inferred from its specific heats and its isobaric expansion coefficient. This means that from thermal measurements alone (specific heats and expansion coefficient) one can predict the value of a purely mechanical property, compressibility. Furthermore, many equations relating thermodynamic properties are the same for all substances—no matter how complex! It is not suggested that such equations can be translated directly into societal risk equations. However, with the help of clever choices of fundamental variables one may be able to discover risk-management relationships involving societal responses to various forms of risk stresses.

COMPARING THE EMERGENCE OF THE CONCEPT OF QUANTITATIVE RISK WITH THAT OF HEAT

The word risk is used with many nuances in different contexts, often leading to confusion and misunderstanding. It is used to imply hazard, danger, harm, uncertainty, probability, or surprise (singly or in various combinations). Douglas (1985, p. 20) quotes *two* divergent definitions recommended by the United Nations for the risk of toxicity in chemicals:

(*a*) Risk is a statistical concept—defined as the expected frequency of undesirable effects arising from exposure to a pollutant and (*b*) Risk, *R*, can be estimated by some sort of product of the probability of the event, *P*, times the severity of the harm, *H*:

$$R = P \times H . \tag{1}$$

These definitions sanction the word risk for two different and incommensurate quantities, probability, *P*, and utility, *R*.

We find that risk analysts also differ from one another in their use of the word risk. The Hazard Assessment Group at Clark University includes only the hazard *effects* in its definition of risk (Kates, Hohenemser, & Kasperson, 1985). This is a narrower definition than that given by Fischhoff, Lichtenstein, Slovic, Derby, and Keeney (1981): "The existence of a threat to life or health." Douglas and Wildavsky (1982) consider risk to be "a collective construct" (p. 187) encompassing dangers in four kinds of policy groupings: "foreign affairs . . . , crime . . . , pollution . . . , economic failure" (p. 2).

A somewhat similar situation existed in physics when the nature of heat was being debated in the days of the caloric theory. Quantity of heat and the degree or intensity of heat were not distinguished from each other until the middle of the 18th Century (Cardwell, 1971, p. 31). The word heat gained the well-defined modern meaning of energy only in the middle of the 19th Century after it was established conclusively that heat is a form of energy. In current physics usage, the word heat is reserved for describing the energy *flow* between two systems caused by their differences in temperature. In everyday language heat is still used with a much wider range of connotations, including that of heat as stored energy and as a synonym for temperature.

The precise differences in meaning of heat and temperature have first to be explicated before communicating a physical idea involving heat or energy. Take the simple case of an ice-water mixture. The temperature of this mixture remains constant even when heat is flowing in from the surroundings. The heat input produces a change in state. It melts some of the ice. Not until all the ice is melted will the heat flow produce a temperature change. Furthermore, even when all the ice is liquified, the amount by which the temperature will change as a result of a fixed amount of heat input will depend very much on the initial temperature. (For water the greatest temperature change will occur when the initial temperature is 35 degrees Celsius.) There is a unique relationship (which differs for every substance) between the amount of heat added to a substance and the temperature rise that is produced. It is the existence of this commonly observed relationship that was, and often still is, the source of the confusion between the meaning of heat and temperature. Most of the time, when heat is added to a substance, its temperature increases. But, as is evident in our ice-water mixture example, there are also common situations where the temperature does not increase!

Similarly, we use the word risk to describe both the probability of the occurrence of the harmful event, *P*, and its actual occurrence, *R*. But these quantities are by no means the same or even necessarily proportional to each other. An interchange of these two meanings of risk, of *P* and *R*, can lead to the same kind of confusion that is caused by an interchange of temperature and heat. This confusion is compounded in situations where Eq. (1) becomes more complex. Although some relationship exists between probability of harm and expected adverse effect, $R=R(P)$, this is, in general, not as simple as expressed in Eq. (1), $R=P \times H$. For example, we cannot necessarily assume that an increase in probability of harm will inevitably result in more injuries or deaths. If a traffic intersection becomes more dangerous (by increased traffic or faster driving), the probability of harm for the pedestrians crossing the intersection increases; but it does not necessarily follow that more people will be killed. Awareness of the increased danger may make pedestrians more cautious so that actually fewer people are killed than before.

P and R must be clearly distinguished not only because they are different entities, but because they are of a dissimilar nature. P is analogous to an intensive variable of a thermodynamic state (e.g., temperature) and R is analogous to an extensive variable of state (e.g., energy). (An extensive variable is one whose value varies with the size of the system. The value of the intensive variable does not vary with size.) In order to describe a specific risk situation *both* P and R are needed. The precision of knowledge of one is often greater than that of the other. The equation relating these two must involve another extensive variable, such as the number of people exposed or the geographical area affected, etc.

Comprehensive overviews of formal definitions of risk and risk-preference relationships have been published by Coombs (1972), Libby & Fishburn (1977), Schaefer (1978), Fischhoff, Watson, & Hope (1984) and others. The need for agreement on a common risk vocabulary was recognized by the Society for Risk Analysis when it established its Committee for Definitions (Gratt, 1984).

At a recent meeting of this Committee a general ambivalence about the definition of the word risk permeated the discussion. Thirteen definitions for risk were considered (see Table 1) (Gratt, 1987). Comments ranged from a proposal not to adopt any definition (because of the wide variations of meaning that the word risk assumes in different disciplines) to a proposal to restrict the word risk to mean only the conditional probability of an adverse event, definition 5. Our attempt to characterize these definitions succinctly is shown in Table 2. From this tabulation we readily see that seven of the 13 definitions of risk pertain to probability (P), two to consequences of harmful events, R (which we label C in Table 2) two to a combination of P and C, and two to a combination of P and H. Proposals defining risk merely as a probability can be compared to measuring only the temperature, when one estimates the possibility of getting burned by a hot object. However, what really counts in determining the harm to be expected from touching a hot object is, in addition to the high temperature, the heat capacity of the object and the duration of the contact. Similarly, in societal risk situations, in addition to the probability of the harmful event, P, it is the severity of the potential harm, H, and the length of time of exposure to it, t, that determines the expected harm. Indeed, a societal risk situation is never fully described by merely specifying the probability of the event occurring per unit time. Table 2 illustrates that in discussing risk situations, people are as likely to be talking about the risk probability, P, as they are to be talking about the risk consequences, C.

SOCIETAL RISK — THERMODYNAMICS PARALLELS

Societal risk and thermodynamics have some striking common features. We briefly discuss five of these in order to lend credence to our proposal for using the discipline of thermodynamics as a template for structuring our thinking about risk. These features are tabulated in Table 3.

Global Relevance

Both subjects occupy "global" domains. The principles of thermodynamics are relevant to all the natural sciences and apply to every form and constitution of matter. The concepts of societal risk are likewise relevant to the social sciences, that is, to the lives and activities of all human beings and all societies.

Table 1. Definitions of Risk Considered by the SRA Committee on Definitions

1. Risk is the possibility of loss, injury, disadvantage or destruction.

2. Risk is an expression of possible loss over a specific period of time or number of operational cycles.

3. Risk (consequence/unit time) = Frequency (events/unit time) × Magnitude (consequence/event).

4. Risk is a measure of the probability and severity of adverse effects.

5. Risk is the conditional probability of an adverse event (given that the causative events necessary have occurred).

6. Risk is the potential for unwanted negative consequences of an event or activity.

7. Risk is the probability that a substance will produce harm under specified conditions.

8. Risk is the probability of loss or injury to people and property.

9. Risk is the potential for realization of unwanted, negative consequences to human life, health, or the environment.

10. Risk is the product for a probability of an adverse event times the consequences of that event were it to occur (dimensions of consequence × time).

11. Risk is a function of two major factors: (*a*) the probability that an event, or series of events of various magnitudes, will occur, and (*b*) the consequences of the event(s).

12. Risk is a probability distribution over all possible consequences of a specific cause which can have an adverse effect on human health, property or the environment.

13. Risk is a measure of the occurrence and severity of an adverse effect to health, property or the environment.

Micro- and Macroscopic Perspectives

To understand either subject thoroughly we need two very different but complementary approaches, the macroscopic view and the microscopic one. In thermodynamics, the macroscopic description is called classical (or phenomenological) thermodynamics and the approach starting from the microcosm is known as statistical mechanics. It was only after the discovery of discrete energy states at the turn of the century that the older subject of classical thermodynamics was integrated in a consistent way with statistical considerations. A natural scientist must study both of these topics for an in-depth understanding of the properties of any particular substance.

Similarly, in societal risk management the macrorisk assessment must be supplemented by an understanding of the (individual) microrisks if a risk-reduction management strategy is to be effective in a given situation. While coping with microrisk is a concept innate to human beings, macrorisk management is a fairly new idea. (Macrorisks concerning health have emerged from the growth of the relatively young science of

Table 2. Characterizations in Terms of R, P, H of the 13 Selected Definitions of Risk Proposed to the Definitions Committee of SRA

1. $WR = P$	8. $WR = P$
2. $WR = C$	9. $WR = P$
3. $WR = C$	10. $WR = P \times C$
4. $WR = P + H$	11. $WR = P + C$
5. $WR = P$	12. $WR = P[C]$
6. $WR = P$	13. $WR = P + H$
7. $WR = P$	

Legend: WR = the word risk; P = possibility, probability, potential or frequency; H = severity; C = consequences of harmful event.

Table 3. Parallels Between Thermodynamics and Risk Systems

FEATURE	THERMODYNAMICS	RISK
1. Global Relevance	All Matter	All Lives
2. Dual Perspective		
a) Macroview	Classical Thermodynamics	Societal Risk
b) Microview	Statistical Mechanics	Individual Risk
3. New Dimension	Temperature, T	Probability of Harm, P
4. Unifying State Variable	Internal Energy, U	Societal Resources, Re
5. Derived Variable	Entropy, S	Severity, H

epidemiology.) The distinction between these two vantage points is often blurred in risk discussions and in much of the risk literature. However, this distinction has recently been appropriately emphasized in an insightful article by Sharlin (1986). Reporting on a study of EPA's effectiveness in communicating risks associated with ethylene dibromide (EDB), Sharlin points out that macrorisk assessment and management (societal risk) constitutes EPA's regulatory role and microrisk assessment (individual risk) constitutes EPA's public information role.

We cite a second example of a macro- and a microview of societal risk: the data for accidental deaths of coal miners from 1950-1970, as presented by Wilson and Crouch (1982). The coal-mining "risk" over this 20-year period is graphed there in two different ways: (1) in fatalities per million tons of coal mined per year, and (2) in fatalities per thousand employed miners per year. The first graph shows a drop greater than 50%, while the second shows a widely fluctuating, but nevertheless marked, increase. The authors pose the rhetorical question: Is coal mining getting safer? They comment that it would be possible to use each measure to support one of the two opposing views on the safety of coal mining!

But this could happen only if we fail to distinguish between a macro- and a microview of risk. These risk measures make it quite clear that for society as a whole the coal-mining risk has decreased — the macroview — but for the individual miner the risk probability has increased — the microview. This example illustrates the importance of classifying "risk measures" according to a scheme that clearly identifies the values to be considered. Are decisions on the management of coal-mining risks to be based on society's need for coal or on the safety of the miners? We are not looking merely at two different measures of risk (the raw data are identical!), but, much more fundamentally, at two different values: the value of the life of a miner versus society's demand for energy.

Although the two risk views must be clearly distinguished, they are interdependent. Risks cannot be realistically evaluated for a society without considering the risk perceptions of the individuals within that society. Conversely, individuals cannot fully understand their own risks without an appreciation of the related societal risk factors.

Quantitative Variables: Temperature, Energy and Entropy

In addition to the two rather straightforward parallels presented above, we propose three, more specific, comparisons between thermodynamics and societal risk. They relate to the roles of temperature, energy and entropy in thermodynamics. Temperature, T, we have already compared with the probability of harm, P; (internal) energy, U, will be compared with societal resources, Re, and entropy, S, will be compared with the "severity" of harm, H defined by Eq. (1). We may describe H also as the potential adverse consequences, or the magnitude of the quantity that is subject to damage.

New Dimensions: Temperature, T, and Probability of Harm, P

One of the basics of thermodynamics, both experimental and theoretical, is the concept of temperature. Temperature is defined as the property which determines the direction of net thermal-energy flow between two systems when they are placed into thermal contact with each other. Theoretically, temperature is defined as the partial derivative of the internal energy, U, with respect to the entropy, S.[a]

Likewise, the assessment of risk is based on a measured and/or theoretically predicted probability of harm. A measured probability can be derived from risk data based on past experience and a theoretical probability can be derived from some assumed model for the process causing the harm.

The temperature of a system is experimentally definable only in relation to another system with which it can exchange internal energy. It might be argued that the "riskiness" (risk probability) of a societal activity such as travel by air is meaningful, strictly speaking,

---

a. For a simple thermodynamic system the internal energy, U, is a function of the entropy, S, the volume, V, and the number of identical constituents of the system, N. The temperature is then defined in terms of the partial derivative of U with respect to S, keeping the other variables constant.

only in relation to the "riskiness" of another activity of the same kind, such as travel by automobile. Between the two modes of travel societal resources may be "exchanged."

In a specification of a thermodynamic system, one value of the temperature, T, is sufficient only if the system is in equilibrium. Likewise, one probability distribution for societal harm is satisfactory only for a well-specified situation. As a new societal risk is projected into public awareness, society generally responds with mitigating safety measures. Some time elapses before the societal risk is tolerated at some level. It is when toleration sets in that one can assign a set of time independent P values to that risk. An example would be the tolerated highway fatalities associated with various countries (50,000 for the U.S., 6,000 for Britain, 10,000 for France, etc.).

In the development of thermodynamics, temperature constituted a new "independent" experimental variable. This variable required the definition of a new universal unit of physical measurement, the Kelvin. Temperature also signified, for practical purposes of measurement, a new "dimension" in the world of physics.[b] For societal risk it is the probability of harm, P, which is the "new" measurement variable and it likewise signifies a new "dimension" in public "decision" space. No new unit is required because P is a pure number (dimensionless). This characteristic of P is also responsible for some of the ambiguity in common risk situations. The expression "new dimension" is meant to evoke appreciation of the concept of higher dimensions that emerges in analytic geometry when, after considering all the possible positions on a line, one advances to consider all the intersections on a planar grid and after that all the points in a three-dimensional array. At each of these conceptual steps the old possibilities are MULTIPLIED manyfold. Furthermore, to identify the position of a point in a space with a new dimension requires knowledge of the value of the additional coordinate. Translated into public decision space, this means that a full description of alternative risk-management policies requires knowledge of numerous values of P according to the number of different types of harmful effects associated with each policy.

"Ascending" Dimension: Probability of Harm, P

Thermodynamics is traditionally introduced into the physics curriculum after the study of mechanics and motion. There is a compelling reason for this order. Students first have to understand the description of an idealized physical world in a four-dimensional space-time coordinate system before they are prepared to conceptualize additional physical dimensions such as temperature and pressure. Temperature and pressure are always required for describing the thermodynamic state of substances of the real world. As we "ascend" from the material world of the natural sciences into the "risk world" of the social sciences we need to add another dimension to those employed by physical sciences. This is the probability of harm to humans, P. Just as the meaning of temperature had to be sorted out in terms of energy and entropy, $(T=dU/dS)$, so the confusing multiple meanings of P need to be sorted out if risk analysis is to become a more trustworthy and influential tool for explaining societal dangers.

Unifying State Variables: Internal Energy, U, and
Societal Resources, Re

We draw the fourth parallel between the concept of internal energy (of thermodynamics) and the notion of "resources" (economic and human resources) in societal risk management (see Table 4).

---

[b]. There are five fundamental standards for physical measurements: mass, length, time, temperature, and electric current. In discussions of the consistency of physical units they often are referred to as different "dimensions."

Table 4. Unifying Variables and Asymmetries of Thermodynamics and Risk

UNIFYING VARIABLES: Internal Energy, U; Societal Resources, Re

Thermodynamics:	$\Delta U_{of} =$	$Q_{into} +$	W_{on}
	Internal Energy	Heat	Work
		Lives	Money
Risk:	$\Delta Re_{of} =$	$L_{into} +$	M_{into}
	Resources	Labor	Capital

$$L = L\ (\text{Physical}) + L\ (\text{Mental})$$

Asymmetries

Thermodynamics:	$W \rightarrow Q$	not symmetrical with	$Q \rightarrow W$
Risk:	$M \rightarrow L$	not symmetrical with	$L \rightarrow M$

In macroscopic thermodynamics the concept of energy includes, besides the kinetic and potential energy of the system, the internal energy associated with the microscopic and submicroscopic motions of the constituents of the system. Similarly, the resource riches in a risk-conscious society include, in addition to its material wealth, the value of individual human labor and human creativity (human resources).

Thermodynamics was born (in part) as a result of two major conceptual advances: (1) the clear differentiation between temperature, T, and heat, Q, and (2) the experimental proof that led to the realization that heat is a form of energy.

The internal energy state of any system, U, can be increased either by heat, Q, flowing into the system, or work, W, being done on the system. The realization that both of these methods of energy transfer can have the same effect on the energy state of the system led to the formulation of the first law of thermodynamics: The change (Δ) *of* the internal energy (U) is equal to the heat (Q) flowing *into* the system plus the work (W) done *on the system*.

$$\Delta U_{of} = Q_{into} + W_{on}. \tag{2}$$

Similarly, two important advances are occurring in the development of societal risk management:

1.   Recognition is growing that there is a need for a clearer distinction between the probability of occurrence of a harmful event, P (during a specified interval of time), and the harmful consequences that might result in a given situation, R.

2.   More resources are being invested in safety measures to minimize risk exposure.

We shall now use these two ideas to formulate an equation for quantitative risk that is analogous to the first law of thermodynamics.

To emphasize that quantitative risk has the dimensions of an extensive variable (one whose value varies with the size of the system considered), we take as a specific case the harm represented by human fatalities. Rather than using R for consequences, we now use the symbol L for lives which may be lost or gained. Let L be positive when lives are "added" to a society or are saved (i.e., when fewer than the average number of deaths occur in the next "appropriate" interval of time). Then L is negative when a greater than average number of fatalities occur.

In analogy with the internal energy, U, we shall construct a societal-state variable, which we shall call Societal-Resource-Wealth, Re. This will include all the resources — human as well as material — of the society.

In thermodynamics, there is no unique reference level for U. Only *changes* in energy can be measured. Similarly, Re has no unique reference level. The effectiveness of societal risk management can be determined only by the *changes* produced in the total value of Societal-Resource-Wealth, Re.

In our analogy, positive L is the inflow and negative L is the outflow of lives or human resources. Economic resources are represented by the symbol M for monies. A statement of the first "law" of societal risk, then, is: A change in resources of a society (Re) equals the flow of lives into society (L) plus the flow of economic resources into the society (M).

$$\Delta Re_{of} = L_{into} + M_{into}.$$ (3)

Re, the resources of society, can increase either by adding lives, L (saved, born or added on from outside), or by developing economic resources, M (natural, financial, etc.). Thus, in the same way that internal energy became the unifying variable for keeping track of the qualitatively different forms of energy transfer to a system represented by heat and work, Re becomes the unifying variable for the flow of the two disparate forms of societal resources: lives and money. An investment in safety measures, represented by a negative value of M (because monies are flowing out) would be compensated for by a positive value of L (number of lives saved). Depending on the conversion factor (i.e., the monetary equivalence assumed for a life saved), the total resource state of society, Re, would increase, decrease or remain the same.

This formulation of societal risk highlights a number of sensitive issues. For example, it may focus on the limits to the amount of money society can spend on safety measures, that is: the total monies flowing out cannot be greater than the total available economic resource, M (out) $\leq M$ (total). It may also draw attention to the conversion factor required for the value of life (monies per life M/L). In contrast to thermal energy, which has a specified conversion factor of 4.186 Joule/Calorie, the M/L conversion factor (which enters real societal decision making, whether one likes it or not), is highly variable and dependent (among many other factors) on the economic wealth of the society.

The process can also function in reverse. Money can flow in (i.e., by not investing in safety measures) and lives can flow out. Again depending on the money/life conversion factor, there could be a net gain or net loss for the resource wealth of the society.

Such an idea came to mind recently when a radio announcer proclaimed (prior to the 1986 Superbowl football match between the New York Giants and the Denver Broncos) that the Superbowl stadium was not safe! What does that statement really mean? Utilizing the suggested thermodynamics analogy, it means that the expectation of a mishap evaluated in terms of number of lives potentially lost multiplied by the value of each life is greater than the amount of money that those responsible for the Superbowl were willing to invest to prevent a mishap. If a mishap had occurred, then American society would have been poorer not only as reckoned by the dollar cost of the physical damage but by the cost of "human resources," lives that would have tragically gone to waste.

In thermodynamics it is the minimization of internal energy which determines the equilibrium state of an (isolated) system. In risk management it is some optimization principles for the resource state of the community to be managed which ideally would determine the temporary equilibrium state. To define the meaning of optimum, one would have to specify the variables that are to be minimized, maximized or "compromised." The optimum conditions depend on the goals of the society and these will vary with the specific risk application.[c]

Our economic and political language has already adjusted to this idea of the "equivalence" of human labor and economic resources. Instead of Departments of Labor we have Departments of Human Resources. Irrespective of our ethical sensibilities, we are often forced to evaluate lives saved in terms of costs. These valuations apply, for example, to technology choices and to medical treatment decisions. How many lives will be saved by installing a safety device and what is the implicit worth of each life? or, what public medical costs are justified?

There is another and even more significant point to be made here. The L value really consists of two parts: the L (physical) and the L (mental = intellectual, emotional, "human," etc.). It is the L (mental) which is even more difficult to measure than the "physical" value of human life. Indeed, it is emotional arguments about these types of measurements that tend to polarize modern societies into the "materially oriented" and the "spiritually oriented." The latter measure risks to human life and life's value in a much more complex, many dimensional, domain. In this more complex domain, only one of many dimensions of human existence is addressed by the utility concept of economics. In the eyes of a spiritually oriented society, a formulation of a risk problem is incomplete if it includes only the resources that are under human control, even in the presence of virtually infinite human resources. Ignoring this perception may sometimes prevent the resolution of a risk controversy or the acquiescence to a risk.

Another thermodynamics idea which applies to risk is the asymmetry of converting one type of energy flow, Q (heat), into the other type, W (work). It is possible to convert all of W into Q but not the other way around. Only a portion of Q can be turned into W. In considering societal resource conversions, it is possible to turn M (money) into L (life) but

---

c. An example of "optimization" criteria in an area spanning the physical and social sciences appears in a recent paper on solar energy conversion (Clark, 1986). In this paper, the economic/technical "optimum" configuration is shown to be derived from considerations of life-cycle costs, annual cash flow, maximum average annual saving or return on investment, payback period, and the economics of inflated or discounted value of money.

it is never possible to turn all of L into M. When a life disappears, valuable societal resources other than those that can be expressed by M are also depleted (creativity, kindness, etc.).

Derived Variables: Entropy, S, and Severity, H

We have previously drawn the analogies between risk probability, P, and temperature, T, and between risk consequence, R (alternatively expressed as lives, L), and heat, Q. We now propose a third analog between severity, H, and thermodynamic entropy, S. The severity for society depends on how many people will be affected and on the number of different types of consequences which may result from a given hazard. We confine our discussion here to the meaning of severity as used in Eq. (1): $R = P \times H$. We consider four aspects of H that are analogous to aspects of S.

(1) Both H and S depend on the number of participating units: people in a societal risk situation and particles in a substance, respectively. In the language of thermodynamics, they are both extensive variables. This dependence on the number of participating units is illustrated by Okrent's example (Okrent, 1980) of an ocean liner crossing the Atlantic. If it sinks with three hundred people aboard, the societal risk consequence is 100 times that of a rowboat sinking with only 3 people aboard. Although the risk probability for the ocean-liner tragedy, Po, is much less than for that of the rowboat, Pr, the statement about the magnitude of the societal-risk consequence is true because of the liner's large H. Societal risk (in contrast to individual risk) always involves a societal size factor, H. In thermodynamics the entropy also depends on the size of the substance, with the large sample possessing more entropy than the small sample of the same substance.

(2) The values of both H and S may be derived from measurements of R and Q, respectively, if P and T are known for the particular situations. Thus, one could estimate the size of the population using intravenous drugs, H, from the number of diagnosed AIDS cases in the population, R, and a known or estimated probability of contracting AIDS in that population, P. Expressed in an equation:

$$H = R/P . \tag{4}$$

We write a similar equation for entropy but we use delta S instead of S to emphasize that it is the changes in S which are measured by any given flow of heat:

$$\Delta S = Q/T . \tag{5}$$

For example, when 1 g of ice melts at 273 K upon the addition of 80 calories of heat, the 1 g of water has 0.293 cal/K more entropy than the ice. This leads us to the third comparison.

(3) H increases with the number of types of harmful consequences in a risk situation and S increases with the number of degrees of freedom into which heat can flow in a substance. Let i represent a running index for the consequences being considered. Then the total severity of the risk situation is the sum over i of Ri divided by Pi, where Ri is the ith consequence and Pi is the probability for that consequence to happen.

$$H = \sum_i Hi = \sum_i \frac{Ri}{Pi} . \tag{6}$$

This aspect is exemplified by the "severity" associated with a nuclear-power accident. A nuclear accident is considered to be of high severity (even when the probability of its occurrence is very low) because of the many different types of consequences associated

Table 5. Major Differences Between Thermodynamic and Societal Systems

Characteristics	Thermodynamic Systems	Societal Systems
a) Number of Different Types of Forces	Few	Many
b) Experimental Goals	Isolation	Intervention
c) Data Banks	Large	Small
d) Degrees of Freedom	Small, Predictable	Large, Unpredictable
e) Aggregation of Constituents	Easy	Hard
f) Fluctuations of Values	Can't Exceed Original Value	Can Exceed Value By Orders of Magnitude

with it. In the Chernobyl accident, although the number of fatalities was low, there were many other harmful societal effects (economic, psychological, agricultural, social). (Parenthetically, it is interesting to note that we tend to ignore the economic, psychological and social harm that ensue when large numbers of lives are lost in other high-visibility accidents such as air disasters.)

(4) Values for H and S can be obtained by theoretical calculations as well as by experimental measures. H may be calculated from exposure models or by postulating scenarios or constructing simulations. For theoretical predictions of S, the techniques of statistical mechanics are employed and the known properties (such as the energy-level scheme) of the constituents are utilized, including those properties manifested as a result of the constituents' interactions with each other. Comparison of theoretical and experimental values of S may lead to the discovery of new degrees of freedom of a thermodynamic system. Similarly, calculations of H and experimental verification may lead to the discovery of new sources (or consequences) of the risk.

CAVEATS FOR THE APPLICATION OF THE THERMODYNAMIC PARADIGM

To stimulate exploration of the conceptual transfer from thermodynamics to risk, we have highlighted a few common features of risk and thermodynamics: the global relevance, the complementary vantage points, the importance of a new kind of intensive variable, and the utilization of a unifying and a derived extensive variable. In examining these features and in searching for others, it would be prudent not to carry this analogy too far (as has sometimes been done in applying natural science concepts to the social sciences) and to be aware of some of the fundamental differences between thermodynamic systems and societal systems. Five of these differences are summarized in Table 5 and briefly discussed below.

(a) In thermodynamics there are only a very few and rather well-known types of forces. But the societal risk state is affected by many different and sometimes unexpected types of forces: social, economic, political, psychological.

When thermodynamics (dealing with matter and energy) makes predictions about the properties and behavior of the physical world, it assumes that there are no unrevealed forces at work which would interfere with the known natural laws.

In making societal risk predictions, on the other hand, we must remember that it is not only possible, but often very likely, that new forces will emerge during the period covered by the predictions. In fact, a prediction itself creates some new forces which may—if they have enough power—interfere with the "course of nature" and thus change the predicted outcome.

A striking example is provided by the result of the Multiple Risk Factor Intervention Trial (MrFit, 1976). The trial was motivated by the hypothesis that reducing major risk factors (particularly smoking, dietary fat and lack of exercise) would decrease the incidence of cardiovascular disease. This hoped-for result was so widely accepted as a true prediction by journalists and health-care professionals that they widely disseminated this information to the public. As a result, individuals in the control group changed their behavior in a manner similar to the recommendations made to the experimental group. This led to an equivocal result. Although there was a marked improvement in the cardiovascular state of the experimental group, the differences from the control group all but disappeared.

(b) Theories in both thermodynamics and risk emerge from data. But risk data are, in many domains, very sparse and also, in general, much harder to obtain than thermodynamic data.

(c) The number and kinds of degrees of freedom available to a molecule are much smaller than the behavioral options of a human being.

(d) There also is a qualitative difference between the ease of aggregating molecules into macroscopic structures and the ease of creating societal units, legal codes and institutions in society which determine the conditions under which the management of societal risk operates.

(e) The economic values of resources (such as oil) may vary with time by orders of magnitude. While the energy delivered by a tank of gasoline may also vary, such fluctuations are relatively small. (They depend only on the energy conversion efficiency of the engine and on the chemical and mechanical purity of the fuel.)

SUMMARY

The importance, the diversity, and the complexity of societal risk management are well recognized. Most risk professionals are also aware of the difficulties involved in sorting out important considerations from the less important ones and in communicating available scientific knowledge on risks to policy makers, risk managers and citizens at large. In this paper we considered the structure of thermodynamics as a template for clarifying some of these complexities. By referring to this well-developed scientific discipline we might sharpen our formulation and communication of risk concepts and we might find new directions for worthwhile societal experiments and data collection. In order to lend credence to this rather novel proposal for looking at risk management, we discussed ideas which were developed in thermodynamics, such as the usefulness of the concept of an idealized, isolated generalized system, and of the macro and micro approach to understanding the physical properties of substances. We also conjectured that the concepts of temperature, energy and entropy of thermodynamics have some similarity to the risk concepts of probability, consequences and severity.

The great difficulties and limitations imposed by the much greater complexity of social behavior (compared to physical behavior) have been portrayed in the section on caveats.

In companion papers we plan to develop the thermodynamic paradigm along three lines: (1) presenting significant risk generalizations that are evoked by the basic laws of thermodynamics, (2) distinguishing among risk-probability estimates according to their methods of derivation, and (3) proposing response coefficients for risk-management strategies similar to response coefficients for changes in thermodynamic variables.

ACKNOWLEDGMENTS

Support for this work was provided by a Temple University Grant-in-Aid of Research and by National Science Foundation Grant SES-8796182 to Decision Research. We wish to thank Charles Vlek, Sarah Lichtenstein and Esther Offenbacher for their helpful comments on an earlier draft of the manuscript.

REFERENCES

Campbell, T. C., 1980, "Chemical Carcinogens and Human Risk Assent?" *Federation Proceedings* **39**:2467-2484.

Cardwell, D. S. L., 1971, *From Watt to Clausius*, Cornell University Press.

Clark, J. A., 1986, "Thermodynamic Optimization: An Interface with Economic Analysis," *Journal of Non-Equilibrium Thermodynamics* **10**:85-122.

Coombs, C. H., 1972, "A Review of the Mathematical Psychology of Risk and Risk-Taking," Report No. MMPP 72-6, Michigan Mathematical Psychology Program.

Douglas, M. T., 1985, *Risk Acceptability According to the Social Sciences*, New York: Russell Sage Foundation.

Douglas, M., and Wildavsky, A., 1982, *Risk and Culture*, University of California Press.

Fischhoff, B., Lichtenstein, S., Slovic, P., Derby, S., and Keeney, R., 1981, *Acceptable Risk*, New York: Cambridge University Press.

Fischhoff, B., Watson, S., and Hope, C., 1984, "Defining Risk," *Policy Sciences* **17**:123-139.

Gratt, L. B., 1987, "A Proposal for Consistent Definitions," *Proceedings of the Society for Risk Analysis International Workshop on Uncertainty in Risk Assessment, Risk Management, and Decision Making, September 30– October 3, 1984, Knoxville, TN*, V. T. Covello, L. B. Lave, A. Moghissi, and V. R. R. Uppuluri, Eds., pp. 241-249, Plenum Press, NY.

Gratt, L. B., Chairman, "The Definition of Risk and Associated Terminology for Risk Analysis," *Proceedings of the Society for Risk Analysis 1987 Annual Meeting, November 1-4, 1987, Houston, TX* (to be published).

Kates, R. W., Hohenemser, C., and Kasperson, J. X. (Eds.), 1985, *Perilous Progress: Technology as Hazard*, Boulder, CO: Westview.

Libby, R., and Fishburn, P. C., 1977, Autumn, "Behavioral Models of Risk Taking in Business Decisions: A Survey and Evaluation," *Journal of Accounting Research*, 272-292.

MrFit, 1976, "Multiple Risk Factor Intervention Trial: A National Study of Primary Prevention of Coronary Heart Disease," *Journal of the American Medical Association* **235**:825.

Okrent, D., 1980, "Comment on Societal Risk," *Science* **208**:372-375.

Schaefer, R., 1978, "What Are We Talking About When We Talk About "Risk?" A Critical Survey of Risk and Risk Preference Theories, RM-78-69, International Institute for Applied Systems Analysis.

Sharlin, H. I., 1986, "EDB: A Case Study in Communicating Risk," *Risk Analysis* **6**:61-68.

Wilson, R., and Crouch, E. A. C., 1982, *Risk/Benefit Analysis*, Ballinger Publishing.
World Health Organization, 1978, *Environmental Health Criteria: Vol. 6 — Principles and Methods for Evaluating the Toxicity of Chemicals: Pt. 1 — Geneva and Campbell.*

Risk in Defense Policy Decisions

M. Elisabeth Paté-Cornell

Stanford University
Stanford, CA

ABSTRACT

This paper shows how the use of risk analysis and decision theory can improve the strategic debate and the choice of some defense policies. Three different levels of abstraction and quantification in the use of these methods are presented and illustrated. In all three cases, Bayesian techniques permit reasonings and conclusions that are not easily achievable, if at all reachable, by classical methods of strategic analysis.

KEYWORDS: Risk, uncertainty, defense policy, strategic decision, nuclear conflict

INTRODUCTION

The choice of a defense policy often involves uncertainties, either strategic or tactical. The traditional approach to such decisions is to try to reduce or possibly eliminate these uncertainties, for example, through extensive contingency planning. Recognizing that some uncertainties always remain in matters of defense and must be treated explicitly is a relatively new way of thinking (see, for example, Estes, 1983). This paper describes how probabilistic methods of risk analysis and decision theory can be used for that purpose.

Classical strategic reasoning (e.g., the analysis of deterrence factors by George, 1974) relies implicitly on concepts of risk and uncertainty but treats these concepts in a qualitative manner. In particular, this approach does not allow the analyst to weight quantitatively the relative importance of two elements of a problem that have potentially opposite effects on the conclusion. In addition, the debate often focuses on a limited number of possible scenarios because the complexity of most strategic questions can make an exhaustive reasoning impossible in practice. Game theory (Luce and Raiffa, 1957), which in its development stage was often applied to the resolution of strategic problems, can provide some valuable insights. This method, however, has some major drawbacks; for example, it is not adequate to treat sequential decisions, nor does it account for uncertainties about the opponents' preferences such as the losses that they may be willing to accept for political gains. Yet, these trade-offs are critical in many strategic issues (Strait, 1987).

The explicit treatment of uncertainty in strategic reasoning has become still more crucial with the existence of nuclear weapons because of the magnitude of the potential destruction and the costs of possible miscalculations. This paper shows how the use of quantitative analytical methods based on probability and utility can improve the strategic debate and the choice of some defense policies. Examples of such analyses for particular issues can be found in recent literature (e.g., Parnell, 1985; Judd *et al.*, 1986; Strait, 1987). In some cases, it is not even necessary to actually quantify the elements of such models to obtain valuable insights (Starr and Whipple, 1987).

Consider, for example, the following questions of defense policy, all of which involve uncertainty and value judgments in decision making:

- What constitutes a stabilizing policy? Is uncertainty, in general, a stabilizing factor?

- Does the decoupling of nuclear and conventional forces in Europe increase or decrease the probability of a generalized nuclear conflict?

- Does the increasing vulnerability of the American land-based Intercontinental Ballistic Missiles (ICBMs) justify a change in the United States policy of missile launch in response to attack signals?

For each of these questions, there are different possible answers according to the weight allocated to different factors and to assumptions about technical performances and human behaviors. These illustrative examples show how risk analysis methods can be used at least at three different levels. The first one involves formal models based on concepts of Bayesian probability, utility, and rationality as defined by von Neumann (1947) and Savage (1972). The second level is based on the relative ranking of scenario probabilities and relies on the concept of conditional probability. The third level results in the quantification of the outcomes of possible scenarios under the form of probability distributions and requires the full development of risk analysis models using the tools of reliability and decision analysis such as fault trees, event trees (Paté, 1984), decision trees (Raiffa, 1968), and influence diagrams (Shachter, 1986). In all three cases, the use of Bayesian methods permits reasonings and conclusions that are not easily achievable, or are simply unreachable, by classical methods of strategic analysis.

UNCERTAINTY AND STABILITY

Conflicting Effects

At the first level, risk analysis and decision theory can be used in the strategic debate on the basis of concepts and formal models rather than actual quantification. This type of reasoning is illustrated by a discussion of how uncertainty affects stability.

As a working definition of crisis stability, assume that a policy is stabilizing if it reduces globally the attractiveness of a first strike for the parties involved in a potential conflict. To simplify the reasoning, assume that there are only two parties involved: the United States and the Soviet Union. For the independent observer, a policy is defined as globally stabilizing if it reduces the sum of the probabilities of first strike by the United States ($p(AA)$) and by the Soviet Union ($p(SA)$). According to this definition, a given policy is globally stabilizing if it reduces the overall probability of nuclear war, even though it may increase the probability of attack by one of the two parties.

Many policies have a dual potential to be either stabilizing or destabilizing according to which term prevails [variation of $p(AA)$ or variation of $p(SA)$], and quantification of probabilities is necessary to reach a conclusion. The debate about the Strategic Defense Initiative (SDI) provides a clear example of this phenomenon. On one hand, SDI is considered destabilizing by those who think that it provides an incentive for the United States to strike first with a higher probability of absorbing a weakened retaliatory attack on the part of the Soviet Union. Therefore, SDI increases $p(AA)$. On the other hand, SDI is considered stabilizing by those who think that, everything else being equal, it reduces the effects and, therefore, the attractiveness, of a Soviet attack on the United States, thus reinforcing deterrence and reducing $p(SA)$. Both arguments are correct, but it is impossible to reach a global conclusion without weighting these two types of contradictory effects by quantifying a measure of propensity to attack involving information and risk attitudes. Because decision theory is used in the rest of this paper as a descriptive approach to strategic actions, the first issue is the question of rationality in the propensity to attack.

Rationality

Whether or not it can be assumed that the Soviets (or the Americans for that matter) are "rational" in defense decisions is a question that is often debated in strategic discussions. Most of the time, however, what is actually meant by rational is that the trade-offs that either side is willing to accept are "reasonable." "Reasonableness," in this context, is often vaguely defined as a value system similar either to the current Western way of thinking or to preferences of the particular group assessing rationality. This notion differs from rationality in its axiomatic definition as discussed below. For example, the behavior of a particular country determined to exterminate a specific group from its population may be perfectly rational in the axiomatic sense of the term, yet incompatible with Western values and unacceptable in a civilized world.

The axioms of decision theory as defined by von Neumann (1947) have been formulated under several different equivalent forms (e.g., Savage, 1972). The form of the axioms used here is a set of five properties of the utility function that characterize rational choices under uncertainty (Howard, 1984). The question is to examine their descriptive applicability to decisions of strategic defense.

- Orderability: In general, everything else being equal, one can safely assume the logical preference of less losses to more, and of more political control to less. Circularity in preferences does occur over time as political thinking evolves and values change, but at the time of a given decision there is little doubt that there is an ordering of preferences.

- Monotonicity: In strategic decisions, it seems reasonable to assume, everything else being equal, preference for gambles involving a higher probability of success as estimated at the time of the decision.

- Continuity and substitutability: The existence of certain equivalents of uncertain gambles, not as desirable as the highest outcome but not as undesirable as the lowest one, is the essence of bargaining in cease-fires and peace treaties.

- Decomposability: This is the most questionable of the rationality axioms from a descriptive perspective because it implies that what matters is the outcome (e.g., a level of losses) and its probability, and not the sequence of events leading to it. In times of conflict, some elements of the process are often relevant to the decision. These elements, however, can often be considered integral parts of the outcomes. For example, it may be that by restricting the description of the outcomes to actual losses and political control, one misses important attributes such as national honor or flexibility for future moves. If such elements are critical, they can be included as part of the description of the outcomes and thence in the utility function. The question then is: what immediate losses are nations willing to incur as a price to pay for such attributes?

Decision theory is best suited as a normative approach to individual decisions. It has at least two major limitations when used as a descriptive tool for the analysis of strategic decisions. First, some psychological factors may introduce biases and errors of logic in the reasoning of decision makers (Kinder and Weiss, 1978; Kahneman, Slovic, and Tversky,

1982). Second, the method does not account for the organizational nature of the decision; therefore, it ignores the effects of the procedures used in sequential or collective decision making. A simple process such as a majority vote can result in clear violations of the axioms even if each of the participants acts rationally, so that the organization as a whole cannot be regarded as one rational decision maker. If, however, one considers the decision of a National Command Authority acting together in full concert, the axioms are not a bad approximation of the underlying principles of choice. In some cases, individual decisions by key actors such as political leaders or military commanders may have to be analyzed separately using individual preference structures.

The author thus concurs with others, such as Bueno de Mesquita (1981), to conclude that the von Neumann axioms of rationality can be used as an approximation of the principles underlying strategic decision making. In particular, one may assume that the matter is important enough that the nations involved have designed procedures that do not systematically favor suboptimal options. Yet, the reservations expressed by others, such as Steinbrunner (1982), regarding the relevance of the process are valid and their implications for the results of stochastic models must be carefully considered.

Propensity to Attack

Consider the decision of the Soviet Union to launch a preemptive nuclear strike on the United States when war seems imminent and unavoidable. There is a small chance, however, that war will not occur after all. Given the stakes in an all-out nuclear conflict, this small chance may actually play an important role in the decision. A critical issue in such a situation is the possibility of misperception and miscalculation. For example, either side may be wrong in its estimation of its own actual strategic potential (capabilities of forces actually unknown) given objective military means (level of forces assumed to be known). Either side may also be wrong in its assessment of the opponent's forces and intentions. Concentrating on the miscalculation problem, the variables of the model are defined as follows:

O: Soviet objective military means as estimated by the Soviets (assumed known by them).

O':   Soviet subjective assessment of the U.S. military means.

M: Subjective assessment by the Soviets of their own military potential.

M':   Subjective assessment by the Soviets of the U.S. assessment of the U.S. military potential.

W: Soviet propensity to attack the U.S. ($W>0 = >SA$).

W':   American propensity to attack the Soviet Union ($W'>0 = >AA$).

SA: Soviet attack on the U.S. ($\underline{SA}$: no attack).

AA: American attack on the Soviet Union ($\underline{AA}$: no attack).

f: Probability density function as assessed by the Soviets.

p_{SU}:   Probability of events as assessed by the Soviets.

p: Probability of events as assessed by the observer.

Δ: Variation of a quantity (variable or parameter).

Multiattribute utility theory (Keeney and Raiffa, 1976) is used here to express preferences given trade-offs, for instance, between potential losses of different types and magnitudes and different levels of political control. Assume that the attributes of the utility function for the Soviet Union are limited to the following:

L: Losses to the Soviet Union (civilian, military, industrial, environmental).

PC: Political control [gain or loss, e.g., from 0 (total loss of control) to 1 (total control of NATO)]. This assumes that losses inflicted to NATO matter to the Soviet Union only to the extent that they imply political control over NATO).

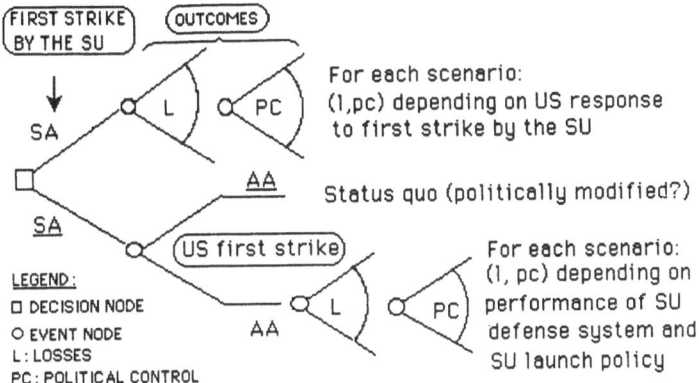

Fig. 1. Simplified Decision Tree for a First Strike by the Soviet Union.

From the Soviet side, the decision to attack can be analyzed using the decision tree shown in Fig. 1.

The expected utilities for each of the alternatives are the following:

$$Eu(SA) = \int\int\int_{m,pc,l} f(m \mid o;SA) \times f(l,pc \mid m;SA) \times u(l,pc) \times dm\ dpc\ dl\ ,$$

$$Eu(\underline{SA}) = p_{SU}(\underline{AA}) \times u(\text{status quo}) + p_{SU}(AA) \times \int\int_{pc,l} f(l,pc;\ AA,\underline{SA}) \times u(l,pc) \times dpc\ dl\ .$$

In these formulas, the probability of a preemptive strike on the Soviet Union by the United States as assessed by the Soviets is

$$p_{SU}(AA) = \int\int_{o',m'} p_{SU}(W'>0 \mid m',o') \times f(m' \mid o') \times f(o') \times do'\ dm'\ .$$

The axioms of rationality discussed above imply that the Soviet Union will decide to attack if the expected utility of attacking is greater than the expected utility of not attacking. One can thus define the *propensity to attack* on the part of the Soviet Union as the difference between these two values:

$$W = Eu(SA) - Eu(\underline{SA})\ .$$

No attack takes place when W is negative. The decision is reversed when W becomes positive. A similar computation can be done for the decision of the United States to launch a preemptive strike against the Soviet Union. For the outside observer, however, all of the variables above, including both sides' attitudes towards risk and estimations of the states of the world, become state variables in his own estimation of the probability of SA or AA. All the variables involved, including preferences, can vary over time, particularly in times of crisis. For this observer, the probability of a nuclear attack over a given time horizon T is

$$p(T)=p(SA,T)+p(AA,T)=p(W>0,T)+p(W'>0,T)\ ,$$

in which $p(SA,T)$ is the probability of a Soviet preemptive attack on the United States during the time interval T, and $p(W>0, T)$ is the probability that W becomes positive during the same time interval. $p(AA,T)$ and $p(W'>0,T)$ are defined in a similar way.

The concept of propensity to attack and the method for its measurement permit the definition of stability in quantitative terms and allow weighting potential conflicting effects of some policies. One can now define as stabilizing, over a given time horizon, a policy that reduces the probability of an intentional preemptive nuclear attack as assessed by the independent observer. Therefore, a policy is stabilizing for the time interval T if the variation $\Delta\{p(SA,T) + p(AA,T)\}$ is negative. Note that this particular model does not include the possibility of an accidental strike, although its probability increases with the severity of a crisis situation. Accidental strikes, however, can be included as illustrated below (see "Launch Policy and Reliability of the C^3I").

Obviously, different observers may have different assessments of the variables or different opinions as to what variables are relevant to the issue. This formulation allows introduction of additional variables when desired, discussion of values and probabilities for each of them, and separation of their effects on the result. Note also that this model can accommodate a large spectrum of preferences, for example, from "better red than dead" to "better dead than red." It also permits the analyst to dispel some misconceptions that have obscured past debates about crisis stability, for example, the idea that uncertainty in itself is always a source of stability.

Is Uncertainty Stabilizing?

The notion that increasing uncertainty about the outcome of an attack is a deterrent in itself has been pervasive in European as well as American strategic thinking. As a general principle this idea is wrong, because it relies on two assumptions that do not always hold: that more uncertainty always implies greater probability of higher losses, and that the potential assailant is necessarily risk averse. Whether or not these assumptions are correct depends on the circumstances.

Uncertainty, in fact, can be increased by introducing the possibility of less losses to the nation that considers striking first, therefore making the option to attack more attractive. In other terms, the distribution of losses can be spread exclusively towards outcomes that are less costly for the assailant. There may be complete dominance of the more certain and more costly situation by the more uncertain but always less hazardous one. In an extreme case, the option to strike first looks more attractive in a situation where there is a small chance of "winning" (uncertainty) rather than an absolute certainty of defeat. For this reason alone, increasing uncertainty does not necessarily increase deterrence.

Consider, as a illustrative example of this dominance effect, the role of France in the Western alliance after that country decided to leave the NATO Integrated Military Command. It has been argued since then that the French position is stabilizing and deterring in essence because it introduces an unpredictable element in the Western defense. This argument, however, does not hold. Indeed, facing a divided NATO rather than a united alliance does introduce more uncertainty in the Soviet military planning of an attack on Western Europe, but mainly in the sense of spreading the probability distribution of potential Soviet losses towards lower values. According to the model proposed above, it therefore increases the Soviet propensity to attack Western Europe regardless of the risk attitude of the Soviet leaders. France may have other valid reasons to want its independence from the NATO military wing. Her position, however, does not increase crisis stability with respect to a position of military unity within NATO because the additional uncertainty that it introduces in a potential decision to attack by the Soviet Union mainly increases the attractiveness of an attack option, and is thus counter to deterrence.

Even when there is no such dominance effect, and when uncertainty increases the probability of high levels of losses, the propensity to attack may decrease for the risk-averse but not for the risk-prone. The risk-taker may find reassurance in the fact that the increased uncertainty that spreads the probability distribution of potential losses may introduce a slim probability of lower losses at the same time as it increases the probability of higher losses. If one considers past military decisions, risk aversion has certainly not always been the norm, in particular in high-level crisis situations. In fact, Ellsberg (1987) shows that during the Vietnam war, the decisions of the U.S. military were often risk-prone. One can argue in the same way that many of the decisions that led to the start of World War I in Europe were probably risk-prone unless the information on which they relied was extremely biased.

It is true that, in some cases, introducing more uncertainty by increasing the probability of higher losses will deter the risk-averse. Thus for the independent observer, in order to come to a conclusion about the stabilizing or destabilizing effects of a defense policy, it will often be necessary to quantify his own assessment of the risk attitudes and the quality of the information used by both sides in reaching their final decisions.

DECOUPLING IN EUROPE

When concepts alone no longer suffice, risk analysis techniques can be used to rank the probabilities of different scenarios. This second level involves some quantification through simple event trees. The illustrative example used in this section is whether the decoupling of conventional and nuclear strategic forces in Europe increases or decreases the probability of a generalized nuclear conflict.

Deterrence in Europe

Deterrence in Europe is currently based on the doctrine of flexible response involving a continuum of options from conventional forces, through nuclear tactical weapons, through nuclear intermediate range weapons, to strategic nuclear weapons. Elimination of intermediate, and possibly of short-range tactical weapons, would thus leave a gap between the use of conventional forces and the use of strategic nuclear weapons.

In the United States, the media often convey the impression that the mere reduction of the number of existing nuclear weapons will reduce *de facto* the probability of a nuclear conflict. In Europe, however, there is the uneasy feeling that decoupling increases the attractiveness to the Soviet Union of a conventional attack on Western Europe (WEU, 1987), that such a conflict may easily escalate into a full-scale nuclear conflict if the losing side decides to use strategic forces as a last resort, and, therefore, that the overall probability of a nuclear war may well be increased rather than decreased by decoupling as defined above.

Decoupling and Probability of a Nuclear Conflict

There are several contradictory elements in this problem that cannot be weighted by a simple qualitative reasoning. They include the following:

- Decoupling may increase the probability of a Soviet conventional attack on Western Europe.

- It may decrease the probability of a U.S. nuclear intervention in the case in which NATO is losing control.

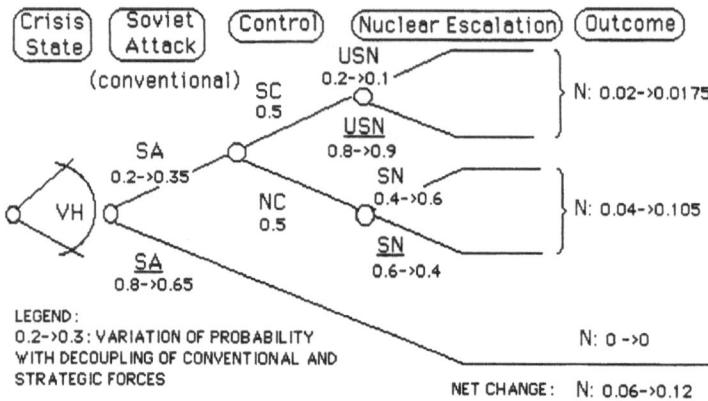

Fig. 2. Probability of Escalation of a Conventional Conflict in Europe.

- It may increase the probability that the Soviets use nuclear weapons in the case where they lose control on conventional ground because it reduces the probability of a U.S. nuclear retaliation to such a move.

Altogether, it is, therefore, impossible to tell *a priori* if decoupling increases or decreases the probability of nuclear conflict. In order to come to a conclusion, it is necessary to assess the relative variations of the probabilities of the two main scenarios leading to nuclear war. The questions are thus: (1) Does decoupling increase the probability of a conventional conflict in Western Europe? If so, to what degree? (2) What is the probability of a nuclear escalation with and without decoupling, given the outcome of the initial conventional conflict? The following risk analysis model can be used to integrate the different parts of the problem.

Variables:

CS:	Crisis state.
VH:	Highest value of the crisis state (e.g., DEFCON 1 in the U.S.).
SA:	Conventional Soviet attack on Western Europe (*SA*: no attack).
SC:	Soviet control in the conventional conflict.
NPC:	NATO control in the conventional conflict.
USN:	Nuclear escalation of the conflict by use of strategic weapons by the U.S. or by nations of Western Europe (*USN*: no escalation on NATO side).
SN:	Nuclear escalation of the conflict by the Soviet Union (*SN*: no escalation on the Soviet side).
N:	Nuclear escalation of the conflict by either side.
p(x):	Probability of *x*.
p(x|y):	Conditional probability of *x* given *y*.

Consider in a severe crisis situation, the different possible scenarios represented on the event tree of Fig. 2.

Conditional on a severe ("very high") crisis state, there are five scenarios in this analysis, two of which end in a nuclear conflict (critical scenarios): (1) Soviet attack, NATO control, and Soviet use of strategic nuclear weapons and (2) Soviet attack, Soviet control, and NATO use of strategic nuclear weapons. [It is assumed here that nuclear escalation will be triggered by the side that is losing control on conventional ground. Other assumptions can be incorporated in the analysis using the same method.] The probability of a nuclear conflict (N), given a very high crisis state, is the sum of the probabilities of these two critical scenarios.

Quantification and Sensitivity

The probability of each critical scenario can be computed by multiplying marginal and conditional probabilities for the considered values of the variables:

$$p(SA,NC,SN \mid VH) = p(SA \mid VH) \times p(NC \mid SA,VH) \times p(SN \mid SA,NC,VH) \ ,$$

$$p(SA,SC,USN \mid VH) = p(SA \mid VH) \times p(SC \mid SA,VH) \times p(USN \mid SA,SC,VH) \ ,$$

$$p(N \mid VH) = p(SA,NC,SN \mid VH) + p(SA,SC,USN \mid VH) \ .$$

In the following table a set of illustrative data is proposed and the results of a very simple risk analysis model are displayed.

Table 1. Variation of the Probability of Escalation with Decoupling

	Before elimination of intermediate and tactical weapons		After decoupling
$p(SA)$	0.2	->	0.35
$p(SC) = p(NC)$	0.5	->	0.5 (unchanged)
$p(USN \mid SC)$	0.2	->	0.1
$p(SN \mid NC)$	0.4	->	0.6
Net result: $p(N \mid VH)$	0.06	->	0.12
But if assessor believes that $p(SC \mid SA)$ = 0.9, then, using the same values for the other data,			
Net Result: $p(N \mid VH)$	0.044	->	0.0525

In this illustrative example, it is assumed that decoupling does not change the probabilities of the possible outcomes of a conventional conflict, but that it increases the probability of a Soviet attack from 0.2 to 0.35, decreases the probability of nuclear escalation in case of the defeat of NATO from 0.2 to 0.1, and increases the probability of nuclear escalation in case of defeat of the Soviet Union from 0.4 to 0.6. The result is that the probability of scenario 1 (nuclear exchange initiated by NATO) decreases slightly from 0.02 to 0.0175, but the probability of scenario (2) increases sharply from 0.04 to 0.105, resulting in a global increase by a factor of two of the overall probability of a nuclear conflict given a severe crisis situation. If the external observer believes that the probability of victory by Soviet conventional forces is higher than 0.5 (e.g., 0.9), the probability of a global nuclear conflict is only slightly increased by decoupling. This simple analysis may need to be refined but is sufficient to illustrate clearly the trade-offs that exist between deterrence, political independence, and the existence of nuclear weapons.

This model can be modified to introduce other factors. For example, one may want to analyze further the process of escalation. One possible scenario, in the present situation, is that nuclear escalation may start with the use of intermediate-range or short-range tactical nuclear weapons, eventually leading to the launch of strategic weapons. The removal of intermediate nuclear forces and tactical weapons may indeed reduce the probability of this particular escalation scenario but not necessarily the overall probability of nuclearization of a conflict if it increases sufficiently the probability that the defeated side will threaten to launch its strategic missiles.

LAUNCH POLICY AND RELIABILITY OF THE C^3I

Some complex issues may involve a large number of variables, each of which may take a large number of values. In classical strategic discussion of resulting scenarios, it is extremely difficult not to lose track of the different elements and their relative importance. It is at this third level that risk analysis and full-scale probabilistic models can be most useful because there is simply no other way to reach such results. At the same time, the models become more complicated, more intimidating, more difficult to communicate, and the results may be less easy to interpret for the layperson. When it really matters, however, the understanding of these results and of the way by which they were obtained is worth the effort.

Consider the question of whether the increase in the vulnerability of the U.S. ICBMs to a Soviet first strikes justifies a change in the U.S. missile launch policy (Garwin, 1980). Assume that the initial situation is one of launch on impact and that the considered alternative is launch under attack. The dilemma results from the trade-off between the probability of a false alert, possibly followed by an accidental nuclear attack, and the probability of not being able to respond to either because of a failure in the system of command and control or because of destruction of the retaliatory forces. There is thus a critical link between the choice of a launch policy and the reliability of the system of Command, Control, Communication, and Information (C^3I).

Probability of Error

In a study by Paté and Neu (1985), a method was presented to compute the probabilities of type 1 error (false negative), type 2 error (false positive), and action as planned, given a particular launch policy and the reliability of the C^3I under this policy. The two types of errors were further subdivided between error of *information* (type 1: missed signal; type 2: false alert) and error of *action* (type 1: failure to launch when planned; type 2: accidental attack). The method is first to compute the probability of the different scenarios leading to the three types of situations (type 1 errors, type 2 errors, no error), then to compute the overall probability of each type of situation for each launch policy, and finally to compute the variation of these probabilities in a shift of policy.

For each launch policy, the configuration of the C^3I is represented by its functional diagram (Henley and Kumamoto, 1981). Fault-tree analysis based on the system configuration described by the functional diagram then allows identification of the failure modes (i.e., the conjunctions of events leading to failure) for each type of error and for each launch policy. An event tree is used to construct an exhaustive set of scenarios and to identify those corresponding to no error, a type 1 error, or a type 2 error. The total probability of a type 1 (or, respectively, a type 2) error is the sum of the probabilities of all scenarios leading to a type 1 (or, respectively, a type 2) error. The result of this model is the probability of a type 1 error and the probability of a type 2 error for each launch policy and thus a description of trade-offs in a shift of policy measured by variations of error probabilities.

Note that the probabilities involved depend in a large part on the circumstances as anticipated at the time of the analysis. For example, the probability of a Soviet attack varies with the crisis state, and the probability of an intended retaliatory attack on the part of the U.S. depends on many elements, including the personality of the president. Therefore, these probabilities vary with time and the model itself has to be adapted to the circumstances at the time when a decision is made. The numbers that were computed were displayed for illustrative purposes only and on the basis of unclassified information.

The use of an event tree in this study illustrates clearly the power of a systematic analysis of events and random variables. It would be extremely difficult to combine mentally the different possibilities leading to different types of errors without the help of a tool that allows exhaustive treatment of the ten scenarios considered. In this form, risk analysis allows construction of the relevant scenarios, computation of the probability of each of them, and computation of the probabilities of errors by grouping the scenarios by categories and adding the relevant probabilities.

Consequence Analysis

Because the consequences of type 1 and type 2 errors are quite different, the next step is a complete analysis of the outcomes. In the following phase of the work described above, this is done by computing the probability distribution of potential losses in the first exchange of a nuclear conflict between the United States and the Soviet Union. In this part of the study, only two launch policies are considered: launch on impact and launch under attack. The losses are described, respectively for the United States and for the Soviet Union, by three attributes: the number of civilian casualties, the loss of military potential (a proxy for the short-term development of the conflict), and the loss of industrial capacity (a proxy for the long-term development of the conflict). A more complete consequence model should also include other attributes such as losses of civilian property, damage to the environment, and a measure of political control. The scope is limited to a first exchange of strategic nuclear weapons (ICBMs, SLBMs, and weapons on bombers), assuming no intentional first strike on the part of the United States. The two major scenarios are thus: (1) a Soviet first strike followed either by U.S. retaliatory attack or no response in case of a type 1 error, and (2) a U.S. accidental attack followed by a Soviet retaliatory attack. The results are a shift in the probability distribution of the different types of losses. By discretization of these distributions, one can then show the variation of the probability of different levels of losses (e.g., massive losses) either in terms of marginal probability or conditional on a high level of crisis.

The tool that is used in this part of the study is called an influence diagram: it is equivalent to a decision tree in its fundamental properties and the results that it provides, but it also displays explicitly the probabilistic dependencies among random variables (Howard and Matheson, 1984; Shachter, 1986). Because it does not display in its graphic form the different values of the decision and state variables, the influence diagram is much more compact that the corresponding decision tree. As an example, Fig. 3 shows the influence diagram that was used to compute the probability distribution for civilian losses in the United States.

The variables involved are the following:

CS: Crisis state (the starting node).
SAA: Soviet attack action. Four different scales of attack, each of which can be intentional or accidental, and no attack at all.
ASD: American signal detection: scale of the Soviet attack as perceived by the U.S. C^3I.
ALT: Lead time for American reaction to signal of attack (i.e., time as perceived by the U.S. between signal interpretation and nuclear explosion).
ALP: American launch policy.

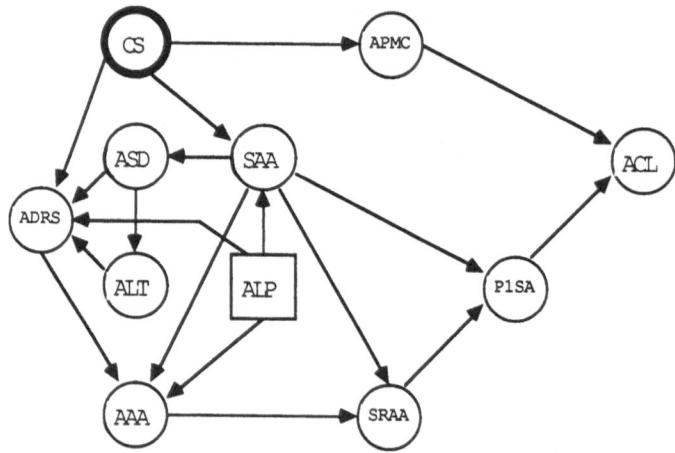

Fig. 3. Influence Diagram for the Analysis of U.S. Civilian Losses in the First Phase of a Nuclear Conflict.

ADRS: American decision to respond to signal: scale of the intended response.
AAA: American attack action: actual attack, given intentions of the National Command Authority, military means, and performance of the C^3I.
SRAA: Soviet response to American accidental attack.
P1SA: Phase 1 Soviet action: scope of Soviet attack; combination of potential first strike and response to a U.S. accidental attack.
APMC: American measures for the protection of civilians. Potential evacuation due to severe crisis situation in high-risk areas. Does not assume any protective measure during the short-lead time provided by the C^3I.
ACL: American civilian losses in the first phase of a nuclear exchange.

Figure 3 shows the probabilistic dependencies among these variables. For example, the American decision to respond to a signal of attack is assumed to depend on the crisis state, the American signal detection, the lead time, and the U.S. launch policy. By varying the American Launch Policy option and using the probabilities obtained in the previous part of the study, one can thus relate the choice of a launch policy to the consequences that may follow in terms of potential losses and the corresponding uncertainties. This kind of model can also be used, for instance, to assess the effects of a risk reduction center, which, among other things, may reduce the probability of response to an accidental attack both from the American and the Soviet side.

In this model, some of the data is circumstantial; most of it is controversial. The goal of the method is to allow consideration of the probability of error in the decision and, in doing so, to test the sensitivity of the results to divergences of opinions among the experts, the politicians, and the public. The numbers used are thus not presented as definitive estimates. Instead, for each variable, some of the key issues involved in the assessment of probabilities are identified. The following example shows how the link is drawn between information found in literature and illustrative probabilities. In the influence diagram of Fig. 3, the probabilities of different levels of Soviet attack action (*SAA*) are assessed conditional on different levels of international crisis (*CS*). Some of the issues discussed in the literature are critical in the choice of illustrative values for the probabilities $p(SAA \mid CS)$.

Examples of such issues include the following: (1) an accidental launch by the Soviet Union in case of high crisis may be the result of a technical accident but may also be unintentional, e.g., an unauthorized launch due to a breakdown in command and control; (2) the probability of an intentional launch by the Soviet Union, and in particular of a massive attack, may not be highest at the peak of a crisis but afterwards, in particular in, traditional times of low activity; (3) the "bolt out of the blue" seems unlikely; and (4) a policy of launch under attack may well be a deterrent to a Soviet first strike compared to a policy of launch on impact. An important characteristic of the risk assessment model is that it provides a means of measuring the sensitivity of the result (here an evaluation *a priori* of the probability distribution of potential losses in a nuclear attack on the U.S.) to different beliefs and expert opinions. Note that in the model described in this section, there is no attempt to introduce values and utilities but rather to describe the risks involved through probability distributions.

CONCLUSION

Methods of risk analysis and decision theory seem to be regaining ground in strategic reasonings, but there is some reluctance to use them in actual policy making for two reasons. The first is the difficulty in quantifying Bayesian probabilities which are rarely based on long statistical time series but often represent expert opinions. These numbers are "soft" by nature but represent an improvement over the simple statement of an opinion without any attempt to quantify a degree of certainty. The second reason is that it forces decision makers to face undesirable outcomes and unpleasant trade-offs which, given a chance, they would rather avoid (Steinbruner, 1982). Yet, there are clear advantages to the proposed method. The first is that it provides a logical framework for the combination of subjective opinions for which there are no better substitutes. The second advantage is that this method puts limits on reasoning by wishful thinking, which can be particularly dangerous when the stakes involve the possibility of nuclear destruction.

The probabilities involved are seldom "hard numbers" because many uncertainties are not related to frequencies in repeated experiments, but rather to lack of knowledge about a phenomenon. The results are most meaningful to the people who have to make the decision anyway and, therefore, whose implicit or explicit assessments of probabilities are the bases of the decision. The goal of stochastic models, in which this treatment is explicit, is to avoid errors of logic in the reasoning.

Stochastic methods can be used at several levels of generality, abstraction, and quantification. The first level illustrated here relies on concepts and on formal models. The second level calls for comparison of probabilities and requires only relatively simple manipulations of marginal, conditional, and joint probabilities. The third level involves full quantification of often complex sets of variables, but provides more information about the consequences, as well as the probabilities, of the different scenarios. In this last case, the results can seem obscure to the layman and one of the challenges is to present them in a clear and understandable manner. This can be done, for example, through discretization of the final probability distributions, rankings, and comparisons. The second and the third levels of analysis are particularly useful when conflicting results could be obtained, depending on the choice of parameters and of the values given to them at the time of the discussion.

The techniques of risk analysis and stochastic modeling can be most useful in the analysis of defense policies for two purposes. The first one is to clarify some concepts, such as uncertainty itself, and to facilitate the debate by providing a common language and a framework to think more exhaustively about complex scenarios. The second purpose is to provide information to actually make specific policy decisions when elements such as variations of the probability of attack, or the probability of high levels of potential losses, are critical to the policy making. In both cases, risk analysis and decision theory provide tools to pose the problems more sharply than verbal descriptions of classical strategic thinking. When needed, these models permit the introduction of additional variables and the assessment of their effects on the results. The utility of these results depends on the analyst's ability to define the relevant parameters.

REFERENCES

Bueno de Mesquita, B., *The War Trap*, Yale University Press, New Haven and London, 1981.

Ellsberg, D., Seminar at the Graduate School of Business, Stanford University, 1987.

Estes, H. M., "On Strategic Uncertainty," *Strategic Review* 11:36-43, 1983.

Garwin, R. L., "Launch Under Attack to Redress Minute Man Vulnerability," *International Security* 4(3):117-139, Winter 1979-1980.

George, A. L., *Deterrence in American Foreign Policy: Theory and Practice*, Columbia University Press, New York, 1974.

Henley, E. J., and Kumamoto, H., *Reliability Engineering and Risk Assessment*, Prentice Hall Pub., Englewood Cliffs, New Jersey, 1981.

Howard, R. A., "Risk Preference," in *Readings in the Principles and Applications of Decision Analysis*, 2:627-664, Howard and Matheson, Eds., Strategic Decision Group, Menlo Park, California, 1984.

Howard, R. A. and Matheson, J. E., "Influence Diagrams," in *Readings in the Principles and Applications of Decision Analysis*, 2:719-762, Howard and Matheson, Eds., Strategic Decision Group, Menlo Park, California, 1984.

Judd, B., Younker, L. W., and Strait, R. S., *Decision Analysis Framework for Evaluating CTBT Seismic Verification Options*, Lawrence Livermore Laboratory, Livermore, California, UCID 20853, 1986.

Keeney, R. L. and Raiffa, H., *Decisions With Multiple Objectives*, Wiley, New York, 1976.

Kahneman, D., Slovic, P., and Tversky, A., *Judgment Under Uncertainty: Heuristics and Biases*, Cambridge University Press, Cambridge, United Kingdom, 1982.

Kinder, D. R., and Weiss, J. A., "In Lieu of Rationality, Psychological Perspectives on Foreign Policy Decision-Making," *Journal of Conflict Resolution* 22(4), 1978.

Luce, R. D., and Raiffa, H., *Games and Decisions*, Wiley, New York, 1957.

Parnell, G. S., "Large Bilateral Reductions in Superpower Nuclear Weapons," Doctoral Thesis, Department of Engineering-Economic Systems, Stanford University, July, 1985.

Paté, M. E., "Fault Trees vs Event Trees in Reliability Analysis," *Risk Analysis* 4(3):177-186, 1984.

Paté, M. E., and Neu, J. E., "Warning Systems and Defense Policy: A Reliability Model for the Command and Control of Nuclear Forces," *Risk Analysis* 5(2):121-138, 1985.

Raiffa, H., *Decision Analysis*, Addison-Wesley Pub., Reading, Massachusetts, 1968.

Savage, L. J., *The Foundation of Statistics*, Dover Pub., New York, 1972.

Shachter, R. D., "Evaluating Influence Diagrams," *Operations Research* 34:871-882, 1986.

Strait, R. S., "Decision Analysis and Strategic Interaction," Doctoral Thesis, Department of Engineering-Economic Systems, Stanford University, Stanford, California, 1987.

Starr, C., and Whipple, C., "The Strategic Defense Initiative and Nuclear Proliferation from a Risk Analysis Perspective," *Risk Assessment in Setting National Priorities*, Plenum Pub., Proceedings of the Annual Meeting of the Society for Risk Analysis, Houston, Texas, November, 1987 (see page 151).

Steinbrunner, J. D., "The Doubtful Presumption of Rationality," *American Defense Policy*, 5th Edition, J. F. Reichart and S. R. Sturm, Eds., Johns Hopkins University Press, Baltimore, MD, 1982.

von Neumann, J., and Morgenstern, O., *Theory of Games and Economic Behavior*, Wiley Pub., New York, 1947.

Western Europe Union, "Déclaration Commune des Ministres des Affaires Etrangères et de la Défense des Pays de l'Union de l'Europe Occidentale," The Hague, the Netherlands, October, 1987.

The Strategic Defense Initiative and Nuclear Proliferation from a Risk Analysis Perspective

Chauncey Starr and Chris Whipple
Electric Power Research Institute
Palo Alto, CA

ABSTRACT

The risks from nuclear weapons and the effect of the strategic defense initiative (SDI) on these risks are examined from perspectives commonly used in risk analysis. Two primary issues addressed are how risks and responses to risks of nuclear war are perceived differently depending on whether a short-term or long-term viewpoint is taken, and on whether SDI enhances or replaces the traditional nuclear arms philosophy of mutually assured destruction. Special attention is given to the distinction between defensive systems which protect nuclear weapons from a first strike, and thereby assure the capacity for a retaliatory response, and defensive systems meant to protect people and cities.

KEYWORDS: Nuclear weapons, nuclear proliferation, SDI

THE RISK ANALYSIS PERSPECTIVE

Few, if any, risks rival nuclear weapons in the potential severity of consequences. Extreme geophysical events may have comparable capacity for damage, but these events have a historically low frequency which is beyond our control. The probability of nuclear war cannot be estimated from the historical frequency record the way geophysical events can. No comparable experience appears to be available to provide a frame of reference for thinking about a risk of this magnitude. Standard theories for evaluating risk, utility theory, for example, can be of some value, but under any approach it is difficult to think about a risk with near-infinite cost and finite probability, hence a risk of near-infinite expectation of loss.

Risk evaluation requires consideration of both probability and consequence. For nuclear weapons, consequences are crudely calculable, although there are sizable uncertainties in the many consequences of large-scale use, for example, in the effect on climate (nuclear winter), or from late effects (hunger, disease). Such uncertainties regarding the consequences of nuclear war may be of limited significance, because the large inventories of weapons currently in place mean that consequences of a nuclear war would be so severe that knowing precisely what effects would occur has little decision-making usefulness. (This viewpoint is not unanimously held; some argue that incremental reductions in weapons inventories are not only symbolic but also provide meaningful risk reductions; others argue that civil defense is feasible.)

The other half of risk is probability. The effect of alternative strategic military policies on the probability of nuclear war cannot be estimated with the confidence of consequence estimates. But if actions which reduce consequences have only limited practical value, then the major avenue available for reducing risks from nuclear war lies in reducing its probability. However, because single-event probabilities cannot be proved or demonstrated, public discussion on the risk tends to emphasize consequence and its mitigation, e.g., nuclear winter and the Strategic Defense Initiative (SDI). The situation is complicated because most national policies regarding nuclear weapons influence both probability and consequence.

This paper applies a qualitative risk analysis framework to public issues concerning nuclear war and the SDI. The objective is to clarify the nature of the problem and to distinguish the more important issues from those that are less important. In our view, the public debate relating to the risk from nuclear war is not organized around a coherent framework. The SDI has been described by the administration as a system for defending civilian populations, and, to a large extent, its feasibility and utility have been publicly considered on this basis. While our intention is not to provide a technical review of the SDI issue, the analytical framework for considering the potential for an SDI system to alter nuclear weapons risk is addressed as an example of a useful qualitative risk analysis.

It is important to recognize the limited nature as well as the usefulness of such a qualitative risk analysis. As in this study, it does present the basic issues, the main factors involved, the alternative premises and assumptions behind the several national policy perspectives, and the main interactions that create the perceived risks. This background might then permit a similar dissection of the alternatives to management of these risks. In contrast to a quantitative risk analysis, this qualitative approach does not attempt to assess the probabilities of events, or the realistic magnitudes of the consequences, or the priorities of national policy in risk management. A qualitative risk analysis is most useful when a consensus does not exist on the definition and structure of the problem. The arms control debate, and particularly nuclear weapons proliferation and the SDI, are certainly examples of qualitative disagreement on whether proposed policies reduce or increase risk. This paper will attempt to disclose the structure of this debate by a first-cut qualitative risk analysis.

THE STATEMENT OF THE PROBLEM

The very horror of a nuclear war is so intimidating to all sane peoples that many nations have valued the possession of a nuclear weapons arsenal as the ultimate defense for preventing a major military attack. For nuclear powers, this is the concept of "Mutually Assured Destruction"—popularly known as MAD—which assumes that a first strike with nuclear weapons would result in an immediate retaliation of the same kind, so that both antagonists would be equally devastated. Thus, a nuclear arsenal in the hands of both antagonists might (and perhaps does) prevent a war too horrible to contemplate.

More recently, the debate in the U.S. on the merits of the SDI has raised the novel question of whether an improved defense may increase the risk of a war by destabilizing the MAD balance.

MIND-SETS ON NUCLEAR WEAPONS RISK

As a first step in considering the opposing positions in the U.S. debate on arms control, it appears that the pro-military "hawks" consistently seek assurance of a retaliatory capability, whereas the anti-military "doves" consistently seek an international consensus on the unacceptability of nuclear weapons. Both are seeking to avoid the horrors of a nuclear war. We suggest that their difference in approach arises from a basic difference in personal values and perceptions of societal motivations. We refer to this combination of values and assumed motivations as a mind-set. In our simplified description, pragmatists view the world as it is in the short-run, and idealists view the world as it should be in the long-run.

We suggest in Table 1 the major aspects of these differing mind-sets. As consideration of Table 1 might suggest, national policies for the risk management of arms control issues might well incorporate near-term and long-term programs that balance the objectives of both pragmatists and idealists. Although MAD appears to have been an effective deterrent to nuclear war, the small but finite annual probability of such a war occurring for any reason suggests that MAD is not an acceptable long-term solution. MAD thus stabilizes the risk, but does not remove it.

Table 1. Mind-Sets on Nuclear Weapons Risks

Subject	Pragmatists (Hawks)	Idealists (Doves)
Human behavior	Historically consistent	Perfectible
Societal values	Observed preferences	Better world ideas
Alternatives	Politically feasible incremental changes	Fresh concepts
How chosen	Popular opinion	Public education
Goals	Short-Term Stability (defer risks)	Long-term removal of threats (reduce risks)
Morality	Self-defense is justified	Killing is abhorrent
Risk	Some risk is appropriate for end objectives	Avoid responsibility for catastrophe

VERTICAL PROLIFERATION TRENDS

The historical milestones in the growth of vertical proliferation issues are summarized in Table 2. These milestones highlight the shift in the role of nuclear weapons which followed the shift in delivery systems from aircraft to missiles. The concepts of military offense and defense associated with the airborne bombers of the 1950s were made obsolete by the intercontinental ballistic missile (ICBM). Geographical defense arrangements no longer were adequate. ICBMs could be launched from submarines as well as from land. The time of flight was so short that available response systems could not provide a defense. The premise of "no defense" drove the superpowers to the MAD concept of an offensive balance. More recently, the development of a space launch capability for maneuvering devices in outer space has opened the possibility for an

Table 2. Historical Trend of Vertical Nuclear Proliferation

- U.S. sole nuclear weapons power in 1946.

    - Bomber delivery restricts ranges.
    - U.S.S.R. vulnerable only from european bases.

- U.S.S.R. becomes nuclear weapons power in 1949.

    - Europe is vulnerable.
    - Long-range bombers needed to threaten U.S.

- U.K., France, China develop nuclear weapons inventory (1960s).

- Intercontinental ballistic missile (ICBM) removes geographical
  barriers to vulnerability (1960s).

- Antiballistic missile defense (ABM) and strategic defense initiative
  (SDI) challenge the "no defense" axiom.

Table 3. Vertical Proliferation—Traditional Superpower Nuclear Weapons Concepts

- Ultimate defense and threat are the same—MAD.

- Sufficient strength is needed to counter any threat.

- Maintain large inventory and delivery systems to assure equality.

- Primary function of nuclear weapons is prevention and intimidation, not
  use (Korea, Vietnam, Falklands, Near East, Afghanistan).

- Provide public reassurance of controlling and coping with potential nuclear
  weapons threats.

    - Military objective: Retaliatory capability.
    - Civilian objective: Physical security.

antiballistic missile (ABM) defense system, which in its most advanced speculative mix has been termed a Strategic Defense Initiative (SDI).

At the present time, these changes have resulted in several characteristic premises in the strategic thinking of the super powers. These are listed in Table 3. Most interesting is the realization that nuclear weapons would not be used to resolve international conflicts between the superpowers, because of MAD, and are unlikely to be used between a nuclear power and a non-nuclear nation, because of world-wide opprobrium. Thus, their primary function is intimidation—like the "beware of the dog" sign in front of a house. This situation places the military establishments of the superpowers in the awkward posture of maintaining their nuclear weapons capability without a reasonable probability of its use, or a clear consensus on its social acceptability.

This then emphasizes the last item in Table 3, the difference between protecting a nation's civilian structure and protecting its military capability. In order to make MAD a viable preventive concept, the military needs to maintain a nuclear arsenal large enough to destroy the functional and social structure of an antagonist, even after an initiating offensive first strike has damaged much of its inventory. So the military concentrates on protecting its inventory of thousands of weapons. In contrast, the civilian sector faces the realization that a few hundred weapons could destroy most cities in their country. So, while the average citizen worries about incoming nuclear weapons, even in small numbers, the military concentrates on the hundreds of weapons surviving after a first-strike attack of thousands. This lack of common objective is the cause of much of the rancorous debate between the pro- and anti-disarmament sectors of our society. The "doves" see the large nuclear arsenal as a threat to civilians and the "hawks" see it as a necessary self-defense of the military (Table 1).

SDI AND THE MAINTENANCE OF RETALIATORY CAPACITY

Whether the SDI increases or decreases risk depends on many factors. To explore the role of SDI on risk, we start with premises about the technical capabilities an SDI system might have. The Reagan Administration has supported the SDI as an alternative to MAD; it is described as a defensive system for protecting people against nuclear attack. Opponents of the SDI have been happy to argue against the SDI on this basis; the criticism often encountered is that protection of civilian targets from nuclear missile attack requires a level of reliability that is realistically unachievable, especially considering possible countermeasures against the SDI and the inability to conduct realistic tests of the full system. When one takes the diverse delivery systems available to the U.S. and the U.S.S.R., e.g., submarines and cruise missiles, the prospects for protecting cities seem infeasible.

However, the SDI could be a stabilizing factor by helping a country maintain a retaliatory capacity by mitigating a first-strike destruction of the military target. This capability does not require the near-perfect reliability of a system for protecting cities; as long as a significant fraction of the targeted nuclear missiles would survive a first strike, retaliatory capacity is assured. In this role an SDI is potentially beneficial in countering the offensive capacities of accurate MIRVs and of measures to detect and target nuclear missile submarines. Thus the SDI is consistent with past technical changes made to reduce vulnerability of nuclear weapons to a first strike.

UNCERTAINTY AND STABILITY

But there are conditions under which an SDI system can also be destabilizing, as shown in Table 4. One case of concern is that a highly effective SDI could make it possible for a nation with superior nuclear weapons strength to make a first strike and avoid unacceptable retaliation. A second concern is that an SDI could spur a new arms race, because the SDI requires a greater offensive force to apply the same threat. A 75% effective SDI system could be effectively neutralized by increasing offensive stockpiles four-fold.

The present MAD balance depends on the mutual perception that either side can and would retaliate to a nuclear attack, and that the consequences of such a retaliation would be unacceptably severe. This condition for stability does not require that the weapons inventories of hostile nations be in balance. As long as each side's retaliatory response carries unacceptable costs, neither side will make a first strike. Under this framework, an important source of risk is the potential that one or both sides perceive that they could make or suffer a first strike without an assured, unacceptable response. Whether through superior

Table 4. Implications of Hypothetical Example

- If no mutual SDI capability:

  - Emphasizes first strike military value.
  - Multiple nuclear weapons systems needed to prevent inventory
        destruction (submarines, mobile, etc.).
  - No inherent stability margin.

- If partial mutual SDI capability:

  - Preserves MAD, with large inventory.
  - Preserves retaliatory capability.
  - Military value of first strike is reduced.

- If perfect mutual SDI capability:

  - Removes value of ballistic missile inventory.
  - Shifts offense to alternative delivery systems (cruise missiles,
        Bombers, etc.).

- Security implications of a perceived SDI imbalance:

  - First-strike balance destabilized.
  - MAD destabilized.
  - Nuclear weapons inventory would be increased simultaneously
        with SDI catch-up.
  - Stabilization would be reestablished at higher inventory levels.

numbers of weapons or through technical changes such as in the accuracy of weapons delivery, this circumstance would increase the destabilization of MAD.

The unilateral approach to assure a capability to respond to an attack is to seek weapons delivery systems that are difficult to destroy prior to their use and difficult to defend against if used. The shift many years ago from long-range bombers to ICBMs removed the possibility that air defenses could prevent successful counterattack. The deployment of MIRVs, threatening as a first-strike weapon against land-based missiles, has led to expanded reliance on submarine-based missiles and on cruise missiles. The large inventories of nuclear weapons in the U.S. and U.S.S.R. illustrate the perceived benefit of projecting a strong retaliatory capability.

Unpredictability of the willingness or ability of a country to retaliate is stabilizing to the extent that it prevents a first strike against it. However, if a country is unsure of its capacity to retaliate if attacked, it may be prone to launch its weapons earlier (i.e., on less evidence that an attack is occurring) than if it were certain of its retaliatory ability. This means that superiority in weapons and delivery systems can increase risk even for the side with superior forces. Mutual equality in both capability and perceived determination to reply is needed to preserve the MAD balance.

TRADE-OFFS IN RISK — NUCLEAR AND CONVENTIONAL WEAPONS

The international opprobrium associated with nuclear weapons has been a factor in preventing their use since World War II. Nations possessing nuclear weapons have been involved in numerous military conflicts, e.g., the U.S. in Korea and Vietnam, the U.S.S.R. in Afghanistan, the U.K. in the Falklands, but nuclear weapons were not used. The barrier to use nuclear weapons is apparently large. The measure against this barrier is a stated policy of "no first use." A commitment by nuclear weapon states not to introduce nuclear weapons into a conflict is a helpful step in reducing the risk from nuclear weapons. The greater the distinction between conventional and nuclear weapons, the less likely it is that nuclear weapons will be used in any conflict short of a war in which national survival is at stake. Conversely, as the gap between conventional and nuclear weapons widens, the barrier to nuclear weapons use increases and they lose their power to deter conventional war. This of course has been the stated reason for U.S. unwillingness to announce a no-first-use policy for Europe.

CONCLUSIONS

This very simplified qualitative risk analysis of the potential role of SDI, in the context of the more general framework of nuclear weapons proliferation, discloses several basic relationships pertinent to national policy. Although SDI has been usually presented to the public as intended to defend civilians, it is evident that only a near-perfect and presently implausible system could do so. Such a hypothetically perfect SDI would also destabilize MAD as a deterrent to nuclear war. On the other hand, a realistic SDI would be only partially effective, but could aid the stability of the MAD balance by providing some protection to retaliatory capability after a first strike. It would also provide some defense to the stray inadvertent launch accident from any source. However, a one-sided SDI capability would reduce confidence that unacceptable retaliation would occur, and thus destabilize MAD. In general, an evaluation of SDI should be based on the total system for nuclear weapons use, including inventories, mix of delivery systems, and their vulnerabilities. In this perspective, SDI may be considered as the most recent of the technical sequence of military subsystems (e.g., submarines, cruise missiles) intended to assure retaliatory capability. The dilemma created by the multifaceted aspects of SDI is illustrated in Table 5.

Although a mutually equal SDI effectiveness would support the MAD balance, a verification of such a balance between antagonists would clearly be needed to prevent the escalation of inventories arising from uncertainty and suspicion. Thus an arms control agreement, properly verifiable, is pragmatically desirable to control inventories and is idealistically desirable as a basic step to eventually reducing the residual threat of a nuclear war.

In this somewhat complex matrix of military and geopolitical issues, this risk analysis approach, although qualitative, discloses some of the chief factors affecting the dynamics of the situation. This presentation is intended to illustrate in summary form these principal factors in nuclear weapons proliferation and to provide an example of an analytical framework for disclosing the risk management options which might, over time, reduce the risks of a nuclear war. This qualitative analysis is purposely simplified to clarify the issues and does not discuss the caveats, subtleties, and long-term side effects of these risk management options.

Table 5. The Dilemma

- SDI appears to substitute defense for offense, but unbalanced feedback process might increase nuclear weapons inventories.

- MAD has apparently stabilized superpower peace for four decades.

> To preserve MAD:
>
> - Undertake partial SDI (or equivalent).
> - Avoid unilateral arms control reductions if SDI exists.
> - Maintain an inventory sufficient for assured retaliation capability.
>
> SDI may increase inventory goals.

- To minimize nuclear weapons potential catastrophe.

> Eliminate MAD:
>
> - Perfect SDI.
> - Verifiable nuclear weapons reductions (as a cost-effective alternative, successful SDI may stimulate verifiable nuclear weapons reductions).
>
> Develop military and political alternatives for mutual intimidation.

The Hazardous Air Pollutant Prioritization System

Ila L. Cote
U.S. Environmental Protection Agency
Research Triangle Park, NC

Alice S. Pelland
Radian Corp.
Research Triangle Park, NC

Stanton P. Coerr
Mandarin Systems, Inc.
Research Triangle Park, NC

ABSTRACT

The Hazardous Air Pollution Prioritization System (HAPPS) is used to rank toxic air pollutants as the first step in the Environmental Protection Agency's regulatory assessment under the Clean Air Act. The purpose of this exercise is to assist in setting the priorities for assessment and focus resources on the chemicals with the greatest potential for public health risks. The HAPPS has evolved over several years based on a system developed at the Oak Ridge National Laboratory and later modified at Argonne National Laboratory. The current system can be run on a personal computer and focuses on air pollutants rather than multimedia exposures. It is designed to rank large numbers of chemicals based on limited, readily available information. Ranking is based on health information and surrogates for population exposure. More specifically, substances are scored on eight factors: oncogenicity, mutagenicity, reproductive and developmental toxicity, acute lethality, other toxic effects, the potential for airborne release (production volume and volatility), bioaccumulation and existing standards. The system possesses a somewhat flexible ranking scheme that can be easily modified to reflect different issues such as concern for acute exposures and the impact of combustion sources on exposure.

KEYWORDS: Air pollutants, toxics, ranking, priority setting, Clean Air Act

INTRODUCTION

The purpose of this paper is to describe the approach used by the Environmental Protection Agency's (EPA) Office of Air Quality Planning and Standards (OAQPS) to select toxic substances for regulatory assessment under the Clean Air Act. Ideally, a full range of toxicological and epidemiological information coupled with detailed current emissions and human exposure data would be used in this priority setting process. Such complete information, however, is seldom available. The problem is further complicated by

159

the large number of substances that are potential candidates for study. Thus, a procedure which can initially rank many substances on the basis of limited, readily available information is useful so that resources for detailed review might be allocated efficiently. The procedure used by this office, the Hazardous Air Pollutant Prioritization System (HAPPS), is described in this paper.

SELECTION

Selecting a subset of chemicals for further evaluation from the universe of all chemicals is perhaps the most difficult step in the overall evaluation process. This is due to the large number of chemicals and the numerous sources of information to be considered. As a consequence, the selection step is less formalized and requires less information per chemical than the remainder of the process.

Throughout the process, the concern is greater that a chemical associated with significant public health risks will be overlooked or excluded from evaluation than that a chemical will be evaluated for which there is little public health concern. As a consequence, a bias exists, particularly in the early stages of this process, toward further evaluation. As a chemical moves through the process, the information and apparent risk required to proceed to the next step become greater. Eventually, for some chemicals, the process ends in the very detailed risk assessments necessary to support federal regulations.

To select chemicals to be ranked, an extensive review of the primary literature is periodically conducted. Almost any indication of the potential for substantial exposure and adverse health or environmental effects can cause a chemical to be added to the list of chemicals to be ranked. For chemicals already on the list, the data used for ranking are also updated. In the interim, to keep the list updated, the OAQPS relies on large-scale review or data collection efforts conducted by other organizations. For example, the Toxic Substances Control Act requires that new information indicative of significant chemical hazard be reported to the EPA's Office of Toxic Substances. These submissions may provide information that results in a chemical being added to the list of ranked chemicals. Other efforts utilized include data from the National Toxicology Program (NTP), as well as the research laboratories at the Environmental Protection Agency and the National Institute of Environment Health Sciences. Also, the International Agency for Research on Cancer may note information that results in inclusion of a chemical in the HAPPS list. Subsequent to compilation of the list of chemicals to be ranked, the priorities are assigned using the procedure described below.

RANKING

The ranking procedure was first developed by the Oak Ridge National Laboratory for EPA's Office of Toxic Substances in 1980.[1] Two years later, A. E. Smith and D. J. Fingleton of the Argonne National Laboratory modified this system to emphasize inhalation as a route of exposure (as opposed to the multimedia approach of the Oak Ridge system).[2] This system was subsequently further modified and computerized.[3] The current version reflects increased concerns for potential health effects resulting from short-term exposures and reproductive/developmental effects and is designed to allow more flexibility by providing several different user-selected ranking schemes. The various options will be discussed later in this article.

Several guidelines were followed in the development of the most recent version of the HAPPS ranking:

- Generally, readily available summary documents or computerized data bases were used because searches of primary sources were too expensive for what is a preliminary prioritization step.

- The procedure was set up to be as objective as possible and to use uniform sources of data. Thus, a particular set of substances receives the same score when prioritized by different persons.

- A user should be able to rank substances quickly and to easily update data; hence the system was computerized.

- The procedure is only intended to produce reasonable, initial rankings and was purposely designed not to draw upon expert judgment. Decisions on data interpretation, the validity of data and similar technical items are left for the review step described subsequently.

Sources of Data for Ranking

These guidelines place considerable constraints upon the procedure by restricting the sources of data to be used and by limiting the effort expended to rank a single compound. The *Registry of Toxic Effects of Chemical Substances* (RTECS) was decided to be the most suitable source of health data.[4] The RTECS is a concise, easily used summary of toxic effects which is kept current by continual updates in a computerized format and by quarterly updates in microfiche copy. The major drawback to the use of RTECS is that the data, while taken from peer-reviewed journals, are of variable quality. The lowest concentration reported in the open literature to produce a given effect is used for RTECS. The lowest concentration reported in the literature may not represent the lowest-observed-effect level that might be chosen after extensive review of the data. Hence, use of RTECS creates a bias toward false positives. Nevertheless, RTECS reports health data for more chemicals than other data bases, contains data related to most of the criteria used for ranking substances, and was the most easily accessible reference available.

This office uses the on-line versions of RTECS and of the Hazardous Substances Data Base (HSDB) for physical properties and production volume. The HSDB production is part of the National Library of Medicine's TOXNET system. The HSDB data are extracted from data compiled by the Stanford Research Institute.[5] Like RTECS, HSDB is a concise, easily accessed data base that is kept relatively current on-line. In addition to RTECS and HSDB, HAPPS requires users to check documents prepared by the National Toxicology Program (NTP) which describe the status with respect to NTP testing.[6,7]

HAPPS Factors, Groups and Alternative Ranking Schemes

Substances are ranked by scoring them in eight areas. These eight areas, referred to as "factors," are: oncogenicity, mutagenicity, reproductive and development toxicity, acute lethality, effects other than acute lethality, potential for airborne release, bioaccumulation, and existing standards. The airborne release factor is further divided into subfactors: production volume and volatility. Substances with the potential to enter the environment as particles are given a moderately high weight in the volatility subfactor. The option was chosen after analysis of the relative ranking of chemicals for which exposure was known to occur via particles as opposed to gases. For weighting in various ranking schemes the eight factors are combined into five groups: (1) carcinogenicity, (2) reproductive and developmental toxicity, (3) general toxicity, (4) exposure, and (5) standards. Table 1 presents the factors, groups, and subfactors used in HAPPS.

Different weights are associated with various criteria within each of the eight HAPPS factors. "Criteria" refer to specific data characteristics that receive a certain weight or

Table 1. Subfactors, Factors and Groups Used in HAPPS

Group	Factors
I. Carcinogenicity	A. Oncogenicity B. Mutagenicity
II. Reproductive and Developmental Toxicity	A. Reproductive and Developmental Toxicity
III. General Toxicity	A. Acute Lethality B. Effects Other than Acute Lethality
IV. Exposure	A. Potential for Airborne Release - Production volume - Vapor pressure and physical state B. Bioaccumulation
V. Existing Standards	A. Existing Standards

score. For example, the oncogenicity and reproductive developmental factors have as their weighted criteria, in descending order: (1) evidence in humans by the inhalation route, (2) evidence in humans by other routes of exposure, (3) evidence in two or more species, (4) evidence in one animal species, and (5) scheduled for testing. The impact of animal data by inhalation versus other routes can also be incorporated. Weights assigned in the carcinogenicity and reproductive/developmental groups are based on these qualitative weight-of-evidence type criteria. Additional weight is given for satisfying more than one criteria category, as opposed to scoring in only one category. For toxicity and exposure potential, quantitative criteria are used (see example for Acute Lethality, Table 2).

In the scoring process, a substance first receives a score for each factor based on the criteria that are relevant for the chemical in question. Then, the eight factor scores are combined according to specific weighting schemes and a score is calculated for each of the five groups. Finally, group scores are combined to arrive at one score for the substance. This final score is the basis for the ranking.

The HAPPS has several alternative weighting schemes that users can select for special purposes. The alternatives are created by varying the way certain factor and group scores are calculated and by varying how the groups are weighted, relative to each other. One example of an alternative weighting scheme is an emphasis on health effects resulting from short-term exposure. This is accomplished by increasing the weight of the acute lethality factor within the toxicity group, reducing the emphasis on the reproductive and developmental toxicity groups, and omitting carcinogenicity data from consideration in calculating the final score for each substance. Another alternative is consideration of combustion products. A number of compounds are known to enter the air as combustion products rather than or in addition to emissions from intentional production. Dioxin is an example of such a chemical. The HAPPS allows the user to select an alternative ranking

Table 2. HAPPS Criteria for Acute Lethality
(in descending order of weight)

| Species | Exposure Route and Concentration | | |
	Inhalation (mg/m^3)	Oral (mg/kg)	Dermal (mg/kg)
Human	≤ 50	---	---
Human	---	≤ 5	≤ 5
Animal	≤ 50	≤ 5	≤ 5
Human	50 - 500	---	---
Human	---	5 - 50	5 - 200
Animal	50 - 500	5 - 50	5 - 200
Human/Animal	500 - 5000	50 - 500	200 - 500
Human/Animal	5000 - 10000	500 - 5000	500 - 5000
Human/Animal	>10000	>5000	>5000

procedure that emphasizes the potential contribution to exposure of known combustion products. The availability of these alternative weighting schemes increases the flexibility of the system and its use in consideration of a wider variety of special problems.

REVIEW OF RANKED LIST

As noted earlier, selection and ranking are biased toward false positives, as opposed to false negatives. Hence the review of the ranking prior to regulatory evaluation is critical if resources are to be expended efficiently. The ranking process allows this review effort to be focused on a more limited number of chemicals. For each highly ranked chemical, abstracts of current publications and secondary sources of information are examined. In addition, a few critical papers may be studied. The purpose of this review is to validate the ranking by reviewing sources of information other than those used in the ranking. Also, at this time, other relevant factors can be considered, e.g., pending impact of regulatory action by other governmental groups. Subsequent to review of the highly ranked chemicals, the ranking may be modified to reflect the additional information. Chemicals are then chosen from this modified list for regulatory assessment. Regulatory risk assessments take anywhere from 8 months to several years to complete. Historically, EPA has regulated or intends to regulate one-third to one-half of the chemicals chosen from the HAPPS list for regulatory assessment.

SUMMARY

This paper briefly reviews the procedure by which the EPA's Office of Air Quality Planning and Standards selects chemicals for regulatory assessment under the Clean Air Act. Chemicals are first selected for the purposes of ranking using many diverse sources of information. The chemicals are ranked primarily on the basis of their potential for causing several health effects, production volume, and volatility. Additional information on the most highly ranked chemicals is then reviewed and chemicals are chosen for more detailed regulatory risk assessment.

Although it is anticipated that this system will continue to be used as described above, efforts currently are under way to incorporate more detailed source information into the ranking process.

REFERENCES

1. R. H. Ross and J. Welch, *Proceedings of the EPA Workshop on the Environmental Scoring of Chemicals*, ORNL/EIS-158, EPA-560/11-80-010, Oak Ridge National Laboratory, Oak Ridge, Tennessee (1980).
2. A. E. Smith and D. J. Fingleton, *Hazardous Air Pollutant Prioritization System*, U.S. Environmental Protection Agency, Research Triangle Park, North Carolina (EPA 450/5-82-008) (1982).
3. Radian Corporation, *Modified Hazardous Air Pollutant Prioritization System*, U.S. Environmental Protection Agency, Research Triangle Park, North Carolina (EPA Contract No. 68-02-4330) (1987).
4. R. J. Lewis and R. L. Tatken, Eds., *Registry of Toxic Effects of Chemical Substances*, Vols. I and II, U.S. Department of Health and Human Services, National Institute for Occupational Safety and Health, Cincinnati, Ohio (1979).
5. Stanford Research Institute, *Chemical Economics Handbook*, Stanford Research Institute, Menlo Park, California (1982).
6. National Toxicology Program, *Chemicals on Standard Protocol*, Carcinogenesis Testing Program, National Toxicology Program, Bethesda, Maryland (1987).
7. National Toxicology Program, *Annual Plan*, National Toxicology Program, Bethesda, Maryland (1987).

Air Emission Risk Assessment Sensitivity Analysis for a Coal-Fired Power Plant

Gregory S. Kowalczyk
Northeast Utilities Service Co.
Hartford, CT

Lawrence B. Gratt and Bruce W. Perry
IWG Corp.
San Diego, CA

Evyonne Yazdzik
Northeast Utilities Service Co.
Hartford, CT

ABSTRACT

The Electric Power Research Institute's Air Emissions Risk Assessment Model (AERAM), which assesses the risk of toxic air emissions from coal-fired power plants, was used to study the excess cancer risk due to inhalation of airborne emissions from the Mt. Tom Power Plant in Holyoke, Massachusetts, and the key sensitivities of the risk assessment process were investigated as a guide for future analyses. With arsenic used as the pollutant, source-term descriptions, air-dispersion parameters, population exposure parameters, and dose-response models were varied to assess the effect on the risk result. It was found that eliminating pollutant enrichment on fly-ash particles reduced the risk by 50%. Changing from complete ground reflection to no reflection in the air-dispersion model reduced the risk by 57%; changing the settling velocity had no effect on the risk. Using the urban mode 2 stability pattern increased the risk by 124%, while using different stability array data changed the risk by about $\pm 30\%$. The effect of using age-specific breathing rates in the exposure module was small (an 11% decrease in the risk), as was the impact of considering only respirable particles (a 4% decrease). The largest variation in the risk was observed when the Environmental Protection Agency's unit risk factor was replaced with one of four different low-dose extrapolation models utilized with three dose-response data sets; variations of up to 10 orders of magnitude were found. These results show that careful selection of dose-response data and low-dose extrapolation models should be a fundamental concern in risk assessment.

KEY WORDS: Air toxic risk assessment, coal-fired power plant, air emission cancer risk

INTRODUCTION

Generation of electricity by coal-fired power plants causes the emission of a variety of air pollutants, some of which are carcinogens. Although control technology can be used

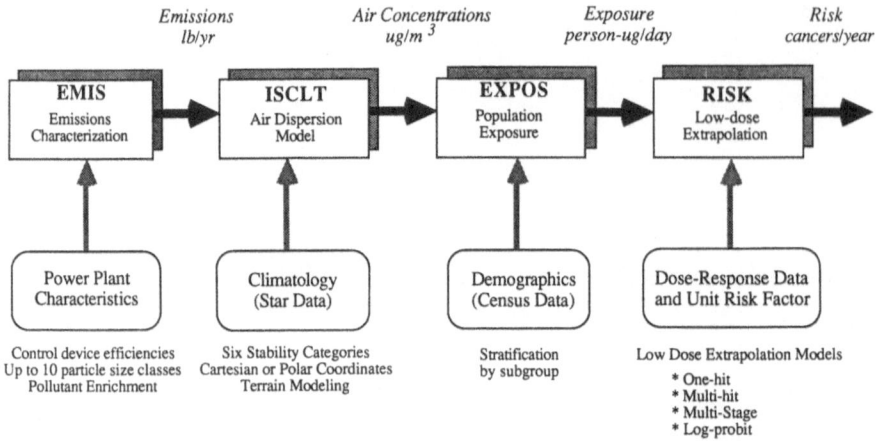

Fig. 1. Air Emissions Risk Assessment Model (AERAM) Methodology.

to reduce these emissions, some pollutants will be released into the atmosphere. It is difficult to determine the exposed population's actual cancer risk due to the pollutants; however, risk analysis can be used to estimate the health risks and to evaluate the risk reduction associated with various control strategies. In the process, addressing the uncertainty of risk estimates is an important step, since the results are sensitive to both model options and data selection. A proper understanding of these sensitivities will aid in the estimation of risk uncertainty and the decision-making processes involving health risks.

This paper describes a risk assessment sensitivity study which used the Electric Power Research Institute's Air Emissions Risk Assessment Model (AERAM) to revise and extend an earlier analysis of cancer risks due to arsenic emissions from Northeast Utilities' (NU) Mt. Tom Power Plant in Holyoke, Massachusetts. The sensitivity of cancer risk due to variations in data input values and program options were examined. The resulting sensitivities can be used to focus attention on the most important data needs for risk analysis.

BACKGROUND

The development of AERAM, a computerized model for estimating human cancer risk associated with toxic air pollutants emitted from coal-fired power plants[1,2] was sponsored by the Electric Power Research Institute (EPRI) to aid utility decision makers in the evaluation of power plant control strategies. AERAM is an integrated four-module FORTRAN computer code with each module addressing one of the four components of the risk analysis process: pollutant emission, atmospheric dispersion, population exposure, and health risk estimation. A flow-chart of these modules with an overview of the AERAM methodology is shown in Fig. 1. The recently developed AERAM manager was used as an aid for handling and modifying the extensive input and output data files for the sensitivity analysis. The AERAM manager is a top-level, menu-driven program which performs file handling, allows editing of AERAM data sets, checks input data prior to execution, executes the four AERAM modules, allows browsing of output, and provides context-specific help.[3]

The emission module (EMIS) uses coal properties, power-plant operating parameters, and removal efficiency of the pollution control devices to calculate the stack pollutant emission rate for the pollutant of interest. The program can account for the pollutant's enrichment on the fly ash as a function of particle size.

The air-dispersion module, which uses EPA's Industrial Source Complex – Long Term (ISCLT) disperson model, determines ambient air concentrations of the pollutant from the plant. The primary input data required for this module are meteorological data in the form of a stability array (STAR) summary, which is a tabulation of the joint frequency of occurrence of wind speed and direction stratified according to Pasquill atmospheric stability categories. The program can model terrain, dry deposition, and surface reflection of particles. This module calculates the ambient air pollutant concentration at each receptor in a user-defined exposure grid.

The exposure module, EXPOS, estimates population exposure using the modeled pollutant concentrations, demographic data, and breathing rates. Exposure is calculated for each receptor in the exposure grid. The module has the capability to account for population subgroups.

The risk module, RISK, calculates cancer risks based on a unit risk factor or dose-response data from human exposure or animal studies. It extrapolates risk from high to low doses and, when necessary, from animal species to humans. AERAM currently includes four low-dose extrapolation models: the one-hit, the multihit, the multistage, and the log-probit models. The results include an estimate of the excess human lifetime cancer risk for the total study population.

NU was the first utility to apply AERAM to one of its facilities, using it to evaluate the lung cancer risk associated with exposure to two carcinogens, benzo(a)pyrene (BaP) and arsenic, from its Mt. Tom Power Plant.[4] Mt. Tom was built in 1960 as a coal-fired electrical power generating facility, but in 1970, it was converted to burn oil and, except for about six months during the oil embargo in 1974, remained on oil until December 1981, when it began burning coal again. This generating station has one front-fired boiler designed to burn pulverized bituminous coal. The station has a peak capacity of 155 mW(e) normal gross output.

NU's calculations with AERAM showed that the excess cancer risk for arsenic emissions to the surrounding population in a 20 by 20 km study area was a very small percentage of the risk from the background arsenic level. Details of the calculation, including the fuel pollutant concentrations and stack emissions, are described elsewhere.[4,5]

METHODS

AERAM needs diverse data to perform risk calculations, some of which can be based on measurements while others must be estimated. Since modeling with AERAM involves selection of input data and various model options, the results are very dependent on these choices, although some may affect the results more than others. The following procedure was used to investigate the effect of input data selection and model options on the results of the baseline risk analysis of arsenic emissions from the Mt. Tom Power Plant.

Ten AERAM options/parameters were chosen for the sensitivity analysis: particle enrichment, ground reflection, settling velocity, adjustment of stability for surface type, modeling surface terrain, the STAR data, the breathing rate of subpopulations, respirable particles, the low-dose extrapolation model, and the dose-response data. Figure 2 presents these parameters as a branching tree of possible choices.

The baseline Mt. Tom Power Plant analysis is shown in Fig. 2 as a dashed line taking one route at each branch of the tree. The sensitivity of the risk estimate to changes in

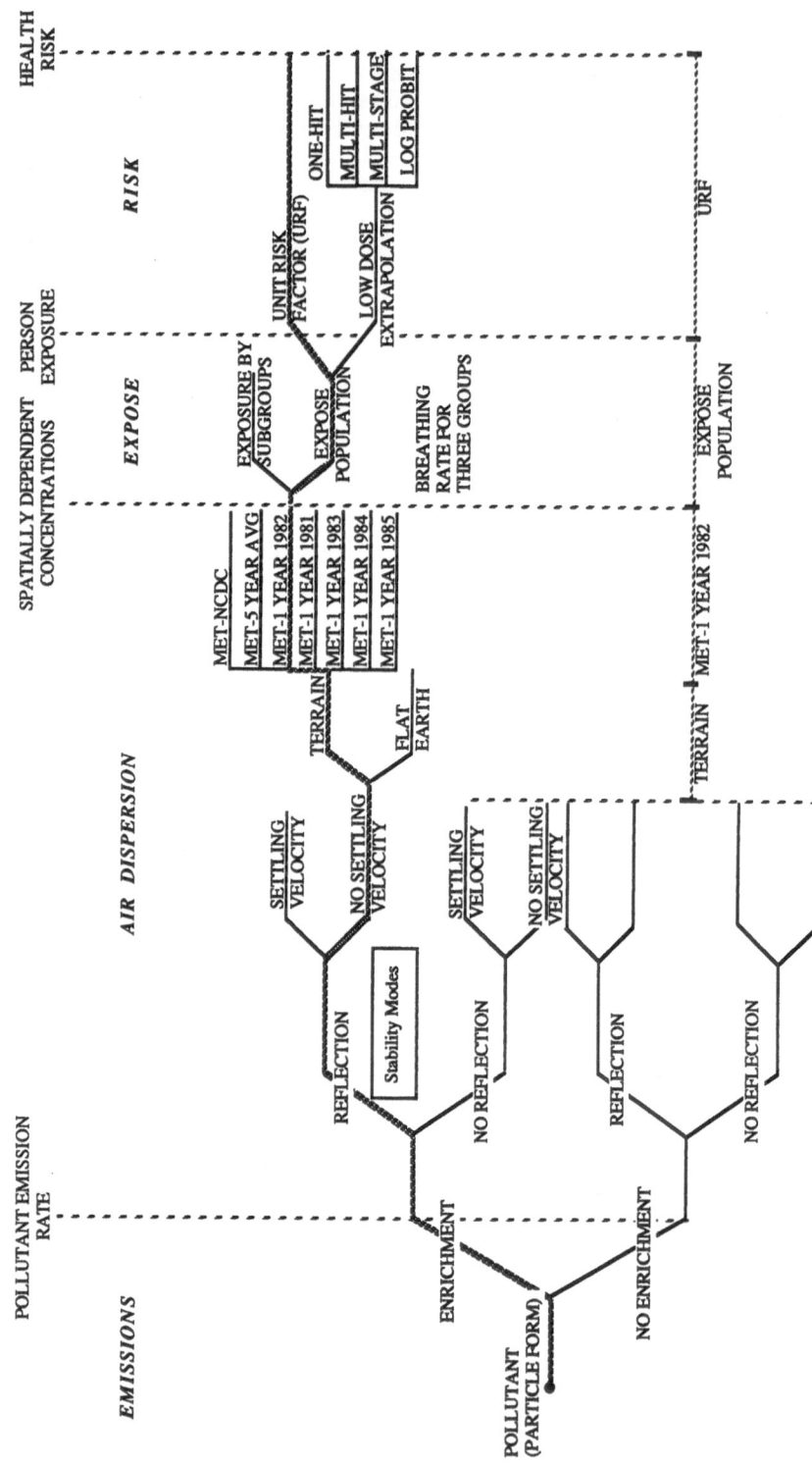

Fig. 2. Basic Risk Calculation Options and Variations for Sensitivity Study.

parameter values was investigated relative to this baseline case. The baseline calculation is a reasonable best-estimate except for the use of the unit risk factor, which is a plausible upper bound. When the unit risk factor is used, as in the baseline case, changes in the risk estimate are proportional to the changes in exposure.

In the sensitivity study, values were selected for each parameter to represent a range. In some cases, the values were simply a dichotomous choice to use or not use an option. Other parameters involved discrete options such as the stability data adjustment, which has three possible modes: rural mode, urban mode 1, and urban mode 2. In the rural mode, rural mixing heights and the Pasquill-Gifford (PG) standard deviations for the indicated stability category are used. In the urban modes, urban mixing heights are used. In urban mode 1, the stable E and F categories are redefined as neutral (D) stability, and the PG standard deviations are used. In urban mode 2, the E and F stability categories are combined and the PG deviations for the stability category one step more unstable than the indicated category are used. Appropriate values were used for other parameters to represent the analysis variations. For example, three different dose-response data sets obtained from the literature were used for the calculation of risk. This analysis involved 20 separate runs of AERAM.

RESULTS

The settings or values for each of the ten options/parameters and the corresponding risk sensitivities are summarized in Table 1. The sensitivity is expressed as the ratio of the risk with the given test input over the risk in the baseline case. By examining these risk ratios, the relative sensitivity of the different parameters is found.

The sensitivity of the risk result to source-term parameters in the emission module considered pollutant enrichment, which occurs as a function of particle size. The emission rate for the study pollutant, arsenic, is based on its concentration of 8.8 µg/g in the coal. Particulate arsenic was modeled with lower particle diameter bounds of 0.1, 0.5, 2.0 and 10 µm, with associated control efficiencies[6] of 98.5, 95.0, 99.1, and 99.7%, respectively. The choice of particle enrichment or nonenrichment impacts the resulting emission by a factor of 0.5, with a similar change in the risk. Enrichment refers to the change in trace element concentration from the bulk material to the fly ash as a function of particle size. Arsenic has been shown to undergo significant enrichment with decreasing particle size. Enrichment factors for arsenic were based on in-stack samples collected at a coal-fired power plant.[7]

Additional model choices can impact the risk concern options that affect the pollutant concentration at the receptor sites. These include the ground reflection and settling velocity of the pollutant, dispersion coefficients, terrain elevation, and meteorology.

The removal of airborne emissions by dry deposition is included to improve the dispersion model accuracy.[8] Larger particulates can be brought to the surface by the combined process of atmospheric turbulence and gravitational settling. Smaller particulates and gases tend to be reflected from the surface. ISCLT allows for both gravitational settling and dry deposition using a settling velocity. Reflection can be accounted for by an input parameter that specifies the fraction of material reaching the ground surface by the combined process that is reflected from the surface. The sensitivity to both the reflection coefficient and the settling velocity can be important.[9] The results in Table 1 indicate that removal of ground reflection decreased the concentrations and risk by 57%. The use of 100% reflection is an upper bound, as some particles are trapped at the surface. The effect of using the calculated settling velocites was nil. On the other hand, the addition of settling velocities at an extreme value of 0.01 m/sec increased the concentrations and risk by 53%.

Table 1. Summary of Air Emissions Risk Assessment Model Sensitivity Results

AERAM Module	Description of Variant	Baseline[a]	Variation from Baseline	Risk Ratio
EMIS	Particle Enhancement	Enrichment[b]	No Enrichment	0.51
ISCLT	Ground Reflection	100%	0%	0.43
	Settling Velocity	0	Calculated Velocities[c]	1.00
	Dispersion Coefficients	Rural Mode	Urban Mode 1	1.19
			Urban Mode 2	2.24
	Receptor Terrain	Elevations	Flat-earth	0.68
	Stability Array (STAR) data	1982 Measured	5-year Ave. Measured	0.91
			1981 Measured	0.94
			1983 Measured	0.98
			1984 Measured	0.81
			1985 Measured	0.83
			NCDC STAR[d]	1.30
EXPOSE	Exposure Groups	One Group	Three Age Groups	0.89
	Respirable Particles	All sizes	0.1 to 10 um only	0.96
RISK	Dose-Response Data Set	Unit Risk Factor	Tseng[13,14]	
	Low-Dose Extrapolation Model		One-Hit	2.4
			Multi-Hit	6E-03
			Multi-Stage	1.49
			Log-Probit	2E-05
	Dose-Response Data Set	Unit Risk Factor	Mabuchi[15]	
	Low-Dose Extrapolation Model		One-Hit	1.59
			Multi-Hit	1.17
			Multi-Stage	0.26
			Log-Probit	5E-06
	Dose-Response Data Set	Unit Risk Factor	Ishinishi[16]	
	Low-Dose Extrapolation Model		One-Hit	0.18
			Multi-Hit	3E-10
			Multi-Stage	3E-09
			Log-Probit	6E-09

a. Based on an annual excess lifetime per capita cancer risk of 3.87E-8 calculated for a population of 129,082 using a unit risk factor of 4.3E-3 lifetime cancers per microgram of arsenic per cubic meter.
b. Based on enrichment factors of 5, 5.3, 2.2 and 1 for the particle size groups, respectively.
c. Based on settling velocities of 3.4E-6, 5.7E-5, 1.4E-3, and 8.3E-2 m/sec for the respective particle size groups using equations for the aerodynamic mean diameter and settling velocity from reference 8. If an upper bound settling velocity of 1E-2 m/sec is used for all groups, the relative risk is 153%.
d. Based on National Climatic Data Center data for Chicopee Falls, Mass. locted 16.8 km from Mt. Tom.

Meteorological data used in the dispersion module were obtained from a tower located on site. This tower is equipped with instruments that provide real-time wind speed and direction, ambient temperature and stability class data. The base year chosen was 1982, which turned out to give the highest health risk of a five-year set of measurements. Results indicative of changes in the exposure and risk under different meteorological conditions are shown in Table 1. With respect to the base year, the five-year average produces a risk which is about 10% less. On the other hand, if National Climatic Data Center (NCDC) data are used for the closest site to Mt. Tom, the risks are 30% higher. This site, Chicopee Falls, is about 17 km south of Mt. Tom and, although those data may not be representative of the site topography, this study indicates that use of best available STAR data should not cause major discrepancies in the results.

The impact of assuming an elevated terrain versus the "flat-earth" approximation normally used to simplify risk analyses was considered in the sensitivity study. Receptor elevations were taken from a topographic map. The results shown in Table 1 indicate that the flat-earth approximation can result in an underestimation of the unit-risk-factor estimated risk by about one third, which is consistent with previous results.[10]

The population in each receptor zone was estimated using 1980 census data. A key assumption of the exposure module and risk assessment is that the entire population in the study area is continuously exposed for a 70-year lifetime at a constant exposure level. This is a static approach because the population in the study area is assumed to be fixed. Sensitivity of the risk results to the addition of age-specific breathing rates in the exposure analysis was investigated by stratifying the baseline population data into three age groups: under 5; 5 to 17; and 18 years and older. The age distributions from the 1980 census[11] for Northampton and Holyoke were assigned to the northwestern and southeastern halves of the population grid, respectively. The Northampton age distribution was 4.3%, 15.7%, and 80.0%, while the Holyoke age distribution was 7.1%, 20.5%, and 72.4% for the under 5, 5 to 17, and 18 and older age groups, respectively. The under 5 age group exposure calculation assumed a breathing rate of 7 m^3/day,[12] the 5 to 17 age group a rate of 15.0 m^3/day, and the 18 and older group a rate of 22.4 m^3/day breathing rate.[12] The resulting risk was 89% of the baseline risk; the refinement of population age distribution apparently has relatively little effect on the risk.

The health hazard from the inhalation of particles depends on the concentration of deposits in regions of the human lung. The size of the particles largely determines their fate in the respiratory system. The majority of particles with an aerodynamic diameter larger than 10 μm are trapped in the nasal passages and thus prevented from entering the lung. Finer particles (about 1 μm) easily penetrate to all parts of the respiratory system. Very fine particles (less than 0.1 μm) are exhaled if not chemically reacted in the lung. The respirable particles were considered to be in the range of 0.1 to 10 μm for investigating the effect of expected health risk by considering all particles versus only respirable particles. Because of arsenic's high enrichment on small particles, the mass of arsenic in particulates is primarily in the respirable range (96%), so the risk is reduced by only 4% when only respirable particles are considered.

Arsenic has been implicated as a human carcinogen. The uncertainty of the carcinogenic response to arsenic exposure was considered in the sensitivity analysis. The EPA unit risk factor was chosen for the basic risk estimate. Low-dose extrapolation models with different dose-response data sets were a key part of the sensitivity study. Dose-response data used in the risk module for arsenic were obtained from the literature. The first data set is based on epidemiological studies by Tseng *et al.* of arsenic ingestion in drinking water.[13,14] Tseng *et al.* found increased incidence of skin cancer in Taiwanese villagers exposed to arsenic-contaminated drinking water. Within a population of 40,421 individuals, the overall prevalence rate of skin cancer was 10.6 per 1000. The second and third data sets for arsenic are more directly applicable, being based on inhalation exposures.

Ishinishi *et al.* studied the tumorigenicity of arsenic trioxide to the lung in Syrian golden hamsters by intermittent installations.[15] The third arsenic data set is based on the investigation of cancer mortality by Mabuchi *et al.* among 1,393 workers exposed to high concentrations of inorganic arsenicals for varying lengths of time during the manufacture of pesticides at a plant in Baltimore, Maryland.[16] Dose-response data were derived using the tabulated data for the observed and expected number of deaths from lung cancer and other causes, and from Standardized Mortality Ratios by high exposure duration to arsenicals and nonarsenicals among male production workers first employed before 1955. The use of three dose-response data sets and four low-dose extrapolation models were compared to the use of the unit risk factor for arsenic as part of the sensitivity study and the results are shown in Table 1.

Results for the health risks based on the three data sets for arsenic differ by a factor of 13 for the one-hit model, a factor of 10 million for the multihit model, a factor of 1 billion for the multistage model, and a factor of 10 thousand for the log-probit model. The variation from the unit risk factor baseline case ranges from a factor of 5 for the one-hit model to 16 or 17 orders of magnitude for the other models. Results for the health risks are based on point risk estimates from each of the four dose-response functions. The range of results is indicative of the large uncertainty in dose-response modeling. The results for the different dose-response data sets indicate that the model choice between unit risk and low-dose extrapolation presents the greatest variability in the results. If the low-dose extrapolation scheme is selected, the greatest variability in the results appears to be due to the choice of data sets.[17]

Since many assumptions were made in the air emission risk estimation process, it is instructive to look at overall sensitivities to understand the uncertainty in risk estimates. The results in Table 1 show that an analyst could compute excess risks that vary from a factor of 10 higher than the baseline results to at least 10 orders of magnitude lower. In general, the lower uncertainty end point is zero; that is, there is no induced cancer from the expected levels of exposure to the pollutant. The upper uncertainty end point given by a unit risk factor is likely to yield a conservative estimate. It should be understood that the use of an upper bound at any point in the analysis does not allow for a best estimate of the risk with its associated uncertainty bounds.

Another issue is the impact of including the background level of the pollutant in risk calculations because the nonlinearity of the low-dose extrapolation models means that the total exposure is needed for an accurate estimate of risk. The background levels of arsenic in the Mt. Tom area are very high relative to the calculated concentrations from plant emissions. The sensitivity results reported herein have not included background and the reported risks are for excess cases from Mt. Tom emissions only.

CONCLUSIONS AND RECOMMENDATIONS

The AERAM code provides a useful and convenient means to analyze the sensitivity analysis of health risk related to toxic air emissions from coal-fired power plants. The development of a user-friendly interface for AERAM has allowed for flexibility and efficiency in performing air emission risk assessment sensitivity studies.

This sensitivity analysis was performed to examine the effects of changing some of the analysis options. Changes in the source term usually result in proportionate changes in the risk. The effects of changes in model options and data that impact the source and dispersion calculations are all relatively small, being of the order of less than a factor of 3. However, the risk results were very sensitive to the choice of dose-response data and, as had been observed in an earlier study,[17] to the choice of low-dose extrapolation model.

This means that careful consideration of dose-response data and model selection are paramount in the calculation of risk.

ACKNOWLEDGEMENTS

This paper was based partially on work performed in support of the Electric Power Research Institute (Research Project 2141-2). This paper represents the views of the authors and does not necessarily reflect the views or policies of Northeast Utilities Service Company, IWG Corp., or the Electric Power Research Institute.

REFERENCES

1. Arthur D. Little, Inc., "Assessing the Health Risks of Airborne Carcinogens," EPRI Research Project 1946-1, EA-4021, Electric Power Research Institute, Palo Alto, CA, 1985.
2. Mindware Corporation, "Air Emissions Risk Assessment Model (AERAM), Programmer's Guide," (Draft), Columbus, Ohio, 1987.
3. IWG Corp., "Air Emissions Risk Assessment Model (AERAM) Manager User's Guide" (Draft), San Diego, CA, 1987.
4. Kowalczyk, G. S., L. B. Gratt and P. F. Ricci, "An Air Emission Risk Assessment for Benzo(a)pyrene and Arsenic from the Mt. Tom Power Plant," *JAPCA* **37**:361-369 (1987).
5. Kowalczyk, G. S., "Emission and Atmospheric Impact of Trace Elements from a Reconverted Coal-Fired Power Plant," Paper No. 84-12.6, 77th APCA Annual Meeting, San Francisco, June 1984.
6. Revised from previous analysis in Reference 4.
7. Gladney, E. S., J. A. Small, G. E. Gordon, W. H. Zoller, "Composition and Size Distribution of In-Stack Particulate Material at a Coal-Fired Power Plant," *Atmos. Environ.* **10**:1071 (1976).
8. H. E. Cramer Co., Inc., "Industrial Source Complex (ISC) Dispersion Model User's Guide," U.S. Environmental Protection Agency, Research Triangle Park, North Carolina, 1979.
9. Tesche, T. W., R. Kapahi, R. Honrath, and W. Dietrich, "Improved Dry Deposition Estimates for Air Toxics Risk Assessment," Paper No. 87-73 B.1, 80th APCA Annual Meeting, New York, June 1987.
10. Gratt, L. B. and M. Dusetzina, "A Numerical Comparison of the Human Exposure Model and the Air Emissions Risk Assessment Model," Paper 87-98.3, 80th APCA Annual Meeting, New York, June 1987.
11. U.S. Bureau of the Census, *County and City Data Book, 1983*, U.S. Government Printing Office, Washington, D.C., 1983.
12. International Commission on Radiological Protection, *Report of the Task Group on Reference Man*, Pergamon Press, New York, New York, 1974.
13. Tseng, W. P., H. M. Chu, S. W. How, J. M. Fong, C. S. Lin, and S. Yeh, "Prevalence of Skin Cancer in an Endemic Area of Chronic Arsenicalism in Taiwan," *J. Natl. Cancer Inst.* **40**: 453-463 (1968).
14. Tseng, W. P., "Effects and Dose-Response Relationships of Skin Cancer and Blackfoot Disease with Arsenic," *Environ. Health Persp.* **19**:109-119 (1977).
15. Ishinishi, N., A. Yamamoto, A. Hisanaga, and T. Inamasu, "Tumorigenicity of Arsenic Trioxide to the Lung in Syrian Golden Hamsters by Intermittent Installations," *Cancer Letters* **21**:141-147 (1983).
16. Mabuchi, K., A. Lilienfeld, L. M. Snell, "Lung Cancer Among Pesticide Workers Exposed to Inorganic Arsenicals," *Arch. Environ. Health* **34(5)**:312-320 (1979).

17. Perry, B. W. and L. B. Gratt, "Multiple Low-Dose Extrapolation Models in Cancer Risk Assessment of Power Plant Air Emissions," Paper No. 87-41.6, 80th APCA Annual Meeting, New York, June 1987.

A Systematic Approach for Environmental Risk Assessment

Sheldon S. Lande
3M Company
St. Paul, MN

ABSTRACT

This paper discusses the approach used at 3M Company to assess environmental risk for the manufacture and use of existing and new products. Risk can be assessed based upon two options: (1) rate of environmental release or (2) estimated environmental concentrations. Information required for the first option consists of release rate estimates and intrinsic properties for the materials released to the environment during product manufacture and use. The second option requires information for estimating material dispersion in the environment. Some example applications of the approach are discussed.

KEYWORDS: Environmental risk assessment, hazard assessment, material balance, exposure assessment, environmental release

INTRODUCTION

About 15 years ago 3M Company started a corporate program to assess environmental risk as part of an overall corporate product responsibility strategy for newly introduced products. The program was designed to identify potential risks both for manufacturing operations and for any intended uses and anticipated misuses of the product. It was also to recommend risk management where necessary. The program has evolved with the expanding knowledge bases in environmental sciences and risk assessment, with changes in regulations, and with experience.

This paper describes the practical aspects of doing environmental risk assessment at 3M. The first part describes what information is used and presents an idealized model of how risk is assessed. The second part discusses the current risk assessment program and how the idealized model can be used as a framework to modify it.

ENVIRONMENTAL RISK ASSESSMENT APPROACH

The 3M risk assessment program uses conventional ideas on risk and hazard. Risk assessment combines assessments of exposure and hazard for materials released into the environment from product manufacture and use.[1] Hazard assessment derives No Observable Adverse Effects Levels (NOAEL) for materials released into the environment during manufacture and use of the product.[2,3] While the elements of the exposure

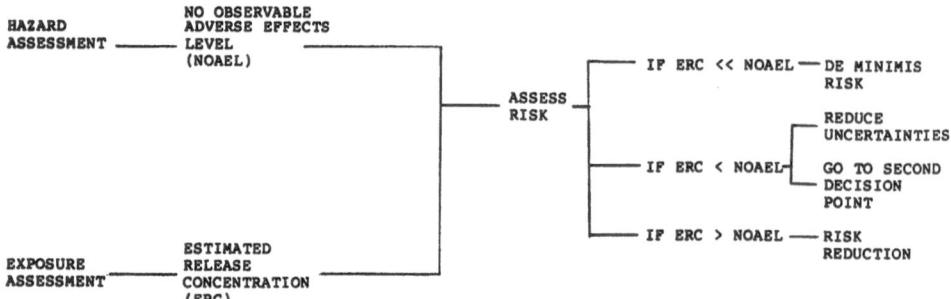

Fig. 1. First Decision Point: Environmental Release Concentration.

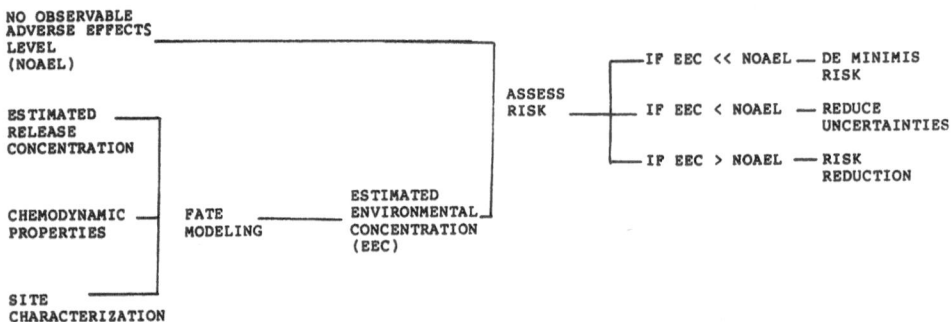

Fig. 2. Second Decision Point: Estimated Environmental Concentration.

assessment are not unusual,[4] the approach to gather and apply the information does appear unique. This section describes the risk assessment process and how the necessary information is gathered and applied.

Assessment Process. The risk can be assessed on the basis of either the material concentrations at the environmental release point (Fig. 1) or the projected environment concentrations (Fig. 2). The environmental release decision point assigns risk by comparing the NOAEL with the environmental release rate expressed in appropriate concentration units [environmental release concentration (ERC)]. A *de minimis* risk is assigned if the ERC is well below the NOAEL.[5] If the *de minimis* risk can not be assigned, the risk assessor can opt to reduce uncertainties in the release assessment process of the first decision point, go on to the second decision point, or recommend risk reduction.

Estimated environmental concentration (EEC) values are estimated from the environmental release data either through simple fate estimations or formal environmental fate modeling.[6] Computerized environmental models are available at 3M or through consultants for calculating the environmental concentration distribution of the released materials and their degradation products. The models require information on environmental conditions for the release points and surrounding area and chemodynamic properties for the released materials and their degradation products. Risk is assigned by comparing EEC to NOAEL. *De minimis* risk is assigned if EEC is well below NOAEL. If EEC is near NOAEL, the options are to continue reducing uncertainties in the assessment process or to suggest risk reduction. When EEC is greater than or equal to NOAEL, risk reduction alternatives are evaluated with the assessment background information as a framework.

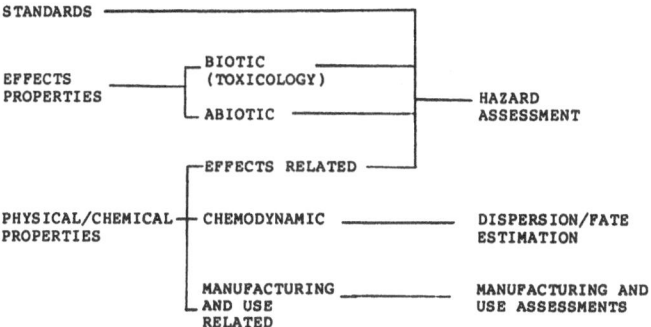

Fig. 3. Materials Evaluation.

Assessment Information. A materials evaluation gathers and evaluates environmental standards and the intrinsic properties needed for the risk assessment (Fig. 3). These intrinsic properties are effects, physical/chemical properties related to manufacture and use, and chemodynamic properties. The effects are the standards, toxicological and ecotoxicological properties applied in the hazard assessment. Chemodynamic properties are those which are required to estimate environmental concentration through appropriate environmental fate estimation. The manufacturing/use related properties are those applied in estimating material balance and release during the exposure assessment.

The hazard assessment follows the established approach of gathering standards, toxicological data for humans and ecotoxicological data for other target organisms, and abiotic effects data. The NOAEL for a material is assigned by applying an appropriate safety factor to the lowest concentration causing an adverse environmental effect.

In concept, exposure assessment quantifies material release, characterizes the release and describes the environment of the release, site, and gathers intrinsic property information. In practice, the product manufacturing and use exposure assessments have different approaches to gathering information. While specific information can be obtained for the manufacturing assessment, product use assessment will require more speculation.

The exposure assessment for product manufacturing, which is described in Fig. 4, consists of environmental release and fate analyses. The release assessment delineates manufacturing into a sequence of unit operations. An engineering analysis on each unit operation evaluates chemical transformations and distribution of the raw, intermediate, product and byproduct materials. The release assessment continues with analysis of control technologies applied to the initial waste stream. For example, wastewater treatment at the manufacturing site alters composition of the wastewater but causes atmospheric emissions of volatiles and solid waste disposal of sludges. Release characterization describes how the components of the waste stream enter the environment. For example, wastewater could have components discharged into a publicly owned sewage treatment plant or into a receiving stream, components released into the atmosphere as fugitive emissions, and components in a sludge disposed on land.

If the EEC decision point is used for risk assignment, information must be gathered for modeling environmental degradation and distribution of the released materials. This includes details on the physical conditions of the environment (on the geology, atmosphere, and hydrology and on the population and ecology) as well as the chemodynamic properties. Site-specific data can be obtained for estimating dispersion of materials released from product manufacturing.[5]

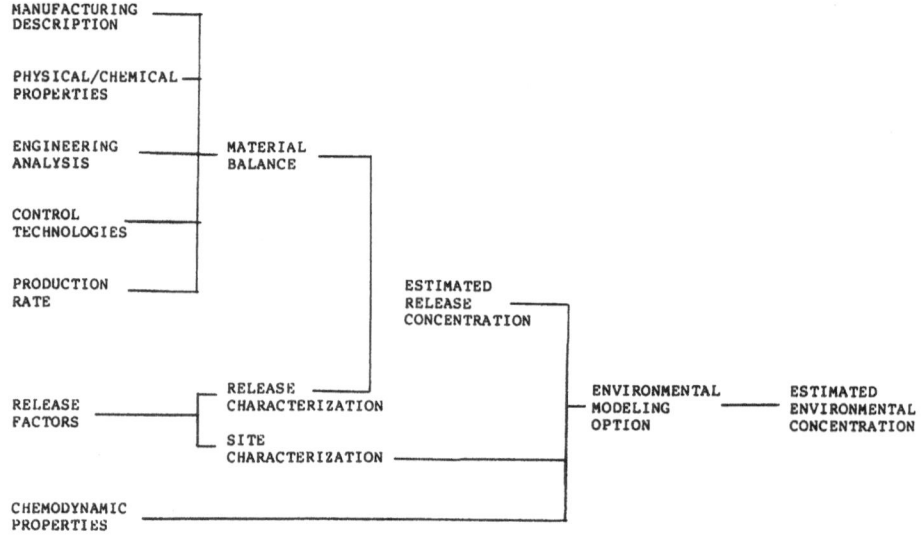

Fig. 4. Manufacturing Exposure Assessment.

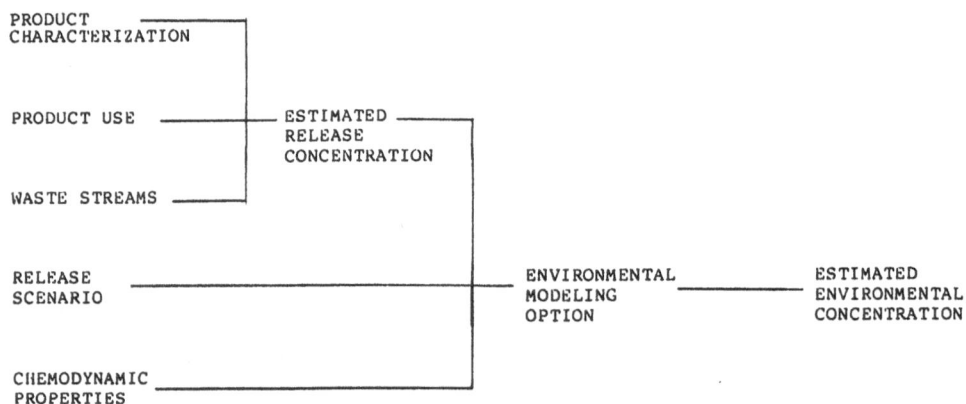

Fig. 5. Product Use Exposure Assessment.

Since specific details are not available for product use, information for environmental release assessment and environmental concentrations must be estimated on the basis of experience and conjecture (Fig. 5). The usual approach estimates release (ERC) for anticipated use and "worst case" situations, including blatant misuse. The EEC process must develop appropriate scenarios for release points, which adopts both "average" and "worst case" situations. Obviously, uncertainties are higher for information coming from hypothetical scenarios than from site-specific data. The general approach is applied for specialty chemicals and formulated products (for example, commercial coatings) as well as for evaluating hard goods (such as tapes and mechanical or electrical devices.)

PROGRAM OPERATIONS

Background. A description of 3M Company and its products will facilitate understanding how the model helps optimize the environmental risk assessment program. The 3M Company manufactures and sells over 45,000 products in 10 major market areas. Their manufacture utilizes more than 8,000 purchased raw materials, which are composed

Table 1. Category Approaches to Optimize Assessment

Approach	Information	Application
• Product Line Assessment	• Risk Assessment	• Reference Document
	- Specific Material Categories	• Guidelines for Future Risk Assessment
	- Specific Manufacture	
	- Specific Use	
• Material Category Evaluation	• Material Evaluation	• Reference Document
- Critical Chemical (3M Unique)	- Intrinsic Properties	
- Other Categories (Not 3M Unique)	- 3M Uses	
• Technology Assessment	• Manufacturing Exposure	• Reference Document

of more than 5,000 different chemicals. Products are sold to consumer, industrial, and service markets. Most products are finished articles (such as tapes) and formulated products (such as encapsulating resins). Some low-volume speciality chemicals are sold, but 3M does not sell commodity chemicals. New and modified products are developed and introduced through more than 40 operating divisions. Technology transfer between operating divisions results in expanded uses and variations of existing products. Many requests for environmental risk assessments arise from slight modifications or extensions of existent product lines or new applications of existing products.

Application. Since so many individual assessments utilize similar information, several approaches have been implemented to optimize information utilization (Table 1). The approach described above provides a framework to most effectively gather and apply information necessary for risk assessment.

The product line assessment gathers and evaluates the information for products which have common raw material categories, manufacturing technologies, and product uses. A full environmental risk assessment on this information yields a background document, which can be implemented for new assessment requests either directly or through screening guidelines. Guidelines establish an assessment procedure to assign *de minimis* risk based upon (1) product use limits, (2) product manufacturing limits, and (3) specific raw material limits. An expert system based on this approach is now being evaluated for existing risk assessment guidelines.

Material category evaluations develop background information on similar raw materials, intermediates, and products. The material category approach includes the "critical chemical" program for chemicals unique to 3M. The approach is also applied to materials which are not unique to 3M but have broad use throughout the company. Information gathered for the materials categories consists of intrinsic properties, major 3M related uses, and known environmental releases. A background reference document which is developed with this information is applied to subsequent product assessments.

The technology assessment approach gathers and interprets environmental release and exposure information and known environmental problems common to a specific manufacturing process or series of unit operations routinely used in manufacturing. Examples of technologies are solution coating and drying, extrusion, and batch reactions. Technology assessment results supply background information which is applied in subsequent product risk assessments. Product line assessments usually contain *de facto* technology assessments.

The environmental risk assessment can also utilize the model to optimize information-retrieval capabilities. Information developed during the risk assessment is accessible in hard copy files which are searched by alpha-numeric cataloging. The currently available computerized systems can search product risk assessments based on product name and division of origin and can identify raw material and intermediate material information in previous product risk assessment reports based upon name or code number assigned to the materials. No retrieval system exists to search for information derived from technology assessments. Efforts to upgrade the information searching capability and access can use the model as a framework for analyzing program needs and determining how well alternative search and file retrieval systems will meet these needs.

In general, risk is assigned at the first decision point; that is, it is based upon the ERC. When the assessor is unable to assign risk at the first decision point, the model facilitates choosing between the alternatives of reducing uncertainties in the first decision point assessment or going into the second decision point (EEC based). If the assessor is unable to assign risk at the second decision point, the model provides insight into the most efficient approach for reducing uncertainties throughout the assessment.

Uncertainties for intrinsic property data depend primarily on the parameter and its source. Uncertainty depends upon the quality and description of the property's measurement and error analysis. Uncertainty is least with values from in-house measurements or from primary or secondary literature and commercial data bases with documented error analysis. Estimated intrinsic properties from theoretical relationships and correlation equations or in the literature without documented error have the most uncertainty. The NOAEL uncertainty depends upon the uncertainties in intrinsic effects properties combined with issues of "acceptable" risk.[3]

The manufacturing assessment is affected by uncertainties in the release analysis and fate modeling. Environmental release accuracy is limited by variations allowed within product standards (which are usually performance based) and normal variations in equipment operation, operator practices, and maintenance. More detailed product description information, engineering analysis, and release characterization will reduce uncertainties in ERC. Monitoring both validates estimated release values and provides a value with reduced uncertainty for the risk assessment.

Uncertainties in the estimated environmental release in a product use assessment vary with the type of product and how well we can anticipate the customer's practices. The first approximation will be based upon intended application of the product combined with estimates that the consumer will use "reasonable" practices in any processing and will follow all existing regulations. Potential unintended misuses of the product and poor disposal practices for the spent product and related wastes are applied as "worst case" situations.

OTHER UTILIZATION

The material balance and environmental release portions of the manufacturing release assessment have potential applications in response to other regulatory initiatives. The material balance and release assessment approach could facilitate estimation of information required by the Superfund Amendments and Reauthorization Act (SARA), Title III, Section 313, for responding to manufacturing site requirements on environmental release of listed chemicals. The material balance approach also provides insight into waste minimization issues.

SUMMARY

The development and application of the model has helped to improve the operation and quality of the current environmental risk assessment program. The model highlights the environmental release aspects of the overall risk assessment process. The major applications so far include developing an explicit description of a standard risk assessment performance and delineating how various approaches to gathering and evaluating categories information (product line assessment, technology assessment, and material category evaluation) can optimize risk assessment operations. Future applications are planned for developing improved information retrieval capabilities, for exploring rule-based approaches to risk assessment, and for developing artificial intelligence capabilities.

REFERENCES

1. *Risk Assessment in Federal Government, Managing the Process*, National Research Council, National Academy Press, Washington, D.C., (1983).
2. P. W. Preuss and A. M. Ehrlich, "The Environmental Protection Agency Risk Assessment Guidelines," *Journal of the Air Pollution Control Association* **37**:784-791 (1987).
3. J. V. Rodricks and R. G. Tardiff, "Conceptual Basis for Risk Assessment," Assessment and Management of Chemical Risks, ACS Symposium Series 239, American Chemical Society, Washington, D.C. (1984).
4. "Proposed Guidelines for Exposure Assessment," U.S. Environmental Protection Agency, *Federal Register* **49**:46304 (1985).
5. J. Mumpower, "An Analysis of the *de minimis* Strategy for Risk Management," *Risk Analysis* **6**:437-446 (1986).
6. C. Travis, "Exposure Assessment in Risk Analysis," in *Risk Analysis in the Chemical Industry*, Chemical Manufacturers Association, 46-60 (1985).

The Use of Risk Index Systems to Evaluate Risk

Gary R. Rosenblum
Atlantic Richfield Co.
Los Angeles, CA

Steven A. Lapp
Design Sciences, Inc.
Pittsburgh, PA

ABSTRACT

Risk indexing is a useful and powerful tool that can provide valuable information on the risks associated with chemicals and chemical processes. As defined here, risk indexing is the process of modeling and scoring chemical hazard and exposure parameters to produce a rapid and simple estimate of relative risk. The principles of risk indexing can be applied to a variety of risk assessment projects to set priorities and help manage resources. Within the framework of basic risk assessment principles, however, different levels of risk index depth can be developed. A simple risk index can be considered to be at one end of a continuous risk analysis spectrum, where hazard and exposure are quickly estimated through the use of simple models. A full quantitative risk analysis would be at the other end of this risk analysis spectrum, where hazard and exposure are tested, measured, and assessed as rigorously as possible. Choosing the depth of the risk analysis is a critical decision that depends on such factors as the time and resource commitment to the project, the number of materials or processes to be analyzed, and the desired endpoint of the risk index analysis. This report demonstrates three levels of risk index analysis for a hypothetical ammonia storage system and analyzes the relative merits of each. The analysis and discussion offer guidance for appropriately selecting the depth of risk index analysis based on the desired risk analysis endpoint.

KEYWORDS: Risk indexing, prioritization, risk management, fault tree, relative risk

INTRODUCTION

Risk indexing, a process of estimating and scoring hazard and exposure parameters to give a rapid and simple estimate of relative risk, has gained widespread acceptance as a cost-effective prioritization and screening tool for health risk assessment programs. Experience with health risk assessment has shown that it is an expensive and labor-intensive process, and much time and money can be wasted if the chemicals with the greatest potential for health risks and associated liability are not identified and assessed first. Without a prioritization plan it will not be known whether a risk was worth assessing until the time and money have been spent. A prioritization system helps defend difficult decisions for choosing among chemicals competing for limited health risk assessment resources.

In order to be cost-effective, a prioritization system must be simple, rapid, and accurate. The appeal of risk indexing is that it can be all of these. Risk indexing also has appeal to management charged with risk management decision-making responsibilities who may be unfamiliar with the details and mechanics of the risk assessment process. Because risk indexing for health risk prioritization simplifies basic risk assessment principles, risk indexing can be an effective way to acquire a global grasp of the issues. Excessive simplicity can, however, lead to problems, and some of these will also be examined.

Cost-effectiveness and defensibility are two strong advantages of risk indexing. A logical and well-designed risk index system will rapidly and consistently screen for company operations having the greatest potential for causing harm on a large scale.

Defensibility is one of the strongest assets of a scientific risk index system, both internally and externally. Internal defensibility provides the company management with justification for expending resources for a health risk assessment program, and can provide the means for most effectively dividing limited resources among risks competing for analysis resources. External justification of priority setting could be important for defending company activities in a courtroom, or to a regulatory agency, where it might be necessary to explain why risk assessments are being conducted on certain chemicals or processes and not on others.

Risk indexing is not a substitute for a detailed quantitative risk analysis. It is a planning tool useful for screening, ranking, and setting priorities. As such, risk management decisions should not be based solely on a risk index or other prioritization system. Risk management should be based on a much deeper analysis of chemical risks.

The practical necessity of trying to assess hundreds or even thousands of risks for chemicals and industrial operations with limited resources has led to the creation of highly detailed risk index systems. These are systematic risk evaluation tools that have the features of a rapid and simple risk index and some of the depth and accuracy of a full quantitative risk analysis. These detailed risk index systems are virtually limitless in their possible variations. This report briefly reviews some risk index systems recently presented in the literature as examples of what a risk index system would consist of.

In the second section of this report, a hypothetical chemical risk situation is analyzed by two risk index systems of different depths and by a fault-tree analysis, which together illustrate the utility of simple, detailed, and highly detailed analyses, respectively. The three-way analysis of the hypothetical risk situation can provide some guidance in selecting the depth of analysis necessary for a given task.

RISK INDEXING SYSTEMS

At a basic level, a risk index has some similarities with a risk analysis. They both define a relationship between the inherent hazard of a material and the exposure potential. For a risk analysis, the characterization of exposure and the assessment of the hazards are as detailed and well defined as possible. The risk analysis usually quantifies unit risk at a specific exposure level. It may rely on many hundreds of thousands or even millions of dollars worth of toxicology studies, often carried out specifically for the risk analysis. The exposure data characterize exposed populations as comprehensively as possible, frequently utilizing exposure monitoring measurements in specific locations or statistically modeled recreations of the exposure patterns. Exposure potential can be modeled by fault-tree or similar analyses. In short, the risk analysis is complex, time-consuming and usually costly.

It is necessary when particular risks must be known as accurately as possible, or when legal issues demand that an exhaustive analysis must be performed.

For many situations, an in-depth analysis is not appropriate. It is for these situations that a risk index is developed. A risk index, in contrast to the full analysis, defines the hazard-exposure relationship by more simplistic models requiring less data and less analysis. As it is difficult to describe a typical risk index, three representative examples of risk index systems selected from the literature are summarized here to provide some idea of the variations involved with modeling and quantifying inherent hazard and potential exposure.

The three systems selected are the Integrated Risk Index System,[1] the Chemical Scoring System for Hazard and Exposure Assessment,[2] and the Priority Selection Method.[3]

The Integrated Risk Index System (IRIS) is the most simplified system of the three. It multiplies combined scores of three kinds of hazards (health, fire, and environmental) with a simplified exposure estimate. The health hazards considered are acute, subchronic, carcinogenicity, mutagenicity,reproductive and teratogenic effects. Physical (fire) hazard scores are based on flammability and explosivity, and environmental hazard scores are based on bioaccumulation and ecological effects. Expert scorers are expected to use their professional judgment and are provided with written criteria for consistent scoring. Exposure potential is based on the population type potentially exposed (occupational, or general public/consumer) and the volume of production and environmental fate of the chemical. This exposure estimate assumes that more stringent control over exposure is exercised in the occupational setting than in the public/general consumer sector, and that potential for exposure increases as production volume increases.

The Chemical Scoring for Hazard Assessment System is similar to the Integrated Risk Index System in that it includes objective criteria to score chemicals for hazards (oncogenicity, genotoxicity, developmental toxicity, acute and chronic mammalian toxicity, and aquatic toxicity), and for exposure (bioconcentration, chemical production volume, occupational exposure, consumer exposure, environmental exposure, and environmental fate). Some differences are that the criteria are more detailed and complex than those in IRIS, and the scores are not combined into a final single score for each chemical, but instead are listed on a matrix. While the highly detailed criteria give improved accuracy of scoring, the investment in time for data collection and analysis is increased, and the impact of data gaps is larger.

The Priority Selection and Risk Assessment System assesses four aspects of chemical substances: intrinsic physiochemical properties and hazards, data gaps for hazard assessment, differing exposure parameters, and priority assessment for further risk assessment and filling data gaps. The system breaks each of these four aspects into smaller units that are scored according to specified criteria. The physiochemical parameter combines scores on such areas as vapor pressure, water solubility, flammability, explosivity, etc. Toxicological properties scored include skin irritation, carcinogenicity, reproductive and teratogenic effects, and various ecotoxicological properties. Exposure is scored by criteria-defined assessment of professional, personal, or occupational exposure, as well as environmental spread and persistence. The parameters can be combined in different ways to provide risk indices for personal risk assessment, environmental risk assessment, and data-gap priority. The criteria are quite detailed, thus leading to significant time investment in the prioritization stage.

Other risk index systems exist and are currently in use by major industrial corporations for cost-effectively auditing their operations for catastrophic risks, and by U.S. government agencies, such as the Environmental Protection Agency (EPA), for prioritizing chemical risks at superfund sites.

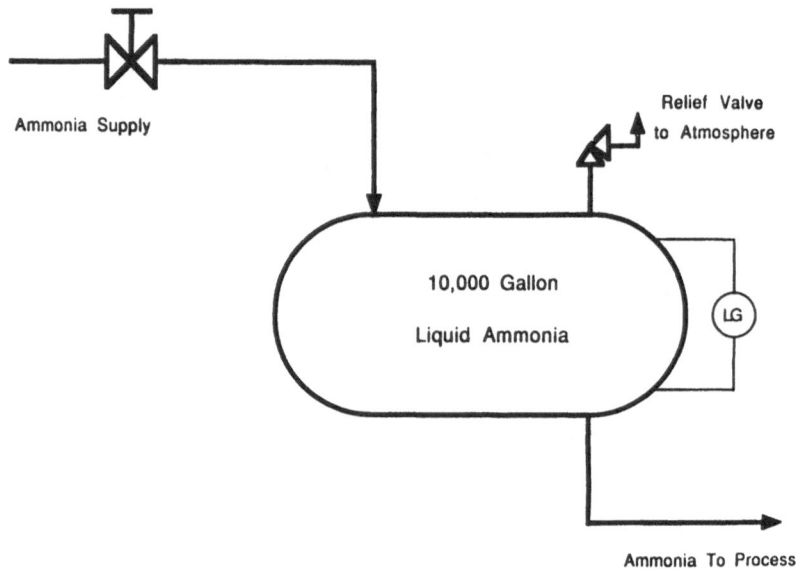

Fig. 1. An Ammonia Storage Tank.

RISK INDEX UTILITY ANALYSIS

The small sample of risk index systems presented above suggests that systems can be quite varied in their approach and that each approach may have certain advantages or trade-offs when utilized for specific tasks. One trade-off concerns the increase in time and resource expenditure as the depth of analysis is increased. This creates a situation in which it may be difficult to determine what level of index depth is appropriate for a particular risk prioritization task, given a certain limited level of resources. In an age in which resources are scarce and efficiency is prized, maximizing the utility of the risk index is clearly desirable.

In order to assess how the various levels of risk index depth can affect the utility of the system, a hypothetical risk situation was created and three different depths of risk analysis were applied. The hypothetical risk situation was a series of three ammonia storage tanks (Figs. 1-3) and the risk of ammonia overflow was assessed for each.

. The three storage tanks possess identical storage capacity but have different levels of technology to prevent overflow. The first tank (process 1) has a level gauge and is manually controlled (Fig. 1). The second tank (process 2) has some automatic level control through a level transmitter (LT) that activates a level recorder controller (LRC) (Fig. 2). The LRC modulates the signal from the LT and controls a pneumatic valve, which reduces and then shuts off the flow into the tank as the tank reaches a preset capacity. The third tank (process 3) possesses automatic level control, an alarm, and an interlock (Fig. 3). The system is similar to that shown in Fig. 2, except that it adds a level switch high (LSH) that activates at a preset high level, setting off an alarm to alert the operator to slow the flow into the tank. It also has a level switch high high (LSHH) that activates at the highest level before overflow and triggers the interlock to vent the pneumatic valve and immediately stop all flow into the tank.

Two risk index systems based on the Integrated Risk Index System (IRIS) and a fault-tree analysis were created to assess the relative risks of the hypothetical ammonia tank

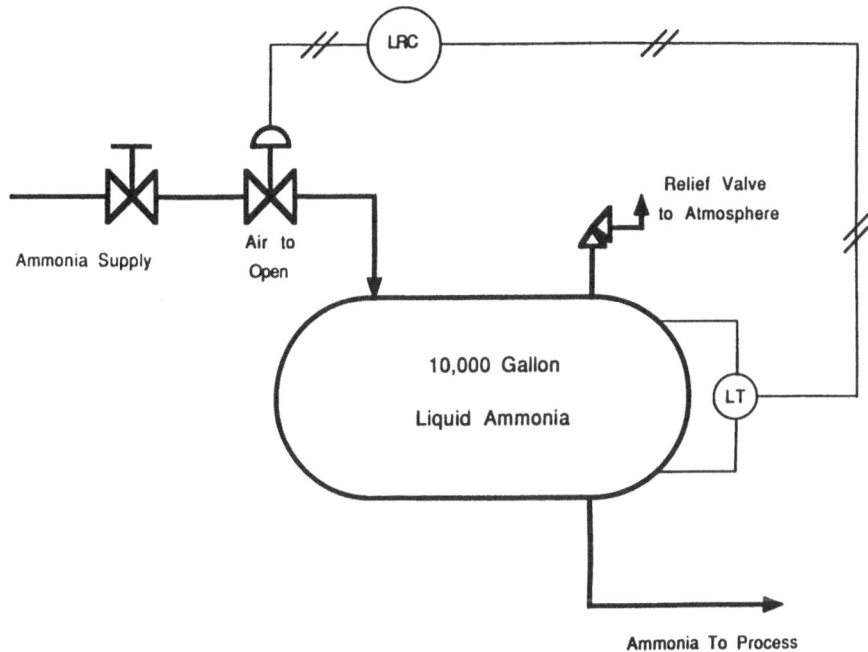

Fig. 2. Ammonia Storage Tank with Automatic Level Control.

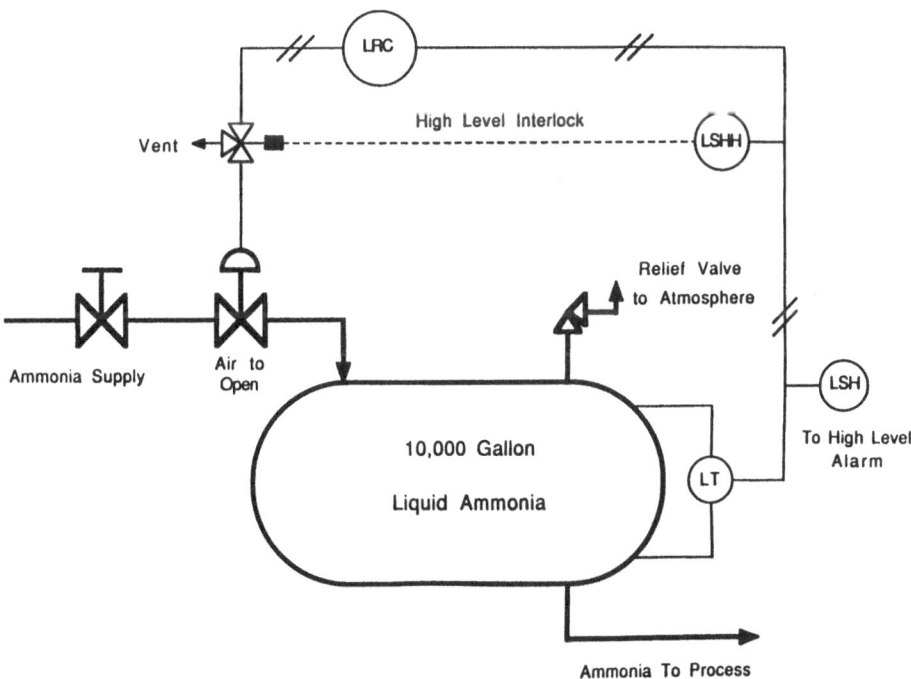

Fig. 3. Ammonia Storage Tank with Control, Alarm, and Interlock.

farm. They are termed the "simple risk index," the "detailed risk index," and the "fault-tree quantitative analysis." All three methods assess the inherent hazard of the material stored through use of the IRIS hazard scoring model. Because the chemical in all three processes is ammonia, the hazard score can be considered constant.

The simple risk index relies on the volume stored to quantify the exposure potential. The detailed risk index incorporates not only the volume of the material into the exposure potential assessment, but also a variety of factors for assessing different levels of engineering protection against accidental release. This approach utilizes specific algorithms to generate a prediction for a rate of accidental releases. Details of this risk index system are found in Roth and Lapp.[4] This risk index system would take more time and effort to collect the data on the automatic flow control and overflow protection devices. For example, an assessor might have to travel to the location to properly collect the data.

The fault-tree analysis quantitatively determines probability of failure for every step of each potential failure mode. This assessment must account for such things as the engineering logic for each protection system, construction quality, maintenance, and lifetime failure rate for each item. This obviously is a resource-intensive method that would require on-site inspection and engineering analysis.

Figure 4 demonstrates the results from using the simple risk index system. Because it utilizes volume stored to estimate exposure, and all three processes are of equal capacity, this system is obviously not suitable to define the relative risks of the three ammonia tanks. If, however, the hypothetical tank farm were storing water, ammonia, and hydrogen cyanide, this system would then be able to differentiate relative risks. This is because the simple risk index is consequence oriented, which means that it is based on differentiating risks of events assumed to happen (i.e. the risks resulting from complete leakage of each tank). This simple system could therefore be used to prioritize risks where the differences were based on inherent hazard and gross volume differences. This might be useful, for example, for auditing and prioritizing the potential for catastrophic releases from a thousand different chemical storage sites around the world. The simple system could not differentiate the more subtle risks, such as those resulting from process controls, but a relative ranking could be accomplished rapidly and probably would not require the time and resources to visit each site.

Figure 5 shows the outcome from using the detailed risk index. The detailed risk index estimates the probability of overflow in process 1 to be once in 4 years, process 2 once in 10 years, and process 3 once in 40 years. By integrating an assessment of the overflow control equipment into the detailed risk index model, it is possible to differentiate relative risk between the processes based on process control. Based on this assessment, it would seem that the interlock system in process 3 greatly reduces the risk of overflow. But the results of the fault-tree analysis (Fig.6) indicate that this supposition is faulty and point out a pitfall of basing risk management decisions (e.g. upgrading process 2 to have the same interlock as process 3) on a risk index system.

Whereas both the detailed risk index model and the quantitative fault-tree analysis predict an overflow to occur once in 4 years for process 1 and once in 10 years for process 2, the fault-tree analysis predicts an overflow once in 15 years for process 3, which is different from the estimate derived from the detailed risk index. The quantitative fault-tree analysis, which was able to assess the relative risks with greater precision for the more subtle risk, as would be expected from the level of effort required to produce the assessment, indicates that process 3 actually has a greater potential for overflow than was previously suspected, and the addition of the high-level interlock and alarm provides only marginal improvement.

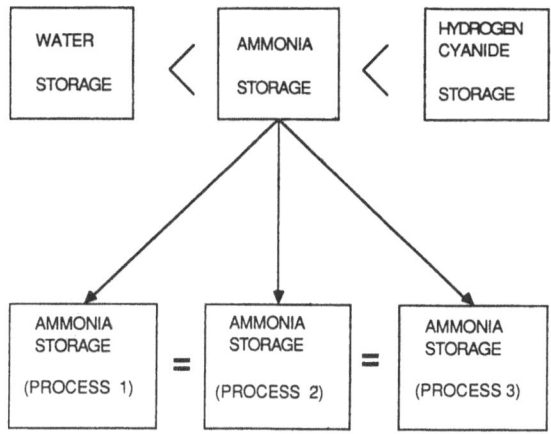

Fig. 4. Ranking Relative Risks: Simple Risk Index System.

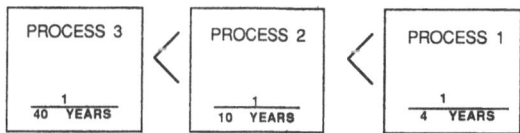

Fig. 5. Ranking Relative Risks: Detailed Risk Index System.

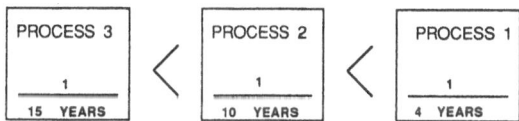

Fig. 6. Ranking Relative Risks: Fault-Tree Risk Analysis.

The fault-tree analysis is outlined in Figs. 7-9. An examination of the process 3 tree (Fig. 9) shows that the reason for only marginal improvement from adding the alarm and interlock is that the failure of the level sensor is a common-cause failure of all three protective loops (control, alarm, interlock). The integrity of the process could be improved considerably by using a separate level sensor for the alarm and the interlock. Note that the detailed index system does not account for this, because it assesses whether or not the process controls are there, and does not assess the interaction between components. This shows why risk index systems are not substitutes for quantitative risk analyses and why caution should be exercised before risk management options are based on a risk index.

However, if the time and resources for fault-tree analyses of one thousand storage sites across the country are not available, an appropriate risk index system could be designed. Note that the detailed risk index and the fault-tree analysis provided the same qualitative ranking of the three processes. The detailed risk index could be suitable for prioritizing the thousand sites and selecting the riskiest locations for which fault-tree analyses could subsequently be conducted. Common-cause failure situations could then be found in a cost-effective manner.

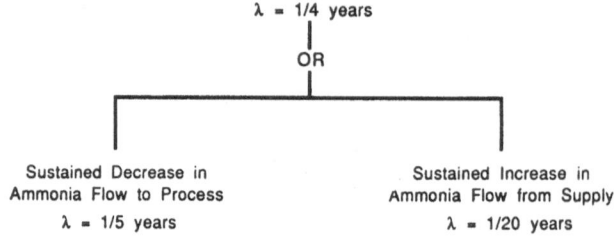

λ = failure rate

Fig. 7. Fault-Tree Analysis of Process 1.

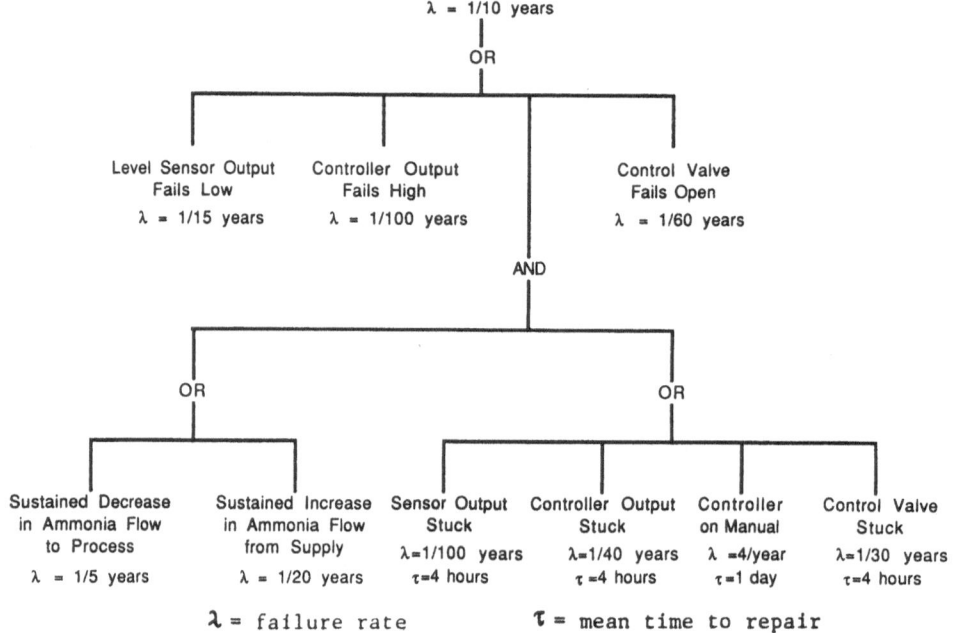

λ = failure rate   τ = mean time to repair

Fig. 8. Fault-Tree Analysis of Process 2.

CONCLUSIONS

The comparison of the three depths of analysis of the hypothetical risk situation can be presented graphically. Figure 10 provides a graphic view of the relative power and limitations of three broad levels of risk quantification. Curves A, B, and C do not represent actual data points, but instead are representative curves selected from a continuum of risk analysis possibilities for demonstration purposes.

Risk indexing with simplistic exposure and hazard modeling (C) provides the ability to screen for high-risk catastrophic-type situations and chemicals for which the analysis can be consequence oriented. However, as the relative risk decreases, the ability of the more simplistic screening system to differentiate between two more subtle risks, such as those arising from differences in safety engineering, diminishes as well.

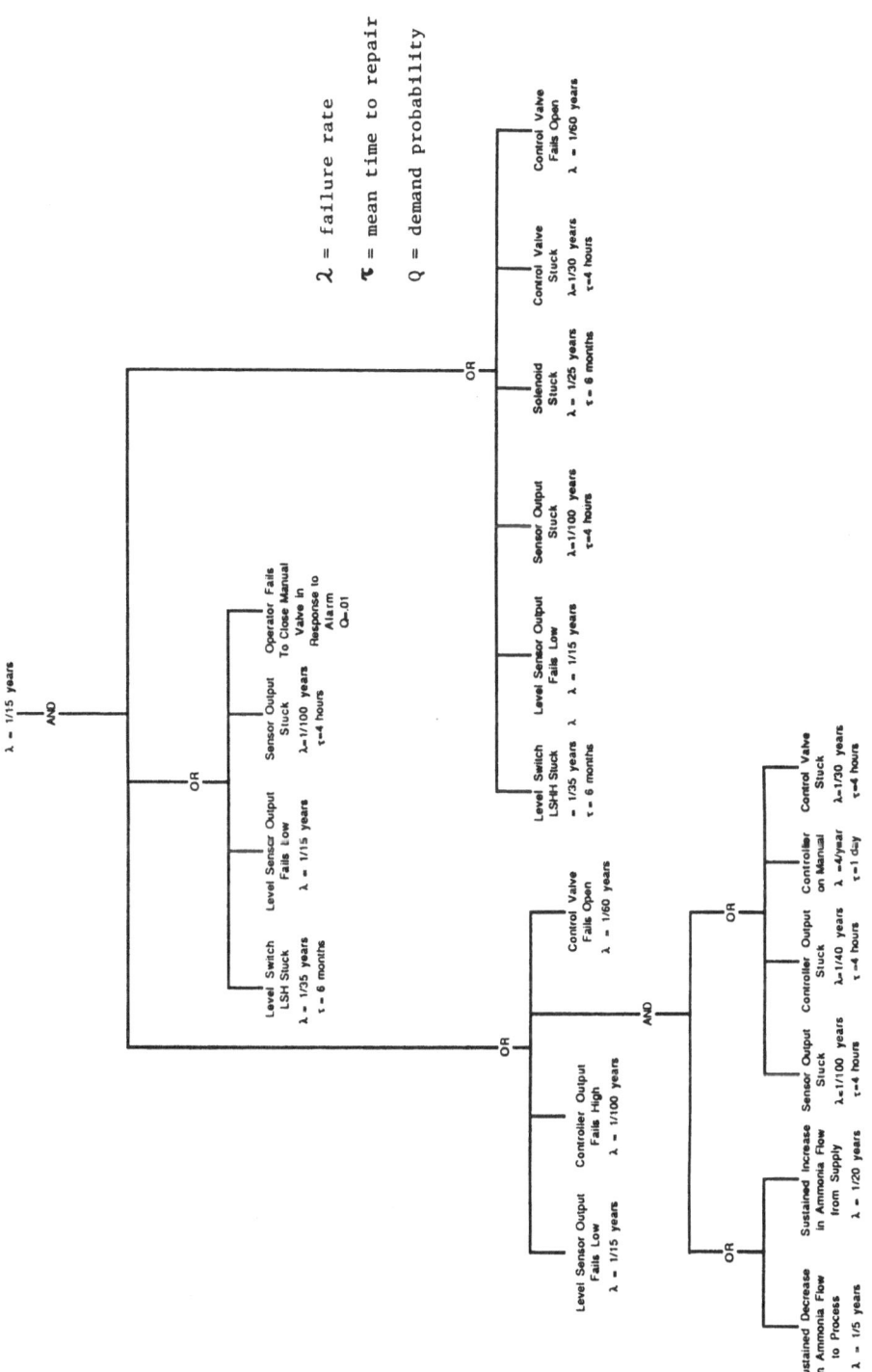

Fig. 9. Fault-Tree Analysis of Process 3.

λ = failure rate

τ = mean time to repair

Q = demand probability

Level Switch
LSH Stuck
λ = 1/35 years
τ = 6 months

Level Sensor Output
Fails Low
λ = 1/15 years

Sensor Output
Stuck
λ=1/100 years
τ=4 hours

Operator Fails
To Close Manual
Valve in
Response to
Alarm
Q=.01

λ = 1/15 years

Level Sensor Output
Fails Low
λ = 1/15 years

Controller Output
Fails High
λ = 1/100 years

Control Valve
Fails Open
λ = 1/60 years

Sustained Decrease
in Ammonia Flow
to Process
λ = 1/5 years

Sustained Increase
in Ammonia Flow
from Supply
λ = 1/20 years

Sensor Output
Stuck
λ=1/100 years
τ=4 hours

Controller Output
Stuck
λ=1/40 years
τ=4 hours

Controller
on Manual
λ = 4/year
τ=1 day

Control Valve
Stuck
λ=1/30 years
τ=4 hours

Level Switch
LSH Stuck
λ = 1/35 years
τ = 6 months

Level Sensor Output
Fails Low
λ = 1/15 years

Sensor Output
Stuck
λ=1/100 years
τ=4 hours

Solenoid
Stuck
λ = 1/25 years
τ = 6 months

Control Valve
Stuck
λ=1/30 years
τ=4 hours

Control Valve
Fails Open
λ = 1/60 years

191

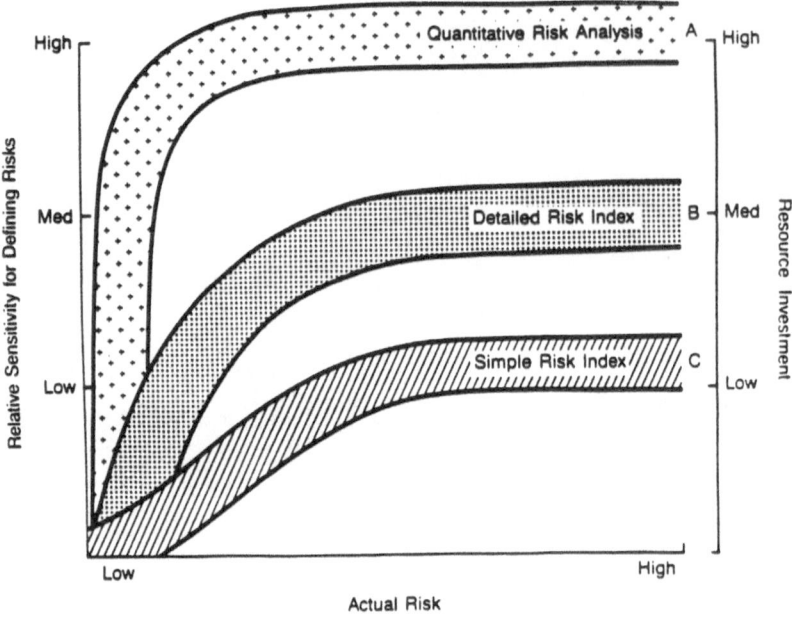

Fig. 10. Risk Index Systems and Relative Sensitivity for Defining Actual Risk.

A risk index using a more complex and accurate exposure and/or hazard assessment model (B) will provide much greater differentiation between lesser risks and improved overall accuracy. It maintains some ability to differentiate between risks as the actual risk decreases. The trade-off for this approach is the increased time and resources expended for model development, implementation, and data collection. The advantage is that it can provide a useful analysis with far less resource expenditure than a quantitative risk analysis.

Curve A is representative of a quantitative risk analysis in which the hazard and exposure are analyzed through toxicology testing and hazard assessment, rigorous analysis of actual exposure monitoring data, and full quantitative analyses of the probabilities of exposure and adverse effect. It is clear that this is the best approach to defining the risk of a chemical or operation, especially for relatively low risks, but it is also clear that a large resource investment is necessary to accomplish this task. This approach maintains its ability to distinguish between risks down to the lowest levels of actual risk.

These curves can also highlight the pitfalls of improperly using risk index systems. The most basic error is mistaking curve C for curve A, or using a relatively simple risk index system when a more rigorous analysis is appropriate for the desired task. The problems that can stem from misuses of risk indexing are often not apparent until it is too late. A misused risk index can produce disastrous consequences, such as unwarranted complacency leading to unforeseen catastrophe, or lost opportunity through misapplied resources.

These problems are avoidable by emphasizing the preliminary nature of the risk index system analysis. It should be made clear that models of hazard and exposure can only provide dimensionless numbers of relative risk, which are useful only for priority setting and resource management, not risk management.

The false sense of precision and certainty that the quantitative nature of a risk index may imply to a decision-maker should be dispelled. Because a risk index produces

192

dimensionless numbers, any risk/benefit calculations will result in meaningless analyses, and any calculations concerning the cost of incremental risk reduction should be treated with extreme caution. Unless the exposure analysis is very deep, there will not be sufficient detail and accuracy to allow cost per risk-reduced calculation. This subtle concept may get lost in the translation of the benefits of the risk index analysis to upper management. A risk index is likely to be misused if it is sold as an answer to a problem rather than as a tool to cost-effectively guide further questioning.

When these pitfalls are known and avoided, risk indexing is a useful and potentially powerful tool for a variety of risk assessment projects to help manage resources. How the exposure and hazard are estimated will greatly influence what questions the risk index can answer and what analyses can be performed from the data. Examples of some different but related prioritization activities include identifying and then prioritizing risky operations for further assessment, or identifying and then prioritizing risky chemicals produced or handled by the company regardless of the site. The risk index may also be used to help determine priorities for audit teams to inspect operations either for risk identification or for checking on the progress of risk management program implementation. A risk index can also assist in the prioritization of toxicity testing.

All these activities can be coordinated through use of similar data sets, which provides a unified and consistent approach. The hazard index concept is useful for designing risk index systems, because the hazards of a chemical can be described as "inherent."[5] Once the hazards of a chemical are modeled through use of a sound approach, they can be considered constants for all the different risk indexing activities outlined above.

Therefore, when a hazard index is established, the exposure estimation element of the risk analysis will become the ultimate influence on the structure of the risk index and on the questions it will be able to raise and answer. While a risk index is not an ultimate answer to risk assessment problems, it is a versatile tool to cost-effectively guide and structure the resource-intensive risk assessment process.

DISCLAIMER

The opinions expressed in this paper are those of the authors and do not necessarily represent those of ARCO (Atlantic Richfield Co.)

REFERENCES

1. Rosenblum, G. R., W. S. Effron, J. L. Siva, E. R. Mancini, R. N. Roth, "Integrated Risk Index System," *Risk Analysis in the Private Sector, Proceedings of the Society for Risk Analysis International Workshop, 1983*, Edited by C. Whipple and V. Covello, Plenum Press, New York (1985).
2. O'Brien, T., R. H. Ross, P. Y. Lu, "Chemical Scoring for Hazard Assessment," *The Toxicologist* 6(1):281 (1986); Proceedings of the Society of Toxicology, New Orleans (1986).
3. Sampaolo, A., R. Binetti, "Elaboration of a Practical Method for Priority Selections and Risk Assessment Among Existing Chemicals," *Regulatory Toxicology and Pharmacology* 6:129-154 (1986).
4. Roth, R. N., and S. A. Lapp, "A Relative Risk Indexing System for Industrial Operations," presented at the Society for Risk Analysis 1987 Annual Conference, Houston, TX (not submitted for inclusion in this proceedings volume).
5. Rosenblum, G. R., "Prioritizing Health Risk Assessments," *Risk Analysis in the Chemical Industry*, Edited by T. O'Leary, Chemical Manufacturers Association, Washington, D.C. (1985).

Probabilistic Seismic Risk to Non-Nuclear Facilities

B. G. Cassidy, W. S. LaPay and D. F. Paddelford

Westinghouse Electric Corporation
Pittsburgh, PA

ABSTRACT

This paper presents the overall methodology and results of a seismic risk assessment performed on two facilities in California. The purpose of this study was to assess the financial risks associated with earthquakes at non-nuclear facilities. Analytical tools, assumptions and models applicable to the assessment of seismic risk to nuclear power plants were utilized in this study. Structural failures that could potentially result in loss of the buildings were evaluated and the ground acceleration levels that could induce failures of critical structures were calculated. A frequency distribution of earthquakes was next constructed based upon exceeding ground acceleration levels as a function of distance from identified seismic events and the estimated intensity of each seismic event. The ground accelerations associated with the structural failure data were then combined with the constructed frequency distribution to determine the frequency of seismic events that equaled or exceeded the ground acceleration levels that would induce structural failures on the two sites. The resulting distribution estimates the potential risk of earthquakes inducing structural failures as data input for the consequential financial risk associated with these failures. The methodology application discussed in this paper can be a valuable tool used by insurance companies and by commercial and industrial firms to identify areas of high seismic risk that can then be used to evaluate financial risk, adequacy and value of insurance coverage, and the need to perform a seismic upgrade program.

KEYWORDS: Earthquakes, seismic risk, non-nuclear

INTRODUCTION

We present the results of an analysis to assess probable seismic-induced damage to two non-nuclear facilities (Facility 1 and Facility 2) in northern California. The intent of this study was to develop a methodology which could be completed with minimal time and cost. This study concentrated on the seismic adequacy of the building structures housing capital equipment, since significant loss of contents can occur with the loss of structural integrity of the building. An evaluation of the equipment within the buildings would require a detailed study of the equipment capacity as well as the localized structural integrity of the building structure in the vicinity of the equipment. It is possible that the structures would remain intact, but equipment within the structures could be damaged if not anchored or protected. Therefore, to limit the scope of this analysis, only structural failures are addressed and these failures are assumed to induce maximum damage.

METHODOLOGY

The analysis of the seismic risk consists of the following steps.

Assess the critical failure modes of building structures and their associated seismic ground acceleration capacities (fragility).

Calculate the ground acceleration at each seismic intensity as a function of distance from the epicenter of the seismic event (attenuation).

Construct the distribution of earthquakes as a function of distance from the epicenter of each earthquake at each ground acceleration level and compare to the seismic ground acceleration capacities associated with the critical failure modes of building structures at each facility site.

Construct the frequency distribution of earthquakes for each facility.

Determine the integrated area around the site that includes all epicenters that would cause ground acceleration values exceeding the capacity estimated for the postulated failure modes of the structures at each facility site.

This method was used to calculate the probability of the event occurring close enough to the facility to cause the specified damage by ratio of the damage area to the total area over which frequencies had been determined. Each probability was then multiplied by the frequency of each intensity. The frequency contributions of each seismic intensity were then summed to calculate the frequency at each ground acceleration level.

Fragility

The quantitative measurement of the performance of a structure under stress is important in defining its seismic capacity and probabilistic risk of failure. This is accomplished by calculating its capability (reserve strength) to perform its functions, given the seismic events.

Fragility limits describe the quantitative measurement at which something will cease to perform its functions. In defining structural seismic capability, the first step is to define the critical performance functions; we concentrated on the seismic adequacy of the building structures housing capital equipment.

Buildings may have several fragility limits associated with their failure modes, as follows:

    buckle and collapse
    yield and loss of load capacity
    formation of a plastic hinge, resulting in collapse
    loss of anchorage load carrying capacity
    shear failure
    bearing capacity of soil exceeded
    structural bearing failure
    bursting
    fatigue

The fragility limit is determined by analysis or testing. Sources of data used to define fragility limits can be past experience in qualification programs, documented data associated with past earthquakes, or data associated with dynamic loadings that can be related to the defined seismic events. The use of experience data is a recognized and acceptable qualification procedure as evidenced by its use in Reference 1.

Since fragility limits represent the maximum capacity of a structure, they are calculated with the design margin removed so that there is no reserve strength in the value established. This requires that the analyst be knowledgeable in the design margins associated with the various analytical or testing procedures used to show qualifications. Design margins associated with the following stress and strength factors may have to be reflected in the calculation of fragility limits:

Stress
    a. damping
    b. broadening of response spectra
    c. analytical approach (e.g., response spectra, time history)

Strength
    a. stress criteria used to define failure
    b. material strength
    c. material behavior

Uncertainty is also defined for each fragility limit. This factor is used in the probability analysis to reflect variations from the norm. Such factors are expressed in terms of standard deviations, based on:

a. Known statistical data, in which the standard deviation is determined from statistical data based on tests, such as tests on material strength.

b. Expected behavior, in which the standard deviation is obtained based on expected behavior established by experience by assuming that the conservatism is set at the nominal value of 1.0 plus a set number of standard deviations (e.g. 3).

c. Zero variation, in which a standard deviation of zero is set for cases where the variation is known to be zero, such as strength factors associated with stress criteria where the design margin is known exactly.

In Reference 2, design margin and standard deviations associated with stress and strength factors are discussed in more detail, along with the mathematical approach used to calculate the probability of stress exceeding strength. This paper also demonstrates how the reserve strength of structures determined, using fragility limits, can be used to achieve a reduction in the scope of a probabilistic risk assessment program, prioritize structures in a seismic upgrade program, justify postponement or elimination of non-critical items in a seismic evaluation program, and obtain a quantitative measure of the reserve strength associated with equipment or structural systems.

In the studies performed, in addition to all safety factors associated with loading and stress limits having been removed, the minimum material properties were used. The added capacity due to increased material strength and other potential factors was reflected by an uncertainty factor provided for each of the failure modes. An uncertainty factor of 1.2, for example, represents a potential increase in the acceleration limits of twenty percent.

Facility 1 has a predominantly wood superstructure with steel trusses and concrete reinforced footings. Based on the design and construction of Facility 1, the horizontal lateral seismic loads that the building structures are assumed to be subjected to are based on a response spectrum analysis using a Housner spectrum. Six critical areas associated with the building structure were evaluated. Most of the critical areas are localized near one critical equipment position (called position C). The six critical areas are: (1) strut columns

Table 1. Critical Failure Modes of Facility 1

Failure Area	Minimum Seismic Ground Acceleration Capacity	Uncertainty Factor
1. Strut Columns Near Position C	0.8g	1.2
2. Strut Columns Away From Position C	1.1g	1.2
3. Strut Footing Soil Pressure Near Position C	0.8g	1.0
4. Strut Footing Soil Pressure Away From Position C	1.1g	1.0
5. Roof Columns	0.5g	1.1
6. Soil Under Center Columns	1.1g	1.0

near position C; (2) strut columns away from position C; (3) strut footing soil pressure near position C, (4) strut footing soil pressure away from position C, (5) roof columns, and (6) soil under center columns. The critical failures and their associated seismic ground acceleration capacities are summarized in Table 1. These acceleration limits represent the point at which failure would occur. Note that the roof columns (item 5) of this facility have been identified as failing first. Failure of these columns will result in failure of the roof and, consequently, failure of equipment below roof. The seismic margin can be increased by strengthening these columns or protecting the equipment.

Facility 2 is in a reinforced concrete structure. This building will exhibit a combined shear- and cantilever-beam bending response during a seismic event. The building displacement shape can be adequately described following a shear-beam formulation. The acceptable seismic loads are taken from the Uniform Building Code. Six critical areas associated with this building structure were evaluated. The critical failures and their associated seismic ground acceleration capacities are summarized in Table 2. As seen in Table 2, area 5 will most likely fail, but failure of this element will not result in significant damage if the equipment is adequately protected from "architectural" ceiling material. Similarly, failure in area 2 can result in localized damage, but not in complete loss of all capital equipment.

Earthquake Magnitude as a Function of Distance (Attenuation)

Decrease in intensity of ground shaking with distance from the epicenter region is called attenuation. Attenuation relationships are developed to estimate the intensity of seismic-induced ground motion (peak ground acceleration) at the site. The equation for attenuation used in this study (Reference 3) is

$$a = b1*\exp(b2*m)/(R + 25)^{b3} \ ,$$

Table 2. Critical Failure Modes of Facility 2

Failure Area	Minimum Seismic Ground Acceleration Capacity	Uncertainty Factor
1. Floor Columns	0.9g	1.3
2. Roof Steel Columns	1.3g	1.2
3. Lateral Bracing	1.4g	1.2
4. Foundation Footings	14.5g	1.2
5. Strut Brace for Tee Bar Suspended Ceiling	0.3g	1.2
6. Soil Pressure	2.7g	1.0

where a is the peak ground acceleration at the site (cm/sec^2), m is the magnitude of the earthquake, R is the distance to the energy center and $b1$, $b2$, and $b3$ are coefficients evaluated by using recorded strong-motion data, with

$$b1 = 2,154,000/(R)^{2.1},$$

$$b2 = 0.046 + 0.445(\log R),$$

$$b3 = 2.515 - 0.486(\log R).$$

The attenuation, a, divided by 980 cm/sec^2 yields the g-level for each magnitude m and distance R.

Earthquakes in California of magnitude V or greater in the Modified Mercalli intensity scale are listed in Reference 4. The distance from each epicenter was estimated (R) and the attenuation was calculated for all earthquakes within one hundred miles. Based on the expected ground acceleration (g-level) at each intensity and possible ground acceleration capacity of the critical structures, only the largest earthquakes would have an impact in a radius greater than 60 km. Therefore, the data base was reduced to include only those events that could impact either site.

Figure 1 is a plot of calculated g-levels as a function of distance R for each of the earthquake intensities.

Frequency of Seismic Events

A frequency distribution as a function of magnitude was developed for each site. The distribution of earthquakes according to their magnitudes is given by the recurrence relationship (Reference 5):

$$\log N = a - bm_i ,$$

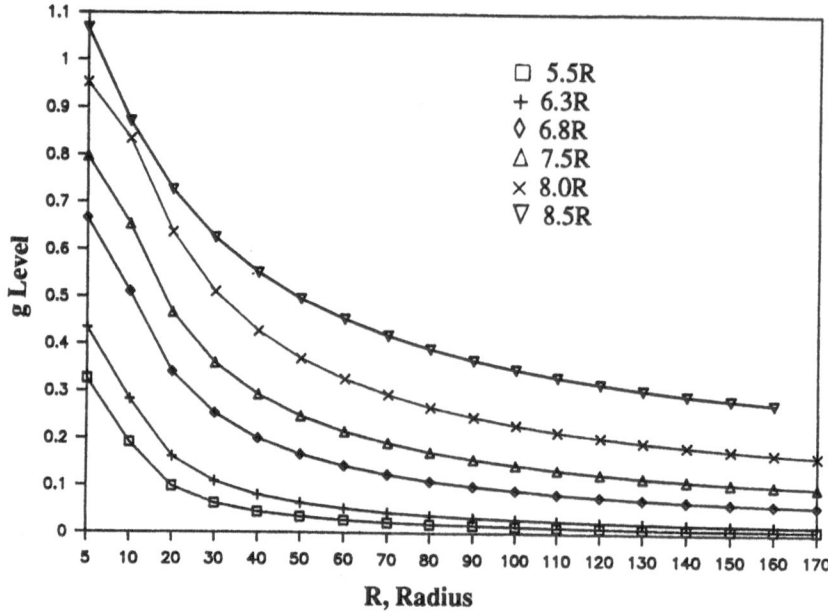

Fig. 1. g-Level vs Distance.

where N is the number of earthquakes with Richter magnitude equal to or greater than m_i for a given source and over a given interval of time. A linear regression analysis of the frequency of exceeding each earthquake magnitude was performed; this indicated that the frequency distribution is essentially a straight line on a logarithmic scale. An allowance for uncertainty is made by increasing the frequency by 20 percent. The frequency distribution for each facility is given in Fig. 2, which shows frequency (seismic events/year) versus the magnitude of the earthquake. Curves are shown for both facilities (FAC1 and FAC2).

Probability of Earthquake Damage

As can be seen in Fig. 1, not all earthquakes will result in g-levels equivalent to the ground acceleration capacity of the critical failures. In order to account for the possible damage from any earthquake of magnitude greater than 5 (Richter Scale), the probability of each earthquake magnitude over an area of radius 60 km was assessed as occurring anywhere within that area. As an example, an earthquake of magnitude 8 anywhere within this area would result in an increasing g-level, depending on distance from the epicenter, but would not induce maximum damage everywhere within the area. For each magnitude, the probability of an unacceptable g-level was found by

$$\frac{\text{Area causing damage}}{\text{Total area in which frequencies were estimated}} = \frac{r^2}{60^2},$$

where r is the radius around the facility within which an earthquake of given magnitude must have its epicenter in order to cause the postulated damage. This assumes that earthquake sources are randomly located within the defined area. The critical radius was found for each intensity and the probability of damage calculated.

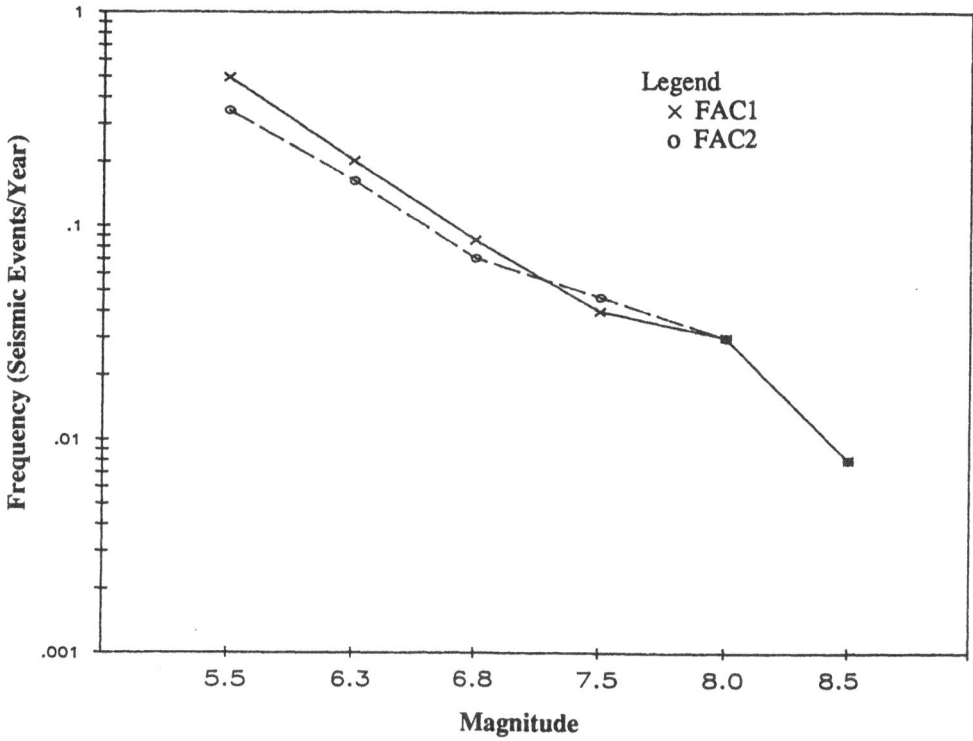

Fig. 2. Frequency of Exceedence vs. Magnitude.

Frequency of Earthquakes Inducing Damage

The probability of each earthquake magnitude resulting in g-levels of 0.1 to 0.9 was calculated and multiplied by the frequency of each magnitude at each g-level. The contributions from each earthquake magnitude were then summed to obtain the total frequency of reaching each g-level. Figure 3 shows the frequency (events per year) of exceeding each g-level (from 0.1g to 1.0g) for each site facility (FAC1 and FAC2).

RESULTS

The frequency per year for the lowest ground acceleration capacity (g-level) is reported for each site. The frequencies are equivalent to the random chance in x years that one earthquake of any intensity equal to or greater than a Richter intensity of about 5.5 could cause the listed failure mode. This is a random chance and as such does not predict when such an event could occur. It should be noted that this analysis is based on peak ground accelerations and no analyses of resonance earthquake frequencies were addressed.

Area 5 in Table 1, roof columns failure, has the lowest seismic ground acceleration capacity for Facility 1. This is postulated to cause major damage to the building and to be the controlling failure mode for major capital investment loss. Total destruction is postulated at a 0.8 g-level. The estimated frequency of these two failures are:

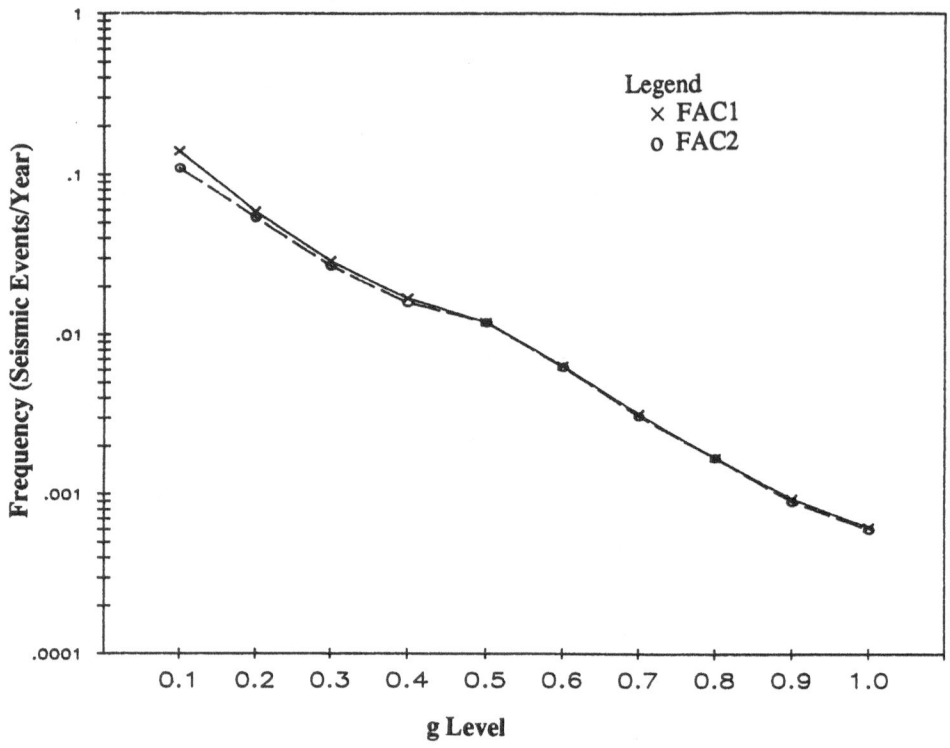

Fig. 3. Frequency of Exceedence vs. g Level.

Facility 1

Failure Mode	g-level	Frequency	Chance
Roof columns fail	0.5g	1.2E–2/year	1/83 years
Total destruction	0.8g	1.70E–3/year	1/588 years

Area 5 in Table 2, strut brace for tee bar suspended ceiling fails, has the lowest g-level. This failure would not result in significant damage if the equipment within the area were adequately protected from "architectural" ceiling material. Failure of the floor columns is postulated to cause major damage to the facility.

Facility 2

Failure Mode	g-level	Frequency	Chance
Suspended ceiling fails	0.3g	2.7E–2/year	1/37 years
Floor columns fail	0.9g	9.0E–4/year	1/1000 years

The results of this study will be utilized to determine the financial risk at these facilities and any major structural changes to further protect the building structures. This type of analysis was also applied as part of a risk study to a chemical facility. The seismic-induced failure of chemical holding tanks was determined to be a dominant contributor to public risk.

REFERENCES

1. "IEEE Recommended Practice for Seismic Qualification of Class 1E Equipment for Nuclear Power Generating Stations," IEEE Std 344-1987.
2. W. S. LaPay, B. A. Bishop, and S. C. Chay, "Reserve Strength as a Measure of Fragility," *Proceedings of the Workshop on Seismic and Dynamic Fragility of Nuclear Power Plant Components*, NUREG/C-0070, BNL-NUREG-51924, August 1985.
3. "PRA Procedures Guide, Section 11.2 Seismic Risk Analysis," Vol. 2, NUREG/CR-2300, U.S. NRC, Washington, DC, January 1983.
4. "Earthquake History of the United States," Publication 41-1, U.S. Department of Commerce, Washington, DC, 1973.
5. Gutenberg, B., and C. F. Richter, "Earthquake Magnitude, Intensity, Energy and Acceleration," *Bulletin of the Seismological Society of America* 32:163-191 (1942).

Catastrophic Damage from Dam-Break Floods

L. Douglas James and Melanie L. Bengtson

Utah Water Research Laboratory
Logan, UT

ABSTRACT

Technological advances that improve social welfare also expose large numbers of people to disastrous technological failures. The impacts of historical natural disasters show a scale effect that magnifies losses from a disaster (property scale) to a catastrophe (community scale) to a super catastrophe (regional scale). For example, destruction of a community causes much more damage than can be accounted for by the additional houses destroyed. One reason is that larger events lessen society's capacity for emergency response. Damage to construction industry capacity delays repair, disrupts markets, and adds to the cost of recovery. A second and more telling reason is that a community itself is an entity of value. Harms to community institutions, functions, and spirit cause continuing economic and social losses. A literature review was used to conceptualize the dimensions of disaster loss magnification, identify threshold indicators of magnification, and approximate the order of magnitude of magnification from the worst conceivable dam-break scenario.

KEYWORDS: Flood damage, dam failure, natural disasters, social impact, flood loss

PROBLEM STATEMENT

Technologies that in normal times enhance social welfare may expose people to catastrophic losses during extreme events. As an example of a technology with a low-probability/high-consequence impact, a large dam that has supplied water, power, and flood benefits to many people for years might suddenly fail and cause far larger losses than any catastrophe yet experienced in the United States.

Flood losses would accrue as rising waters destroy property, damage buildings, and disrupt land uses. Procedures are well established (Water Resources Council, 1980) to estimate flood losses based on property replacement, building repair and land use substitution from market costs subject to maximums of equaling values in use (Milliman, 1984). However, the results can grossly underestimate losses when markets, the repair industry, and transportation systems are disrupted. The reason is that when a few properties are flooded, locally available capacities are adequate to replace, repair, and relocate. Large floods exceed these capacities. As a result, a dam break flooding 10,000 homes inflicts more than 1000 times the loss caused by a thunderstorm flooding 10 homes.

Relatively little work has gone into estimating this "damage magnification." We do not understand the causative processes, let alone how to estimate their magnitude during a given catastrophe. The authors undertook an exploratory study (James *et al.*, 1986) to review magnification during historical catastrophes, search the literature for useful constructs, identify indicators of magnification situations, approximate the magnification likely during the worst conceivable dam-break scenario, and suggest useful research directions.

LOSS ESTIMATES FROM IMPACT-RECOVERY PROCESSES

Flood wetting causes damage during a relatively short inundation period. The losses continue through negatively reinforcing feedback mechanisms (Forrester, 1969) over an aftermath period. Wetted goods become unsalvageable, and wetted facilities deteriorate, requiring more costly repair. Recovery begins as positively reinforcing feedback mechanisms stop further losses (salvage goods, dry and clean facilities, make arrangements for dislodged activities) and eventually restore or even improve the community.

For quantifying losses to goods, facilities, and land uses, one can consider the relationship

$$L = S_c^* A_s + U_c^* R_c^* A_p + \int_0^{T_R} L_c^* A_u \ dt \ , \tag{1}$$

where

S_c = Unit cost of replacing stored goods.
A_s = Amount of stored goods to be replaced.
U_c = Unit cost of repair work.
R_c = Damage per facility.
A_p = Number of damaged facilities.
L_c = Cost of relocating activities and extra cost of conducting them in their temporary environment.
A_u = Amount of land use activity disrupted.
T_R = Time until the repairs are completed.

All three terms on the right side of Eq. 1 contain a unit value (S_c, U_c, and L_c), an index of the extent of flooding (A_s, A_p, and A_u), and a measure of the extent of damage (1, R_c, 1). Unity is used where complete loss is expected. A_s, A_p, A_u, and R_c would obviously be much larger during a dam-break flood but amounts can be estimated by known methods. The unknown is the increase to expect in unit values. The three terms will increase by different amounts and for different reasons.

First, S_c would increase when the goods that have to be purchased exceed the amount the local market can supply within an acceptable time frame. The replacement goods would have to be specially made or shipped at extra cost.

Second, repair cost increases when labor, material, and equipment requirements exceed what the local repair industry can provide. Two factors apply. With respect to the amount of repair needed, R_c has an initial value of R_i (Fig. 1) immediately after the flood (dependent on flood depth, velocity and sediment load) and increases over time at a deterioration rate D_r caused by rust and decay. With respect to unit repair cost (U_c), the urgent repair demand and limited supply of labor and materials raise prices and attract less capable people doing inferior work.

The time required to effect the repairs includes a time to mobilize the repair forces (M_t) and a time to complete the repairs at rate R_r. The repairs will be completed at time T_R (Fig. 1) at cost:

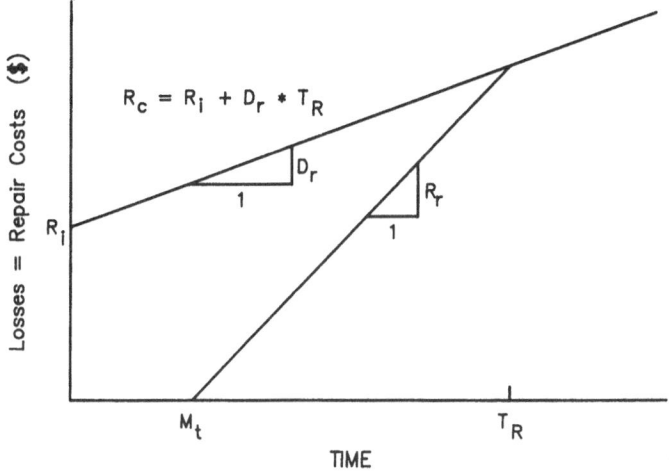

Fig. 1. Flood Losses Caused by Repair Costs.

$$R_c = R_i + D_r * T_R. \qquad (2)$$

Consequently, R_c is increased by forces that increase D_r or M_t or decrease R_r, all attributes of a very large flood.

Third, the cost of substituted uses (D_c) tends to increase over time as longer term provisions become more expensive. Accordingly,

$$D_c = f(M_a, t), \qquad (3)$$

where M_a is the land-use activity mobility. With this formulation, the loss L_c is the area under the D_c v. t curve to time T_R and thus also increases with factors that increase R_c.

In summary, the unit losses are increased by four factors:

1. Greater demands for replacement goods and repair labor and materials that bid up costs in a flood disrupted market (S_c and U_c).

2. Slower mobilization and repair by a flood-disrupted repair industry (M_t and R_r).

3. The reduced mobility of land-use activities with a flood-disrupted transportation system (M_a).

4. A longer recovery time (T_R) exacerbated by greater damages (larger R_i and D_r).

All four factors have greater effects during larger floods.

PROCESS SCALES

Yancey *et al.* (1976) found the cost and duration of repair to depend on a "repair capacity." Local repair forces have set amounts of labor, equipment, and materials. If the damages are small compared to this capacity, the repair can begin quickly (M_t) and proceed efficiently (R_r). If the damages exceed the capacity, prices (U_c) are bid up and a longer duration (T_R) is required to complete the job. A quantum increase in unit damages thus occurs and can be considered a scale effect with repair capacity the indicator of movement to the next larger scale.

Scale effect analogies occur in physics, meteorology, and hydrology. In physics, for example, subatomic, atomic, pore, basin, continental, and planetary studies consider different processes while recognizing that other scales are operating at larger and smaller levels. Socio-economically, the damage processes that affect the functioning of an individual or building are set within processes that affect the functioning of a community, an industry, or a market. Both of these are set within damage processes over a larger region that obstruct restoration forces from moving into damaged areas.

The physical sciences suggest that larger scale processes are slower. For example, in physics, electrons orbit atoms in infinitesimal fractions of seconds and stars hardly change positions in millennia. With respect to floods, this longer duration at larger scales increases the losses. In addition, the longer time frames extend the dynamic feedback mechanisms (negative and positive), increase uncertainty, and spread the probability distributions of potential impacts. The largest scale losses are the most uncertain, and aversion to major uncertainty intensifies utility (if not damage) magnification.

In this exploratory study, the property, community, and regional scale events are termed flood disasters, flood catastrophes, and flood super catastrophes. A magnification factor is defined as a multiplier of normal (disaster) damages to estimate catastrophe damages. A super magnification factor is multiplied by the normal damages to estimate the damages caused by a super catastrophe.

Use of this concept requires an index to judge the scale of a particular event. The ability of a community, industry, or market to respond to flood damage depends on its capacity to provide emergency services, supporting infrastructure, and repair goods and services. A major flood simultaneously burdens and reduces this capacity. The burden can be estimated from the "normal" losses, and the reduction can be estimated from the losses to the repair industry. The ratio of the burden to the reduced capacity offers a starting index.

DATA BASE

Empirical observations of natural catastrophes were used as a source of ideas to identify magnification causes and amounts; the principles and standards for water resources planning (Water Resources Council, 1980) provided a check list for the search. Contemporary descriptions and later objective assessments of dam failures and other natural disasters were reviewed for physical impacts and their resultant consequences (James et al., 1986). The reviews emphasized catastrophes occurring in the United States in recent years to match present technological and socio-economic conditions and to minimize cultural, economic, and transportation network differences across international boundaries.

Altogether, about 3500 lives have been lost in the 12 historical dam breaks large enough to take lives in the United States (Baldewicz, 1984). The largest, in Johnstown, PA, took 2209 lives. The 12 dam breaks divide between six older ones between 1874 and 1928, 35 years without a life-taking dam break, and six recent ones between 1963 and 1977. The institutional factors contributing to the grouping in two distinct historical periods deserve consideration.

INDIVIDUAL   COMMUNITY   REGION

Fig. 2. Process Scales.

A variety of major physical disasters that included dam-break floods gave useful information for understanding larger scale processes. Twelve events probed in detail for this study were:

1. The Johnstown, Pennsylvania, Dam-Break Flood of May 31, 1889.
2. The Galveston, Texas, Hurricane, September 8, 1900.
3. The San Francisco, California, Earthquake and Fire, April 18, 1906.
4. Failure of Saint Francis Dam, California, March 13, 1928.
5. Lynmouth, England, Cloudburst Flood, August 15, 1952.
6. Hurricane Camille, Mississippi Coast, August 18, 1969.
7. Floods in Bristol England, July 11, 1970.
8. Hurricane Agnes, Pennsylvania, June 20-26, 1972.
9. Buffalo Creek Dam-Break Flood, West Virginia, February 26, 1972.
10. Rapid City, South Dakota, Cloudburst and Dam-Break Flood, June 9, 1972.
11. Australian Floods of 1974.
12. San Francisco Bay Region, California, Floods and Debris Flows, January, 1982.

DIRECT IMPACT PROCESSES

The collected evidence showed that scale magnification occurs for many impacts besides economic damages. The initial review (James *et al.*, 1986) of the historical data suggested nine principal processes with impact magnification, two of which (disturbing experiences and mental health) were later joined. Environmental processes are also discussed here because of their current importance, even though environmental impacts received little note during the historical disasters.

These processes fall into three groups. The first consists of disruptions to emergency services that would reduce subsequent losses if executed efficiently. In the second group, flood events harm capital investment, community infrastructure, health care services, and environmental systems. Each harm must be repaired following the general pattern of impact and recovery outlined for economic losses in Figs. 1 and 2, and each repair has its own T_R. The third group includes social and psychological processes whose impacts continue for durations determined by recovery of the second group processes but at magnitudes determined by independent factors. All nine processes in the three groups are discussed below.

Group 1: Provision of Emergency Services

Prompt emergency actions can substantially reduce losses. People are helped in reaching places of safety, retrieving property, and securing food and shelter until normal living conditions are restored. Major floods disrupt communications and transportation and slow these services. Important rescues may be missed. The management structure and supplies for emergency action may be disrupted. The impact of these inefficiencies is to increase R_i and D_r (Eq. 2), extend T_R, and increase S_c (unless property is salvaged). Market stresses increase unit costs.

Group 2: Processes Involving Direct Repair

A smoothly functioning economy produces goods and services in the private sector to satisfy human needs. A flood disrupts the economy by directly destroying some goods, reducing capacity to produce other goods and services, and forcing some activities to cease or move elsewhere. The local economy, through market processes, mobilizes and imports construction capacity to replace goods, repair facilities, and restore activities. At some threshold impact level, costs increase in disrupted markets, and disruptions in the repair industry increase mobilization time (M_t) and slow repair work (R_r). The threshold for this scale change can be indexed by the ratio of disaster-induced repair demand to disaster-reduced repair capacity. The normal capacity can be indexed by the annual construction activity in the economic trade area (Yancey *et al.*, 1976). The capacity reduction can be estimated from flood impacts on construction equipment and building materials in the economic trade area.

The community infrastructure provides public services such as electricity, phones, water, access, stores, etc. A flood disrupts this infrastructure, forcing local governments and utilities to mobilize and import means to provide temporary service and restore regular service. Their M_t and R_r are slower, expenses increase, and the longer T_R for infrastructure repair also increases the cost of dependent economic restoration. The threshold at which magnification begins can be indexed by the ratio of infrastructure repair demand to capacity in the area where local utilities can draw for help.

A community provides a different sort of service in filling health and safety needs with rescue teams, hospitals, doctors, medical supplies, etc. A flood leaves people injured, trapped and requiring rescue, sick from prolonged exposure, and dying. Afterwards, threats to health linger (D_r) from contaminated water supplies and insects. Floods that exceed a threshold magnitude divert the first attention of local health departments to mobilizing and importing temporary help to restore services. During this period, their M_t and R_r are slower, and expenses are larger. The longer T_R also slows economic and infrastructure restorations. The threshold for the scale change can be indexed by the ratio of health care demand to capacity in the local health care region.

A flood disrupts the community's environment by destroying habitats, especially those concentrated in riverine environments, and creates ecological imbalances whose impacts continue after the waters recede. Ecological processes will begin restoration, but their ability to do so depends on how the damage taxes the natural restorative capacity. Analogous techniques based on Eq. 1 can be applied to estimate these impacts too.

Group 3: Processes with Dependent Durations

Recovery from the social and psychological impacts of major flooding depends on the rate of direct repairs of the Group 2 processes and on interactions with all the other processes. Four major sources of social and psychological impact were identified.

People form social groups, including family and friends, for help and fellowship throughout their lives. Rural communities emphasize neighbors; urban social groups are more scattered spatially. A flood causing people to vacate a disaster area separates them from social support when they particularly need it. The amount of loss depends on the extent of the social disruption, and a scale effect emerges when significant numbers of people begin to vacate. The duration of the loss depends upon the repair times (T_R) for buildings, infrastructure, health services, and environment.

Mental health is disturbed by flood experiences, including seeing death or injury, losing homes and possessions, exposure to dampness and cold, and the discomforts of makeshift arrangements and restoration responsibilities. The loss rate depends on the experiences, and a scale effect emerges with the disruption of health care services and social group support.

Another process that can disturb mental health is loss of personal time. People order their personal time to fulfill obligations and enjoy leisure. A flood diverts countless hours to making temporary arrangements for work and living, cleaning and restoring damaged property, dealing with repair people, and making financial arrangements. The unfamiliar activities during a period of general disruption add frustrations and costs. Disrupted infrastructure and community services increase and extend these difficulties.

Loss of personal property can also impact mental health and personal time. Post-flood interviews have shown that people often feel losses of personal treasures and records more deeply than their monetary losses. Possessions have unique values to their owners even when worthless to others. These losses are greatly reduced by quick recovery and rehabilitation of the items in question. The suddenness of a dam break event makes it more difficult to flee with valued items, the extent of the devastation makes it more difficult to find them, and losses to infrastructure slow the return and search.

INDIRECT IMPACTS

In addition to direct impacts, a major dam failure may indirectly affect water resources planning. Such a disaster is bound to make designs more costly and conservative, slow approvals, and constrain water resources management programs. Funds will be harder to raise on bond markets. Environmental opposition will increase, people living downstream from proposed dams will express their fears more vocally. Political opposition will increase, and stronger liability laws may further constrain development.

COMMUNITY-SCALE REPAIR

The above analysis examines the relationship between the people impacted by flooding and damages impairing the functioning of community-level institutions. At a still larger scale, an analysis could examine the relationship between impacted communities and regional-level institutions. Example factors include availability of regional help to provide emergency services in flood-ravaged communities; regional assistance in supporting damaged or destroyed local repair markets, infrastructures, or health services; and restoring the community as an efficient entity serving the needs of its citizens.

FIRST APPROXIMATION OF DIRECT MAGNIFICATION MAGNITUDE

Based on rough data, a dam failure with an impact on the local repair industry slightly larger than that of Hurricane Agnes (Yancey *et al.*, 1976) would have a magnification factor on economic losses of about 6. Because of governmental control, one would expect less magnification for the other processes involving direct structural restorations. Greater magnification is expected in the four social and psychological processes (perhaps as high as 40) due to the volatility of reinforcing negative impacts and slow recovery of these processes. Indirect effects add a new and potentially much larger dimension of loss.

LOCATION ANALYSIS

The magnification associated with failure of a specific dam could be estimated by mapping the inundated area and determining the damage to property and to community markets, repair industry, infrastructure, and health services. This information could then be used to assess the probability of a scale effect and employed in an analytic approach to estimating losses utilizing Eq. 1.

SUMMARY AND RESEARCH RECOMMENDATIONS

The above constructs, hypothesized from rough data, match empirical historical evidence and suggest that impacts at large scales will add substantially to flood losses during a major dam-break catastrophe. More and more carefully collected data are needed. Whenever large physical disasters occur, information should be systematically collected and analyzed. Research is also recommended on how markets disrupted by catastrophic events recover afterwards. The planners' ideal would be a multiple-feedback simulation model that represents event, aftermath, and recovery interactions among the nine processes to estimate the consequences of events reaching catastrophic and super catastrophic levels.

ACKNOWLEDGMENTS

The study that provided the basis for this paper was supported by the Hydrologic Engineering Center, U.S. Army Corps of Engineers, in cooperation with ASCE Task Committee on Spillway Design Flood Selection. Much of the empirical data was compiled from original records by Al Hussan Sumani.

REFERENCES

Baldewicz, W. L., 1984, "Dam Failures: Insights to Nuclear Power Risks," in *Low-Probability/High-Consequence Risk Analysis*, Waller, Ray A. and Vincent T. Covello, Eds., pp. 81-90, Plenum Press, New York.

Forrester, Jay W., 1969, *Urban Dynamics*, M.I.T. Press, Cambridge, MA.

James, L. Douglas, Sumani, Al Hussan, and Bengtson, Melanie L., 1986, "Catastrophic Damage from Dam-Break Floods," Utah Water Research Laboratory, Utah State University, Logan, UT.

Milliman, J. W., 1984, "A Needed Economic Framework for Floodplain Management," *Water International* 9(3).

Water Resources Council, Part II, 1980, "Proposed Rules: Principles, Standards, and Procedures for Planning Water and Related Land Resources," Federal Register, Monday, April 14.

Yancey, T. N. Jr., L. Douglas James, D. Earl Jones, and Jeanne Goerdert, 1976, "Disaster-Caused Increases in Unit Repair Costs," Proceedings of the ASCE, *Journal of the Water Resources Planning and Management Division* 102(WR2):265-282.

Risk Reduction Strategies for Setting Remedial Priorities: Hazardous Waste

Alfred Levinson

New Jersey Institute of Technology
Newark, NJ

ABSTRACT

The usual approach to overcoming community opposition to the siting of a hazardous waste facility is to offer compensation to the host community. Most communities have refused to serve as hosts to hazardous waste facilities regardless of the amount of compensation. A risk-risk approach is proposed as an alternative where the increased risk to the community as a result of the new hazardous waste facility is offset by a reduction in risk by the hazardous waste generators in the community. This could be accomplished by a reduction in the generation of hazardous waste or toxic emissions by these firms. A multiobjective programming model is proposed to evaluate this alternative policy.

KEY WORDS: Risk-risk, hazardous waste, multiobjective.

INTRODUCTION

Increased concern over the risks from environmental pollutants has led communities to oppose the siting of various types of waste facilities within their boundaries. Various compensation schemes have been proposed to make communities more amenable to accepting waste facilities, but these usually have meant trying to identify the amount of compensation the host community would require to drop opposition. In most cases, the communities have refused to accept these facilities regardless of the amount of compensation offered.

This paper proposes an alternative to the usual compensation schemes which may make hazardous waste facilities more acceptable to host communities. Called the risk-risk approach, it avoids forcing the communities to place a monetary value on an increase in the risk of death or illness when involuntarily accepting a hazardous waste facility. The risk-risk strategy can also take various forms and be applied to radioactive waste storage facilities.

HAZARDOUS WASTE RISK TRADE-OFF STRATEGIES

Since hazardous waste reduction has recently come to the forefront as a dominant hazardous waste policy, it can now be integrated into a risk-risk strategy combining it with

213

an earlier strategy which was proposed and largely rejected—the siting of hazardous waste facilities in the municipalities where the waste was generated.

During the hearings for hazardous waste siting legislation in some states, there were proposals to require the treatment and disposal of hazardous waste in the municipality where it was generated [Morell and Magorian, 1982]. This was opposed by representatives from industrial communities on the grounds that the risks associated with the production of the products that we all use should be shared by all the people. They claimed that the industrial communities were already bearing most of the risk from the production of these products in the form of increased air and water pollution. This was the "fair share" debate. It was a question of political equity.

These objections can be addressed by combining waste reduction policies with policies requiring the treatment and disposal of the waste in the communities where it is generated. The goal is to offset the increase in risk from the operation of a hazardous waste disposal or treatment facility in the host community with a reduction in the generation of hazardous waste by companies in that community, which would reduce the current risk from the generation of hazardous waste faced by the people in the community.

This combination of policies means that when a hazardous waste facility is sited in a municipality there will be no overall increase in risk to the people in that community from the hazardous waste. Since this will force some of the hazardous waste generators to reduce their waste, which might increase their costs of operation, they might be compensated by the owner of the hazardous waste facility. The cost of compensating the firms that reduce their waste would be passed on to the companies that ship their waste to the hazardous waste facility. Any companies that find their costs of doing business in that municipality have become too high may choose to relocate elsewhere. The community would thus lose the jobs, income and taxes generated by that business. These are the trade-offs that would have to be considered by that community when they are deciding the extent to which hazardous waste must be reduced by the community's firms. This gives the community full control over the generation, treatment and disposal of hazardous waste, the risks associated with that waste and the trade-offs between the risks from the waste and the benefits derived from the hazardous waste generators doing business in that community.

This is similar to the bubble policy used by the Environmental Protection Agency (EPA) in the control of air pollution. If it would give the community greater flexibility and make it easier to control pollution and hazardous waste, the bubble policy might be expanded to include water pollution and hazardous waste. Thus, firms could elect to reduce their emission of air pollutants instead of their hazardous waste if it would cost less.

This risk-risk approach can be applied to other types of wastes. An example where this was applied was the acceptance by Oak Ridge, Tennessee, of the Department of Energy's (DOE's) monitored retrievable storage facility (MRS) to consolidate, package, and temporarily store spent nuclear fuel until a permanent storage repository is constructed. In addition to monetary compensation, acceptance was predicated upon the cleanup of various environmental problems at existing DOE facilities in Oak Ridge [Sigmon, 1987]. Although Oak Ridge did not view the siting of the MRS there as posing a major health or safety risk, this still represents a risk-risk trade-off—a reduction in the risk from the environmental problems at other DOE facilities there and an increase in risk from the MRS. There is another major difference—the nuclear waste was generated outside Oak Ridge.

This risk-risk approach could also be applied to the problem of siting a permanent nuclear waste storage repository. One possibility is to link the policies governing the permanent storage of nuclear waste with those banning underground nuclear tests. One might include in a treaty banning underground nuclear tests a provision requiring the conversion of the nuclear test grounds into permanent nuclear waste storage repositories.

For the United States, this would mandate the conversion of the atomic test grounds in Nevada into a nuclear waste repository. This should make the siting of the nuclear waste repository in Nevada less objectionable to the people of Nevada. The increase in the risk from the nuclear waste storage repository would be offset by the elimination of the risk from underground nuclear tests.

The hazardous waste risk trade-off policies suggest two types of models. The first, which was introduced above, assumes that all hazardous waste generated in a region will be treated and disposed of in that region and that there will be no increase in risk to the region as a result of the siting of a hazardous waste facility. The second model takes a shared risk approach. Communities or regions can share in the treatment and disposal of the hazardous wastes generated in their regions. Two regions, A and B, for example, may form a compact to share in the incineration and landfilling of the hazardous wastes generated in their regions. Assume each of the two regions generates hazardous wastes which require both incineration and landfilling. Under the compact, Region A would house the incinerator and accept hazardous waste for incineration from both regions A and B. In turn, region B would contain the landfill and accept hazardous waste for landfilling from both regions A and B. This model can easily be integrated with the first model.

THE REGIONAL SELF-CONTAINED HAZARDOUS WASTE DISPOSAL MODEL

This is a multiobjective programming model where risk minimization and cost minimization (or profit maximization) are the objectives. All hazardous waste is treated and disposed of in the region. It is assumed that there will not be a net increase in risk to the region from the siting of the hazardous waste facilities, i.e., the reduction in risk from the reduction in the generation of hazardous waste will be at least as great as the increase in risk from the disposal or treatment of the hazardous waste in the region (see Fig. 1).

There are T treatment or disposal methods which include landfilling, incineration and solvent recovery. There are M major generators of hazardous waste. There are three sizes of facilities for each type of treatment or disposal method. A simple version of the model is as follows:

Let:

$$X_i^t \quad = \quad \text{the amount of waste of type } t \text{ generated by firm } i, \ i\epsilon M, \ t\epsilon T,$$

$$W_i^t \quad = \quad \text{the amount of waste of type } t \text{ reduced by firm } i, \ i\epsilon M, \ t\epsilon T,$$

$$F_j^t \quad = \quad \text{facility of type } t, \text{ size } j, \ t\epsilon T, \ j = 1,2,3,$$

$$L_j^t \quad = \quad \text{minimum operating level of } F_j^t,$$

$$U_j^t \quad = \quad \text{maximum operating capacity of } F_j^t,$$

$$R_i^t \quad = \quad \text{cost of reducing waste of type } t \text{ at firm } i,$$

$$C_j^t \quad = \quad \text{cost of treating/disposing waste type } t \text{ at facility size } j,$$

$$g_i^t(Z^t) \quad = \quad \text{risk from the generation of waste type } t \text{ at firm } i,$$

$$h_i^t(Z^t) \quad = \quad \text{risk from the treatment/disposal of waste type } t \text{ at facility size } j,$$

$$Z^t \quad = \quad \text{waste of type } t.$$

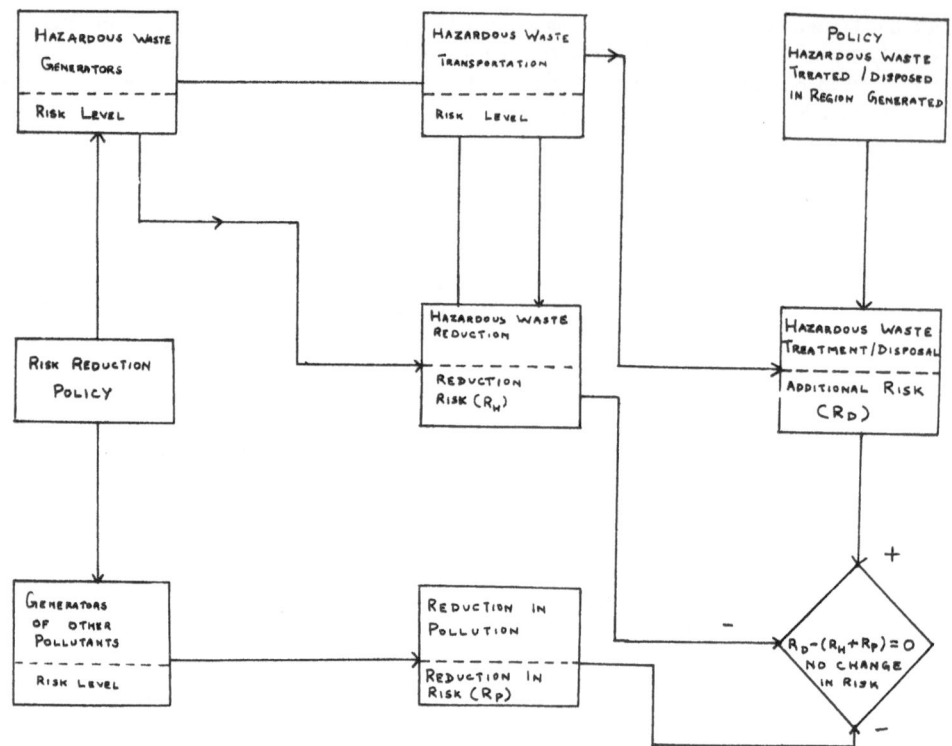

Fig. 1. Zero Risk Increase Regional Self-Contained Hazardous Waste Treatment/ Disposal System.

Objectives:

Cost minimization,

$$\min \left[\sum_j \sum_t \left(C_j^t \sum \bar{X}_i^t \right) + \sum_t \sum_i R_i^t W_i^t \right] .$$

Risk minimization,

$$\min \left[\sum_t \sum_i g_i^t (\bar{X}_i^t) + \sum_t \sum_j h_j^t \left(\sum_i \bar{X}_i^t \right) \right] .$$

Basic Constraints:

$$X_i^t - W_i^t = \bar{X}_i^t. \tag{1}$$

$$\sum_i \bar{X}_i^t \leq F_j^t , \; j = 1, 2, 3. \tag{2}$$

$$L_j^t \leq F_j^t \leq U_j^t , \; j = 1, 2, 3. \tag{3}$$

$$\sum_t \sum_j h_j^t (\bar{X}_j^t) - \sum_t \sum_i g_i^t (W_i^t) \leq 0. \tag{4}$$

Additional constraints:

    Transportation constraints
    Job loss constraints
    Siting constraints
    Environmental constraints

Note: Job loss minimization can also be added as an objective.

Three trade-off schemes will be used to test the sensitivity of the model. They will be presented as alternative policy options facing the region: (1) health risk aversion relative to jobs and income, (2) greater concern over potential loss of jobs and income relative to possible health risks, (3) balance between health risks, and jobs and income.

This model could be expanded to include the minimization of the risk from air pollutants and water pollutants, along with the minimization of the risk from hazardous waste. This would be an expansion of the bubble principle.

If sufficient data on the risk from hazardous wastes exist, it may be possible to develop a more sophisticated model using the partitioned multiobjective risk method [Asbeck and Haimes, 1984].

SHARED RISK HAZARDOUS WASTE FACILITY SITING MODEL

This is a multiobjective program with a bargaining game. The objectives are cost minimization and risk minimization for the whole state. There are N regions in the state. Each region contains firms which generate T types of hazardous waste. Each region can choose to site T hazardous waste facilities or to form coalitions with other regions to share the treatment and disposal of hazardous waste generated in their regions (see Fig. 2).

This can be taken as an extension of the Regional Self-Contained Hazardous Waste Treatment/Disposal Model where the objective for the coalition is cost minimization and risk minimization with the requirement that there be no net increase in risk to the coalition as a result of the treatment or disposal of hazardous waste in the regions represented in the coalition.

This is similar to the Coal Logistics and Environmental Control (CLEC) model, which was used to develop noninferior solutions for utility coal conversions in several regions in New England and to determine optimal coalitions which would optimize the cost of conversion and the environmental impact [Ratick, 1983].

CONCLUSION

Two models have been presented here to evaluate alternative risk reduction strategies for setting remedial priorities. They were applied to the specific case of hazardous waste treatment and disposal. The Regional Self-Contained Hazardous Waste Treatment/Disposal Model enables policymakers to examine the trade offs between the risks from the treatment and disposal of hazardous waste and possible losses in jobs and income within a community. The Shared Treatment Model, which is an extension of the Regional Self-Contained Model, allows policymakers to evaluate the possibilities of various communities forming compacts to share the risks from the treatment and disposal of hazardous waste and the potential loss of jobs and income.

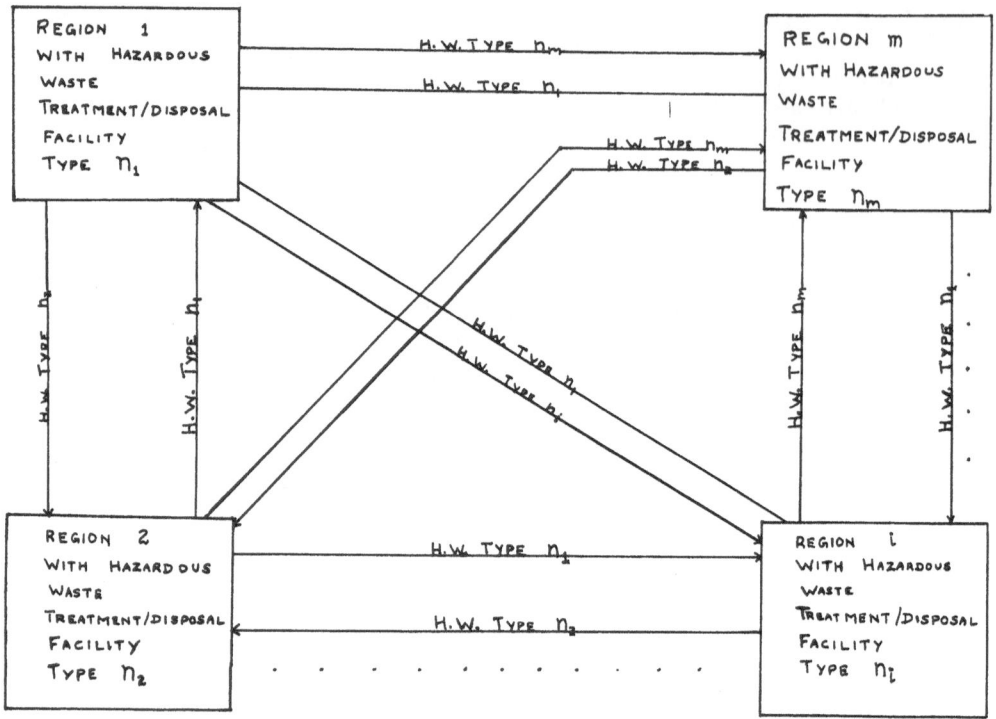

Fig. 2. Shared Treatment/Disposal of Hazardous Waste Among Regions.

The application of these risk-risk models may make it possible to resolve the impasse that faces most hazardous waste policy decisionmakers. It also demonstrates the need to link what may appear to be unrelated policies to produce innovative solutions to some of the complex and apparently unsolvable problems facing policymakers.

REFERENCES

Asbeck, E., and Haimes, Y., 1984, "The Partitioned Multiobjective Risk Method," *Large Scale Systems* **6**:13-38.

Morell and Magorian, 1982, *Siting Hazardous Waste Facilities*, Ballinger, Cambridge, MA, pp. 154-158.

Ratick, Samuel J., 1983, "Multiobjective Programming with Related Bargaining Games," *Regional Science and Urban Economics* **13**:55-76.

Sigmon, E. Brent, 1987, "Achieving a Negotiated Compensation Agreement in Siting: The MRS Case," *Journal of Policy Analysis and Management* **6(2)**:170-179.

Educating the Public About Toxicological Risk

Michael A. Kamrin

Michigan State University
East Lansing, MI

ABSTRACT

In view of the increasing public awareness of environmental contaminants and the increasing concern about the risks to health that they pose, it is essential that the level of public education about the assessment and management of such risks be greatly expanded. This education can be achieved by providing programs for intermediaries, e.g., reporters or local health personnel, as well as for the general public. During the past five years, the Center for Environmental Toxicology at Michigan State University has developed a number of approaches to provide such education to this variety of audiences. These include written materials, workshops, symposia, courses, computer materials, an inquiry-response system and a community assistance program. This paper will describe these efforts and discuss the way they contribute to increasing public understanding of toxicological risk.

KEYWORDS: Risk communication, risk education, risk information

INTRODUCTION

In the years since the end of World War II, the public has become aware of a succession of environmental problems that have posed potential and actual risks to health. These include the air pollution incidents at Donora, Pennsylvania, and London, England, which led to many deaths and illnesses and the contamination of the food chain by halogenated hydrocarbons, such as PBBs and PCBs, which may yet result in long-term human health effects. The number and diversity of these contamination incidents has increased as more and more effort has been applied to discovering sites of past contamination and monitoring the present environment, especially drinking water and indoor air. Unfortunately, the increased awareness of the public of such events has not been matched by an increased understanding of the magnitude of the risks associated with such events. As a result, there is a great deal of fear and it is difficult to arrive at public policy which reflects the severity of real situations rather than the perceptions that people have.

The Center for Environmental Toxicology at Michigan State University was established in 1978 in response to a number of contamination incidents in Michigan in the 1970s, especially the PBB contamination of the food chain. The Center was created to coordinate the response of the scientific community to such incidents. As with other university units, it is responsible for research, education and service. However, in distinction to many other units, the service function is an integral part of the overall program. This function is carried out through an outreach program which has several components, including publications, presentations, and projects. The communication of risk is an essential feature of all of these components. This article will describe how the outreach programs are structured and how they educate the public about toxicological risk.

INQUIRY-RESPONSE SYSTEM

One of the most fundamental aspects of outreach is the inquiry-response system. Staff respond daily to telephone inquiries and letters about myriad environmental concerns. These have changed over time, but generally involve a limited number of different problems and types of chemicals. Pesticides are of most concern, and the problem of most interest is the possible health effects of exposure to such chemicals. Related questions deal with removal of unwanted pesticides that may have been misapplied and the proper disposal of household pesticides so that they do not move through the air, soil, or water and result in human exposures. Health effects of groundwater contamination and chemicals in indoor air are also subjects of frequent questions.

In responding to all of these inquiries, staff members try to provide some education as well as an answer. It is important for the inquirer to understand why the response is being given as well as what the answer is. For example, the risk may be put into some perspective by comparison of ambient levels with workplace levels and frequencies of exposure or with doses which were administered to laboratory animals. Thus, each person who contacts the Center provides an opportunity for education about toxicological risk.

COMMUNITY ASSISTANCE PROGRAM

On a broader level, the Center has recently initiated the Community Assistance Program in Environmental Toxicology (CAPET), which is designed to provide assistance to communities that are experiencing environmental toxicology problems and do not have the resources to address them. These might include groundwater contamination problems or air pollution. In these situations, it is usually a group that is being educated, at least initially. This group often consists of members of the local health department, Cooperative Extension Service, Planning Commission, and Board of Commissioners. Education of this local group is provided in the context of the assistance being rendered, and often is passed on to the public in that community in the course of resolving the environmental concern. In addition, as part of CAPET, the public may receive a balanced view of the local situation at open meetings organized by the Center. Education about risk and risk assessment are integral parts of any assistance rendered through the CAPET program.

EDUCATIONAL PROGRAMS

Through contacts with the public resulting from either the inquiry-response system or the CAPET program, Center staff receive a variety of inputs on the issues which are of most importance to citizens in Michigan. These inputs are utilized in developing publications and presentations which can address these critical concerns. Although public concerns are focused on specific issues, Center staff also develop materials that address the background concepts which are critical to increasing citizens' abilities to understand and respond to both current and future issues. Examples of these are the pamphlet entitled *Toxicology for the Citizen*[1] and presentations on "Basics of Toxicology." The former provides an overview of toxicology, including some discussion about how risks are assessed and standards determined. The presentations provide similar information, but usually in somewhat more depth.

Although the foregoing efforts can reach a large number of citizens, it is necessary to provide education for intermediaries if a significant percentage of the population is to be reached. These intermediaries include members of the media, Cooperative Extension Agents, local health department personnel, state agency personnel, medical professionals and environmental groups. The Center has utilized a variety of avenues in order to reach these groups. One has been to organize conferences and workshops of both limited and broad scope. An example of the former is an annual media workshop that is open only to working reporters. Toxicological risk has been addressed to some degree in each of these workshops and was the focus of the one held in 1987.

An illustration of the broader type of event is the symposium and workshop on Reducing Uncertainty in Risk Assessment, which was held in May, 1987. This brought together scientists, environmental group members, state and local agency personnel, and industry representatives to listen to expert presentations and to discuss the strengths and limitations of current risk assessment procedures. The proceedings of previous conferences on PCBs[2] and dioxins[3] have been published to enlarge the potential audience. In addition to organizing such events, Center staff have assisted in the development of workshops relating to environmental toxicology sponsored by other units, both inside and outside the university.

Center staff have also taken advantage of available opportunities to appear at meetings of groups which are in constant contact with the public and whose members could provide a vital link to citizens. An example is the Michigan Environmental Health Association, which is composed mainly of representatives from local health departments and the state health department. Center staff provided talks on basic toxicology at regional meetings of this group during the past year. Center staff have also addressed the annual meetings of the Michigan State Medical Society and informed physicians about environmental toxicology issues of concern to their patients and how they might best deal with questions relating to risk of environmental contaminants. Center staff have also appeared at numerous meetings of local organizations and provided similar educational messages.

In addition to these less traditional forms of education, the Center has organized basic courses which were presented through the Lifelong Education Programs of Michigan State University. One, entitled "Toxic Chemicals and Human Health," attracted a range of students, including high school teachers, local business people, local health staff, environmental groups' representatives, and state governmental employees. Another, entitled "Toxic Chemicals and the Environment," attracted a similar mix of attendees. The group of participants represented the sort of people who could carry the message to the general public. Thus, more formal education related to risk has also provided an opportunity to educate the public through intermediaries.

INTERACTIVE PROGRAMS

One unique cooperative effort is deserving of attention because it represents a mechanism for public education which has not yet been tapped. This is coordination among universities, local and state government, industry, and environmental groups. In Michigan, a Home Chemical Awareness Coalition was formed about two years ago with just this type of representation. The focus is on the proper use and disposal of household toxic substances and the Coalition has served as a resource for communities which are interested in programs for recycling or disposing of such substances. The work of the Coalition has included the preparation of materials to help in planning such activities and workshops for local officials who might be involved in such efforts. More recently, the group has developed some educational materials aimed at fourth to sixth graders and these will soon be available for distribution to teachers in Michigan. These materials, while practical in nature, also introduce some important concepts which are needed in understanding toxic substances. Thus, the Coalition has provided a way to educate both citizens and potential citizens about risks and how to control them. It also illustrates that cooperation is possible among groups which have a common interest in educating the public about the true risks of toxic chemicals in our society.

A second unique type of interaction has stemmed from the Center's affiliation with Michigan State University's Cooperative Extension Service. Center staff have provided an Extension Toxicology Program in Michigan, a program which is present in only a few other state extension services. Recognizing the commonality of problems across the country, the extension toxicologists in four states, Michigan, New York, California and Oregon, have joined together in cooperative, public education efforts. The main focus of these efforts at present is the production and distribution of Pesticide Information Profiles (PIPs) and associated Toxicology Information Briefs (TIBs). The Profiles provide easily understandable overviews of the human and environmental toxicology of a number of pesticides in common use. The Briefs explain basic toxicological concepts, such as dose-response, in simple language and are designed to help in understanding of the Profiles. Once completed, these materials will be available for distribution in each state through the Cooperative Extension Service in that state. Interstate cooperation represents another largely unexplored potential for public education about toxicological risk.

SUMMARY

Overall, it is clear that there are a number of mechanisms available for communicating toxicological risk education to the public. The Center for Environmental Toxicology has utilized a wide variety of these approaches and has developed materials, both internally and in conjunction with others, which may be useful to others engaged in the same pursuit. It is clear that much more needs to be done and that public policy with respect to toxic chemicals can be improved if toxicological risk education is pursued more vigorously by the scientific community.

REFERENCES

1. A. Marczewski and M. Kamrin, *Toxicology for the Citizen*, 2nd Edition, Center for Environmental Toxicology, Michigan State University, E. Lansing, Michigan (1987).
2. F.M. D'Itri and M. Kamrin, Eds., *PCBs: Human and Environmental Hazards*, Ann Arbor Science (Butterworth), Boston, MA (1983).
3. M. Kamrin and P. Rodgers, Eds., *Dioxins in the Environment*, Hemisphere Publishing Corp., New York (1985).

Upper-Bound Estimates of Carcinogenic Risk: More Policy Than Science

Joseph P. Rieth and Thomas B. Starr

Chemical Industry Institute of Toxicology
Research Triangle Park, NC

ABSTRACT

A major obstacle to the development of reliable quantitative risk estimation models which describe the potential human carcinogenicity of chemicals is the problem of extrapolation between species. One approach has been to define a standardized parameter, carcinogenic potency, which could be used to compare the risks associated with exposure to different chemicals. The Environmental Protection Agency uses a conservative estimate of carcinogenic potency, the upper-bound, rather than the corresponding maximum likelihood estimate. The data presently used to define both the maximum likelihood and upper-bound potency estimates are obtained primarily from high dose, chronic rodent bioassays. These tests provide little data regarding mechanism of action, and the dose range per chemical is usually narrow. Others have demonstrated that a strong correlation exists between maximum likelihood potency estimates for rats and mice which developed tumors during chronic exposure to the same chemical. This has been used to justify the extrapolation from these data to human risk. The high correlation, however, is an artifact caused by two features of the experimental design: (*i*) the small number of animals per dose group in each study and (*ii*) the high correlation between the maximum doses tested in rodents for each chemical. We have illustrated this by showing that the correlation between the maximum likelihood estimates of carcinogenic potency for rats and mice which did not develop tumors is as large as the correlation for those that did ($r=0.79$, $n=82$ and $r=0.83$, $n=83$, respectively). The correlation is also strong between rats developing tumors and mice not developing tumors ($r=0.73$, $n=76$), and mice developing tumors and rats not developing tumors ($r=0.76$, $n=93$). Correlations of upper-bound carcinogenic potency estimates for positive and negative experiments are of similar magnitudes. These high correlations are inevitable because the maximum likelihood and upper-bound risk estimates are inversely related to the maximum dose tested. Rather than being plausible estimates of carcinogenic risk, these indices are merely convoluted safety factors. The use of chronic bioassay data to determine such carcinogenic potency indices should be replaced with risk estimation models that can utilize mechanistic information.

KEYWORDS: Carcinogenic potency, upper-bound potency estimate, maximum likelihood carcinogenic potency estimate, chronic rodent bioassay, cancer risk

INTRODUCTION

Numerous authors have reported large correlations between estimates of carcinogenic potency for chemicals tested in chronic rodent bioassays on rats and mice. These studies have included examinations of maximum likelihood potency estimates (MLE)[1,2,3,4] and estimates of virtually safe doses.[5] The interest in these correlations stemmed from the need to evaluate the potential human carcinogenicity of chemicals. Because the extrapolation from rodent data to human risk is uncertain, it seemed logical to investigate the degree of similarity between potency estimates from two rodent species. If the rodent carcinogenic potency estimates were not similar, then the interspecies extrapolation of these parameters from rodents to humans would not be plausible. The opposite occurred, however; the high correlations which were found between estimates of carcinogenic potency for rats and mice were used as evidence that such indices could legitimately be used for human risk assessment.

It was shown recently that the high correlations between rat and mouse potency estimates were artifactual.[6] This was because (1) the maximum doses tested (MDT) were highly correlated between the two species and (2) the maximum likelihood carcinogenic potency estimates were essentially equivalent to the inverse MDT. The high correlations were thus inevitable. Nevertheless, reports on strong correlations between potency estimates continued to be published[3,4,5] and the Environmental Protection Agency (EPA) still continues to use the upper-bound potency estimate for human risk extrapolation.[7]

The relationship between the MLE and the true carcinogenic potency is unknown because (1) the MLE are essentially equivalent to the inverse MDT, (2) the MDT are measures of subchronic toxicity, and (3) no standardized relationship between subchronic toxicity and carcinogenicity has been defined which holds for all chemicals. Whenever the doses tested are similar, the MLE from the linearized multistage model calculated from experiments in which the rodents developed no tumors are essentially the same as those from experiments producing a statistically significant treatment-related carcinogenic response. Chemicals have similar maximum likelihood potency estimates not because they cause similar numbers and/or types of tumors but because their subchronic toxicities are similar.

Upper-bound estimates of carcinogenic potency also mirror the chronic bioassay maximum dose. The EPA uses an upper-bound rather than maximum likelihood carcinogenic potency estimate because the former is more conservative; the upper-bound implies a higher risk for a given dose. We show here that upper-bound potency estimates are little influenced by the experimental carcinogenicity of the chemical. As with the MLE, the upper-bound potency estimate measures the inverse maximum dose tested, not carcinogenicity. In practical terms, the cancer risks predicted by the upper-bound estimates are so enormous, and the doses calculated using a virtually safe level of 10^{-6} are so minute, that regulatory agencies can feel confident that use of upper-bound risk levels will certainly ensure human safety. Any safety imparted by these risk estimates is due to their extreme conservatism, however, and not to any underlying scientific understanding of carcinogenesis. Because upper-bound potency estimates are not measures of the true carcinogenicity of chemicals, their use for meaningful risk-benefit analysis is delusive. The use of such carcinogenic potency estimates for human risk assessment must thus be viewed as a matter of policy, not science.

METHODS

Estimates of rodent carcinogenic potency and information on the chronic bioassays from which they were derived were taken from the Carcinogenic Potency Database (CPD).[8] The maximum likelihood potency estimate provided in the CPD was in the form of a TD_{50};

this was the dose corresponding to a treatment-related tumor incidence of 50% relative to controls. The TD_{50} is related to another commonly used maximum likelihood carcinogenic potency index, the low dose slope (β) of the one-hit model that passes through the TD_{50}, by $\beta = \ln(2)/TD_{50}$.[1,6] Each TD_{50} in the CPD was accompanied by a lower 99% confidence bound; these may be transformed into upper-bound potency estimates (similar to those used by U.S. regulatory agencies) by division into the natural logarithm of 2. The potency estimates we used were βs and their corresponding upper-bounds. For a more complete description of the CPD see Reference 8.

Correlations between upper-bound potency estimates were calculated for chemicals tested on both rats and mice. Chemicals were sorted into four subgroups on the basis of their experimental carcinogenicity: (1) positive (carcinogenic) in both species, (2) negative in both species, (3) positive in the rat but not the mouse, and (4) positive in the mouse but not the rat. For groups 1 and 2, the most potent upper-bound (largest) was selected from among the studies on each chemical to be consistent with the EPA convention of using the experimental results from the most sensitive animal strain in the risk analysis.[7] For groups 3 and 4, the most potent experimental result was chosen for the positive group and the least potent was selected for the negative to make the estimates as different as possible.

Maximum likelihood and upper-bound potency estimates, and maximum doses tested, were selected for those chemicals tested on the mouse in group 1. These were plotted to illustrate the close inverse relationship between these potency estimates and their respective maximum doses.

RESULTS

Figure 1 illustrates the relationship between upper-bound potency estimates calculated for chemicals which were experimental carcinogens and those which were not. In the series Figs. 1A to 1C, chemicals which were positive carcinogens in both rats and mice are plotted as circles. These are overlaid for comparison with chemicals, each shown as an "x," which were, respectively, noncarcinogenic in both species (Fig. 1A); positive in the rat and negative in the mouse (Fig. 1B); and negative in the rat and positive in the mouse (Fig. 1C). The correlation coefficients for the four sets of upper-bound potency estimates ranged from 0.76 to 0.84. Thus, the rat and mouse carcinogenic potency estimates were highly correlated irrespective of whether or not the chemicals were experimental carcinogens.

Figures 1A to 1C show extensive overlap in the distributions of the upper-bound potency estimates from carcinogenic and noncarcinogenic chemicals. This demonstrates that upper-bound potency estimates often indicate that noncarcinogenic chemicals pose a greater carcinogenic risk than experimental carcinogens. Regions of extremely high potency and very low potency may be seen in Fig. 1A, in which no overlap occurs between the chemicals which were carcinogenic and those which were noncarcinogenic in two species. Most chemicals are found in the central overlapping region, which comprises approximately four orders of magnitude.

Figures 1B and 1C illustrate that the rat and mouse upper-bound carcinogenic potency estimates are highly correlated even when the most potent estimate for a carcinogen is compared to the least potent estimate for a noncarcinogen. This result demonstrates that the high correlation between carcinogenic potency estimates must be due to something other than similarities in the carcinogenicity of the chemicals.

Figure 2 shows the tight inverse relationship between upper-bound carcinogenic potency estimates and maximum doses tested. Also shown is the maximum likelihood potency estimate. The upper-bound on carcinogenic risk was within a factor of 10 of the

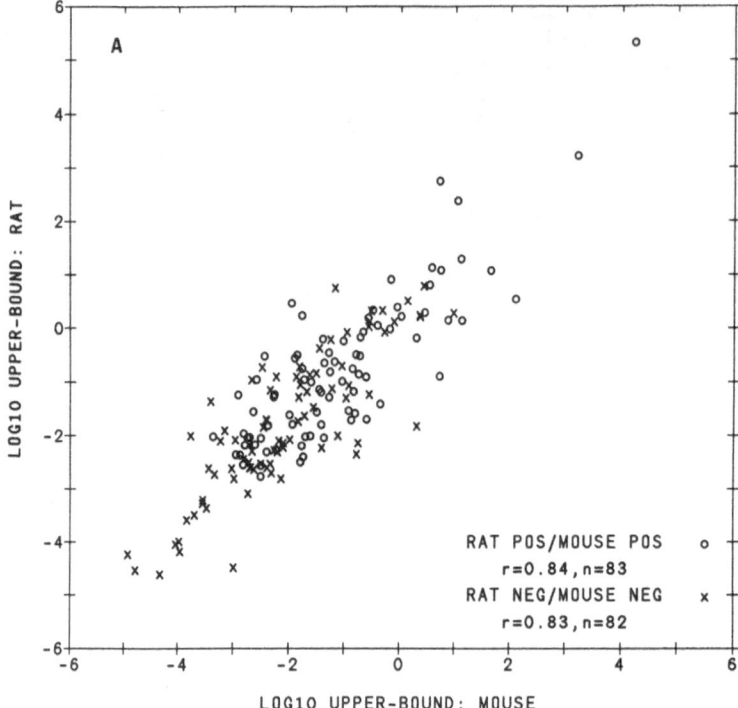

Fig. 1. Upper-bound carcinogenic potency estimates (on logarithmic scales) for chemicals tested on rats and mice during chronic carcinogen bioassays. In each of Figs. 1A, 1B, and 1C, the upper-bound estimates for chemicals which produced a positive carcinogenic response in both species are plotted as circles. These are overlaid, for comparison, with upper-bound estimates from experiments in which one or both of the species did not develop tumors (each chemical represented by an "x").

inverse maximum dose tested for 94% of the chemicals and was always within a factor of 35. This empirical observation explains the high correlations seen in Fig. 1.

DISCUSSION

Figure 1 illustrates that a serious problem exists with the use of upper-bound carcinogenic potency estimates based solely on conventional chronic rodent bioassays. The upper-bound estimates for each chemical were all highly correlated between rats and mice. The correlations were as strong for chemicals which were positive carcinogens in one species and negative in another as they were for chemicals which were carcinogenic in both species. This occurred even though the most potent and least potent estimates for the positive and negative groups, respectively, were used. Upper-bound carcinogenic potency estimates for noncarcinogens are derived from essentially no information beyond the dose levels tested. Upper-bounds for experimental carcinogens, on the other hand, are produced using full tumor incidence data from the chronic bioassay. One would thus expect the latter potency estimates to differ from the former strictly on the basis of the latter's more robust database. Because the upper-bound estimates are essentially equivalent for carcinogens and noncarcinogens, it is apparent that differences in tumor incidence are not reflected in the potency estimate to any significant degree.

226

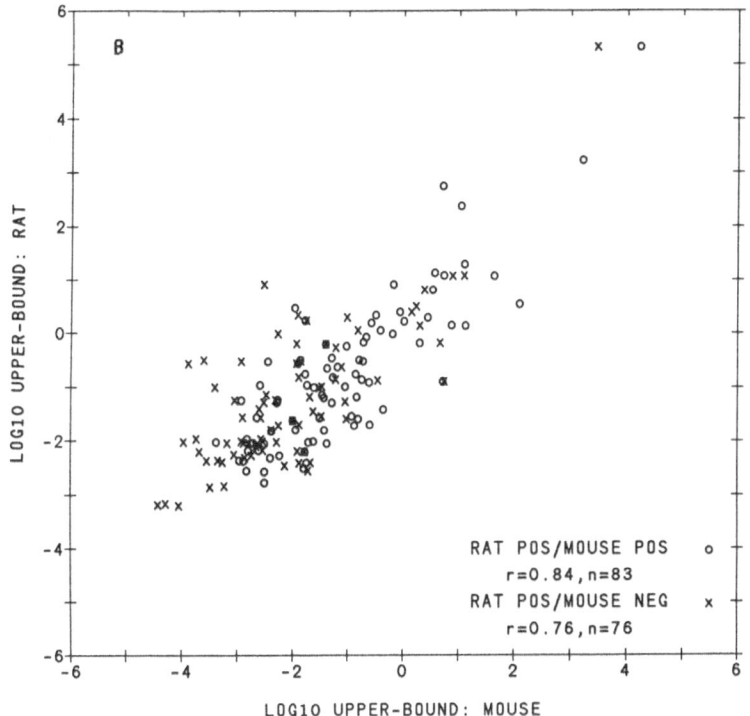

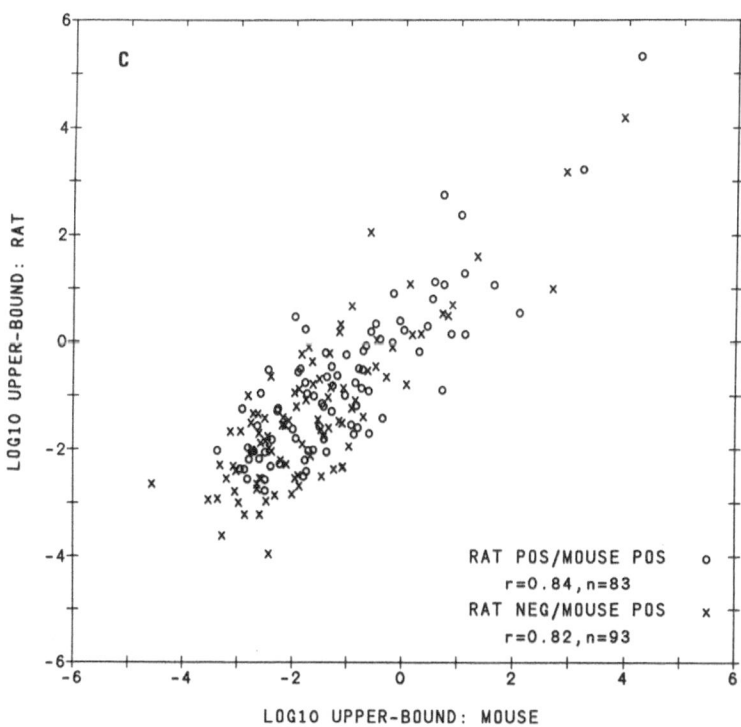

Fig. 1.

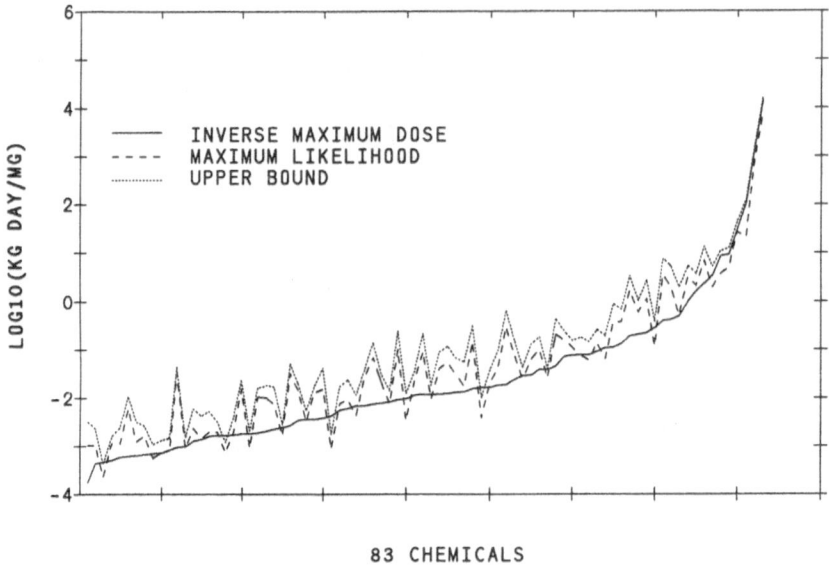

83 CHEMICALS

Fig. 2. Comparison of the inverse maximum dose tested with maximum likelihood and upper-bound carcinogenic potency estimates for 83 chemicals found to be carcinogenic in the mouse. The maximum doses were sorted in ascending order. The upper-bound estimates were within a factor of 10 of the inverse maximum dose tested in 94% of the cases.

Figure 2 shows that the upper-bound potency estimates are approximately equal to the inverse maximum dose tested. Because the maximum doses tested in chronic bioassays are highly correlated between rats and mice,[6] it follows that the upper-bound potency estimates must be correlated also, irrespective of whether or not the chemicals are carcinogenic. Since the maximum dose tested is almost always the maximum tolerated dose (MTD),[7] the "carcinogenic" potency estimate actually measures the inverse MTD, not carcinogenicity. Thus the high correlations between rat and mouse potency estimates are artifactual and are not indicative of similar carcinogenic mechanisms in the two species.

If the correlations between the upper-bound estimates for the experimental carcinogens were significantly larger than those for the noncarcinogens, the case might be made that the extra correlation above that caused by artifact was due to real similarities in carcinogenic response between species. This was not observed, however; the correlation coefficient for the positive group was essentially equivalent to that for the negative group ($r=0.84$ and $r=0.83$, respectively). These high correlations can therefore not be used to infer the plausibility of using upper-bound carcinogenic potency estimates for human risk estimation. These high correlations do show, although rather clumsily, that the subchronic toxicities of many chemicals are highly correlated between rats and mice.

Figure 2 also shows that one does not need to conduct a chronic bioassay to calculate an upper-bound potency estimate. Because the upper-bound is essentially the inverse maximum dose tested, one can simply find the maximum tolerated dose with a subchronic study and use its inverse as the potency estimate. It seems reasonable to expect an estimate of carcinogenic risk to be more dependent on the induction of a carcinogenic response than on experimental design parameters. Present regulatory cancer risk estimates do not meet this criterion.

Some have argued that the good correlations between carcinogenic potency and inverse maximum dose tested were not artifactual but instead were true biological phenomena.[9,10] This line of reasoning leads to the conclusion that the chronic rodent bioassay is redundant (rather than inadequate) for quantitative carcinogen risk assessment because all that is necessary to derive a carcinogenic potency estimate is some measure of acute toxicity. We do not agree with this opinion for two reasons. (1) Bernstein *et al.* showed that the possible values of potency estimates derived from the one-hit model had a limited range about the inverse MDT.[6] This relationship is shown empirically in Fig. 2 to hold for the multistage model. The potential range of the potency estimates relative to 1/MDT is defined by the sample size used in the study and the level of statistical significance desired. Also, (2) we have found no biological evidence which indicates that carcinogenicity is necessarily and quantitatively linked to acute toxicity for every chemical. The conclusion that the strong inverse relationship between carcinogenic potency and maximum dose tested is a biological observation implies that, in effect, every chemical is a carcinogen. Although we readily concur that chronic toxicity may, through such mechanisms as cell proliferation, be manifested in a carcinogenic response, it is certainly not evident to us that such toxicity is a necessary prerequisite to carcinogenesis. In addition, the evidence demonstrating that the inextricable linkage between upper-bound potency and inverse maximum dose tested is artifactual is, to us, overwhelming.

It should be recognized that U.S. regulatory agencies are often in the undesirable position of having been mandated to perform carcinogen risk assessments for chemicals most of which do not have a database adequate for the task. In such situations, the traditional toxicological approach for endpoints other than cancer has been to rely on safety factors. The EPA is reluctant to use safety factors for a variety of good reasons,[11] but in practical terms the current usage of upper-bound risk estimates is little different from the safety factor approach. The EPA is careful to qualify their risk estimates so that the uncertainty associated with them is clear.[7] With this philosophy in mind, we feel that the use of upper-bound risk estimates and their associated virtually safe dose levels should be described more as policy and less as science. These parameters impart a measure of safety because the former is huge and the latter is vanishingly small, not because they are plausible measures of the true carcinogenic risk. To characterize them as meaningful scientific parameters is more than a misrepresentation because it impedes the introduction of newer, and hopefully more meaningful, human cancer risk estimation methods into the regulatory process.

REFERENCES

1. E. Crouch and R. Wilson, "Interspecies Comparison of Carcinogenic Potency," *J. Toxicol. Environ. Health* **5**:1095-1118 (1979).
2. E. A. C. Crouch, "Uncertainties in Interspecies Extrapolations of Carcinogenicity," *Environ. Health Perspec.* **50**:321-327 (1987).
3. L. S. Gold, C. Wright, L. Bernstein, and M. deVeciana, "Reproducibility of Results in 'Near-Replicate' Carcinogenesis Bioassays," *J. Nat. Cancer Inst.* **78**:1149-1158 (1987).
4. D. W. Gaylor and J. J. Chen, "Relative Potency of Chemical Carcinogens in Rodents," *Risk Anal.* **6**:283-290 (1986).
5. J. J. Chen and D. W. Gaylor, "Carcinogenic Risk Assessment: Comparison of Estimated Safe Doses for Rats and Mice," *Environ. Health Perspec.* **72**:305-309 (1987).
6. L. Bernstein, L. S. Gold, B. N. Ames, M. C. Pike, and D. G. Hoel, "Some Tautologous Aspects of the Comparison of Carcinogenic Potency in Rats and Mice," *Fundam. Appl. Toxicol.* **5**:79-86 (1985).
7. Environmental Protection Agency, "Guidelines for Carcinogen Risk Assessment," *Fed. Regist.* **51(185)**:33992-34003 (1986).

8. L. S. Gold, C. B. Sawyer, M. Magaw, G. M. Backman, M. deVeciana, R. Levinson, N. K. Hooper, W. R. Havender, L. Bernstein, R. Peto, M. C. Pike, and B. N. Ames, "A Carcinogenic Potency Database of the Standardized Results of Animal Bioassays," *Environ. Health Perspect.* **58**:9-319 (1984).

9. L. Zeise, R. Wilson, and E. Crouch, "Use of Acute Toxicity to Estimate Carcinogenic Risk," *Risk Anal.* **4**:187-199 (1984).

10. E. Crouch, R. Wilson, and L. Zeise, "Tautology or Not Tautology?" *J. Tox. Environ. Health* **20**:1-10 (1987).

11. Office of Science and Technology Policy, "Chemical Carcinogens: A Review of the Science and Its Associated Principles," *Fed. Regist.* **50(50)**:10372-10442 (1985).

Risk Communication: The Need for Incentives

E. A. C. Crouch, E. D. Klema, and Richard Wilson

Harvard University
Cambridge, MA

ABSTRACT

In this paper we emphasize that a major problem in public understanding of risk is a failure of the various actors to communicate. We believe that this is largely due to a lack of desire to communicate rather than to an ignorance of *how* to communicate. We show this by four examples and discuss ways this can be changed within the framework of U.S. society.

KEYWORDS: Risk communication, trichloroethylene, chloroform, nuclear wastes, radon gas, aflatoxin

INTRODUCTION

All over the federal government and the risk analysis community the "buzz words" for 1988 are "Risk Communication." It seems to be felt by scientists that the public does not understand risk and demands from the government unreasonable actions and precautions. It also seems to be felt by some members of the public that scientists and regulators do not understand their concerns. There is a failure to communicate. Much of the present discussion about risk communication seems to assume that the failure is due to inability of the actors to communicate. By going through some examples—or case studies—in a preliminary way, we suggest that there is often a lack of *desire* to communicate, even among those whose duty it would seem to be. We analyze this a bit further and suggest some ways in which society might create incentives for better communication.

ASSUMPTIONS

The authors are in the business of teaching; two for 40 years each. In our profession it is axiomatic to consider knowledge to be "a good thing" and the more knowledge the better. This may not always be the case and in the examples we illustrate how some people in some circumstances may consider some knowledge to be bad. We will also touch upon the question of the quality and validity of the information that is communicated.

The American view of democracy holds that an informed citizenry will, in the long run, make the best possible decisions. We accept that, but show that in the short run, a manager may choose not to inform the citizenry so that the decisions he deems right can be implemented quickly, without waiting for the slow educational process.

Without trying to be accusatory and certainly not being complete, we suggest several reasons why actors in the risk analysis process may wish to withhold information.

a. A government manager may want to implement a decision quickly and indeed may be ordered to do so by a court and not want possible interference by someone who might criticize his decision.

b. Industry might wish to delay decision by government, so that they may continue unabated for several years to make profits or at least have time to institute control measures calmly.

c. Unions, suspicious perhaps of industry, might prefer to withhold favorable information and, in particular, any information that suggests that workers can reduce their own risk.

d. Academics may wish to withhold information to protect their future research and may not want federal sponsors to know that the hazard they are spending $1 million a year on is trivial, whereas a more important one their rival is working on is ignored.

e. Environmental groups, and in particular their Washington lawyers, may be content with their political power and not feel the need to bolster their case.

f. Any one of the above may actively spread erroneous ideas through inadvertence.

All of these postulated reasons have one feature in common; in the *short term* there is a perceived advantage to withholding information. However, in the *long term* the advantage tends to disappear, suggesting that our axiom may be correct. They all possess another common feature: the sooner or more reliably widespread communication and understanding occurs, the sooner the advantage disappears.

EXAMPLES

In this section we go briefly over a number of examples where it is obvious that information is partially withheld.

TCE vs Chloroform

In the last century parasitic and infectious disease has been eliminated as a major cause of death in the western world (Doll, 1979). In large part this has been due to two engineering triumphs: the installation of main drainage to keep detritus out of the water supply and the chlorination of drinking water to kill bacteria. The reduction of risk by chlorination is large and evident.

In the 1970's it was shown that chlorination of water supplies produces chloroform as the chlorine interacts with organic matter (Rook, 1974; Morris, 1975). The amount of chloroform is roughly proportional to the amount of organic matter originally present, and is as high as 100 μg/l in some water supplies and 100 μg/l in the EPA regulatory limit (Federal Register, 1977). Chloroform has been shown, when given by gavage, to cause liver cancer in mice and kidney cancer in rats, and, when calculated by what are now conventional methods (Anderson *et al.*, 1983), 100 μg/l gives a risk of about one in ten thousand per life, 100 times the level at which EPA often regulates (Crouch *et al.*, 1987).

Trichloroethylene (or TCE) is an organic solvent used in both domestic and industrial processes. It is often found as a containment of well water at levels of 10 μg/l and occasionally at 500 μg/l. It also causes liver cancer in mice, but it takes 20 times *more* TCE than chloroform to induce the same tumor rate. It has not definitively been shown to cause tumors in rats. The present EPA regulatory limit is 5 μg/l, with a risk 400 times *less* than the regulatory limit for chloroform.

The regulatory authority, the Office of Drinking Water of the EPA, knows and understands this comparison, as do many toxicologists. However, they refuse to educate the public directly about this. They were challenged to do so at a toxicology meeting in 1980; at least one junior member of the office tried to get comparisons in the public statements, only to have them removed by his superior (Anon., 1987). In the regulation of "chlorinated hydrocarbons" (under which TCE is regulated), chloroform—an obvious chlorinated hydrocarbon—is omitted (Federal Register, 1984) even though this omission was explicitly pointed out in public comment.

Thus the *risk assessment* document is incomplete, and the *risk management* process of the EPA was altered by incomplete information. The reason for this omission is (presumably) a desire to save trouble and forestall public pressure to reduce chloroform levels, which cannot easily be done. It would have required explicit recognition of the tradeoff between chlorination and the risks of not chlorinating or the costs of alternative methods (with their own risks), something the EPA was apparently not prepared to do.

The present situation is such that junior persons in regulatory agencies are misinformed. We know of two cases in California where TCE was measured in well water at levels several times the (low) EPA standard, and it was recommended that users switch to city water with a calculated hazard due to chloroform 50 times as great.

One remedy for this situation is for scientists and others to continually point out this omission, particularly at any meeting where the press is present. But what is needed is *an incentive* for the EPA to do the job thoroughly. One way might be for the Science Advisory Board, or subset thereof, to explicitly review all risk assessments presented to management *with this particular point in mind*.

Nuclear Wastes

One could write a whole book on nuclear power, but we confine ourselves to one aspect: the long term disposal of high level radioactive wastes.

It is worth stressing one aspect of nuclear energy. It arose from the work of a relatively small number of people who agreed that its hazards be understood and not be concealed. This contrasts with almost all previous technologies where the long term hazard was not known, or if known, ignored, and few attempts were made to discover them.

From the point of view of perception, this is crucial. On one side, it appears to be a technology with possibly large long-term hazards. On the other hand, scientists claim that the long-term hazard is less than that of competing technologies. Scientists who cannot now conceal what they previously decided to openly admit, often state that it is the most easily solved waste problem in our society, and while refusing to predict the future absolutely, point to the fact that radionuclides *do* decay, whereas arsenic, for example, *does not*.

In this case, it is (partially as a result of the past decision of openness) the scientist who wants to communicate, and someone else, who for any reason wants to inhibit nuclear power, who wants to delay communication.

The political cry against waste, "Not in my own backyard," is often heard. However, there are many who would willingly have the waste in their own backyard—for a reasonable fee. The more accurate (but less morally acceptable) cry would be "not in my *neighbor's* backyard." We can make a parallel here. Each of us can say: "I don't want heavy delivery trucks on my street, *except when they deliver to me.*" Because I like my neighbors, I don't mind trucks delivering to them provided it is not a nuisance object like some noisy 'musical' instrument. Society has learned to cope with this. On some streets interstate truck travel is allowed; but then the federal government pays a large fraction of the cost of upkeep of these roads. Nonetheless, expanding their number is a hassle.

Again, we see most clearly the steps that those in our profession can take to reduce the problem. We must ensure that reports on nuclear waste are written clearly, stating risk comparisons insofar as possible and calling attention to those cases where risks are not properly described or compared.

Radon Gas

Radon gas is a chemically inert gas, arising in nature from the α particle decay of radium, which in turn, arises from the decay of ^{238}U. Radon decays with a half life of 3.8 days into a series of daughter products, radium A, radium B, radium C', radium D, radium E, and finally lead. Since it is a gas, radon can be easily breathed into the esophagus and lungs where a few atoms decay, depositing by recoil atoms of the daughter products. The α particles from these can then cause lung cancer.

In a home, radon comes up from the ground, particularly when the barometric pressure drops, leaving the ground to outgas. It is trapped for some hours by the house, exposing its occupants.

Lung cancer has been observed among uranium miners for many years. Epidemiological studies relate the risk to measured radon levels. It is tempting, then, to assume that the same risk is undergone by anyone with the same exposure.

The hazard was not fully recognized until about 1975. Even then, Professor Dade Moeller wrote about the gas coming from building walls. The EPA, considering the problem uninteresting, would not renew his $10,000 contract. The matter came to attention in 1984 when Mr. Watros of Pennsylvania set off radiation alarms as he *entered* a nuclear power plant. He was exposed to over 200 times the U.S. average concentration of radon gas.

Now EPA has a recommended level for houses of 4 pCi/l, and 20% of the U.S. housing stock has levels above this. Pronuclear advocates are ecstatic, pointing out that this radiation hazard exceeds all nuclear power hazards, and the yearly cancer toll by conservative EPA estimates exceeds the toll from Chernobyl. But is this real?

Several studies have been made of the lung cancer rate of women who died before 1970 and were presumably non-smokers. In several counties and countries, the cancer rate in high radon areas is no higher than the rate in low radon areas, whereas the EPA formula would suggest 5 times the risk. This indicates that, at least for non-smokers, the risk is greatly overestimated—perhaps twentyfold. At low concentrations, less than 20 pCi/l, the risk for non-smokers could be *zero*. Yet in recent research on risk communication by EPA scientists and contractors (Desvousges *et al.*, 1987) 2,300 N.Y. state homeowners were told that the lower limit of risk for concentrations in this region was *finite*. In this example, we can criticize the EPA from both directions.

The EPA, who ignored the problem for so long, is now *reluctant to state* their possible overestimate. Presumably they fear that an on-again, off-again approach will confuse the public. But is it not possible that the relevant EPA administrators, having acquired a large staff to cope with the problem, act in the way C. Northcote Parkinson described in his books (1955, 1960) and prefer not to have the staff cut when public alarm is reduced by the truth being told? Or is it that once the juggernaut of EPA has started moving in one direction, it is hard to stop?

Aflatoxin in Nuts and Milk

Aflatoxin B1, one of the most potent and acutely toxic carcinogens known to man, grows in a mold on corn and nut products. Therefore it can be found in trace quantities in many foodstuffs, particularly if the nuts are not dried quickly.

The FDA permits peanut butter to be sold with 20 ppb of aflatoxin therein, and in 1977 the average was about 2 ppb. Even at the average level, 16 oz. of peanut butter gives a one in a million risk according to the linear dose relationship considered proper for a prudent public policy. The risk associated with eating 16 oz. in a year (a modest allowance) is seventy times the 10^{-8} lifetime risk the FDA considers appropriate for food additives.

While this may be proper as a management policy, it seems to us that the FDA could explain it better. When describing regulations about additives, the levels at which natural carcinogens occur could be explained and the public would be reassured by the *extra* care taken for food additives. However, the FDA has missed many public opportunities for explaining the hazards of aflatoxin and other natural carcinogens, as in their public statement about safety of carcinogenic residues (Federal Register, 1985a). It is left up to academics such as Bruce Ames (Ames *et al.*, 1987) and myself (Wilson, 1986) to do so.

There are adverse consequences of this refusal. Bruce Ames emphasizes the waste of public effort on trivia. But we point out here a direct example where attention of public health authorities is misdirected. At a meeting of public health officials in San Antonio, Texas, on April 2, 1986, there was a panel discussion of recent spillage of carcinogens in the mid-west. High levels of chlordane were found in milk over much of the mid-west. It was over 100 times the FDA standard for continuous consumption. However, high levels of aflatoxin were also found in corn; corn seems to convert it into the less toxic aflatoxin M. Assuming the carcinogenic potency of aflatoxin M is one-tenth that of aflatoxin B, it posed a far more serious hazard than chlordane. Yet no public health official had alerted the public to this, and the comment of one of us (RW) went almost unnoticed by the press, although one risk analyst, widely known for his opposition to misuse of pesticides, agreed.

Worse still, the federal government does not seem to want to know the hazard. There has been no animal bioassay for aflatoxin M (although the National Toxicology program has tested over 300 chemicals), and so our estimate of carcinogenic potency is indirect.

In our view, if there were better public information, some part of the attention would be redirected from chlordane to aflatoxin.

CAUSES OF THE PROBLEM

With these examples in front of us, how can information transfer and public education be improved? It is easy to find scapegoats.

1. Victor (1987) argues that scientists should communicate more often.

2. Robinson (1987) responds that scientists talk often and freely but too few people listen.

3. Westheimer (1987), writing in *Science*, takes as a blameworthy example the new "core curriculum" at Harvard University. This, compared with the previous

"General Education Program" at Harvard, allows students to take fewer courses in quantitative science.

4. Cohen (1985) blames journalists for not being interested in propagating the "truth." He wishes they would get their information from the "main line" scientists rather than from "fringe" scientists. However, members of the former group are often very difficult to get for an interview and are usually careful not to discuss matters outside their area of personal competence, while the latter usually are glad to make themselves readily available.

5. Scientists are unwilling to "prostitute themselves" by entering the public arena, particularly where the action is (a) appearing as expert witnesses in tort cases; (b) giving testimony to legislatures; and (c) writing letters and OP-ED articles to newspapers.

While recognizing that such criticisms have at least partial validity, we wish to accept American society largely as it is, with all the societal constraints on college deans and newspaper editors and consider possible changes *within this framework*.

POSSIBLE SOLUTIONS

Kantrowitz (1967, 1975) repeatedly suggested the formation of an "Institute for Scientific Judgment" for these questions (popularly paraphrased—probably inaccurately—into "Science Court"). Recently, "Centers for Risk Assessment" have been proposed and even started. Such institutions can easily become paralyzed.

INCENTIVES

One feature of U.S. society seems to be distinct from all other societies: the number of lawyers; that is, the ubiquitous nature of the legal adversary system and in particular the tort law.[a]

The decision of a jury is considered by many to be the ultimate expression of democracy. Anyone who refuses to appear before a jury is opting out of an important part of the (U.S.) process.

We can see small changes in the habits of our own profession that could alter the situation considerably. Scientists are reluctant to enter the public arena and appear as expert witnesses before juries because

a. colleagues consider it "prostitution,"

b. funding agencies for research say that "you are spreading yourself too thin,"

c. it takes time from the exciting, self-absorbing process of scientific research, and

d. the prospect of being cross examined by an opposing attorney who is not interested in searching out facts but is interested in damaging the credibility of the witness, discourages many scientists.

---

a. In debate one of us has tried to use this as proof that there is no absolute safety. "There is no number of lawyers so small that they cannot do harm." However, this is obviously wrong. While two lawyers in a town can sue each other, if there is only one lawyer he will starve to death.

The opinions indicated by a and b can be changed, and the financial rewards of appropriate fees can often overcome c.

PEER REVIEW

Scientists rely upon external review committees to review their work. So, we believe, should public agencies. For example, in the EPA it is the EPA that decides on the hazard to be addressed, the EPA staff that carries out the risk assessment, the EPA staff that passes it (or, as we noted before, a part of it) up to the management, and the management that makes the decision. At what stage the pubic is informed is unclear.

The Science Advisory Board is a fine addition to this; but it tends to confine itself to questions of whether the science has been introduced at the start of the process, rather than whether the information presented upwards to the risk manager is complete and properly summarized with the uncertainties highlighted. We suggest that all EPA statements in the Federal Register should be *peer reviewed*.

EXAMPLES OF MODIFICATION

As a simple example, we show a statement (see box) sent out by the California Department of Health about contamination of ground water. As noted before, this resulted in a switch to surface water with an increased calculated risk. We suggest the following modified letter:

"Based on limited test results finding trichloroethylene in ground water, the Sacramento County Health Department and the Central Valley Water Control Board are sampling wells around....to determine the levels of potentially injurious chemicals."

"The area of concern is....."

"We emphasize that the levels are well below the levels at which toxic effects are known, but that a lifetime exposure to these chemicals may slightly increase the risk of chronic ailments. There is no undue danger from short term exposure at these levels, and we are expediting the measurements so that we may shortly know any long term danger."

"Meanwhile, we note that all food and drink poses *some* risk, including bottled water or city water, and we will give recommendations shortly."

"If you have questions you may wish to read California Department of Health pamphlet number ??? and the scientific references therein, or call the Central Valley Regional Control Board at 445-0270 or County Health Department at 366-2101."

PROPOSITION 65

This year proposition 65 in California has drawn a lot of attention. We note, with amusement, that two distinguished cancer experts, Dr. Arthur Upton and Dr. Bruce Ames differed when voters were considering the proposition. Now that it has been enacted, we argue that it can help avoid the problems we discuss here and become a huge education program about cancer and carcinogensis.

As the law presently seems to be administered, anything emitting a carcinogen posing a risk level of 10^{-6}/life or greater must have this fact advertised. Exceptions include natural carcinogens and those regulated by public agencies. It appears that restaurants, producing benzopyrene by broiling steaks are *not* excluded, and every menu must state the fact that cooking food produces carcinogens. They can, of course, add a statement that cooking food is an age old practice which often prevents spreading of disease and should not be lightly abandoned.

We suggest that public agencies that have statutory exemptions should set the tone by abandoning them. With every water bill can come a brief statement of why water is chlorinated and the unavoidable carcinogenic consequences of chlorination. The risk of using asbestos water pipes could also be stated. Other interesting and important educational examples will no doubt come to mind.

ACKNOWLEDGEMENTS

The ideas in this paper have been derived from many discussions with scientists, lawyers, reporters and (in a mostly one-way communication) juries. Their help is gratefully acknowledged.

REFERENCES

Ames, B., *et al.*, 1987, *Science* **236**:271.

Anderson, E. L., *et al.*, 1983, *Risk Analysis* **3**:277.

Cohen, B. L., 1985, "Journalism and Nuclear Power," *Columbia Journalism Review.*

Crouch, E. A. C., *et al.*, 1987, "Cancer Risk Management: An Overview of Federal Regulatory Decision Making," *Eng. Sci. Tech.* **21**:415.

Desvousges, W. H., Smith, V. K., and Fisher, A., 1987, "Communicating Risks Effectively," paper presented at the Air Pollution Control Association Meeting, Oct. 26-27, 1987

Doll, R., 1979, "The Pattern of Disease in the Post Infection Era, National Trends," *Proc. Roy. Soc.* **B205**:47.

Federal Register, 1979, "National Interim Primary Regulation. Control of Trihalomethanes in Drinking Water, Final Rule," **44**:6824.

Federal Register, 1985a, "Sponsored Compounds in Food Producing Animals: Criteria and Procedures for Evaluating the Safety of Carcinogenic Residues," October 31.

Federal Register, 1985b, "National Primary Drinking Water Regulations: Volatile Synthetic Organic Chemicals: Final Rule," FR46880, Nov. 13.

Kantrowitz, A., 1967, "Proposal for an Institute for Scientific Judgment," *Science* **156**:763.

Kantrowitz, A., 1975, "Controlling Technology Democratically," *American Scientist* **63**:505.

Morris, J. C., *et al.*, 1975, "Formation of Halogenated Chemicals by Chlorination of Water Supplies: A Review," PB-241511.

Parkinson, C. Northcote, 1955, "Parkinson's Law."

Parkinson, C. Northcote, 1960, "The Law and the Profits."

Robinson, J. D., 1987, *Science* **237**:119.

Rook, J. J., 1975, "Formation of Haloforms During Chlorination of Natural Waters," *Water Treatment Exam* **23**:234.

Victor, D., 1987, *Science* **236**:897.

Westheimer, F., 1987, *Science* **237**:1165.

Wilson, R., 1986, "What Level of Risk Is Acceptable," Presented at SOM Workshop, Crystal City, VA, Jan. 14.

An International Approach to Health, Safety and Environmental Risk Management for a Diversified Corporation

Lorraine T. Daigle[a] and Ann C. Smith
Allied-Signal Inc.
Morristown, NJ

ABSTRACT

A program of risk analysis is presented which has been shown to be an effective tool in the management of health, safety and environmental risks in a diversified multinational corporation. This three-step program involves informal plant visits designed to assess the compliance status and degree of sophistication of health, safety and environmental programs. These assessments are followed by more formal reviews which include consultation, training and program support. Finally, formal reviews are conducted to verify compliance with respect to company and legal requirements. Critical elements of this three-step approach include top management support, knowledge of applicable criteria and communication of review results. Several tiers of reporting occur which are designed to communicate relevant information to management from line supervision through the Board of Directors. Corrective action is planned by local management with assistance from other organizational groups as required. Plan implementation is tracked at several levels. The program works in conjunction with other mechanisms of health, safety and environmental protection to assure effective management of these risks.

KEYWORDS: Health, safety and environmental management, regulatory compliance, risk management, international, environmental audit, occupational health and medicine, industrial hygiene

INTRODUCTION

Over the years, Allied-Signal has developed a strong program to manage health, safety and environmental risks. The program basis is understanding of applicable requirements, which include law, regulation, standards of good practice and the Corporation's own health, safety and environmental policy and other internal procedures. The program includes guidance, training and oversight from various levels of the Corporation. As one of the tools to manage these risks, this diversified, multi-national corporation uses a plant review system geared to increasing degrees of sophistication of plant programs.

---

a. Present address: Johnson & Johnson, New Brunswick, NJ

Table 1. Measured Risks

Compliance with:

- Laws and Regulations
- Corporate, Sector and Company Policies and Procedures
- Standards of Good Practice

Table 2. Review Objectives

Types of Reviews	Risk Objective
• Assessment	• Measurement
• Assurance	• Management
• Surveillance	• Compliance Verification

The entire program began, of course, domestically. In recent years, all program elements have gradually been introduced internationally as well. Before continuing, it is helpful to describe the scope of the program.

Health, safety and environmental programs which come under this review system include: Occupational Health (Industrial Hygiene), Occupational Medicine, Safety and Loss Prevention, Product Safety and Integrity, Air Pollution Control, Water Pollution Control, Oil and Chemical Spill Prevention and Solid and Hazardous Waste Management.

Compliance is measured or evaluated relative to: laws and regulations, internal policies and procedures, and acceptable standards of good practice within the professional discipline (Table 1). This program does not include a self-evaluation mechanism, although it could be expanded so that a plant could determine its own health, safety or environmental status at any time.

The program is comprised of three types of reviews: assessment reviews, assurance reviews and surveillance reviews. Initially, an informal plant visit is performed which is designed to assess the compliance status and degree of program development within the corporation, or to *measure* risks. These visits are called "assessments." The second visit is more formal and may include consultation, training, and program support in order to assist the plant in reaching compliance, or in *managing* risks. These are termed "assurance" reviews. Finally, formal surveillance reviews (elsewhere sometimes called "environmental audits") are conducted for the purpose of *verifying compliance* with rules, regulations, policies, and procedures; that is, that risk management systems are in place and working (Table 2).

The discussion of each of these three reviews will include the planning of the review, the tool used, the actual visit description, the review process or system used, reporting and the corrective action mechanisms (Table 3).

Table 3. Review Content

All Reviews Involve:
- Planning
- Tool to be used by the evaluator
- Visit to the operation
- Consistent process or system of evaluation
- Reporting mechanism
- Corrective action

	Responses
39. If sampling is performed, are there written procedures for:	
A. Sampling?	[ ] Yes [ ] No
B. Calibration?	[ ] Yes [ ] No
C. Analysis?	[ ] Yes [ ] No
D. Handling of samples?	[ ] Yes [ ] No
If yes to any of the above attach copy.	
40. Does the sampling record include a calibration log, collection and analytical methods?	[ ] Yes [ ] No
41. If an outside analytical laboratory is employed, has it been certified by AIHA?	[ ] Yes [ ] No

Fig. 1. Assessment and Assurance Reviews: Sample Occupational Health Questionnaire.

ASSESSMENT REVIEWS

Planning

The planning for assessment visits is initiated by a corporate professional in a specific discipline who lists those operations where basic information is not available (e.g., a recent acquisition) or which warrant a reevaluation. Changes that could result in a revisit include new products, new or changed processes, substantial increase in area and/or census of the operation or plant relocation. Management is contacted for agreement with the list of the specific operations that the professional wishes to visit. It is the responsibility of the management of the operations to set up the necessary invitations and dates. Assessment reviews are planned to require usually no more than a two-day visit and frequently can be accomplished in one day.

The tools used during this visit are specific questionnaires and check lists (Figs. 1 and 2). They are used to record findings and provide space for program and facility description. The reviewer completes the questionnaire and retains it as part of the reference file on the operation. These questionnaires are available to professionals in other disciplines who may be visiting the location in the future. The purpose of the tool is to afford the visitor a degree of understanding of the status of the plant's programs in the specific discipline. By using the same tool during each review, the process remains consistent from one plant to another, thus serving as an internal quality control device.

```
                                                    Responses

RESPIRATORY PROTECTION  [ ] Not applicable
                        Go to question 56

43.  Do plant workers use repiratory
     protection?                                    [ ] Yes [ ] No
     For US only: if yes, are only
     NIOSH approved respirators used?               [ ] Yes [ ] No

44.  Is there a written program
     detailing selection, areas of
     use, fitting, maintenance,
     storage and employee training?                 [ ] Yes [ ] No
     If yes, attach copy.

45.  List respirators used and exposure
     for which they are being worn.
```

Fig. 2. Assessment and Assurance Reviews: Sample Occupational Health Questionnaire.

Process

An informal opening meeting is held with management. The plant manager and the individual responsible for the discipline under review meet with the visiting professional, who describes the reason for the visit and the reports and documents which will be generated as a result of the visit. Plant management is usually asked to describe elements of the operation which have impact on the program under review.

Following the opening meeting a plant tour is conducted. The questionnaire is reviewed with the person responsible for the discipline at the plant. At no time is the individual requested to *prove* any of his/her statements, but rather to provide a narrative of how the programs or systems operate in the plant. Any documents that would better define the program are made part of the reviewer's files if the reviewer finds it necessary. For example, since respirator programs are required to be in writing and because we have noted difficulties in this area in past reviews, reviewers have begun to ask for a copy of the written respirator program during assessment visits.

Reporting System

Prior to leaving the plant, an informal meeting is held with plant management to review the findings and recommendations, if any. Proposed distribution of the written report to be prepared later is discussed during the closing meeting as well. Upon returning to his or her office, the professional writes a descriptive report and directs it as described. Distribution usually includes sector, division, or company professionals who are responsible for the discipline being reviewed. The plant manager and the individual at the plant who is responsible for the discipline are the addressees of the report. Generally, the review report is submitted in draft to the plant for comments and questions before being finalized.

Corrective Action Required

Plant management is requested to respond to deficiencies noted in the report. If a response is made, a copy is requested for the professional's file in order to close the loop.

The plant may opt to develop a plan of action with estimated dates when each action will be implemented. This is not a necessary step, but plant management is notified that if the plant is visited in the future for an assurance review, the report will be used as one of the documents against which the assurance system is tested. Therefore, it is beneficial for the plant manager to have documentation of his/her response to the assessment report.

ASSURANCE REVIEWS

The second tier of this program is the assurance review. These visits are more formal and may include consultation with the plant, assistance in training, and program support with the primary purpose of helping the plant reach compliance. These reviews are thorough inspections by the professionals in the five basic health, safety and environmental disciplines: Pollution Control, Safety and Loss Prevention, Product Safety, Occupational Health and Occupational Medicine.

Planning the Review

These reviews are usually performed at plants that either have prior recorded assessments or are planned for acquisition or divestiture. The same questionnaires are used for assurance and assessment reviews, but more backup documentation is gathered in an assurance review. Reports generated by prior insurance company visits, regulatory inspections, or internal reviews are some examples. The findings, recommendations and plan of action that the plant developed in response to previous reviews are used as additional evaluation tools.

Visit Description

A formal meeting is held with plant management, which is informed of the in-depth review that will occur, what subjects will be covered and what documents will be used for the review. Plant management is asked to describe plant operations with particular reference to operations that have changed or new processes or products that are planned for the very near future. An in depth tour is conducted during which the reviewer notes areas of interest for later follow-up.

Process

The individual responsible for the discipline to be reviewed is interviewed and asked to provide documents for the program elements discussed. These documents are reviewed as an aid in evaluating the programs.

Formal Reporting System

A close-out interview is held with plant management where findings and recommendations are discussed.

A formal report is directed to the plant manager with distribution to his/her management in the division, company or sector. The report requests a written corrective action plan. These types of reviews may uncover items which are sensitive to the corporation. Although this is rarely the case, the review report may be edited by a lawyer from the sector, company or division which is being reviewed. If legal advice is sought based on the facts disclosed in the draft report, it is possible that the report may be modified and addressed to counsel of the reviewed business group as a privileged communication. However, that is the rare exception and not the rule.

	Yes	No	N/A
1. Are employees exposed to air contaminants for which exposure limits have been established by either regulatory agencies or internal policy?	—	—	—
2. Are there physical agents (such as noise, ionizing radiation, microwave radiation, heat stress) to which employees are exposed?	—	—	—
3. Have baseline measurements of employee exposures to air contaminants and physical agents been made?	—	—	—

Fig. 3. Surveillance Review: Sample Occupational Health Questionnaire.

Such reviews assist each operating sector president to understand his/her compliance with the health, safety and environmental policy of the corporation and with applicable laws and regulations.

SURVEILLANCE REVIEWS

In 1978 Allied-Signal adopted an internal environmental surveillance program. The objective of this program is to provide top management, including the Board of Directors, with independent verification that the corporation's operations are in compliance with Allied-Signal's Health, Safety and Environmental Policy and Procedures as well as with applicable laws and regulations. One of the current staff of three surveillance professionals leads each review. This helps to provide consistency over time. These reviews, frequently termed "environmental audits" by other firms, are very thorough reviews of the plant's compliance status in the health, safety and environmental area over a specified period—usually at least a year.

The reviews are organized in six subject areas: Occupational Health and Medicine, Safety and Loss Prevention, Product Safety and Integrity, Air Pollution Control, Water Pollution Control and Spill Prevention, and Solid and Hazardous Waste Management.

Planning

All plants and operations including US and foreign operations are within the scope of the program. For domestic operations, usually only one discipline is reviewed per visit. Overseas, several disciplines may be reviewed during a visit to reduce travel time and costs to the corporation. Operations (primarily manufacturing facilities) are randomly selected from a pool of plants that the Surveillance Director, the various Corporate Health, Safety and Environmental Directors and the consultant believe might pose significant environmental risks. Therefore, low environmental risk operations receive considerably less attention than high risk ones. The program consists of about 50 reviews per year.

Each review is conducted by a team. This team is comprised of a full-time surveillance professional as the team leader, a corporate specialist in the discipline under review and a representative of a consulting firm. The consultant acts as an external reviewer, an independent third party, thus providing the advantages of an external "auditor."

Visit and Process

The tool used in surveillance reviews is a formal protocol for each review subject and includes an introductory questionnaire (Fig. 3).

5. Obtain an understanding of the location's Occupational
   Health Program for the following:

   a. Exposure of personnel to chemical hazards
      (See Step 6).

   b. Physical agent hazards
      (See Step 7).

   c. Training and awareness programs
      (See Step 8).

   d. Engineering controls
      (See Step 9).

   e. Respiratory protection program
      (See Step 10).

   f. Hearing conservation and noise control
      (See Step 11).

   Document your understanding in narrative or flow
   chart form. Based upon your understanding of the
   program and plant operations, designate the scope
   of the review, explaining additions and deletions
   from the general review program.

Fig. 4. Surveillance Review: Sample Occupational Health Protocol.

11. Confirm that the following have been done relative
    to hearing conservation and noise control based on
    the understanding developed in step 5.

    a. Sound level surveys have been conducted.

    b. Methods of control have been evaluated and applied.

    c. Employees are included in an effective HCP.

    d. Hearing protection equipment and measures are
       approved by a qualified individual.

Fig. 5. Surveillance Review: Sample Occupational Health Protocol.

Elements of a sample occupational health protocol are shown in Figs. 4 and 5. Each protocol directs the reviewer to describe applicable requirements, the systems and programs at the facility to meet those requirements, and the test steps to verify compliance. These protocol steps require the reviewer to document his/her understanding, using appropriate exhibits to exemplify plant practices. When all review steps are completed, the team has generally compiled 50-100 pages of writing and a series of exhibits about an inch thick. These "working papers" constitute the official record of the review.

The visit is very formal, beginning with an opening meeting with plant management and generally lasting 3-5 days. It involves an orientation tour, interviews, review of pertinent documents, and compliance tours. Frequent status reports are made by the team leader to plant management and other personnel.

Table 4. International Review Program Issues

• Culture	• Regulation/Law
• Language	• Technical sophistication

Reporting

Team members continuously feed back impressions to the responsible plant personnel with regard to the compliance of any specific system. This helps to eliminate misconceptions or false trails during the review. On the last day of the visit, a complete detailed oral exit interview and a written list of findings are provided to the plant manager. Appropriate plant personnel usually participate in this meeting and are free to comment on findings.

A draft report is sent to the plant manager and other involved personnel for comments and corrections. The final written report is sent to the company president with distribution to plant management and other affected parties. It is written by the team leader and indicates the scope of the review as well as compliance deficiencies relative to federal, state and local regulatory standards; applicable policies and procedures; and standards of good practice. Plant management is requested to develop a formal response to the review report to describe corrective action anticipated or taken in response to the findings. The responsibility for tracking implementation of action plans lies with business management.

An overview report of the results of these reviews is made by the surveillance director at all regular meetings of the Corporate Responsibility Committee of the Board of Directors. A representative of the external consultant group is also present and is asked to provide an independent report in private to the Committee.

About 20% of the corrective action plans developed in response to these reviews are subjected to follow-up review by the surveillance group. Follow-up reviews are conducted one to two years after the initial review. They provide an additional verification step and relieve some of the concerns that legal minds of the corporation may have regarding "smoking guns" of outstanding documented review findings.

INTERNATIONAL ISSUES

The extension of this review program into the international arena is an ongoing activity. The major difficulties encountered have centered on cultural differences, language, regulatory development and technical sophistication (Table 4). Nevertheless, Allied-Signal continues to transport this program of tiered reviews to its overseas locations, working in a stepwise fashion to *assess* risks, *manage* them, and then *verify* the ongoing functioning of appropriate control systems.

SUMMARY

All these reviews—assessment, assurance and surveillance—examine past and present health, safety and environmental systems. The purpose of the reviews is to provide protection to the Corporation—its employees, Board of Directors, and shareholders—by identifying deficiencies and weaknesses relative to laws, regulations, policies and procedures, so that they can be promptly corrected. In addition, they develop information which can be fed back into ongoing management program activities: guidance, training, program oversight, capital project review, etc. The monetary benefits are real and substantial, although difficult to measure.

This Corporation views Health, Safety and Environmental compliance as a risk management activity, i.e., continuously identifying, estimating, assessing, evaluating, intervening and controlling risks. Through this system of reviews, operations are safer places to work, offer safer environments for our neighbors, provide safer products for our customers, and maintain and enhance the health of the Corporation.

Risk Perception of Technologies: The Perspective of Similar Groups in Portugal and the United States

M. Peter Hoefer, S. Basheer Ahmed,
Elayn Bernay, and Colin DuSaire

Pace University
New York, NY

ABSTRACT

The perception of risk is affected by demographic biases. This paper studies how those biases have caused segments of the population of two countries to perceive risk associated with contemporary technologies. Portugal and the United States are both western but culturally different. A questionnaire was distributed to college-level business students (mostly graduate students) at universities in Lisbon and New York. Statistical results show similarities and differences between the Portugal sample and the American sample. What was consistent in both countries was the way the sexes differed in their perception of risk.

KEYWORDS: Risk perception, technologies, Portugal and the United States

INTRODUCTION

This study was conducted to determine differences in perception of risk of high technology within demographic groups and across cultures. A study of perceived risk of technology by business college students in New York City was conducted by M. P. Hoefer and S. B. Ahmed in 1985.[1,2] In that study significant differences in perception were noted within certain demographic groups; in particular, strong differences were noted between sexes. Women almost universally saw more risk associated with technologies than men. This current study, thus, also serves to ascertain changes in attitude that have occurred within this group during the past two years.

Studies of risk perception conducted by P. Slovik[3] and others have shown that sharp differences may exist among countries. A. Wildavsky commented as early as 1979 that the United States was "on its way to becoming the most frightened."[4] What happens when comparing groups from similar backgrounds but from countries with different cultures?

METHODOLOGY

For this study, MBA candidates and college business majors from Portugal and the United States are compared. Many of the MBA candidates in both countries are already responsible middle managers in their companies; all are potential business leaders representing a group that will play an important part in the political process in their countries.

In Portugal, a sample of 52 was drawn from MBA candidates and college business majors at the Universidade Nova de Lisboa, Faculdade de Economia; in the United States, the sample consisted of 50 MBA candidates at Pace University's Lubin Graduate School of Business.

Although neither sample may be representative of their countries as a whole, they do represent the population of educated, potential business leaders in their respective cities: Lisbon and New York.

A questionnaire was developed using an eleven-point scale to measure the extent of perceived risk in the following areas of high technology:

(1) Nuclear power generation (NPG)
(2) Robotics (R)
(3) Genetic engineering (GE)
(4) Information technologies (IT)
(5) Space defense technologies (SDT)
(6) Nuclear weaponry (NW).

For each technology, multiple statements were developed and the respondent was asked to indicate the degree of risk associated with that statement, on a scale of 0 to 10, wherein a response of 0 represented no perceived risk and a response of 10 indicated the highest degree of perceived risk.

Averages were calculated for each statement and for each area. These were then cross-tabulated by:

(1) Country (U.S. vs. Portugal)
(2) Place of residence (urban vs. suburban)
(3) Sex
(4) Age
(5) Annual family income
(6) Religion
(7) Education.

The data were subjected to analysis of variance (ANOVA), a robust technique in determining significant differences between groups when the following theoretical requirements are met:[5]

(1) All samples are random and independently selected
(2) All populations are normally distributed
(3) All population variances are equal (homoskedasticity)

The data in this study meet these requirements. In the analysis that follows, important differences are reported as "observed significance levels" or "p-values."[6] A p-value is defined as the lowest level of significance at which a null hypothesis can be rejected. Within this context, a p-value may be interpreted as being the highest probability of making an error (type I) when a significant difference in the perception of risk is observed within a demographic or cultural group. All differences achieving a p-value of less than 0.20 are reported in the analysis that follows. We used the more sophisticated ANOVA instead of the simpler t-test because some of the demographic areas contained more than two subgroups (such as religion). The differences noted by ANOVA for two groups are the same as those noted by a *t*-test.

STATISTICAL RESULTS: LISBON VERSUS NEW YORK

In this section we compare the mean responses to the blocks of questions about each technology by country. Table 1 shows the mean scores associated with the six technologies. Significant differences (*p*<.20) are noted by an asterisk next to the technology; the *p*-value is in the last column.

Table 1. Mean Scores Associated with the Six Technologies

		Lisbon, Port.	New York City, U.S.	*p*-value
(1) NPG		6.73	6.81	
(2) R	(*)	4.28	4.82	.16
(3) GE	(*)	3.69	4.29	.04
(4) IT	(*)	5.05	5.53	.06
(5) SDT		5.64	5.79	
(6) NW		6.44	6.78	

Of interest is the similarity across the two cultures of these measures of perceived risk. For both samples, nuclear power generation and nuclear weaponry had the highest perceived risk. The ranking from highest to lowest is also identical for both countries.

The findings also confirmed the observation by Wildavsky that Americans are the most frightened; in each case the U.S.A. had a higher perceived risk than Portugal.

Each of these measures of perceived risk represents a composite of responses to a series of statements. Examination of the responses to individual statements within each group revealed some interesting differences between the two cultures.

Nuclear power generation (NPG) — For both groups the problem of nuclear waste disposal elicited the highest perceived risk. However, on those statements where we asked about their willingness to "trade-off" the usage of nuclear power for economic gains, the Americans were much more willing than the Portuguese to compromise, even though they had a higher perception of risk of nuclear danger and of nuclear accidents. This is an interesting result in the values of the two cultures, especially since Portugal is one of the poorest countries in all of Western Europe, and the cost of electrical power in Lisbon is relatively higher than in New York City. In discussion with the Portuguese students, it was apparent that they were knowledgeable and proud of their cultural heritage and values and committed to preserving their "way of life" even if it is at the cost of economic gains.

Nuclear weaponry (NW) — Both groups perceived a danger of war as a result of the Nuclear Arms race. The Portuguese, however, were almost equally divided between agreement and disagreement on the statement: "An accident that starts a nuclear war could happen tomorrow." In contrast, the Americans had a higher perception of risk of such an accident.

Space defense technology (SDT) — There were eight statements presented to measure perceived risk of space defense technology, or "StarWars." No statistically significant differences were found between the two groups on any of the individual statements. For both groups, the strongest agreement was on the statement: "A space based defense system will increase the 'price' of war."

Information technology (IT) — On eight of the nine statements used to measure perceived risk of information technology, the American students had a higher perceived risk than the Portuguese students. The highest perceived risk in both groups was the influence of the media on public opinion, that those who controlled the media had the greatest influence over the people and that mass communication did NOT lead to a well-informed public. The Americans, however, were significantly more concerned than the Portuguese with the loss of privacy. The Portuguese were more concerned than the Americans, however, that "Mass communications will eventually create a race thinking alike."

Robotics (R) — In general, the Portuguese exhibited much more favorable attitudes towards robotics and were less apt than the Americans to see it as a forerunner of job loss and less likely than the Americans to view robots as dehumanizing.

Genetic engineering (GE) — While genetic engineering overall had the lowest perceived risk for both groups, significantly more Americans were concerned with a catastrophe due to genetic research. Portugal, an agrarian country, was much more in favor of genetic engineering for every element that dealt with food production, whether animal or crop.

STATISTICAL RESULTS: MALE VERSUS FEMALE

Consistent with the findings in our earlier studies,[1,2] both in Portugal and the U.S.A., with very few exceptions, women evinced a higher degree of perceived risk with technology than men.

Table 2 shows the mean responses for each technology by sex for Portugal and Table 3 shows the same data for the U.S.A. Significant results are noted by an asterisk, with the significant p-value appearing in the last column of the table.

Table 2. Portugal

	Male	Female	p-value
(1) NPG (*)	6.44	7.45	.10
(2) R (*)	3.96	5.08	.07
(3) GE	3.67	3.75	
(4) IT	5.02	5.13	
(5) SDT (*)	5.36	6.31	.09
(6) NW	6.51	6.27	

In Portugal, women indicated a higher perceived risk than men in every area except nuclear weaponry (NW). On this measure, although the average is slightly higher for males, the difference is not statistically significant. Significant differences, however, are found in the areas of nuclear power generation (NPG), robotics (R), and space defense technologies (SDT).

254

Table 3. U.S.A.

	Male	Female	*p*-value
(1) NPG (*)	6.40	7.58	.04
(2) R	4.60	5.19	
(3) GE	4.36	4.17	
(4) IT	5.67	5.31	
(5) SDT	5.62	6.08	
(6) NW (*)	6.32	7.57	.03

For U.S.A. respondents, women, on the average, had a higher perceived risk than men in all technologies but genetic engineering (GE) and information technologies (IT). While the pattern for U.S. respondents by sex is interesting and consistent with our previous study, the only differences between men and women that prove to be statistically significant are for nuclear power generation (NPG) and nuclear weaponry (NW).

To arrive at a measure of perceived risk of nuclear power generation, three types of statements were developed:

    a) Fear of accidents (like at Chernobyl)
    b) Problems of waste disposal
    c) Willingness/unwillingness to trade economic gains for nuclear power.

In both Lisbon and New York, women had a higher perceived risk of accidents and were significantly more unwilling to trade economic gains for nuclear power. The Portuguese women had a higher perceived risk of problems of waste disposal than either the Portuguese men or the American women.

Comparing the New York and Lisbon cross tabulations by sex, it is interesting to note the similarities and dissimilarities in patterns. As shown in Tables 2 and 3, the averages for perceived risk of nuclear power generation (NPG) are virtually identical.

On nuclear weaponry (NW), American women, for each statement, significantly perceived a higher risk than did the Portuguese women or the men in either group.

The patterns of perceived risk of space defense technology (SDT) in the two countries were similar. There was a significant difference between the Portuguese men and women, with the women perceiving a higher risk. This was also true for the American sample, but the difference was not statistically significant.

There was a statistically significant difference in the perception of risk of space technology between the Portuguese men and women, with the women perceiving a higher risk. This pattern also pertained to the American sample, but the difference was not statistically significant. Of interest, too, was the difference between the Portuguese women and the American women. The Portuguese women were much more concerned with the "possibility of war" with a space-based defense system, and the American women were more concerned about "catastrophic errors."

The statements used to measure perception of risk of information technology (IT) essentially fell into two categories: infringement of privacy and influence of public opinion. Although there were no statistically significant differences between men and women in either country, both the Portuguese and the American men were more concerned than the women about the influence of the media. And, as has been previously stated, the Americans were more concerned than the Portuguese about the loss of privacy and confidentiality.

On every one of the seven statements about robotics (R), women had a higher perceived risk than men. This difference was statistically significant for the Portuguese sample but not for the U.S.A. sample. And, for most measures, the American men perceived a higher degree of risk and fear of the intrusion of robotics into the workplace than did the Portuguese men.

Although there was a statistically significant difference between Portugal and the U.S.A. on the perceived risk of genetic engineering (GE), there was very little difference between men and women within each country.

STATISTICAL RESULTS: URBAN VERSUS SUBURBAN

We divided the residence demographic into "urban" and "suburban." For the Portuguese sample, respondents living in urban areas perceived a higher degree of risk than the non-urban group for all six technologies. In contrast, for the New York sample, no differences in perception, based on place of residence, were found.

Table 4 below lists the mean response to the block of questions for each technology for the Portuguese sample.

Table 4. Portugal: Urban versus Suburban

	Urban	Suburban	p-value
(1) NPG (*)	6.99	5.40	.03
(2) R (*)	4.50	3.10	.07
(3) GE (*)	3.87	2.74	.03
(4) IT	5.12	4.68	
(5) SDT (*)	5.85	4.58	.07
(6) NW (*)	6.72	4.88	.05

With the exception of information technology (IT), all of the differences are statistically significant with a reliability of over 90%.

No differences were found in the New York sample on the basis of residence in this study. The urban "spread" in the Northeastern U.S.A. is much greater than in Portugal, where one encounters non-industrial, essentially agricultural areas just outside the center city. Thus, the sharp difference found in the Lisbon sample based on urban versus suburban residence may more correctly reflect the difference between urban versus rural perceptions.

STATISTICAL RESULTS: ANALYSIS BY RELIGION

Analysis of the data by religion within the Portuguese sample, provided a number of surprises. The first unanticipated result was that 17% of the Lisbon students classified themselves as "atheist," as compared to only 6% of the New York students. Thus for Portugal on the basis of religion, there were only two segments: Christian and atheist. The New York sample was too diversified to provide any meaningful analysis based on religion.

Table 5 provides the cross tabulations by religion for the Portuguese sample. Christians perceived a higher average risk than atheists for each of the six technologies.

Table 5. Portugal: Analysis by Religion

	Christian	Atheist	p-value
(1) NPG	6.81	6.42	
(2) R (*)	4.52	3.27	.08
(3) GE (*)	3.82	3.17	.18
(4) IT (*)	5.12	4.53	.13
(5) SDT	5.76	5.43	
(6) NW	6.58	5.87	

Portugal is a Catholic country so one must assume that virtually all of the Christians in the sample were, in fact, Catholic. This may explain the statistically significant difference in attitude between the two groups on genetic engineering (GE), which has been criticized by the Catholic Church. It does not explain, however, the other significant differences nor the pattern of higher perceived risk throughout. Can one perhaps assume that the atheist, without religion to fall back on, vests greater reliance on science?

SUMMARY

The comparison of risk perception of college business students by demographic areas in Portugal (Lisbon) and the U.S.A. (New York) provides some interesting parallels and differences. First, the ranking of technologies according to perception of associated risk was the same for both samples with the two nuclear technologies causing the most concern. Second, in almost all cases, women perceived more risk than men. These two findings are in agreement with the results of previously cited studies.[1,2] Finally, for all technologies, the American sample, overall, perceived more risk than did the Portuguese sample, overall. This seems to corroborate the statement attributed to Wildavsky[4] about Americans being the most concerned about new technologies.

REFERENCES

1. S. B. Ahmed and M. P. Hoefer, "Public Perception of Risk in Developing Technologies: A Case Study of a Business Community in Manhattan," in: *Risk Assessment and Management*, Vol. 5, pp. 325-335, L. B. Lave, Ed., Plenum Press, New York (1985).

2. M. P. Hoefer and S. B. Ahmed, "How a Business Community Perceives Risk in Modern Technologies," published by the Center for Applied Research (Working Paper #41), Pace University, New York, NY, March 1985.

3. T. Englander, K. Farago, P. Slovik, and B. Fischoff, "A Comparative Analysis of Risk Perception in Hungary and the United States," *Social Behavior* I:55-56 (1986).

4. A. Wildavsky, *American Scientist* **67**:32 (1979).

5. G. W. Snedecor and W. G. Cochran, *Statistical Methods*, 6th ed., Iowa State University Press, Ames, IA, 1967.

6. H. Kohler, *Statistics for Business and Economics*, Scott-Foresman and Co., Glenview, Ill, pp. 596-597, 1985.

Policy Lessons from Risk Communications Practice[a]

Rob Coppock

National Academy of Sciences/National Research Council
Washington, DC

ABSTRACT

This paper briefly addresses the difficulties of risk communication, and describes three lessons concerning the activities. First, only in the rarest instances can effects be observed that follow directly from specific communication efforts; the decision arena and the context within which the communication takes place are of great importance. Second, a variety of different kinds of risk messages usually need to be received before an effect may be observed; reinforcement is essential. Third, the risk communicator faces a fundamental dilemma between clarity and completeness. Sacrificing completeness for clarity will offend scientists and may undermine the legitimacy of the analysis; sacrificing clarity for completeness will confuse the lay public and may render the message ineffectual.

KEYWORDS: Risk communication, policy lesson, clarity, completeness

Several months ago, when I agreed to make a presentation on policy lessons from risk communications practice, I expected to be able to draw upon several case studies prepared for the National Academy of Sciences study on risk perception and communication. However, that study has proceeded in different directions than I had imagined, so that my comments today rely on materials drawn from the open literature rather than from sharply focused case studies of the practice of risk communications. I apologize for any false advertising in the announced title of my talk.

I want to start by conducting a little thought experiment. Imagine that somehow we were able to circumvent the human subjects committee and force a set of intelligent, well educated lay people to read the literature on risk communication. I think that one response would be surprise at the degree to which careful scientific studies have produced conclusions about how to communicate risks that correspond to common sense. For example, let me paraphrase just a couple of examples from the literature.

---

a. The views expressed here are those of the author, and do not necessarily reflect the policies or positions of the National Academy of Science or the National Research Council, or any units associated with them.

1. Avoid technical and bureaucratic language and address people's concerns; provide information that people can understand and relate to their own experience.

2. There is not a single, homogeneous public; listen to people, understand their interests and attitudes, and speak to their concerns.

3. Be clear and open; state your assumptions, your uncertainties, and the methods used to obtain your conclusions.

How can it be that risk communicators need to be told to do what any responsible adult would do when communicating with his or her neighbors? I think, at least in most cases, that risk communicators are neither ignorant nor do they lack good will. Rather, risk communication is so complicated that even the simplest tasks cannot be accomplished.

This means that risk communication is exceedingly difficult. Consider the record. If there is one thing that experience shows, it is that no matter how well you communicate, it probably won't work.

1. Bill Ruckelshaus did an excellent job in the press conferences about the EDB crisis; his presentation flawless, the scientific basis explained, the arguments supported with visual material. But it still wasn't enough to deal with the crisis.

2. The basic data about the extent to which wearing seat belts lessens the likelihood of injury has been forcefully communicated to the public. The communications were successful and were understood. But seat belt laws have been repealed in several states anyway, for reasons that have nothing to do with the degree of risk.

There are, however, a few success stories. The Stanford Heart Disease Prevention Program Three Communities Study is one. It demonstrated the ability to influence people's behavior in ways that reduced the likelihood of their contracting heart disease. But it also had a number of factors unavailable to most risk communicators. Not only was there a mass media communications program, but also an active program of face-to-face instruction targeted at the high-risk population.

I suppose the anti-smoking campaign is another example of success. Again, it is unusual. Perhaps the best summary of the reasons for its success is that smokers heard the same message from the family physician, the high school biology teacher, newspapers, and television — and after twenty years we begin to observe effects.

The reason for going through this depressing litany is to suggest that risk communication is not something that can be judged in isolation. In each of the examples I mentioned, the most important factors influencing the outcome were not the extent to which the communications were clear, forceful, and so on, but rather the groundswell of fear, the dominance of motivations for personal control over individual action, the power of peer pressure, or the difficulty of changing habits involving addictive substances.

So I believe the first policy lesson from risk communication practice is that in only the rarest instances can you expect to observe effects that follow directly from specific communications efforts. The outcome may depend upon entirely different factors than the quality of the risk communications effort. What really matters is the decision arena and the context within which the decision takes place.

Of course, this does not mean that you should not pay careful attention to your communications activities. For it will be difficult enough to do a good job of communicating risks. But there will be many things in addition to your communications that affect outcomes.

One reason risk communication is so difficult is the complexity of the knowledge to be transmitted. In fact, risk communication probably needs to include a broader range of information than is generally incorporated in risk assessments.

One purpose of risk communication is to transmit information for an informed decision. The decision maker sitting at the apex of an organization, be it a government agency, a private enterprise, or a public interest group, will want to be knowledgeable about four different kinds of information with respect to a particular choice: technical information, economic information, legal information, and political information. Technical and economic information pertain to the hazard, its likely consequences, and possible interventions aimed at preventing, alleviating, compensating, etc. Legal and political information describe prescriptions and proscriptions on action, as well as the extent of support or opposition.

Risk assessments usually contain quite a lot of technical information, and usually some economic information. But they are generally sparse when it comes to legal and political information. This is probably as it should be. The political and administrative traditions in this country call for decisions to be based on the facts in so far as possible. The thrust of administrative law is to provide a written record of the facts, of the procedures used to develop and check them, and of the way the decision is devised. In this context, it is useful to have a clear statement of the facts as a part of the policy making process. Risk assessments usually strive to serve this purpose.

Yet it is clear that an informed decision will draw upon a broader spectrum of information than is usually found in risk assessments. Thus we may conclude that risk communications will need to be broader than risk assessments. I think that risk communication to the public should present the same spectrum of information that is required by decision makers. I think that public communications programs should also refer to technical, economic, legal, and political factors as appropriate.

Thus, the second policy lesson I would draw from risk communications practice is that a variety of different kinds of information must be combined to be effective. Of course, the process of combining them is neither straightforward nor simple.

Now this brings us to a basic dilemma: in order to communicate effectively, you want a relatively simple, straightforward message; but in order to be professional and build credibility, you want to rely on a comprehensive and sophisticated understanding of the phenomenon under question. This dilemma between clarity and completeness is imbedded in most risk communications situations.

It is a dilemma within the technical and economic components, as well as within the other information upon which decision makers will draw. Let me illustrate some of the difficulties with an example from risk assessment.

The way we usually characterize uncertainty is incomplete in a fundamental sense. Neither point estimates nor worst case scenarios present all the uncertainties involved. Confidence intervals express only the arithmetic uncertainties associated with the mathematical and statistical calculations. We simply have no systematic and rigorous way to express other kinds of methodological and scientific uncertainties.

I think this is a problem of characterizing scientific judgment in general. Let me give you an illustration. At the National Academy of Sciences we recently conducted a conference on the valuation of health effects in environmental decisions. We invited both practitioners of risk assessment and cost/benefit or cost/effectiveness in federal agencies, as well as a few national scientists actively involved in some aspect of the regulatory process. One case that was discussed at length, regulation of arsenic from copper smelters, led scientists to express concern about the rationalistic assessment models under discussion. As I read the transcript, they felt that the rationalistic models fail to capture the concerns that scientists have in mind when thinking about the issue, and that something very important is lost in the process. Recently developed information that arsenic is a carcinogen dominated the rationalistic assessment, but the scientists saw that new data in terms of decades of knowledge that arsenic is a neurotoxin. Concern was expressed that the analytical model failed to incorporate adequately all the factors involved in the scientific judgment. We have trouble enough communicating to lay people the mathematical uncertainty, which we can characterize with vigor. We have only the most rudimentary methods for characterizing conceptual and methodological uncertainty.

This is the last policy lesson I wish to draw from the practice of risk communication. The risk communicator faces a fundamental dilemma: between clarity and completeness of communication. Sacrificing completeness for clarity will offend scientists and may undermine the legitimacy of the analysis; sacrificing clarity for completeness will confuse the lay public and may render the analysis ineffectual. There is no fine line to follow. Rather, a choice must be made in each case. And each choice will have both advantages and disadvantages.

A final comment: I have not addressed the practice of risk communication in detail, so that my lessons for policy, if they are lessons at all, are of a fairly general nature. There are also conclusions to be drawn of a much more detailed and differential sort. Some approaches and methods are most likely to work best in certain circumstances. Some ways of presenting complex information are clearer and more easily understood than others. These more detailed conclusions will be of considerable help to risk communicators and others. And they are beginning to emerge. The work that Vince Covello is describing in this session is one example, and the study on risk perception and communication currently underway at the Academy is another. But I believe constructive skepticism of the sort I have used to derive my policy lessons will help keep expectations in proper perspective.

History of the Impurities Policy: Risk Assessment for Food and Color Additives

L. Robert Lake and Kennon M. Smith

U.S. Food and Drug Administration
Washington, DC

ABSTRACT

The FDA impurities policy embodies a regulatory approach which distinguishes between a food or color additive as a whole and its constituents for the purpose of determining when the Delaney clause is triggered. Under the impurities policy, the Delaney clause does not apply to a carcinogenic impurity in a food or color additive unless there is a finding that the additive as a whole induces cancer. This interpretation of the Delaney clause enables the agency to use risk assessment to determine whether, under the general safety provisions for food and color additives, there is a reasonable certainty that no harm will result from the proposed use of an additive. The use of risk assessment in the implementation of the impurities policy permits the FDA to manage insignificant risk from low-level carcinogens consistent with the anti-cancer prohibitions of the Delaney clause.

KEYWORDS: Delaney clause, general safety (provisions) clause, impurities policy, risk assessment

INTRODUCTION

Generally speaking, before most food and color additives may be used in the food supply, they must receive pre-market approval from the U. S. Food and Drug Administration (FDA). The approval requirements are set forth in the Federal Food, Drug, and Cosmetic Act (FD&C Act),[1] which was amended, in part, by Congress in the Food Additives Amendment of 1958,[2] and again in the Color Additive Amendments of 1960.[3] The safety of food and color additives is assessed by FDA according to the provisions of the so-called "general safety"[4] and "Delaney"[5] clauses of the FD&C Act applicable to such additives.

In an effort to infuse some flexibility and practicability into the stringent anti-cancer prohibitions of the respective Delaney clauses, the FDA developed its impurities policy. Today, the impurities policy is implemented by the agency as a mechanism by which it can regulate food and color additives that contain insignificant amounts of carcinogenic substances. In order to appreciate fully the context in which this regulatory policy was developed, it is necessary to have some understanding of the historical development of risk assessment and its role in the formulation of the impurities policy.

THE GENERAL SAFETY CLAUSE

The general safety clauses applicable to food[6] and color[7] additives, respectively, provide that such additives cannot be approved for use unless data presented to the FDA establish that the additives are safe. The concept of safety has been codified in the federal food[8] and color[9] additive regulations. The regulations provide that "safe" means that there is a reasonable certainty that the substance is not harmful under the intended conditions of use. Current scientific knowledge is inadequate to establish with complete certainty the absolute harmlessness of the use of any substance. Thus, absolute safety is not required. Safety may be determined by scientific procedures or by a general recognition of safety. In its safety assessment of a given additive, the FDA will also consider the probable consumption of the substance, the cumulative effect of the substance in the diet, and any other safety factors deemed appropriate.

THE DELANEY CLAUSE

The Delaney clause is an anticancer provision. Simply stated, it provides that no additive shall be deemed to be safe if it is found to induce cancer when ingested by man or animal, or if it is found, after tests which are appropriate for the evaluation of the safety of such additives, to induce cancer in man or animals.

The Delaney clause was first enacted in 1958 as part of the Food Additives Amendment of the FD&C Act.[10] It was later included in the 1960 Color Additive Amendments of the Act.[11] In ostensibly unequivocal language, the Delaney clause prohibits the use of carcinogenic additives in the food supply. This provision, which has been the focus of continual controversy since its inception, is today viewed by many as a legislative dinosaur caught up in an era of rapidly advancing science and technology.

During the lifetime of the Delaney clause, we have experienced tremendous improvements in our methods of detecting substances in foods. Because of these highly-advanced improvements in our methods of analysis, we are now capable of detecting carcinogens in our food supply in minute amounts.[12] Not surprisingly, therefore, as more food constituents have been investigated for potential carcinogenicity, more have been determined to be carcinogenic. Thus, we are no longer as naive about our potential exposure to carcinogens as we once were. Similarly, toxicological methodology has also improved, although not with equal advancement.[13] Such improvements, however, have nevertheless led to an increased certainty with which we can assure the safety of our food supply.

Advancements in science and technology are not always accompanied by reformations in the law which governs that to which the science and technology are applied. The enactment of the Delaney clause nearly three decades ago provided additional governmental assurance as to the safety of the food supply. However, because we are today regulating substances previously undetectable in foods and still using the Delaney clause of yesterday, rulemaking has become increasingly awkward and public confidence increasingly skeptical.

While the public often expects the FDA to ensure absolute safety through its implementation of the Delaney clause, the agency has recognized for some time that absolute safety is an unattainable goal since quantification of actual risk is imprecise. The FDA's recognition and acceptance of some uncertainty in the process of risk assessment has provided the necessary flexibility to allow the agency to continue its regulatory decision-making in accordance with (and in spite of) the Delaney clause.

The FDA encountered great difficulty in its attempts to make regulatory decisions about low-level carcinogens now being detected in some food additives. Furthermore, it was faced with the troublesome task of determining the magnitude of potential risk intrinsic in such substances. The agency reluctantly acknowledged the inevitability of some exposure to low-level cancer risk, and then took up the challenge of developing a mechanism by which it could regulate low-level carcinogens consistent with the Delaney clause.

Through its use of agency discretion and statutory interpretation, the FDA first developed a risk assessment approach to regulating animal drug residues in edible tissues in 1973. Subsequently, the agency developed the "constituents policy," now known as the "impurities policy." In implementing its impurities policy, the FDA modified its earlier interpretation of both the Delaney clause and the general safety clause provisions of the food and color additive amendments.

MONSANTO CO. v. KENNEDY

A significant "stepping stone" along the road to the FDA's adoption and implementation of the impurities policy was established in 1979 when the United States Court of Appeals decided the case now simply referred to as *Monsanto Co. v. Kennedy*.[14] This case played an important role in the later development of the FDA impurities policy because it provided judicial underscoring of broad administrative discretionary powers that the agency ultimately exercised in the implementation of that policy.

The *Monsanto* Case dealt with the specific issue of whether a minute amount of a substance which might migrate from food packaging materials into food constituted a food additive under the FD&C Act.[15] The Court acknowledged that the FDA has discretionary authority, inherent in the statutory scheme, to avoid literal application of the statutory definition of "food additive" where negligible migration clearly presents no public health or safety concerns. Thus, the *Monsanto* decision allowed the FDA to determine that the application of the Delaney clause is not triggered by the presence of a carcinogenic substance which occurs in a non-carcinogenic food additive if the amount of the substance is so small that it presents a reasonable certainty of no harm. The *Monsanto* decision does not state, however, that an additive which has been determined, as a whole, to cause cancer could escape the application of the Delaney clause.

Another important distinction is worthy of note: The Court was concerned with the issue of *de minimis* migration, not *de minimis* risk. The Court did not affirm the use of risk assessment procedures, but rather focused upon the issue of what amount of a migrating substance present in food is sufficiently significant to constitute a "food additive."

The reasoning of this decision supported a regulatory approach later suggested by the FDA in an Advance Notice of Proposed Rulemaking.

1982 ADVANCE NOTICE OF PROPOSED RULEMAKING — POLICY FOR REGULATING CARCINOGENIC CHEMICALS IN FOOD AND COLOR ADDITIVES

In 1982, the FDA issued an Advance Notice of Proposed Rulemaking which proffered a new policy for assuring the safe use of food and color additives containing minor amounts of carcinogenic chemicals.[16] It was suggested that this policy would replace the previous case-by-case approach to regulating such additives with an approach that applied consistent standards.

The prevailing scientific conviction in 1958 was that the state of the art would not permit scientists to establish a tolerance for a carcinogen. Thus, it was impossible to determine reliably a level of exposure to a carcinogen below which one could be assured that there was no significant increase in the risk of developing cancer. At the time, scientists lacked the technology to assess adequately the risk presented by a particular chemical.

During the two decades that followed, the FDA, with few exceptions, interpreted the amendments to the FD&C Act to ban the use of any additive that was found to contain or was suspected of containing minor amounts of carcinogenic chemicals, even if the additive as a whole had not been found to cause cancer. The agency generally concluded that a safe level could not be established for a carcinogenic chemical in a food additive and consistently took regulatory action against additives that contained carcinogenic chemicals.

During the same twenty years, however, there were rapid developments in analytical capabilities that made it increasingly possible to detect and identify components of substances (such as food and color additives) at ever lower levels. Coupled with this development was a great increase in the number of substances studied for carcinogenicity in animal bioassays.

As the number of chemicals found to cause cancer in animals grew, and as the ability to detect the components of a substance became more acute, the chances that a food or color additive would be found to contain a carcinogenic component increased. The agency reasoned, however, that not all additives found to contain carcinogenic components would themselves be shown to induce cancer. Yet, if the FDA continued to implement the regulatory approach it had followed in the past, it would be forced to refuse approval of each such additive, even though the additive itself was safe.

It had become clear that the FDA would have to reconsider how it determined whether there had been an adequate demonstration of the safety of a food or color additive that contained a carcinogenic component but that was not itself carcinogenic. The agency thereafter began to distinguish between carcinogenic food and color additives and those food and color additives containing a carcinogenic component, where such additives had not been shown to be carcinogenic.

It was in this regulatory context that the FDA introduced its impurities policy. Citing the *Monsanto* decision as support for this new regulatory approach, the agency reasoned that if the FDA has discretion to disregard low-level migration into food of substances from additives because such migration presents no public health concern, then the FDA may also disregard, after appropriate tests, a carcinogenic chemical in a non-carcinogenic food additive, provided the agency determines that there is a reasonable certainty of no harm from the chemical. Furthermore, the agency expressed confidence that it now possessed the capacity, through the use of extrapolation procedures, to assess adequately the upper level of risk presented by the use of a non-carcinogenic additive that contained a carcinogenic chemical.

1982 FINAL RULE — D&C GREEN NO. 6

Simultaneous with the publication of the 1982 Advance Notice of Proposed Rulemaking previously discussed, the FDA issued a final rule permanently listing the color additive D&C Green No. 6 for use in externally applied drugs and cosmetics.[17]

D&C Green No. 6 is a dye containing residual amounts of unreacted chemicals used in the manufacture of the color additive. One of the residual chemicals surviving the manufacturing process of D&C Green No. 6 has been demonstrated to be an animal carcinogen. The unavoidable carcinogenic constituent detected in D&C Green No. 6 did not contribute any color to the additive, nor did it impart any color to drugs, cosmetics, or the human body. Consequently, the FDA concluded that the residual chemical was not a color additive within the meaning of the FD&C Act, but rather was merely an impurity in the color additive. The agency recognizes that residual amounts of reactants and manufacturing aids are commonly found among the constituents of many color additives. The presence of such constituents is not unique to color additives; in fact, numerous contaminants are unavoidably present in all chemical products, including highly purified chemicals.

Under the general safety clause for color additives, a color additive cannot be approved for a particular use unless the data presented to the FDA establish that the color is safe for that use. In earlier actions, the FDA had terminated approval of several color additives that contained or were suspected of containing carcinogenic impurities or constituents. The agency had been unable to resolve questions about the safety of use of such additives. However, by 1982, the FDA no longer believed it must refuse to approve a color additive simply because it contained or was suspected of containing a carcinogenic impurity or constituent. The agency reasoned that by exercising agency discretion previously endorsed by the *Monsanto* Court, it could find the presence of a carcinogenic constituent or impurity in a color additive to be so insignificant as to present no public health or safety concern. Furthermore, the agency explained that through the use of improved risk assessment procedures, it could now, with confidence, adequately estimate the upper limit of risk presented by use of such a color additive. Relying on conservative extrapolation models and risk assessment procedures inherent in the agency's implementation of its impurities policy, the FDA evaluated the safety of D&C Green No. 6 and thereafter approved its use as a color additive.

SCOTT v. FDA

Later in 1982, the FDA announced its approval of the color additive D&C Green No. 5 for use in drugs and cosmetics.[18] D&C Green No. 5 contains minute amounts of the same unwanted but unavoidable carcinogenic impurity as does D&C Green No. 6. Once again using its constituents policy to conclude that the impurity was not the color additive subject to the Delaney Clause, the agency likewise approved D&C Green No. 5. The FDA found that the color additive as a whole was safe under the general safety clause after quantitative risk assessment procedures had established that the color additive presented an insignificant risk under conditions of use.

The case of *Scott v. FDA*[19] involved the judicial review of the FDA's decision to approve the use of D&C Green No. 5. Relying heavily on the reasoning underlying the agency's approval of this color additive, the United States Court of Appeals rejected the challenge to the FDA's action and affirmed the regulation approving the color additive.

The *Scott* decision represents a significant triumph for the FDA. The Court accorded great deference to the agency in upholding its interpretation of the Delaney Clause. The Court thus ratified the impurities policy as an application of a reasonable statutory interpretation appropriately within the administrative discretion of the agency. Moreover, the Court held that the FDA acted lawfully in its use of risk assessment procedures to conclude that the use of the additive posed an insignificant risk of human cancer.

THE IMPURITIES (CONSTITUENTS) POLICY

The impurities policy embodies a regulatory approach which distinguishes between the additive as a whole and its constituents for the purpose of determining when the Delaney clause is triggered.[20] The policy consists of three elements:

1. clarification of exactly what a "food additive" is;

2. interpretation of the Delaney clause to apply only when the additive itself has been shown to cause cancer; and

3. use of risk assessment to determine whether the additive is safe under the general safety clause.

Using this approach, the food additive would be the substance that is actually intended for use in food or for food contact. All nonfunctional chemicals present in that substance are characterized as "impurities" of the additive. Impurities include chemicals such as residual reactants, intermediates, and manufacturing aids, as well as products of side reactions and chemical degradation. An impurity, although part of the additive as a whole, would not itself be considered to be an additive for purposes of the Delaney clause because it is not intended for use in food or for food contact.

As discussed earlier, the Delaney clause requires the disapproval of any food or color additive that has been shown by appropriate testing to be a carcinogen. However, the Delaney clause does *not* expressly require the disapproval of a food additive containing a carcinogenic impurity where the additive itself, taken as a whole, is non-carcinogenic. Under the FDA impurities policy, the agency's statutory interpretation establishes that the Delaney clause does not apply to a carcinogenic impurity in a food additive unless there is a finding, after appropriate tests, that the additive as a whole induces cancer. This interpretation of the Delaney clause enables the agency to use risk assessment appropriately to determine whether, under the general safety provisions for food and color additives, there is a reasonable certainty that no harm will result from the proposed use of the additive.

During recent years, it has become necessary for agency scientists to develop quantitative extrapolation procedures to estimate human risk from low-level exposure to carcinogenic chemicals, both because carcinogenic potency is so widely variable and because human exposure to carcinogenic chemicals in food additives would occur at low levels. Quantitative risk assessment models can be used to estimate the upper limit of potential risk created by exposure to such a chemical. After relevant animal data have been thoroughly evaluated, an appropriate risk assessment procedure can then be used to calculate an acceptable maximum exposure level to the chemical based on a suitable risk standard. Once the FDA has determined a maximum acceptable level of exposure to a carcinogenic chemical, the agency would then consider whether the amount of actual exposure to the chemical from food additives is within the acceptable level. If necessary, the agency would develop specifications for such chemicals in individual additive regulations, taking into account total estimated exposures.

The general safety provisions for food and color additives dictate that such additives must meet the FDA's toxicological requirements before they can be approved. The use of risk assessment procedures under the impurities policy does not weaken the requirements of the general safety clause, but rather provides a regulatory approach for determining whether a non-carcinogenic additive that contains minor amounts of a proven carcinogenic chemical is safe. Thus, risk assessment procedures provide a method for estimating the levels of such chemicals that meet the general safety clause standard of safety.

The impurities policy is today implemented by the FDA as a practicable, scientifically sound, and prudent means of limiting those instances in which it would have been otherwise necessary for the agency to disapprove the use of a food additive, without compromising the public health protection afforded by the FD&C Act. This policy is implemented solely in those instances where data demonstrate that there is a reasonable certainty that no harm will result from the use of a food or color additive that contains a carcinogenic chemical.

RISK ASSESSMENT

Throughout the recent conceptual development of risk assessment at the FDA, one principal has remained axiomatic: knowledge of the actual risk to humans from relatively low-level exposures to animal carcinogens is elusive.[21] Partly because of this uncertainty, FDA had earlier been reluctant to use risk assessment procedures for regulatory decision-making. Many theoretical models have been developed to extrapolate from animal experimental data to the relatively low levels of possible human exposure, but they can vary widely in the risk values that they predict.

The primary weakness of risk assessment derives from the fact that it is predicated upon substantial assumptions. Such assumptions do little to bolster confidence in the accuracy of predictions of actual risk using these procedures. Nevertheless, the agency is confident that by using certain conservative extrapolation models it is possible to estimate an upper limit of risk. While the estimate of the risk may be exaggerated by conservative extrapolation models, the estimated risk will not be understated. Risk assessment can therefore be used to determine whether, under the general safety provisions for food and color additives, it has been established with reasonable certainty that no harm will result from the intended use of a particular additive.

CONCLUSION

Invention is oft purported to have been mothered by necessity. Likewise, both risk assessment and the impurities policy were conceived of necessity — in particular, an agency's need for a practicable approach which would allow regulatory decision-making to continue despite the increased detection of low-level carcinogenic impurities in food and color additives.

The use of risk assessment procedures and the implementation of the impurities policy have evolved concurrently at FDA, being ultimately combined to form a mechanism for regulating low-level carcinogens consistent with the Delaney clause and its anti-cancer prohibitions. Applied sensibly, both concepts are today utilized by the FDA as effective tools with which to manage insignificant risk of cancer to humans from widespread and innovative uses of food and color additives.

REFERENCES

1. Federal Food, Drug, and Cosmetic Act (1938), Pub. L. 717, ch. 675, 52 Stat. 1040, as amended; codified at 21 U.S.C. sec. 301 *et seq*.
2. Food Additives Amendment of 1958, Pub. L. No. 85-929, 72 Stat. 1784 (1958).
3. Color Additive Amendments of 1960, Pub. L. No. 86-618, 74 Stat. 397 (1960).
4. 21 U.S.C. sec. 348(c)(3) [food additives]; 21 U.S.C. sec. 376(b)(4) [color additives].
5. 21 U.S.C. sec. 348(c)(3)(A) [food additives]; 21 U.S.C. sec. 376(b)(5)(B) [color additives].
6. 21 U.S.C. sec. 348(c)(3)(A).

7. 21 U.S.C. sec. 376(b)(4).
8. 21 C.F.R. sec. 170.3(i), definition of "safe" with respect to food additives.
9. 21 C.F.R. sec. 70.3(i), definition of "safe" with respect to color additives.
10. See *supra* note 2; Delaney clause applicable to food additives codified at 21 U.S.C. sec. 348(c)(3)(A).
11. See *supra* note 3; Delaney clause applicable to color additives codified at 21 U.S.C. sec. 376(b)(5)(B).
12. Scheuplein, "Risk Assessment and Food Safety: A Scientist's and Regulator's View," *Food Drug Cosm. L. J.* **42**:237 (1987).
13. *Id.*
14. *Monsanto Co. v. Kennedy*, 613 F.2d 947 (D.C. Cir. 1979).
15. 21 U.S.C. sec. 321(s), definition of "food additive".
16. 47 Fed. Reg. 14,463 (1982).
17. 47 Fed. Reg. 14,138 (1982).
18. 47 Fed. Reg. 24,278 (1982). See also final rule; termination of stay and confirmation of effective date, 47 Fed. Reg. 49,628 (1982).
19. *Scott v. Food and Drug Administration*, 728 F.2d 322 (6th Cir. 1984).
20. See *supra* note 16.
21. See *supra* note 12.

Establishing a Threshold of Regulation

Alan M. Rulis

Food and Drug Administration
Washington, DC

ABSTRACT

The 1958 Food Additives Amendment to the Federal Food, Drug and Cosmetic Act defines a food additive as "...any substance, the intended use of which results or may reasonably be expected to result, directly or indirectly, in its becoming a component ... of ... food...." Thus, even materials which may migrate to food in very small quantities from packaging materials have been subjected to regulation. With advances in analytical methods of detection over the years, ever lower levels of migrating substances are technically subject to food additive regulation. Is there some level below which a substance need not be considered a food additive? The agency has grappled with this "Threshold-of-Regulation" (T/R) question ever since 1958 and generally has handled such situations on a case-by-case basis. Increasingly, however, there have been calls for the agency to develop a T/R policy. To develop such a policy, the agency must consider a number of important issues such as the reliable measurement of low levels of migration; low dose risk assessment; the likely potencies of possible carcinogens; and an array of legal and regulatory constraints. The presentation will center on possible ways the agency might weave together this array of considerations to arrive at a viable policy for the Threshold of Regulation.

KEYWORDS: Food additive, *de minimis*, Threshold of Regulation

INTRODUCTION

The Food, Drug, and Cosmetic Act (the FD&C Act, or the Act) defines a food additive as:

> "...any substance the intended use of which results or may reasonably be expected to result, directly or indirectly, in its becoming a component or otherwise affecting the characteristics of any food (including any substance intended for use in producing, manufacturing, packing, processing, preparing, treating, packaging, transporting, or holding food...."[1]

Congress added this definition to the Act in 1958, at the same time it passed the Food Additives Amendment requiring that any new food additive be shown to be "safe" and that it be the subject of an approved food additive petition before use in food. The development of the petition containing the required safety data is the responsibility of the petitioner.

The above definition is very broad. Although it specifically excludes certain special categories of substances such as those that are "generally recognized as safe" (GRAS) and those whose use was sanctioned prior to 1958, it nevertheless captures within the premarket petition process a wide range of substances. These include not only chemical substances added directly to food to accomplish an intended technical effect in the food itself, but also food packaging and other food-contact materials and their components that might migrate unintentionally into food. These additives are referred to as "indirect additives." Over the years the FDA has reviewed hundreds of petitions for indirect food additives, and the present inventory of such additives totals thousands of substances.

It should not be surprising that ever since 1958 there has been debate about whether the FD&C Act ought to be interpreted generally to require premarket safety evaluation of indirect additives at all. There has also been discussion about whether, and to what degree, the agency may consider certain food-contact situations to be outside the food additive definition and thus beyond the scope of application of the premarket safety evaluation requirements of the Act; this may occur when the level of migration to food of a given substance is very low.[2,3] Indeed, the Food and Drug Administration has needed to decide, case-by-case, over the years whether petitions should be required for certain indirect situations, particularly when the probable migration to food of a given chemical substance was exceedingly small or nonexistent. Such situations could be considered to be "below the Threshold of Regulation."

The agency's traditional scheme of premarket petition requirements for food-contact materials is fairly simple in principle.[4] When materials migrate into food in amounts that correspond to probable levels of exposure of 1 ppm or more in the human diet, the usual regimen of testing would generally require a full battery of animal feeding studies, including studies of chronic duration. Between 50 parts per billion (ppb) and 1.0 ppm, only subchronic (90 days in rodents) studies are usually required. Below 50 ppb exposure is said to be "virtually nil," but a petition that includes acute toxicity data is still generally required. The focus for application of a threshold-of-regulation policy is not, of course, those levels of exposure that are too low to be of potential concern. For example, situations based on theoretical calculations with dietary exposures of only parts per quadrillion (ppq) or lower usually can be resolved with little or no effort. However, food-contact exposures below 50 ppb but not so low as to be below any reasonable level of concern (e.g., the 1.0 ppb range) present the greatest challenges. For situations in this so-called "gray zone" the need for premarket safety evaluations may be justified, especially if potential risks could exceed a trivial level. Often, however, a petition may not be needed. In these situations a new threshold-of-regulation policy could be helpful in aiding decision making.

In recent years, threshold-of-regulation decisions have become a more pressing problem as a result of a number of trends. Over the last three decades analytical chemical methods of measurement have become more sophisticated and powerful. Today the analyst may be able to measure in food (or in food-simulating solvents) the presence of migrating additives at levels as low as one ppb or even lower. Furthermore, with the continual growth of the inventory of chemical substances tested in carcinogenicity bioassays, the list of carcinogens has been continually expanding. (Carcinogenic food additives would, of course, be proscribed under the Delaney Clause of the Food Additives Amendment.)

Another driving force is the legal doctrine of *de minimis*. This doctrine, usually expressed in Latin as *de minimis non curat lex* (the law does not concern itself with trifles), states the idea that the law need not be literally applied in certain situations involving trivial infractions. (This doctrine is intrinsic to most statutory schemes.) In *de minimis* situations it would serve no useful purpose to enforce the law literally because the net societal benefit would be negligible or perhaps even negative, and to do so might not be helpful in "implementing the legislative design."[5]

Still another concept giving impetus to development of a threshold-of-regulation policy is the so-called "principle of commensurate effort." According to this idea, a regulatory agency should distribute its limited resources according to priority, based on informed estimates of probable hazard. That is to say, for any group of similar regulatory situations, the agency's use of resources should roughly increase with the presumptive threat to public health. In the case of indirect food additives, however, we have found at the FDA that the principle of commensurate effort has not always been applied. We find that trivial or near-trivial issues can at times consume resources out of all proportion to their potential public health consequences while more important issues wait for resolution.

The trends outlined above demonstrate why the agency should develop a consistent policy under which it can better handle potential threshold-of-regulation situations. The challenge to the FDA is to provide simultaneously: a) a means to waive regulation of noncarcinogenic food-contact substances with predicted levels of migration to food that are so low as to not satisfy the definition of a food additive; b) a means to protect public health when the migrating substance turns out to be a carcinogen or other potent toxin; c) some means of monitoring its own threshold-of-regulation decisions over time to permit consistent decisions and any needed reevaluation in light of new information; d) a means of setting priorities so that problems of greater concern for public health protection receive a greater share of the available resources.

For the past year-and-a-half an FDA task force has been analyzing these issues with the goal of developing a set of recommendations for a threshold-of-regulation policy. This paper will describe some of the issues that the task force has dealt with, some of the risk assessment concepts applicable to the problem, and ways to weave together administrative, legal, and scientific perspectives into a workable policy.

REQUIREMENTS FOR A THRESHOLD-OF-REGULATION POLICY

A threshold-of-regulation policy, to be effective, ought to meet certain criteria. First of all, it must provide the agency with criteria to judge consistently whether given situations with a proposed food-contact material are indeed below the threshold-of-regulation. These criteria must be grounded, to the extent possible, on a firm scientific foundation. Only then will they possess credibility and be defensible. Specifically such criteria should contain: (1) the probable level of migration of the chemical substance to food and the resulting probable human dietary exposure; (2) the "use conditions" of the chemical substance, i.e., whether it is intended for repeated use or single service, and whether it is part of a complicated food package design such as a multiple-layer laminate construction presenting issues of "functional barriers"; (3) a probable upper-bound risk estimate, if a potential migrant should turn out to be a carcinogen; (4) other factors such as the temperature and the length of time that food will be in contact with the substance; the amount of food contacting a given amount of the food-contact material containing the chemical of interest; and any applicable factors that relate to the total diet, the amount of food receiving the migrant.

The policy will also need to specify procedures to maintain appropriate records, to handle all situations consistently, and to carry out the policy in a cost-effective manner. Specifically, such procedures will need to consider: (1) "who" will make threshold-of-regulation decisions, industry independent of the FDA (a "do-it-yourself" policy), the FDA itself, or some neutral expert panel of scientific peers; (2) the FDA makes the decisions whether it should use an internal panel, some "authorized" person, a full-blown review committee, or, possibly, even some type of "expert-based" computer program. Furthermore, some mechanism must assure that the execution of the policy does not run counter to the principle of commensurate effort. In principle, the effort expended by the agency in deciding on the threshold-of-regulation status of a specific situation should not exceed the effort alternatively expended by the agency in processing a petition for the same substance at a comparable level of exposure. The "effort equation" is not quite so neat in reality, however, because the former situation results in the agency not asserting regulatory control over a specific use of a substance by rulemaking, while the latter results in the agency retaining such control by written regulation. Such disparate outcomes may warrant expending differing levels of effort. Finally, the policy must discriminate sufficiently between the two alternatives. That is, in its day-to-day application, some chemicals must pass and others must fail. If all candidates were to fail or all to pass, it is then, of course, not a very useful policy.

RISK ASSESSMENT CONSIDERATIONS

In constructing a threshold-of-regulation policy for chemicals that are candidates for exemption from the petition process, it would first seem easiest to base the policy solely on exposure. Substances migrating to food below some arbitrary fixed level (say, for example, 1 ppb in the total diet) could be deemed to be, by definition, "below the threshold-of-regulation." This is the approach which has most often been discussed by both agency and industry representatives since the first threshold-of-regulation proposals were put forward in about 1967. It appeals to our toxicological "intuition" ("the dose makes the poison") and would be quite simple and convenient. In reality, however, this choice does not seem appropriate. The problem with such an approach lies in the choice of the level. Chemicals present an enormous range of intrinsic toxicities; thus, a threshold-of-regulation level low enough to protect the public from exposure to unwarranted risk from any and all substances would require such a low level as to preclude a policy of any usefulness—virtually everything would fail. It is necessary to go beyond the realm of only exposure considerations and to base a policy on the realization that risks of adverse health consequences from chemical substances in the diet result from the combined effects of dietary exposure *and* potency (risk per unit level of exposure, or intrinsic toxicity). Both variables are crucial to an estimation of these risks, and food-contact substances near the threshold of regulation may vary in both over many orders of magnitude.

Exposures to food chemicals may vary over many orders of magnitude. Some chemicals added to food may be in the range of several percent of the diet. At the other end of the spectrum in the realm of threshold-of-regulation situations, eight orders of magnitude lower, are exposures on the order of one part per billion (ppb).[6] Toxicities of chemicals may vary over many orders of magnitude as well. For example, the lowest-effect-levels of typical food additives vary from about 1 mg/kg b.w.(body weight)/day up to about 1×10^4 mg/kg b.w./day.[6] For carcinogens, the range of possible toxicities is even greater. Potencies of carcinogens range from 1×10^{-5} (mg/kg b.w./d)$^{-1}$ up to 1×10^3 or even 1×10^4 (mg/kg b.w./d)$^{-1}$, a range of at least eight orders of magnitude.[7] Clearly, when statutory safety standards are applied to such potentially disparate levels of risk, some situations will arise in which it can be logically argued that the risks presented are *de minimis*,[8] and thus below the threshold-of-regulation. But that determination depends upon an assessment of both exposure and potency.

A POSSIBLE APPROACH

Need to Consider Potential Carcinogenic Risks

In developing a threshold-of-regulation policy that operates at the low end of the exposure range of chemicals in food, carcinogenic risks must be distinguished from classical toxicological endpoints. Aside from a relatively small handful of substances such as biochemically "specific" toxins or highly reactive chemicals possessing exquisite toxicity, most low-dose risks tend to be carcinogenic toxic effects rather than "classical" toxic effects associated with threshold doses (doses below which the risk of harm is actually zero). This fact is portrayed in Fig. 1. The figure shows probability density distributions of toxic effect doses for three groupings of different chemical substances displayed along the horizontal axis.[2] Curve "1" shows a probability density distribution of LD_{50} doses for a subset of 18,000 chemicals listed in the Registry of Toxic Effects of Chemical Substances (RTECS). Curve "2" shows a probability density distribution of lowest-effect-levels for subchronic or chronic noncarcinogenic toxic responses resulting from a study of food additive toxicity profiles carried out by FDA.[6] Curve "3" is a nonlinear least squares best-fit Gaussian distribution to a histogram of collected TD_{50}'s from NTP carcinogen bioassays (published previously).[2,8] Note that the TD_{50} doses of the carcinogens distribute generally at orders-of-magnitude lower doses (generally further to the right in the figure) than the classical toxic responses. This is the reason that it is necessary to include concern for the carcinogenesis endpoint in developing a threshold-of-regulation policy, and why a threshold-of-regulation policy must, to some extent, be driven by concern for potential carcinogenic risk.

Using a Probabilistic Approach

The chemicals that will be the subject of this policy are, of course, not known to be carcinogens in either man or in animals. If they were carcinogens their exemption from any regulatory control would be difficult to justify; thus, any procedure for handling such substances would need to be tailored to the specific situation. However, to have a threshold-of-regulation policy for food additives that will preclude unwarranted potential carcinogenic risks, if substances are found to be carcinogens, the presumptive carcinogenic risks of these substances must be considered. One convenient way to address this concern is to examine the probability density distribution of potencies for the subset of chemicals shown to be carcinogenic in animal bioassays by the oral route of administration [in essence the histogram distribution in (Curve "3") Fig. 1].

Because the potencies of carcinogens seem to distribute in a Gaussian fashion, it may be possible to create a threshold-of-regulation policy that can be applied at the lowest levels of likely human exposure, and still to assure, in a probabilistic sense at least, that the risk of harm from carcinogenesis will be confined below some level of upper-bound risk (for example, 1×10^{-6} per lifetime).

It is possible to do this under the (crude but widely used) assumption of linear proportionality of carcinogenic risk as a function of dose. Under this assumption, carcinogenic potency is defined simply as the slope of a straight line connecting the TD_{50} (the effective dose that is toxic to 50 percent of the test animals) to the origin of the dose-response graph (risk being just potency times dose). Then a "10^{-6}-risk-equivalent" exposure distribution can easily be created by sliding curve "3" in Fig. 1 to the right by the required amount (a distance corresponding to the logarithm of the ratio of 0.5 (from the TD_{50}) to 1×10^{-6}). The result is shown in Fig. 1 as curve "4."

This demonstrates why the threshold-of-regulation problem has always seemed so intractable in light of concerns about carcinogenesis. If a dietary level of 1.0 ppb is chosen as a threshold-of-regulation level, nonpetitioned exposures to substances at levels higher than that (to the left of that on Fig. 1) would be precluded. But because curve "4" is bisected by a human dietary exposure of about 1 ppb, there is about a 50 percent probability that a substance allowed at levels of up to 1.0 ppb, which was found to be a carcinogen, could present an upper-bound risk of greater than (and occasionally much greater than) 1×10^{-6} per lifetime. That is because nearly half of the area under curve "4" lies to the right of the 1.0 ppb line. In order to protect at a level sufficient to assure less than a 1×10^{-6} upper-bound risk per lifetime from carcinogens as potent as aflatoxin B-1 or 2,3,7,8-tetrachlorodibenzo-p-dioxin (TCDD) are presumed to be (potencies corresponding to abscissa values of 6 to 8 in Fig. 1), the threshold-of-regulation level would have to be so

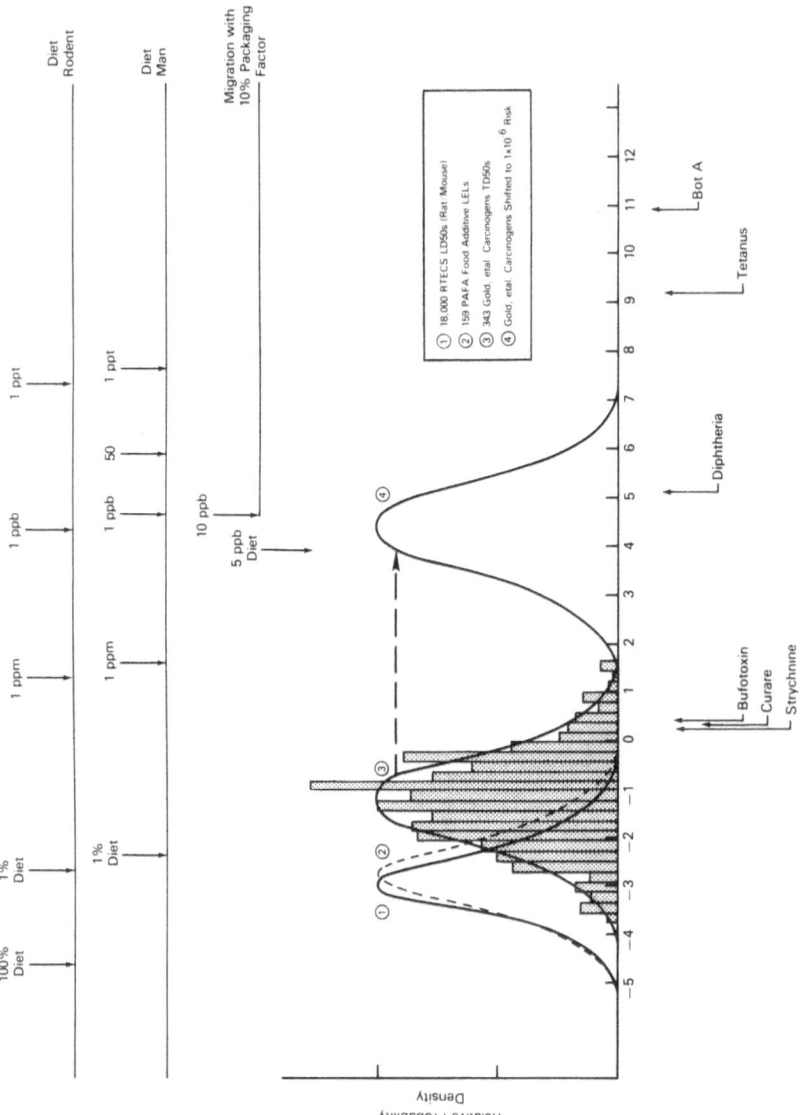

**Fig. 1. Probability Distributions of Toxicity for Three Groupings of Chemical Substances: (1)
18,000 Rat- or Mouse-LD$_{50}$s Obtained from the Registry of Toxic Effects of Chemical Substances; (2)
Lowest-Effect-Level Data from 159 Food Additives Studied in FDA's Priority-Based Assessment of Food
Additives; and (3) TD$_{50}$s from 343 Rodent Carcinogens by the Oral Route Compiled by Gold, *et al.***

low (lower than one part per trillion (ppt) in the diet) that virtually nothing would be able to pass the threshold-of-regulation criterion.

Here it must be remembered that analytical capabilities, although vastly improved over those of yesteryear, rarely deliver both quantitation and confirmation of identity for migrating species at levels of less than 1.0 ppb in food or food simulating solvents. Because not all food is packaged in any one given food-contact substance, a so-called "packaging factor" (the simple ratio of the amount of the diet packaged in the packaging material of interest, to the amount of all other packaged food) may help mitigate this problem. For example, if it is found that a candidate for "threshold-of-regulation" is present in a material that contacts only 5 percent of the packaged food in the diet, then a 1 ppb migration level corresponds to a 50-ppt dietary intake. From curve "4" of Fig. 1 we see that such a level would be sufficient to assure an upper-bound risk of 1×10^{-6} or less, with about 85 percent probability should the substance in fact turn out to be a carcinogen.

It should be reemphasized here that this model is crude and is based on some very bold assumptions. For example, it has been assumed: (1) that the results of the animal carcinogen bioassays may be related directly to potential human risk; (2) that human risk at low doses may be estimated reliably by using a simple linear proportional extrapolation model on the published TD_{50} values derived from the bioassay results; and (3) the carcinogens used to create the probability density distribution of potencies comprise a representative set of chemicals that could also reasonably be expected to be components of food-contact materials.[8]

CONCLUSION

Clearly, the development of a threshold-of-regulation policy demands much more than just "picking a number." A framework to efficiently aid the agency in consistently and fairly deciding when any given situation involving food-contact materials may be exempted from premarket safety review must combine administrative, legal and scientific considerations. Thus, in addition to considerations of likely upper-bound levels of risk, the policy must be cognizant of the limits of analytical chemistry, the requirements of record keeping and the impact of the Freedom of Information Act on those records. The implications of the *Monsanto v. Kennedy* Court of Appeals decision of 1979[9] must be considered, and also the thirty years of debate in the legal and scientific communities about the boundaries of the "food additive" definition. The framework must function efficiently without creating too many "exceptions to the rule," and should not use more resources in its implementation than would a "full-blown" petition review process. Examination of the available data on the universe of known potencies as described above may provide some basis for the choice of a threshold-of-regulation level or range of levels. Using such a probabilistic approach on the data may help to resolve whether to "simply set a threshold level and be done with it," or to ensure public health protection (from potential carcinogenic risks) in the absence of a petition without setting a single level as a threshold. The present approach, of course, is only one of a spectrum of options available to the FDA in its quest for a solution to the threshold-of-regulation challenge. In coming months, the Center for Food Safety and Applied Nutrition will continue to examine these approaches, as well as other possible solutions to the problem, attempting to resolve the issue in a practical, cost-effective, fair, and internally consistent manner.

REFERENCES

1. Federal Food, Drug, and Cosmetic Act, as amended, (Title 21 U.S. Code), 1958, U.S. Government Printing Office, Washington, D.C.

2. A. M. Rulis, "*De Minimis* and the Threshold of Regulation," Food Protection Technology, Proceedings of the 1986 Conference for Food Protection, Charles Felix, Ed., Lewis Publ., Inc. and references cited therein (1987).

3. J. H. Heckman, "Fathoming Food Packaging Regulation: A Guide to Independent Industry Action," *Food Drug Cosmet. Law J.* **42**:38-49 (1987).

4. C. J. Kokoski, "Regulatory Food Additive Toxicology," Second International Conference on Safety Evaluation and Regulation of Chemicals, and Course on Safety Regulations in the USA, Boston University School of Medicine, Cambridge, MA, October 24, 1983.

5. *Public Citizen v. Young*, United States Court of Appeals for the District of Columbia, October 23, 1987.

6. A. M. Rulis, "Safety Assurance Margins for Food Additives Currently in Use," *Regulatory Toxicol. and Pharmacol.* **7**:160-168 (1987).

7. L. S. Gold *et al.*, "A Carcinogenesis Potency Data Base of the Standardized Results of Animal Bioassays," *Environ. Health Pers.* **58**:9-314 (1984).

8. W. G. Flamm, L. R. Lake, R. J. Lorentzen, A. M. Rulis, P. S. Schwartz, and T. C. Troxell, "Carcinogenic Potencies and Establishment of a Threshold-of-Regulation for Food Contact Substances," in *De Minimis Risk*, Whipple, Ed. Plenum, New York, NY.

9. *Monsanto v. Kennedy*, D.C. Cir. 1979; 613 F. 2d 947.

The Use of Pharmacokinetics in Food Safety Evaluation

Robert J. Scheuplein

Center for Food Safety and Applied Nutrition
Food and Drug Administration
Washington, DC

ABSTRACT

The use of risk assessment, i.e., permitting acceptable risk, in food safety evaluation creates as a practical matter the need to predict adverse effects in humans from those observed in animal studies with greater accuracy and precision. In trying to establish the qualitative and quantitative relevance of animal studies, toxicologists are forced to make two fundamental assumptions. The first is that there is a qualitative similarity in the effect produced by a chemical in animals and humans, and the second is that the intensity of biological response in both species is proportional to the appropriate tissue dose. To convert these assumptions into useful procedures, we need to answer two pharmacokinetic questions: (1) What is the appropriate interspecies scaling factor? and (2) What is the impact of non-linear kinetics on the required high-to-low dose extrapolation? These issues are discussed very generally in the paper.

KEYWORDS: Pharmacokinetics, food safety, blood levels, effective dose, interspecies scaling

INTRODUCTION

The fundamental scientific task of the CFSAN (Center for Food Safety and Applied Nutrition) as far as the regulation of food and color additives is concerned is how to establish the safety of these substances, typically ingested by people at low levels, from animal studies at much higher doses. Regulatory agencies like the FDA spend a great deal of time determining the appropriate studies to require, recommending protocols for the conduct of these studies and finally evaluating the submitted studies to assure that the safety issues have been adequately addressed. The process is complex and lengthy primarily because the questions are not amenable to ready answers on the basis of the data that are provided or that can reasonably be provided. The law requires that the FDA establish safety to a reasonable certainty on the basis of the scientifically most reliable animal data. The two fundamental questions—the relevance of animal response to likely human response and the fidelity of high-to-low dose extrapolation—cannot now be answered with certainty or precision. The law anticipates this and expects the FDA to make an informed judgment on the human relevance of these studies. And so we rely on past experience and the judgment of qualified experts to help us interpret the data. The process usually unfolds in the following way. We assure that the submitted animal studies are complete, of good

1. Incidence	Tumor response in target organ in animals of the lifetime oral administration of agent.	$- \mathrm{I}\,a$
2. Intake Rate	Average amount of agent administered daily to animals for a lifetime	$- DR$ (mg/day)
3. Effective Dose	Average daily "body concentration" administered to animals over a lifetime.	$- C$ (mg/(kgbw)^n/day)

$$C = \frac{DR}{W^n} = \mathrm{mg}/(\mathrm{kgbw})^n \cdot \mathrm{day}$$

Fig. 1. The Three Measured (or Derived) Experimental Quantities Typically Measured in Carcinogen Bioassays That Form the Basis for Estimates of Risk.

quality and that they tell a self-consistent story. Any additional questions raised by the experimental observations are carefully considered and sometimes resolved by requesting specific further studies, sometimes by getting independent expert opinion. The most significant toxicological end point is determined and ultimately a tolerance or an ADI for a thresholdable substance or a risk number for a non-Delaney carcinogen is calculated.

It is important to emphasize that a large amount of effort and judgment is involved in this "hazard evaluation" phase of the process because the amount of work involved is usually underestimated. There are two major reasons for this. First, all this information is compressed in the end into a single critical study—the one with the most reliable and usually most sensitive end point from which the NOEL (no effect level) or risk is determined. One often tends to lose site of all the prior work and analysis that were necessary to test and exclude other possible adverse effects. Second, the current emphasis, as our session this morning illustrates, is on more quantitation of risk and more understanding of mechanism. We support this emerging trend, but I want to point out that a good deal of toxicological study is usually necessary before one reaches this stage.

When the target organ is identified and dose-response data are in hand, there are typically only three experimentally measured quantities explicitly given, and the process seems simple—deceptively so. To take a specific example, let us consider a carcinogen, rodent bioassay. But everything will at least in principle apply as well to any toxicological effect resulting from a chemical.

EFFECTIVE vs. ADMINISTERED DOSE

Figure 1 shows the three quantities typically measured and reported in carcinogen bioassays: the tumor incidence—I, intake rate; DR (mg/day); and "effective dose," c (mg/(kg.bw)^n day). The intake rate is the daily administered dose and the "effective dose" is an approximation to the actual effective dose. The symbol c is used to indicate that an attempt is made to account for animal size and to produce an effective concentration from an administered quantity. This is a very crude attempt since no account of metabolism, clearance, sequestration, or distribution is taken. This is one area where the insight from pharmacokinetics and PB-PK (physiologically-based pharmacokinetic) modelling should prove most helpful.[1]

In trying to establish human relevance, toxicologists are forced to make two fundamental assumptions. The first is that there is a qualitative similarity in effect produced by a chemical in rodents and in humans. The second is that the intensity of biological response in both species is proportional to the appropriate tissue dose. But what is the most appropriate measure of tissue dose? It is not likely to bear so simple a relation to the administered dose as routinely assumed. Most might agree that the true effective concentration is the concentration of biologically active, unbound form(s) of a substance at the involved receptors. If concentration at the receptor sites can safely be assumed to be proportional to the concentration in the blood or blood plasma, then a practical measure of effective internal exposure from a single dose can be expressed as the integral of concentration in blood or blood plasma over time (i.e., the area under the concentration-time curve-AUC)

$$\int_{t=0}^{t=\infty} C(t)dt = AUC = \frac{F \cdot D}{V_D k_e} = \frac{F \cdot D}{C_L} \quad,$$

where F is the fraction systemically absorbed, k_e is the elimination rate constant, V_D is the apparent volume of distribution and C_L is the total body clearance. The last two relationships hold when all absorptive, distributive and eliminative processes are linear, but not necessarily otherwise.[2] And, of course, the simplicity of this expression is due to our assumption that modelling the distribution and elimination processes of the body by single compartment kinetics is sufficiently accurate. As the administered dose is increased into the toxic range, it becomes increasingly less likely that these individual processes remain linear and the proportionality between effective dose AUC and administered dose—D can no longer be expected.

For a chronic study like the rodent bioassay in our example, the steady state concentration (or the average concentration around which the daily ingestion levels oscillate) bears a similar relation to dose rate, DR

$$C_{ss} = \frac{F \cdot DR}{C_L} \quad,$$

with the caveat that all elimination processes are again linear.

SCALING

After consideration of "effective dose," the next most important use of pharmacokinetics in safety evaluation based on animal studies is in interspecies scaling. The FDA and the EPA typically used different scaling factors to correct dose or dose rate for interspecies differences. The FDA uses amount per unit body weight, and the EPA uses amount per unit surface area. To go from x mg in the mouse to the desired toxicologically equivalent dose x'mg in the human via mg/kg equivalency yields a scale factor of 2,030, e.g.,

if:

$$\chi_{mg}/kg \cdot bw(mouse) = \chi'_{mg}/kg \cdot bw(human) \quad,$$

then:

$$\chi'_{mg} = \frac{70kg}{0.0345kg} \cdot \chi_{mg} = 2,030\chi_{mg} \quad.$$

Whereas via $[mg/kg/bw]^{2/3}$ equivalency the scale factor is 160.4, e.g.,

$$\chi''_{mg} = \left(\frac{70}{0.0345}\right)^{2/3} = 1.60.4\chi_{mg} \; .$$

The ratio of these factors is 12.7, and this is the inherent difference in these scaled NOEL's and risk numbers when mouse data are used.

Pharmacokinetics offers a bit more insight into a possible physiologically based scaling factor. If the steady state blood level is a reliable indicator of effective concentration, then the equivalent blood levels of active agent in the animal (A) and in humans (H) (species sensitivity being equal) should produce equivalent toxic effects.[3] The basis for interspecies scaling would then be a normalization to equivalent blood levels, i.e.,

$$C^H_{ss} = C^A_{ss} \; .$$

Obviously, for many substances, dioxin, for example, the susceptibility of different species is not the same, but this is not a scaling problem. An adjustment for pharmacodynamic differences or for susceptibility differences between strains or species is certainly not expected to track anything so simple as proportional dose scaling in terms of $[bw]^n$, nor would one expect a single generic interspecies or interstrain factor, independent of the specific chemical, to provide such an adjustment. But for many substances, assuming that the steady state blood levels of the active substances in both species are the same, the dose rate in animals required to produce the same effects in humans is:

$$DR_A = \frac{F_H}{F_A} \cdot \frac{C_{L,A}}{C_{L,H}} \cdot DR_H \; .$$

Thus, one would need a higher dose rate in animals to balance, say, more efficient absorption or less efficient clearance in humans. Since clearances are found generally to scale as approximately $bw^{0.75}$, one is led on this basis to expect a generic interspecies scaling factor somewhere between the FDA's $bw^{1.0}$ and the EPA's $bw^{0.67}$. Current data on the comparison of animal bioassay data with human epidemiological cancer data are simply not accurate enough to decide which scaling procedure is generally better. The pharmacokinetic approach illustrates that the quest for a generic scaling factor is probably not worth too much effort. PB-PK modelling of the distribution, metabolism and elimination of the same substances in various species shows that not only clearance but other physiologic processes scale to different powers of body weight. The final steady state concentration depends on a functional average of these individual scale factors which are very likely to vary with the way the body handles different chemicals.

COMPARATIVE PHARMACOKINETICS

The scaling example illustrates that the use of pharmacokinetics in food safety evaluation is an exercise in *comparative* toxicology, physiology, etc. CFSAN needs to *compare* doses, blood levels, important metabolites, and the concentrations of important metabolites reaching receptor cells in both the test animals and in humans. Figure 2 illustrates the major issues that we are concerned with when extrapolating between species. The figure indicates in a heuristic way how the human risk (e.g., tumor incidence) depends on the ratio of comparable quantities that may be different between animals and humans.

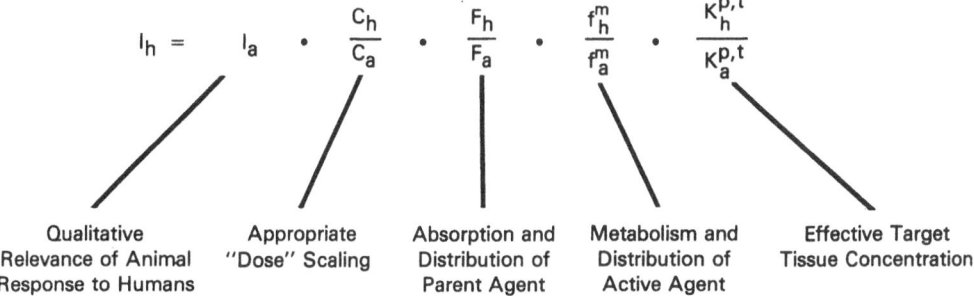

$$I_h = I_a \cdot \frac{C_h}{C_a} \cdot \frac{F_h}{F_a} \cdot \frac{f_h^m}{f_a^m} \cdot \frac{K_h^{p,t}}{K_a^{p,t}}$$

| Qualitative Relevance of Animal Response to Humans | Appropriate "Dose" Scaling | Absorption and Distribution of Parent Agent | Metabolism and Distribution of Active Agent | Effective Target Tissue Concentration |

Fig. 2. The Use of Comparative Pharmacokinetics in Animal-to-Human Extrapolation. The estimate of human tumor incidence (human risk) is directly dependent on a knowledge of pharmacokinetic quantities or their ratios in animals and in humans.

The figure itself represents an enormous oversimplification. I've put in only the major factors, and even left out an important term correcting for the fraction of the active substance in unbound form. The toxicity of a substance is usually proportional to the concentration of "free" form; the extent of plasma protein binding can vary between species and may affect toxicity as much as or more than any of the factors given.[4] Other factors that may need consideration are listed below:

Factors Altering Clearance and Toxicity

Diet
Interaction with other chemicals
Variation in plasma and tissue protein binding
Genetic variation in drug metabolism (polymorphism)
Decline in protein binding in aged animals
Pathological changes in the eliminating organs
Species differences in bile and urine flow
Species differences in intrinsic rate of metabolism
    (e.g., cytochrome P_{450} in rat liver × 10 human)
Reduction in body size (= increased relative liver size)

IMPACT OF NONLINEAR KINETICS

When linear kinetics are observed, as we have assumed, the values of the fraction absorbed, F, and the clearance, C_L, remain constant in the equation AUC = F.D/C_L for a range of doses. And if AUC can be taken as the measure of internal exposure during a toxicity test, then the effective concentration will be proportional to dose throughout the dose range. When elimination mechanisms, like biotransformation, renal tubular secretion or biliary excretion, themselves depend on dose, the AUC dependence on dose becomes more complex. Figure 3 shows a quadratic increase in exposure (AUC) with dose when the rate limiting elimination process obeys Michelis Menton kinetics.[2] It also illustrates the dependence of these elimination processes on the metabolic parameters (V_{max}, K_m, etc.). When the parent molecule is the active agent one could expect disproportionately high toxicity with increased dose. Linear extrapolation of this result to lower doses could vastly overstate the toxicity at the lower doses because no account is taken of the change in clearance with dose. When the metabolite is the active agent, high-dose-saturation of the biotransformation processes that produce the metabolite will result in disproportionately

$$AUC_\infty = \int_0^\infty C)t) \, dt$$

Linear

Nonlinear

$$\frac{dc}{dt} = -k_e C$$

$$\frac{dc}{dt} = -\frac{V_m C}{K_m + C}$$

$$C = C_o e^{-k_e t}$$

$$C = C_o - V_m t + k_m \ln\left(\frac{C_o}{C}\right)$$

$$AUC_\infty = \frac{C_o}{k_e} = \frac{D}{V_D k_e}$$

$$AUC_\infty = \frac{C_o}{V_m}\left(\frac{C_o}{2} + K_m\right) = \frac{D}{V_D V_m}\left(\frac{D}{2V_D} + K_m\right)$$

Fig. 3. The Effective Dose, i.e., AUC, Is No Longer Directly Proportional to Intake Dose (D) When Elimination Mechanisms (and Others) Are Saturable. As in the example above, AUC can increase disproportionately with D and can be dependent on metabolic parameters (V_{max}, K_m, etc.).

low toxicity; and linear extrapolation can underestimate the risk.[5] It is clear that metabolites can play a part in toxicity, particularly when they are formed at substantial rates and when they have slower elimination rates than the parent compound. This is likely to be true for at least some of those carcinogenic compounds that require metabolic activation.

CONCLUSION

The two major extrapolations required in order to predict human risk from animal data—animal-to-man and high-to-low dose—are both critically dependent on reliable pharmacokinetic data.

Animal-to-Man

The fundamental problem introduced by animal to human extrapolation (independent of high-to-low dose issues) is that the ultimate human relevance of animal studies and the values of the critical parameters, defining e.g., scaling, absorption, distribution, biotransformation, binding, etc., require detailed understanding of the pharmacokinetic behavior of the chemical in *both species*, the test animal and the human.

High-to-Low-Dose

When nonlinear kinetics occur (the usual case), distribution and elimination processes particularly may be highly nonlinearly dependent on dose. Over the large dose range typically involved, nonlinear absorptive or elimination processes may become important over different portions of the dose range. Extrapolation models that fail to include such mechanisms when they occur are not credible.

Both the amount of animal data we ask for and the amount of human data available severely limit the full range of this pharmacokinetic comparison. We rarely have enough information to go to the second step in Fig. 2. How often have you seen a table of plasma or blood concentrations as a function of administered dose in a carcinogen bioassay report? And if we did have the animal data, would we know the human values? Not knowing this information can be judged to be justifiable in food safety regulation if appropriate safety factors are used for noncarcinogens and if all carcinogens are banned. Safety factors in effect compensate for our lack of precise information by assuming the worst and providing an ample added margin of safety. But it is not self-evident how our lack of information would be compensated for if we were to permit widely disseminated risk from carcinogens under an acceptable risk cancer policy. The use of risk assessment in food safety evaluation creates as a practical matter a need for better pharmacokinetic comparisons. The need may not be acute as long as our high-to-low dose risk assessments remain conservative as they now are. But it is clear that the trend to reduce conservatisms in high-to-low dose extrapolation places an increased need to understand the errors we may be making in the animal-human extrapolation.

REFERENCES

1. *Pharmacokinetics in Risk Assessment, Drinking Water and Health*, Vol. 8, National Academy Press, Washington, DC (1987).
2. E. J. O'Flaherty, "Differences in Metabolism at Different Dose Levels," in *Toxicological Risk Assessment, Volume I, Biological and Statistical Criteria*, D. B. Clayson, D. Krewski, and I. Munro, Eds., CRC Press, Inc., Boca Raton (1985).
3. B. Clark and D. A. Smith, "Pharmacokinetics and Toxicity Testing," *Crit. Rev. Toxical.* **12**:343 (1984).
4. J. R. Gillette, "Dose, Species and Route Extrapolation: General Aspects," in *Pharmacokinetics in Risk Assessment, Drinking Water and Health*, Vol. 8, ibid.
5. M. E. Anderson, "Saturable Metabolism and Its Relationship to Toxicity," *Crit. Rev. Toxical.* **9**:105 (1981)

Consideration of Cell Proliferation and Adduct Formation as Parameters Affecting Risk Outcome

Ronald J. Lorentzen and Robert N. Brown

Center for Food Safety and Applied Nutrition
Food and Drug Administration
Washington, DC

ABSTRACT

For the most part, the quantitative aspects of risk assessment for carcinogens currently involve using standard animal bioassay data and making assumptions about dose-response such as linearity at low-dosage levels and no threshold for the effect. These procedures are not considered to predict actual risk but to reflect an upper bound on risk. The true, but indeterminate, risk could be as high as this upper bound value and as low as zero. As we learn more about mechanisms of carcinogenesis, we anticipate that more biological information will lead to refinement of predicted dose-response curves and, hence, predicted risk. Two types of biological information currently being explored for their value in this regard are cell proliferation and covalent adduct formation with cellular macromolecules. Both of these types of data have the potential to allow reliable characterization of the dose-response curve to lower doses than is possible with bioassay data alone. This may allow in certain cases for the predictions of the upper bound risk to be closer to the actuarial risk at the low exposure levels usually encountered. Much work and careful interpretation need to be done before these approaches become routinely viable. However, the time has come to take seriously the potential that this kind of biological information can have on quantitative risk assessment.

KEYWORDS: Threshold, cellular proliferation, DNA adducts, pharmacokinetic models, pharmacodynamic models

INTRODUCTION

The actual risk for cancer from exposure to a known carcinogen at low doses where the risks are low cannot currently be determined. Animal bioassays and human epidemiology studies can, at best, define the dose-response curve down to a level corresponding to an actual risk of 10^{-1} to 10^{-2}, the latter value being overly optimistic about the capability except under the most optimal circumstances. While not a strictly scientific issue, an increase of 10% in the lifetime incidence of human cancer is generally agreed to be unacceptable under almost any conceivable circumstances, particularly for an avoidable situation such as the addition to or contamination of food by carcinogens. Therefore, there is the need to extrapolate the dose-response curve to a region where the implied risks are much lower in order to determine an acceptable human exposure. The dilemma is this need to extrapolate to low risk/dose with the knowledge that it cannot be performed with any semblance of relative accuracy. There are those who say that, given the uncertainty associated with low dose risk assessment, it should not be used.[1] However, given the increasingly sensitive methods of detection being developed by analytical chemists and the proliferation of the number of known carcinogens, there does not seem to be any realistic alternative to evaluating risk as some function of carcinogenic potency and exposure. It is reasonably safe to conclude that virtually every substance is either carcinogenic itself, or contains carcinogens as impurities at *some* level, although, fortunately, the level of contamination is low in the majority of cases. Therefore, a method is needed to distinguish between "safe" and "unsafe" exposures to carcinogens in the safety evaluation of food and color additives and contaminants of food. Risk assessment is such a method.

Nevertheless, how do we deal with the dilemma of not being able to predict risk accurately at low dose? If there is anything approaching a consensus in the field of carcinogen risk assessment, it is that linearity of the dose-response curve at low doses defines the reasonable upper bound on risk at low doses.[2] The true risk is thought to be potentially any value ranging from this upper bound to zero, the latter if the principal mechanism of action involves some thresholdable phenomenon. Any number of non-threshold models, mathematical or otherwise, can produce essentially any value within this range and none can really claim to have any inherent superiority in ability to make predictions at low dose. And we are not speaking of small differences between models, but differences often of many orders of magnitude at human exposure levels. There are reasons for expecting a linear association between dose and response at low doses, excepting genuine thresholdable circumstances. The only question—and this is the pivotal question—is where on the dose-response curve does or should linearity be presumed to begin? The FDA, other regulatory agencies and other practitioners of cancer risk assessment have adopted a philosophy of linear-at-low-dose extrapolation for carcinogens. This is considered by most to be conservative in favor of the public health, and the risks derived by such an approach are expressed not as actuarial but as upper bound risks. Whether this extrapolation is carried out by using sophisticated computer software or computationally less intense techniques is of little matter, insofar as estimated values rarely vary by more than a factor of two if conservatively and consistently implemented—a trivial difference in the context of much larger uncertainties in the typical risk assessment process. Unfortunately, when data are restricted to tumor incidences, which is most often the case, the curve readily defined by those data is for relatively high dose and response (or risk). The data do not allow valid inferences of curve shape below these relatively high levels and linearity is imposed at the relatively high end of the dose-response curve. In some cases, this may be entirely appropriate. In other cases where considerable non-linearity exists in the observable dose-response curve, there arises suspicion that the non-linearity may continue to lower doses even though this is not verifiable by incidence data alone.

Are there any ways to improve risk assessments, particularly for data from studies showing steep dose-response curves based upon tumor incidences? The answer is probably yes. But it is not generally easy, and scientific advancement in some of the approaches appears to be necessary before significant progress will be made. Three potentially significant approaches are: (1) establishing a threshold for an indirect carcinogenic effect associated with a steep dose response, (2) using DNA adducts as biomarkers of the dose-response curve to much lower levels than is possible with tumor incidence data, and (3) establishing quantitatively the role that cellular proliferation can play in enhancing a neoplastic process, hence, affecting the dose-response characteristics.

INDIRECT MECHANISMS

There are often a number of reasons for suspecting that a substance may be inducing neoplasia by an indirect or "secondary" mechanism of action and that a threshold may exist. Establishing this to some reasonable consensus is not generally easy and regulatory activity cannot be based upon mere scientific speculation about mechanism. Nevertheless, we expect in the future that research will be able in some cases to demonstrate to a reasonable certainty indirect mechanisms of carcinogenesis. A case where this has already been done is with melamine.

Melamine administration by the oral route is associated with an increased incidence of transitional cell papillomas and carcinomas of the urinary bladder in male F344 rats. The administration is also associated with an increased incidence of urinary bladder calculi. A committee of FDA scientists concluded after reviewing all of the evidence that the bladder tumors were not a reflection of chemical carcinogenesis by melamine *per se*, but were caused secondarily by the bladder calculi produced by melamine administration. Without elaborating in great detail, the main reasons that lead to this conclusion are:

- the ability of bladder calculi to induce bladder tumors in rodents has been well established;

- a highly significant association was found to exist between bladder tumors and bladder stones (p. 0.001);

- melamine has a structure not likely to have carcinogenic activity unless previously metabolized, yet virtually all, if not all, melamine is excreted unchanged in the urine;

- the calculi have been identified as phosphate salts of melamine;

- when levels of melamine are administered where no calculi are produced, no tumors are observed;

- melamine is not genotoxic.

Of course, bladder calculi formation could have serious toxicological consequences. However, the cancer risk is predicted to be zero at lower doses where calculi are not formed.

DNA ADDUCTS

For carcinogens that are genotoxic, evidence has been accumulating for many years that covalent binding to DNA is a pre-initiation event that should bear some relationship to the ultimate result, namely tumor formation. Because of the potential for detecting DNA adducts at low levels of applied dose, the hope has been that the dose-response curve for incidences (risk) of cancer would be reflected by the level of adduct formation at low dose. If this were possible, the estimate of low levels of risk, or alternatively the estimate of dose corresponding to low levels of risk, could be performed with greater accuracy than can be inferred from tumor incidence data alone. This has been coined "molecular dosimetry" by Swenberg and colleagues at CIIT.[3] It seems likely that ultimately such an approach will offer a viable improvement to the linear-at-low-dose assumption now used on the basis of tumor incidence. However, at present, attempts to validate an approach where DNA adduct formation could be used to define low dose response in carcinogenesis quantitatively have not been entirely successful. Some of the reasons for this are obvious, while others are not altogether clear. Many DNA adduct formation studies have been performed over the years.

But very few have been done under the conditions reflecting the lifetime carcinogen bioassay from which lifetime cancer risks are estimated. It would be difficult indeed to have confidence in a correlation between tumor incidence derived from a lifetime study and DNA adduct formation measured in short term and single dosing studies. Moreover, carcinogens generally produce several different adducts with DNA nucleotides and, as Swenberg has recently so aptly put it, "not all DNA adducts are created equal."[4] In other words, the specific adduct(s) associated with the development of neoplasia needs to be identified before we can have confidence that a real correlation exists. This can be done. But it is not easy and requires considerable sophisticated experimentation to accomplish. From our perspective, the most elegant research in this area is being performed by Swenberg and co-workers at CIIT. They are using diethylnitrosamine (DEN) as one of their model compounds and have essentially demonstrated that ethylation of the 0-4 position of thymidine is the principal adduct associated with developing hepatocellular neoplasia.[5,6] They have also measured essentially the steady state levels of this adduct in the livers of male F344 rats continuously administered DEN over a wide range of concentrations in the drinking water.[7,8] This continuous exposure is important in that it mimics the conditions of carcinogen bioassays. Whether or not the dose-response curve of adduct formation versus administered dose correlates in some way with the dose-response curve for liver tumor incidence requires some validation before it could become potentially useful in risk assessment. They chose to use the large BIBRA study on DEN which employed sixteen dosage levels over a broad range.[9] This large bioassay design is important because the use of data from a typical two or three dose bioassay probably would not allow for any confidence in whatever correlation was found. In Fig. 1, the log-log plots of liver tumor incidence versus administered dose and the concentration of 0-4 ethylthymidine adducts (compared to total thymidine) in liver cell DNA versus administered dose are given (taken from ref. 8). From the outset, it must be noted that the two plots are derived from different rat strains. In particular, in the BIBRA study in England, the Colworth strain of rat was used, whereas the measurement of adducts was performed using Fischer 344 rats. This is a confounding factor, the importance of which is difficult to determine. As can be seen, although the plots for both tumor incidence and adduct formation are monotone, increasing with increasing administered dose, there is a lack of parallelism, one that persists even after subtraction of background tumors. The DNA adducts could not be used *per se* as a reliable dosimeter for tumor incidence or risk. This is somewhat disappointing on the surface, and, undoubtedly, the reasons for the poor correlation will become clear with further experimentation. The good news may be that strict parallelism may not be necessary, depending upon the model used for tumor prediction. A brief mathematical discussion of why strict parallelism between a dose surrogate, such as adduct formation, and tumor response is not always necessary, within the context of time and dose dependent multistage models, is relegated to the Appendix. However, to quote Swenberg *et al.*, "a much greater understanding of scientific principles surrounding molecular dosimetry of DNA adducts and carcinogenesis will be required if quantitative risk assessment is to be placed on a firm scientific basis."[8] It may have been too optimistic to expect a straightforward relationship between a presumed early event in carcinogenesis and the final outcome of a long process. Hopefully, further research will reveal useful relationships that will allow reliable extrapolations to very low doses based on molecular biomarkers.

CELLULAR PROLIFERATION

There seems to be little doubt that cellular proliferation can play a role in the carcinogenic process. Early theories of carcinogenesis, such as the irritation theory, at least implied that cell turnover from an irritant stimulus played some pivotal role. Liver regeneration after partial hepatectomy is known to enhance carcinogenesis by hepatocarcinogens.[10] If one visualizes carcinogenesis as a multistage process, it is easy to imagine that cellular proliferation could affect virtually every stage from the cell division

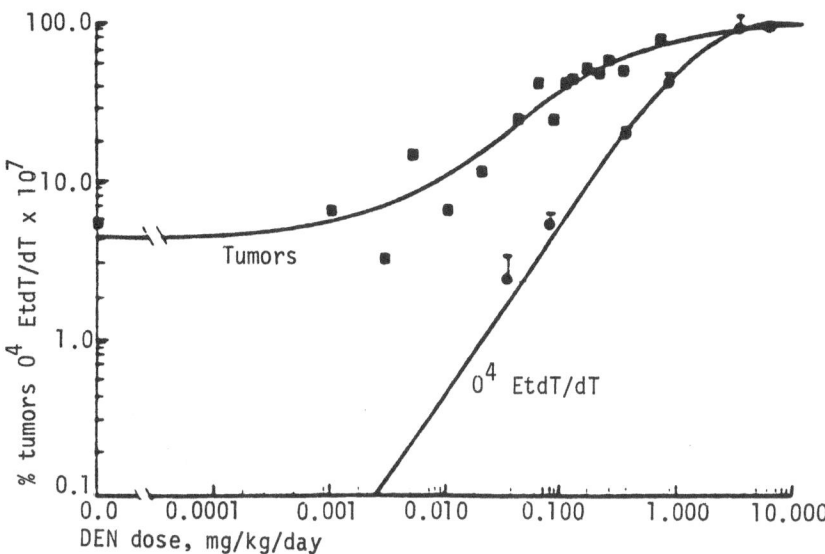

Fig. 1. O^4EtdT **Concentrations in Livers of Fischer Rats After 49 Days Continuous Exposure and Liver Tumor Incidences in Colworth Rats After Continuous Exposure for 2.5 years (from Ref. 8).**

needed to fix the "mutation" caused by the DNA adduct in the initiation stage to the increase in the rate of growth of a tumor in late stages. It needs to be pointed out that the mechanism by which a substance induces cancer is not known in any case to this detail. The mechanisms remain as theories. Nevertheless, it seems intuitive that some incorporation of cellular proliferation into dose-response or risk assessment should be possible for those substances which not only induce cancer but affect the rate of cell proliferation. Can this be done in a credible manner and still give a reasonable certainty that the risk predicted at low dose is not egregiously underestimated? Probably, if models and assumptions are carefully scrutinized and implemented in a computationally conservative manner, especially where model assumptions are not easily validated yet result in large risk changes compared to classical conservative extrapolation procedures.

It is probably a good idea for the toxicologists and the mathematicians to work together, yet from their own perspectives, to examine mathematical models that include cellular proliferation (or adduct formation) as part of their mechanism. The model that appears to be receiving the most attention of late is the so-called M-V-K model after its developers, Moolgavkar, Venzon and Knudson.[11,12] In Fig. 2, this is the underlying scheme for the model taken from a recent paper by Thorslund *et al.*[12] It is a two-stage model where two events (mutations to the authors), represented by M_0 and M_1, at the cellular level are obligatory to produce a malignant cell, C_2. During this process, normal or stem cells, C_0, and first stage transformed cells, C_1, can proliferate or die, represented by functions B and D, respectively. The number of C_0 and C_1 cells present, governed by proliferation and death, affects the probability of producing a malignant cell.

Biologically, this is a very appealing but highly simplistic view of the process of carcinogenesis. This is not necessarily a disadvantage. Moolgavkar and Knudson claim the model explains a number of neoplastic phenomena, mostly in humans.[12] We have no reason to dispute these claims. However, these presumed validations of the model are based on high incidence situations. We are interested in low incidence (risk) situations and

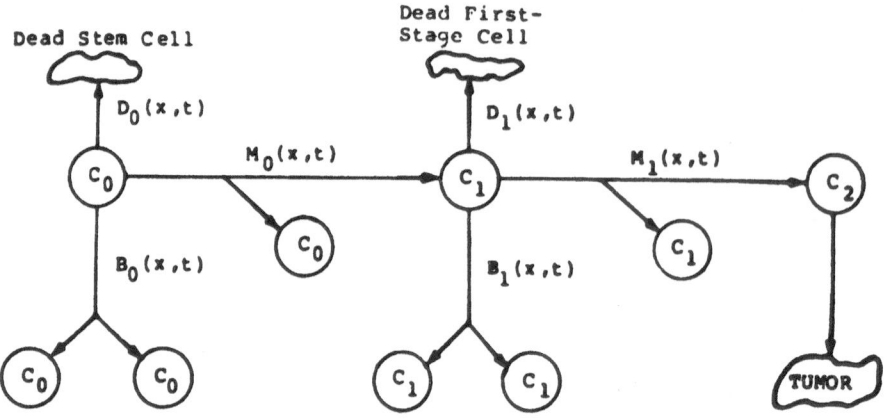

Fig. 2. Scheme of Exposure- and Time-Dependent M-V-K Model (from Ref. 1).

the model has not yet been explored sufficiently to decide whether it can be of use for low dose risk assessment.

The model is simplistic in that it collapses what is best described as an n-stage multistage process into two stages, yet there are $n-1$ ways of implementing such a collapse. Thorslund et al., for example, describe C_1 as a "pre-neoplastic, first-stage cell, which can proliferate into a premalignant clone."[13] If premalignant means benign, yet neoplastic, then progression from preneoplastic to neoplastic takes place due to cell proliferation alone without any events or stages. This may not be a bad approximation. However, that remains to be shown.

As Moolgavkar and Knudson have shown, cancer incidence predicted by the model is very sensitive to small changes in the net rate of cell proliferation.[12] This places a heavy burden on how one describes the cell proliferation as a function of dose. It is reasonably clear that both cell groups C_0 and C_1 should be modelled as functions of both dose and time to properly predict tumor incidence or risk. In other words, there has to be a curve-fitting exercise, not on tumor incidence and dose, as is the case with most other models, but with cell proliferation and dose and time. Our ability to do this may be better at lower doses than for classical tumor endpoints. However, for low dose extrapolation there is good reason to question the reliability of risk predictions at low dose, especially under nonconservative assumptions on the rate of cell proliferation at low dose. The choice of modelling function for cell proliferation may provide as much variation in low dose risk as we get from the various models used to predict risk on the basis of tumor incidence alone. Nevertheless, this needs to be explored both experimentally and mathematically to determine the feasibility of using not necessarily just the M-V-K model but any that takes into account cellular proliferation (or adduct formation).

APPENDIX

Pharmacodynamic (PD) models, such as the M-V-K model, can be thought of as modelling the cancer process at the cellular or subcellular level as a function of organ or tissue concentration x, taking into account transition rates from one preneoplastic or neoplastic cellular state to the next and taking into account cellular growth or proliferation rates within each stage. Physiologically based pharmacokinetic (PB-PK) models express the tissue concentration x of the proximate carcinogen as a function $x(d)$ of nominal administered dose d, where the function $x(d)$ generally takes into account the physiological absorption, distribution, elimination and metabolism of the parent compound. Mathematically, the composition of these two, frequently independent, submodels of the cancer process can be expressed as a composite PB-PK-PD model as follows (treating most time-dependent processes as time-independent for purposes of simplicity):

$$P(x(d))=1-\exp(-M_{00}[1+f_1(x(d))]M_{10}[1+f_2(x(d))]C_0[1+f_3(x(d))]H(d,t)),$$

where:

M_{00} and M_{10} are background, time-independent, transition (or mutation) rates from states 0 to 1 and 1 to 2;

f_1 and f_2 are S-shaped functions of relative increase over background transition rates;

C_0 is the background, time-independent, number of normal stem cells;

f_3 is an S-shaped function of relative increase over background number of normal stem cells;

$H(d,t)$ is a time (t) and dose (d) dependent function of first stage or preneoplastic stage cellular growth given by:

$$H(d,t)=(\exp(G_0[1+f_4(x(d))]t)-1-G_0[1+f_4(x(d))]t)/G_0^2[1+f_4(x(d))]^2,$$

where:

G_0 is the background, time-independent, growth rate for preneoplastic cells;

f_4 is an S-shaped function of relative increase over background preneoplastic growth rate.

Expressing dose response relationships for biomarkers as apparent relative (rather than absolute) increases over background biomarker levels may not be conceptually optimal for conservative extrapolation purposes. However, one can clearly see from the above equation that nonlinearities (and perhaps even linearities at comparable dose levels d) in two or more of the four biomarker fitting functions—f_1, f_2, f_3 and f_4—could easily lead to nonparallelism between predicted tumor response $P(x(d))$ and any one of these four dose surrogates even at low dose and perhaps for low-background tumor rates where $P(x(d))$ tends to be linear if one uses linear-at-low-dose fitting functions. Although many compounds may cause cancer predominantly by acting on only one subprocess within one stage, it is also quite plausible that many compounds may significantly affect more than one stage of the cancer process and/or may significantly affect several subprocesses (e.g., both transition rates and cell growth rates) within one or more stages. This potential lack of observable parallelism between nominal dose response and the surrogate biomarker dose response is even more likely to occur when the relationship between nominal dose d of the parent compound and internal tissue dose $x(d)$ of the parent or active metabolite is nonlinear in the observable ranges, as might occur in many cases of saturable (nonlinear) Michaelis-Menton kinetics.

Although the above models are probably too simplistic to describe the actual cancer process in most instances, eventual model validation will require that a realistic composite dose response model show the ability to predict reasonably well, qualitatively and even quantitatively, both any expected parallelisms as well as possibly expected progressively divergent nonparallelisms between tumor response and progressively earlier-stage biomarkers of the carcinogenic process. Next, model validation would seem to require a confidence limit (standard error) analysis of the many parameters simultaneously estimated by the composite model, a concomitant sensitivity analysis of their potential dose response implications, and a stable, conservative, algorithmic implementation of low dose linearity, based upon that sensitivity analysis. Lastly, before believing we have truly progressed beyond mere curve fitting approaches to conservative low dose tumor extrapolation, a substantial data base, under chronic bioassay conditions, needs to be developed whereby we can compare the estimated biomarker subfunctions against actual laboratory measurements for those biomarkers (e.g., adduct formation or cell proliferation measurements within the context of a chronic animal cancer bioassay). This will yield clues as to whether and under what conditions curve fitting exercises with these new more complex composite models will be sufficiently credible for conservative regulatory risk assessment purposes, in the presence and, perhaps even in the absence, of extensive biomarker measurements for routine chemicals.

REFERENCES

1. L. J. Carter, "How to Assess Cancer Risks," *Science* **204**:811-816 (1979).
2. D. M. Hughes, R. D. Bruce, R. W. Hart, L. Fishbein, D. W. Gaylor, J. M. Smith and W. W. Carlton, "A Report on the Workshop on Biological and Statistical Implications of the ED_{01} Study and Related Data Bases," *Fund. Appl. Toxicol.* **3**:129 (1983).
3. J. A. Swenberg, F. C. Richardson, J. A. Boucheron and M. C. Dyroff, "Relationships Between DNA Adduct Formation and Carcinogenesis," *Environ. Health Perspect.* **62**:177-183 (1985).
4. J. A. Swenberg, Transcript: Seminar Series on Scientific Issues Related to Risk Assessment, January 8, 1987, ILSI Risk Science Institute, Brookings Institution, Society for Risk Analysis.
5. J. A. Swenberg, M. C. Dyroff, M. A. Bedell, J. A. Popp, N. Muh, U. Kirstein and M. F. Rajewsky, "O^4-Ethyldeoxythymidine, But Not O^6-Ethyldeoxyguanosine, Accumulates in Hepatocyte DNA of Rats Exposed Continuously to Diethylnitrosamine," *Proc. Natl. Acad. Sci.* **81**:1692-1695 (1984).
6. M. C. Dyroff, F. C. Richardson, J. A. Popp, M. A. Bedell and J. A. Swenberg, "Correlation of O^4-Ethyldeoxythymidine Accumulation, Hepatic Initiation and Hepatocellular Carcinoma Induction in Rats Continuously Administered Diethylnitrosamine," *Carcinogenesis* **7**:241-246 (1986)
7. J. A. Boucheron, F. C. Richardson, P. H. Morgan and J. A. Swenberg, "Molecular Dosimetry of O^4-Ethyldeoxythymidine in Rats Continuously Exposed to Diethylnitrosamine," *Cancer Res.* **47**:1577-1583 (1987).
8. J. A. Swenberg, F. C. Richardson, L. Tyeryar, F. Deal and J. Boucheron, "The Molecular Dosimetry of DNA Adducts Formed by Continuous Exposure of Rats to Alkylating Hepatocarcinogens," *Prog. Exp. Tumor Res.* **31**:42-51 (1987).
9. R. Peto, R. Gray, P. Brantom and P. Grasso, "Nitrosamine Carcinogenesis in 5120 Rodents: Chronic Administration of Sixteen Different Concentrations of NDEA, NDMA, NPYR, and NPIP in the Water of 4440 Inbred Rats, with Parallel Studies on NDEA Alone of the Effect of Age of Starting (3, 6 or 20 Weeks) and of Species (Rats, Mice or Hamsters)," in O'Neill, VonBorstel, Miller, Long, Bartsch, N-Nitroso Compounds, Occurrence, Biological Effects and Relevance to Human Cancer, IARC Scientific Publication No. 57, International Agency for Research on Cancer, Lyon (1985).
10. R. Schulte-Hermann, "Reactions of the Liver to Injury," in *Toxic Injury of the Liver Part A*, E. Farber and M. M. Fisher, eds., Marcel Dekker, New York (1979).
11. S. H. Moolgavkar and D. J. Venzon, "Two-Event Models for Carcinogenesis: Incidence Curves for Childhood and Adult Tumors," *Math. Biosci.* **47**:55-77 (1979).

12. S. H. Moolgavkar and A. G. Knudson, "Mutation and Cancer: A Model for Human Carcinogenesis," *J. Natl. Cancer Inst.* **66**:1037-1062, (1981).
13. T. W. Thorslund, C. C. Brown and G. Charnley, "Biologically Motivated Cancer Risk Models," *Risk Analysis* **7**:109-119 (1987).

Global Risk Assessment[a]

Michael H. Tiller

Tiller Consulting Group, Inc.
St. Louis, MO

ABSTRACT

This paper discusses the need for global risk assessment and suggests a general philosophical design approach for such a monumental task. One key to success of such an endeavor lies in the integration of global risk assessment data needs into the overall framework of global change study recently initiated by several organizations, including the International Council of Scientific Unions.

We feel there is an acute need for global risk assessment. Anthropogenic activities are garnering larger and larger percentages of the total emissions of a variety of chemicals into the environmental media and greatly speeding up existing and creating new biogeochemical cycles. The effects of such activity on natural earth systems are largely unknown, particularly as spatiotemporal scales of potential impact challenge our analytical capability.

Additionally, current retrospective study of the sun, earth, and other terrestrial planet system factors such as solar flux, planet paleoclimate, orbital dynamics, and atmospheric composition suggest a much greater rate and intensity of change of our environment than previously thought. In itself, such natural change can present a potential threat to humans.

Ideally, global risk assessment will identify and analyze potential human and natural event scenarios, their probability of occurrence and potential level of damage to humans and earth systems. Such a risk assessment may be useful in suggesting policy changes in regards to chemical handling, agricultural practice and education.

This paper is exploratory in nature and perhaps will serve as a point of departure for more definitive design of a global risk assessment.

KEYWORDS: Risk assessment, global risk assessment, earth system

---

a. A revised and updated version of this paper titled "A Tentative Global Risk Assessment Framework," can be found in Liptak, S. C., J. W. Atwater, and D. S. Mavinic, *Environmental Engineering Proceedings of the 1988 Joint CSCE-ASCE National Conference, Vancouver, British Columbia, July 13-15, 1988,* Montreal: Canadian Society for Civil Engineering, pp. 46-53, copyright Canadian Society for Civil Engineering, 1988.

OVERVIEW

Introduction

Almost twenty years ago, Chauncey Starr suggested that technological risk can be quantified (Starr, 1969). Risk analysts in the nuclear power industry began doing just that in the 1970s. Then along came Bhopal, and the use of risk assessment technology in the chemical and related industries accelerated greatly (J. C. Consultancy, 1986; American Institute of Chemical Engineers, 1985; Garrick in SRA, 1987). While chemical engineers were beginning to assimilate risk assessment, the shuttle disaster occurred. Again, the rapid migration of risk assessment technology took place, this time into space (see, for example, the Risk Analysis and Space Policies Session in SRA, 1987). While in space, the risk analysts will look out their portholes and see earth. Perhaps this change in perspective will catalyze synoptic application of risk assessment technology to Buckminster Fuller's Spaceship Earth.

This "perspectives" revolution began in the ecologic community when the first photos of the earth from space became available in the 1960s. Our past cultural attitudes toward the "infinite" resources of the environment are now being questioned with greater frequency (Richards, in Clark, 1986).

There is an urgent need for global risk assessment. Anthropogenic (human-related) activities are capturing larger and larger percentages of the total emissions of a variety of chemicals into the environmental media and greatly speeding up existing biogeochemical cycles and creating new ones. The effects of such activity on natural earth systems are largely unknown, particularly as space-time scales of potential impact challenge our analytic capability (Bolin, 1983; Goody, 1982; Goudie, 1986; Moore, 1984; National Research Council, 1984, 1986). Recent retrospective study of the sun, earth and other terrestrial planet system factors, such as solar flux, planet paleoclimate, orbital dynamics, and atmospheric composition, suggest a much greater rate and intensity of "natural" change of our environment than previously thought (Haberle, 1986; Klemes, 1985; Williams, 1986). Natural change alone can present a potential threat to humans.

Ultimately, each environmental medium (atmosphere, biosphere, lithosphere, hydro/cryosphere, solarsphere, and magneto/electrosphere) contributes and is affected by natural and anthropogenic change. The number of major sources and scale of change to earth systems are enormous. Hypotheses about previously unexpected or poorly understood relationships between media and their contents are being put forth (Lovelock, 1979, National Research Council, 1986, Rosenzweig, 1986). For example, relationships between oceanic phytoplankton, atmospheric sulphur, cloud albedo, and climate have been hypothesized (Charlson, 1987; Bates, 1987). The number of feedback loops and emergent properties of each subsystem makes the earth system very difficult to study. The goal of a global risk assessment is to provide answers to questions such as:

- What initiating events threaten current global environment?

- What are the probabilities of their occurrence?

- What level of damage would be sustained?

Recent international cooperation has already started a global change study effort (ICSU, 1986; Malone, 1985; McCauley, 187; NASA, 1986). One of the goals of this study is to provide a body of knowledge needed to assess the future of the earth (National Research Council, 1986).

This paper will discuss general aspects of a global risk assessment and suggest one organizing template based on matrix theory formalism which may be appropriate in organizing the study. This paper is exploratory in nature and may serve as a point of departure for a more erudite discussion of a global risk assessment endeavor.

The Risk Triplet

Risk assessments constitute a formal method for describing various potentially adverse initiating events, probabilities of their occurrence, and expected level of damage for a system. One definition of risk is the measure of the probability and severity of adverse effects. Risk assessment can be thought of as the determination of the risk triplet:

(1) identification of the initiating event,

(2) the probability of the event's occurring, and

(3) the measure of damage.

A risk assessment is essentially a listing of the various initiating events, along with their associated probabilities and consequences. Conceptually, the list of initiating events is infinite. The fact that any risk assessment is finite means that risk assessments are inherently incomplete. So, in a sense, the risk analyst must first explicitly list all known initiating events and then attempt to describe the size of the residual risk not explicitly listed. Generally speaking, a family of risk curves is generated, displaying an expected level of damage over some probability range. An introduction to risk assessment can be found in Kaplan (1981) with a summary of applications in Garrick (1984).

The following sections discuss issues affecting the global risk assessment. These issues include the concepts of change, space-time scaling, boundaries, models, and the interfaces of different kinds of systems. These issues have a fundamental impact on the risk assessment. Once these issues are discussed, a matrix theory application for the assessment of global risk is briefly proposed.

ISSUES AFFECTING THE GLOBAL RISK ASSESSMENT

Change, Space-time and Boundary Issues

A risk assessment attempts to identify and quantify potential adverse change to a system. Change is indeed the central issue surrounding risk assessments. In the *I Ching*, a distinction is made between three types of change: nonchange, cyclical change, and sequential change (Baynes, 1984).

Nonchange is the background against which change is made apparent. Cyclical change is the rotation of phenomena, each succeeding the other until the starting point is reached again. Examples of cyclical change include seasonal changes such as carbon dioxide concentrations in the earth's atmosphere related to photosynthetic activity. Sequential change is an onward moving process that never returns to its starting point. Sequential change incorporates time's arrow, which points the way from past to future.

Chaisson, 1987, believes that the Eastern mind has been dominated by cyclical change, while the Western mind has been dominated by sequential change. A synthesis of the cyclical and sequential views must be effected to give an adequate account of earth systems (Gould, 1987).

Cyclical and sequential change may only be an artifact of the space-time frame used in observing the particular system under study. Consider observing the changes in behavior of a system over some time T, every t seconds. The frequency of observable behaviors can be divided into three ranges, depending upon the relative length of time spent observing the system:

- low frequencies, much less than $1/t$,

- middle-range frequencies, and

- high frequencies, greater than $1/t$.

Behavior of the system determined by the low frequency modes will be so slow that this behavior is perceived as a constant. The high frequency behavior is not directly observable; we see only an "average" of this change. Thus, the middle band of frequencies will be the only easily observable interactions of the subsystems under study (Pattee, 1973).

The space-time scale used is crucial in the identification of the initiating event. For example, it has been known for some years that change in the concentration of CO_2 in the atmosphere has a seasonal (cyclical) pattern related to photosynthetic activity. Yet only after years of monitoring CO_2 concentrations on Mona Loa, Hawaii, was it realized that this seasonal change is superimposed on an apparently longer currently increasing component of change.

Whether the longer term change is another cyclical component of natural change over a much longer period or an anthropogenically caused change remains to be seen. In any case, the longer time perspective suggests that the rise in CO_2 concentration should be considered as a possible initiating event of global proportions.

Thus the comprehension and analysis of change in any system are highly dependent on the space-time frame being used to examine the system. It follows, then, that the measure of damage of the change is also highly dependent on the observational window we are using. Global risk assessment must incorporate the concept of change during the analysis of an initiating event: Over what space-time frame and with what observational frequency is the potential change in one of the system state variables being analyzed?

Analyzing a system using different space-time frames can give divergent perspectives on the measure of damage of an already identified initiating event. For instance, on one space-time scale a risk assessment of several acres of a forest may show that an initiating event such as a fire would cause major debilitating change to this system. Paradoxically, another risk assessment of the forest, done over a relatively larger space-time scale, may show that the change brought about by the fire is indeed crucial for the survivability of the entire forest.

How large should the space-time scale of a global risk assessment be? Consider Milankovich-type orbital variations which scale from tens of thousands to hundreds of thousands of years (Pisias *et al.* in Ryan, 1986). Chinese civilization has lasted for thousands of years (Temple, 1987). Should our risk assessment cover such a time period or go with a human generational time frame and consider changes in human gene pools caused by the founder effect (Diamond, 1987a,b; Lewin, 1987)? Or are we satisfied with a window of a few hundred years, which Americans are probably more comfortable with?

Should global risk assessments be limited to the upper boundary of the atmosphere (Malone, 1985)? Should they include the impact of solar variability on climate, which would implicitly extend the spatial analysis by about 93 million miles (Kerr, 1987; Holton, 1982; National Academy of Sciences, 1985)? Should they include interactions of the earth with a cloud of interstellar gas that is now streaming through the solar system and other similar events, suggesting a cosmic scale for the assessment (Paresce, 1987)?

Fortunately, a component of systems theory called hierarchy theory can aid in the assignment of the risk assessment space-time boundaries. Hierarchy theory integrates the concept of change into the analysis of systems by relating members of hierarchical systems according to frequency or rate of change, not size. Hierarchy theory is beginning to be used successfully in modeling various components of the earth system. A rapidly developing literature describes hierarchy theory and various environmental applications (Bertalanffy, 1968; Grene, 1987; O'Neill, 1986; Pattee, 1973; Tiller in Dietz, 1986; Whyte, 1969).

Unfortunately, the risk assessment literature contains little explicit analysis utilizing hierarchy theory. Yet some limited attention has been given to the setting of boundary conditions in the systems analysis community (Jones, 1982). Only recently have initiating events of a regional scale, external to the engineered system under study, been considered (Budnitz, 1984).

Models

Models are an important tool in all branches of science and will play a crucial role in a global risk assessment. Conceptually, a model is built by using three universals: deletion, distortion and generalization. The process of deletion entails excluding portions of the world to render the analysis simpler. The process of distortion changes the actual relationships among parts of the system. Generalization is the process by which a specific experience comes to represent the entire category of which the experience is a member (Bandler, 1975).

These universals prevent a complete description of the earth system. A full description would eventually require complete self-reference, which is impossible. This condition is analogous to Kurt Godel's proof that a mathematical system cannot contain all the rules which create it (Nagel, 1958; Hofstadter, 1979).

Philosophically we have been faced for several decades with a loss of certainty, as Kline (1980) describes. Our mathematics, once thought to be the epitome of determinism, have enigmatically "shrunk to statistical truths." Nonlinear science suggests that some non-deterministic natural processes such as weather cannot be predicted for more than two weeks into the future, due to the underlying chaotic nature of the system being studied (National Academy of Sciences, 1987, 1985; Lorentz, 1984; Schneider, 1987; Tribbia, 1987).

Paradoxically, this same nonlinear behavior of some systems is helping us to understand them (Prigogine, 1984; Sander, 1987). For example, medical research has revealed that many physiological parameters vary chaotically in the healthy individual, while more regularity can be a sign of pathology (National Academy of Sciences, 1987; West, 1987). The risk assessor is already using fractals, one of three major paradigms of nonlinearity, to describe the environmental dilution of potentially toxic effluents in river basins (Seiler, 1986). The nonlinear science contribution to global risk assessment may be that the earth system can be described by some set of rules. Yet this set of rules may not lead us closer to prediction of change within the system.

Modeling on a world scale is, fortunately, not new. Such modeling is being done in prospective and retrospective categories. Prospective models include, for example, World 3, World Integrated Model, Latin American World Model, United Nations Input-Output World Model, Global 2000, global circulation models, global biogeochemical models and various climate impact models (Stumm, 1977; Trabaka, 1986; Schneider, 1987; Bolin, 1983, 1986; Council on Environmental Quality, 1979; Kates, 1985; Linder, 1987; Mills,

1982; National Aeronautics and Space Administration, 1986; Sperber, 1987). These prospective models attempt to predict change on a global scale in such areas as economics, population, energy growth, and flow of mass, energy and information within various earth subsystems.

Retrospective modeling of paleoclimatic and other past features of terrestrial planets are also being done (Haberle, 1986; Hecht, 1984; Williams, 1986). The strength in retrospective models lies in their ability to impart to the risk assessor an appreciation of potential space-time scales and types of change possible within a terrestrial planet system. For example, modeling of past change of carbon dioxide concentration in paleoatmospheres suggests that the current increasing change in concentration discussed earlier is unusual both in the rate and the quantity of change.

Some of both kinds of models address some aspects of global risk; none address a synoptic risk assessment of the earth as a whole. The risk assessor will have to sort through these models, perhaps incorporating some of them within the larger framework of a matrix-based risk assessment. The matrix-based risk assessment will be discussed later.

Initiating Event Selection and Systems Interfacing

Ironically, we do not yet have a robust model which identifies potentially adverse initiating events. Nor is there a model which aids in the selection of which initiating event to analyze, assuming that it can be identified. If we do not know the initiating event, we certainly cannot analyze it. This uncomfortable issue, first raised by Douglas (1982) has received little attention.

Douglas points out that the selection of risks to analyze has a strong social component. This implies that human systems are very important throughout risk assessment. Yet human systems are different from engineered or natural systems and may prove to be the most intractable for integration into a formal global risk assessment (Vickers, 1984). The full dimension of risk cannot be captured solely by the probability of an adverse initiating event because this misses the human element (Covello, 1982).

The issues of social responsibility and the fuzzy decision models for collective global resource management will factor greatly in any global risk assessment. Models need to be developed describing how humans adapt (Ortner, 1983; World Meteorological Organization, 1984). One of the major philosophical issues we need to face is: what do we owe our future generations (Rayner in SRA, 1987)?

In brief, a global risk assessment will require sound analysis and interfacing of three types of systems: natural, human-engineered and human. Our current method of analysis by use of the risk triplet may undergo extensive revision during this process.

MATRIX THEORY FORMALISM FOR GLOBAL RISK ASSESSMENT

The Global Assembly Equation

A global risk assessment will include dozens of scientific disciplines, very large, diverse data sets, and scales of study ranging from atomic to global. To maximize effectiveness there must be a simple organizing template for such an endeavor.

Kates (1985) describes the framework for an impact model for analyzing the effects of climate change on human populations. This model assumes a change in a state variable describing some aspect of climate, such as temperature. This variable acts upon the second part of the model, a population in a given area on earth. The goal is an accurate description of a second state variable, such as Gross National Product, resulting from the climate change on the population.

Kates' model can be generalized into three components. The first component is the Earth System, which in turn consists of interconnected media subsystems describing the atmosphere, biosphere, lithosphere, hydro/cryosphere, solarsphere, and magneto/electrosphere.

The second, the Response System, delineates the response of humans to the perceived change in one or more state variables describing the Earth System. The output of this second component is one or more state variables describing an impact category. These state variables, in turn, are inputs for the Social System component, whose output is one or more state variables describing final damage states.

The Earth System component transforms initiating events into earth states. The Response System component uses the earth state as input to develop the frequency of occurrence of different impact categories. With defined impact categories, the Social System component translates impact categories into final damage states to earth inhabitants.

Four pinch points are used in this global risk matrix model. These are:

- the initiating event (i.e., an asteroid collision, continued release of "green house" gases, etc.),

- the earth state (i.e., a change in climate, etc.),

- impact category (i.e., full, grazing or no collision of asteroid, based on human response to the threat), and

- final damage state (i.e., carrying capacity of the earth system drastically reduced).

Each pinch point represents a result of a "lower level" risk assessment. By use of a matrix formalism developed by Kaplan, 1982, we can assemble the three component assessments into a full statement of risk. This statement of risk can be called the global assembly equation. The end result of this Kaplan assembly equation can take the form of a series of risk curves which relate initiating events to final damage states.

As there may be hundreds of initiating events, thousands of state variables and dozens of "lower level" risk assessments, matrix algebra should be used to relate which initiating event affects which state variable. Disassembly of the overall statement of risk is relatively easy, as each pinch point would be described as a algebraic vector. Various diagnostic matrices can be developed and important initiating events isolated from the overall study. An overview of the general methodology can be found in Garrick (1984) and Sancaktar (1982).

CONCLUSION

Kairos

The Greek language is much richer than English in describing time. Kairos, meaning the time is ripe, is a good example. Today is kairos for starting a global risk assessment. The recently begun international cooperation for global change study is already providing important data which can be used for a synoptic risk assessment of the earth. This risk assessment can provide important feedback to the global change study. For instance, the risk assessment matrix will organize and force use of interchangeable data sets and will guide portions of the change studies towards the gathering of data relating to crucial adverse initiating events.

ACKNOWLEDGMENTS

We are indebted to M. J. Berry, S. Hawkins, W. C. Clark, A. Dreimanis, V. Krapivin, T. F. Malone, N. Polunin, W. G. Sombroek, T. Wilson, L. Worrel and many others not mentioned here for suggesting and providing reference material and ideas. Responsibility for this paper lies solely with the author.

REFERENCES

American Institute of Chemical Engineers, 1985, *Guidelines for Hazard Evaluation Procedures*, American Institute of Chemical Engineers, New York.

Bandler, R., Grinder, J., 1975, *The Structure of Magic*, Science and Behavior Books, Inc., Palo Alto

Bates, T. S., Charlson, R. J., and Gammon, R. H., 1987, "Evidence for the Climatic Role of Marine Biogenic Sulphur," *Nature* **329(6137)**:319-321.

Baynes, C. F., and Wilhelm, R., translators, 1984, *The I Ching*, Princeton: Princeton University Press.

Bertalanffy, Von Ludwig, 1968, *General System Theory*, George Braziller, New York.

Bolin, B., Doos, B. R., Jager, J., and Warrick, R. A., Eds., 1986, *SCOPE 29: The Greenhouse Effect, Climatic Change, and Ecosystems*, John Wiley & Sons, New York.

Bolin, B., Ed., 1983, *SCOPE 24: The Major Biogeochemical Cycles*, John Wiley & Sons, New York.

Budnitz, R. J., 1984, "External Initiators in Probabilistic Reactor Accident Analysis — Earthquakes, Fires, Floods, Winds," *Risk Analysis* **4(4)**:323-335.

Chaisson, E., 1987, *The Life Era*, The Atlantic Monthly Press, New York.

Charlson, R. J., Lovelock, J. E., Meinrat, O. A., and Warren, S. G., 1987, "Oceanic Phytoplankton, Atmospheric Sulphur, Cloud Albedo and Climate," *Nature* **326**:655-661.

Chivas, A. R., Barnes, I., Evans, W. C., Lupton, J. E., and Stone, J. O., 1987, "Liquid Carbon Dioxide of Magmatic Origin and Its Role in Volcanic Eruptions," *Nature* **326(6113)**:587-589.

Clark, W. C., and Munn, R. E., Eds., 1986, *Sustainable Development of the Biosphere*, Cambridge University Press, Cambridge.

Council on Environmental Quality, 1979, *The Global 2000 Report to the President*, Vol. I and II, U.S. Government Printing Office, Washington.

Covello, V. T., Menkes, J., and Nehnevajsa, J., 1982, "Risk Analysis, Philosophy, and the Social and Behavioral Sciences: Reflections on the Scope of Risk Analysis Research," *Risk Analysis* **2(2)**:53-58.

Diamond, J. M., 1987, "Who Were the First Americans?" *Nature* **329(6140)**:580-581.

Diamond, J. M., and Rotter, J. I., 1987, "Observing the Founder Effect in Human Evolution," *Nature* **329(6137)**:105-106.

Dietz, J. D., Ed., 1987, *Proceedings of the 1987 Specialty Conference Environmental Engineering*, American Society of Civil Engineers, New York.

Douglas, M., Wildavsky, 1982, *Risk and Culture*, University of California Press, Berkeley.

Gaffeny, J. S., Streit, G. E., Spall, W. D., and Hall, J. H., 1987, "Beyond Acid Rain," *Environmental Science and Technology* **21(6)**:521-524.

Garrick, B. J., 1984, "Recent Case Studies and Advancements in Probabilistic Risk Assessment," *Risk Analysis* **4(4)**:267-279.

Goodey, R., 1982, *Global Change: Impacts on Habitability, A Scientific Basis for Assessment JPL D-95*, Jet Propulsion Laboratory, Pasadena.

Goudie, A., 1986, *The Human Impact on the Natural Environment*, MIT Press, Cambridge.

Gould, S. J., 1987, *Time's Arrow, Time's Cycle*, Harvard University Press, Cambridge.

Grene, M., 1987, "Hierarchies in Biology," *American Scientist* **75**:504-510, September-October.

Haberle, R. M., 1986, "The Climate of Mars," *Scientific American* **254(5)**:54-62.

Hecht, A. D., Ed., 1985, *Paleoclimate Analysis and Modeling*, John Wiley & Sons, New York.

Hofstadter, D. R., 1979, *Godel, Escher, Bach: An Eternal Golden Braid*, Basic Books, Inc., New York.

Holton, J. R., Ed., 1982, *Solar Variability, Weather & Climate*, National Academy Press, Washington.

J. C. Consultancy, 1986, *Risk Assessment for Hazardous Installations*, Pergamon Press, London.

Jones, L. M., 1982, "Defining Systems Boundaries in Practice: Some Proposals and Guidelines," *Journal of Applied Systems Analysis* **9**:41-55.

Jouzel, J., Lorius, C., Petit, J. R., Genthon, C., Barkov, N. I., Kotlyakov, V. M., and Petrov, V. M., 1987, "Vostok Ice Core: A Continuous Isotope Temperature Record Over the Last Climate Cycle (160,000 years)," *Nature* **329**:403-408.

Kaplan, S., 1982, "Matrix Theory Formalism for Event Tree Analysis: Application to Nuclear Risk Analysis," *Risk Analysis* **2(1)**:9-18.

Kaplan, S., and Garrick, B. J., 1981, "On the Quantitative Definition of Risk," *Risk Analysis* **1(1)**:11-27.

Kates, R. W., Ausubel, J. H., and Berberian, M., Eds., 1985, *Climate Impact Assessment: Studies of the Interaction of Climate and Society (SCOPE 27)*, John Wiley & Sons, New York.

Kerr, R. A., 1987, "Sunspot-Weather Correlation Found," *Science* **238**:479-480.

Kerr, R. A., 1987, "Winds, Pollutants Drive Ozone Hole," *Science* **238**:156-158.

Klemes, V., 1985, *Sensitivity of Water Resource Systems to Climate Variations*, World Meteorological Organization, Geneva.

Kline, M., 1980, *Mathematics: The Loss of Certainty*, Oxford University Press, New York.

Lewin, R., 1987, "Africa: Cradle of Modern Humans," *Science* **237**:1292-1295.

Linder, K. P., and Gibbs, M. J., 1987, "Potential Impacts of Climate Change on Electrical Utilities: Project Summary," Air Pollution Control Association Annual Meeting and Exhibition, New York, NY, June 21-26.

Lorentz, E. N., 1984, "Irregularity: A Fundamental Property of Atmosphere," *Tellus* **36A**:98-110.

Lovelock, J. E., 1979, *Gaia: A New Look at Life on Earth*, Oxford University Press, Oxford.

Malone, T. F., and Roederer, J. G., 1985, *Global Change*, Cambridge University Press, London.

McCauley, L. L., and Warner, L., Eds., 1987, *Earthquest Boulder*, University Corporation for Atmospheric Research.

Mills, W., 1982, *Global Models, World Futures, and Public Policy: A Critique*, Office of Technology Assessment, Congress, Washington.

Moore, B., and Dastoor, M. N., Eds., 1984, *The Interaction of Global Biochemical Cycles*, Jet Propulsion Laboratory, JPL Pub 84-21, Pasadena.

Nagel, E., and Newman, J. R., 1958, *Godel's Proof*, New York University Press, New York.

National Academy of Sciences, 1985, "Panel on Weather Prediction Technologies," in Research Briefings, National Academy Press, Washington.

National Academy of Sciences, 1987, "Report of the Research Briefing Panel on Order, Chaos, and Patterns," in Research Briefings, National Academy Press, Washington.

National Aeronautics and Space Administration, 1986, *Earth System Science Overview: A Program for Global Change*, National Aeronautics and Space Administration, Washington.

National Research Council, 1984, *Global Tropospheric Chemistry: A Plan for Action*, National Academy Press, Washington.

National Research Council, 1986, *Global Change in the Geosphere-Biosphere: Initial Priorities for an IGBP*, National Academy Press, Washington.

National Research Council, 1986, *The Earth's Electrical Environment*, National Academy Press, Washington.

O'Neill, R. V., Deangelis, D. L., Waide, J. B., and Allen, T. F. H., 1986, *A Hierarchical Concept of Ecosystems*, Princeton University Press, Princeton.

Ortner, D. J., Ed., 1983, *How Humans Adapt a Biocultural Odyssey*, Smithsonian Institution Press, Washington.

Paresce, F., and Bowyer, S., 1986, "The Sun and the Interstellar Medium," *Scientific American* **255(3)**:92-99.

Pattee, H. H., Ed., 1973, *Hierarchy Theory*, Braziller, New York.

Prigogine, I., and Stengers, I., 1984, *Order Out of Chaos*, Bantam Books, New York.

Rosenzweig, C., and Dickinson, R., Ed., 1986, "Climate-Vegetation Interactions," Office for Interdisciplinary Earth Studies Report OIES-2, Boulder.

Ryan, P., Ed., 1986, "Changing Climate and the Oceans," *Oceanus* **29(4)**:1-100.

Sancaktar, S., 1982, "An Illustration of Matrix Formulation for a Probabilistic Risk-Assessment Study," *Risk Analysis* **2(3)**:137-147.

Sander, L. M., 1986, "Fractal Growth Processes," *Nature* **322**:789-793.

Schneider, S. H., 1987, "Climate Modeling," *Scientific American* **256(5)**:72-80.

Seiler, F. A., 1986, "Use of Fractals to Estimate Environmental Dilution Factors in River Basins," *Risk Analysis* **6(1)**:15-25.

Society for Risk Analysis, 1987, "Risk Assessment in Setting National Priorities," Society for Risk Analysis, Proceedings of the 1987 Annual Conference, McLean.

Sperber, K. R., Hameed, S., Gates, W. L., and Potter, G. L., 1987, "Southern Oscillation Simulated in a Global Climate Model," *Nature* **329(6137)**:140-142.

Starr, C., 1969, "Social Benefit Versus Technological Risk," *Science* **165**:1232-1238.

Stumm, W., Ed., 1977, *Global Chemical Cycles and Their Alterations by Man*, Abakon-Verlagsgesellschaft, Berlin.

Temple, R., 1987, *The Genius of China: 3,000 Years of Science, Discovery, and Invention*, Simon & Schuster, New York.

Trabacka, J. R. *et al.*, Eds., 1986, *The Changing Carbon Cycle: A Global Analysis*, Springer Verlag, New York.

Tribbia, J. J., and Anthes, R. A., 1987, "Scientific Basis of Modern Weather Prediction," *Science* **237**:493:499.

Vickers, G., 1983, "Human Systems Are Different," *Journal of Applied Systems Analysis* **10**:3-13.

Wallen, C. C., 1984, "Present Century Climate Fluctuations in the Northern Hemisphere and Examples of Their Impact," World Meteorological Organization, Geneva.

West, B. J., and Goldberger, A. L., 1987, "Physiology in Fractal Dimensions," *American Scientist* **75**:354-365, July-August.

Whyte, L. L., Wilson, A. G., and Wilson, D., Eds., 1969, *Hierarchical Structures*, Elsevier Publishing Company, Inc., New York.

Williams, G. E., 1986, "The Solar Cycle in Precambrian Time," *Scientific American* **255(2)**:88-96.

World Meteorological Organization, 1984, "Report of the Study Conference on Sensitivity of Ecosystems and Society to Climate Change," World Meteorological Organization, Geneva.

Cost Benefit and Environmental Risk Aspects in the Context of European Air Pollution Control Strategies

Otto Rentz, Thomas Morgenstern, and Georg Schons

Institute for Industrial Production
University of Karlsruhe (TH)
Federal Republic of Germany

ABSTRACT

Approaches to develop and evaluate control strategies for air pollution control have been implemented and validated by the authors. Time-phased energy supply models have proved to be meaningful working tools for the assessment of different pollution control strategies. Cost optimization is applied as the usual objective function; however, other objectives, as, for example, supply reliability and emission minimization, have been considered. One shortcoming of this approach, however, is the lack of a comparison of the resulting economic situation with the public benefit.

In this paper, methodologies are described which are based on the optimization and harmonization of air pollution control strategies in all member countries of the European Communities. The results are mainly from a study being carried out by the authors on behalf of the Commission of the European Communities. In the study the national energy supply structure has been analyzed to achieve the optimal allocation of emission reduction measures. The cost-effective reduction of environmental risks related to air pollution has been determined by weighting measures in different sectors, e.g., utilities and industry, for a defined reduction level.

KEYWORDS: Cost benefit, European air pollution, energy supply structure, environment, cost-effectiveness analysis, systems analysis, Europe

INTRODUCTION: AIR POLLUTION CONTROL AND TECHNOLOGY RISK MANAGEMENT

What is the relation between air pollution control and the management of technological risks?

This is a question which you might often hear in discussion with European scientists or politicians in the field of air pollution. Just in the past few years it has become more and more obvious that the research work in the field of air pollution control, i.e., assessment of environmental damage and evaluation of abatement measures, should be understood as part of risk analysis. Similarly, launching emission control technologies should be understood as part of risk management. An example of the new orientation is the international colloquium "La Maitrise des Risques Technologiques" in Paris, organized by ACADI[a] on behalf of the UNESCO in December 1987.

The reasons why this closeness to the field of risk analysis has often not been realized by those experts who are involved in the development of emission reduction measures are easy to understand and include the following:

- The lack of clarity about the chain of cause and effects, i.e.,

    - nonlinearity of the relation of cause and effects,
    - the existence of synergism,
    - cumulative damages (the damage remains hidden until it becomes explosive).

- The difficulty in proving the effectiveness of emission abatement measures.

- The difficulty in quantifying the impacts of damage to the environment, as well as the chance of its rescue.

In Europe many investigations have been launched in order to optimize the energy supply structure with regard to environmental restrictions (see, e.g., Refs. 1, 2, 3, and 4). The criterion for a decision has been mostly to minimize the costs of the energy supply system inclusive of the equipment with flue gas treatment. These models are able to treat scenarios with different emission reduction levels (e.g., 30%, 50%, or 70% reduction SO_2 and/or NO_x). However, one drawback of these models is that nothing is included to aid the politician in making a choice between the different scenarios and the conclusive strategies. That is, the benefit for the national economy is not taken into account. This disadvantage could be one reason why these models do not find their way into decision making on the political level.

Some attempts to remove this drawback by comparing the different scenarios with their corresponding benefits are described in this paper.

CONCEPTS OF THE MONETARY EVALUATION OF THE COST OF AIR POLLUTION

The usual starting point of economic considerations in the field of cost-benefit analysis is as simple as it is plausible (see Fig. 1). The utility curve and cost curve are compared. The optimal allocation of emission reduction measures is achieved when the marginal control costs meet the marginal damage costs.

As is well known, the major drawback in comparing marginal cost and marginal benefit lies in an exact definition and determination of the utility curve.[5,6] Noteworthy investigations in the Federal Republic of Germany are based on the assumption that a Pareto-Optimum, where marginal cost and utility correspond, should be sought. For private goods the utility curve can be described as the willingness-to-pay. The concept of these investigations was to determine the total benefit as the sum of the marginal willingness-to-pay.[5,10,11] Even though reasonable arguments support this approach, it has serious disadvantages, as noted below:

---

a. Association des Cadres Dirigeants de l'Industrie pour le Pregrès Sociàl et Economique.

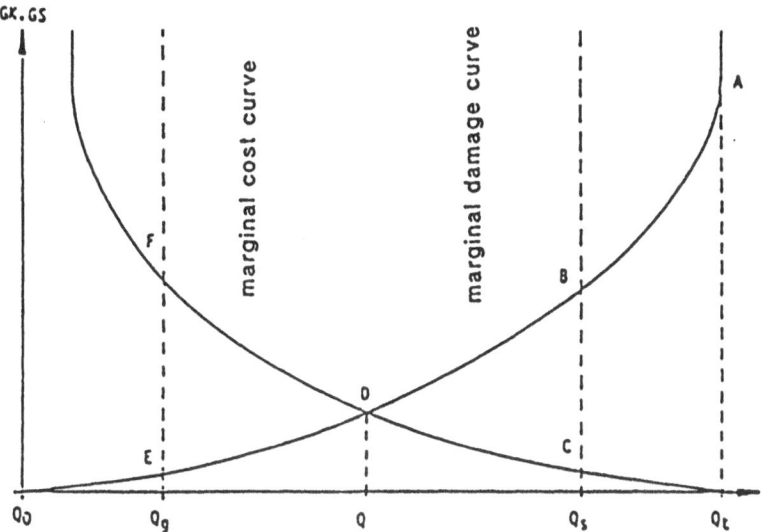

Fig. 1. Scheme of Cost-Benefit Analysis for Air Pollution Control Policy.[5]

- A correct individual assessment would have to be based on complete information on the subject; in reality, this does not occur.

- The incomplete information available to both the experts and the general population leads to strategies that vary with time. That is, strategies derived from results of such a cost-benefit analysis have to be revised as additional information becomes available.

- The environment has a limited assimilative capacity which is independent of any assessment. That is, infinite marginal costs would not always ensure removal of the damages.

- In the case of highly toxic emissions, e.g., dioxin, the consequences of careless policies are well known. However, for most of the pollutants discussed, cumulative effects are still being observed, and this time lag delays correct evaluations.

- The "willingness-to-pay" approach correlates to a large degree with the income of the persons asked; i.e., it tends to favor the rich, so that clean air could become a luxury.

On the other hand, the following advantages are against a rejection of this approach:

- The monetary evaluation of environmental damage stimulates a more rational discussion of the problem and makes grading possible.

- Material and immaterial damages are taken into account in the calculations; i.e., a "willingness-to-pay" analysis provides a better assessment of the utility than balancing the cost of removing the damage, since the consumers' surplus is considered.

- The effects of meaningless answers by uninformed persons can be avoided by simple testing.

- The neglect of the effects of income distribution can be estimated and compensated for by econometric methods.

- The overestimation of the damage costs by using the "willingness-to-sell" approach is avoided.

THE EVALUATION OF THE COSTS OF AIR POLLUTION CONTROL

As mentioned above, the authors are mainly dealing with instruments supporting a cost-effectiveness analysis of emission control strategies on both a national and a European level. The Commission of European Communities provided the working tool for this application, the dynamic and optimizing EFOM 12 C Model. In a first phase of the study, this model was adjusted to the new requirements. It was then modified and extended so that the following elements of emission reduction measures could be modeled and placed in competition:

- Ability to implement less extensive policies such as fuel switching or combustion modifications.

- Explicit modeling of emission reduction technologies for conventional fossil-fired energy conversion technologies as secondary measures.

- Implementation of new "integrated" technologies, e.g., combined cycle power plant with and without coal gasification, pressurized fluidized bed combustion, etc.

- Consideration of mobile sources.

The subsystems of EFOM are concatenated from primary energy to final or utilization energy (Fig. 2). All sectors causing environmental problems are now equipped with an environmental module. The measures in the different subsystems are weighed against each other with respect to cost effectiveness. (In public discussion it is implied that SO_2 and NO_x are reduced and that the emissions of particulates are balanced. Actually the problem of dust emissions receives little attention in the FRG.) General economic assumptions have been modeled according to the results of econometric studies on behalf of the Commission of the European Communities.[7]

Oriented to these basic figures, different scenarios have been defined in order to cover a wide scope of control strategies:

- "Doing-nothing" Case. This scenario is used as the reference scenario—compulsory for a European study—but somewhat unrealistic for the German investigation.

- "Legal" Case. The simulation of the current legal situation, which for some European countries is synonymous with the first scenario.

- "Cost-effectiveness" Cases.

  - 30% SO_2/NO_x-reduction.
  - 50% SO_2/NO_x-reduction.
  - 70% SO_2/NO_x-reduction
    (simultaneously referred to 1980-emission-levels).

In this context, the "30%-reduction" case can be seen as an extended interpretation of the 30% reduction agreement of most of the members of the Economic Commission for Europe (ECE).[8]

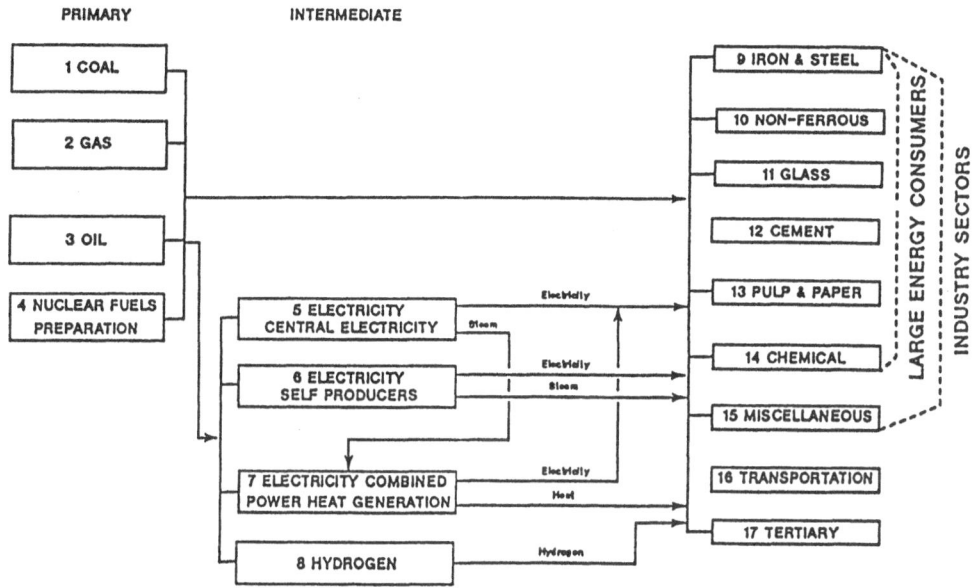

Fig. 2. The Modular Structure of EFOM.

In conclusion, the major results of this application should be pointed out:

- Systems analysis of the environmental problems of the energy supply.

- Computation of annual average costs, investments, fixed and variable costs, sector by sector (i.e., determination of the total cost in relation to different emission levels).

- As a result of linear programming, the marginal cost of the environmental restriction reflects the gradient of the total cost curve. The lack of a marginal cost curve has been complained about in a lot of publications (see, e.g., Refs. 5, 6, 9, and 13).

COMPARISON OF THE COST AND BENEFIT OF EMISSION CONTROL OF THE POLLUTANTS SO$_2$ AND NO$_x$

As mentioned above, the total utility curve reflects the issues of a large questionnaire in the Federal Republic of Germany published in 1985.[5,10,11] The people interviewed evaluated five different levels of air quality—from pure air (Level I) to smog (Level V). These five quality levels correlate with SO$_2$ deposition-concentrations (see Fig. 3). However, for a comparison of utility and cost we should make the following assumptions:

- The entire region (FGR) is to be interpreted as one emission source.

- This region is also the valley of depositions; i.e., no local differences in the distribution of pollutants are considered.

- In evaluating distribution models (e.g., MESOS), it has been shown that the import and export of emissions are balanced in this region.[12]

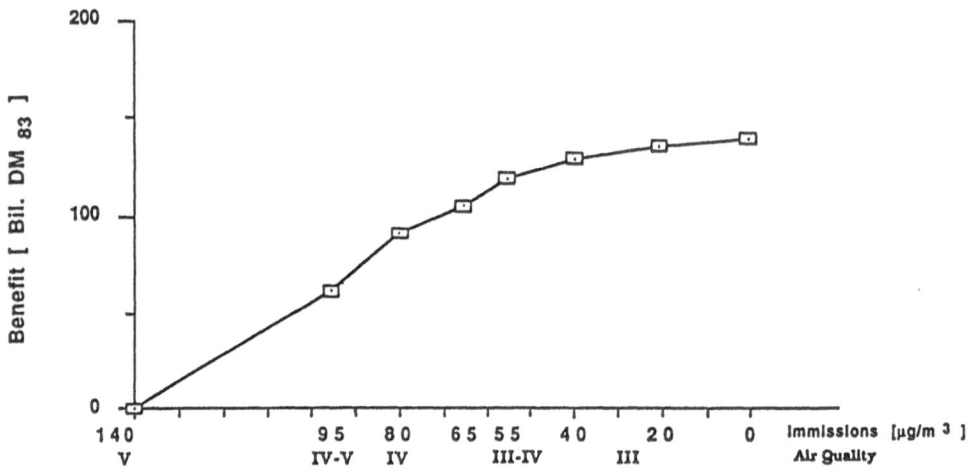

Fig. 3. Total Utility Curve of Clean Air (Federal Republic of Germany, 1984).[5]

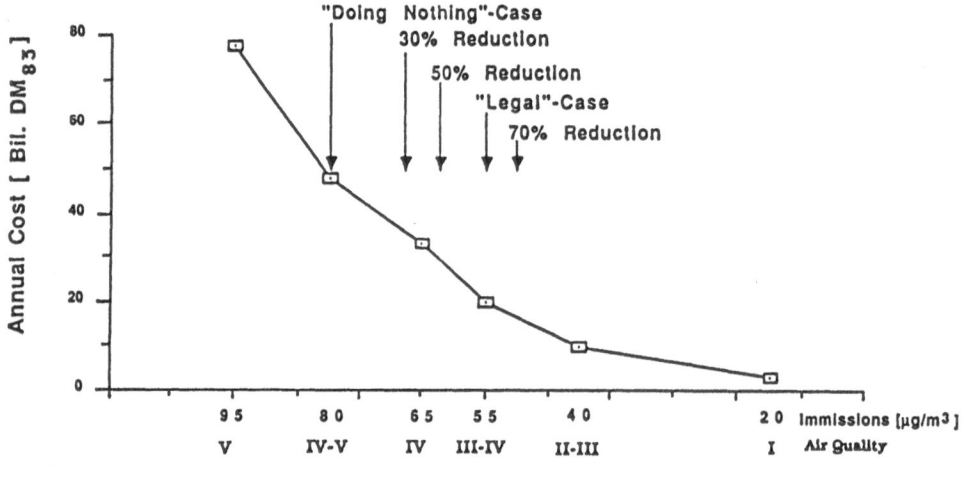

Fig. 4. Damage Cost Curve.

- Because of these border-crossing effects, the effectiveness of emission control measures in the Federal Republic of Germany is reduced by approximately one-half.

Moreover, we assume that emission reduction measures are put into effect without any major delay; i.e., time-lag effects are neglected.

The damage costs are shown in Fig. 4. The reference scenario corresponds to a damage of about 48 billion DM in 1983. By enforcing restrictive standards for SO_2 and NO_x emissions ("Legal" Case, flue gas desulphurization, selective catalytic reduction for power plants, catalytic converters for new cars), the achieved air quality level is associated with damage costs of approximately 20 billion DM.

As emphasized above, the actual application of the energy-environment model EFOM is oriented toward SO_2 and NO_x emissions. Figures 5 and 6 demonstrate the

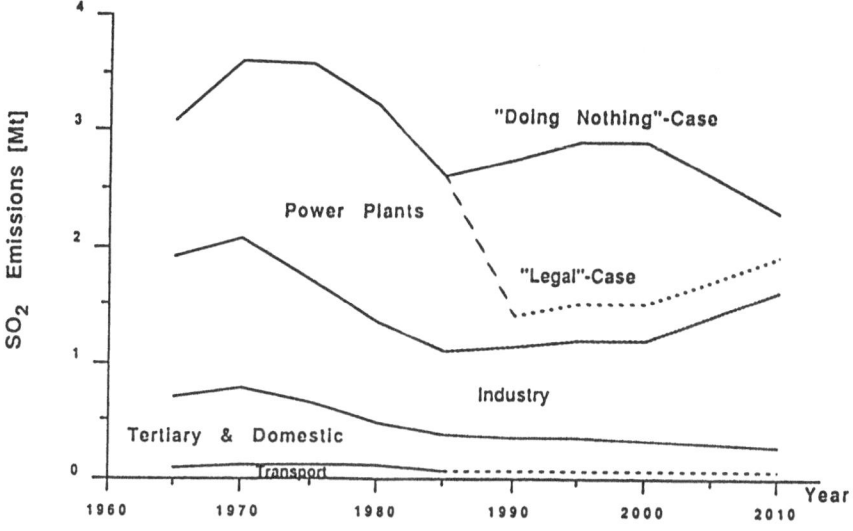

Fig. 5. Development of SO₂ Emissions.

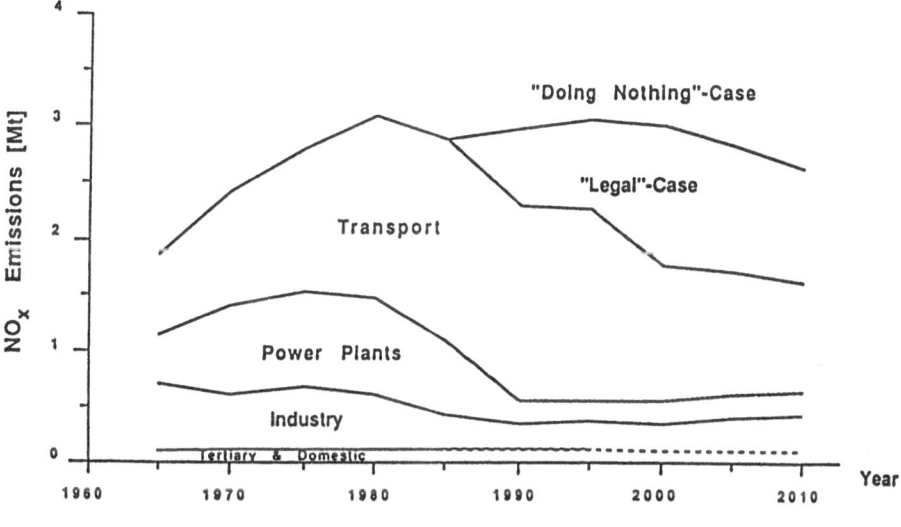

Fig. 6. Development of NOₓ Emissions.

development of both pollutants in the "Doing-nothing" Case and "Legal" Case. The effectiveness of the current legislation is clearly indicated. Moreover, the contribution of the energy-intensive sectors to the overall emissions is reflected.

For more detailed information about this model application, especially in the European context, see Ref. 4. To obtain economic results, all additional costs (interest rate five percent) were determined (see Fig. 7).

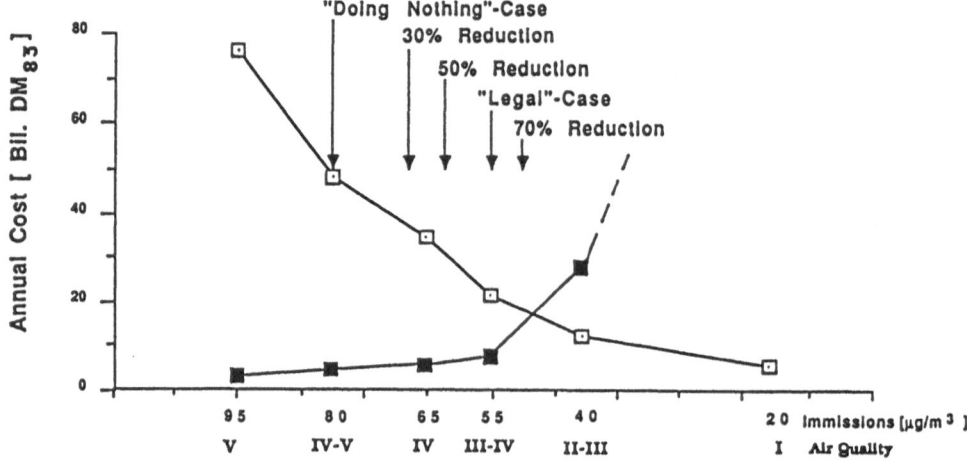

Fig. 7. Cost and Benefit of Environmental Protection.

If it is assumed that by 1990 environmental legislation will put the expected emission reduction into effect (60% SO_2, 25% NO_x), additional costs of about 7 billion DM/year will arise. Both benefit and reduction cost curves intersect with a price of about 18 billion DM.

A more effective reduction of NO_x can be obtained only by consistently equipping cars with three-way catalytic converters. This reflects the increase of the cost curve beyond the intersection point.

CONCLUSIONS

One should beware of interpreting the point of intersection as "the market price of clean air." However, it can provide an idea of the order of magnitude where environmental protection is still reasonable in the eyes of the population. The following can also be concluded:

- The questionnaire was based on a general assessment of the air quality as opposed to the reduction cost curve, referring only to the main single pollutants in the actual public discussion, SO_2 and NO_x. The assumption that the damage is mainly caused by NO_x and SO_2 leads to a misinterpretation in the assessment of human health, since other pollutants (e.g., CO, volatile organic compounds, lead, etc.) have been neglected.

- The state of information about single pollutants will be improved, and evaluations of the benefit by reducing these pollutants will be made.

- There will always be a lack of information available to experts and the public. That is, there will always be an "after-effect" ("after Seveso," "after Harrisburg," "after Tschernobyl") which will shift the public assessment immediately. Such hazards displace the public discussion of continuous damage to nature, e.g., the "Waldsterben."

- The assimilative capacity of nature is limited. It does not rely on the preferences of the consumer. It is an objective fact of the ecological system. Only a few "harmless" pollutants require no standards beyond which major impacts to the ecological systems are expected.

- The low effectiveness of the actual standards in the Federal Republic of Germany demonstrates drastically the international character of the environmental problem in Europe. The "cost-effectiveness" study of the Commission of the European Communities[4] is only one of the initial steps in the harmonization and improvement of the emission and deposition situation.

- Environmental protection amply repays the effort. The current legislation even in the FRG does not yet correspond with the public assessment of the benefit. More stringent legislative restrictions could be justified.

REFERENCES

1. Rentz, O., Th. Hanicke, and R. Hempelmann, "Practical Experience with a Dynamic LP-Model for the Planning of Optimal Energy Supply Strategies Taking into Account Detailed Technologies for Pollution Control," *Operations Research Verfahren* **44**:653-664 (1981).
2. Hanicke, Th., "Wirtschaftlich-technische Optimierung des Energieversorgungs-systems der Bundesrepublik Deutschland anhand eines linearen multikriteriellen Optimierungsmodells," Dissertation, Universität Karlsruhe, 1985.
3. Rentz, O. *et al.*, "Entwicklung von technisch-wirtschaftlichen Strategien für Emissionsminderungsmaßnahmen für Schwefeldioxid und Stickoxide aus stationären Produktionsanlagen in Baden-Württemberg," Forschungsbericht KfK-PEF 13, April 1987.
4. Commission of the European Communities, Ed., "Optimal Control Strategies for Reducing Emissions from Energy Production and Energy Use," to be released in March 1988.
5. Schulz, M., *Der monetäre Wert besserer Luft*, Frankfurt, Bern, N.Y., 1985.
6. OECD, Ed., *The Costs and Benefits of Sulphur Oxide Control: A Methodological Study*, Paris, 1981.
7. CISI–WHARTON, Ed., "L`Europe d`ici 2005 selon quatre scenarios contrastes de l`economic mondiale," Contract 86B0709-11-008-17-N with the Commission of the European Communities, 1987.
8. Economic Commission for Europe (ECE), Report of the third session of the Executive Body ECE/EB.AIR/7, Geneva, Annex I, Article 2, 1985.
9. Environmental Resources Limited, *Acid Rain — A Review of the Phenomenon in the EEC and Europe*, London, 1983.
10. Wicke, L., *Die ökonomischen Milliarden*, Kösel, München, 1986.
11. Umweltbundesamt, Ed., *Kosten der Umweltverschmutzung*, Berichte 7/86, Berlin.
12. Halbritter, G. *et al.*, "Regionale Immissionsanalysen für pflanzentoxische und säurebindende Luftverunreinigungen, Projekt Europäisches Forschungszentrum für Maßnahmen zur Luftreinhaltung (PEF)," 1. Statuskolloquium, Proceedings March 1985, Karlsruhe (FRG).
13. OECD, Environment and Economics, Vol. II, Paris, 1984.
14. Umweltbundesamt, Ed., *Daten zur Umwelt 1986, 87*, Erich Schmidt Verlag Berlin, 1986.

Health Effects of Occupational Exposures of Welders: Multicenter Study Using a Uniform Core Protocol

R. M. Stern, F. LaFerla, and F. Gyntelberg
World Health Organization Regional Office for Europe
Copenhagen Ø, Denmark

ABSTRACT

Because of the large number of welders (2% of the working population), worldwide, even a small occupationally related excess risk of cancer would represent an important public health hazard. Responding to the repeated observation of a small respiratory tract cancer overincidence in many independent measurements, The World Health Organization, Regional Office for Europe, Copenhagen, has initiated a multicenter historical-prospective study of the effects of welding, compared to non-welding populations, and the effects of welding on stainless steel (which provides a unique exposure to the putative carcinogens Ni and Cr) compared to non-stainless steel welding populations. It is hoped that by the use of a uniform core epidemiological protocol, the data from a number of national or local studies, when pooled under the auspices of the International Agency for Research on Cancer, Lyon, will permit a statistically significant determination of the magnitude of the overall effect: anticipated range of SMR=1.3-1.4. The study anticipates a total cohort of 14,000 welders of both categories, with a total of 100 expected cases at SMR=1.0, and is considerably more powerful than any single study designed around a local cohort. It is also large enough to provide important but almost unavailable information concerning cumulative dose, latency, and age on the effect of occupational exposure to metals.

KEYWORDS: Epidemiology, multicenter studies, welders' health, WHO (World Health Organization)

A multicenter collaborative study of the possible health effects of welding has been organized under the WHO/Regional Office for Europe (F.LF.), coordinated by the Danish Welding Institute (R.M.S.) and the Clinic Of Occupational Medicine, Copenhagen University Hospital (F.G.). Studies of cancer incidence and mortality are coordinated by IARC, Lyon.

Welders have been chosen as a study group for several reasons: they make up from 0.2-2% of working populations in most of the industrialized countries, they are known to have high occupational exposures to toxic fumes and gases the nature of which depends to a great deal on the welding process and application, many current technologies have been in use for over 40 years, and, in spite of an exceptionally large volume of literature, little is known about the absolute health effects of the occupation.

A common core protocol is in use,[1] and at present, studies of mortality, morbidity and occupational hygiene have been initiated. Mortality studies focus on cancer of the respiratory tract: programs are being designed or are underway at, among others: INRS, Vandoevre-Les-Nancy; Institute of Cancer Research, Surry; Odense University Hospital; Telemark Central Hospital; National Board Of Occupational Safety and Health, Solna; Institute for Documentation Information and Statistics, Heidelberg; Institute for Tumour Research, Genoa; Institute of Occupational Health, Helsinki. In an attempt to improve the power of the study, incidence and mortality data are to be pooled under the coordination of IARC. The major aim of the study is to determine if the apparent excess of respiratory tract cancer among welders is related to occupation by comparison with non-welding populations, and to identify the causes, if possible, by comparing age, exposure and latency stratified data from welders exposed to stainless steel welding fumes with those without such exposure. Clinical studies of lung function, accompanied by exposure measurements and biological monitoring, are being undertaken at the Institute of Occupational Health, Budapest; the Institute of Preventive Medicine, Bratislava; the University of California School of Medicine, Los Angeles; and the Central Institute for Occupational Medicine, East Berlin.

One of the first activities has been to convene an international conference to assemble the state-of-the-art-knowledge concerning constituents of welding fumes and gases, the nature and extent of occupational exposures, laboratory evidence for toxicity of fumes, gases and their constituents, clinical evidence for adverse health effects, and a review of morbidity and mortality studies.[2] The proceedings provide a guideline for current and future investigators in terms of planning the details of research programs. In particular, the difficulty with which delayed health effects could be identified or measured in any single cohort study, or excess risk could be associated with specific processes or applications, emphasizes the need for collaborative studies. A major effort is being made to identify separately the effects of exposures to mild steel (MS) and to stainless steel (SS) welding fumes. The interest with respect to SS fumes arises because they contain high concentrations of Cr(VI) and Ni, both of which are putative human carcinogens, and welders appear to suffer a 40% excess incidence of respiratory tract cancer, as illustrated by almost all incidence or mortality studies. This excess could have a life-style (non-occupational) origin, or it could arise from either a universal exposure to welding fumes among all welders, or it could be localized to several particular technologies at high risk (e.g., SS welding).[3] Such differences can only be studied with the help of large collective cohorts, since SS welders make up only 10% of the welding population. The current program envisions a collective cohort of 14,000 welders, and anticipates approximately 100 cases divided almost equally between SS and MS cohorts. Because there are very different strategies for reducing a universal small risk or a localized high risk, the outcome of the collaborative study will have a significant effect on priority setting for risk reduction within the industry. In particular, large research programs are under way to try to identify the nature of the toxic risk associated with the fumes from a wide variety of processes and applications so as to provide early guidelines for risk reduction when and if the extent of human health risk has been established.

A first quality control exercise in chemical analysis of welding fumes has been completed, and a second round is underway in an effort to harmonize hygienic monitoring and epidemiology by ensuring a uniform method for dose assessment, and to develop exposure estimates based on job descriptions. In addition, the core protocol contains models of health status and exposure questionnaires, and minimum programs for biological monitoring, lung function testing, study of non-respiratory effects, and exposure monitoring. An experimental non-invasive technique for determining the quantity and quality of magnetic dust burden of the lungs (magnetopneumography) is introduced as a complement to biological monitoring, together with other suggestions for innovative studies. The activity is intended as a model exercise and could easily be extended to other

Summary Of National Cohorts

| Country | Cohort Size | | | Expected Cases, SS/MS | Type of Cohort |
	Stainless Steel	Mild Steel	Reference		
Denmark	3288	2040	3711	20/13	Welders and other workers: 90 sites >1 year employed not left by 1964
England	(600)	(600)	--	7/7	Welders at 18-25 sites employed >5 years before 1970
Finland		1689	11004	-/34	Welders in 4 shops 5 shipyards > 1 year controls:pipefitters platers,machinists
France	440	600	2560	1.6/2.7	Welders at 4 sites employed in 1976 3 matched controls per welder
Germany	1221	--	1624	6.3/-	Welders in 25 sites employed <1970
Italy	(600)	(250)	--	3.0/1.0	Welders at 2 sites
Norway		(600)	(1200)		shipyard welders
Sweden	234	208	--	2.0/3.0	Welders at 8 sites employed >5 years
Total	6323	5987	20099	40/60	Followup 1981-5

industrial sectors, especially those involving mixed exposure to metals, with the ultimate aim of establishing quantitative estimates of dose-response for Cr, Ni, etc.

Additional participation in this program is hereby solicited on a worldwide basis, and interested parties, including trade unions and employers' associations, are welcome to contact any of the senior investigators to obtain information on the current state of the activity.

REFERENCES

1. Stern, R. M. et al., "Health Effects of Occupational Exposure of Welders," in *Studies in Epidemiology Part 1*, Health Aspects of Chemical Safety, Interim Document 15, WHO/Euro, Copenhagen 1984.

2. Stern, R. M., Berlin, A., Fletcher A. C., Hemminki, K., Järvisalo, J., and Peto, J., International Conference on Health Hazards and Biological Effects of Welding Fumes and Gases: Summary Report., *Int. Arc. Occup. Environ. Health* **57**:237-246 (1986).
3. Stern, R. M., "Process Dependent Risk of Delayed Health Effects for Welders," *Env. Health Perspect.* **41**:235-253 (1981).

Comparative Risk Assessment of Transportation of Hazardous Materials in Urban Areas

Ashok B. Boghani[a] and Krishna S. Mudan[b]
Arthur D. Little, Inc.
Cambridge, MA

ABSTRACT

The transportation of hazardous materials (HAZMAT) through urban areas is one of the most challenging risk-related issues faced by modern society. Society cannot function without this activity, yet the concentration of people and property in urban areas makes the potential for damage due to a HAZMAT release very high. Clearly, a methodology is required to systematically identify and quantify risk in this situation, help develop mitigation measures, and evaluate the effectiveness of such measures.

This paper describes a methodology to perform a microlevel risk analysis at street level in an urban area, thereby achieving the above objectives. The methodology requires studying each transportation movement through the urban areas in terms of material transported, quantity transported, vehicle characteristics, characteristics of the route (land usage, length, speed, type of road), and potential release scenarios. When the transportation pattern of a particular commodity is too complex to analyze, typical movements can be used as substitutes.

Then, using historic accident frequency and release consequence databases, a comparative risk index is obtained for each movement. (A spreadsheet program is used for calculations.) The effects of mitigation measures, such as additional regulations requiring changes in route and vehicle characteristics, can then be studied by comparing the risk index numbers for various possibilities. The paper illustrates the methodology, using an example.

KEYWORDS: Hazardous materials, transportation risk, risk assessment, urban areas

INTRODUCTION

The use and transportation of hazardous materials create risks that are of great concern to the public, the government and the industry. Bulk transportation of hazardous materials by highway involves the use of tank trucks or trailers and certain types of more specialized bulk cargo vehicles. In all, trucks transport more than sixty percent of the

---

a. Manager, Transportation Consulting Group

b. Now with Technica, Inc., Columbus, OH

hazardous materials not carried by pipelines, with just under fifty percent of this material being gasoline.[1] Since trucks carry hazardous materials the greatest number of miles and carry the largest number of shipments, it is not surprising that this mode of transportation is also responsible for the greatest number of accidents.

Tank trucks are usually tractor-semitrailer vehicles or smaller bobtail-type units. The tanks themselves are usually constructed of steel or an aluminum alloy, but may also be stainless steel, nickel and other materials. Capacities are usually in the range of 3000-9000 gallons, although slightly smaller, and larger units are available. Intermodal tanks which consist of a tank within a protective rigid framework, one-ton tanks which are lifted on and off the transporting vehicle, and large gas cylinder bundles are also commonly used for bulk transport by highway.

Commodity breakdowns for trucks, as described in various data sources, are not considered very accurate and vary widely. However, a one-month survey of cargoes in Virginia found a fairly close match (by percent) to the average distribution of commodities involved in accidents. The comparison for commodities involved in accidents in Virginia also matched national accident breakdowns fairly well.[2] National involvement in accidents by type of commodity for the time period July 1973-December 1978 was:

Flammable liquids	60.5%
Combustible liquids	16.3%
Corrosives	11.6%
Flammable compressed gases	3.2%
Oxidizers	2.1%
Poisons (liquid or solid)	2.1%
Nonflammable compressed gases	1.9%
Explosives	1.5%
Radioactive materials	0.5%
Flammable solids	0.3%

Truck accidents on roadways, regardless of the cargo involved, are generally due to:

- Collisions with other vehicles

- Collisions with fixed objects (such as bridges or overpass supports)

- Running off the road

- Overturns

These four events are most likely to result in a release of large quantities of hazardous materials. However, smaller releases may arise due to defective or loose valves, fittings or couplings, weld failures, and various other structural defects.

The impacts of truck accidents (with releases) can vary widely. A set of recent accidents is described below to illustrate the types of events and consequences which may occur.

- A tank truck carrying 11,000 gallons of gasoline blew a tire, struck a cement barrier which ripped open the side of the tank, and then burst into flames. The accident took place on Interstate 95 near Peabody, Massachusetts. State troopers closed down the highway while a crash crew from Logan International Airport spread foam on the wreckage. The highway remained partially closed for several days, since one section had melted and needed to be replaced. There were no injuries or deaths. (December 3, 1985)

- The town of Littleton, New Hampshire, was spared a potentially catastrophic accident when a tank trailer loaded with 9,200 gallons of propane jack-knifed on an icy hill, and then tipped on its side about 75 yards from a large storage tank of liquid propane and less than 100 yards from several large fuel oil storage tanks. No propane leaked from the truck, but a diesel fuel tank was ruptured. A half mile radius of the accident was evacuated of 1500 people until the propane was safely transferred to another vehicle the next day. (February 11, 1982)

- A truck pulling two tank trailers loaded with molten sulphur collided with a highway barrier on a toll bridge and burst into flames, taking two lives and injuring twenty-six. Fire fighters encountered difficulty extinguishing the fire and rescuing victims. Visibility at the time was poor due to a fog inversion, and the spilled sulfur had burned through water supply lines. (January 19, 1986)

- In Houston, a tank truck carrying liquefied anhydrous ammonia collided with a car and fell from an elevated highway to a busy freeway. It exploded violently on impact, releasing billowing clouds of ammonia. Four persons (including the truck driver) were killed, dozens of motorists were overcome by the fumes in a three-mile area, and at least 100 were treated at area hospitals. The fumes were so thick that police helicopters were initially repelled. The city was forced to use all available ambulances and privately-owned hearses to transport the injured. (May 12, 1976)

- Although certain details are unclear, a tank truck carrying liquid propylene sprang a leak in the vicinity of a crowded campsite in Spain. Flammable gases spread from the truck, encountered a source of ignition, flashed back to the vehicle, and caused a boiling liquid expanding vapor explosion (BLEVE) with a large fireball. The death toll from burns was approximately 170. Numerous other people suffered moderate to severe burns but otherwise survived. (July 11, 1978)

Clearly, tank accidents involving hazardous materials can have significant impact. In the following section, we have outlined a relatively simple approach to combine the accident frequency with accident impact and to determine a comparative risk level.

DESCRIPTION OF APPROACH

The first step in performing the risk assessment is to establish what is meant by risk and what variables quantify it. We have assumed that the risk in truck transportation of hazardous materials lies in a truck becoming involved in an accident which results in a reasonable-size spill. Such a spill can then lead to a more severe consequence, such as pool fire, BLEVE, unconfined vapor cloud explosion (UVCE), or confined vapor cloud explosion (CVCE).

In transportation activities, the probability of a very large spill (dumping the complete content of the truck) is usually quite small. Depending upon the probable frequency of these small and large spills and the damage they cause, we can estimate the average damage per year, which is a good measure of risk. Two additional measures of risk which may be of interest are the probable frequency and magnitude of a catastrophic spill. A catastrophic spill is defined as a spill of the complete content of a truck in a heavy residential or heavy commercial area. The magnitude and frequency of a catastrophic spill will have a direct bearing on the extent of preparation required of the emergency response teams, and therefore these two quantities can also be used for expressing risk analysis results.

Damage can be expressed in terms of dollars, injuries or fatalities. To do so, however, requires a detailed evaluation of all possible accident scenarios, which is not generally needed for comparing various alternatives for hazardous transportation. What is of interest is a comparative measure of risk (e.g., risk of gasoline vs. risk of fuel oil; risk without tank size limitation vs. risk with tank size limitation). Therefore, damage has been expressed in terms of a dimensionless number which can be used to compare risks for various alternatives but will not give any absolute measure of risk.

Figure 1 shows the procedure used to arrive at these measures of transportation risk. The first step involves using past experience and judgment to determine the comparative consequences of large or small spills involving each hazardous material and land use type (heavy commercial or residential, light commercial or residential, industrial, vacant, elevated roadway, or bridge). As a measure of comparative consequences, a weighted "impact factor", a non-dimensional number is derived for each combination of material, land use, and spill size (either a full tankload or a partial tankload). These impact factors take into account both the relative damage caused by various scenarios (pool fire, boiling liquid expanding vapor explosion, confined or unconfined vapor cloud explosion, or no ignition) and the relative likelihood of these scenarios, given that a spill occurs. The mathematical expression for calculating the impact factor is provided below.

For each material n, spill size m, and land use type j,

$$I_{njm} = \frac{1}{100}\left[\frac{P_c}{100} \sum_i P_{aijmn}\, d_{ijmn} \right.$$
$$\left. + \frac{100-P_c}{100} \sum_i P_{bijmn}\, d_{ijmn} \right] \tag{1}$$

Where

I_{njm} = Impact factor for material n, spill size m and land use type j

P_c = Conditional probability for collision given an accident (in %)

P_{aijmn} = Conditional probability for scenario i given a collision type of accident, land use type j, spill size m and material n (in %)

P_{bijmn} = The above for a non-collision type accident (in %)

d_{ijmn} = Damage score for scenario i, in land use type j, spill size m, and material n.

$\sum_i P_{aijmn}$ = 100 for each j, m and n

$\sum_i P_{bijmn}$ = 100 for each j, m and n

Next, the impact factors are combined with estimates of the number of trips made by transporters of hazardous materials and the average spill frequency of a typical trip to arrive at an overall risk for each commodity transported. These estimates are based on the total volume of each commodity transported, the characteristics of typical routes (gathered from a route survey), and historical data on accident and spill frequencies. The calculations for overall risk are provided below.

For each material, n:

Accident frequency (per year)/n $= 2 \times \sum_o \sum_j \sum_r L_{jron} f_r T_{on}$ $\tag{2}$

Spill frequency (per year)/n $= \sum_m \sum_o \sum_j \sum_r L_{jron} f_r T_{on} P_{pm}$ $\tag{3}$

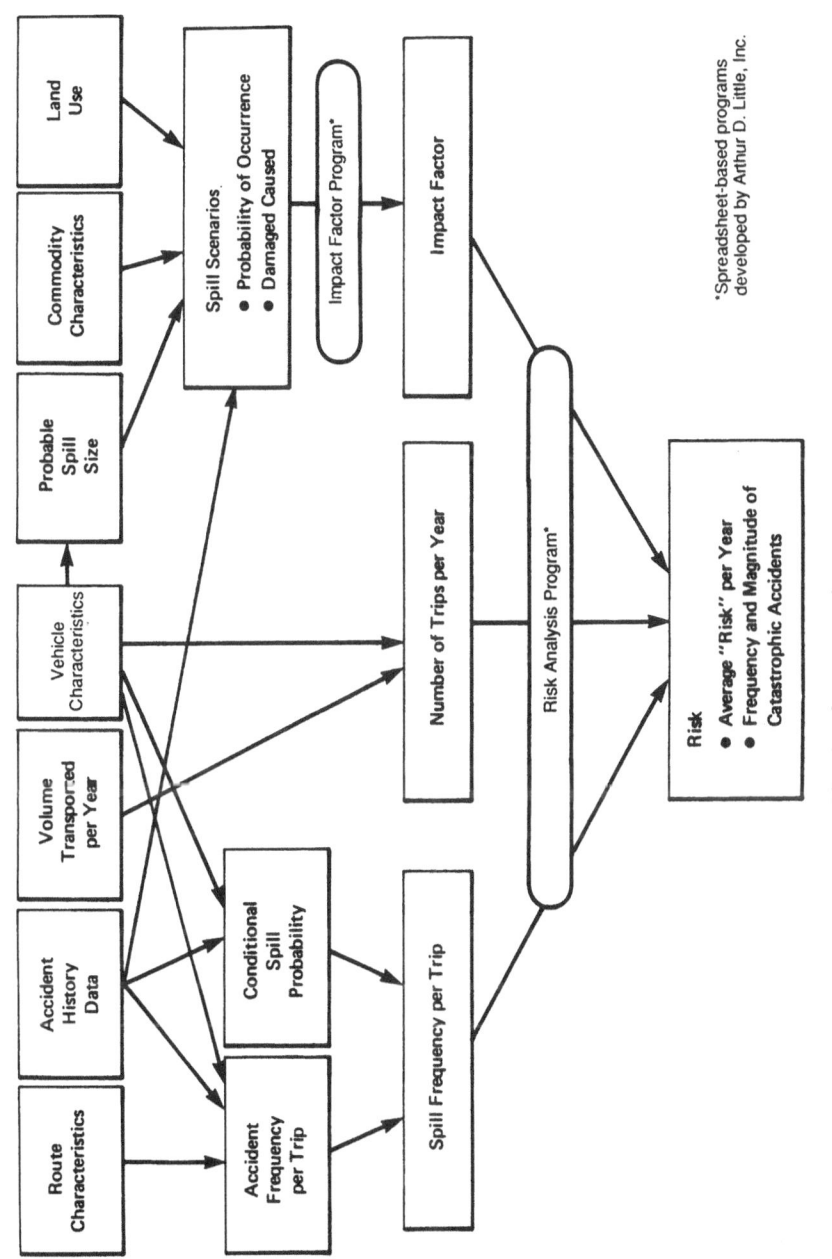

Fig. 1. Risk Analysis Approach.

*Spreadsheet-based programs developed by Arthur D. Little, Inc.

Damages per year/n $= \sum_m \sum_o \sum_j \sum_r L_{jron} f_r T_{on} P_{pm} I_{njm}$ (4)

Catastrophic spill frequency/n $= \sum_{j=HR/HC} \sum_o \sum_r L_{jron} f_r P_{pm}$

m=large spill (5)

where

L_{jron} = Distance (in miles) of road type r through land use j, for route o of material n (one way distance)

f_r = Accident frequency (accidents/truck miles) for road type r

T_{on} = Trips per year on route o of trucks carrying material n

P_{pm} = Fraction of accidents which result in a release of amount m

I_{njm} = Impact factor [from Eq. (1)]

HR/HC = Heavy residential/heavy commercial

Note that the accident frequency has a factor 2 associated with it to account for empty return trip. Since a spill can take place only for accidents to loaded trucks, no such factor is needed.

EXAMPLE

We used this methodology to compare risk of transporting various hazardous commodities through a major metropolis. We also compared the risks involved in conducting transportation activities under various regulations. In this section we briefly discuss some examples of the results obtained.

Considerable literature is available on truck accidents. From a review of this literature, we selected the following rates as appropriate for accident frequency and conditional probabilities of release.

Road Type	Accident/Truck Miles	Conditional Probability of Release	
		Small Spill	Large Spill
Turnpike	1.4×10^{-6} (Ref.3)	0.2 (Ref.5,6,7)	0.02 (Ref.5)
City	8.2×10^{-6} (Ref.4)	0.2	0.02
Ramp	3.5×10^{-6} (Ref.3)	0.2	0.02
Bridge	1.4×10^{-6}	0.2	0.02

The risk associated with hazardous material transportation was assessed in terms of two typical spill sizes. Based on past experience, it was assumed that the spill which occurs once every five accidents is usually about 10% of the total content of the tank truck, while the large spill which happens only once every fifty accidents is equal to the full load of the tank truck.

The commodities selected for the study are gasoline and fuel oil. It was assumed that the gasoline is transported in 4000 gallon trucks, and the fuel oil in 6500 gallon trucks. The amount spilled can then be estimated as:

Commodity	Small Spill (10% of Full Load)		Large Spill (Full Load)	
	Amount	Fraction	Amount	Fraction
Gasoline	400g	0.1	4000g	1.0
Fuel Oil	650g	0.1	6500g	1.0

The calculation of impact factors involves estimating both the relative likelihood of the different accident scenarios and the relative damage caused by each of these scenarios for spills occurring on various types of land use. The following assumptions are used in estimating impact factors:

- *Heavy Commercial/Heavy Residential (HC/HR) Areas*: Here the road surfaces are relatively impermeable. Therefore the spilled fuel may spread evenly to form a large pool fire. In the absence of immediate ignition, a large cloud of flammable vapor may form. Because of partial confinement (buildings, houses and structures), the delayed ignition of the vapor cloud could lead to large overpressures.

- *Light Commercial/Light Residential/Industrial Areas*: Here the road surfaces are similar to those of the first classification. However, the population density is much lower. In addition, since there are fewer confinements, an unconfined vapor cloud explosion is equally likely.

- *Vacant Areas*: These may or may not have paved, impermeable road surfaces. Therefore, the extent of pool spreading, hence the flame sizes, is small. Confined vapor cloud explosion is highly unlikely. Since the population density is likely to be very small, the impact of any consequence will be relatively low.

- *Two Level (Double Deck) Bridges on Water Without Grating*: An accident involving fuel spill on the top level of a bridge is likely to lead to formation of a pool fire. Since the confinements (even on the lower deck of the bridge) are only partial, the explosion overpressure is smaller than comparable numbers for HC/HR area. If the spill is confined completely to the top deck of the bridge, there is a potential for an UVCE.

- *Single Level Bridge on Water with Grating*: In this case the BLEVE and confined vapor cloud hazards are zero. There is no liquid pool formed on the grating. There will be some hazard due to a pool fire on water. A delayed ignition is likely to lead to an UVCE.

- *Single Level Bridge on Water Without Grating*: Spills of flammable liquid on the bridge surface may lead to a large pool fire. However, because of relatively open atmosphere, unconfined vapor cloud explosions are more likely than CVCE.

- *Elevated Road*: Here we have assumed that pool fire hazards are comparable to HC/HR area. The impact of BLEVE will be small based on the assumption that fewer people would be exposed. Similarly, both UVCE and CVCE are possible events, but with lower impacts.

To quantify relative damage, each situation was assigned a damage score of 1 to 400. For example, if a small (500 gallon) spill of gasoline in a vacant lot which does not ignite causes damage equal to 1, it was assumed that a pool fire caused by a large (full tank) spill of gasoline on a ramp or elevated road will cause damage equal to 150, and a confined vapor cloud explosion caused by a large spill of gasoline in a heavy residential area will cause damage equal to 400. These numbers are, of course, based on judgment, but they take into account the vapor dispersion and thermal radiation intensity which can be calculated for the commodities and spill sizes of interest.

In assigning these scores, we assumed that the damage score for a no-ignition situation is proportional to volume spilled, for a pool fire it is proportion to the square root of the volume spilled, and for a CVCE or UVCE it is proportional to the cube root of the volume spilled.

The relative likelihood of the various scenarios depends in part on whether or not a given accident involves a collision. The principal difference between collision and non-collision accidents is the probability of having an ignition source nearby. Generally, the collision type of accident is more likely to lead to a pool fire because of the greater likelihood of an ignition source in the vicinity. On the other hand, a non-collision accident may cause a spill which is not ignited immediately and therefore forms a vapor cloud. If the cloud subsequently reaches an ignition source, there will be a vapor cloud explosion — a *confined* explosion if there are high-rise buildings nearby, and an *unconfined* explosion in areas with less concentrated development.

An additional assumption is that a boiling liquid expanding vapor explosion cannot occur if the entire truckload of material is spilled (i.e., if spill size is "large").

For the vehicles used, we assumed that 70% of all accidents are collisions involving more than one vehicle, and, of the remaining 30%, one third are single vehicle collisions and the rest are rollovers. Thus, 80% of all accidents are collisions, 20% are non-collisions. The value assumed for percentage of rollovers is somewhat higher than that reported by HSRI.[4] However, another reference[8] reports 34% of all accidents to be single-vehicle and 25.4% of all accidents to be non-collision, which are similar to the values used in this analysis.

The impact factors were developed using a spreadsheet based program. An example of impact factor calculations is shown in Table 1. The impact factors for gasoline and fuel oil for various spill sizes and land use are shown in Figures 2 and 3. We used the above method for assessing risks involved in the transportation of gasoline and fuel oil along two routes. We segmented each route according to type of road and type of land use, and measured length of each segment. Finally, knowing the amount of gasoline and fuel oil transported through the city, we calculated the number of trips.

All this information was combined using a spreadsheet-based program to generate risk analysis results shown in Table 2.

Table 1. An Example of Impact Factor Calculations

Commodity: Gasoline
Land use: HC/HR
Spill size: 500

Scenario	Conditional Probability Collision	No Collision	Damage Score
BLEVE	5	5	100
Pool Fire	90	45	50
CVCE	4	40	200
UVCE	0	0	0
No Ignition	1	10	5

Conditional Probability of Collision   80

WEIGHTED IMPACT - 68.0

As can be seen, the second route for either commodity is two and a half times less risky than the first route. Also, transportation of fuel oil is almost twice as risky as that of gasoline, primarily because:

- More fuel oil is transported (almost two times) than gasoline.

- The transportation routes of fuel oil lie in much more heavy commercial/heavy residential areas.

- The capacity, and therefore the average spill size, of a fuel oil truck is higher than that of gasoline.

Similar analysis was repeated for various other cases in order to evaluate the current risk management strategy of the city.

CONCLUSIONS AND RECOMMENDATIONS

This paper describes a methodology for performing a comparable risk assessment of hazardous transportation in urban areas and illustrates it by an example in which the risks posed by two different commodities transported along two different routes in a metropolis are compared. The principal advantage of the proposed methodology is that it combines various items involved in a risk analysis of this type (i.e., land use, road type, route miles, commodity characteristics, volume transported, tank size, and vehicle characteristics) into a risk number which can be used to make decisions on:

- Effectiveness of various regulations,

- Commodities posing maximum risk to the city,

- Route restrictions,

- Tank volume restrictions,

- Commodity restrictions, and so on.

This methodology can thus assist urban areas in developing a framework for making decisions regarding transportation of hazardous materials through their boundaries.

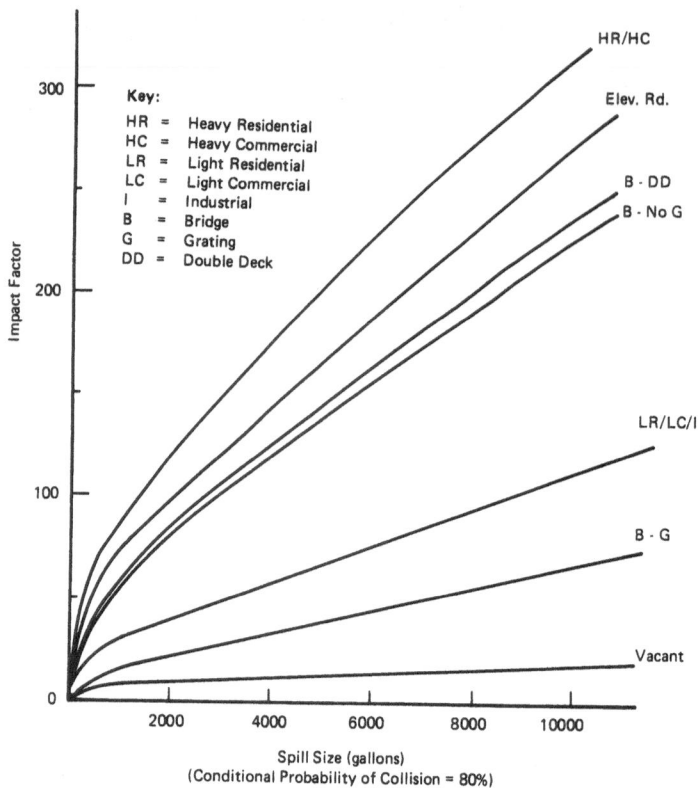

Fig. 2. The Impact Factors for Gasoline.

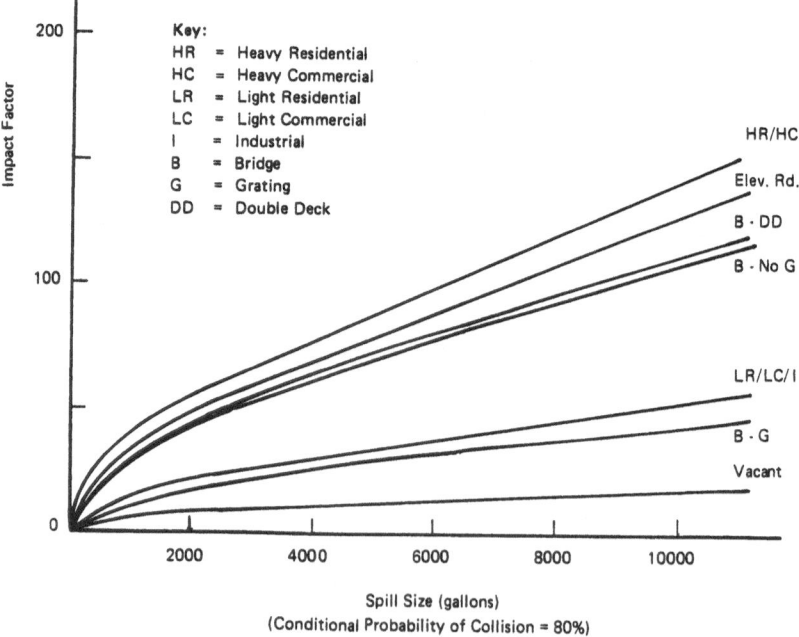

Fig. 3. The Impact Factors for Fuel Oil.

Table 2. Typical Risk Analysis Results

A. Total Risk/Yr.

Commodity	Route Alternative	Risk/Yr* Risk/Yr	Total
Gasoline	1	22.9	
"	2	8.0	30.9
Fuel Oil	1	40.1	
"	2	16.4	56.5
			87.4

* An index related to damage caused by spills.

B. Predicted Frequency of a Catastrophic Spill**

	Route Alternative 1	Route Alternative 2	Total	Spill Size
Gasoline	0.0045/yr	0	0.0045/yr (once very 220 yrs)	4000g
Fuel Oil	0.088/yr	0.016/yr	0.104/yr (once every 10 yrs)	6500g

** A release of total content in Heavy Residential or Heavy Commercial Area

REFERENCES

1. Anon., "Transportation of Hazardous Materials—State and Local Activities," prepared by OTA, SET-301, Washington, DC, Government Printing Office, March 1986.
2. Urbanek, G. L., and Barber, E. J., "Development of Criteria to Designate Routes for Transporting Hazardous Materials," prepared by Peat, Marwick & Mitchell Company for FHWA, NTIS PB81-164725, September 1980.
3. Saccomanno, F. F., and Chan, A.Y.- W, "Economic Evaluation of Routing Strategies for Hazardous Road Shipment," Transportation Research Record No. 1020, December 1985.
4. Ervin, R. D. *et al.*, "Future Configuration of Tank Vehicles Hauling Flammable Liquids in Michigan," PB81-143281, prepared by Michigan University for Michigan Department of State Highways & Transportation, December 1980.
5. Geffen, C. A., "An Assessment of the Risk of Transporting Propane by Truck and Train," prepared for the U.S. Department of Energy, Report No. PNL-3308, Pacific Northwest Laboratory, operated by Battelle Memorial Institute, March 1980.
6. Abkowitz, M., Eiger, A., and Srinivasan, S., "Assessing the Releases and Costs Associated with Truck Transport of Hazardous Wastes," prepared by Rensselaer Polytechnical Institute for the EPA, January 1984.
7. Pijawka, K. D., and Radwan, A. E., "The Transportation of Hazardous Materials," Dangerous Properties of Industrial Materials Report, September/October 1985.
8. Kloeber, G. *et al.*, "Risk Assessment of Air Versus Other Transportation Modes for Explosives and Flammable Cryogenic Liquids, Vols. I and II," prepared by ORI, Inc., prepared for the U.S. Department of Transportation, Materials Transportation Bureau, DOT/RSPA/MTB-79/14, PB80-138480, December 1979.

Drinking-Age Laws: An Evaluation Synthesis of Their Impact on Highway Safety

Thomas Laetz, Roy Jones, and Phil Travers
U.S. General Accounting Office
Denver, CO

ABSTRACT

Controversy surrounds the concept of minimum drinking age that legally restricts alcoholic beverages to a specific age group and the effects of such a law on highway safety. Since passage of federal legislation promoting a "national minimum drinking age" of 21 in July 1984, more than 20 studies have examined the effects of raising the drinking age. Those opposed to this legislation take issue with the efficacy of the law, its fairness, and the constitutionality of its sanctions. Because policymakers are often faced with conflicting opinions over the assessments of risk and the appropriateness of alternative mitigation options, General Accounting Office (GAO) conducted an evaluation synthesis of all existing evaluations of drinking-age laws to determine the extent to which they provide empirical support for federal and state initiatives to raise the legal drinking age. After eliminating studies that did not meet our minimum criteria for acceptable research, we synthesized results of the remaining studies in order to determine if there was enough evidence to support generalizations regarding the isolated effects of raising the drinking age on various factors, such as traffic accidents and alcohol consumption. Among our findings we were able to support the conclusion that raising the drinking age reduced alcohol-related traffic accidents for age groups affected. The evaluation synthesis approach proved a useful technique for analyzing a body of evaluative literature to determine the extent of remaining uncertainty surrounding a policy determination.

KEYWORDS: Minimum drinking age, highway safety, synthesis

INTRODUCTION

In 1986, the U.S. General Accounting Office (GAO) was asked by the Congress to evaluate existing evidence pertaining to whether the risk of young driver involvements in alcohol-related traffic accidents was reduced by raising state drinking ages to conform with national legislation encouraging a uniform drinking age of 21 years of age.[a] As both an accounting and evaluation staff to congressional committees and their subcommittees,

---

a. This paper is based on the report, *Drinking-Age Laws: An Evaluation Synthesis of Their Impact on Highway Safety* (GAO/PEMD-87-10), March 1987; requested by the Chairman, Subcommittee on Investigations and Oversight, Committee on Public Works and Transportation, House of Representatives.

GAO is often requested to evaluate the effectiveness of legislative policy and program implementation. In this case, we were asked to analyze the technical and methodological soundness of existing evaluations of minimum drinking-age laws and to assess the credibility of claims based on their findings. More specifically, we assessed the available evidence concerning the effect of raising the legal drinking age on (1) traffic accidents; (2) beverage alcohol consumption; and (3) other effects, such as the spillover effect on underage youth (typically 16- and 17-year olds), the effect of border crossing to states with lower minimum ages, and the long-term effect of the law change.

RESULTS IN BRIEF

The available evidence for the outcomes we examined varied considerably; however, there was enough evidence to conclude that raising the drinking age has, on average, a direct effect on reducing alcohol-related traffic accidents among affected age groups (typically 18-, 19-, and 20-year-olds) across states. Each state can generally expect fewer traffic accidents, but how many will depend on the particular outcome measured[b] and the characteristics of the state.

We also found that raising the drinking age may result in a decline in the consumption of alcohol and in driving after drinking for the affected age group; however, the limited quantity and quality of evaluations[c] for these outcomes impede making generalizations.

Finally, we found insufficient evidence to draw firm conclusions on the spillover effects, border crossing effects, and the permanence of effects over time. However, we found some evidence that there was no effect on the crash experience of underage youth, but some evidence that the short-term effects of the law change may hold up over time.

BACKGROUND OF THE ISSUE

Congressional concern over the disproportionate involvement of young drivers in alcohol-related traffic accidents prompted the passage of national legislation (Public Law 98-363) in July 1984. The Congress became involved in this otherwise state issue because of concern for the border crossing problem--that is, the risk posed to young drivers crossing state lines in order to obtain alcohol not legally available to them in the state where they reside.[d] The legislation provides for withholding a portion of federal highway funds from states that continue to allow the purchase or public possession of alcoholic beverages after October 1986 by persons younger than 21 years of age. Crossover sanctions, which require compliance with the rules of one federal program as a condition for receiving funds for

---

b. All but one of the traffic accident studies that we synthesized to reach our conclusions focused on the law's effect on accidents, with varying degrees of seriousness, where drivers were from the affected age group. Cross-sectional and time series analyses were used to identify an effect of the law on six outcomes that included drivers involved in crashes that resulted in a fatality, a fatality and an injury, property damage, or just an injury. The one exception was a study which looked at the law's effects on youth among the affected age group who were victims of fatal crashes.

c. Alcohol consumption and driving after drinking studies primarily rely on survey data and, in the case of consumption, aggregate alcohol sales data. Both sources of data are not as reliable as other data on the involvement of alcohol in traffic accidents.

d. During 1984 hearings, it was estimated that 56 percent of the borders in this country separated states that had different legal drinking ages. Therefore, the Congress encouraged the establishment of a uniform drinking age nationwide as a way to reduce the incidence of driving between states after drinking among those affected by the law.

another program, were previously used in 1974 to encourage the states to adopt a 55-mile-per-hour speed limit.

At the time of our study, 23 states increased their drinking age to 21 years of age in response to the uniform drinking age law and growing empirical evidence that raising the drinking age reduced traffic fatalities. However, in spite of this growing pressure and the potential loss of federal funds, 8 states and Puerto Rico had not complied with the federal requirement by October 1986. Reasons given by opponents of this legislation, other than the insufficiency of evidence supporting its efficacy, were that it would (1) have negative consequences, such as reducing state alcohol sales-tax revenue, (2) unfairly penalize most youths for the excesses of a few, (3) jeopardize the right of the states to control the availability of alcohol, and (4) not work as effectively as other deterrents, such as stricter enforcement of existing laws.

Subsequent to a congressional hearing that was held to examine this study and other related research activities, most of the remaining 8 states raised their minimum drinking-age laws in compliance with the federal legislation. In addition, on June 23, 1987, the U.S. Supreme Court ruled favorably on the constitutionality of the law. The federal legislation had been challenged by the State of South Dakota on the grounds that either the Tenth Amendment or Twenty-first Amendment bars the Congress from conditioning a grant of federal highway funds to a state upon the state's adoption of a minimum drinking age of 21. Although the results of our evaluation synthesis were referenced in two legal briefs presented to this Court, the oral arguments did not address any empirical evidence regarding the efficacy of the law.

STUDY SCOPE AND METHODOLOGY

In this study, we applied the evaluation synthesis methodology[e] to the existing body of literature on the relation between minimum drinking-age laws and highway safety. This methodology basically involves (1) identifying the appropriate research questions; (2) collecting studies and other information; (3) developing minimum threshold criteria to judge the acceptability of the study; (4) eliminating studies that failed to meet minimum threshold criteria; (5) synthesizing study results; and (6) identifying gaps in the evaluative knowledge.

Identifying the Questions

Research questions were identified through a general literature review and then negotiated with the congressional requester. Because it is generally acknowledged that drinking-age laws do not affect traffic accidents directly but are mediated by a variety of intervening variables, we also decided to assess studies of alcohol consumption and driving after drinking. A simplified conceptual model of the potential intermediate and long-term effects of the legislative change is presented in Fig. 1.

The model depicts how changes in the legal drinking age interact with other factors, such as marketing practices and changes in the availability of alcohol, to influence drinking-and-driving behavior. The evaluation literature on the subject focuses on traffic accidents as an indicator of this behavior and, to a lesser extent, on patterns in alcohol consumption. Few of the authors whose work we reviewed discussed any theoretical premise upon which to base their studies of the drinking age. Using a variety of measures,

---

e. A general review of the evaluation synthesis methodology is presented in GAO's The Evaluation Synthesis (Institute for Program Evaluation, Methods Paper 1, April 1983).

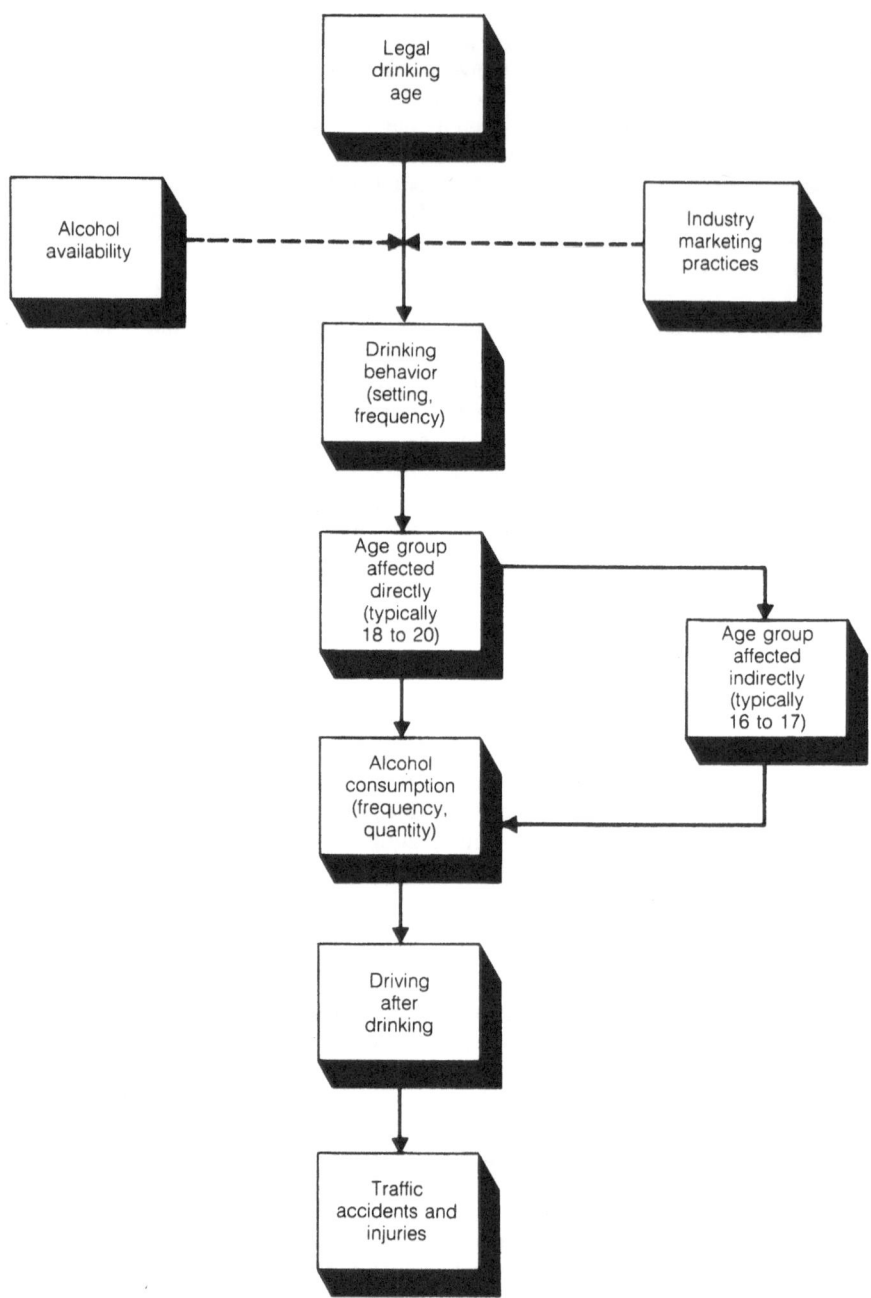

Fig. 1. Conceptual Model Linking a Minimum Drinking-Age Law With Highway Safety Outcomes.

most tested directly for a relationship between the legal drinking age and crash experience or alcohol consumption.

Collecting Studies and Other Information

Because our objective was to identify all available documents on the effects of drinking-age laws, we cast a broad net in an attempt to find not only the most frequently cited published work but also unpublished evaluations conducted by state and local governments, independent researchers, and other research organizations.

Our approach to identifying relevant documents was three-pronged and was made up of an examination of computerized bibliographic files, surveys of alcohol and highway safety officials, and personal interviews with experts in the field. In order to minimize publication bias and maximize the likelihood of collecting as complete a compilation of evaluations as possible, we sent an initial questionnaire to 114 state officials, asking them to identify evaluations and reports that had been completed in their state. We also sent a bibliography of the evaluations we had identified to researchers and knowledgeable experts to uncover work that we might have missed by other means. The results of the surveys yielded more than 80 documents of relevance, including 22 evaluations not previously identified.

Although we uncovered over 2000 citations on the subject, 400 were considered documents on youth and drinking, and only 82 of these documents addressed the minimum drinking age issue. Because we were asked to focus our review on the effects of raising the drinking age, 33 of the studies, which addressed the effects of lowering the drinking age, were eliminated from more in-depth analysis. The remaining 49 studies were evaluated against our minimum threshold criteria in order to identify those studies that were suitable for synthesis. We did, however, cross-index critiques and summaries of evaluations to all studies of drinking-age laws as an independent source of information for rating purposes. Figure 2 reconciles our research questions with the evaluation literature.

Developing the Rating Criteria

A review panel of GAO staff and independent experts was formed to develop rating criteria and review studies of direct relevance. Because no universally agreed-upon rating criteria existed, we developed the criteria, basing them on a preliminary review of the literature and prior evaluation syntheses. We considered the characteristics of the studies-- for example, measures used, questions examined, the nature of the law change, and designs employed--in refining existing criteria for purposes of examining the specific literature we were reviewing.

The panel developed criteria for two generic types of studies: cross sectional studies, comparing two or more defined groups for a single point in time, and before-and-after studies, comparing groups at two or more points in time. We rated all studies in terms of (1) the existence and adequacy of comparison groups, (2) the source of data used, (3) the appropriateness and comparability of measures used, (4) the appropriateness of methods for taking chance into account, and (5) the extent to which a study controlled for other factors and provided quantitative measures of differences. For before-and-after studies, we also looked at (6) data that were comparable and (7) controls for the non-independence of measures.

Eliminating Unacceptable Studies

To critically assess the methodological quality of the 49 evaluations, three raters reviewed each study independently. They were asked to identify the study questions

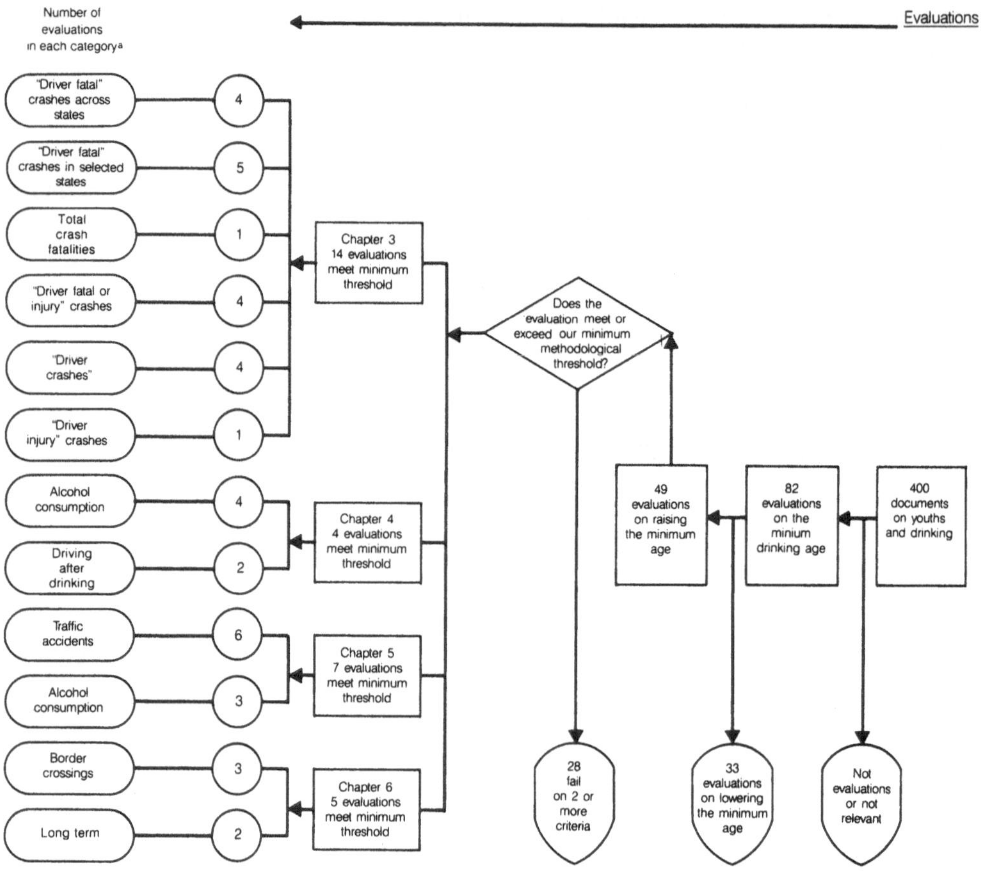

Evaluations

"Driver fatal"
crashes across
states — 4

"Driver fatal"
crashes in selected
states — 5

Total
crash
fatalities — 1

"Driver fatal or
injury" crashes — 4

"Driver
crashes" — 4

"Driver
injury" crashes — 1

Alcohol
consumption — 4

Driving
after
drinking — 2

Traffic
accidents — 6

Alcohol
consumption — 3

Border
crossings — 3

Long term — 2

Chapter 3
14 evaluations
meet minimum
threshold

Chapter 4
4 evaluations
meet minimum
threshold

Chapter 5
7 evaluations
meet minimum
threshold

Chapter 6
5 evaluations
meet minimum
threshold

Does the
evaluation meet or
exceed our minimum
methodological
threshold?

49
evaluations
on raising
the minimum
age

82
evaluations
on the
minium
drinking age

400
documents
on youths
and drinking

28
fail
on 2 or
more
criteria

33
evaluations
on lowering
the minimum
age

Not
evaluations
or not
relevant

[a]These numbers do not always equal the total number of studies within or between chapters, since
some evaluations considered more than one question.

Fig. 2. Reconciliation of Our Synthesis Questions and the Evaluation Literature.

addressed, and, for each question, to rate the study against appropriate criteria[f]. After independently rating each study, the panel met to discuss its strengths and weaknesses and to reconcile differences in individual ratings. The studies that contained no serious flaws or were flawed but of sufficient quality to contribute to policy formulation were grouped by study question for more in-depth reviews. Among the 49 studies we reviewed, 28 did not meet our minimum threshold criteria. Table 1 summarizes the ratings for these studies against the seven criteria.[g]

---

f. The raters gave an overall rating of acceptable, questionable, or unacceptable for each research question addressed in our evaluation. An unacceptable rating was typically given to studies failing to meet two or more criteria.

g. Many of these studies did not adequately take chance into account by employing appropriate statistical tests, which is a prerequisite for linking changes in measures of effect to a change in the law. In addition, many inappropriate comparisons were cited when, for example, studies merged data from age groups not directly affected by the purchase-age policy with data for those directly affected (the experimental group) by the law.

Table 1. Reasons for Unacceptable Study Ratings

Criterion	Traffic accidents	Consumption and driving after drinking	Effects on other youths	Other effects	Total
Comparison group comparability	14	4	0	7	25
Description of source data	7	0	0	0	7
Comparable measures	8	5	0	3	16
Test for significance	14	1	2	5	22
Quantitative measure of difference	18	5	2	8	33
Comparable before- and-after data	5	0	1	1	7
Account for nonindependent observations	4	0	0	0	4
Total[a]	70	15	5	24	114

[a]Totals do not equal the 28 studies judged unacceptable, since most of these studies failed to meet two or more criteria and some studies dealt with more than one outcome.

Synthesis of Results

Once the rating process was completed, the panel members reviewed the studies in groups by study question, in order to assess (1) what was known concerning the question, (2) how confident we were about the available evidence, (3) how adequate the information was, and (4) what knowledge gaps remained. While the initial phases of the review process focused on the strengths and weaknesses of individual studies, during this phase we focused on the quantity and quality of evidence across studies. Only studies that met our minimum threshold criteria were used to assess what was known about the effects of the law change.

In our synthesis, we looked for patterns in the study findings, possible limitations in measures used and comparisons made, and the ability to generalize the results. We also considered the quantity of the evidence and whether it accumulated from study to study. In this way we assessed both quality and quantity in order to determine the strength of evidence for each question. For example, we found support for our conclusion regarding the effects of raising the drinking age on traffic accidents in the number of multiple observations of similar direction and often similar magnitude that were obtained in studies that used alternative approaches to analyze various measures[h] of traffic accidents. Further support for our conclusion comes from the knowledge that such consistent findings as we found for traffic accidents rarely occur in reviews of this sort.

Evaluation of Knowledge Gaps

We found that there was insufficient evidence to address some of the research questions; however, we found support for the law's efficacy in reducing alcohol-related traffic accidents. We recommended to the Congress that, although we could study this issue further, it was clear to us that the law was having its desired effect.

---

h. Two methods were primarily used to measure the influence of alcohol on traffic accident statistics. The direct method relies on police reports on the impairment of the drivers involved in a crash. The indirect method relies on selective characteristics of a crash, such as time of day, to serve as a predictor or surrogate indicator of alcohol involvement in a crash. All studies relied on crash data attained through either the federal fatal-accident system or state records.

STRENGTHS AND LIMITATIONS

Some evaluation questions can be answered only by looking across several studies, and one strength of our method is that it supplies a systematic way of doing this. In considering the findings of different studies, while accounting for the quality and quantity of evidence for each specific question, we were able to provide an indicator of what is known, what is unclear, and what questions remain unanswered. An additional advantage of the evaluation synthesis method is that it establishes an easily accessible base of knowledge which can be used in assessing future evaluation questions.

An evaluation synthesis necessarily depends on the amount and quality of the information available. Because we primarily relied on information as it was reported in the published and unpublished sources we examined, our information on each study was limited. We made every attempt to obtain the most information possible, but time constraints restricted our ability to contact all authors to clarify ambiguities, request additional information, or obtain primary data.

IMPLICATIONS FOR RISK MANAGEMENT

The evaluation synthesis approach is both timely, in that it makes use of existing evaluations, and integrative, in that it brings together evaluations that have fragmented information over time. We believe that the evaluation synthesis method has application in other areas where policymakers are confronted with conflicting assessments of potential hazards and their mitigation. This method allows a study team, with the assistance of experts, to cut through the methodological underbrush that often impedes resolution of controversial issues. Unfortunately, subsequent debates in state legislatures made little use of our study results. Those who were opposed to raising the minimum drinking age avoided reference to the empirical evidence and, instead, focused public attention on other issues, such as states' rights and federal blackmail.

Fatality Incidence Stratified by Driver Behavior: An Exploratory Analysis with FARS Data

James S. Licholat and Richard C. Schwing

General Motors Research Laboratories
Warren, MI

ABSTRACT

Driver behavior indicators were used to stratify FARS (Fatal Accident Reporting System) data on occupant fatalities. Data for automobiles, vans, and light trucks were stratified by:

- single or multiple vehicle accident
- driver use of alcohol
- driver use of safety belt

The weekly pattern of occupant fatality incidence was examined for all strata except the sparse categories of fatal crashes wherein the driver wore a safety belt.

The temporal patterns of three of the non-belted groups were remarkably similar, with daily peaks shortly after midnight and with higher incidence throughout the weekend. The fourth group, multiple vehicle crashes not involving alcohol, had a quite uniform daily pattern peaking at afternoon rush hour. When crudely controlling for exposure, even this group, recorded as non-alcohol involved, again revealed the after-midnight peaks, an indication of substantial underreporting of alcohol involvement.

This study also shows that belted drivers are less than proportionately involved in crashes involving fatalities. Whereas 25% to 40% of surveyed drivers and passengers[1] always wear their safety belts, drivers wearing belts are represented in only 5.6% of fatal occupant crashes. Similarly, this study shows that alcohol-involved drivers are over-represented in vehicle crashes producing occupant fatalities.

These factors indicate that relatively few drivers are exposed to the "average risk" and that the distribution of risk is skewed. Specifically, non-drinking belted drivers are so under-represented in crash statistics that it is misleading to indicate that their risk is the average risk of driving, which is 1.0×10^{-8} deaths per person mile of travel. On the other hand, when dealing with those who drink and drive while unbelted, we are dealing with risks which greatly exceed this average risk.

KEYWORDS: Driver behavior, safety belt use, alcohol use, temporal patterns, occupant fatality

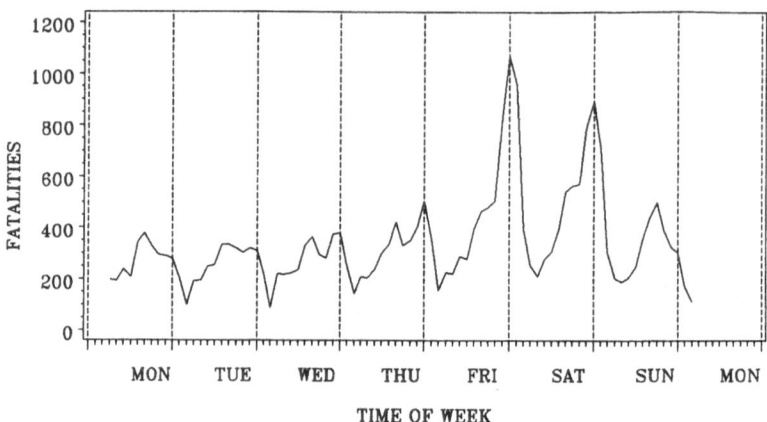

Fig. 1. Total Occupant Fatalities in Cars, Vans, and Light Trucks (1983 FARS Data). (Vertical lines indicate midnight to 2 a.m. interval.)

INTRODUCTION

In the 19th century, a rash of horse-kick fatalities in the Prussian cavalry led Von Bortkiewicz to study kicking accident records. He wanted to determine whether the accidents were random events or whether there had been a change in circumstances involving carelessness by soldiers or an increase in the number of wild horses. He concluded that the occurrences over a 10-year period of observation were indeed random events and that no special interventions were required.[2]

This study is similarly motivated; however, the focus is to better understand the human behavior components which lead to the clearly non-random motor vehicle fatality incidence shown in Fig. 1. Schwing and Kamerud[3] controlled these fatality incidence data for passenger miles traveled at different times of the week. They found that the risk of travel varied from a low of 0.32×10^{-8} fatalities per person mile of travel to a high of 43×10^{-8} fatalities per person mile of travel (a ratio of 134:1).

The perspective or emphasis of the present study is on the human behavior components in fatal accident causation. We are not concerned with accident severity as a function of alcohol or safety belt use, given a crash.

DATA AND METHOD

The present analysis considers a very specific subset of highway fatalities. We considered car, van, and light truck accidents (either single or multiple vehicle) in which one or more occupants were killed. Excluded, therefore, among highway fatalities, are pedestrians, cyclists, and medium to heavy truck occupants.

Our measure of risk in this study is taken as the incidence (number of occurrences) of fatalities. The fatality data were examined both with and without controlling for passenger miles travelled as a crude measure of exposure.

The fatal accident data were stratified by indicators of driver risk-taking propensity. The stratifications were done at three sequential levels. The first level was stratified by whether the accident involved one or more vehicles. The second level was stratified by whether a driver in the accident was using alcohol. For this stratification, the FARS indicator variable "drinking driver" was used to stratify the sample.

The third stratification was by the driver behavior criterion as to whether or not a driver was belted. In the case of multiple car accidents, if any driver was using a safety belt, the accident was classified in the "belted" category.

As in the earlier time-of-week study, groups of data were subsequently partitioned over time throughout the week. In this study we used 84 two-hour periods of the week. Again, the interest was to determine whether or not each of the strata was randomly distributed over the week. Because 94% of the fatalities were in crashes involving unbelted drivers, these data represented the largest sample and were the only ones selected for partitioning by the time of week in this study. However, the remaining victims, riding in vehicles driven by relatively prudent drivers, could provide an interesting sample for subsequent analysis.

Fatality Data

Our source of fatality data is the 1983 Fatal Accident Reporting System (FARS) study.[4] Operated by the U. S. National Highway Traffic Safety Administration, FARS is an annual census of fatal motor vehicle accidents that occurred on the public roads of the United States (NHTSA, 1984). For each hour of the week, we compiled the number of fatally-injured vehicle occupants within each group stratified as above.

When controlling for exposure within each group, the numerator in the "risk" term is the variable "FATALS," where FATALS(i) is the percentage of all occupant fatalities that resulted from travel in each two-hour period i.

Travel Data

Our index of exposure is PMT(i), the percentage of passenger miles of travel that can be ascribed to each two-hour interval i. We estimate this quantity using data from the GMR Automobile Usage Study, which surveyed the travel activities of 2000 households during a one-week period in 1979.[5] One disadvantage of such household travel studies is that actual exposure results from combined household, commercial, and government travel, which may have a weekly pattern different from that of household travel alone.

For each of the approximately 70,000 trips in the auto usage data, we know the time it started, its duration, the number of passenger-miles it represents, and its household weighting factor. The vast majority (96.0%) of trips were less than one hour long, 2.6% were between one and two hours long, and only 1.4% were over two hours long. This justifies the following simplification: all the weighted passenger-miles of travel for each trip were assigned to just one two-hour interval in the week, namely, the one containing the trip's midpoint time. This defines a unique period for each trip, allowing travel estimates for each hour of the week to be extracted using routine data processing procedures.

Incidence in Categories

A total of 28,665 victims, identified in FARS, were coded with sufficient information to identify the time of week and perform each stratification.

Stratifying in the order:

- single or multiple vehicle
- driver use of alcohol
- driver use of safety belt

343

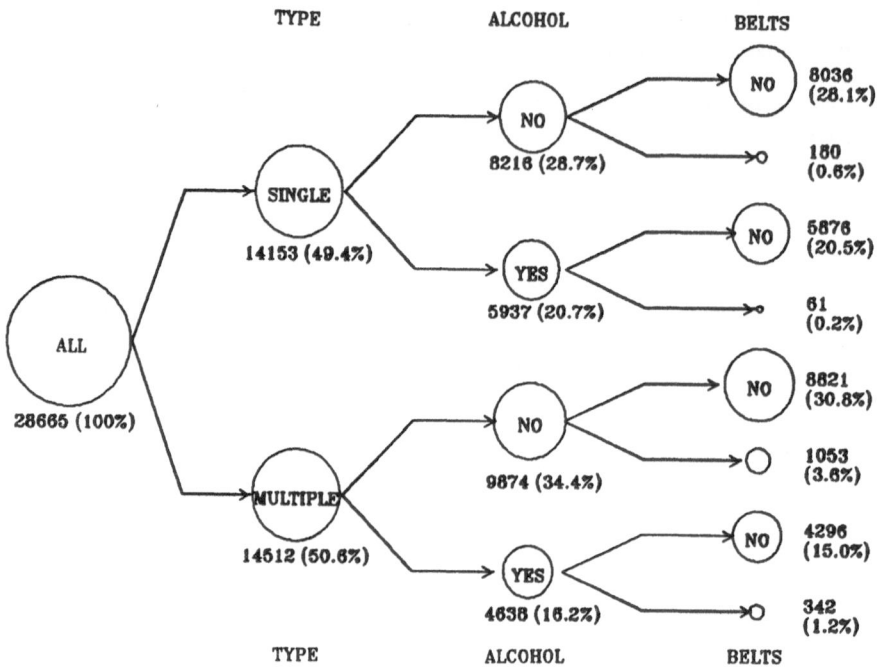

Fig. 2. Automobile, Van, and Light Truck Occupant Fatalities Stratified by Number of Vehicle, Alcohol Use, and Belt Use.

provides the category sizes depicted in Fig. 2.

Occupant fatalities in car, van, and light trucks are nearly equally distributed between single and multiple car crashes. Single vehicle crashes, often characterized by excess fatigue, speed, or alcohol, account for 14,153 occupant fatalities or 49.4% of the total. Alcohol use among drivers in these single vehicle crashes was identified in cases involving 5,937 fatalities (20.7% of the total) in the FARS sample. Alcohol use among drivers was not identified in the reports on the balance of these single vehicle crashes, accounting for 8,216 or 28.7% of the total occupant fatalities in cars, vans, and light trucks.

Note in Fig. 2 that on subsequent stratification of the occupant fatality data, the odds are very low for drivers being involved in a fatal collision while using a safety belt (whether or not they were identified with alcohol use).

In the case of multiple vehicle crashes, the largest group of fatalities, 9,874 (or 34.4% of the occupant fatalities) were in crashes in which neither driver was identified as using alcohol. Again, crashes causing occupant fatalities while the drivers of either vehicle were wearing safety belts were relatively rare.

RESULTING PATTERNS OF INCIDENCE

Of the eight groups resulting from the stratifications, the four largest involve accidents in which none of the drivers were belted. The incidence patterns of these larger groups, subsequently partitioned over the 84 two-hour periods of the week, are presented in

Figs. 3a through 3d. The midpoint of the midnight to 2:00 a.m. period is designated by the vertical dashed lines.

Very similar patterns are generated by three of the four groups of data in Fig. 3. The three groups are:

- single vehicle - no alcohol involved - (3a)
- single vehicle - alcohol involved - (3b)
- multiple vehicle - alcohol involved - (3d)

The similarities involve periodic peaks in the daily incidence after midnight with generally higher incidence on Friday through Sunday.

It is also noted that when alcohol is involved (Figs. 3b and 3d) the variation in fatalities among the different hours of the day is larger than when alcohol is not involved (Figs. 3a and 3c). Furthermore, we note that, although a substantial number of fatalities occur during morning rush hours, virtually none of the fatalities were recorded with alcohol involvement.

The atypical pattern was generated by the multiple vehicle — no alcohol group (3c). In this case, the daily peaks occurred between 2:00 and 4:00 p.m. and the daily pattern was nearly the same throughout the week. Since this pattern resembles the pattern of travel represented in Fig. 4, a calculation was performed to see if travel exposure provides an adequate control for exposure for this group of fatalities.

As in the earlier study,[3] we have calculated "relative risk" as the incidence (number of occurrences) in a two-hour period divided by exposure. Note that, in contrast to the Schwing and Kamerud study, we cannot control as precisely for exposure since we do not know passenger miles traveled conditioned on belt use and alcohol. We are therefore controlling for exposure in a crude manner, assuming the non-alcohol involved, non-belted group exposure is adequately controlled for by the pattern of travel shown in Fig. 4.

The relative risk ratio and logarithm of this ratio are represented in Figs. 5a and 5b, respectively, for each two-hour time interval in the week for multiple vehicle accidents when alcohol is not involved. Having controlled for exposure or traffic density, a strong periodic component remains in these figures.

For completeness, we have performed this "control for exposure" calculation for all four categories represented in Figs. 3a through 3d. The results look similar. We would be very cautious, however, in trying to interpret these results as Figs. 3b and 3d involve alcohol, and surely the exposure data derived from diary data do not account for this parameter. The results of these calculations are presented in Figs. 6a-d. The logarithm of relative risk is presented in Figs. 7a-d.

DISCUSSION

Alcohol Stratification

The now-familiar pattern of alcohol use is once again evident in the patterns produced in this study. Other researchers have consistently found after midnight to be the riskiest time for serious accidents. Stratifying the data by single vehicle crashes and multiple vehicle crashes and by alcohol use has not changed this general pattern in three of the four strata.

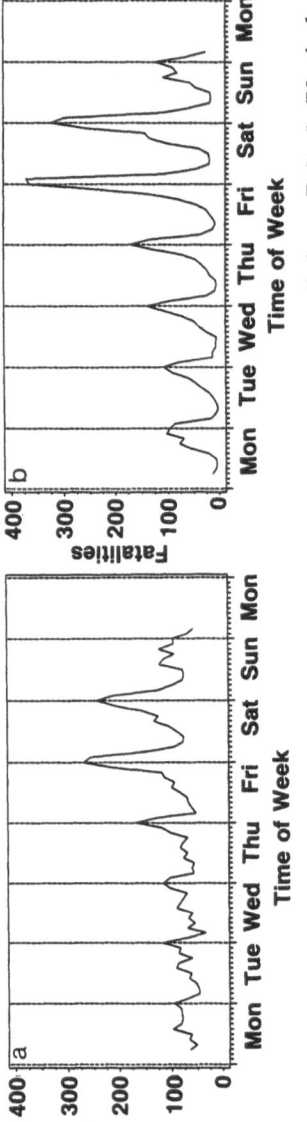

Fig. 3a-b. Fatalities for Single Vehicle Accidents with None of the Drivers Belted. (Vertical lines indicate midnight to 2 a.m. interval.) (a) Alcohol Not Involved. (b) Alcohol Indicated for Driver.

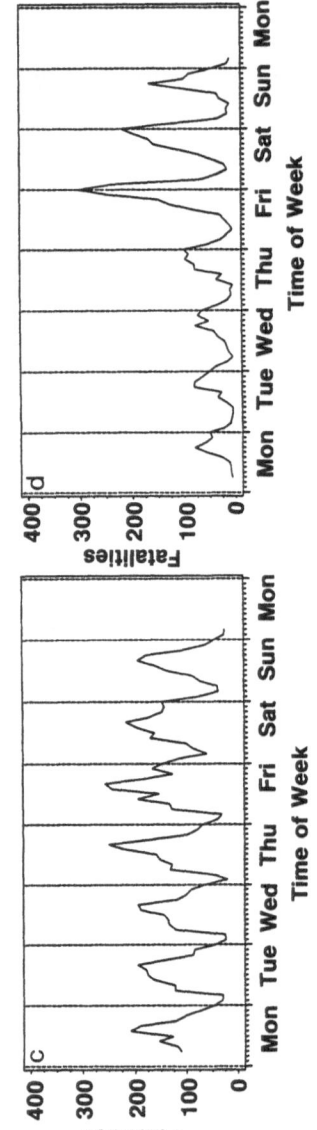

Fig. 3c-d. Fatalities for Multiple Vehicle Accidents with None of the Drivers Belted. (Vertical lines indicate midnight to 2 a.m. interval.) (c) Alcohol Not Involved. (b) Alcohol Indicated for at Least One Driver.

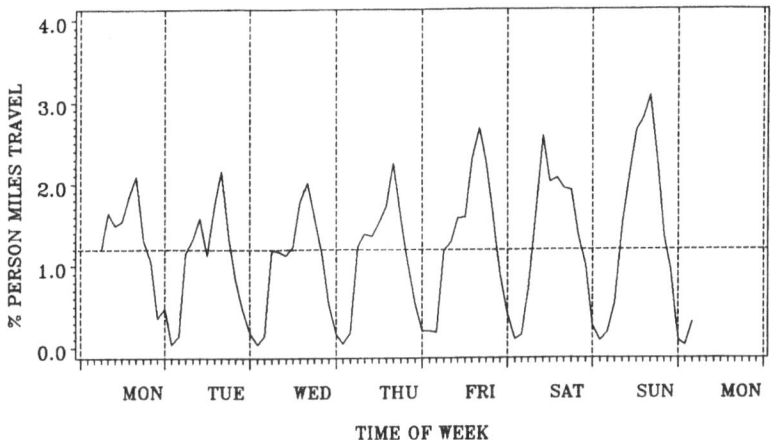

Fig. 4. Pattern of Travel for Cars, Vans, and Light Trucks. (Vertical lines indicate Midnight to 2 a.m. interval.)

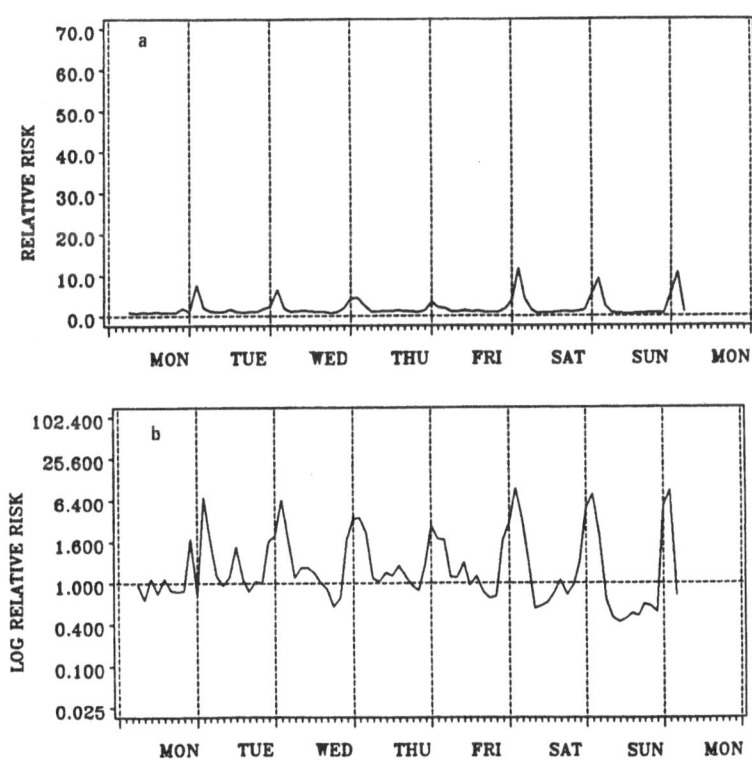

Fig. 5a-b. Relative Risk for Multiple Vehicle Accidents with None of the Drivers Belted and Alcohol Not Involved. (Vertical lines indicate midnight to 2 a.m. interval.) (a) Linear Plot. (b) Log Plot.

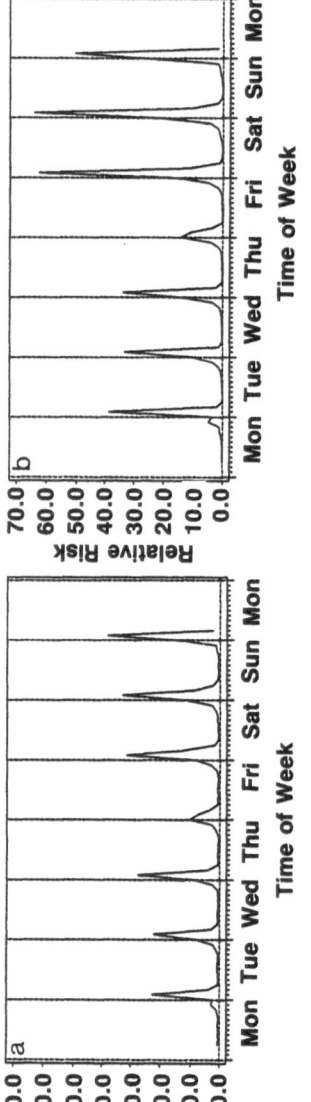

Fig. 6a-b. Fatalities for Single Vehicle Accidents with None of the Drivers Belted. (Vertical lines indicate midnight to 2 a.m. interval.) (a) Alcohol Not Involved. (b) Alcohol Indicated for Driver.

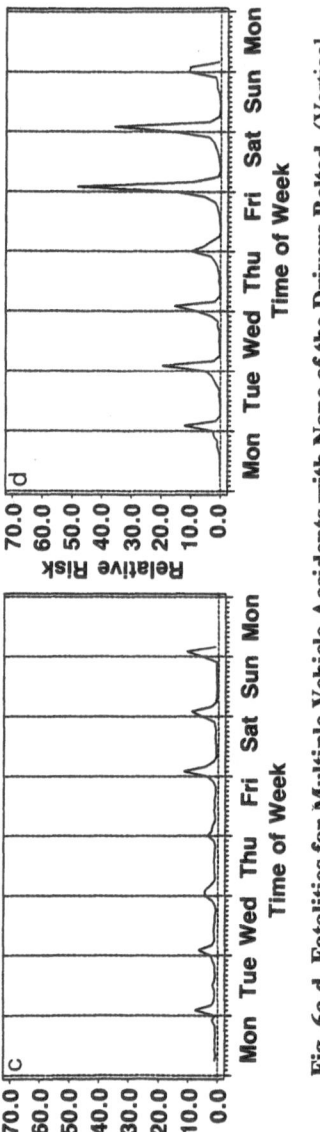

Fig. 6c-d. Fatalities for Multiple Vehicle Accidents with None of the Drivers Belted. (Vertical lines indicate midnight to 2 a.m. interval.) (c) Alcohol Not Involved. (b) Alcohol Indicated for at Least One Driver.

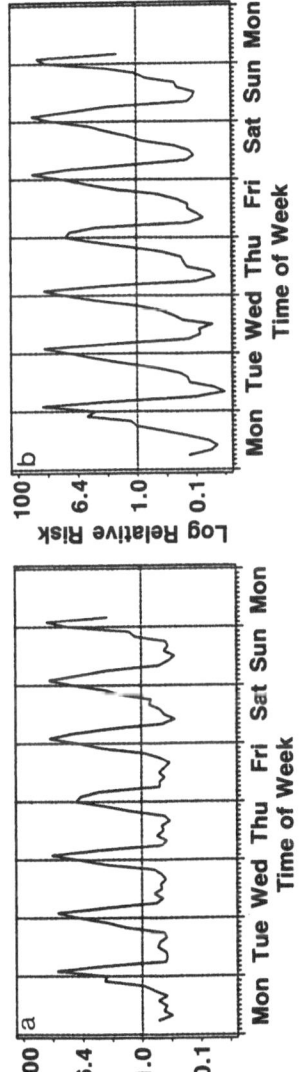

Fig. 7a-b. Fatalities for Single Vehicle Accidents with None of the Drivers Belted. (Vertical lines indicate midnight to 2 a.m. interval.) (a) Alcohol Not Involved. (b) Alcohol Indicated for Driver.

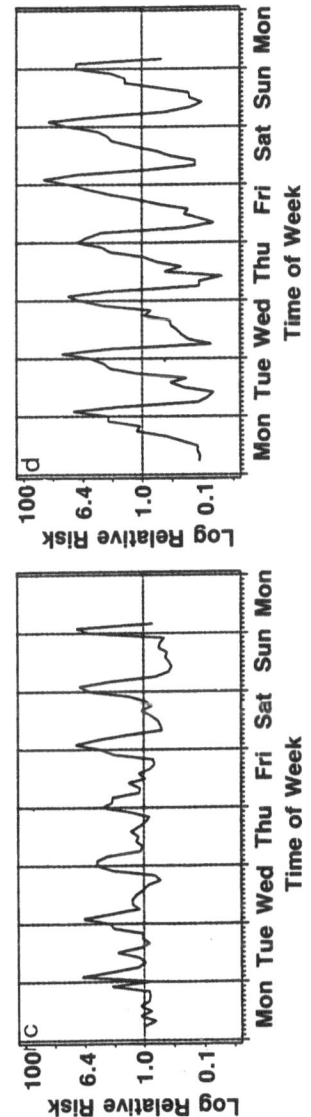

Fig. 7c-d. Fatalities for Multiple Vehicle Accidents with None of the Drivers Belted. (Vertical lines indicate midnight to 2 a.m. interval.) (c) Alcohol Not Involved. (b) Alcohol Indicated for at Least One Driver.

349

The multiple vehicle crash, not-alcohol-involved drivers' stratum generated a time of week pattern more consistent with traffic volume patterns, with peaks during afternoon rush hour. This pattern persisted throughout every day in the week.

Surprisingly, however, when controlling this stratum for exposure, the familiar after-midnight peaks again appear in the daily pattern. Assuming that the person-mile of exposure is an adequate control for this group of accidents, the after-midnight peaks are approximately 6 times more risky than the average risk throughout the week.

Safety Belt Stratification

Relatively few fatal accidents were generated by drivers wearing safety belts. In the stratum, multiple vehicle crash, not-alcohol-involved, there were 1,053 fatalities (3.6% of occupant fatalities) generated where one or more drivers wore a safety belt. To determine whether or not 3.6% is expected, knowing that 30.8% of the occupant fatalities were generated in the similar stratum by drivers not wearing belts, one would have to know the following:

- percentage of belted drivers,
- a factor to allow for the obvious yet unidentified alcohol involvement,
- a factor to account for seat belt effectiveness, and
- a factor to account for a belt user's potential to be more risk averse.

If only the first factor was important, the ratio 3.6 over 30.8 indicates that slightly more than 10% of the driver population was belted in 1983. Since observation surveys indicate that over 13% of the drivers wore belts in 1983,[6] it is possible that the other factors are significant. These factors have been explicitly studied by Evans and Frick.[7]

CONCLUSIONS

All strata examined in the occupant fatality data have a component which is strongly associated with alcohol involvement. Even the stratum identified as alcohol-not-involved contains a strong hint of the pattern of alcohol involvement, indicating, as others have argued[8] (e.g., Texas A&M Study, 1986), that there is great need for better quality reporting of alcohol involvement in traffic accidents.

ACKNOWLEDGMENT

Useful suggestions to improve accuracy, perspective and clarity were provided by Dana B. Kamerud and Abraham D. Horowitz.

REFERENCES

1. Lieb, R. C., F. Wiseman and T. E. Moore, "Automobile Safety Programs: The Public Viewpoint," *Automobile Safety*, Summer 1986, pp. 22-30.
2. Covello, V. T. and J. Mumpower, "Risk Analysis and Risk Management: An Historical Perspective," *Risk Analysis* 5:103-120 (1985).
3. Schwing, R. C. and D. B. Kamerud, "The Distribution of Risks: Vehicle Occupant Fatalities and Time of the Week," Research Publication GMR-5686, General Motors Research Laboratories, Warren, Michigan, January 1987.
4. National Highway Traffic Safety Administration, *Fatal Accident Reporting System—1983*, DOT-HS-806-705, March 1984.

5. Horowitz, A. D., "Automobile Usage: A Factbook on Trips and Weekly Travel," Research Publication GMR-5351, General Motors Research Laboratories, Warren, Michigan, April 1986.

6. Hedlund, H. H., "Recent U. S. Traffic Fatality Trends," in *Human Behavior and Traffic Safety*, L. Evans and R. C. Schwing, Eds., p. 13, Plenum Press, New York, 1985.

7. Evans, L. and M. C. Frick, "Safety Belt Effectiveness in Preventing Driver Fatalities Versus a Number of Vehicular, Accident, Roadway and Environmental Factors," *Journal of Safety Research* (in press, 1987).

8. Pendleton, O. J., Hatfield, N. J. and Bremer, R. "Alcohol Involvement in Texas Driver Fatalities: Accident Reports Versus Blood Alcohol Concentration," SAE-8600371, in *Alcohol, Accidents, and Injuries, Society of Automotive Engineers*, Warrendale, Pennsylvania, 1986.

An Assessment of the Risks of Stratospheric Modification

John S. Hoffman[a]

Environmental Protection Agency
Washington, DC

John B. Wells

The Bruce Company
Washington, DC

ABSTRACT

The large spatial scale, long-time scale, irreversibility, and magnitude of potential effects of stratospheric ozone depletion distinguish it from other environmental risks faced by the U.S. Environmental Protection Agency. The *Montreal Protocol on Substances that Deplete the Ozone Layer*, which requires a 50% global reduction in chlorofluorocarbons and a freeze on halons, was developed to address these risks. Recent scientific evidence, however, suggests that our estimates of ozone depletion may have been underestimated and may necessitate a review of the Protocol's stringency.

KEYWORDS: Ozone layer, stratospheric ozone depletion, global, ultraviolet radiation, Montreal Protocol

GLOBAL SCALE

Most environmental problems that EPA considers are local or regional in scope. Particulates, for example, are essentially a local problem that can be ameliorated by controls on local sources. Acid rain has a wider geographic scale, but remains regional. Ozone depletion, however, is truly a global problem. CFCs are stable in the lower atmosphere, and within two years after release are uniformly distributed throughout the global atmosphere (World Meteorological Organization, 1986). CFCs are used by all nations (the U.S. currently produces one-third of the world's CFCs), and effective ozone protection requires that emissions from all nations be considered.

---

a. The opinions expressed in this article are the authors' and do not necessarily reflect the opinions of any organization.

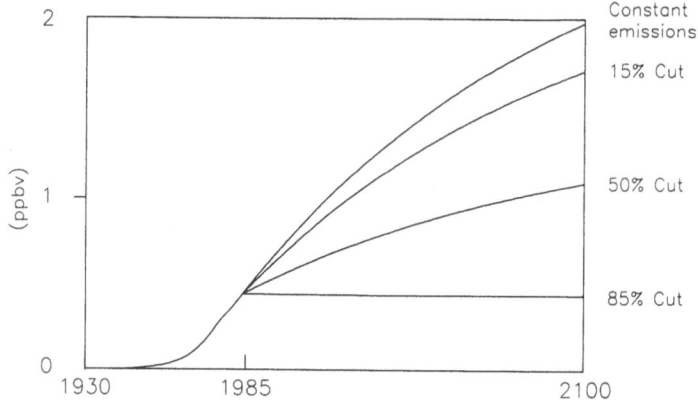

Fig. 1. Atmospheric Concentrations of CFC-12 from Different Emission Trajectories. Atmospheric concentrations of CFC-12 would continue to rise if emissions were not cut. The Montreal Protocol, which calls for a 50 percent reduction in emissions, will still allow atmospheric concentrations to grow. Only a cut of 85 percent or more could stabilize atmospheric concentrations. Source: Hoffman, (1986)

LONG TIME SCALE

Most environmental problems have a shorter time scale. The precursors to acid rain, for example, have atmospheric lifetimes that are usually measured in terms of weeks or months. Thus, rather quickly after curtailing their emissions, their atmospheric concentrations would fall.

CFCs, however, have atmospheric lifetimes of over one hundred years (World Meteorological Organization, 1986). They accumulate in the atmosphere, and concentrations would continue to rise even if their emissions were held constant. Figure 1 shows that to stabilize the concentration of CFC-12, for example, would require a reduction in global emissions of 85 percent (Hoffman, 1986).

IRREVERSIBLE

The problem of stratospheric ozone depletion is thus semi-irreversible — emissions today will deplete ozone at some point in the future, by which time emission controls could not prevent damage to the ozone layer. In 1986, for example, the world emitted 376.5 million kilograms of CFC-12 into the atmosphere (Chemical Manufacturers Association, 1987). Given the 150 year atmospheric lifetime of CFC-12, more than one-third of these emissions, 138.5 million kilograms, will remain in the atmosphere in the year 2135 and pose a potential threat to stratospheric ozone.

MAJOR EFFECTS

The magnitude of the damages that could result if significant depletion of stratospheric ozone occurs also distinguishes it from other environmental problems. Increases in harmful ultraviolet radiation (UV-B) due to stratospheric ozone depletion would damage human health (skin cancer, cataracts and systemic immune suppression), welfare (accelerated degradation of outdoor plastics and earlier formation of ground level oxidants), and the environment (perturbations to terrestrial and aquatic ecosystems). Table 1

Table 1. Summary of Benefits of the Montreal Protocol
(Projected Effects Through 2075)

	Without Montreal Protocol	With Montreal Protocol
Ozone Depletion (%)	39.9	1.3
Skin Cancer Cases (including melanoma)	154 million	3.7 million
Skin Cancer Deaths (including melanoma)	3.2 million	66 thousand
Cataract Cases	18 million	610 thousand
Reduction in Essex Soybean Yield (%)	> 7.5	0.4
Loss of anchovy larvae (%)	> 25	0.0
Increase in Tropospheric Ozone (%)	> 30.9	1.1
Cost of UV-B stabilizers ($ bill)	4.69	1.57
Increase in Equilibrium Global Temperature (degrees C)	5.8	4.3
Sea Level Rise (centimeters)	97.8	86.7

ASSUMPTIONS: Based on projected ozone depletion for the no controls case (approximately 2.5 percent annual growth in CFC emissions and historic growth in other trace gases) and implementation of the Montreal Protocol provisions (assuming participation rates of 100% for U.S., 94% for other developed nations, and 65% for developing nations). Assumes that the Antarctic ozone hole has no global implications; i.e. that current one-dimensional models adequately project depletion. Effects were computed based on dose-response models developed for EPA's stratospheric ozone risk assessment (U.S. EPA, 1987a), and are reported in its regulatory impact analysis (U.S. EPA, 1987b).

shows the damages estimated for continued growth in use of CFCs at approximately 2.5 percent per year and for the reduced growth achieved by global participation in the Montreal Protocol (U.S. EPA, 1987a). These estimates were prepared for EPA's Regulatory Impact Analysis (U.S. EPA, 1987b), and served as part of the technical basis for its proposed regulation.

TROUBLING UNCERTAINTIES

EPA's baseline estimates of stratospheric ozone depletion and its effects assume that current atmospheric models can correctly project depletion caused by CFCs. In May 1985, our confidence in the current generation of such models was shaken when researchers from the British Antarctic Survey reported that total ozone above Halley Bay, Antarctica, during September and October had fallen 40 percent from the 1960s to 1984 (Farman, Gardiner, and Shanklin, 1985). These data were soon confirmed by measurements from the Nimbus-7 satellite, and it was learned that the phenomenon extended to a region larger than the entire Antarctic continent (World Meteorological Organization, 1986). Preliminary data from the same set of satellite instruments indicate that global ozone may have dropped by five percent over the last six years (Kerr, 1987).

Neither the Antarctic ozone hole nor the potential global ozone losses can be predicted or replicated by current atmospheric models. If the emerging scientific evidence shows that the global atmosphere may be more sensitive than previously believed to chlorine, our estimates of future depletion and effects may be seriously underestimated.

CONCLUSION

The large spatial scale, long time scale, and the magnitude of potential effects of stratospheric ozone depletion distinguish it from other environmental problems faced by EPA. The Agency's current risk analysis concluded that continued emissions of CFCs and Halons would lead to significant damage to human health, welfare and the environment. These findings supported the negotiations culminating in the Montreal Protocol, which requires a 50 percent global reduction in CFCs and a freeze on Halons. The protocol contains an assessment and revision framework which will need to consider future possibilities, especially the possible implications of the Antarctic ozone hole.

REFERENCES

Chemical Manufacturers Association, 1987, *Production, Sales, and Calculated Release of CFC-11 and CFC-12 through 1986*, Chemical Manufacturers Association, Washington, DC.

Farman, J. C., B. G. Gardiner, and J. D. Shanklin, 1985, "Large Losses of Total Ozone in Antarctica Reveal Seasonal ClOx/NOx Interaction," *Nature* **315**:207-210.

Hoffman, J. S., 1986, "The Importance of Knowing Sooner," in *Effects of Stratospheric Ozone and Global Climate, Volume 1: Overview*, J. G. Titus, Ed., United States Environmental Protection Agency and United Nations Environment Programme, Washington, DC.

Kerr, R. A., 1987, "Has Stratospheric Ozone Started to Disappear?" *Science* **237**:131-132.

U.S. Environmental Protection Agency (EPA), 1987a, *Assessing the Risks of Trace Gases that Can Modify the Stratosphere*, U.S. EPA, Washington, DC.

U.S. EPA, 1987b, *Regulatory Impact Analysis: Protection of Stratospheric Ozone*, U.S. EPA, Washington, DC.

World Meteorological Organization, (WMO), 1986, *Atmospheric Ozone 1985. Assessment of Our Understanding of the Processes Controlling Its Present Distribution and Change*, WMO Report No. 16, WMO, Geneva, Switzerland.

Quality Assurance in Risk Models for Regulatory Decision Making

Geraldine V. Cox

Chemical Manufacturers Association
Washington, DC

ABSTRACT

Public health policy demands conservative decisions to allow an adequate margin of safety. Risk assessment is the tool policymakers use when they decide to set controls to a level of public safety. If the science of risk assessment is confused with policy decisions in risk management, we are no longer managing risks. We must insist on the most accurate risk model and resist the temptation to introduce bias from exposure assumptions, poor data, improper biological assumptions, types of models used, and misleading presentation of model results. Although conservative policy is appropriate for risk management (the policy decision-making phase), conservatism in the risk assessment phase leads to serious errors—whether they be introduced consciously or unconsciously.

Risk models should be as accurate as possible, and the results should reflect the most plausible conditions. Additional results should give the maximum and minimum plausible values and results of sensitivity analysis as support to the most plausible estimate. This data presentation is not common practice, and, currently, the policymaker is often misled into conclusions that are not supported by the data.

Conservative assumptions often lead to results so inaccurate that, if the practice remains uncorrected, the entire field of risk assessment and risk management will lack credibility. Safety factors should be applied to the results of risk modelling—not to the input data.

Errors of small proportion magnify into inaccurate conclusions if sensitivity analysis and corrections for error propagation are not incorporated into routine quality assurance procedures for all risk models.

Risk modelling is an evolving art and it will take many years before we are able to achieve the same rigor of quality control/quality assurance as analytical chemistry (or even toxicology). But, if we do not improve the current estimates immediately, modelling will have such poor acceptability that it will not survive to develop the precision and accuracy that it can achieve. With time and diligent care, risk assessment/risk management approaches can, and should, emerge as a respected discipline.

This paper addresses data input errors in hazard assessments, considers the problems with input data in exposure assessments, discusses errors in numerical manipulation, and provides suggestions for quality control in risk assessment methods.

KEYWORDS: Errors, hazards, risk assessment, numerical manipulation

INTRODUCTION

The United States Environmental Protection Agency adopted a two-step approach to risk assessment. Risk assessment was defined as a process that would answer two questions: (1) How likely is an agent to be a human carcinogen? and (2) If an agent is a human carcinogen, what is the magnitude of its public health impact given current and projected exposures? Anderson (1983) points out that at one time the Agency assigned positive evidence to one of three broad categories: "(1) strongest evidence—for positive epidemiological data supported by animal data, (2) substantial evidence—for the broad range of positive results from animal bioassay tests, and (3) suggestive evidence—for positive short-term test results, or for borderline animal or human results." Anderson reports that the names for the categories were dropped because of the general misunderstanding of the names, but the categories remained. "The upper bound is calculated using reasonably conservative exposure estimates and the linear non-threshold dose-response curve is regarded by scientists working in risk assessment in the United States as usually placing a plausible upper bound on risks—recognition that the lower bound may approach zero or be indistinguishable from zero stems from the uncertainties associated with the mechanisms of carcinogenesis, including the possibility of detoxification and repair mechanisms, differences in metabolic pathways, and the role of the agent in the cancer process."

Yet this conservative approach is not necessarily in the best public interest. Nichols and Zeckhauser (1985) observed, "Conservatism in risk assessment is the standard approach in a wide range of regulatory decisions, in fields ranging from control of toxic wastes to approval of new drugs. Motivated by a desire to emphasize safety over other considerations, the conservative approach employs assumptions that tend to overstate risk. This bias in assessment leads to [inappropriate diversions of] limited risk reduction resources. The outcome may be less overall safety gains than would result from decisions based on more realistic assumptions. The fundamental problem is that conservatism confuses risk assessment with risk management."

While the discussion that follows addresses the errors in data, manipulation and presentation, along with the need for quality assurance/quality control discipline in risk models, it does not discuss the risk models. This paper may appear unduly harsh toward the Environmental Protection Agency (EPA). That is not the intent. Most risk models were performed for EPA, and hence EPA is the richest source of examples. Some examples are old, and the mistakes have been corrected, but they are used here to illustrate what can go wrong if the modeller isn't vigilant in control of input data. Most examples relate to cancer assessment because more work has been done in that area than in others, but the principles identified can be used to evaluate any type of environmental risk model.

DATA INPUT ERRORS IN HAZARD ASSESSMENTS

Hazard assessments must assess data from the chemical, physical and biological properties of the material under study. Traditional toxicology has generated reams of data, much of which are virtually irrelevant to risk assessment. Many data used in risk assessment are based on bioassays from animals dosed at such high levels "to get a result" that the measured outcomes bear little resemblance to real world exposures. If testing protocol does not properly specify administered dose and provide some insight to delivered dose, relevant pharmacokinetics and proper test species, the data are misleading and can give the wrong result. Risk modellers must learn to view data critically and reject much of the currently available information since the data were gathered in such a way as to make them inapplicable for risk assessment techniques. Some of the problems, but, by no means all, are presented below.

Use of Maximum Tolerated Dose

The greatest error in input data is probably the use of maximum tolerated dose (MTD) results. While high doses of test materials are necessary to obtain statistically significant biological endpoints, the results from these tests should not be weighted as heavily as they are in hazard assessment models. MTD is old toxicology and should be replaced by more relevant types of assays such as pharmacokinetics or in vivo cell culture. For the same investment of research dollars, more relevant data than MTD can be obtained for hazard assessment using different tests altogether. The famed megamouse study clearly showed that the answer was not simply to test at more levels. MTD tests are not only cruel to the test animals, they frequently do not reflect normal metabolism at environmental exposures. MTD is based on high exposures for long periods of time that saturate biochemical pathways. Therefore, test results often are based on metabolic pathways that would not be seen normally—a metabolism that is often irrelevant to the desired test. Greim *et al.* (1981) observed that high doses of a chemical may have other biological effects than lower doses by overcoming metabolic inactivation mechanisms.

If the metabolism in test animals does not follow the same pattern as in humans, what is the value of the number? Maximum tolerated doses should be replaced by more accurate measures that allow more meaningful interpretation, or at least have less prominence in risk models.

If the administered dose is too high, it may cause underreporting. The high dose may kill the receptor cells, and lower dose testing might have shown different results.

In addition, the following erroneous assumptions have entered the practice of risk assessment:

Assumption 1: All Mutations Cause Cancer. Use of high dose testing to predict mutagenicity (and therefore carcinogenicity, as one theory goes) is founded on the assumption that once a mutagenic event happens, it is irreversible and will lead to formation of neoplasia. Scientists know this to be untrue, but often models fail to allow for repair of damaged chromosomes. When compounded by upper bound estimates, the two assumptions lead to unreasonably high estimates of probable risk.

Single changes in genetic material do not necessarily result in neoplasia, as Greim *et al.* (1981) observed. We are exposed to small amounts of mutagenic material in just about everything we eat—much of it of natural origin (Ames, 1984). If no repair mechanism existed, we would never reach adulthood, because these natural mutagens would have exacted their toll early in our lives. The American Council on Science and Health (1985) issued a valuable lay publication, "Natural Carcinogens in American Food," that details the natural "sea of carcinogens" in which we swim.

Models of carcinogenicity, however, use data from high levels, usually the maximum tolerated dose, and extrapolate to low levels, without considering repair mechanisms. This high-to-low extrapolation is not only without scientific support because it doesn't allow for natural recovery, it also encourages error magnification as the model is processed in the computer. This unnecessary error (derived from using only conservative estimates and later compounded by numerical manipulation) is what government modellers call "cascading of conservatism"—technically error propagation.

Assumption 2: Man is More Sensitive Than a Mouse or Rat. Greim *et al.* (1981) observe that the translation formula is

$$\text{Linear } P = \beta \times d,$$

where P is the fractional incidence of excess tumors, d the average dose in mg/kg, and β a parameter (in units of kg/day/mg) to be estimated from the experimental data. Additionally, a species correction factor which has been chosen as the cube root of the ratio of the body weights can be applied to β to calculate the dose per surface area. This always implies that man is more sensitive than rats or mice. This is not true because many materials that cause cancer in mice have not done so in man. Man has a more complex physiology. As a higher species on the evolutionary scale, man has many more repair mechanisms and other defenses than rodents. To assume that man is more sensitive is wrong, because data do not support the assumption. In fact, test mice, such as the standard B6C3F1 strain, have been bred to have high rates of spontaneous tumors from viral changes in the DNA. Irons *et al.* (1987) have shown a presumptive role for endogenous retrovirus sequences in 1,3-butadiene-induced lymphoma in B6C3F1 mice.

If test animals with an extremely high rate of spontaneous tumors are used, such as the B6C3F1 mice, the test material's ability to cause mutation (hence cancer) is not even being measured. What is measured is merely the test material's ability to promote changes in already altered (by previous precancerous mutation) test animals. If these results from defective animals are used as if the results represented de novo cancer, and it is then assumed that the test animals under-represented man as a surrogate, the significance of the data would be grossly magnified. Manipulation of these biased data in subsequent models will add to error propagation (especially the cubic function of the mouse to man calculation).

Reynolds *et al.* (1987) suggest that the B6C3F1 mouse liver might provide a sensitive assay system to detect various classes of proto-oncogenes that are susceptible to activation by carcinogenic insult.

Assumption 3: Mechanisms of Carcinogenicity Can Be Ignored in Modelling. Chism and Rickert (1985) observed that 2-nitrotoluene induces DNA repair in the in vivo–in vitro hepatocyte unscheduled DNA synthesis assay in male, but not female, Fischer-344 rats. The structurally related 2,6-dinitrotoluene, which also displays a sex-specific toxicity, requires biliary excretion for bioactivation. Their results indicate that biliary bioactivation is necessary for the material to produce an effect. If this mechanism is not relevant in human physiology, for example, data from these studies would not predict human effects. To use the data in predictive modelling without testing for their relevance to humans not only would distort the results but also would give the wrong result.

Assumption 4: Administered Dose Is the Same as Delivered Dose. Starr (1987) observed " . . . [T]he actual quantitative relationship between administered and delivered doses is a reflection of the entire spectrum of biological responses to exposure ranging from physiological responses of the whole organism to intra-cellular biochemical responses in target tissues. Thus the administered dose provides at best an indirect surrogate measure of the delivered dose, and the relationship between these two measures of exposure need not be a simple linear one. . . .

"Various mathematical models of carcinogenesis are conceptualized and formulated in terms of interactions between the biologically active forms of chemical agent, i.e., delivered dose, and cellular macromolecules in target tissues. These models lack the structure necessary to characterize the many physiological and pharmacokinetic factors that determine the actual relationship between administered and delivered doses. They characterize only the dependence of the carcinogenic process on the delivered dose.

"It is now known that low-dose risk extrapolations based upon the linear proportionality assumption will yield risk assumptions that are either excessively conservative (too high) or anti-conservative (too low) when the true administered/delivered dose relationship is non-linear (Hoel *et al.* 1983)."

Starr (1987) illustrates this difference between administered and delivered dose with an example of the work done by Chang *et al.* (1983). Both rats and mice were exposed to 15 ppm formaldehyde. Mice have a greater respiratory depression reflex than rats to irritating gases; therefore, they receive about one-half of the delivered dose of formaldehyde as do similarly exposed rats after correcting for the size of the nasal cavity. This reflex can account for the tremendous difference in high-dose tumor incidence (50% in rats, 1% in mice). This factor is not true for all inhaled gases. Rats receive only one-half of the delivered dose that mice receive with another gas, acrylic acid. The respiratory depression reflex is chemical-specific.

Starr (1987) describes other factors that affect delivered dose, such as mucus secretion, a major factor in formaldehyde toxicology. He also observes that formaldehyde dose delivered to DNA of replicating cells is non-linearly related to airborne concentration. Significantly less binding was observed with airborne formaldehyde concentrations of 2 ppm than would be expected from the linear data at 6 ppm or greater. Starr notes that initial factors of ten for an overestimate do not seem great, but when applied to the multistage model, favored by the government, the risk measure is raised to an integer power. Thus, if an exposure is overestimated by a factor of 5, the risk of tumor is overestimated by a power of 125. (See Fig. 1.)

Starr (1987) presents maximum likelihood estimates of risks associated with exposure to 1.0 ppm airborne formaldehyde, as determined with four commonly used quantal response models that were fit to the formaldehyde bioassay data of the Chemical Industry Institute of Toxicology (CIIT) for Fischer-344 rats. The first row in Table 1 shows results obtained by using administered dose—airborne formaldehyde concentration—as the measure of exposure. The second row shows corresponding estimates obtained by using covalent binding data as a measure of the delivered target tissues. The third row provides the ratios of the administered dose-based risk estimates to the corresponding delivered dose-based risk estimates.

Table 1. Maximum Likelihood Risk Estimates for 1.0 ppm Airborne Formaldehyde

Dose Measure	Probit	Logit	Weibull	Multi-stage
Administered	2.6×10^{-5}	2.9	5.9	251
Delivered	4.0×10^{-14}	0.02	0.07	4.7
Reduction Factor	6.6×10^{8}	133	83.4	53.4

Of course, none of this is directly related to man, who may have another mechanism of action. Perhaps inhalation is not as significant as dermal absorption or ingestion. This point is raised simply to show that the risk modeller must decide if the animal model has any relevance to man.

Assumption 5: Test Results Always Indicate the Effects of the Chemical Tested.
Some early studies by Tannenbaum (1940) documented the effect of simple diet on rates of neoplastic tumor development. Mice fed a normal balanced diet—sufficient to maintain the normal body weight—exhibited a lower incidence of lung and other cancers than mice fed

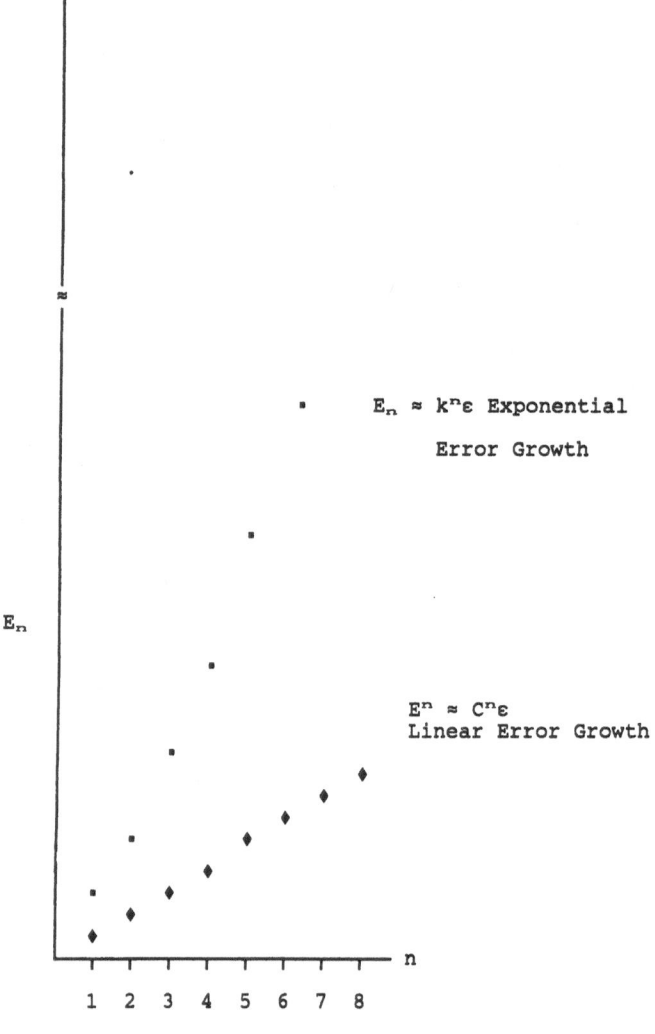

Fig. 1. Linear and Exponential Growth of Errors from Burden and Faires (1985).

the same diet, but in slightly higher quantities—enough to create obesity. Others [Jose (1979), Roe and Tucker (1974), Tucker (1979), Conybeare (1980), Roe (1981) and Casarett and Doull (1980)] have confirmed the carcinogenic effect of obesity in rats and mice. In addition, these authors report that underweight rats also had higher tumor incidence than the rats fed the normal balanced diet.

Crowding and other stresses may also increase the incidence of tumors independent of the materials being tested.

Interspecies Variability

Starr (1987) observed, ". . . [I]t would appear that the cookbook-like approach to interspecies extrapolation currently in use by the regulatory agencies yields little more than imprecise estimates of the reciprocals of maximum doses tested, no matter what the actual tumor incidence data are! . . . Bioassay data by themselves, no matter how they are analyzed, are simply inadequate to the task of accurately and quantitatively predicting human cancer risk. At best, they permit an identification of those chemical compounds that are carcinogenic in test species under the specific conditions of the studies." [NOTE: EPA has begun to recognize delivered dose concepts in the hazard assessment document on methylene chloride.]

Calabrese (1987) discussed the problems with interspecies variation. He observed that chemically induced hepatomas in mice and their extrapolative relevance to humans is a major point of controversy in the field of predictive toxicology. The standard cancer bioassay performed by the National Toxicology Program uses the B6C3F1 mouse and Fisher 344 rat. As previously discussed, the B6C3F1 mouse has a retrovirus oncogene which makes the majority of chemicals test positive. (This is due to the assay which de facto measures promotion rather than initiation in the carcinogenic process.) Calabrese observed that in 85 chronic exposure studies, the B6C3F1 mice developed hepatomas in 45 studies, whereas the rats developed tumors in only 15 studies. Yet many times the 30 chemicals which showed activity in only the B6C3F1 mice are classified as probable carcinogens based on the single mouse bioassay.

Epidemiological Information

Perhaps the most perplexing issue is the reluctance of modellers to use sound epidemiology data when they are available. The tendency is to use epidemiology data when the results show an adverse effect, and to ignore them when the data do not show an effect. The feeling seems to be that the findings of harm have more substance than findings that are inconclusive or don't show an adverse effect. These studies should be given more weight than animal studies on systems that are not reflective of the real world.

Unfortunately, epidemiology can be riddled with as many errors as risk models, and before these studies are used, they must be carefully reviewed for quality. Most exposed populations are small and exposure data are weak. All too often the studies are conducted with biased sampling, improper controls and insufficient exposure assessment, which makes them invalid. Neighborhood surveys by activist groups are prime suspects for poorly planned studies (Grisham, 1986).

Risk managers are always asked to make sure the action is safe. Safe implies the absence of harm, and the truth of the matter is that one can never prove a negative. It is always easier to say there is a problem rather than there is not a problem. This bias against negative epidemiology is a major weakness in risk analysis and a serious impediment to understanding differences between humans and laboratory specimens. Human data should be the strongest data, rather than the weakest. Risk assessment technologies must find a way to incorporate both laboratory and well-developed epidemiology data to provide the most accurate health assessment for human risk.

Hazard assessments are made with incomplete data. We will never have enough data to make decisions with absolute certainty. If insufficient data are available to do a model, best professional judgment should replace the model. Models should not be a substitute for good judgment, and the results should always be compared with actual experience.

DATA INPUT ERRORS IN EXPOSURE ASSESSMENT

The Chemical Manufacturers Association (CMA) commissioned a study by Picard, Lowe and Garrick, Inc. (1987) to evaluate exposure assessment models. They evaluated fifteen models and concluded that six were exemplary. The evaluations were based on the following eight criteria:

- Usefulness	- Accuracy
- Completeness	- Feasibility
- Adequacy	- Certainty
- Objectivity	- Verifiability

The authors present a detailed listing of the factors underlying exposure assessment in Section C of their report. This listing is far too comprehensive for this paper, but anyone who is planning exposure assessment would do well to use the factors list as a checkoff for a properly done study.

Improper exposure assessment—both present and predicted—can yield major sources of error in risk modelling. Some of these sources of error include the following:

Assumptions About Production

All too frequently emission assumptions are based on full production and twenty-four hour per day production. While this is an admirable goal, it is not reality. At best, a plant will run somewhere around 80% capacity. Often the estimates are based on models of emissions rather than measured data, and these will vary widely.

Many of the estimates used in regulatory analysis have been found to be based on obsolete manufacturing technology or on a process that is not used by most of the industry. For example, EPA singled out maleic anhydride plants for priority based on their calculations of the total benzene emissions from eight maleic anhydride plants, when, in fact, only six of the eight were operating at the time of analysis (CMA, 1980), and one plant that was operating at the time of the model development was missed.

The conversion rate of 94.5% that EPA used at one plant was inaccurate; the plant's conversion rate was 97%, which greatly reduced actual benzene emissions. (This resulted in a 3% total overestimation of emissions at this plant.) At the time, reduction equipment was being installed that would reduce the emissions from another plant by more than 91% and the national total by 55%. A third plant had work in progress to reduce its emissions by 92%.

In short, owing to industry developments, actual benzene emissions from maleic anhydride plants were going to be 40% lower than estimated within a short time—without regulatory action by EPA. On the correct production basis alone, estimated risk from benzene exposure was one-sixth of that estimated by the agency.

Continuing with the maleic anhydride example, CMA (1980) reported that EPA used "model plants" to calculate the population at risk. When compared to actual plant operations, the "model plant" had little relevance. When corrected to actual conditions, the exposure values were 60% less than what EPA calculated would be the result after regulation with new standards.

Assumptions on Patterns of Exposure

Many work place exposure analyses are based on an assumption of an 8-hour exposure and a 5-day/week exposure. "Normal" work patterns allow for a certain amount of purging and recovery over the weekend. Workers may work 5 days with 8 hours as an average, but this exposure can extend from 24 hours a day for 3 months (e.g., on ships at sea), or vary widely for swing shift workers or those whose daily pattern varies (e.g., mechanics or electricians in a factory when they are assigned to multiproduct operations).

Use of the Theoretical Maximally Exposed Individual

Agency regulators use a fictitious character, the maximally exposed individual. This is a source of bias, because no such individual exists. The agency would be well served to use, instead, the most probably exposed individual, and then find the range of the exposure. This will greatly reduce compounded error.

EPA's benzene exposure model (for maleic anhydride) assumed that exposed individuals would sit on their front doorstep 24 hours a day, 365 days a year for 70 years. This alone introduced tremendous conservative data bias.

Vollmerhausen (1987) reports that EPA used 365 days/year to determine exposure values for children in the PCB risk model for Wide Beach. Wide Beach is southeast of Lake Erie and receives an average of 72 inches of snow per year. Obviously, the children will not be in contact with the soil every day of the year owing to snow, rain, illness and school. The age of the child is a major factor in assessing soil ingestion as it is much more likely to occur in younger children than in infants. By reducing the number of days of exposure to a more realistic 125 for children 1 to 6 years of age and 113 for 6-year-old children, a very different exposure pattern develops.

Selection of High Values Rather than Representative Values

A recent example of ambient monitoring bias was documented by the American Petroleum Institute (1987) in their analysis of EPA ozone monitoring data. EPA used only the monitor with the highest value to determine if the study area exceeds the ozone standard. API found that in three of the four areas studied, the readings for the average of all monitors are most like the readings of the lowest monitor, not the highest. In Detroit, a single, high ozone value monitor forced the area into noncompliance. By using the single, high-value monitoring station only, EPA badly biased the results. If the average of all stations in a study area were used, approximately 85% of the study areas would meet attainment standards. The gap in data, based on using the high-value monitor, represents billions of dollars that would have to be spent unnecessarily.

The desire to be conservative in input data had major implications in a soil ingestion model in Wide Beach, New York, as Vollmerhausen (1987) reported. A single value for PCB concentration was used in the risk model. This datum was obtained from a catch basin—a very improbable play (hence exposure) area for children. This high value was 5,300 mg/kg PCBs in soil. The geometric mean of PCBs in soil of the 59 homes tested was 6.6 mg/kg. Based on the range of data, <0.05 to 600 mg/kg, the 95% confidence interval is 0.183 to 238 mg/kg. No matter how one looks at the weight of the data, 5,300 mg/kg, the number used by EPA, is simply unrepresentative.

Sampling Bias

It is all too easy to place monitoring systems in areas where we think the highest values might be recorded. Yet this is not representative of the ambient conditions in the area, and the biasing of sample collection locations may cause unduly high and nonrepresentative values. When later used by modellers who are unfamiliar with the site location, this error can be magnified well beyond the ambient conditions in the study area. This yields the same type of error as selecting the highest values, as detailed above, without consideration of the most probable values.

Inadequate Sampling Protocol

The American Chemical Society (1983) observes, "One source of uncertainty in estimating chemical exposure is in determining the location and number of samples to be taken for analysis, for these samples must accurately represent the exposure levels at the place and time of exposure. For example, it may make a difference whether the soil is sampled at the surface or at a depth of 6 inches. Proper scientific methods must be observed, as an error in sampling will be propagated throughout the entire analysis. Furthermore, enough samples must be taken in order to perform statistical analyses critical to ensure the reliability of the data."

Sampling and Analysis Error

Some methods of sample collection, sample preservation, transportation and analysis are subject to errors, and improper collection, preservation and analysis yield invalid data. Unfortunately, many data in government data banks could not stand scrutiny of quality assurance and quality control (QA/QC), and the use of data—even published data—without some form of quality check can lead to serious errors if they are used directly for modelling. This QA/QC data problem is especially true of the older environmental data bases such as STORET. Back in the days of the mercury scare, many persons used a technique for mercury analysis which later proved to yield incorrect data and some of these old (and incorrect) data are still used today.

Before sampling data are used in models, the risk modeller should establish that the analyses were done within the time limitations for the analytical procedure and proper QA/QC programs were in place. If samples were not processed within the specified time for the test, the data should not be used.

Matrix effects are a perplexing problem in environmental analyses. Far too many of the environmental sampling techniques are developed in pure matrices such as distilled water. When analysts try to use the "cookbook" methods, many do not work because other materials in the environmental sample interfere. Unfortunately, some of these data find their way into data banks and are used for models. Matrix interferences can yield higher or lower values. Evaluation of the internal standards, and recovery of both the analyte and the entire extraction process are essential if the numbers are to be relevant. If the recovery data are poor, the results should not be used in the model. Matrix effects create especially difficult problems when trace analysis is performed. The error introduced from the workup procedure of a difficult natural matrix can be greater than the measured value.

All too often modellers say, "We have to use these data—they're the only ones we have." The response should be "Bad data are bad data. They should not be used even if they are the only data available. When in doubt, throw it out!"

In summary, risk modellers should not accept monitoring data at face value simply because confidence limits are available. These data need the same rigorous scrutiny as biological data.

NUMERICAL AND MODELING ERRORS

Three common errors in computer computations are truncation errors, round-off errors and error propagation. These are, of course, exclusive of the human errors of data input and programming which are all too common.

The most significant error in risk analysis is exponential propagation. Burden and Faires (1985) define error propagation as $|E_n| \approx C_n \varepsilon$, where C is a constant independent of n, ε is the input error, and the growth of error is linear. If $|E_n| \approx k^n \varepsilon$ for some $k > 1$, the growth is exponential. Linear growth of error is usually unavoidable. Exponential growth should be avoided at all costs since k^n becomes large rapidly from very small values of n.

This will lead to unacceptable sizes of error, regardless of the size of ε. This is illustrated in Fig. 1.

Inadequate Sensitivity Analysis in the Models

Munro and Krewski (1981), as reported by Shu et al. (1987), show that different models can estimate risks differing by three or four orders of magnitude based on the same bioassay data set. Shu et al. (1987) also report that Sielken (1988) and Sielken et al. (1986) found that under certain conditions these models can be insensitive to data. For example, significant alterations in the animal data did not appreciably change the estimated risk.

Sensitivity analysis, both relative and absolute, should be calculated for each risk assessment model. Relative calculations should be based on a 1% change in each input datum, and the change in the result noted. In linear relationships, this change should be insignificant. Then the absolute sensitivity should be calculated by using the most reasonable and the two extremes of the values in the model. First, the lower-bound values are substituted for each input value, one at a time, and then the same process is repeated for the upper-bound value. Finally, the average value is substituted for each value in the model. In this manner, the most sensitive value of the input parameter can be determined. Once the sensitive parameters have been determined, the extreme values may be used to approximate an uncertainty range.

In Vollmerhausen's (1987) analysis of soil ingestion, the concentration of the soil was the most sensitive parameter. These values caused the estimate to vary by three orders of magnitude. When all of the worst case assumptions were made, the model overcalculated risk by 7 to 10 orders of magnitude over the best case assumption. If lowest assumptions were made, the exposure was insignificant.

Verification

Often it seems that the risk modeller has lost touch with the system being modelled. For example, in the risk assessment on ethylene dibromide, EPA overestimated the values for farm workers to the point that (if they had been accurate) many cases of illness should have been seen. This was not the case. These conclusions were never field verified before the numbers were released to the media.

When data are available, results from models should be tested against these data.

EPA's benzene assessment for maleic anhydride facilities overestimated the cases of leukemia attributable to benzene emissions from maleic anhydride plants in the area of the seven plants (CMA, 1980). They estimated a current incidence rate of 600 cases per year. In actuality, the number was closer to 400. Then-current maleic anhydride operations might have contributed 0.0037 cases/year (one case every 270 years). If the proposed new standard had been adopted, the cases avoided would have been only 0.0012 cases/year (one case every 830 years). In either case, the contribution from plant operations was negligible relative to the normal incidence in the population. (EPA did drop the regulation in response to industry comments, but many resources were wasted by industry in reviewing a flawed analysis.)

Plausibility is a factor all too often missing from many risk assessments. Many risk practitioners can be more interested in the model than in the information being modelled. It is important to have the assumptions and the results checked against observable field data—especially epidemiological data, when available. Cause and effect predictions must be assessed for biological plausibility as part of the reality check as well.

INCOMPLETE PRESENTATION OF MODEL RESULTS

We tend to lose sight of the purpose of risk models. Risk models are designed to assist decision makers with health-based decisions. All too often, the policymaker is given single number (maximum plausible) estimates with no range of variability.

This practice should stop. The most plausible value should be presented first, with minimum and maximum plausible levels to bracket the range of uncertainty. Policymakers can understand such a simple concept. If the sensitivity analysis and the 95% confidence limits—along with the most probable value—are not given, policymakers should send the results back to the modeller for completion. Upper and lower levels of uncertainty must become routine submissions along with the model results.

Different models yield different 95% confidence limits. Sielken (1985) demonstrates that the fitted multistage model and the Weibull model are nearly identical, but their 95% confidence limits are very different. The upperbound risk in the linearized multistage model at 1.0 ppm of formaldehyde is 157 times greater than the upper bound corresponding to the Weibull model.

Linear extrapolation to zero is the most conservative model and most widely practiced by the EPA. This assumes no threshold, and may give results that overprotect. This practice can easily divert limited funding from protecting against even greater risks that have not been assessed.

CONCLUSIONS

Conservatism should be part of the policy process and not of the scientific risk assessment. The input data on models should be the most probable values, so as to provide the most accurate estimate of the risk. Using the upper 95% confidence limit means that almost all of the data fall below this value. (This by itself vastly distorts the interpretation.) The policymaker should be given the most likely value, along with upper and lower limits, standard deviation, and sensitivity of the assumption. The actual incidence of the disease (or other modelled characteristic) should be presented to put the risk calculation into perspective. Then, and only then, the policymaker can put a level of conservatism on the most accurate number. As it is now, the conservatism of control is placed on top of risk estimates that are already grossly distorted by biased input data.

Hard numbers are reassuring to the general public because there's a feeling that numbers have scientific validity. Yet EPA's maleic anhydride model was off by a factor of more than 100. This error could have caused serious economic hardship for the communities involved without any health benefits being realized.

Conservative toxicological tests such as the Maximum Tolerated Dose (MTD) may underestimate risk because cells die at such high concentrations rather than exhibit other effects which might be noticed at lower testing levels.

Nichols and Zeckhauser (1985) observed: "In practice, the link between risk assessment and management is often blurred. Fundamental gaps in scientific knowledge and data limitations make risk assessment a highly uncertain endeavor, requiring many choices among competing models and assumptions. There is a strong temptation to have such choices reflect implicit policy judgments rather than science.

"This blur is most apparent in current techniques for estimating the risks associated with carcinogens, which employ conservative assumptions that bias the estimates upward. The intent is to err on the side of safety by minimizing the chance that risks will be underestimated and thus uncontrolled. But this approach, [risk management] intrudes [into] the risk assessment process . . . In deciding how conservative to make their estimates, risk assessors implicitly trade-off risk against other factors. Unless they explicitly acknowledge these trade-offs and quantify them, their assessments will mislead others, including those charged with managing risks . . . and entrusted with public resources to provide the greatest protection for the most people.

"Advocates of the conservative approach are likely to view this tendency towards greater control as a virtue; conservatism is intended to give extra weight to protecting public health and, under conditions of massive uncertainty, to err on the side of safety. In fact, however, conservative risk assessment is a deeply flawed approach to protecting public health. It violates the distinction between risk assessment and risk management, concealing value judgments and policy choices under the cloak of science. It creates capricious differences in the degrees of safety provided across different substances and policy areas, because degrees of conservatism vary widely. Finally, because regulators must make complex trade-offs among different risks, not only between risk and cost, conservatism can lead to less rather than more safety, by misdirecting the public concern and scarce agency and societal resources."

REFERENCES

American Chemical Society, *Chemical Risk: A Primer*, Information Pamphlet, American Chemical Society (1983).

American Chemical Society, "Principles of Environmental Analysis," *Anal. Chem.* **55**:2210-2218 (1983).

American Council on Science and Health, "Does Nature Know Best? Natural Carcinogens in American Food," a report (1985).

American Petroleum Institute, "Ozone Monitor Study," Special Report (1987).

Ames, B. N., "Dietary Carcinogens and Anti-Carcinogens," *Science* **221**:1256-1264 (1984).

Anderson, Elizabeth L. "Quantitative Approaches in Use to Assess Cancer Risk," *Risk Analysis* **3**(4).277-295 (1983).

Burden, R. L. and J. D. Faires, *Numerical Analysis*, 3rd Ed. Prindle, Weber and Schmidt, Boston (1985).

Calabrese, E. J., "Animal Extrapolation, A Look Inside the Toxicologist's Black Box," *Environ. Sci. Technolog.* **21**(7):618-623 (1987).

Casarett and Doull, "Toxicology - The Basic Science of Poisons," 2nd Edition, pp. 119 (1980).

Chang, J. C. F., E. A. Gross, J. S. Swenberg and C. S.Barrow, "Nasal Cavity Deposition, Histopathology, and Cell Proliferation after Single or Repeated Formaldehyde Exposure in B6C3F1 Mice and F-433 Rats," *Appl. Pharmac.* **68**:161-176 (1983).

Chemical Manufacturers Association, Testimony on the Proposed NESHAP for Benzene Emissions from Maleic Anhydride Plants, for EPA (August 21, 1980).

Conybeare, G., "Effect of Quality and Quantity of Diet on Survival and Tumor Incidence in Outbred Swiss Mice," *Food, Cosmet. Tox.* **18**:65-75 (1980)

Grisham, J. W. (Ed.), *Health Aspects of the Disposal of Waste Chemicals*, Pergamon Press, Inc., New York, NY. (1986).

Greim, H., U. Andrae, W. Göggelmann, S. Hesse, L. R. Schwartz and K. H. Summer, "How Relevant are High Doses in Mutagenicity and Carcinogenicity Studies in Animals?" *Prog. in Mutation Res.* **2**:129-147 (1981).

Hoel, D. G., N. L. Kaplan and M. W. Anderson, "Implications of Nonlinear Kinetics on Risk Estimation in Carcinogenesis," *Science* **219**:1032-1037 (1983).

Irons, R. D., W. S. Stillman and M. W. Cloyd, "Selective Activation of Endogenous Ecotropic Retrovirus in Hematopoietic Tissues of B6C3F1 Mice During the Preleukemic Phase of 1,3-Butadiene Exposure," in press, Virology, (1987).

Irons, R. D., W. S. Stillman, and R. S. Shah, "Selective Activation of Endogenous Ecotropic Retrovirus in Tissues of B6C3F1 Mice During the Preleukemic Phase of 1,3-Butidiene Exposure," Abstract, *J. Cell. Biochem. Supp.* **11A**:204 (1987).

Jose, D. G., "Dietary Deficiency of Protein Amino Acids and Total Calories on Development and Growth of Cancer," *Nutr. Cancer* **1**:58-63 (1979).

Nichols, A. L., and R. Zeckhauser, "The Dangers of Caution: Conservatism in Assessment and Mismanagement of Risk," E-85-11 Discussion Paper, John R. Kennedy School of Government, Harvard University (1985).

Picard, Lowe and Garrick, Inc., "Evaluation of Exposure Assessment Methods," PLG-0524 (1987).

Reynolds, S. H., S. J. Stowers, R. M. Patterson, R. P. Maronpot, S. A. Aaronson, and M. W. Anderson, "Activated Oncogenes in B6C3F1 Mouse Liver Tumors: Implications for Risk Assessment," *Science* **237**:1309-1316 (1987).

Roe, F.J., "Are Nutritionists Worried by the Epidemic of Tumors in Laboratory Animals?" Proc. Nutr. Soc. 40:57-65 (1981).

Roe, F. J, and M. J. Tucker, "Recent Developments in the Design of Carcinogenicity Tests in Laboratory Animals," *Proc. Europ. Soc. for the Study of Drug Tox.* **15**:171 (1974).

Shu, H. P., D. J. Paustenback, and F. J. Murray, "A Critical Evaluation of the Use of Mutagenesis, Carcinogenesis, and Tumor Promotion Data in A Cancer Risk Assessment of 2,3,7,8-Tetrachlorodibenzo-p-dioxin," *Reg. Toxicol. and Pharm.* **7**:57-88 (1987).

Sielken, R. L., "Quantitative Cancer Risk Assessments for TCDD," *Food Chem. Toxicol.*, in press (1987).

Sielken, R. L., F. W. Calborg, D. J. Paustenback, H. P. Shu, and F. J. Murray, "Alternative Approaches to Mathematically Analyzing the Bioassay Data for 2,3,7,8-TCDD," *The Toxicologist* **6**:282 (abstract) (1986).

Sielken, R. L., "Some Capabilities, Limitations, and Pitfalls in the Quantitative Risk Assessment of Formaldehyde," in Risk Analysis in the Chemical Industry, CMA, Washington, DC, (1985).

Tannenbaum, A., "Initiation of Growth of Tumors; Introduction; Effects of Underfeeding," *Am. J. Cancer* (1940).

Tucker, M. J., "The Effect of Long-Term Food Restriction on Tumors in Rodents," *Int. J. Cancer* **23**:803-807 (1979).

Vollmerhausen, J., "The Uncertainty in Risk Estimation for a Soil Ingestion Model," Master's Thesis, Johns Hopkins School of Hygiene and Public Health (1987).

Risk Assessment Issues in Implementing Emergency Planning and Community Right-to-Know Legislation

Michael H. Shapiro

U.S. Environmental Protection Agency
Washington, DC

ABSTRACT

In the aftermath of the Bhopal, India tragedy the Environmental Protection Agency (EPA) initiated a program to address the risks posed by the accidental releases of hazardous chemicals. This program promoted voluntary hazard identification and contingency planning by local communities and industry, focusing on a list of extremely hazardous chemicals. In identifying these chemicals, EPA had to develop appropriate criteria for selecting chemicals that could cause serious human health effects on the basis of short term, accidental exposure. After considering a number of approaches, EPA used criteria based on measures of acute toxicity from short term, mammalian testing by the inhalation, oral, or dermal routes of exposure. Applying these criteria to the RTECS data base and EPA's list of chemicals in commerce yielded the list of extremely hazardous chemicals. These chemicals subsequently became the basis for mandatory emergency response planning under Title III of the Superfund Amendments and Reauthorization Act of 1986. In support of this requirement, EPA also developed threshold quantities for on-site storage which trigger notification and planning requirements. EPA's priorities for further work in this area include developing criteria for listing chemicals on the basis of long term health effects that could result from accidental exposure and developing exposure levels of concern for emergency planning purposes.

KEYWORDS: Title III (of the Superfund Amendments and Reauthorization Act of 1986), chemical emergency response planning, extremely hazardous chemicals, criteria for identifying hazardous chemicals, threshold planning quantities

INTRODUCTION

Title III of the Superfund Amendments and Reauthorization Act of 1986,[1] which is also known as the Emergency Planning and Community Right-to-Know Act, will dramatically change the nature and type of information that people have about the presence of hazardous chemicals in their communities. The act represents a blending of two developments with respect to public policy on toxic chemicals: (1) concern over the potential for catastrophic accidental releases of extremely hazardous materials of the type that occurred in Bhopal; and (2) the more general interest in the rights of people to have access to information on those toxic chemicals that are present in facilities in their communities. This paper discusses some of the risk assessment related issues that have been important in implementing the legislation, placing particular emphasis on those parts of the legislation dealing with planning for hazardous chemical emergencies. Much of this discussion will focus on activities that began within the Environmental Protection Agency (EPA) prior to the passage of Title III, and that are being incorporated into our implementation of Title III either through the statutory provisions or through our implementating regulations and guidance materials.

BACKGROUND

In response to the Bhopal tragedy and less serious but still disturbing incidents in the United States, the U. S. Environmental Protection Agency's Air Toxics Strategy began to address planning for chemical emergencies. This was a relatively new area of activity for EPA, although other agencies, particularly the Department of Transportation, had been concerned over this issue for quite some time.

EPA's initial strategy in addressing this concern was to focus on the state and local levels of government as being the primary agencies with responsibility for emergency response planning. This was considered to be a logical extension of the historical roles that these levels of government have played in dealing with both natural and man-made emergencies. EPA's role was viewed as one of providing guidance, technical support and leadership in the area. In addition, the strategy viewed the voluntary cooperation of industry in providing information and sharing in the planning process as crucial to the success of the planning effort.

As part of its strategy, EPA evolved the idea of developing a list of extremely hazardous chemicals to focus nationwide chemical emergency response planning. The decision to develop a list was not made without some serious reservations and internal debate within the Agency. In order to provide some focus and coherence to planning, we recognized that any such list would have to be relatively small in comparison to the numbers of chemicals that could potentially be of concern under a variety of different circumstances. Therefore, we worried that by focusing attention on one particular list of chemicals we would by implication exempt a large number of others that merited consideration in particular situations. In addition, given the perceived urgency of moving forward with a program to address accidental release hazards, there was considerable concern within the Agency as to whether we could do something useful within the time period that we were given. Essentially, the decision to move forward with an air toxics program addressing accidental releases was made in May of 1985 and we were expected to have a guidance document and a list of chemicals available sometime in the fall of that year. Despite these reservations EPA's Administrator, Lee Thomas, made the decision that a list of chemicals was necessary to provide a national planning framework and decided to go ahead with the activity. In addition to identifying the list of chemicals, EPA began development of guidance materials for conducting emergency response planning.

DEVELOPMENT OF THE LIST OF EXTREMELY HAZARDOUS CHEMICALS

The list of extremely hazardous chemicals and the technical components of the Agency's guidance for emergency response planning were developed by an Agency-wide task force. In order to provide for peer review of this effort within the limited timeframe, the Agency enlisted the support of its Science Advisory Board (SAB). The SAB assembled a special panel to review the tentative materials developed by the Agency task force roughly midway through the process of developing the list.

Although the Agency task force had to address a large number of issues, four were paramount:

1. What general criteria would be used to distinguish the most extremely hazardous chemicals? We decided that the primary focus of our efforts would be the selection of general criteria. The list itself would be illustrative of these criteria.

2. Given the criteria, what types of data or studies would be accepted in support of a listing decision?

3. What data bases would be used for identifying the effects?

4. Would we make ad hoc adjustments and additions to the list based on information that was not within our initial criteria or databases?

Criteria for Identifying Hazardous Chemicals

The basic focus of this task was to develop criteria for identifying those chemicals which had the greatest potential for causing wide-spread or catastrophic human health effects as a result of a short-term accidental release. The task was to focus on those measures of chemical toxicity and other effects that would be appropriate for measuring the short-term release exposure scenario. As I mentioned earlier, this was not the first time someone has worried about this problem. There were existing criteria that we could and did draw upon. In particular, it quickly became apparent that whatever the scope of our ultimate criteria, a central component of any set of criteria would have to be based on measures of acute lethality developed from mammalian short-term testing. Such test data are obviously not a perfect indicator of relative toxicity to humans nor are they necessarily indicative of specific types of human health effects. They nevertheless constitute the most broadly available set of basic information that could be correlated with the potential for serious human health effects resulting from short-term exposure. Indeed, everyone who had previously looked at this problem had arrived at the same conclusion. In particular, we find that the most comparable effort was that conducted by the European Economic Community in developing its so called "Seveso" list of chemicals.[2]

The EEC list was based at least in large part on criteria developed for acute lethality based on oral, dermal, and inhalation routes of exposure to mammalian species. The real issue for us was whether criteria based on acute lethality studies could be supplemented by other specific information related to the potential for causing human carcinogenicity, neurotoxicity, reproductive effects, organ effects, or other chronic effects. We recognized that our ability to predict such consequences on the basis of short-term exposure was extremely limited, but we wanted to take at least a brief look at whether it was possible to consider these effects. After an initial examination of possible approaches the SAB recommended, and we agreed, that we could not develop a workable scientific framework for including these effects in the time available. Consequently, the only toxicity criterion we have employed to date is acute toxicity.

In addition to the toxicity criteria, a central question was whether toxicity would be combined with some measure of exposure potential in order to select chemicals. We recognized that for the kinds of accidental releases we were most concerned with, the route of exposure of most concern would be through the air. Therefore, one could argue for considering the potential for release into the air and subsequent dispersion in selecting chemicals. For example, given approximately equal toxicities, a chemical that is a gas at ambient temperature should be ranked of more concern than one that is solid because the potential for an airborne release would be much higher in the former case.

The SAB panel recommended that the possible conditions under which the chemical was used, for example, high temperature and high pressure, had to be considered at least as important as its physical state at ambient temperature. Such factors could only be taken into account on a site specific basis. Consequently, we decided to use criteria based solely on toxicity and leave dispersion considerations for the accompanying technical guidance.

In summary, we decided to use mammalian acute lethality as the sole criterion for selecting extremely hazardous chemicals.

Studies to Support Listing

Once we decided to use mammalian acute lethality data, the next question was what types of studies would be accepted. For example, in the case of the EEC criteria, the explicit indication is that studies would be accepted based on four hour, rat acute lethality studies. However, reported acute lethality tests utilize a variety of mammalian species and a range of exposure times. Moreover, routes of exposure in the reported literature may include oral, inhalation, and dermal. Deciding which studies to include involved a trade-off between our desire to be as comprehensive as possible in identifying potential candidates and our desire to be as consistent as possible. The three areas of most concern were: route of exposure, species, and duration of exposure.

As I mentioned previously, inhalation is the route of exposure that we were most concerned with for accidental releases. However, we reached the same conclusion that the groups developing the EEC list did: appropriately measured dermal and oral acute toxicity studies could be utilized as indicators of relative toxicity. In fact, we based our ultimate recommendations on the same criteria that the EEC and World Bank had developed. In other words, we were willing to use any of the three routes of exposure—oral, dermal or inhalation—as a basis for chemical selection. These criteria are presented in Table 1.

Table 1. Criteria for Acutely Toxic Chemicals

Route	Value
Dermal	$LD50 \leq 50$ mg/kg
Oral	$LD50 \leq 25$ mg/kg
Inhalation	$LC50 \leq 0.5$ mg/l

We did modify the EEC approach in a number of ways. We elected to use the most sensitive mammalian species for which reported test data were available. We also considered studies with exposure durations ranging from 2 to 8 hours. In contrast, the stated EEC criterion is for 4 hour exposures with rats. Our approach allowed us to consider a broader range of available studies, thereby allowing us to evaluate a larger number of commercial chemicals.

Databases

Given the set of criteria and a set of implementing principles, the next step was to develop an illustrative list of chemicals by applying these criteria and principles to appropriate databases. Our major goal was to be as comprehensive as possible in reviewing the reported toxicity literature to identify those chemicals of most significant concern. At the same time, we wanted to focus on chemicals that were actually in commerce in the United States, rather than the entire universe of chemicals for which toxicity data were available. Therefore, we chose to implement our strategy by utilizing as the basic information source for toxicity data the Registry of Toxic Effects of Chemical Substances (RTECS). This is the most comprehensive computerized collection of toxicity information available. We applied our criteria against the RTECS data base and intersected the set of chemicals meeting the criteria with the set of chemicals that were either on the Toxic Substances Control Act Inventory or EPA's list of registered pesticides. These two lists encompass the entire set of chemical substances in commerce in the United States with the exception of chemicals that were subject to regulations solely by the Food and Drug Administration.

The major strength of the RTECS data base is its comprehensive and extensive coverage. On the other hand, its major weakness is that the studies reported in RTECS are of very variable quality and the review given to those studies prior to entry in the data base is somewhat limited. Other data bases that we could have used, such as the National Library of Medicine's Hazardous Substances Database (HSDB), are of higher quality in the sense that more peer review is given to the studies that are included. On the other hand, they are much less complete in their coverage. Therefore, given our objective to be as comprehensive as possible, we felt that the advantages of going with RTECS outweighed the disadvantages. Even with RTECS we recognized that the coverage of the universe of commercial chemicals was very incomplete. The majority of commercial chemicals have no reported acute toxicity studies.[3] So it was necessary to re-emphasize that our list was illustrative and that the criteria were the important end point of our investigation.

Adjustments to the List

The final issue in this exercise, as it is for any similar exercise, was a reality check. Does the list make any sense at all? For example, did the list capture those chemicals which most experts in the field would generally acknowledge to be of concern for emergency response planning? With respect to this question, it became clear that because the toxicity criteria were so restricted, a number of fairly toxic materials could easily be excluded from our list. In particular, we were most concerned with those relatively toxic materials that were in widespread use in commerce. For example, our pure toxicity criteria did not pick up anhydrous ammonia, which is certainly one of the more important chemicals for emergency response planning, given its widespread use in large quantities. We therefore screened an additional set of chemicals meeting two factors: a somewhat relaxed set of toxicity conditions, which are presented in Table 2, combined with an indication that the chemicals were used and produced extensively based on published information on production capacities and volumes. This procedure is somewhat similar to that used by the EEC in identifying additional chemicals for their own list of hazardous materials.

The end results of this activity were a set of criteria for identifying acutely hazardous chemicals and a list of 402 chemical substances which met either the strict criteria or the combined toxicity and production based criteria. The list, criteria, and rationale were published together with guidance materials on emergency response planning in a document called the Chemical Emergency Preparedness Program Interim Guidance. This document was published in November 1985, and public comments were requested.[4] The document also served as the major focus for the Agency's voluntary program to encourage emergency planning at the state and local levels.

Table 2. Criteria for Other Chemicals
(High Production Volume)

Route	Value
Dermal	LD50 ≤ 400 mg/kg
Oral	LD50 ≤ 200 mg/kg
Inhalation	LD50 ≤ 2 mg/l

TITLE III

During 1986, implementation of the Chemical Emergency Preparedness Program continued and we began work on revising the list of chemicals, the criteria, and the guidance documents, based on comments and our own further evaluation of the data at hand. However, Congress got into the act and before we were able to revise our documents they enacted the Superfund Amendments and Reauthorization Act of 1986 (SARA). Sections 302 and 303 of this Act established a required national program of chemical emergency response planning that is built along the lines of the voluntary program EPA had initiated. In particular, it requires the establishment of state and local emergency planning committees and the development of emergency response plans by October of 1988. These plans are to be developed around the list of extremely hazardous chemicals and facilities which handle those chemicals in greater than certain threshold planning quantities. Congress enacted into the legislation our published list of 402 chemicals as the list of chemicals for planning purposes, and initially established threshold planning quantities of 2 pounds for all chemicals on the list unless EPA proposed alternative threshold quantities within 30 days of enactment.

There are many positive aspects to this program and these requirements. However, it is legitimate to point out that in some respects our worst fears, envisioned when we undertook the development of the criteria and the list, were realized. Our list of chemicals, arrived at very quickly using various quality data, were enacted into a statute, including errors that had been subsequently identified. In particular, about 40 chemicals were on the list because of errors in the data base and related problems. Four of these chemicals have subsequently been removed from the list following lawsuits by affected companies and EPA has announced its intention to remove the remaining 36.

In future listing decisions, the new law directs the Agency to consider both short and long term health effects resulting from short term exposure, as well as physical/chemical hazards, in its decision making.

We did succeed within the 30 days that Congress gave us in developing a methodology for revising the threshold planning quantities for notification from 2 pounds to a categorization scheme ranging ultimately from 1 pound to 10,000 pounds. This scheme was arrived at by combining a toxicity rating based on acute toxicity with a measure of potential for exposure as follows:

$$I = \frac{IDLH^*}{V} ,$$

where

I   = Index of hazard potential,

IDLH* = Immediately Dangerous to Life or Health value or surrogate estimate,

V   = 1 for gases, powders <100 microns
         = f(molecular weight) for liquids.

This measure reflects the physical state of the chemical. For example, the method takes into account the volatility of a liquid in a potential spill situation.[5]

In the final regulation published on April 22, 1987,[6] Threshold Planning Quantities (TPQ) were keyed to the index in the following manner:

Index	TPQ (pounds)
<10-3	1
10-3 to 10-2	10
10-2 to 10-1	100
10-1 to 1	500
1 to 10	1,000
10	10,000

We have also developed revised planning guidance to assist local emergency planning committees in their mandated responsibility to develop emergency response plans by October 17, 1988.[7] These guidance materials include some very simplified release and dispersion modeling tables to assist the planning committees in identifying areas of concern within their communities for potential toxic chemical releases.

Where do we go from here? Well, our experience to date and the language of Title III indicate that there are two areas that we need to address as priorities for hazard and risk assessments under the emergency planning provisions: non-acute effects and planning level of concern. First, we clearly need to begin to make some effort at addressing the issue of long-term effects from accidental, short-term exposures. We are going to convene a workshop over the coming year to work on specific areas where we believe there is a potential for making some progress. It will probably be some time, however, before we can modify our criteria to include considerations other than the present acute lethality considerations.

Second, we need to provide more help to communities preparing response plans. In particular, communities are trying to identify zones of concern around facilities. These zones can be considered areas of potential evacuation in the event of an accidental release. In essence, the process of identifying these zones involves combining various release and dispersion modeling methodologies to predict off-site concentrations. These concentrations are then compared with a level of concern that indicates concentration levels above which serious health effects could occur. Frankly, very little appropriate work is available in this area. We have relied primarily on the Immediately Dangerous to Life and Health value of the National Institute of Occupational Safety and Health (NIOSH) or on surrogate estimates of that value for setting levels of concern. These were developed originally by NIOSH for work place situations to identify levels above which workers could not be expected to survive without serious health effects over a 30 minute exposure. We used these numbers and an arbitrary factor of safety primarily because they have been established for a fairly large number of chemicals using a relatively consistent methodology. However, we recognize that this approach is not necessarily well suited to the needs of protecting general populations as opposed to healthy male workers.

There is considerable interest in the area on the part of both industry and government organizations and we will shortly be convening an experts workshop in this area.

REFERENCES

1. U.S. Congress. House. Superfund Amendments and Reauthorization Act of 1986. H. Report 99-962 to accompany H.R. 2005, 99th Congress, 2d sess., 1986.
2. European Economic Community, "On the Major Accident Hazards of Certain Industrial Activities," Directive 82/501/EEC, Official Journal of the European Community, L230, 1982.
3. National Academy of Sciences, *Toxicity Testing: Strategies to Determine Needs and Priorities*, Washington, DC, National Academy of Sciences, 1984.
4. U.S. Environmental Protection Agency, *Chemical Emergency Preparedness Program Interim Guidance*, Washington, DC, 1985.
5. U.S. Environmental Protection Agency, *Threshold Planning Quantities: Technical Support Document for SARA Title III, Section 302 Regulations*, Washington, DC, Office of Toxic Substances, 1987.
6. U.S. Environmental Protection Agency, Federal Register 52(77), pp. 13378-13410, April 22, 1987.
7. U.S. Environmental Protection Agency, *Emergency Planning Technical Guidance Document*, Draft, Washington, DC, 1987.
8. U.S. Environmental Protection Agency, Federal Register 52(107), pp. 21152-21208, June 4, 1987.

System Unavailability Monitoring Study

Mohsen Sharirli

Advanced Technology Engineering Systems, Inc.
Reston, VA

ABSTRACT

The purpose of this study is to analyze the methodologies and applications of various techniques for performing a System Unavailability Monitoring Study. This study includes the development, analysis, and evaluation of some innovative techniques in establishing system unavailability target and trend indicators. The analysis was performed for the Emergency Feedwater System (EFS) of a nuclear power plant. System models, industry-average component unavailabilities, and plant-specific component unavailabilities were used to calculate both the target unavailability and the actual unavailability of the system. This provided information on the marginal risk, reliability, unavailability, and system performance indicators for a system unavailability monitoring program. Some important enhancements contributed by this study include the establishment of unavailability targets and trend indicators, results for evaluation of overall system performance, means for comparisons of various methods of analysis, and identification of areas where changes, if any, can be implemented and the associated marginal change in the system performance and plant risk can be determined.

KEYWORDS: Probabilistic risk assessment (PRA), failure modes and effects analysis (FMEA), criticality analysis (CA), time series analysis (TSA), unavailability analysis

INTRODUCTION

The Probabilistic Risk Assessment (PRA) is a powerful tool for performing system analysis. In system analysis (e.g., system operability studies, reliability analysis, maintainability analysis), a variety of techniques are available to support these tasks. Some of these techniques include Failure Modes and Effects Analysis (FMEA), Criticality Analysis (CA), Time Series Analysis (TSA), and Fault Tree Analysis (FTA).

The FMEA uses an inductive method to perform a qualitative analysis to identify various failure modes of components or systems. This method considers one failure at a time, as opposed to combinations of failures or human factors in the failure cause. If the failure rates of the components are known, the CA follows an FMEA to show the quantitative effect of each component failure on the system. This analysis frequently does not take into account human factors, common cause failures, and system interactions. In addition, the results lack mathematical strength, and the analysis cannot be carried out for large systems.

The TSA uses operating experience for the system and provides qualitative and quantitative information regarding components and system operation. The TSA also provides information to identify the areas in which frequent problems appear and also the corrective actions necessary to control or eliminate the identified problems.

The FTA has been used in areas such as system operations, safety studies, and reliability analysis. This method can provide qualitative and quantitative information about systems and components.

All of these methods pertain to system operation, and the system unavailability monitoring study has been designed to integrate them. The information gained and the results obtained will be used by operators, managers, and engineers in supporting various decision-making processes. They will be used to establish unavailability targets and trend indicators in system unavailability analysis in areas such as system/design modification, replacement of unreliable equipment, testing maintenance optimization, and cost/benefit analysis.

In PRAs, generally using the most probable failure rates, most probable repair times, and various other data available, the system unavailability is determined. The system unavailabilities obtained are the best estimates of the probability that the system will be unavailable at some future time. Such results, although helpful in providing much information, because of their shortcomings cannot be used in other areas of system analysis, e.g., system unavailability monitoring and trending for a certain length of time. The method discussed in this study reflects actual component and system unavailability (using actual plant-specific component unavailability, rather than an estimate of most probable component unavailability) during a specific period of time (for the years 1984, 1985, 1986, and for an average of the three years). The results are used as a performance indicator in order to implement various monitoring and trend analyses in actual system operation.

SYSTEM DESCRIPTION

The Emergency Feedwater System (EFS) of a commercial nuclear power plant was selected for this study. A reduced system model was derived using mathematical judgments and assumptions.

Major Assumptions

Various assumptions were made in this study, primarily in the areas of pipe failure modes, operator actions, dependent or common-cause unavailability, etc. The EFS simplified diagram is shown in Fig. 1.

System Success Criteria

The primary safety function of the EFS is to provide cooling for decay-heat removal following a loss of the Main Feedwater System. The EFS must perform this function until the plant can be safely placed in cold shutdown. The EFS unavailability may be calculated depending on the mode of operation of the system. The success criterion is: The EFS must be able to supply water to at least one of the two steam generators to successfully cool down the plant following a loss of main feedwater. The system analysis has been performed on the basis that the EFS is unavailable when full flow is unavailable to both once-through steam generators, OTSG A and OTSG B.

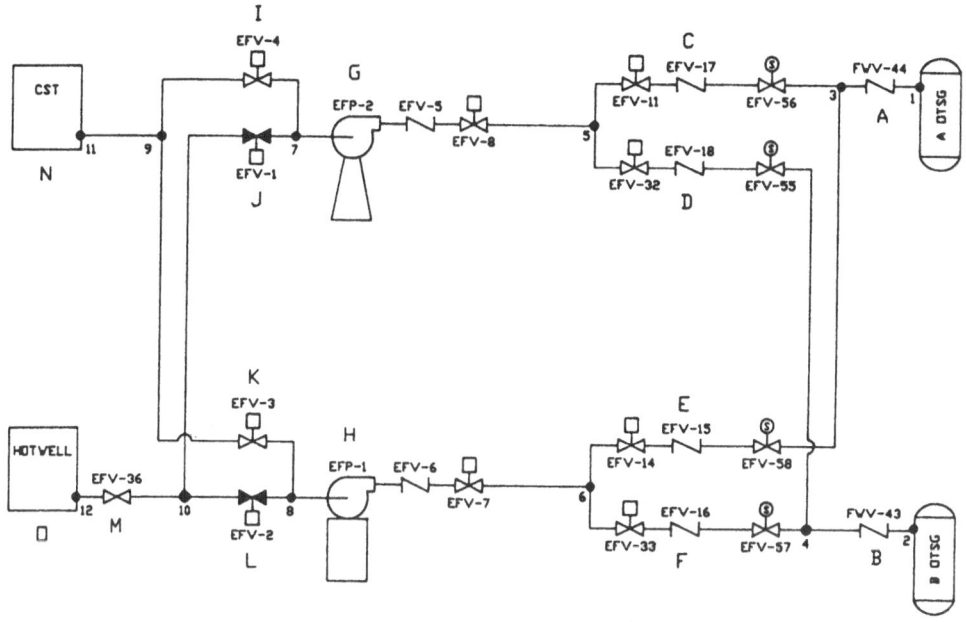

Fig. 1. Simplified Emergency Feedwater System Diagram.

SYSTEM ANALYSIS

A system analysis was performed using system fault tree model and component unavailability data. The system model development is performed using the fault tree analysis. It includes two cases of block model (using the system blocks) and component model (using the system components). The block model serves two purposes. First, it is used to verify the results obtained using the component model. Second, it provides various "lumped" information, such as system trains, when required.

The data sources have been two areas of plant-specific data and the average industry data. Plant-specific data analysis requires the use of many sources of information and records available for a system. Some of these are Work Request Orders, Licensee Event Reports, Preventive Maintenance Control Reports, Field Problem Reports, Surveillance Procedures, and Control Room Logs. In addition, all other information available from various processed records which may significantly aid the analysis, such as reactor shutdowns, is important. The average industry data are selected using the tabularized data analyzed for various components for nuclear power plants. The system analysis and computations were accomplished using hand calculations and computers. The plant-specific data were analyzed for the three years 1984, 1985, and 1986, and for the three-year average. The various areas of calculations include qualitative analysis, quantitative analysis, post-processing study, and other related issues for each case of this study. Qualitative analysis was used to generate various cut sets. Quantitative analysis was used to obtain the cut sets unavailabilities and the system unavailabilities. A sensitivity analysis was performed by extracting various information and generating some useful new areas by which the system unavailability improvements can be performed. This includes design component identification, component high unavailability improvements, and optimization in the unavailability improvements.

There are many areas for sensitivity analysis. For this study, the criticality importance has been selected. This decision was based on various issues. First, the criticality considers the fact that it is more difficult to improve the components with lower unavailability than to improve the components with higher unavailability. This fact is a very important issue in the unavailability analysis. Second, the criticality is closely related and can be followed easily by those familiar with the Vesely-Fussell importance. Also, the results are closely comparable. Finally, the calculations performed are straightforward and do not require complex manipulations, which makes an analysis more manageable and convenient to perform.

There are generally three types of analyses for the system. The *system target unavailability* analysis is performed using the system model and the industry average component unavailabilities. The *system actual averaged unavailability* analysis is done using the system model and the components actual average unavailabilities. This was performed by an averaging process for unavailabilities of similar components. The *system actual unaveraged unavailability* analysis is accomplished by using the system model and the components actual unavailability.

RESULTS

Table 1 provides information regarding the analysis for each of the years 1984, 1985, 1986, and for an average of the three years. This table also provides information for analysis of the target cases. Information from Table 1 was used to plot the cases considered in Fig. 2.

All the results obtained from the criticality importance calculations cannot be reported here, but the analyses of some important items are given as follows:

(1) The actual unaveraged unavailabilities obtained for 1984 (1.365E-5), 1986 (4.208E-6), and the three-year average (7.660E-5) are much less than the target unavailability (1.35E-4). The actual unavailability for 1985 (1.485E-4) was slightly higher than the target unavailability. For 1984, the results revealed that the two components, turbine-driven pump EFP-2 and motor-driven pump EFP-1 (the trains with the two pumps for the case of the block model), have a high importance (each with a value of 0.9997), while others have zero criticality values. For 1985, turbine-driven pump EFP-2 showed the highest importance in the system. The motor-driven pump resulted in a much lower criticality value (trains containing turbine-driven pump and motor-driven pump for the case of the block model). These two pumps were followed by motor-driven valves EFV-8, EFV-11, EFV-32, EFV-33, and EFV-14, and flow control valve EFV-58. For 1986, turbine-driven pump EFP-2 and motor-driven pump EFP-1 (the two pump trains for the block model) showed a high criticality of 0.9999. For the three-year average, turbine-driven pump EFP-2 had the highest criticality value, followed by motor-driven pump EFP-1 and check valve EFV-6. The importance reduced substantially for the remaining components: motor-driven valves EFV-8, EFV-32, EFV-14, and EFV-32; and flow control valves EFV-57 and EFV-58, respectively.

(2) The actual averaged unavailabilities obtained for 1984 (1.413E-5) and 1986 (4.392E-6) are much lower than the target unavailability (1.35E-4). The actual average unavailability for 1985 (4.289E-4) is much greater than the target unavailability. The actual average unavailability for the three-year average (1.502E-4) is slightly higher than the target unavailability.

(3) For 1984, the results showed that turbine-driven pump EFP-2 has the highest criticality, followed by motor-driven pump EFP-1 (the two pump trains for the case of the block model); and motor-driven valves EFV-7, EFV-8, EFV-14, EFV-33, EFV-11, and

Table 1. Emergency Feedwater System Unavailability Results

Case No.	System Model and Unavailabilities	1984	1985	1986	3 Years
3	EFS Unavailability using the component model and actual "unaveraged" component unavailabilities	1.365E-5	1.485E-4	4.208E-6	7.647E-5
4	EFS Unavailability using the system block model and actual "unaveraged" block unavailabilities	1.365E-5	1.485E-4	4.208E-6	7.660E-5
5	EFS Unavailability using the component model and the actual "averaged" component unavailabilities	1.410E-5	4.234E-4	4.401E-6	1.500E-4
6	EFS Unavailability using the system block model and the actual "averaged" block unavailabilities	1.413E-5	4.289E-4	4.392E-6	1.502E-4
1	EFS Unavailability using the component model and industry average component unavailabilities	Target Unavailability = 1.381E-4			
2	EFS Unavailability using the block model and industry average component-block unavailabilities	Target Unavailability = 1.350E-4			

EFV-32. For 1985, the results showed that turbine-driven pump EFP-2, valves EFV-7 and EFV-8, and motor-driven pump EFP-1 have high criticality values (the two pump trains for the block model). For 1986, pumps EFP-1 and EFP-2 (the two pump trains for the block model) showed a high importance, followed by valves EFV-8, EFV-7, EFV-56, EFV-55, EFV-58, EFV-57, EFV-11, EFV-32, EFV-14, and EFV-33. For the three-year average, the highest criticality ranking includes components EFP-2, EFV-7, EFV-8, and EFP-1 (the block containing two pumps for the block model).

CONCLUSIONS

The preliminary steps in this study, such as development of methodologies, collecting information, and gathering the data, provided a broad spectrum of techniques as well as systems for various analyses. Various processing and analyses provided additional information regarding the methods of analyses and the system under consideration (EFS in this case).

A review of the results identifies the following significant items:

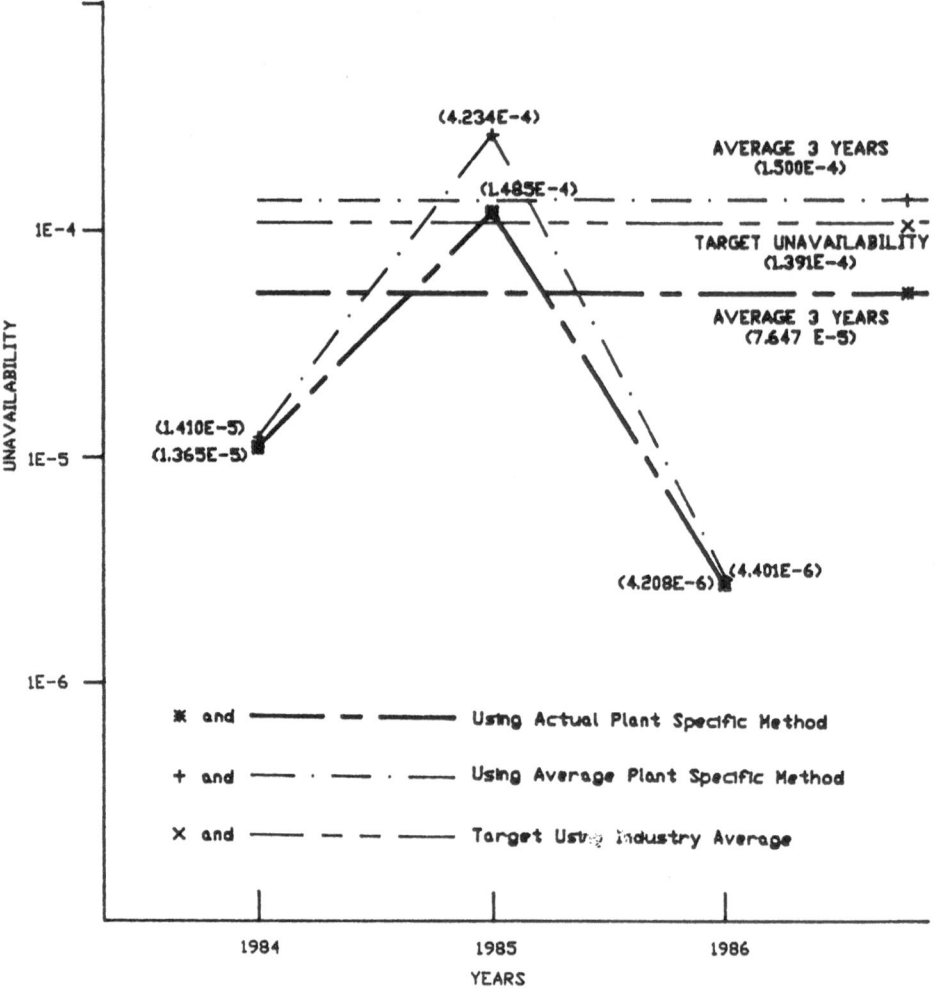

Fig. 2. System Target, Yearly, and Three-Year Unavailability Using Plant Specific, Plant Specific "Average," and Average Industry Component Unavailabilities.

- At two endpoints, 1984 and 1986, the two unavailability values (using the actual unaveraged plant-specific method and using the averaged plant-specific method) are in close agreement and are substantially below the system target unavailability.

- For the three-year average, the unavailability obtained using the actual unaveraged plant-specific method is much lower than the target unavailability, while the unavailability obtained using the averaged plant-specific method is much higher than the target unavailability.

- For the year 1985, both unavailabilities using the two methods are higher than the target unavailability. While using the actual unaveraged plant-specific method resulted in an unavailability very close to the target value, the averaged plant-specific method resulted in unavailability of approximately three orders of magnitude higher than the target unavailability.

A review of the analyses performed and the results obtained leads to the following conclusions:

- For all cases of study, emergency feedwater pumps EFP-2 and EFP-1 have been shown to be the highest unavailability contributors to the system unavailability. Similar conclusions were made when the analyses were performed on the blocks.

- The EFS unavailability obtained using the various methods provided Fig. 2 with interesting information and features. First, using the component average plant-specific method provided higher system unavailability than using the actual component plant-specific method. This unavailability difference (deviation) was higher for 1985 than for other years. This is because more component unavailability data are available for 1985 than for the other years. The justification for the actual component unavailability averaging has been the scarcity of data available. This reveals two items: first, when there is not a great deal of data available for analysis, the results obtained are almost the same as when using actual component "unaveraged" unavailabilities. Therefore, it will not help the results by any means. Second, when there are adequate data available, such averaging would result in non-actual, and, in fact, erroneous, system unavailability. In this study for EFS, the error is by a factor of approximately 3 for the year 1985. This fact is also shown when using the average three-year analysis (error is by a factor of approximately 2). If the objectives are to identify the components that have no unavailability associated with them, this can be performed by various means, such as using other data sources, data treatment, or other processing of the system analysis.

- The results provided various information for the component rankings in sensitivity analysis. For the actual component and block unaveraged unavailabilities, the results identified components or blocks (trains) of importance which would require more attention than others. Also, items in which it is not advisable to invest as much attention as in the previous ones are identified. Similar results are also provided when using the industry average component or block unavailabilities. When making the importance calculations using the actual component or block averaged unavailability method, the information—although interesting—is not as appropriate or accurate as that obtained earlier. Here by averaging we are reducing the unavailability of a component and creating an unavailability for another component which normally has no unavailability. By doing so, we are assigning a criticality importance to this component (with no unavailability) and reducing the actual importance of the other one. Although the results would provide some information to the analyst, they will not contribute more than that information learned previously or that available for further studies. Therefore, this method is not recommended; or one must keep in mind all the items considered in this study.

This study was performed to serve many purposes. It was designed to analyze a system, in this case the Emergency Feedwater System. It also served to identify some new methods in systems analysis. Finally, it served to implement and analyze these methods, their comparisons, and their applications. Depending on the purpose, the objectives, and the cases of the analysis, one or multiple methods selected here may be used for other studies.

The Influence of the Climate of Risk Perception on the Process of Risk Management: Quantification, Evaluation, Decision Making

R. M. Stern

World Health Organization Regional Office For Europe
Copenhagen Ø, Denmark

ABSTRACT

The machinery of risk management (quantification (Q), Evaluation (E), Decision making (D)) is set in motion at the instant of perception of risk, and proceeds in a climate of risk perception which has such strong local variations that various local QED processes can lead to very different management decisions for identical risks, e.g., the *de minimis* approach to carcinogens in Scandinavia, point estimates for carcinogenic risk in the USA, and toxic substance control limits which do not distinguish carcinogens in the UK. Similarly, in any one locality, variation in risk perception between the various self-interest groups involved in the QED process, e.g., scientists, the courts, individual workers, unions, management, regulatory agencies, and the media/public sector, results in very different bargaining positions for each participant. Recently, in Denmark, a cluster of five cases of laryngeal cancer, coupled with positive Ames test results for stainless steel welding fumes (which contain soluble $Cr\ (VI)$) has resulted in rule-making, not open to public debate and largely ignored, identifying stainless steel welding as a potential carcinogenic exposure. The economic recession, however, has subsequently led to agreement on part of the unions, management, and the regulatory bodies, with respect to jointly controlled research funds, only to promote occupational health research which results in reduction of known, unquantified risks, and to discourage investigative studies designed to identify new risks or to quantify old ones.

KEYWORDS: Risk perception, risk management, risk assessment, welders

The process of risk management which consists of a non-structured sequence of overlapping activities (quantification (Q), evaluation (E), decision making (D)) (Fig. 1) is set into action only upon the perception of risk, although a particular risk may have existed for a long period of time before its "discovery."[3] Little attention has been paid to the importance that the local climate of risk perception plays in risk management. Examination of each of the steps will disclose how the initial perception of risk sets the stage for the development of a (sometimes conflicting[6]) perception of the same risk on the part of each of the special interest groups who play specific roles in the management process; it is this collection of perceptions and their interplay which then influences the outcome of each step. The initial perception may arise in many different ways: (1) among workers on the job

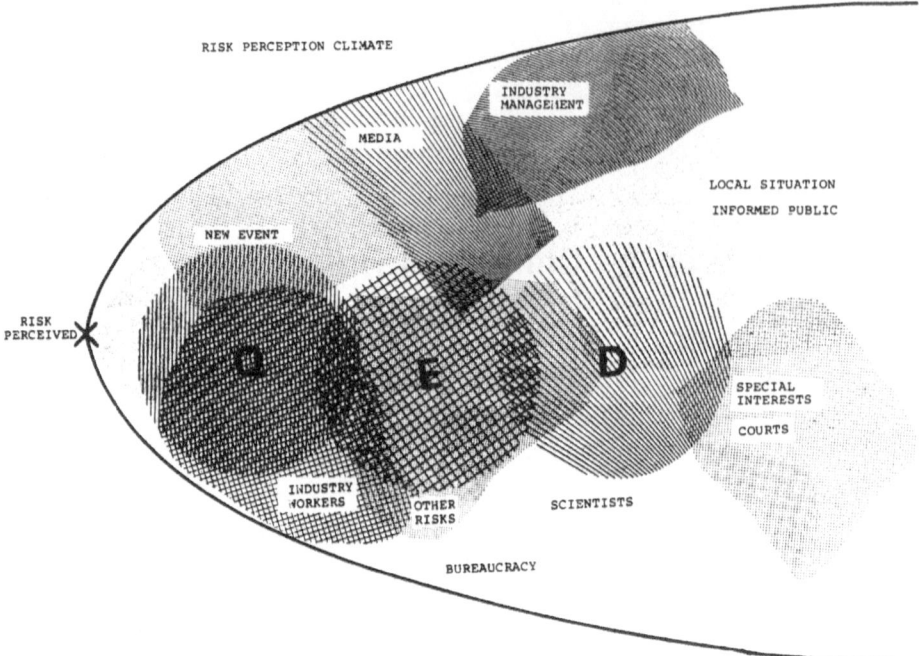

RISK PERCEPTION CLIMATE

INDUSTRY MANAGEMENT

MEDIA

LOCAL SITUATION

INFORMED PUBLIC

NEW EVENT

RISK PERCEIVED

Q E D

SPECIAL INTERESTS

COURTS

INDUSTRY WORKERS

OTHER RISKS

SCIENTISTS

BUREAUCRACY

Fig. 1. The 'QED' Machinery Operates Against a Background of Risk Perception, Which in Turn Alters the Way in Which Each Part Functions. The local details significantly affect the management process and its outcome.

site, e.g., the smell of chlorine, a traumatic accident, etc., in which case it carries the onus of urgency and the need for remedy; (2) among the plant management based either on events off site or on an evaluation of on site loss statistics, whereupon it will be subjected to the local QED machinery without public involvement; (3) in a regulatory body in the course of development of criteria documents, setting in motion the formal QED machinery which includes public hearings; (4) on the part of an activist individual or group, in which case the public climate will be strongly biased towards the protection of the self interest of the constituency; (5) simultaneously at all levels because of some well publicized initiating event, in which case all participants are sensitized to an extent depending on their personal experience or association therewith.

It is assumed that the need for risk management arises from a mandated task of protecting workers' health from unnecessary risks: formally on the part of appropriate regulatory agencies and informally on the part of individual enterprises. The task of the Q machinery is to provide information as to the nature, source and magnitude of risk, the latter at best in absolute terms of risk per unit exposure, and at worst in terms of a relative evaluation with respect to an existing data base of quantitative or semi-quantitative risks for acts or exposures. Individual members of the Q machinery may form special interest groups based on risk perception, e.g., proponents of linear dose-response models or of the utility of *in vitro* tests, etc. The task of the E machinery is to perform an evaluation of the output from the Q step, primarily in terms of establishing priorities for either remedy or legislation. In the absence of uncontested quantitative risk assessment, this evaluation must always contain a strong contribution from risk perception. The D process is one of deciding how to administrate the risk whose magnitude, urgency, and significance have been defined by the E output. The D machinery can range in composition from an individual administrator with

388

absolute powers, through a tripartite committee with members from management, trade unions and government, to a local referendum, each being sensitive to and carrying along specific risk perceptions which will affect the outcome of the management process.

Frequently, only the end product, typically an occupational or environmental exposure limit or concentration, survives after the risk management machinery has completed operation. In such cases it is extremely difficult to introduce new information, e.g., with respect to thresholds and the shape of the dose-response curve, *in vivo* metabolism, new exposure data, etc., or to re-examine the criteria used in priority setting within the risk inventory on hand, so that the scientific and political basis for decision making and the regulatory aspect of risk management are not self renewing. This disfunction of the risk management process is especially unfortunate in the occupational setting, since it can lead to the establishing of exposure limits which do not serve their intended purpose, either permitting workers to remain exposed at risk where the set levels should be reduced, or imposing undue economic burdens by forcing enterprises to reduce exposure levels unnecessarily.

Similarly, with only the legislated limit on hand, it is usually impossible to determine the extent to which perception has played a role in subjective risk management. Such influence only can be analyzed in the case of legal challenge, wherein the QED process can, with unforeseen effect, be replayed in a court of law.[1]

The case history of risk management for welders is instructive since it has been driven by risk perception for at least 40 years. Welders are exposed to high concentrations of potentially toxic fumes and gases, ultraviolet and infrared and electromagnetic radiation, noise, projectiles from grinding operations, electric potentials, uncomfortable, exposed and potentially dangerous working places, etc. (reviewed in Ref. 4) leading to a perception of risk on the part of individual welders, unions, management, and government. While research has provided extensive documentation for exposures to the most obvious source of risk (fumes) and cross-sectional clinical studies of lung function have been made on innumerable welding cohorts, evidence for absolute risk levels is extremely sparse. Very recent studies have suggested that acute and chronic effects of fumes are not as important as accidents and ergonomy in terms of medical wastage,[8] and few absolute data are available on delayed health effects beyond an indication of a small and possibly work related excess incidence of lung cancer.[5]

As a result, a complex industry producing personal protection equipment and ventilation has developed worldwide, industries in many countries have introduced systems of electrode classification according to ventilation requirements, and several companies now offer welding electrodes and gases which replace one toxic substance with another (e.g., hexavalent chromium with lithium, and ozone with nitrogen dioxide), in order to improve effective classification. Thus, the Q machinery has never been brought into wide operation (beyond demonstrations of *in vitro* genotoxicity of stainless steel welding fumes as an indication of potential cancer risk[2]), the E process has used the intensity of perceived risk (e.g., fatal electrical accidents) to set priorities, and the QED machinery has been very active within the industry, responding to market place needs based on perceived risk, while the welding problem has escaped perception at the regulatory level in most countries. Only a few have formal legislation requiring, e.g., eye protection and open-circuit voltage regulators, and specifically list upper exposure levels, or have developed criteria documents for welding fumes.

It is suggested to improve the occupational health risk management process by taking advantage of the sensitivity of the system to risk perception by introducing activities which ensure that appropriate concepts of risk are developed by all participants, and that the QED machinery explicitly consider the importance of perceived risk. Workers can be educated to develop their risk perception along objective lines, and in particular to understand the

389

meaning of variations in absolute risk. Management should be encouraged to develop an inventory of absolute and perceived risks so as to ensure that a feeling of "well-being" is part of workers' health and safety,[7] and local administrations can support public education in terms of personal risk management, so that the public can compare the potential effects of the imposed risk of the workplace with the self-inflicted risks of lifestyle.

REFERENCES

1. Jasinoff, S. and Nelkin, D., "Science and Technology and the Limits of Judicial Competence," *Science* **214**:1211-1215 (1981).
2. Hansen, K. and Stern, R. M., "Nickel and Chromium Compounds and Welding Fumes in Mammalian Cell Transformation Bioassay *in vitro*," in Health Hazards and Biological Effects of Welding Fumes and Gases, Excerpta Medica ICS 676, pp. 305-310, R. M. Stern *et al.*, Eds., Amsterdam, 1986.
3. Stern, R. M., "Analysis of the Decision Making Process in Chemical Safety," *Science of the Total Environ.* **51**:27-62 (1986).
4. Stern, R. M., Berlin, A., Fletcher, A. C. and Järvisalo, J., Eds., "Health Hazards and Biological Activity of Welding Fumes and Gases," Excerpta Medica ICS 676, 596 pp., Amsterdam, 1986.
5. Stern, R. M., "Management of Risk: Application to the Welding Industry," *Risk Analysis* **5**:63-72 (1985).
6. Von Winterfeldt, D. and Edwards, W., "Patterns of Conflict About Risky Technologies," *Risk Analysis* **4**:55-68 (1984).
7. Zielhuis, R. L. an Notten, W. R. F., "Permissible Levels for Occupational Exposure," *Int. Arch. Occup. Environ. Health* **41**:269-281 (1979).
8. Zielhuis, R. L. and Wanders, S.P., "Health Effects and Medical Wastage Due to Combined Exposure in Welding," in Health Hazards and Biological Effects of Welding Fumes and Gases, Excerpta Medica ICS 676, pp. 497-534, R. M. Stern *et al.*, Eds., Amsterdam, 1986.

The Nation-State in Societal Evaluation of Risk: Ambiguity Meets Reality

Ingar Palmlund

University of Stockholm
Stockholm, Sweden

ABSTRACT

Social, cultural and economical dimensions in the politics of societies can be more important than scientific and technological evidence in societal decision-making to protect human health and environment from technological risk.

Three cases are presented to illustrate this thesis. They concern estrogens as growth promotants in animals in the U.S.A., national guidelines for radioactive food intake in Sweden in 1985-1987 after the Chernobyl accident, and national decision-making on the risks of oral contraceptives in the U.S.A. and Sweden in 1960-1980.

KEYWORDS: U.S.A., Sweden, DES, radiation, oral contraceptives, Chernobyl

INTRODUCTION

Nation-states are organizations we all live in. Depending on the type of society we live in and our fundamental values, we have different assumptions of what nations are set up to do, what they should do, how they should do it, and what relationship they should have to society at large.

Our assumptions—and expectations—of the nation are often contradictory. Many of us assume that a nation should act to protect its citizens so that they can live a good life. When we are confronted with threats to human health and environment that we perceive to be part of the way society is organized, we make demands on the nation to institute protection. We also do it when foreign powers threaten our territories. And during recent decades we have increased our expectations on the nation to intervene when we feel threatened by the way technology is used in society.

At the same time, we may have other assumptions on what the nation should be or should do which starkly contradict our expectations concerning the national protection of human health and environment against technological hazards.

There are roughly three types of decisions made in the name of nations:

1. decisions on the form and context of legislation,

2. regulatory decisions on norms that should be observed,

3. decisions in specific cases.

The order reveals something about the stability of the decision. A decision on legislation is difficult to change. Many of the rules in constitutions concern procedures which aim at making it difficult to change laws once they are decided upon. Generally, it is easier to change regulations, and even easier to change decisions in specific cases.

In most countries, laws regarding the protection of human health and environment against technological risk are frames which designate governmental authorities that should issue regulations and take decisions in specific cases within the scope of action stated in the law. The reasons are obvious: scientific and technical knowledge has a great impact on the decisions; the understanding of the causes of technological risks to human health and environment changes rapidly; there is a notion that the precise and rational methods of work used in moving the frontiers of science and technology will provide knowledge of the "correct," the "acceptable," or at least the "appropriate" level of risk. The arguments over risk are presented in scientific terminology. The decisions over risk are legitimated by scientific documentation.

I want to make five propositions concerning the decisions taken in the name of a nation on technological risks to human health and environment:

1. Social and cultural factors may have a greater influence on decisions on technological risk than do science and technology.

2. Regulatory decisions taken in the name of the nation are unstable over time. Not only do they change, they also change in different directions.

3. Decisions on risk are taken in conflictual contexts. The constellations of actors in the decisions on risk essentially define the contents of the decision.

4. The protection of human health and environment is only one set of objectives that influences the decisions.

5. The participation of the public in policy formation is important.

SOCIAL FACTORS IN NATIONAL DECISON-MAKING ON TECHNOLOGICAL RISK: THREE CASES

I will present three cases to illustrate these points. They are drawn from decisions in the U.S.A. and Sweden on technological risks to human health and environment. They are as follows:

1. National decision-making on the risks of estrogens as growth promotants in animals in the U.S.A.

2. National guidelines for radioactive food intake in Sweden 1985-1987 before and after the Chernobyl accident,

3. National decision-making on the risks of oral contraceptives in the U.S.A. and Sweden 1960-1986.

Each case illustrates one facet of the role of nation states in the societal evaluation of risk. For each case I will present a timetable showing some major decisions and point out pertinent implications.

Case I. Estrogens as growth promotants in animals in the U.S.A.

Estrogens are cost-effective growth promotants in animals, since they reduce the amount of food needed to bring the animal up to appropriate slaughter weight. They are also said to make the meat more tender. They have been used in poultry, sheep and cattle.

Diethylstilbestrol (DES) has been the most common estrogen in use for this purpose. In the U.S.A. in the early 1970s, DES was associated with the development of clear-cell adenocarcinoma in young women whose mothers had taken DES during pregnancy. This created a surge of fear and of pity for the victims. It added fuel to the public debate over the hazards of DES as a growth promotant in animals.

Estrogens have been suspected carcinogens since the 1930s. The International Agency for Cancer Research (IARC) in 1974 and 1979 made evaluations of the carcinogenicity of DES. In 1979 the group of experts concluded: "Diethylstilbestrol is casually associated with the occurrence of cancer in humans. There is also *sufficient evidence* for its carcinogenicity in experimental animals." The group of experts also pointed out that DES may be no different from other estrogens as regards carcinogenicity under certain circumstances.

Hormones have been forbidden as growth promotants in animals in several countries.

The national decision-making process in the U.S.A. has taken several decades:

1958	The Delaney amendment ("... no additive shall be deemed safe if it is found to induce cancer when ingested by man or animal...) to the Food, Drug, and Cosmetic Act.
1959	FDA proposed ban of DES in poultry.
1962	The "DES clause" (exempting DES from the Delaney amendment, if the Secretary of the Department of Health, Education, and Welfare was "reasonably certain that no additives would appear in food").
1966	FDA's ban of DES in poultry upheld by court.
1971-1972	Congressional Hearings re DES as growth promotant in cattle and sheep. Bills in Congress to legislate ban of DES.
1971	FDA Commissioner announced action to ban DES. Opposition by pharmaceutical industry and cattle raisers.
1974	FDA ordered by court to prepare guidelines for cattle feeders to resume use of DES.
1976-1979	FDA goes through the procedures to ban DES.
Nov. 1, 1979	The use of DES as a growth promotant in slaughter animals becomes illegal.

1980-1983	Illegal use of DES reported.
1983,1986	FDA approves the use of i.a. estradiol benzoate in implants for promoting growth in cattle.
1986-1987	U.S.A. threatens trade reprisals against the EEC worth up to 120 million British pounds in lost trade between Europe and the U.S.A. if the European Economic Community directive prohibiting the use of hormones as growth promotants in farm animals and import of meat containing hormones after January 1988 goes through.

Case II. National guidelines for radioactive food intake in Sweden in 1985-1987 after the Chernobyl accident

During the 1970s, Sweden went through a social and political crisis over the risks of nuclear energy. The conflicts led to a change of government after some thirty years of rule by the Social Democratic Party. Another result was very stringent public standards on radiation protection.

The Swedish guidelines in 1985 for radioactive food intake stated that sensitive persons should not be exposed to more than 0.1 mSv/year.

On April 28, 1986, elevated levels of radioactivity were measured at two Swedish nuclear plants. Suspicions that the radioactive fall-out came from the U.S.S.R. were soon confirmed. A nuclear reactor at Chernobyl in the U.S.S.R. had exploded.

"No Danger in Sweden" were the newspaper headlines when the news first became known. Within a few days it became apparent, however, that winds and rains had deposited radioactive particles in Sweden. Growing grass and vegetables and milk and meat from animals were identified as potential hazards for humans. The public had to learn the names of new measures: Becquerel (Bq) and milliSievert (mSv). A maximum radiation level was established for the food sold on the market. Food at a higher level of radiation than 300 Bq/kg was not allowed on the market.

The fallout was particularly heavy in some areas in Northern Sweden, where the nomad Lapp population lives. They depend on the raising of reindeer for food and income. Approximately 50,000 reindeer were slaughtered in 1986, but were not allowed on the market because of excessively high levels of Bq. The national government had to compensate the Lapps economically for the loss of income. (Cynical people made jokes: Let's eat the reindeer meat, and we'll spit out the Becquerels.)

In September 1986, the Directors of the Radiation Protection Institutes in the Nordic countries issued a statement saying that it is desirable to assure that an individual in the long run (50 years) receives a dosage of radiation from food of no more than 500 mSv. Actions to avoid a radiation dosage of 50 mSv in the long run should not be undertaken if they were perceived to have major socioeconomic consequences.

In October 1986 a group of medical experts on radiation hazards published an article in the major daily paper *Dagens Nyheter*. "The Lapps are sacrificed without factual reasons," they wrote and argued that the additional radiation the public could receive by eating reindeer meat would not constitute a relevant risk. The 300 Bq/kg for food could be raised to 3,000 Bq/kg without creating a radiation dosage that would be unacceptable from a medical point of view. Moreover, the International Commission for Radiation Protection (ICRP) had recommended that after events like the Chernobyl accident, the added radiation to the background radiation level should not be allowed to be higher than 5-50 mSv/year, corresponding to 200,000 — 2 million Bq/year.

The European community, which previously had been a major export market for Swedish reindeer meat, had established 600 Bq/kg as a maximum acceptable level in food for sale.

In March 1987, the Swedish National Radiation Protection Institute recommended a change in the maximum levels of Bq for food, considerably higher than that enforced. Two months later the Swedish National Food Agency adjusted the level for certain foods (including reindeer meat) to 1,500 Bq/kg. 300 Bq/kg was retained as a maximum level for basic foods. The Lapp Association resisted the change, arguing that people might not want to buy their meat if it had a higher level of Bq than other types of meat.

Official guidelines for the intake of radioactive foods were changed as follows during 1985-1987 (figures within parenthesis are recommended values):

1985	0.1 mSv/year	
1986, May	5 mSv/year	300 Bq/kg (Cs-137). The sale of food > 300 Bq/kg forbidden. The public is advised that it can eat food containing up to 10,000 Bq/kg if it is not consumed often.
(1986, Sept.	<500 mSv/50 years)	
(1986, Oct.	<5-50 mSv/year added to background	< 200,000 - 2,000,000 Bq/year)
(1987, March	1 mSv/year	50,000 - 500,000 Bq/kg (Cs-137) (or 10,000 Bq/kg for sensitive persons)
1987, May		300 Bq/kg basic foods 1,500 Bq/kg other foods

The figures in the table should be compared to the estimated average intake of radioactive food per person in 1986 after the Chernobyl accident, calculated as 0.1-0.3 mSv/year. It was estimated that approximately 90 percent of all Swedes get less than 2,000 Becquerel per year via the food. This can be compared with the estimate that the average Swede takes in 120,000 Bq cesium-137 every year via radiation from the ground and in buildings. It has also been estimated that 8,000 dwellings have a level of radiation comparable to a yearly intake of 2,250,000 Bq.

Case III. National decision-making on the risks of oral contraceptives in the U.S.A. and Sweden in 1960-1980

Oral contraceptives were approved for sale on the U.S. market in 1960 and on the Swedish market in 1964. The premarketing trials would later be denounced as inadequate. The laws to control drugs in the two countries are quite similar. The governmental agencies in charge of the control of drugs in the U.S.A. and Sweden — the Food and Drug Administration (FDA) and Socialstyrelsen (SoS) — have over the years developed a collaboration, not least in the way of exchange of information. The information on which their decisions are based is, moreover, published in medical journals available and read in both countries. Yet the contents and the timing of decisions to protect consumers have been different in the two countries. Generally, the FDA has gone further than the SoS in requirements on information to prescribing physicians regarding the risks of oral contraceptives. The U.S. agency has also gone further than its Swedish counterpart in requiring that patients should be informed of the risks of oral contraceptives via material inserted in the packages sold to them. Decisions to intervene in the market to protect the consumers of oral contraceptives have also been taken earlier in the U.S.A. than in Sweden.

Differences in national politics as regards the same type of risk reveal differences in social cultures and priorities. Agencies set up for the protection of human health do not always behave as rational actors in pursuing their formal goals, reacting in the same way and at the same time to available scientific evidence. One explanation of the difference between U.S. and Swedish regulative action in this instance may be a difference in culture, manifested in politics on fertility control, between the countries. Another may be the national interest in Sweden to support the domestic pharmaceutical industry, a considerable contributor to the important export earnings of the country.

The contents and timing of some of the decisions to restrict the risks of oral contraceptives (OCs) are as follows:

	USA	Sweden
1960	OCs approved. Reports on deaths from thrombosis. FDA reassures public.	Warning against the use of OCs.
1964		OCs approved.
1966	FDA requires labeling to contain warnings of thromboembolic disease.	
1968		SoS informs physicians of definite relationship between OCs and thrombo-embolic disease.
1969	FDA requires smaller estrogen dosage.	
1970	Sequential OCs withdrawn. Only low estrogen dosage brands allowed on the market. Information leaflets for patients required in packages.	
1973		SoS advises physicians that they should preferably prescribe OCs with low estrogen dosage.
1975	FDA requires warning in labeling that women > 40 who smoke run increased risk of heart infarction if they take OCs	Sequential OCs withdrawn from market.
1980		SoS issues explicit warning to physicians that women who smoke more than one packet of cigarettes a day should not take OCs.

CONCLUDING COMMENTS

The three cases presented here illustrate that the social, cultural and economical dimensions in the politics of societies can be more important than scientific and technological evidence in societal decision making to protect human health and environment from technological risk.

There are many goals in national politics. Production and commerce always have high priority. But when the public perceives something as hazardous and out of proportion to benefits, it demands that society should "organize out" the hazard, regardless of whether the risks are high or low in relation to similar threats. A high degree of public emphasis and demand for participation in national policy making on the protection of human health and environment seems an important factor in national decisions to curtail technological risk by imposing restrictive regulations on the forces in the market. If that is true, the education of the public in matters related to health and environment has strategic political importance.

National guidelines to protect human health and environment founded on scientific assessments are adjusted when their implications threaten production and commerce seriously. Thus, other priorities than the protection of human health and environment are manifested in nations' decision-making on risk.

Study of the politics involved in the societal evaluation of risk reveals two basic choices that nations have to make over and over again:

1. How much should individuals be sacrificed to further the interests of the community?

2. How much priority should be given to production before protection of human health and environment?

When the ambiguity in national goals meets the reality of hazards, the "real" national preferences as regards the allocation of benefits and losses in society are revealed. There are many indications suggesting that the welfare of the community and increases in productivity are perennial priorities.

Comparative studies of national decision making on technological risks over time should be encouraged in order to clarify the nature of the political priorities manifested in legislative and regulatory decisions.

REFERENCES

Andersson, Per, 1987, "Höjd gräns för cesium i vilda bär," *Dagens Nyheter*, August 6.

Andersson, Per, Mats Holmberg and Kjell Nyholm, "Dosbidrag från livsmedel, SSI-rapport 87-06, Stockholm, Statens Strålskyddsinstitut.

Douglas, Mary and Aaron Wildavsky, 1983, *Risk and Culture: An Essay on the Selection of Technological and Environmental Dangers*, University of California Press, Berkeley.

Dagens Nyheter, 1960-1987.

Federal Register.

International Agency of the Research on Cancer, 1979, "IARC Monographs on the Evaluation of the Carcinogenic Risk of Chemicals to Humans," *Sex Hormones (II)* **21**, Lyon:IARC.

Läkartidningen, 1960-1986.

"Projekt Tjernobyl — Långsiktig begränsning av cesiumintag via livsmedel," SSI-rapport 87-16, Stockholm, Stratens Strålskyddsinstitut.

Svenskt riksdagstryck.

U.S. Congressional Record and Hearings.

Lind, Magnus G., Stig A. Larsson, Lars Erik Holm and Lars Gunnar Larsson, 1986, "Samerna offras utan sakskäl," *Dagens Nyheter*, October 19.

Lind, Magnus G., Lars Erik Holm, Stig A. Larsson and Lars Gunnar Larsson, 1986, "Gränsvärdet kan höjas," *Dagens Nyheter*, November 5.

Setting Risk Priorities in Environmental Policy: The Role of Women

Susan Turnquist and June Fessenden-Raden

Cornell University
Ithaca, NY

ABSTRACT

In many episodes of environmental contamination, local citizens' groups have been mobilized and led by women who have little or no prior history of civic activism. Some of these "housewife activists" have gone on to participate in, become leaders in, or even establish state and national organizations. This phenomenon has received little attention, yet it raises such questions as: What is it about local environment threats which moves these women to a social activism which has no precedent in their lives? Compared with other communities with similar problems but different (or no) citizens' groups working to remedy or avert the risks, how do communities with housewife-led groups define the issues and take actions and get results? And how are the results different? Do their efforts serve to legitimize the issues they raise in the eyes of their fellow residents and public officials, or do their strategies or social status work against them? If the latter, how do they overcome the barriers to public legitimization of their concerns? This paper discusses the implications of gender differences for risk assessment, communication and management of environmental contamination; considers the differences in local relations with state and federal agencies when local groups are led by such women; and attempts an answer to the question: Does the existence of housewife-led groups suggest the need for reassessment of risk analysis priorities and strategies?

KEYWORDS: Women, emotions, risk assessment and management, environmental risk
policy

A casual look at recent issues of virtually any newspaper would turn up one or more cases of local protest over an environmental issue such as a solid waste incinerator, the siting of a new landfill, or a hazardous waste site. If an article mentions individuals representing local protest groups, frequently the spokesperson is a woman. Quite a lot of these women have never been activists of any sort before. Who are these women, what draws them now into these positions, and how does their participation promise to change the political agenda of environmental issues and the processes by which environmental risk policy emerges? These questions motivated the project described below.

Some writers have sought to establish a connection between environmentalism and feminism, arguing that a commonality of goals has made these movements natural allies (see Geschwender, 1986; Smith and Sands, 1979; Caldecott and Leland, 1983; and King, 1983). However, this potential alliance does not appear have to have developed. Testing for this connection, Geschwender found that only half of her sample of 45 women environmental activists were willing to label themselves feminists. The rest claimed more traditional viewpoints. Feminism is apparently not the inspiration for women behind the new environmentalism.

Was there any other major common characteristic of the women in Geschwender's survey? Thirty-one of 45 were employed outside their homes, 30 of 41 were over 40 years of age, and 17 of 35 had undergraduate degrees. Characteristics of the respondents and their identification with feminism can be seen in Table 1 (adapted from Geschwender, 1986). It should be noted that women in the sample came from all over the country and had been active for at least several years; at the time of their initial concern and involvement they would have been younger and perhaps in a different employment situation.

Table 1. Characteristics of the Respondents and Their Identification with Feminism

	Total	Feminist	Not Feminist
EMPLOYMENT			
Employed outside home	31	20	11
Not employed outside	14	3	11
	(45)	(23)	(22)
AGE			
Under 40	11	7	4
Over 40	30	12	18
	(41)	(19)	(22)
EDUCATION			
With undergrad degree	17	14	3
No undergraduate degree	18	9	9
	(35)	(23)	(12)

From this table it is clear that women active in environmental movements show a wide variation in education, age, and outside employment; activists cannot by typified as "feminists", or as professionals who honed their skills in college, or as "hysterical housewives." What, then, do these women have in common that leads them to dominate in numbers the growing movement of local efforts against toxic contamination?

We wanted to know if women active in local opposition to toxic waste issues shared our hypothesis that women are the movers and shakers in these issues. We hoped to identify common denominators in their concerns, some themes that suggested ways in which many women might view toxic issues differently than most men. And we believed it was likely that their efforts, guided by strong feelings, could change the terms by which environmental risk policies are formed. Events at Love Canal, spearheaded by a homeowners association led by Lois Gibbs, sparked national attention which eventually led to Superfund. At a local level, many groups are trying equally hard to change minds and protect or improve local environmental conditions. Many, perhaps most, of these groups are led by women. Among these leaders are some who believe this may be a movement.

The four sections of this paper which follow the introduction describe our methodology, then discuss characteristics of and themes raised by women who became local environmental activists, women we have called "proto-activists" in recognition of their dedication to an issue rather than to activism, and their self-education in strategies of activism. A third section discusses organizational issues faced by these women. The paper concludes with a short discussion of some implications of this movement for environmental risk policies.

METHODOLOGY

This research was not designed to show that women are the emerging leaders in local environmental movements, although that question remains to be answered. Rather, building on the observation that there are many women leading local groups who have not otherwise been leaders in the community, we sought to understand their perceptions and motivations, their styles of organizing, and the implications of the strength of their efforts for influencing environmental risk policies. We consider this research to be exploratory and, as such, to be an opportunity to identify issues and develop concepts rather than to test hypotheses.

Our methodology was correspondingly open and informal. We called people we knew and asked them for names of women, and we used the membership list of a statewide coalition of environmental groups for additional names. When we interviewed people, usually over the telephone, we asked for additional names of women who might be willing to be interviewed. We interviewed approximately 30 women. In addition, we drew on previous research in other states by ourselves and by others with whom we've worked, and attended a national conference of leaders and members of grassroots groups working to eliminate or prevent local environmental problems.

One problem early on was whether to limit the type of leaders we wished to interview. We had in mind a definite kind of person, a "housewife activist" such as we had met in various communities and as typified by Lois Gibbs of Love Canal. But we found women leaders in such groups as the League of Women Voters and other politically sophisticated groups which have contributed much to promote a safe environment, but which work on other issues as well. Other women had worked since they were college students on issues which were general rather than local environmental problems. Although we did not exclude such women from our study, most of those whom we interviewed were leaders and members of groups working primarily on a local problems. We did not attempt to limit the sample in terms of professional, employment, educational, or income characteristics, or by membership in other groups of any kind.

After developing an interview format, we made preliminary calls to people in our sample to set up telephone appointments for interviews. Potential respondents were dropped if they were not and never had been active in groups; one person declined to be interviewed due to an impending lawsuit involving the local environmental issue. Interviews lasted for 30 to 60 minutes in most cases, sometimes interrupted and later resumed. In the sections which follow, quotations from these interviews are used or paraphrased in order to let women tell their own stories.

CHARACTERISTICS AND THEMES OF WOMEN AS ENVIRONMENTAL PROTO-ACTIVISTS

Characteristics of individuals' involvement varied, of course, according to the unfolding of events in their communities, but in groups where the issue became a lightning rod for community opinion we saw three trajectories of involvement.

"I got into it because of my kids. I *stayed* in it because I got so angry."

The first is typified by a woman who became involved and whose commitment not only survived the experience but deepened to the point that she became active in regional, state, or even national activity relating to policy on toxic contaminants. Her work on the issue became, in some cases, her occupation, paid or unpaid.

> "I had a lot of guilt trips—either not enough time with my kids, or with the issues."

The second trajectory is that of the woman who got involved and, after the group achieved its major objectives, pulled back from heavy involvement, but with a commitment to return to help when she is needed.

> "Your changing alienates you from husband, family, friends, and you lose intimacy."

The third trajectory of involvement concerns women who become active in a local issue and then withdraw completely, usually before the group had achieved its primary goal. One woman pulled out when the issue got "too hot," as a personal lawsuit was threatened. Others cited pressures within their families to withdraw.

Common Characteristics

Characteristics common to many women who became involved, regardless of the outcome of their involvement, included previous apathy to political activity; their motivation by health concerns, particularly concern for their children's health; their annoyance and frustration in initial efforts to get answers from state and local offices; development of a perception that government and industry made great efforts to cover up a situation; their self-education in environmental and health issues and technical terms; their willingness to ask for help; and their belief that they became more knowledgeable about the specifics of issues than did elected or appointed officials.

> "Just a bunch of neighbors who got together and called others."

> "[I was] apathetic before, [became] concerned for my daughter, then annoyed, frustrated, disillusioned, then angry."

> "No degree, no certificate, just J. Q. Public."

> "When it's your child's health, you've got to find out."

> "Don't want to say to my children, 'I'm sorry that happened to you. We should have moved earlier.'"

Shared Perceptions

Many of them shared the perception that women are the major force in local activism against environmental contamination, and gave these explanations:

> "The women may be into professions, but they're still the nurturers."

> "Men don't get involved until the state steps in, or the site becomes a Superfund site."

> "Women are the first to sense a problem, before they can figure out what the problem is."

"Women are just better organizers."

Contrast this with explanations given by others; one woman whose local activism has grown into a deep commitment to the movement noted:

"At a meeting 3-4 years ago, [a Sierra Club official, male] responded to an observation that so many of the leaders in the toxic waste movement are women, by saying, 'Of course, they don't have anything else to do.'"

This reported comment by a leading individual in a national environmental organization gives a glimpse of the depth of incongruence between views from within bureaucracies, be they government, industry, or even established environmental groups, and views from local leaders who are women.

Attitudes of Others

Local leaders struggle against not only resistance to their efforts, but also commonly face attempts to discredit their work because they are women.

"First we were pooh-poohed as women, then as fear mongers."

"'What are you women hollering about; there's nothing wrong.'" (This reported comment was made even though there were men in the group, and present at the time of this comment.)

"Ministers, black and white, are the worst to deal with, especially those over 35. They pat you on the back and say 'You need to be bedded down.'"

Some women felt pressure to take a back seat from men they expected to be allies:

"Young men *in the movement* have ways to make me feel insecure."

"Men in your own organization—you have to deal with machismo. I'll make a point in a meeting and no one will say anything, then 20 minutes later a man will make the same point and people will say, 'great idea—why didn't we think of that?'"

Many women expected the support, if not the active involvement, of other women, especially women with children. They were surprised and disheartened to find opposition:

"What hurts most is the other mothers in the neighborhood who say, 'We don't want you fighting for our children.'"

"The women are the ones I go home and cry about. The rest of them I'll fight."

Several women thought that their sex did not hinder their eventual success in achieving the group's goals, but that it did take longer because they did not have the built-in credibility of men.

"It goes without saying that society doesn't give women the clout, especially early on; if we were men, things wouldn't have taken so long."

Another source of frustration not solely related to sex stereotyping arose from "credibility stereotyping." Respondents thought that their perceptions were given little respect, that technical people reserved for themselves the right to decide if a problem existed:

"If someone was run over by a car, their testimony is believed. If they're run over by radiation or chemicals, they aren't credible unless they have a Ph.D."

Learning to Organize

How does one organize when there is no common workplace, no shared meeting place, no alumni association? How did these women forge consensus, learn to lead, and stand their ground? In a time honored fashion, by telephoning friends and learning by trial and error. Pitfalls included success, which could be followed by attempts by others to take over the leadership of a proven issue. The leader of a group which had a particularly successful time drawing public attention to an issue described her experience and illustrated an assertiveness that many either already possessed or developed as they went along:

"Organizing used to be just phoning friends. Everything got hot, how great it is when everything blows up. After the first meeting, local politicians invited me out for a drink and told me I'd done a great job, and now I'd have to hand leadership over to George. I told him if he valued his family jewels he'd better get away from me. They backed off."

Another woman confirmed that this is not unusual:

"My husband told me once that is *exactly* how to hold your ground. I've used it more times than I can count."

As in many grassroots organizations, an important goal was to involve as many people as possible in active roles. This led to conflicts of style in some cases, and to other problems as well:

"You also come up against differences in approach—executives who snap their fingers and something happens vs. us who work from ground up and want to involve as many people as possible."

"With our group, when men have been on the executive board they've always said the way we do things is wrong. We function better when the executive committee is all women."

"I've gotten men involved, and then the spouses have gotten jealous of *me*, and I lose them both. Then there's the come-ons."

Involving many people demands certain styles of operation, including continual self-education and initiation of new people. Women tended to take an attitude that they had a lot to learn and would only learn it by dogging it:

"We don't mind calling someone up and saying, 'I don't know what to do'."

"Learned the hard way, by doing it."

"Learn to organize by imitating; learn to speak the language of the audience."

Within the organization and on behalf of their organizations, women were vocal about rejecting a traditional tendency to be unassertive about their ideas. They also appreciated working with other women:

"What's wrong with not getting the credit [for your ideas] is that the credit, and sometimes money, doesn't come back to the organization."

"Sometimes, when someone has taken my idea and done it, I was grateful that it was getting done. Other times I took on a male role and shoved it down their throats."

"One of the benefits of being a woman is working with other women. You can cut through the bullshit and get things done."

"Women can see things differently, not just linearly."

Gauging the Opposition and Choosing Tactics

Women had many positive ways to use traditional attitudes that women are nonconfrontational and non-threatening. The feeling that emerged was anger that they were underestimated, but there was recognition that this had tactical advantages:

"Being the token woman, the 'community rep' on a panel was a great opportunity. I could say anything I wanted, refute their lies, and there was nothing they could do."

"You can get a lot of information from officials by playing dumb."

"As a woman, you can be invisible to men. They haven't a clue to this organizing we do."

"I was asked to speak at a meeting [in another area] because people there were too terrified to speak up on their own behalf, afraid they'd get their houses trashed. After the meeting one man said he hoped he could get to his car without being mugged. But I wasn't worried for myself. Who would mug their mother?"

"Men are thrown off guard when you talk intelligently."

Organizers had learned not to expect many allies from among elected officials nor from state agencies:

"They [politicians] only come around during elections, then they ignore us."

"Politicians don't like to fix something they can't see."

While government at all levels came in for some heavy criticism, it was surprising that it was considered as much or more a source of opposition than was industry:

"[State agency name] put more blocks in front of us than [industry name]."

"Best help was Dr. ____ [industry representative]; EPA was no help at all."

"The group had to sue the local government for not following SEQR (State Environmental Quality Review) regulations."

"Can't get laws upheld, especially by state agencies."

"After you've been stung enough by state officials, it can be liberating [from naivete]."

"Government and industry use a strategy of trying to 'wear out' the opposition. They lie."

Consultants also came in for their share of distrust:

"We'll have to hire a Canadian [consulting] firm. Consultants in the U.S. are too contaminated."

"Have you heard of the 'Bouncing Dioxin Theory'? It comes from when an incinerator proposal for New Jersey required a study, which showed the air plume to 'bounce' right out of the community. The same consulting group that prepared the report did the exact same report, with only numbers changed, for a site in Pennsylvania—and the dioxins came down to the ground 120 times faster."

Perhaps because health concerns were a major motivation for most women, and because many uncertainties remain in assessing health impacts of most toxic chemicals, distrust of health studies was mentioned often, particularly for what they didn't examine:

"Don't be taken in by studies of cancer rates. Cancer isn't the only problem. They don't usually look for neurological problems."

"People would talk about rashes. They wouldn't talk about their kids' suicidal problem; they wouldn't talk about their reproductive problems."

"They study cancer at these sites because that's what they're funded for."

"'We have no evidence of...' didn't mean that they looked and found nothing, but that they *didn't look*."

"People shouldn't be asking for health studies, getting involved in that BS. They become tools that are used against you. They're destructive to us."

"We need to be able to do our own monitoring."

Emotion As An Issue

A final theme which stood out concerned women's responses to charges of emotionalism. There were some women who discredited emotional responses on either side:

"You can't react on an emotional level. You *have* to know your stuff."

"I can go in without emotion, with the facts, then officials respond emotionally and on the defensive."

But most women argued with eloquence for the legitimacy of reacting with emotion. Some were uncomfortable that their emotions were somehow used against them; others saw emotion as a tactical tool, and some implied that emotion was an indicator of sincerity.

"I see putting down emotion as political, as dominating, as patriarchal. Emotion is part of our intelligence."

"When they called us hysterical housewives, I said, 'When something is threatening lives, especially mine, I get hysterical!'"

"I don't like people taking advantage of the fact that I have emotions."

"What is so wrong about caring so much about this? 'You should care more' is what I tell them."

"Find their pressure point—they (men) get emotional too."

"Women are usually telling the truth; I can't tell when the men are lying."

"I make no excuses for getting emotional; I try to be rational, but you can't reach everyone that way. Women can do that more easily than men."

"Women do get emotional, after being always verbally patted on the head...makes one want to get up and scream to get attention."

CHARACTERISTICS OF GROUP STYLE

We asked about the types of people in the groups in which our respondents were members or leaders, wondering if perhaps there may be patterns of group organization or membership which corresponded to particular issues or patterns of risk perception or strategies for getting the work done. We saw several types of groups, ranging from those comprised mostly or entirely by women to groups of mostly men. First were those in which women with little or no prior experience do virtually all the work, with a few men offering technical or legal advice. A second type was one comprised of married couples, where women did most of the day-to-day networking but husbands shared the photocopying and took turns babysitting and attending meetings. Both of these types of groups tended to be lower to middle income households where women stayed home to care for children, and where couples had not had previous interest in environmental issues.

A third type of group was a mix of men and women, generally better educated and more affluent than those in the first two groups, who first became interested in the environment during college. A final group we heard about consisted mostly of men. This was seen only in an area with a high concentration of environmental problems and an environmental group in nearly every neighborhood.

Group Development and Strategy

Virtually every group we discussed began with little planned organization or structure. Recruitment of members was informal: women to women contacts, local meeting, letter writing. Group membership shifted as issues became sorted out.

The group members usually taught themselves and each other. They learned to request meetings with officials and/or industry representatives. Those which contacted other environmental groups earliest tended to organize more quickly afterward.

Some groups accepted a political role, moving into politics and lobbying for their issue, bringing lawsuits in order to acquire information being withheld, and using elected representatives to gain their objectives. Other groups rejected activities they saw as political, such as lobbying and litigation.

Most groups relied on a core group to get most work done, but could call on a wide membership when necessary for turnout at rallies, public meetings, and so on. Groups tended to evolve when their goals expanded to more complex problems and adopted a wider scope. Some groups dissolved when their goals were met and group energy dissipated, or when the goals came to be seen as too difficult to achieve before the group lost essential energy or confidence to achieve them.

Common Problems

Groups were forced to "prove" there was a problem; they were challenged as nonprofessionals. One official reportedly stated to members of a group in a public meeting that "You should be home baking cookies." That group did exactly that to raise funds for the group, another perennial problem. A third problem encountered was unfamiliarity and/or frustration with bureaucratic structures. A fourth problem area was the interference of party politics with efforts to keep political activity bipartisan. A fifth problem some groups faced was to avoid becoming undermined by the efforts of industry and government. One strategy in particular was co-optation: leaders became overly familiar with industry and government representatives, in the process of representing the group, to the point where the group felt its efforts were undermined.

Implications of These Groups for Reassessment of Priorities in Risk Assessment and Management

Speaking of a successful political campaign in which one group (50 women, 2 men) got an underdog elected, the leader declared: "When we say we're going to do something, if hell freezes over we're still going to do it." That may sum up the major implication for policymakers: that these grassroots groups are a force to be reckoned with. Their initial involvement is based on a strong need to protect the health and welfare of their families and homes. Continuing involvement is likely among people who become angry at their exclusion from information and decisionmaking. Their tactics are low budget, based on networking, and extremely flexible, all of which give them endurance in their struggle against more established organizations. They do not, for the most part, see themselves as part of a conservationist or other traditional environmental movement, nor do they see their interests being addressed by such groups.

The problems these groups are addressing pose a risk to individuals in ways they perceive as quite personal; the risk is to their families and homes. The local issue initially was compelling to most because of health threats, and eventually became a moral issue for many. Some felt obliged to continue, not by joining established environmental groups but by joining or founding groups centered primarily on toxic waste issues.

"Don't want to leave my daughter a legacy of waste and garbage."

"Don't want to be a crusader as such...can't sit back and not react to the problems."

Their evaluation of actions taken by government and industry in response to the problems is rational as well as emotional, though. Risk managers tend to assume that emotion will be the only field on which women are playing and respond inappropriately to their protests—addressing emotional content while ignoring rational content, or not responding adequately because they feel unequipped to handle the emotional demands.

A strong theme noted is that emotions have an important role in policy decisions and actions—not replacing "rational" responses or judgments, but having similar value. The realm of public emotions in American society traditionally belongs to artists and women. The issue of toxic waste contamination, threatening home and family, also falls into the domain traditionally the responsibility of women. It is not surprising that women from across the social spectrum have become involved, nor that they are making use of the organizational tools available to them to protect their families' welfare. While their success lies partly in understanding the public world they are trying to manipulate, it lies also in using strategies which are not common to that world. Risk managers who wish to communicate more effectively with leaders of local toxic waste action groups—to hear what they are saying, to know what responses are desired, to organize processes of including local groups in decisionmaking activities, to convey information to other decisionmakers, and to convey technical information as completely and accurately as possible to local groups—need to become more comfortable and skilled in giving a place to emotions in environmental risk decisions.

REFERENCES

Caldecott, L. and Leland, S., Eds., 1983, *Reclaim the Earth*, The Women's Press, London.

Geschwender, L., 1986, "Women and the Environmentalist Movement," paper submitted as independent research, Cornell University, Ithaca, NY.

King, Y., 1983, "The Eco-Feminist Imperative," in *Reclaim the Earth*, L. Caldecott and S. Leland, Eds., The Women's Press, London.

Smith, J. and Sands, D., (conference coordinators), 1979, "Women and Technology: Deciding What's Appropriate," Women's Resource Center, Missoula, Montana.

Comparison of Tumor Incidence for Short-Term and Lifetime Exposure

David W. Gaylor

National Center for Toxicological Research
Jefferson, AR

ABSTRACT

Studies to estimate tumor incidence are generally conducted in rodents exposed to a chemical for approximately two years. Exposure of humans to some carcinogens may occur for less than a lifetime from occupations, changing lifestyles, or accidents. A simplifying assumption often is made in risk estimation that risk is proportional to total dose and the age at the time of exposure is ignored. Hence, the estimated cancer risk from an exposure for a fraction ($1/F$) of a lifetime is the lifetime tumor incidence divided by F, where the daily exposure rates are the same. Based on the multistage model, Kodell et al. (1987) show that the risk from a short-term exposure to a carcinogen is less than k/r times the risk predicted from the total dose, where r out of k stages are affected by the carcinogen, whereas at low dosages the transition rate for a stage affected by the carcinogen is assumed to be proportional to the dosage rate. Based on an initiation-growth model, Chen et al. (1988) show that greater discrepancies may exist between the true incidence and the estimate based on average daily lifetime dose. There is a rather limited bioassay database in which tumor incidence at the same age can be compared for different lengths of exposure at the same daily dose rate. From 10 such studies, the observed tumor rates from shorter exposures were from near zero up to 12 times the rates predicted from the average daily dose. Within this very limited database, estimates of short-term tumor incidence were generally less than 10 times the estimate based on average daily lifetime dose.

KEYWORDS: Tumor rates, short-term exposure

INTRODUCTION

The exposure of humans to carcinogens often may occur for less than a lifetime due to changes in lifestyles, changes in the environment, occupational exposure, or accidental exposure. Bioassays from which estimates of tumor incidence are obtained are generally conducted on rodents exposed daily to a chemical for approximately two years. A question arises as to procedures for estimating the risk of tumors from a short-term exposure based upon tumor incidence data for lifetime exposure.

Cancer risk is a function of the dose rate, length of exposure, and age at exposure. A simplifying assumption often made in risk assessment is that risk is proportional to the total dose. That is, dose rate and length of exposure are combined into one measure (total dose = dose rate * length of exposure) and the age at the time of exposure is ignored. Hence, the estimated cancer risk from an exposure for a fraction $(1/F)$ of a lifetime is the lifetime tumor incidence divided by F, where the daily exposure rates are the same. Undoubtedly, this assumption is seldom true, but it may provide a first approximation to risk for short-term exposure.

For the multistage model, the probability (P) of an individual developing a tumor from continuous exposure to a dose rate (d) of a carcinogen up until time (t) is:

$$P(d,t) = 1 - \exp[-(b_0 + b_1 d + \ldots + b_r d^r)(t-w)^k/k!]$$

where k is the number of stages, r is the number of stages affected by the carcinogen, w is the latent period and the b's are constants which are estimated from dose-response data. This model provides a mathematical framework from which risk from short-term exposure at various ages can be estimated from lifetime tumor incidence. Expanding on the work of Whittemore and Keller (1978) and Day and Brown (1980), Crump and Howe (1984) derived results for estimating risks of short-term exposures. Their results depend upon knowing the number of stages in the process, k, and which of one or two stages are affected by the chemical. Kodell *et al.* (1987) showed that the risk from a short-term exposure to a carcinogen is less than k/r times the risk predicted from the total dose where r out of k stages are affected by the carcinogen. For example, risk at any early age to a carcinogen that affects the early stages of the carcinogenic process may be up to k/r times the risk from the same total dose obtained over a lifetime. These results are based upon the premise that at low dosage the transition rate for a stage affected by the carcinogen is proportional to the dosage rate. The results of Kodell *et al.* (1987) provide a basis for establishing a limit on the extent of possible underestimating of short-term risks based upon the assumption that tumor rates are proportional to the total dose or average lifetime daily dose. Still, the value of k must be estimated or assumed.

The multistage model, as it is generally used, does not allow for different sensitivities to a carcinogen at different ages. The different risks arising from these calculations for exposures at different ages are due to the different lengths of time available for the previous and subsequent stages to occur.

When time-to-tumor observance data are available, the proportion of individuals with tumors is a function of the time (age) to the kth power. Also, for low doses the degree of the dose response curve is r. Crump *et al.* (1976), Peto (1978) and Hoel (1980) show that when a carcinogen augments a spontaneously occurring process (e.g., when tumors occur in untreated animals) the dose response is linear $(r=1)$ at low doses. When estimates of k and r cannot be obtained, it has been observed that k is generally 6 or less, and since r may frequently be one, the maximum value of k/r is around 6. Hence, a conservative estimate of risk from an exposure for a fraction of a lifetime $(1/F)$ is less than $6 \times$ (lifetime risk)$/F$ where $1/F$ is less than 1/6. Where $1/F$ is greater than 1/6, the lifetime risk would be used as a conservative upper bound. The significance of this result is that lifetime tumor incidences available from animals exposed continuously from existing chronic bioassays can be used to obtain conservative estimates of cancer risk for any short-term exposure at any age without conducting any additional chronic bioassays for different exposure patterns. Also, it can be shown that the risk from intermittent exposures is the sum of their short-term exposures. For a latent period of w, the risk at time t also needs to be multiplied by $t/(t-w)$ (Kodell *et al.*, 1987). If w is taken to be relatively large, e.g., $w=t/3$, $k=6$, and $r=1$, the maximum risk from a short-term exposure would be expected to be no greater than $kt/r(t-w)$ = 9 times the risk estimated from the exposure averaged over a lifetime.

The purpose of this paper is to examine the available experimental database in the literature to study the validity of using lifetime chronic bioassay data to predict tumor rates for short-term exposures. A second purpose is to check if increasing the short-term risk estimates by a factor of 9 generally results in conservative estimates.

METHODS AND RESULTS

An examination of existing experimental results is performed to determine to what extent the assumption that tumor risk is proportional to total dose (or equivalently the average daily dose) provides a first approximation for tumor risk estimates from short-term exposures. There is a rather limited database in which tumor incidence can be compared at the same age of animals which were exposed to carcinogens at the same dose rate for different lengths of exposure time. The comparison here is limited to those experiments in which the daily dose rate is the same, in order to avoid differences in tumor incidence due to different dose rates. In this way, the differences in tumor rates can be attributed to different lengths of exposure.

The results are given in Table 1 for 10 experiments in which animals were exposed to chemicals for different lengths of time at the same daily dose rate. For example, the first entry in the table gives a comparison of tumor prevalence for animals at 24 months of age where one group of animals was fed 2-AAF for only 9 months. The excess bladder tumor prevalence above background was 0.768 in the group fed 2-AAF continuously for 24 months. Based on total dose proportionality, the estimated prevalence for a 9-month exposure was $(9/24) \times 0.768 = 0.288$. The observed excess bladder tumor prevalence was 0.180 in the group fed 2-AAF for 9 months. Hence, the ratio of the observed to estimated tumor prevalence was $0.180/0.288 = 0.62$. For this case, the observed tumor rate was 62% of the tumor rate predicted by assuming that the tumor rate is proportional to the total lifetime dose.

Examination of the results in Table 1 indicated that the tumor incidence rates from short-term exposures ranged from zero (no GI tract tumors were observed in female rats exposed to NTHP for 6 months) to 11.9 times higher. In the latter group, the tumor incidence for the two month exposure to DES was estimated to be $2/22 = 0.091$ times the 22-month tumor incidence, but the observed incidence for the 2-month exposure was $0.091 \times 11.9 = 1.08$ times the tumor incidence observed in the 22-month exposure groups.

DISCUSSION

From the limited database, it appears that tumor incidences for short-term exposures are generally within a factor of 10 of estimates based upon total lifetime dose. However, many of the short-term exposures were for more than 1/4 of a lifetime. Even if the tumor incidence for the short-term exposure equaled the lifetime tumor incidence, the ratio of the observed to estimated tumor incidence would equal F. Thus, where a chemical was administered for a sizable fraction of the lifetime, a large ratio of the observed to estimated tumor rates would not be achievable. The database only offered a few opportunities to observe large discrepancies between observed and estimated tumor rates in those cases where relatively short exposures were used: vinyl chloride and DES.

For several of the data sets the tumor rates from the short-term exposures were nearly equal to the tumor rates from lifetime exposure: forestomach and GI tract tumors in female rats exposed to NMEA, liver tumors in female rats exposed to NDEA or NTHP, lung and mammary tumors in female mice exposed to vinyl chloride, skin tumors in mice exposed to BaP, and mammary tumors in female mice exposed to DES.

A ratio of one for the observed to estimated tumor rates indicates that the tumor incidence was only a function of the total accumulated dose. Many of the ratios were within a factor of 3 of 1. However, as discussed above, since the short-term exposures in most of the studies covered a sizable fraction of the lifetime, large discrepancies could not be observed. Also missing from the database is information of the tumor incidence for short-term exposures in aged animals. Theory would predict higher tumor rates for late-life

Table 1. Ratio of Observed to Estimated Tumor Incidence for Fractional Lifetime Exposure**

Reference	Chemical	Species/Sex/Site	Exposure Short/ Long	Incidence Observed/ Estimated
Littlefield, N.	2-AAF*	Mouse/F/Bladder	9/24 mos.	0.6
J. Env. Path. Tox.			12/24	0.6
3:17-34, 1980			15/24	0.8
		Mouse/F/Liver	9/24	1.6
			12/24	1.2
			15/24	1.0
Lijinsky, W.	NMEA*	Rat/F/Esoph	7/24	0.9
Food Cos. Tox.		Rat/F/Forstom	7/24	3.4
20:393-399, 1982		Rat/F/GI	7/24	3.1
Lijinsky, W.	NDEA*	Rat/F/Liver	14/24	1.7
Cancer Res.		Rat/F/Esoph	14/24	0.3
41:4997-5003, 1981				
Lijinsky, W.	NTHP*	Rat/F/GI	6/23	0
Ecotox. Env. Safety		Rat/F/Liver	6/23	2.7
6:513-527, 1982				
Kimbrough, R.	Aroclor	Mouse/M/Liver	6/11	0.2
J. Natl. Cancer Instit.				
53:547-549, 1974				
Hong, C.	Vinyl	Mouse/M/Lung	1/6	3.4
J. Tox. Env. Hlth.	Chloride	Mouse/F/Lung	1/6	6.0
7:909-924, 1981		Mouse/F/Mammary	1/6	6.0
Lee, P.	BaP*	Mouse/?/Skin	6/24	3.1
Unpublished			8/24	2.6
Tomatis, L. et al.	DDT*	Mouse/M/Liver	3.5/7.0	0.9
Z. Kreb. Klin. Onk.		Mouse/F/Liver	3.5/7.0	0.7
82:25-35, 1974				
Richardson, F.	DES*	Mouse/F/Mammary	1/22	11.4
J. Natl. Cancer Instit.			2/22	11.9
18:813-822, 1957		Mouse/M/Mammary	1.0/27	0.7
			1.5/27	2.3
			2.0/27	1.4
Richardson, F. & Hall, G.	DES*	Mouse/M/Mammary	1.0/24	1.8
J. Natl. Cancer Instit.			1.5/24	10.2
25:1023-1033, 1960			2.0/24	9.8

* 2-AAF: 2-Acetylaminofluorene
  NMEA : Nitroso-N-methyl-N-(2-phenyl) ethylamine
  NDEA : N-Nitrosodiethylamine
  NTHP : Nitroso-1,2,3,6-tetrahydropyridine

  BaP : Benzo(a)pyrene
  DDT : Dichloro-diphenyl-trichloroethane
  DES : Diethylstilbestrol

**Reprinted from Gaylor, D. W., "Risk Assessments Short-Term Exposure at Various Ages," in *Phenotypic Variation in Populations: Relevance to Risk Assessment*, A. D. Woodhead, M. A. Bender, and R. C. Leonard, Eds., pp. 173-176, Plenum Press, NY, 1988.

exposure to promoters or late acting carcinogens than from exposures early in life. Based upon the initiation-promotion model proposed by Moolgavkar and Venzon (1979), Chen *et al.* (1988) show that greater discrepancies may exist between the true and estimated tumor incidence based upon total dose than is obtained from the multistage model.

It appears from this limited database that predictions based on lifetime tumor rates and average daily exposure of tumor rates for exposures exceeding one-fourth of a lifetime were generally within a factor of 3. Hence, tumor rate predictions between occupational and lifetime exposures or between work-day and continuous inhalation exposure are expected to be nearly as good. Estimates could be improved by studying the pharmacokinetics of intermittent versus continuous exposures. The existing database is inadequate to determine if multiplying short-term (less than one-tenth of a lifetime exposure) tumor risk estimates, based on average lifetime daily exposures, by a factor of 9 will generally provide conservative estimates of tumor risk.

REFERENCES

Chen, J. J., Kodell, R. L., and Gaylor, D. W., 1988, "Using the Biological Two-Stage Model to Assess Risk," *Risk Analysis* 8:223-230.

Crump, K. S., Hoel, D. G., Langley, C. H., and Peto, R., 1976, "Fundamental Carcinogenic Processes and Their Implications for Low Dose Risk Assessment," *Cancer Research* 36:2973-2979.

Crump, K. S., and Howe, R. B., 1984, "The Multistage Model with a Time-Dependent Dose Pattern: Applications to Carcinogenic Risk Assessment," *Risk Analysis* 4:163-176.

Day, N. E., and Brown, C. C., 1980, "Multistage Models and Primary Prevention of Cancer," *J. of the Natl. Cancer Instit.* 64:977-989.

Hoel, D. G., 1980, "Incorporation of Background in Dose-Response Models," *Fed. Proc.* 39:73-75.

Kodell, R. L., Gaylor, D. W., and Chen, J. J., 1987, "Using Average Lifetime Dose Rate for Intermittent Exposures to Carcinogens," *Risk Analysis* 7:339-345.

Moolgavkar, S. H., and Venzon, D. J., 1979, "Two-Event Models for Carcinogenesis: Incidence Curves for Childhood and Adult Tumors," *Mathematical Biosciences* 47:55-77.

Peto, R., 1978, "Carcinogenic Effects of Chronic Exposure to Very Low Levels of Toxic Substances," *Environmental Health Perspectives* 22:155-161.

Whittemore, A. S., and Keller, J. B., 1978, "Quantitative Theories of Carcinogenesis," *SIAM Review* 20:1-30.

Benzene and Leukemia: What Are the Risks and What Do the Data Reveal?

Steven Lamm and Anthony Walters
Consultants in Epidemiology and Occupational Health, Inc.
Washington, DC

Richard Wilson
Harvard University
Cambridge, MA

Hans Grunwald
Long Island Jewish-Hillside Medical Center
SUNY Stony Brook, NY

Daniel Byrd
Consulting Toxicologist
Falls Church, VA

ABSTRACT

Although benzene exposure is recognized as a risk factor for leukemia, there is still no universal consensus regarding which leukemias and what degree of exposure. Definition of benzene-associated leukemia has become more specific as the classification of leukemias is better understood. Review of the studies of benzene-exposed workers finds the excess of leukemia to be generally limited to acute myelogenous leukemia and its variants. Exposure assessment in these analyses has also become more sophisticated. The earliest risk assessments assumed that past exposures were at the legal limit; later analyses considered the exposure data available at the worksite studied. The most recent analyses have focused on the lifetime benzene exposure history of the employees studied, extending beyond the exposure at the studied worksite. Sub-analyses suggest that critical factors included level of exposure, rather than just cumulative exposure. The quality of risk analysis is enhanced when more complete information on exposure is included.

KEYWORDS: Benzene exposure, leukemia classification, rubber hydrochloride manufacturing, leukemia epidemiology

INTRODUCTION

This paper examines the data from the study that forms the epidemiological basis for most quantitative risk assessments for benzene. It fundamentally questions the validity of relating the given cases to the given exposure data. The study, with a recent update published in the *New England Journal of Medicine* (April, 1987) by Rinsky *et al.* of the National Institute for Occupational Safety and Health, presented a quantitative assessment of the association between benzene exposure and leukemia. Historical air sampling data were used where available, and estimates based on existing data were used when no data were available. While the authors stated that "protection from benzene-induced leukemia would increase exponentially with any reduction in the permissible exposure limit," they assumed an exponential formula to fit the Odds Ratio that automatically implies low dose linearity in the dose-response curve.

OBSERVATIONS FROM THE LITERATURE ON BENZENE AND LEUKEMIA

How Leukemias Get Classified

It is important to begin by determining what diseases are of concern with respect to benzene. Table 1 contains the classifications of diseases that fit into the categories of lymphoma and leukemia, traditionally called Neoplasms of the Hematopoietic and Lymphopoietic Tissues (NHLT). The ICD (International Classification of Diseases) numbers reflect distinctions that can rather readily be made both in a medical diagnosis and in differences in embryological development of the cell types. The lymphomas and multiple myelomas are solid tumors of lymphoid cells involving lymph nodes, bone marrow and other organs, whereas the leukemias involve predominantly the bone marrow and peripheral blood, with proliferation of cells in free suspension ("non-solid").

Acute Myeloid Leukemia (AML) Is the Dominant Type of
Leukemia Associated with Benzene Exposure

Within the classification of leukemia, lymphocytic leukemias derive from lymphatic tissue and myelocytic (and monocytic) leukemias derive from hematopoietic tissue. Each cell type further can be classified into two different types of leukemia based on whether the dominant cell is mature (in which case the leukemia is called chronic) or immature (in which case the leukemia is called acute). Thus, the four major types of leukemia are acute lymphoid leukemia (ALL), chronic lymphocytic leukemia (CLL), acute myeloid leukemia (AML), and chronic myelocytic leukemia (CML). The acute myeloid leukemias can be further subclassified as (a) myeloblastic (AML); (b) promyelocytic (APL); (c) myelomonocytic (AMML); (d) monoblastic (AMoL); (e) erythroleukemia (AEL); or (f) megakaryoblastic (AMegL). Hematopoietic tissues are the cells that populate the bone marrow and are the precursors of blood cells. The primary cell is the myeloid line (from myelos, meaning marrow). The myeloid cell lines are also known as myelogenous (because the marrow cells come from them) and as granulocytic (since their cytoplasm shows granules when fixed and stained). It is important to realize that the four major classifications of leukemias are generally regarded as distinct cancerous states, so that it would be contrary to our present understanding of cancer for a particular carcinogen to produce one type of leukemia in some people, and another type in others.

Figures 1, 2, and 3 show the distribution of types of leukemia in cases associated with benzene in several of the larger studies in the literature (Aksoy et al., 1974; Yin et al., 1987; Rinsky et al., 1987). We note that in a review Goldstein (1977) stated that benzene has long been definitely associated with AML and not definitely associated with other types of leukemia. Studies since that time have been consistent with this.

Aksoy's studies of Turkish shoemakers (Aksoy et al., 1974) are very interesting because at the university hematology clinic, where he was Professor in Istanbul, he separated out forty leukemias he found in persons without benzene exposure. For those that were benzene associated, acute myeloid leukemia was most prevalent. For the background leukemias (non-benzene related), Aksoy shows the expected relatively equal representation

Table 1. Classification of Lymphoma/Leukemia Malignant Neoplasms of the Lymphatic and Hematopoietic Tissue

	International Classification of Diseases (ICD)
Lymphoma	
Non-Hodgkins	ICD 200
Reticulosarcoma	200.0
Lymphosarcoma	200.1
Hodgkins	ICD 201
Others and mixed	
(Hairy Cell, Histiocytosis)	ICD 202
Multiple Myeloma	
(Plasma Cell Origin)	ICD 203
Leukemia	
Lymphoid	ICD 204
Acute (ALL)	204.0
Chronic (CLL)	204.1
Myeloid	ICD 205
Acute (AML)	205.0
Chronic (CML)	205.1
Monocytic, etc.	ICD 206
Other	ICD 207
Unspecified	ICD 208

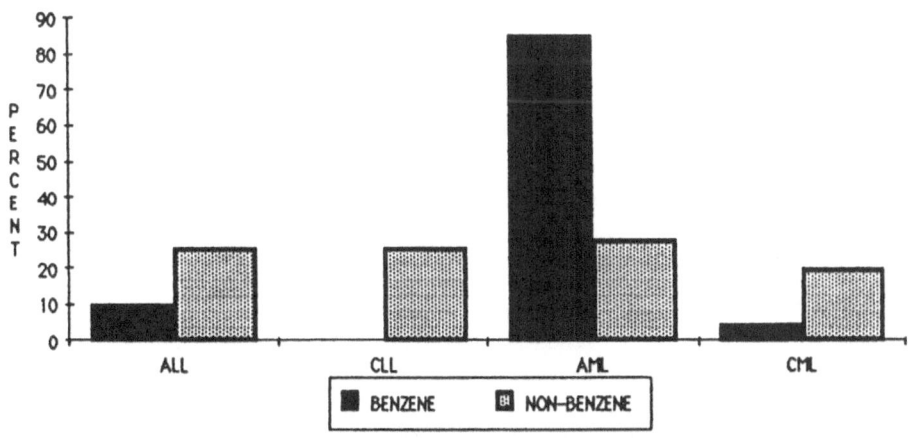

Fig. 1. Distribution of Aksoy's Benzene and Non-Benzene Leukemia Cases.

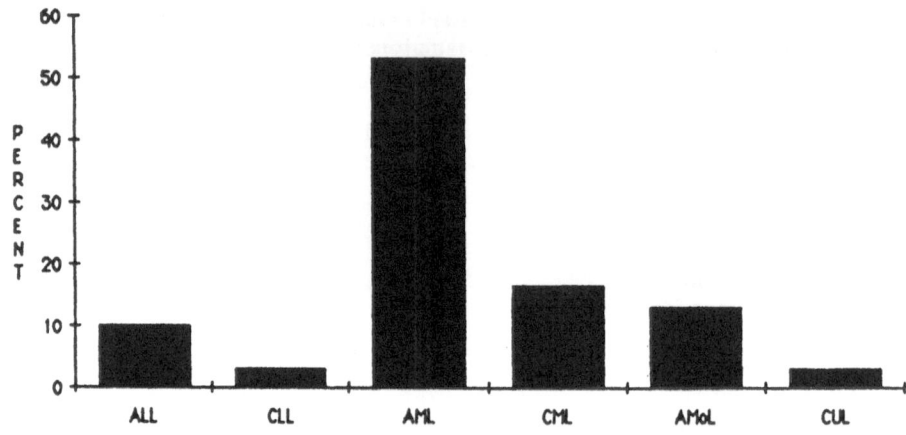

Fig. 2. Distribution of Chinese benzene Leukemia Cases.

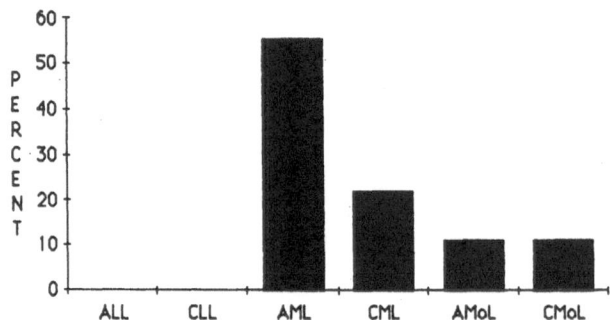

Fig. 3. Distribution of Rinsky's Benzene Leukemia Cases.

of the four classes of leukemia—the acute and chronic lymphocytic leukemias and the acute and chronic myelocytic leukemias. Acute myeloid leukemia is the only type that stood out in excess among the benzene-exposed cases compared to the non-benzene-exposed cases (see Fig. 1). This result seems firm, notwithstanding the fact that many people ignore Aksoy's work because it is not a complete epidemiological study.

Similarly, in the recent Chinese report (Yin *et al.*, 1987) of 30 benzene-associated leukemias, acute myelocytic leukemia stood out in excess above the numbers of other leukemias—ALL, CLL, CML, and unclassified leukemia (CUL) (see Fig. 2).

A preponderance of AML was also seen among a cohort of U.S. chemical workers (Ott *et al.*, 1978) where 5 AML and no other leukemias were seen. In an industry-wide mortality study of chemical workers occupationally exposed to benzene (Wong, 1987), evidence of AML was strikingly absent.

Table 2. Reported Cancers & Leukemias (by Date of Hire)

Location	Dx	Estimated DoH	Cum Benz ppm-yrs
2	AML	1939	478.45
1	MM	1940	0.11
1	MoL	1941	49.99
2	AML	1941	639.84
2	AML	1941	252.66
2	AMoL	1942	98.55
1	PCS	1943	652.66
2	AML	1944	259.50
2	AML	1944	498.23
1	CML	1948	0.10
1	ML	1950	10.16
1	MM	1954	19.50
1	MM	1954	7.75

MM=Multiple Myeloma
MOL=Monocytic Leukemia
PCS=Plasma Cell Sarcoma
ML=Myeloid

Figure 3 presents the distribution by cell type of the leukemia cases reported by NIOSH in its most recent report of the Ohio rubber hydrochloride workers (Rinsky *et al.*, 1987). Again, acute myeloid leukemia is the most common and stands above the rest. Five of the nine cases of leukemia reported in the study were acute myeloblastic leukemia and a sixth was an acute monocytic leukemia. The pattern that again stands out is the dominance of acute myeloid leukemia among the leukemias of benzene-exposed workers.

With respect to benzene, the leukemia that has most often been found in excess is the acute myeloid leukemia (AML). While at various times one study or another has suggested an association with each of the other types of leukemia, across all studies that have been done, it is the acute myeloid leukemias (AMLs) that stand out as benzene associated. Overall, these data suggest that acute myeloid leukemia is the only type of leukemia found consistently in excess among benzene-exposed workers. We further propose that if it is not the only type of benzene leukemia, the risk of other leukemias is much smaller, so that at least its distribution in a benzene-exposed population reflects the distribution of significant exposures to benzene.

RISK ANALYSIS OF BENZENE AND LEUKEMIA

A Brief Description of the Rubber Hydrochloride Study

The study, first reported by NIOSH (Infante *et al.*, 1977), concerned the leukemia mortality of workers first exposed to the wetside of the rubber hydrochloride manufacturing process at three rubber plants (location 1 with Plant 1 and location 2 with Plants 2A and 2B) in Ohio during the 1940s. The most recent 1987 report includes workers assigned to these areas at the three plants through 1965 and reports the deaths from leukemia and lymphomas that have occurred among them through 1981. Table 2 lists these leukemia and lymphoma cases by date of hire as presented by the authors (their Table 4), and the (authors') estimated cumulative benzene exposure calculation.

Table 3. Reported Cancers & Leukemias (by Plant Location)

Location	Dx	Date of Hire	Latency (Years)	Date of Death	Cum Benz ppm-yrs
2	AML	1939	22	1961	478.45
2	AML	1941	20	1961	639.84
2	AML	1941	37	1979	252.66
2	AMoL	1942	15	1957	98.55
2	AML	1944	13.5	1958	259.50
2	AML	1944	15.5	1960	498.23
1	MM	1940	22.5	1963	0.11
1	MoL	1941	17	1958	49.99
1	PCS	1943	24.5	1968	652.66
1	CML	1948	2	1950	0.10
1	ML	1950	3.5	1954	10.16
1	MM	1954	25.5	1980	19.50
1	MM	1954	26.5	1981	7.75

Rubber Hydrochloride Process — Rubber hydrochloride was a thin rubber sheet that was used during WWII to cover tanks and armaments when they crossed the ocean so that the salt water spray would not damage them. The sheets of rubber hydrochloride were made by taking natural rubber, processing it by adding benzene and creating rubber hydrochloride solution, and eventually spreading the resultant solution on a conveyor. The benzene was evaporated and recovered.

Plant History — Plant 1 at location 1 (St. Marys, Ohio) was established in 1939 and continued manufacturing rubber hydrochloride until 1976. Plant 2A at location 2 (Akron, Ohio) commercially produced rubber hydrochloride from 1936-37 through 1949, and Plant 2B, also at location 2, produced rubber hydrochloride from 1949 until 1965. It is asserted that the exposures from the rubber hydrochloride process at each of the three plants were identical. With the exception of one report from Plant 2B, all benzene exposure measurements came from Plant 1. No exposure data are known for Plant 2A. Since exposure data are only presented for the plant at location 1, the equivalency of exposure has to be accepted as a matter of faith.

A Description of AML Cases in the Rubber Hydrochloride Study

Table 3 presents the same data as Table 2, but organized by plant location. These data are quite unusual. Two locations that are considered to be identical in terms of process and benzene exposure have distributions of cases that are entirely different with respect to the only type of leukemia (AML) consistently found in excess among benzene-exposed workers. All of the acute myeloid leukemias (AMLs) came from one location (Plant 2A from which no benzene exposure data are known), while no cases of the acute myeloid leukemias (AMLs) were diagnosed among the workers at the other locations (Plants 1 and 2B) from which the only known benzene exposure data come.

The AML cases at location 2 (Plant 2A) have been estimated to have cumulative benzene doses of 250 to 650 ppm. None of the AML cases have low cumulative benzene exposures. There does not seem to be a gradual increase in the frequency by dose, with some of the cases occurring at low dose and more at high exposure. Furthermore, since there are no known industrial hygiene data for the plant used prior to 1949, all of the benzene exposures assigned to the leukemia cases at location 2 are based on industrial hygiene data from location 1 and from data collected in decades subsequent to that in which these cases were hired. The authors state, "benzene exposure levels measured at location 1 were assumed to be naturally occurring simulations of exposure levels in corresponding areas at location 2, when actual exposure measurements did not exist."

All the cases of acute myeloid leukemia occurred in workers hired at plant 2A prior to 1945, although workers continued to be hired there into 1949, after which hiring occurred at plant 2B into 1965. The group of employees hired at plant 2A prior to 1945 has been followed up through 1981, a period of over 37 years. No member of this group has died of AML since 1961.

Additionally of concern is that the cohort of people from Plant 2A is incomplete. The personnel records from that location prior to 1945 are missing, and the only people whose records of employment exist prior to 1945 are those people who were hired prior to 1945 and continued working past 1945, as they are in the subsequent employment records. Among the cohorts for whom employee rosters are complete (Akron 1945+ and St. Marys 1940+), there is no known case of AML.

The AML cases in this study are reported to have a mean latency of 20.5 years (range 14-37 years) since first exposure. This is surprising, because acute myeloid leukemia from every other exposure is associated with latencies of approximately 2 to 5 years. The latency for AML after chemotherapy is 2-8 years and after radiotherapy is 4-10 years; therefore latency overall is 2-10 years.

The rubber hydrochloride data are unusual for AML and for benzene-associated leukemias. The Chinese data have an average latency of 11 years, and Aksoy reports an average latency of 9 years. Therefore, the latencies might represent a common (probabilistic) process or might confuse *time since first exposure* with *time since critical exposure*. The characteristics of these data raise a question of what, under these circumstances, is the appropriate definition of latency. Latency is usually determined from the time of first exposure to the time of death. It may be that there are exposures subsequent to the initial exposure that are the "inciting" or critical exposures. An analysis of the literature in 1980 showed that 80% of benzene-associated leukemia deaths occurred no more than two years after their last benzene exposure (Lamm, 1980). Observation of such short latency suggests that benzene acts at a late stage in the carcinogenic process, perhaps through action on promotion or progression stages of disease.

The idea that benzene is a promoter rather than an initiator is consistent with two other pieces of data. Benzene has not been found to be mutagenic in appropriate short term tests, and early stage carcinogens are often expected to be initiators. Also, benzene poisoning (pancytopenia) has often been shown to precede leukemia. Thus, in the Chinese Study, those with pancytopenia had a risk ratio of 1500 instead of 5-10.

AMLs in the general population increase in number exponentially with age. Researchers have followed the rubber hydrochloride cohort over time. Given the many years that have passed and consequently the advanced ages of the cohort, one must ask where are the cases of AML that the accepted models would predict.

The industrial hygiene data referenced in the Rinsky analysis (Rinsky *et al.*, 1987) were collected in the 1950s, 60s, and 70s, and then extrapolated back in order to estimate exposures in the 1940s. Most of the samples are area samples with only the data from the mid-1970s being 8 hour time-weighted averages. Nonetheless, they most likely have all been used as 8 hour time-weighted average measurements.

It is important to remember that the benzene exposure standard dropped from a maximum allowable concentration of 100 ppm to a level of 50 ppm (8 hr. time-weighted average) in 1947, and to a level of 35 ppm (8 hr. t.w.a.) in 1948. The standard was further lowered in 1957 to 25 ppm and in 1969 to 10 ppm. These changes in the standard were paralleled by lower exposure of workers, as recognized by the study by Crump and Allen in 1984.

Two research groups (Crump and Allen, 1984 and Rinsky et al., 1987) have presented analyses of the industrial hygiene data in which they have tried to estimate the exposures for the 1940s. Crump and Allen examined these data in a 1984 report for OSHA. They estimated a cumulative benzene exposure amount and a maximum or peak intensity benzene exposure level for each of the employees in rubber hydrochloride production. To calculate or estimate exposure levels for jobs during time periods for which no industrial hygiene data existed, they calculated all exposure data for a particular job as a percentage of the allowable level at that time and then applied that percentage (or proportion) to each time period. Thus, if the data in 1976 indicated that the average exposure to benzene in a particular department was 1.5 times the standard in effect in 1976 (i.e., 15 ppm = 1.5 × 10 ppm), then the benzene exposure in that same department in 1945 was assumed to be 1.5 times the standard in effect in 1945 (i.e., 1.5 × 100 ppm = 150 ppm).

The NIOSH report by Rinsky et al. (1987), handled the industrial hygiene data differently. They limited their calculations to those rubber hydrochloride workers who worked on the "wetside" of the process (the "wetside" was the area where in the 1940s benzene was considered to be perceptible in the air as wet). Rinsky et al. assumed that for a given job the exposure levels were the same both during the periods for which exposure data existed and during other periods when no data existed.

Problems in Patterns of Results

Acute myeloid leukemia is the only type of leukemia consistently found in excess with excess benzene exposure. In the NIOSH rubber hydrochloride study, all industrial hygiene data are from one plant, but all cases of acute myeloid leukemia are from a second plant. This forces one to ask what was different about the exposure at location 2 compared to location 1, even though the study assumes that operations at the two locations were essentially identical.

In the study being analyzed (NIOSH rubber hydrochloride study) no cases of acute myeloid leukemia have been reported for employees hired after 1944, although workers were hired and worked with the rubber hydrochloride process through 1965. This change would lead one to believe some change in exposure occurred after 1944. Some contribution to this positive direction probably should be attributed to the exposure standard for benzene which was lowered to 50 ppm (for an 8 hour time-weighted average) in 1947. The relative risk model would predict very large risk and expected cases that should have occurred between 1961 and 1981, given that exposed workers were generally 20-40 years old in the 1940s.

The data demonstrate non-linearity in the relationship between peak exposure to benzene and occurrence of acute myeloid leukemia. This raises a question as to why regulatory analyses are only based on a linear model; and it also raises the possibility of a threshold for benzene carcinogenicity. For example, Figs. 4, 5, and 6 show different ways of looking at the data. Figure 4 shows the data from Crump and Allen (1984) for benzene exposure and SMR. Figure 5 shows what happens if one simply connects those data points, whereas Fig. 6 (their Fig. 2) connects selected data points by using a linear least squares approximation. It is not difficult to see why a reader would arrive at different conclusions from each presentation of data in isolation.

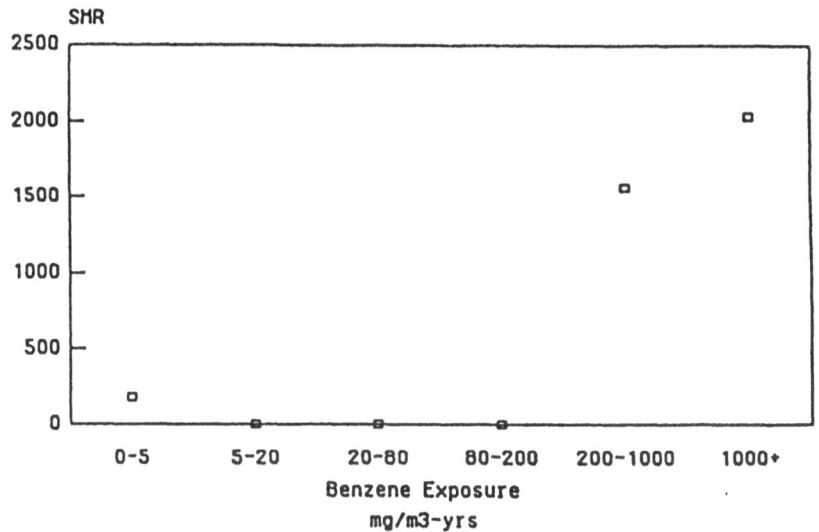

Fig. 4. Scatter Diagram of Leukemia Risk (SMR) by Cumulative Benzene Exposure Based on Data from Rinsky *et al.* and Analysis by Crump *et al.*

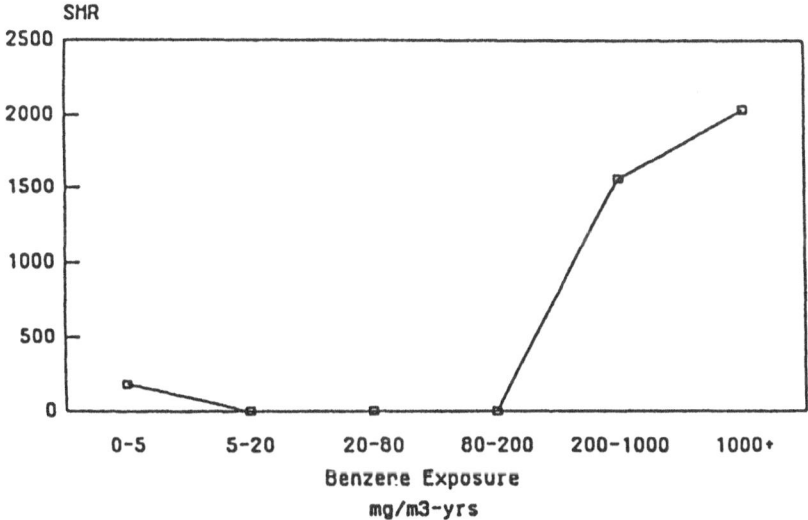

Fig. 5. Line Graph of Leukemia Risk (SMR) by Cumulative Benzene Exposure Based on Data from Rinsky *et al.* and Analysis from Crump *et al.*

Risk Assessments and Issues

We propose that leukemia risk assessments from benzene exposure be based on the distribution pattern of AML alone, as that is the sentinel neoplastic disease associated with significant benzene exposure. This concept is consistent with the assessments of Goldstein (1977) and the studies of Aksoy (1974).

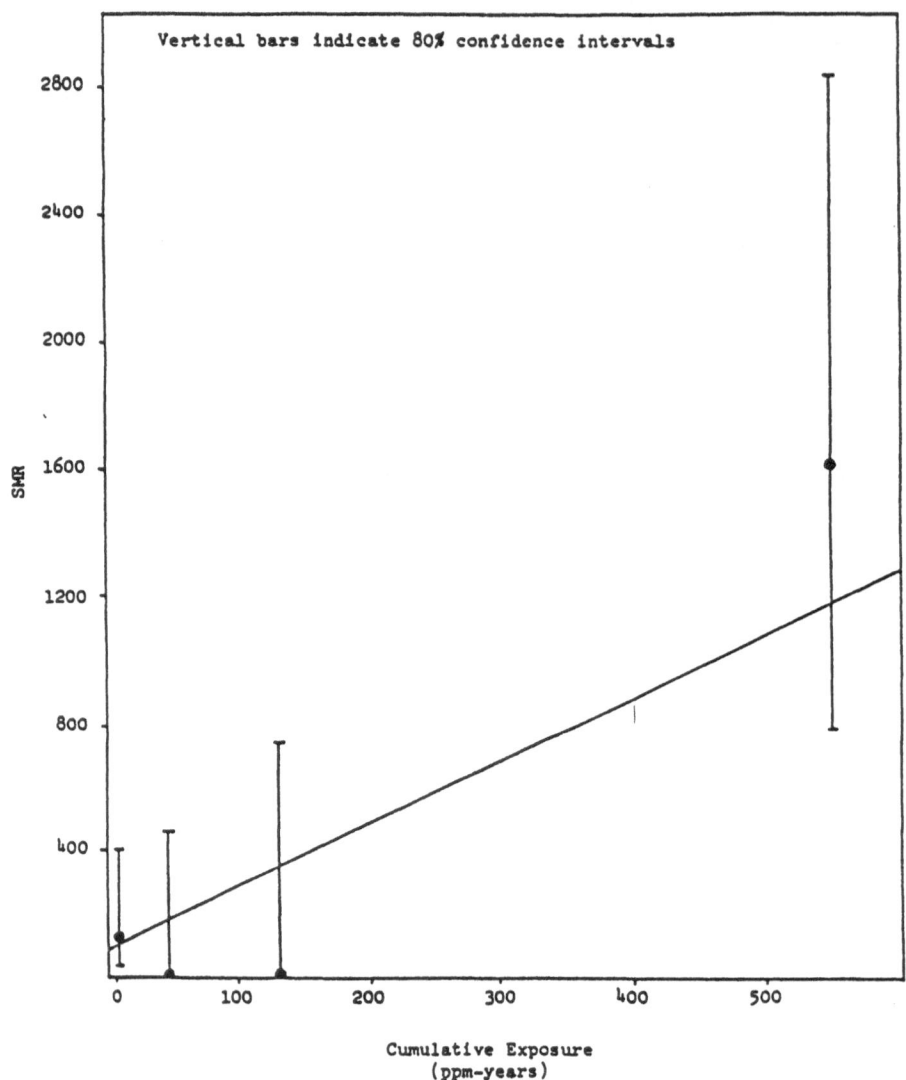

Fig. 6. Linear Least Squares Analysis Graph of Leukemia Risk (SMR) by Cumulative Benzene Exposure (mg/m³-yr) Based on Data from Rinsky *et al.* and Analysis from Crump *et al.*

Risk analysis based on the exposure estimates of Crump and Allen reach the conclusion that a significant excess risk of leukemia was not observed until a cumulative benzene exposure of 450 ppm-years or a peak benzene exposure of 250 ppm had been exceeded. Similar analyses based on the exposure estimates of Rinsky *et al.* reach the conclusion that a significant excess risk of leukemia was observed by the time a cumulative benzene exposure of 250 ppm-years or a peak benzene exposure of 20 ppm had been exceeded.

The exposure circumstances necessary for the development of a significant excess risk of disease mortality can be modelled as in Fig. 7. Here, the cumulative risk from benzene exposure as expressed by peak exposure experienced by workers is observed for

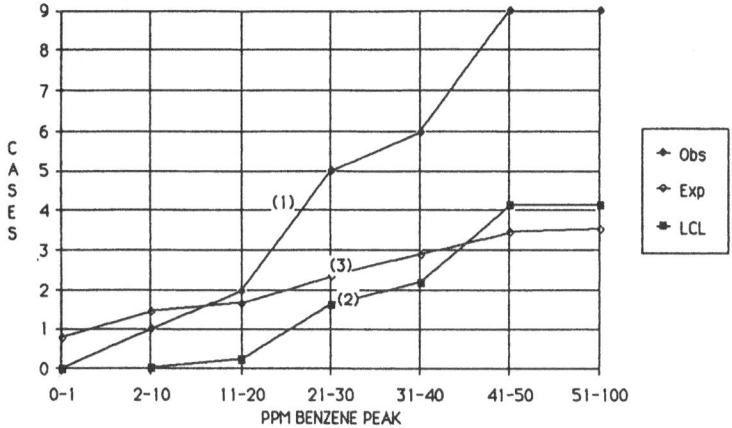

Obs=Observed; Exp=Expected; LCL=95% Lower Confidence Limit

Fig. 7. Benzene-Leukemia Dose Relationship by Peak Benzene Level (ppm) of Rinsky for All 1940-1965 Rubber Hydrochloride Workers.

sequentially higher peak levels of benzene exposure. This graph shows peak benzene exposure levels according to Rinsky's exposure assumptions. Line 1 is the cumulative number of cases of leukemias that have occurred. Line 2 is the 95% lower confidence limit of the observed leukemia cases. Line 3 indicates by peak benzene level the cumulative number of leukemia cases expected among the cohort. It is not until the last two cases are included that there is a significant excess of leukemia occurring among the benzene exposed workers. If the assumptions of Rinsky are used, this occurs at a peak benzene level of 40-100 ppm. If the assumptions of Crump are used, this occurs above 450 ppm.

The determination of which set of assumptions provides the more valid basis for calculating the risk cannot be based on information in these reports alone. Neither Crump and Allen nor Rinsky *et al.* provide independent data that permit determining which of the two exposure assessments is more likely correct. The report by Kipen *et al.* (1987) provides an important key.

Benzene is known to have the toxic hematologic effect of lowering the concentration of white cells in the blood (called lowering the white count.) Hematologic studies of the rubber hydrochloride workers were regularly carried out by their employer. A medical team headed by Goldstein has analyzed the white counts of these workers during the 1940s to determine whether they correlated better with the exposure estimates of Crump and Allen or with those of Rinsky *et al.* Their analysis (Kipen, Cody and Goldstein, 1987) showed for the white counts a correlation coefficient of 0.72 with the Crump and Allen estimates and a correlation coefficient of 0.03 with the Rinsky *et al.* data. Thus, the estimates of Crump and Allen appear to be significantly better than those of Rinsky *et al.*

Previous NIOSH analysis (Rinsky *et al.*, 1982) derived a relative risk for AML of 5-10 for this cohort, based on earlier data. They assumed that this relative risk stayed constant from the end of exposure to death. Since the background risk of leukemia increases markedly with age (see Fig. 8), the relative risk model would predict a markedly greater increase in frequency of AML during the second half of the three decades of observation than during the first half. However, five AML deaths occurred during the period 1950 to 1965 and only one occurred during the period 1966-81. These data suggest a decreasing relative risk (with a short latency period) over time rather than a constant one.

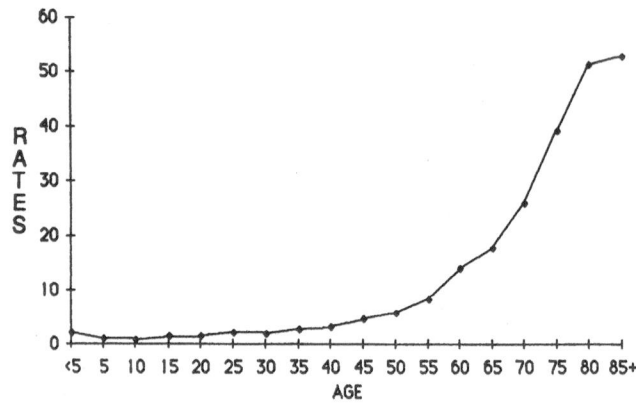

Fig. 8. Age-Specific Leukemia Incidence Rates [Data from National Cancer Institute, Surveillance Epidemiology End Results, Incidence and Mortality Data (1973-77)].

This would accord with an assumption that benzene promotes AML and does not initiate. Risk assessments using a constant risk ratio probably overstate the risk from benzene exposure.

Figures 4-7 all demonstrate that the data show a nonlinear dose-response relationship, as was assumed by Rinsky *et al.* (1987). Auxiliary, indirect, data also suggest a nonlinear dose-response relationship. For example, the likelihood that benzene only follows pancytopenia suggests that the dose-response relationship might be a steep one more like the dose-response expected for an acute toxic effect, rather than the shallow dose-response of a genotoxic carcinogen.

DISCUSSION

Multiple reports have been presented analyzing the frequency of various leukemias as causes of death of the Ohio rubber hydrochloride workers and relating that mortality to the benzene exposure from the rubber hydrochloride process (Byrd and Barfield, 1988). Risk assessments based on these rubber hydrochloride studies have shown the importance of considering all the variables that affect the association between benzene and leukemia. Most basic is the question of definition of the cohort, whether the analysis uses the total cohort as Crump did, or only the wetside workers as Rinsky *et al.* did (see Table 4). The analysis could use the assumption of exposure in the 40s as Crump did (that it was proportional to the standard at the time) or the assumption of Rinsky (that it is linear (constant) over time), remembering that Kipen in his analysis of the white blood count demonstrated that the white count correlates with the Crump estimate of exposures and not with the Rinsky estimate of exposures. Using Rinsky's cohort and exposure as published in the *New England Journal of Medicine*, the analysis results in a significant risk of leukemia occurring with benzene exposures greater than 20 ppm, or cumulative exposure greater than 250 ppm-years. Using Crump's assumptions, significant excess of leukemia does not occur until the peak exceeds 250 ppm or a cumulative exposure greater than 450 ppm-years (see Table 5). The cohort continues under observation, and there will undoubtedly be further reports on them with respect to their mortality. These reports, hopefully, will analyze the information with respect to the exposures and what exposures can be linked to the various cancers.

Table 4. Cohort and Exposure Assumptions of Benzene Leukemia Risk Analysts Studying Rubber Hydrochloride Workers

Cohort Studied

Crump	All Rubber Hydrochloride Workers
Rinsky	Wetside Rubber Hydrochloride Workers

Exposure Estimates for the 1940s

Crump	Proportional to Standard
Rinsky	Linear Over Time

Table 5. Estimates of Benzene Levels Associated with Significant Leukemia Risk, as a Fraction of Various Cohort and Exposure Assumptions for Rubber Hydrochloride Worker Study

RESULTS

	Study Parameters			Significant Excess	
Analyst	Cohort	Exposure		Peak	Cumulative
Rinsky	Wetside	Linear		>20 ppm	>250 ppm-yr
--	Total	Linear		>40 ppm	>250 ppm-yr
--	Wetside	Proportional		>250 ppm	>250 ppm-yr
Crump	Total	Proportional		>250 ppm	>450 ppm-yr

CONCLUSION

Recent data have confirmed the claim of Goldstein (1977) that benzene can cause acute myeloid leukemia (AML). Recent studies fail to confirm the production of any other type of leukemia from benzene exposure.

The following need careful study:

1. There have been suggestions that leukemia has a viral origin; and this partially explains "clustering" of benzene cases, to a greater extent than statistical analysis. Is the distribution of leukemia among leukemia types in the different studies consistent with their being part of the same statistical sample?

2. Does follow-up of the various studies and cohorts confirm the suggestion of a short latency period?

3. Pancytopenia often precedes benzene leukemia. An examination of those with pancytopenia gives, therefore, an almost background-free sample of benzene leukemias. Are any ALL, CLL, or CML preceded by pancytopenia? If not, there is further support for the suggestion here, and earlier (Goldstein, 1977), that AML is the only benzene-caused leukemia. Is some step in the differentiation and clonal expansion to granulocytic cells uniquely sensitive to benzene metabolites?

4. The epidemiological data suggest a nonlinear dose-response relationship. The absence of strong mutagenicity and the presence of pancytopenia suggest a promotional, and possibly nonlinear, behavior. Are there other indirect indications of a nonlinear dose-response relationship?

REFERENCES

Aksoy, M., Erdem, S., and Dincol, G., 1974, *Blood* **44**:837-841.

Byrd, D. M., and Barfield, E. T., 1988, in *Risk Assessment and Risk Management of Industrial and Environmental Chemicals*, C. R. Cothen, M. A. Mehlman, and W. L. Marcus (Eds.), Princeton Scientific, pp. 209-223.

Crump, K. S., and Allen, B. C., 1984, Occupational Safety and Health Administration, docket H-059 B.

Goldstein, B. D., 1977, *Journal of Toxicology and Environmental Health*, Supplement 2, p. 91.

Infante, P. F., Rinsky, R. A., Wagoner, J. K., Young, R. J., 1977, *Lancet* **2**:76-78.

Kipen, H., Cody, R., and Goldstein, B., 1989, "Environmental Health Perspectives," in press.

Lamm, S., 1980, Testimony at OSHA Benzene Hearings.

Ott, G. M., Townsend, J. C., Fishbeck, W. A., and Langner, R. A., 1978, *Archives of Environmental Health* **33**:3-10.

Rinsky, R. A., Smith, A. B., Hornung, R., Filloon, T. G., Young, R. J., Okun, A. H., and Landrigan, P. J., 1987, *New England Journal of Medicine* **316**:1044-1050.

Wong, O., 1987, *British Journal of industrial Medicine* **44**:365-381.

Yin, S-N., Li, G-L., Tain, F-D., Fu, Z-I., Jin, C., Chen, Y-J., Luo, S-J, Ye, P-Z, Zhang, J-Z., Wang, G-C., Zhang, X-C., Wu, H-N, and Zhong, Q-C., 1987, *British Journal of Industrial Medicine* **44**:124-128.

Analysis of Individual Risk Belief Structures

Bruce E. Tonn
Oak Ridge National Laboratory
Oak Ridge, TN

Cheryl B. Travis
University of Tennessee
Knoxville, TN

Lloyd Arrowood and Richard Goeltz
Oak Ridge National Laboratory
Oak Ridge, TN

Chris A. Mann
University of Tennessee
Knoxville, TN

ABSTRACT

An interactive computer program developed at Oak Ridge National Laboratory is presented as a methodology to model individualized belief structures. The logic and general strategy of the model are presented for two risk topics: AIDS and toxic waste. Subjects identified desirable and undesirable consequences for each topic and formulated an associative rule linking topic and consequence in either a causal or a correlational framework. Likelihood estimates, generated by subjects in several formats (probability, odds statements, etc.), constituted one outcome measure. Additionally, source of belief (personal experience, news media, etc.) and perceived personal and societal impact are reviewed. Briefly, subjects believe that AIDS causes significant emotional problems, and to a lesser degree, physical health problems, whereas toxic waste causes significant environmental problems.

KEYWORDS: Risk perceptions, risk beliefs, AIDS, toxic waste

INTRODUCTION

One of the more fascinating discoveries by social scientists in the last thirty years is the tremendous gap between factual accounts of risk and the public's perception of risk (Slovic *et al.*, 1982). The chasm between fact and perception is most notable in the case of nuclear power, but gulfs have also been observed for such diverse risk sources as commercial aviation, living on a flood plain, and common diseases. This state of affairs is not due to irrationality on the part of the public; in fact, people typically desire additional information concerning risk sources (Fischhoff, 1984).

Risk perception research aims to reveal reasons for differences between expert opinion and public risk perceptions. Psychometric work has achieved some success at quantifying and predicting risk perceptions. A typical study might ask respondents to rate the riskiness of a risk source and evaluate the risk source along a number of dimensions (e.g., voluntariness, controllability). Factor analytic work has indicated that unacceptable risk sources rate high in the dreadedness of their consequences and low in what is known about them (Slovic *et al.*, 1984).

Other risk perception research has shown that people exhibit systematically biased decision making when uncertainty is involved (Kahneman *et al.*, 1982). Related research has also revealed serious problems in communicating risk information between experts and the public (Kasperson, 1986) and even in simply defining risks (Fischhoff *et al.*, 1984). For these and other reasons, Keeney and von Winterfeldt (1986) state that dealing with public risk perceptions is still very much of an art.

This paper introduces a new point of view to risk perception research. The intent is to understand the knowledge and beliefs that underlie risk perceptions. In a sense, the goal is to determine people's models of the world because what they believe about how the world works will most assuredly affect what dangers they discern in the world. Thus, for example, in addition to eliciting a risk rating for nuclear power, this approach would also elicit from respondents their knowledge and/or beliefs about nuclear power. For instance, the respondent will be asked what undesirable consequences might be directly caused by, accompany, be associated with, or possibly be related to nuclear power.

To actualize this line of research, ideas and methods have been borrowed from the field of artificial intelligence (AI), or as Fiksel and Covello (1986) would state, from knowledge systems technology. AI offers useful methods to represent risk beliefs and accompanying uncertainties. AI also offers very effective approaches to designing computer software to elicit from respondents their risk beliefs.

The next section of this paper describes the computer software used and its knowledge representation scheme. The following section describes the respondents and the experimental design. The third section presents the results of using the software to elicit beliefs about two diverse risk sources: AIDS and toxic waste. The AI approach yields extremely interesting and useful information about these risk sources which could complement other research approaches.

RISK BELIEF ACQUISITION SOFTWARE

Oak Ridge National Laboratory has developed a computer program known as ARK (Acquiring and Reasoning about Knowledge). Though primarily developed to aid in the development of expert systems, ARK can be easily modified to elicit, through man-machine dialogue, respondents' risk beliefs associated with risk sources. ARK is programmed in Common Lisp for portability and resides on VAX 8600 and Symbolics AI workstation computer systems.

To begin a session with ARK, the respondent is first asked to specify a risk topic, which in this study is either AIDS or toxic waste. Then ARK elicits five basic pieces of information:

1. negative and positive consequences related to risk topic;

2. a propositional primitive most appropriate for each risk topic–consequence pair (i.e., a proposition);

Agent:	AIDS
Propositional rule:	Directly causes
Consequence:	Illness
Likelihood$_1$:	Probability = 0.95
- conditions:	(No effective drug therapy is available)
Likelihood$_2$:	Probability = 0.20
-conditions:	(Drug AZT is effective)
Likelihood$_n$	Nil
-conditions:	Nil
Sources of belief:	News media
Personal salience:	20
Societal salience:	100

Fig. 1. Example of ARK Risk Belief Frame.

3. an estimate of likelihood associated with the consequence;

4. sources of belief about the proposition; and

5. indications of the salience of the consequence from both personal and societal perspectives.

Respondents also have the option of specifying conditions that may change the magnitude of the likelihood estimate. The information is stored in a frame, as illustrated in Fig. 1.

Respondents can specify one of five propositional primitives: directly causes, accompanies, is associated with, possibly relates to, and prevents. The primitives were chosen to coincide with semantics used in everyday cognitive exercises. Tonn and Arrowood (1987) describe how the primitives may be used in reasoning processes. The respondents also enjoyed some flexibility in specifying likelihoods. Five paradigms were provided: probability, % of time, chances, possibility and belief. For example, acceptable answers could be probability .80 or 80% of the time or chances 8 in 10. Help files were on-line to provide explicit definitions of the rules and likelihood paradigms.

Eight sources of belief were provided for selection: (1) personal experience, (2) experiences of acquaintances, (3) the news media, (4) scientific fact, (5) religious doctrine, (6) common knowledge, (7) authoritarian doctrine, and (8) scientific research. Salience ranges from 0 to 100, with the lower bound indicating that the risk source has no significance and the higher bound relating to the highest significance.

ARK is designed for multiple sessions and for the acquisition of extensive risk belief structures. For example, respondents have the opportunity to discuss the consequences (second-order) of each consequent (first order), etc., as long as the discussion proves fruitful. ARK can also incorporate discussions of numerous risk sources. The frame system provides the structure to capture beliefs about inter-relationships between the numerous risk sources and consequences. It is very possible that such inter-relationships, together with secondary and tertiary consequences, may be the most important sources of apprehension about risk sources. Clearly, ARK can collect much information that can be analyzed in a disaggregated fashion, which overcomes a criticism leveled at some psychometric work (Harding and Eiser, 1984).

Representing risk beliefs as frames is new to risk perception research but can complement the factor analytic, repertory grip, and other approaches. However, frames are often used to represent knowledge in cognitive science research (e.g., Cohen and Murphy, 1984; Brachman and Schmolze, 1985). Work by Schank and others (Schank and Riesbeck, 1981) also makes use of a small number of primitives, such as ARK's propositional primitives, to relate agents and consequences. Lastly, simple associations between agents and consequences are consistent both with node-link theories of long-term memory (Barsalom and Bower, 1984; Simon, 1979) and with theories that knowledge can be represented in production systems (Anderson, 1982).

EXPERIMENTAL DESIGN

One hundred and twenty-nine students at the University of Tennessee–Knoxville participated in the study. Sixty-one percent were female, 39% male. Most (80%) were undergraduates, majoring in Business/Engineering (40%), Social Science/Humanities (13%), and Natural Science (6%). Forty percent had other or undecided majors. Seventy-five percent had related courses in logic or computers. Eighty-seven percent expressed moderate to high interest in participating.

A student experimenter booted-up the ARK software and briefly demonstrated the software to each respondent. An instruction sheet was placed next to the terminal and the experimenter was also available to answer questions. The typical session lasted approximately 25 minutes.

RESULTS

The experiment yielded over 500 hundred risk belief frames, as illustrated in Fig. 1. Most entail first order relationships between AIDS or toxic waste and direct consequences, although 68 beliefs about indirect consequences also were elicited. Several aspects of the data necessitated unique data analysis methods.

The most challenging problem was analyzing reported consequences. Because respondents were allowed complete freedom in specifying consequences, means were needed to classify them. Fig. 2 presents a conceptual net that was utilized for the classification. The first distinction is made between concepts that are objects or events. Next, both objects and events may be abstract or physical (i.e., tangible). Specifications continue to be made until the consequences elicited from subjects are distinguishable at the root nodes of the net. Table 1 contains instances of frequently mentioned concepts for some root nodes. For example, concepts such as ill health, discomfort, and infections reside at the object-physical-animate-physical condition node. Concepts such as social outcast, bad name, and innocent victim reside at the object-abstract-societal-social class node. The conceptual net was adapted from a net prepared by Johnson (1987) for use in natural language processing and does not represent a final, unalterable product.

The conceptual net was then used to graphically capture in the aggregate the direct consequences of AIDS (Fig. 3) and toxic waste (Fig. 4). With respect to AIDS, the overwhelming response is its causal relationship to death. Respondents also stated that AIDS causes emotional problems such as depression. The sociological consequences of AIDS possess similar emotional characteristics (e.g., AIDS is associated with being social outcasts, misfits, and alienated). Specific physical-condition problems are mentioned relatively less frequently. Many respondents see advances in economic products (i.e., medical technology) as the only benefit of the AIDS crisis. Secondary consequences mainly continue personal themes related to death, friendships, and physical ailments (Table 2). In summary, the subjects view AIDS as having adverse emotional and physical health consequences.

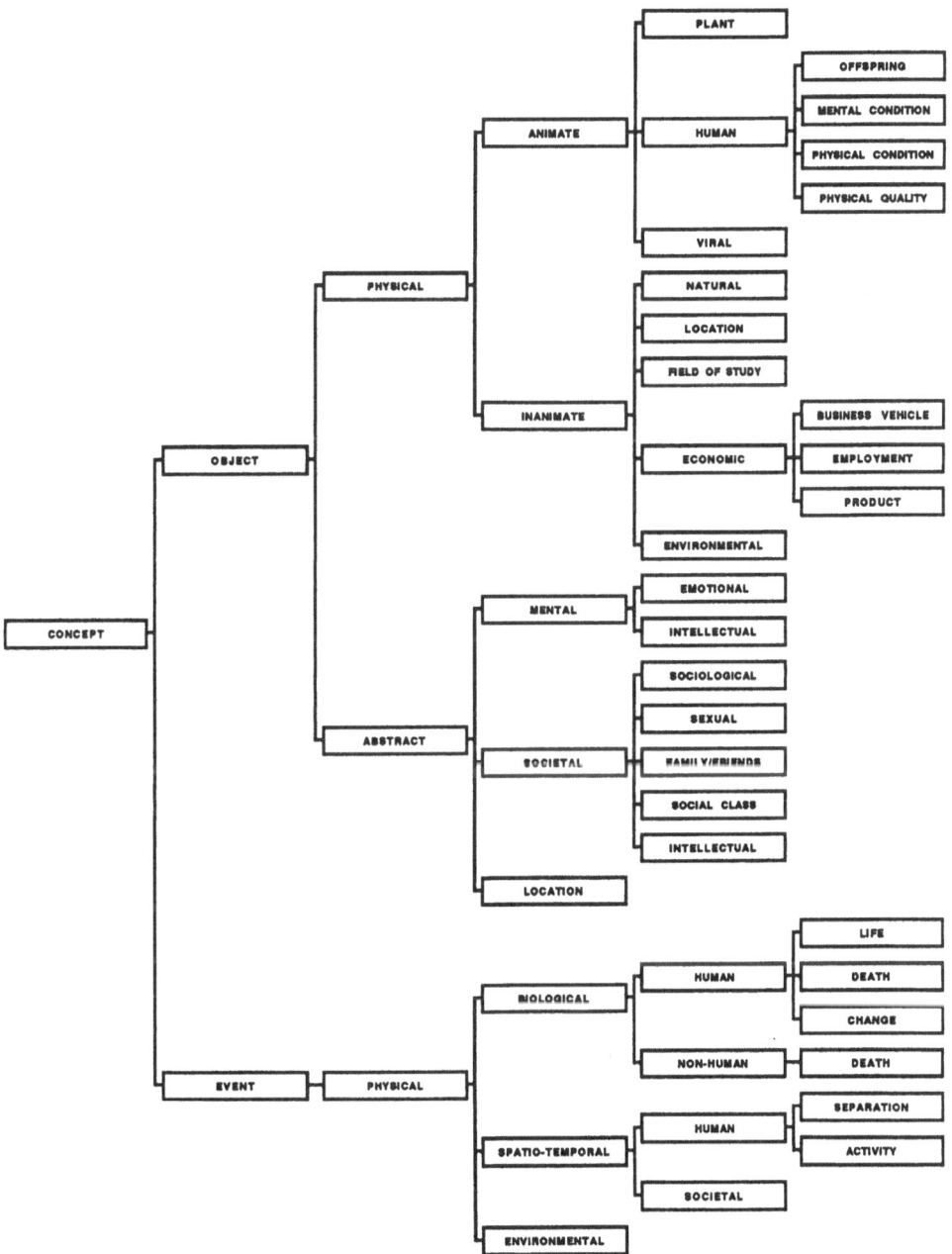

Fig. 2. Conceptual Net.

Table 1. Instances of Frequently Mentioned Concepts

EMOTIONAL ABSTRACTION
- Depression
- Disgrace
- Displacement of anger
- Distress
- Embarrassment
- Emotional issues
- Fear of AIDS
- Fear of blood transfusions
- Fear of unknown
- Grief
- Inconvenience
- Paranoia
- Self-esteem
- Suffering
- Unhappiness
- Rejection
- Relief
- Self satisfaction

SOCIOLOGICAL ABSTRACTION
- Aggression
- Alienation
- Discrimination
- Prejudice
- Ridicule
- Social problems
- Suspicion sexual

SOCIAL CLASS
- Bad name
- Friend
- Homosexuals
- Innocent victim
- Social misfit
- Social outcast

ENVIRONMENTAL OBJECT
- Gas leaks
- Ground pollution
- Hazard to life
- Illegal dump
- No waste storage
- Nuclear waste
- Pollution
- Sanitation problems
- Sludge
- Smell
- Soil erosion
- Waste storage
- Water Pollution

ENVIRONMENTAL EVENTS
- Contamination
- Depletion of minerals
- Destruction of environment
- Destruction of land
- Destruction of life
- Destruction of air

PHYSICAL-CONDITION
- Discomfort
- Fever
- Ill health
- Immune system breakdown
- Infection
- Pain
- Rash
- Unapparent symptoms
- Swelling
- Green hair
- Headache
- Lethargy
- Sleepiness
- Vomit

Inquiries about toxic waste engendered different responses. Most subjects focused on environmental damage likely to be caused by toxic waste. Some mention is made of death of animal and plant life and harm to people. Secondary consequences continue to elaborate on analytical matters associated with environmental problems (Table 2). Missing from the responses are the emotional/sociological components. Toxic waste does not directly affect respondents' (i.e., University of Tennessee students) lives in such an overwhelming manner as is the case with AIDS.

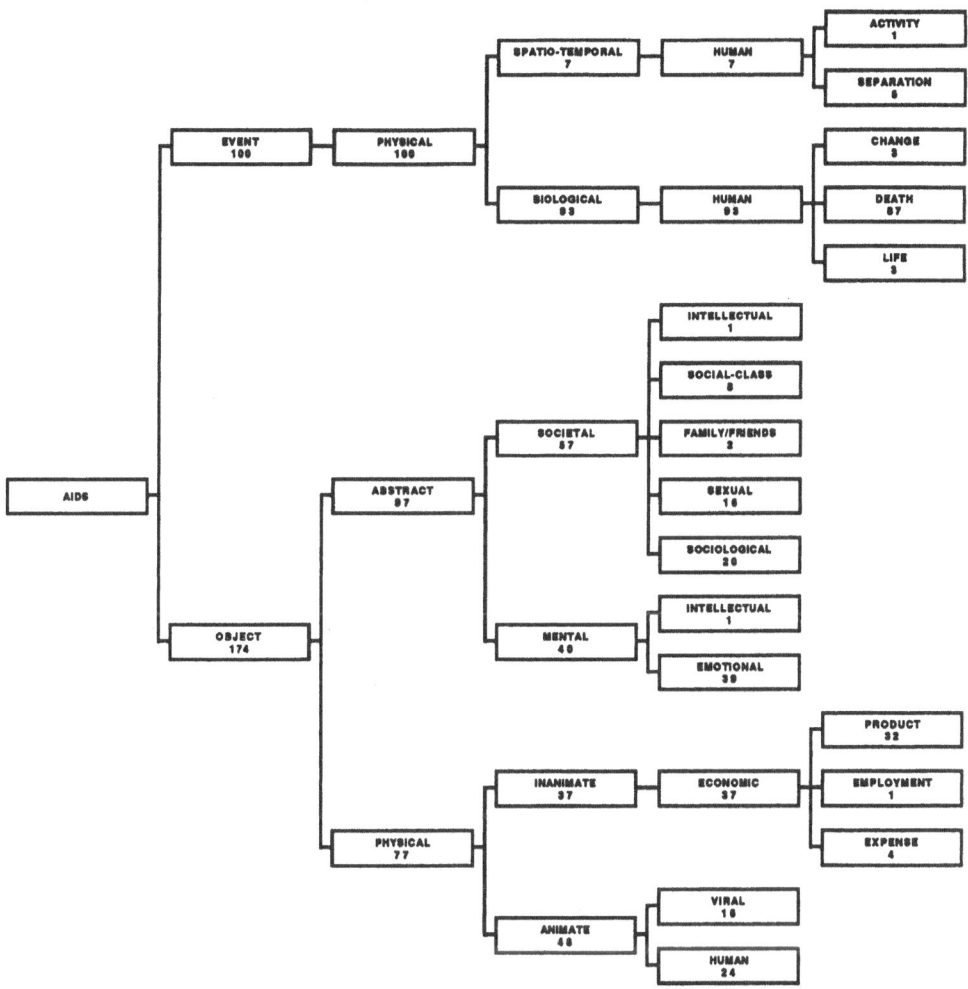

Fig. 3. Direct Consequences of AIDS. (Numbers in boxes are frequencies.)

In addition to the risk beliefs described above, ARK collected from subjects data on the sources and salience of those beliefs (Table 3). For the primary consequences of AIDS and toxic waste, the news media were the most prevalent sources of information, followed by common knowledge. These results dramatically illustrate the importance of the news media. On the other hand, especially with the AIDS data, information about secondary effects is highly influenced by personal and acquaintance experiences. Apparently, beliefs about secondary effects are based on anecdotal evidence. In addition, even though the science-related source of belief categories were often chosen by the respondents, most risk beliefs are not based on scientific evidence directly available to the subjects.

The salience data also suggest that the secondary consequences are more personal because the salience of positive secondary consequences is much higher than the salience of all other primary consequences (Table 3). Thus, only two levels into peoples' belief structures, we appear to find the emergence of the very fundamental concerns of individuals.

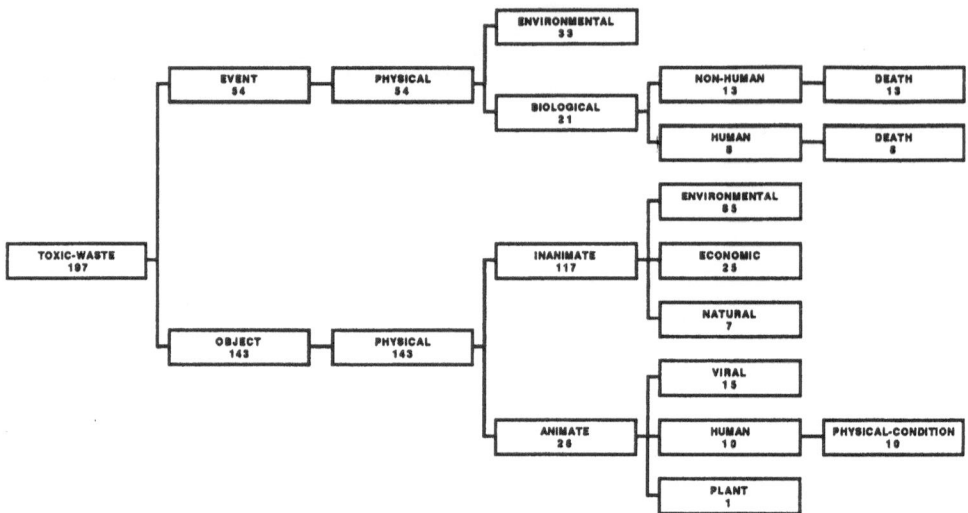

Fig. 4. Direct Consequences of Toxic Waste. (Numbers in boxes are frequencies.)

Apart from ARK, data were collected regarding subjects' age, sex, major, computer background, and interest in participation. Initial data analysis has not revealed any significant correlations between these demographic variables and the salience or source of risk beliefs.

DISCUSSION

The results presented above indicate that the ARK software has been able to collect new types of data which may prove very useful to risk analysis because they capture actual concepts used by subjects when considering risk sources and relationships between concepts. The data indicate, in terms chosen by the subjects, the intense emotional aspects of AIDS and the threat of toxic waste to the environment. The news media strongly influence risk beliefs, although secondary risk beliefs are based on anecdotal evidence. All of the consequences of AIDS and toxic waste rate high in personal and societal salience.

This research can be improved in many ways. Of particular importance, a broader sample of people can be surveyed, from experts to those at different levels of risk from different risk sources. Also, the future research in this area should attempt to create more sophisticated risk belief data bases. For example, subjects could discuss more risk sources and be encouraged to discuss secondary (and beyond) consequences. Richer data bases will provide more detail on how people believe the world works. Complementary research should strive to improve and validate the conceptual net of Fig. 1.

Work also is needed to adapt this type of data to policy analysis purposes. Methods are needed to synthesize the propositional, likelihood, and salience data into an aggregate measure instead of continuing to use separate frequencies. Also, aggregated data, prepared from data bases that encompass all mentioned consequences, should be analyzable using graphical or other techniques. Because everything is computerized, query systems could be designed to provide policy analysts with important information. If the data bases were kept separate, extremely rich ones could be queried as if one were asking the human source. Thus, instead of having to go back to subjects with new and unexpected queries, the data bases could be queried. ARK would need extensive modifications to its knowledge representation scheme and its user interface before this idea could be explored.

Table 2. Secondary Consequences

AIDS AGENT	CONSEQUENCES
Sexual-abstraction	Field of study, Physical inheritance, mental disorder, emotional abstraction, viral object
Economic product	Economic expense, employment, emotional abstraction, human death
Physical condition	Physical condition, viral object, activity human spatio-temporal, death human
Social abstraction	Environmental limitations, emotional abstraction
Social class	Sociological abstraction, emotional abstraction, intellectual societal abstraction, human death
Death human	Change human biological, economic expense, emotional abstraction, separation human
Family friends abstraction	Emotional abstraction
Change human biological	Change human biological, family friends abstraction

TOXIC WASTE AGENT	CONSEQUENCES
Environmental object	Death non-human, physical condition, environmental object, death human, environmental event, employment, economic product
Viral object	Change human biological, death human
Economic product	Environmental object, locational physical
Environmental event	Death non-human, viral object, environmental object
Constructed place	Environmental object
Physical condition	Physical condition, change human biological, emotional abstraction, human offspring
Death human	Location abstraction, viral object, change human biological, physical condition, separation human, life human

Table 3. Mean Values for Sources and Salience of Risk Beliefs

Variable	AIDS Primary Consequences	AIDS Secondary Consequences	Toxic Waste Primary Consequences	Toxic Waste Secondary Consequences
Source of Risk Belief				
Personal Experience	.05	.41	.24	.38
Acquaintance Experience	.11	.53	.16	.23
News Media	.75	.32	.91	.46
Scientific Fact	.27	.12	.20	.28
Religious Doctrine	.04	.03	.01	.18
Common Knowledge	.54	.56	.51	.49
Authoritarian Doctrine	.05	.03	.01	.13
Scientific Research	.37	.35	.27	.46
Mean Number of Sources	2.20	2.35	2.31	2.62
Salience of Consequences				
Negative Personal	−77	−76	−81	−81
Negative Societal	−68	−73	−67	−77
Positive Personal	73	95	75	92
Positive Societal	65	88	58	89

ACKNOWLEDGMENTS

Oak Ridge National Laboratory is operated by Martin Marietta Energy Systems, Inc., under Contract No. DE-AC05-84OR21400 with the U.S. Department of Energy.

REFERENCES

Anderson, J., 1982, "Acquisition of Cognitive Skill," *Psychological Review* **89**:369-406.

Barsalon, L. and Bower, G., 1984, "Discrimination Nets as Psychological Models," *Cognitive Science* **8**:1-26.

Brachman, R., and Schmolze, J., 1985, "An Overview of the KL-ONE Knowledge Representation System," *Cognitive Science* **9**:171-216.

Cohen, B. and Murphy, G., 1984, "Models of Concepts," *Cognitive Science* **8**:27-58.

Fiksel, J. and Covello, V., 1986, "Knowledge Systems for Risk Communication," presented at the NATO Advanced Research Workshop on Expert Systems and Expert Judgment, Porto, Portugal, August.

Fischhoff, B., 1984, "Informed Consent for Transient Nuclear Workers," in R. Kasperson and R. Kates (Eds.), *Equity Issues in Nuclear Waste Disposal*, Oelgeschlager, Gunn, and Hain, Cambridge, MA.

Fischhoff, B., Watson, S., and Hope, C., 1984, "Defining Risk," *Policy Sciences* **17**:123-139.

Harding, C. and Eiser, J. R., 1984, "Characterizing the Perceived Risks and Benefits of Some Health Issues," *Risk Analysis* **4**:131-141.

Johnson, H., 1987, unpublished material, Martin Marietta Data Systems, Denver, CO.

Kahneman, D., Slovic, P., and Tversky, A (Eds.), 1982, *Judgment Under Uncertainty: Heuristics and Biases*, Cambridge University Press, New York.

Kasperson, R., 1986, "Six Propositions on Public Participation and Their Relevance for Risk Communication," *Risk Analysis* **6**:275-281.

Keeney, R. and von Winterfeldt, D., 1986, "Improving Risk Communication," *Risk Analysis* **6**:417-424.

Schank, R. and Riesbeck, C. (Eds.), 1981, *Inside Computer Understanding*, Lawrence Erlbaum, Hillsdale, New Jersey.

Simon, H., 1979, "Information Processing Models of Cognition," *The Annual Review of Psychology* **30**:363-396.

Slovic, P., Fischhoff, B., and Lichtenstein, S., 1982, "Facts Versus Fears: Understanding Perceived Risk," in D. Kahneman, P. Slovic, and A. Tversky (Eds.) *Judgment Under Uncertainty: Heuristics and Biases*, Cambridge University Press, New York.

Slovic, P., Fischhoff, B., and Lichtenstein, S., 1984, "Behavioral Decision Theory Perspectives on Risk and Safety," *Acta Psychologica* **56**:183-203.

Tonn, B. and Arrowood, L., 1987, "ARK – Acquiring and Reasoning About Knowledge," Proceedings of the Third Annual Expert Systems in Government Conference, Washington, D.C., Oct. 21-23.

Sources of Common Cause Failures in Decision Making Involved in Man-Made Catastrophes

Edwin L. Zebroski

Electric Power Research Institute
Palo Alto, CA

ABSTRACT

Four major accidents in large-scale engineering operations are analyzed for the apparent attributes that contributed to key decisions preceding the catastrophic event. The accidents considered include: Bhopal, Challenger Shuttle, Three Mile Island, and Chernobyl. The analysis focuses on the key decisions and assumptions that would have seemed implausible to, or would have been treated differently by, other similarly qualified operations management teams (based on generally available knowledge and practices contemporaneous to the decisions made). The analysis was done to identify measurable attributes that may be causative or associated with common cause failure of perception (CCFP) by a management team.

Ten attributes are identified that were found to be common to all four cases, and an eleventh that was common to three of them. The thesis of Irving Janis that a high degree of common background in the management group leads to CCFP is not necessarily applicable to these accidents. Organizational style and structure, involving multilayered hierarchies with diffuse responsibilities and difficult communications, appear more fundamental in providing the attributes preceding catastrophes.

KEYWORDS: Operations management, man-made catastrophes, common cause failure of perception, engineering operations, safety analysis

INTRODUCTION

A study published in 1973 proposed a thesis for the cause of errors in decision making that contributed to major disasters in foreign policy. Janis correlated such errors with a high degree of commonality of backgrounds of the controlling groups (Janis, 1973, 1983). This was especially striking in cases for which there was a generally high level of relevant skills in a group.

This conjecture was tested in the context of large engineering operations, leading to somewhat different correlations of attributes that appear to be more fundamental than commonality of background experience.

COMMON CAUSE FAILURE OF PERCEPTION IN DICISION-MAKING

For the purposes of this paper, the phrase "Common Cause Failure of Perception" (CCFP) is used to indicate the phenomenon of group acceptance of sets of assumptions and givens that would be likely to be challenged or tested by other groups with more diverse backgrounds.

The approach used to test this conjecture is to consider case histories of four major man-made catastrophes. These are analyzed using published material to determine the major assumptions, explicit or implied by the decisions taken (or omitted) in engineering operations, that were instrumental in producing an unfortunate outcome. The extent to which these basic assumptions were open to substantial question on the basis of generally available contemporaneous knowledge is taken as a possible indication of the effect of CCFP. The attributes that appear to make CCFP likely are derived, and reasonable causative relations can be inferred but not proven. Where the available documents do not provide an explicit record of decisions taken or omitted, the remedies instituted after the accident are used to define the nature of the deficiency.

The four examples selected are: The accident at Bhopal, India, December 1984; the Challenger shuttle accident, January 1986; the accident at Three Mile Island, March 1979; the accident at Chernobyl, April 1986.

Parallelism in some attributes has been noted by others; for example, some similarities of Chernobyl and Challenger were noted by Bernstein and Kushment (1987) as follows:

"Large-scale engineering systems are more than a collection of technological instruments; they are a reflection of their societies and of the management practices and bureaucratic procedures." This quotation was in connection with the apparent parallelisms of deficiencies in the management of design and operation in both the Chernobyl accident and the Challenger accident. The authors remarked, "Chernobyl and the Challenger explosion had a surprising symmetry in some aspects."

The elements of symmetry they remarked upon include:

- A long-term successful program, dating back to the 1950s for Chernobyl, the 1960s for Challenger.

- Major military involvement in the design, objectives, and control of both programs.

- A transition over time from a mission-oriented effort to ordinary repetitive events.

- The National Aeronautics and Space Administration (NASA) began to run the shuttle program as though it were a commercial airliner. Soviet management went into mass production of 1,000-megawatt reactors.

- As technical matters became routine, Soviets politicized their program with questions of party loyalty; adherence to bureaucratic procedure appeared to take precedence over technical ability in the selection of personnel and managers. In NASA, flow of information became very compartmentalized, with safety memos regarding safety features, including "0" rings, having limited circulation. "Bureaucratic compartmentalization in an open society produced results not unlike those in the Soviet program" (Bernstein and Kushment, 1987).

In this paper, a wider range of attributes is considered and compared for the four man-made catastrophes.

THE ACCIDENT AT BHOPAL, INDIA

In December 1984 a chemical plant built by Union Carbide Company, and operated by an Indian affiliate, sustained an operating accident. Tonnage amounts of a toxic volatile substance, methyl isocyanate (MIC), escaped to the environment. Approximately 20,000 people were sickened by the exposure, of whom about 2,000 died within the first two or three weeks. An additional 10 to 15 people died each month for several months after the accident. Some health effects persist, involving respiratory insufficiencies. The accident apparently resulted from the inadvertent introduction of water into a large tank containing 45 tons of methyl isocyanate contaminated with chloroform. The reaction of water with the isocyanate overheated the tank contents, which then vented to the atmosphere through a relief valve. Scrubbers and flares that were supposed to retain vapors of methyl isocyanate did not function.

Key Decision Points

1. The decision to build the plant was made in the context of a large market for the pesticide carbaryl for Indian agriculture. The Indian Government required that there be a majority of local participation in the construction and operation of the plant. A situation of divided responsibilities for managing and monitoring operating policies, personnel selection and performance, training and procedures increasingly developed after the initial startup team was lost. (That team was trained on a similar facility at Institute, West Virginia, under Union Carbide supervision.) The last experienced American supervisor left the plant in 1982.

2. Design and construction of the plant was entirely under Union Carbide control and supervision. The design included many thoughtful features for protection of the system, including overtemperature and overpressure alarms, provision of large vent scrubber capacity, provision for refrigeration of tank contents, and a bunkered concrete covered earth mound to protect the MIC storage tanks from external damage and temperatures. However, the scenario assumptions for protective systems were incomplete (for example, in respect to possible contamination of the tank contents by water, chloroform, or other chloride-containing substances that could attack the stainless steel, or the inadvertent introduction of iron or iron compounds in the form of rust). Particularly hazardous were sneak circuits in the piping that could permit the introduction of water. Water reacts exothermically with MIC, potentially violently at elevated temperatures and pressures. The accident analysis by Union Carbide indicated at least three possible sneak circuits through which water might be introduced into the tank. Plant records indicate that vent filters and a length of process piping were being flushed with water several hours before the accident.

While the initial design of the plant showed many thoughtful protective features and alarms, the absence of a systematic and continuing design review is evident. The analysis of the accident included reproducing the chemical conditions that resulted in the overtemperature, overpressure and venting of methyl isocyanate. The chemical experiments established that, in addition to 45 tons of methyl isocyanate, the tank had to have contained one-half to one ton of water and one to one and a half tons of chloroform. A variety of mechanisms were identified by which such contaminants could enter the tank. A systematic design review could have uncovered the variety of postulated sneak circuits by which water might inadvertently or intentionally be introduced into the tank.

3. Operational supervision and audit was weak, apparently due to confused responsibility for the plant operation between the Indian affiliate company management (Union

Carbide, Ltd.) and the Union Carbide Company that was the majority owner. Routine safety reviews were apparently performed, but not reviews to detect and correct deviations from operating procedures, product specifications, and correct preventive maintenance. A review and audit done by Union Carbide in 1982 noted many deficiencies, but there was apparently no effective followup to ensure that design, procedural, and training deficiencies were corrected.

Absence of an effective on-going operating review and audit process is indicated by the number of conditions in violation of process specifications that had been permitted to occur and persist for considerable periods of time without correction. These included, for example:

a. The storage tank was not leaktight to a required pressure test, indicating the likelihood of sneak circuits for entry of water, for example, from vent lines. This also prevented the transfer of out-of-specification product to a rework tank.

b. A refrigeration system designed to keep tank contents at near zero degrees centigrade was not in operation.

c. The vent gas scrubber had been left in a standby mode for over one month prior to the accident. A flare was not operating.

d. Chloroform contamination of the product was 4 to 5 times higher than the product specification, above which rework by redistillation would be required, but corrective action was not taken.

4. Systematic analysis and training for severe events was present to some extent, but not comprehensive. Emergency procedures and drills for leaks of toxic gases were routinely implemented, since phosgene was routinely handled in the plant in the production of methyl isocyanate. Also, significant attention was paid to fire control measures. However, there was no systematic analysis of low probability–high consequence conditions that could lead to major catastrophes. The procedures and training correspondingly did not sensitize the plant personnel to the importance of various seemingly minor deficiencies that could combine to produce a major disaster. There was no organized system to distinguish alarms and sirens used for routine drills from serious warnings to plant people and local citizenry.

THE CHALLENGER ACCIDENT

The spectacular and tragic explosion of the shuttle booster soon after launching was viewed by hundreds of millions of people. It was also filmed and otherwise recorded from dozens of viewpoints. The precise sequence of events and their root causes remained elusive for many months. There is now general agreement that an "0" ring seal, exposed to lower temperatures at launch time than it was designed and tested for, failed on one of the booster rockets. This led to structural damage of the main propulsion rocket, releasing hydrogen and oxygen and producing a massive explosion.

A Presidential Commission reviewed the circumstances of the accident and of the NASA management systems. In addition, other studies were conducted by Congress, by the General Accounting Office, by the National Academy of Sciences and by NASA itself. A substantial consensus emerged that the management and risk control functions were significantly inferior to those used in other space programs, notably the unmanned space probes and the Apollo moonlander program. There is also some degree of consensus that the program was deeply flawed by the initial acceptance, and continued honoring, of combinations of performance goals and schedules, and of cost goals and constraints, that involved highly implausible assumptions from the outset. Some highly respected and experienced dissenting voices were raised. However, the decision process on the technical specifications and performance goals became increasingly enmeshed in, and dominated by, the politics of the year-to-year and long-term budgeting process.

The contributing and directly causative hardware issues now appear to be reasonably well defined in the available reports (see reference listing).

Several important management reviews continue. The general outlines of the attributes contributing to the catastrophe are now reasonably well-founded. However, more sharply focused delineations of these attributes may still be possible.

Key Decision Points

1. The decision to focus on a manned space program does not of itself appear to be a central problem precluding the achievement of good safety practices with finite resources. However, the further decision to combine the many conflicting specifications for capabilities for launching of both commercial and military satellites, to a variety of low and high orbits, together with the requirements for manned space flight and space station assembly and resupply, was a large step into unknown territory. The further decision to pursue such a combined objectives program to the exclusion of continued development and deployment of expendable launch vehicles was a significant factor in the pressures that ultimately compromised the attainable safety levels of the shuttle program. With one dominant exception, most of the other key decisions were severely constrained by the increasingly negative margin between the cost, the schedule, the performance objectives, and the resources allocated to the project.

2. The decision to proceed with a hydrogen-fueled main booster and strap-on solid fuel boosters, while maintaining target payload size and weight, appeared to preclude launch abort personnel survival features similar in function to those provided in the Apollo manned space program. (These have now been provided for future designs.) It was taken as a working assumption that any of the large variety of potential failures on launch would be so infrequent as to be an acceptable risk.

(Launch failure statistics from considerably simpler systems than the shuttle tend to support estimates of at least one failure in 20 or 30 launches, essentially the level actually experienced with the Challenger accident.)

3. Several elements of the decision making and organizational situation have been faulted, as indicated in the other items. Perhaps the most dominant "common cause failure of perception" was the reluctance to use systematic risk analysis. This is an available and proven technique for recognizing and managing risk exposures. This technique was readily available, it was effectively used in the unmanned space program, and its use to guide risk decisions would not have been appreciably limited by budget or schedule constraints. This implicit decision resulted in policy and philosophical resistance to the use of systematic integrated risk assessment techniques and the associated corrective processes. Some localized systematic risk assessment was done at subsystem levels. Integrated risk assessment and probabilistic risk assessment were (and apparently still are) not highly regarded.

4. Organizational responsibility for systems safety (and the related control authority in respect to design, quality control, and quality assurance) was not adequately integrated and available at decision-making levels. The complex program involved many different contractors for structures, propellants, control systems, instrumentation, computers, life support, etc., each involving design decisions influencing safety, as well requiring intensive quality control and quality assurance to implement, test and maintain the designs. In the

absence of a structured process to integrate safety-related analysis and conformance to specifications, a feature such as the safety margins and temperature limits on "0" rings was perhaps several organizational levels and at least two contractual interfaces removed from people involved in the decisions on schedules and go-ahead for launch.

Memoranda and analyses raising concerns on performance and safety issues were subject to many delays in transmission up the organization chains, as well as being subject to numerous stages of editing and potential vetoes on further transmittals.

Anecdotal information (possibly apocryphal) suggests that NASA management considered that any probabilistic assessment, whether it resulted in 1%, 4%, or 10% risk of launch failure, would be politically unacceptable, and on this basis rejected the use of the technique. This view apparently inhibited its use for the more fundamental purpose of providing orderly tracking of risk exposures indicated by tests, by prior launch experience, and by deterministic analyses. Typically, when such risk analyses are available, a small number of failure scenarios account for a large fraction of the total risk estimated. Development and testing and launch decisions can then be focused on providing the appropriate attention to the dominant risk contributors.

THE ACCIDENT AT THREE MILE ISLAND

Between 4:00 and 8:00 a.m. on March 26, 1979, the unit 2 reactor at Three Mile Island sustained a prolonged period during which a significant part of the inventory of cooling water in the reactor system was lost through an open relief valve. The core overheated, and the metal components in the upper part of the core region melted, forming a layer of resolidified metal and debris near the mid-plane of the core. This layer effectively acted to block cooling from the lower part of the core, and a portion of the fuel as well as metal components melted and flowed to the bottom of the reactor vessel. The lower part of the reactor vessel was still full of water, and the molten material resolidified in the bottom of the vessel. In the course of this, essentially all of the gaseous radioactivity (xenon and krypton) was liberated from the fuel and escaped via the open relief valve to the containment building.

The zirconium cladding of the fuel, when overheated, reacts with water and steam to produce hydrogen. This hydrogen, along with the fission product gases, also leaked into the containment building. The amount of hydrogen was sufficient to exceed the flammability levels for mixtures of hydrogen and oxygen, and a "deflagration" occurred, momentarily raising the pressure in the containment building to about 28 psi, about half of the design pressure of the containment building. Through leakage paths between the containment building and the auxiliary building, some of the water contaminated with the mixture of radioactivity from the damaged fuel, including radioiodine, reached the auxiliary building, along with a large part of the short-lived fission gases, namely xenon and krypton. Eventually, all of the xenon, and a portion of the krypton, escaped to the environment, along with small traces of radioiodine.

Extensive reviews by national commissions and by local health authorities place the environmental impact of the accident as between 0 and 1 additional cases of cancer over the next forty years. (A 360 degree monitoring network did not actually show exposures off site appreciably above natural background.) A much greater impact occurred from public apprehension, due in part to public officials' and media statements of alarm.

Key Decision Points

1. The project was initiated in response to projected load growth in the Pennsylvania-New Jersey area served by General Public Utilities (GPU) and an operating subsidiary

company, Metropolitan Edison. Plans for the project were made in the late 1960s. Construction start was delayed by a change in site from the originally planned location to Three Mile Island. Construction on the two units proceeded uneventfully, with startup of unit 1 in 1974 and of unit 2 in 1978.

2. A key decision was the selection of Babcock & Wilcox (B&W) as the reactor designer and supplier. B&W had less nuclear experience than the two earlier reactor suppliers in the United States. Its initial demonstration project, Indian Point-I, had been troubled by delays in design and construction and substantial overrun in costs. The B&W reactors were generally similar to other pressurized water reactors being built by Westinghouse and Combustion Engineering, but the system was unique in using a once-through steam generator design. This requires more sensitive control of feedwater and more complex and sensitive operations for control of startup and shutdown.

3. The Presidential Commission studies of the Three Mile Accident remarked on the "mindset" evident in the organization. This was, in effect, that a severely damaging event could not happen. Ordinary equipment problems were expected—and had been intensively experienced during the first two years of operation of unit 1 on the same site. Unit 2 had apparently benefited from unit 1 experience. The mindset was not limited to the utility and its operators. The Nuclear Regulatory Commission (NRC) also shared this mindset to some degree, since its regulation did not require many of the actions and procedures that were found to be important after review of the accident.

4. Some of the principal assumptions or "givens" that contributed to the accident were as follows:

a. Compliance with Federal regulations assures safety. Some reviews noted that rigidity of requirements for testing and certifying operators contributed to this mindset.

b. Procedures and training for frequent system upsets was sufficient.

c. Systematic reporting and documenting of minor accidents, component failures, or other observed deficiencies was not necessary for safety or reliability. A similar event that was successfully aborted by operators 18 months previously was not reported adequately.

d. The process of imagining possible severe damage events and scenarios that might lead to accidents involving severe damage, and possible defense measures, was limited to the studies required for plant licensing.

The plant safety analysis report (FSAR), required for licensing, showed that the plant could safely accommodate single failures (structural, electrical, etc.), and also defined "design basis accidents," but not "beyond design basis" accidents, as is now customary. Operators also had some exposure to control room simulators. However, these could only reproduce the scenarios of routine events, not actual or threatened severe accidents.

e. Skill-based training of operators and supervisors with largely fossil power plant experience was appropriate and sufficient for safe operation. Operators typically had high school education plus limited on-the-job courses required to pass licensing examinations. This training rarely would provide overall insight on how some of the plant systems functioned.

f. Control room instrumentation and control knobs, designed for normal routine operation, would be adequate for coping with unusual events or conditions that could lead to severe accidents. Subsequently, "human factors" reviews of the control room have resulted in more ergonomic and comprehensible layouts. Special consoles provide a concise overview of several dozen of the most vital parameters.

THE CHERNOBYL ACCIDENT

The factors of omission and commission include all of the factors present at Three Mile Island as well as several others. The general mindset that severely damaging events could not occur was pervasive at all levels. This is in contrast to U.S. and European design and analysis efforts, which began exploring severe events and rare "beyond design basis" scenarios in the 1970s. For example, the Reactor Safety Study (RSS), issued as report WASH-1400 in 1975, was a comprehensive probabilistic risk analysis of the two major reactor types, PWR and BWR. It came into widespread use in the West after the Three Mile Island accident.

The viewpoint expressed in international conferences by Soviet delegates was that studies of low probability events were useless, and could even increase the risks of plant outage from ordinary hardware problems (by diverting attention from the more probable occurrences that actually happen more frequently). This mindset apparently dominated the Soviet nuclear power establishment involved in research, development, design, testing and operation of the Chernobyl-type or RBMK reactors. In over 300 publications and one major book on this reactor type, there was no significant analysis or discussion of low-probability accidents with potential for severe consequences.

Key Decision Points

1. Goals and objectives of RBMK design: The design is a dual-purpose reactor capable of making weapons-grade plutonium or tritium as well as producing steam for electric power production and district heating. (There is one such dual-purpose reactor in the United States operated by the U.S. Department of Energy. It is currently shut down, although it does not have some of the most hazardous features of the RBMK type.) The requirement for military plutonium production capability was presumably more than satisfied by Soviet reactors operating in the early 1970s. A decision was reached about that time that power-only reactors should be built, rather than dual-purpose, generally similar to western PWRs. However, "Atommash," the factory complex on the Volga that builds large components and pressure vessels, was delayed 5 to 7 years in reaching full production. The RBMK type continued to be built, nominally as an interim measure, since it required less heavy fabrication equipment. However, it imposed an economic penalty (for power-only production) by requiring a much larger physical plant. Also, it included a complex machine for on-stream refueling (a relatively dangerous operation compared to off-line refueling). Finally, a judgment was made that it was impractical to house such reactors in high pressure containment buildings such as are used for Western world PWRs, and also for recent models of Soviet PWRs.

2. A key decision was made to retain the basic design of the dual-purpose reactors. This included the large spacing between tubes that was criticized in the 1977 review of the design done by a British team because it produces the runaway power characteristic. Revising the design for greater safety—for example, by making the tubes in the graphite closer together, as is done with the U.S. Hanford "N" reactor—could eliminate the runaway characteristic that destroyed Chernobyl. It could also make full containment buildings more feasible.

3. Another key decision was to retain on-stream refueling. This was useful for military production that requires a short residence time of fuel in order to make weapon-

grade plutonium. With enriched fuel (2.0% U-235), fuel residence times of nearly 10 years are available in the RBMK design.

4. The decision to omit full containment highlights the split between an old culture favoring weapons plutonium production along with power, and a newer culture favoring power production only. The newer PWRs in the U.S.S.R. have large high-pressure reinforced concrete cylindrical containments. These respond to several common sense safety criteria for keeping steam and radioactivity from reaching the environment in the event of a large pipe break or a severe accident with core damage and release of radioactivity.

5. Continuing review, audit and enforcement of safety practices and procedures was at best superficial. Contrary to practice in the Western world, the test procedure that precipitated the accident was not detailed and subject to review and approval by qualified safety engineers. Many of the steps taken during the test were improvised without review or discussion and included the successive disabling of multiple safety systems over a period of 12 hours.

6. An assumption was made that trained operators could not make extended errors—both conceptual and procedural. U.S.S.R. operators all have a 5-1/2-year engineering degree, contrasted with high school level typical of many U.S. operators. Awareness of this difference from the U.S. reinforced the Soviet overconfidence in their reactor operators. (While a degree-level engineer is now required in U.S. reactors, the factors of "mindset" and the presence or absence of consideration, analysis, and training for infrequent events that might have severe consequences is evidently much more productive of safety than a degree.)

7. Control room instrumentation and control knob layout provided for convenient routine operation, but did not give attention to the ergonomics of severe accidents. Much of the instrumentation was slow in responding, and some of it was available only as output from a teletype printer. Most seriously (and unlike Western reactors) safety systems could be easily bypassed or disabled directly from switches on the control panels.

CONCLUSIONS

A number of common attributes are evident that are shared by all of the examples considered, with only one or two notable exceptions. These attributes are listed in Table 1.

All four cases share at least ten of the attributes listed. One obvious exception is overdependence on compliance with rules, which was not evident in the case of Chernobyl.

The thesis of Irving Janis that excessive commonality of background is a principal factor in accepting implausible assumptions does not appear to account fully for some of the attributes noted. The more fundamental characteristic appears to be the tendency of a controlling group to operate in a rigidly hierarchical style. This apparently can develop for structural reasons (such as the growth of very large and geographically dispersed organizations) rather than primarily from the degree of commonality of experience of the managing group. (It might be more likely, but not inevitable, if the members of the managing group had largely had their primary experience in such environments). A rigid hierarchy tends to foster decisions by committees, diffuse responsibility, promote organization-defensive and turf-protective decision making, and limit communication channels so that the full use of systematic and rigorous learning from experience is deterred. The integration of views and analyses in lower echelons and in support functions appears especially difficult to achieve, since the administrative skills required for large organizations are not commonly accompanied by the depth of physical, technical, and practical insights required to achieve such integration.

Table 1. Frequent/Common Attributes

1. Diffuse responsibility. Rigid procedures and communication channels; large organizational distances between decision makers and those with detailed technical awareness and competence.

2. "Mindset" that success is inevitable or routine; limited, unsystematic consideration of severe risk exposures.

3. Belief that rule compliance is a principal factor in achieving adequate levels of safety.

4. Strong emphasis on commonality of experience and viewpoint; team-player characteristic highly valued.

5. Review of generally relevant experience neglected.

6. Disregard of lessons learned from prior events.

7. Safety analysis and preservation of safety factors subordinate to other performance goals in management and operating priorities.

8. Absence of effective procedures, training, and drills for unusual or severe conditions.

9. Acceptance of design and/or operating features involving hazards that were recognized, controlled or avoided elsewhere.

10. Underutilization of available project management techniques for systematic risk assessment and control.

11. Organizations lacking defined responsibility and authority for integration of safety issues.

REFERENCES

General

Irving Janis, *Victims of Groupthink*, Cambridge: Harvard Univ. Press, 1973, rev. 1983.

Bhopal

Union Carbide Corporation, "Bhopal Methyl Isocyanate Incident Investigation Team Report" (March 1985).

New York Times Investigation Report, New Delhi (January 29, 1985); also, *San Francisco Chronicle* (January 29, 1985).

W. Lepowski, "Bhopal Special Report," *Chemical & Engineering News*, pp. 18-32 (December 2, 1985).

R. E. Taylor, "Carbide Bolstering Methyl Isocyanate Precautions," *Wall Street Journal*, p. 34 (March 25, 1985).

T. Gladwin and I. Walter, "Bhopal and the Multinationals," *Wall Street Journal*, Op. Ed. page (January 16, 1985).

New York Times, "Union Carbide on Bhopal," p. 48 (March 21, 1985).

P. Shrivastava, *Bhopal: Anatomy of a Crisis*, Cambridge: Ballinger, 1987; also, reviewed in *Science* **236**:979 (May 22, 1987).

Challenger

T. Bell and K. Esch, "The Fatal Flaw in Flight 51-L," *IEEE Spectrum*, p. 36 (February 1987).

Report of the Presidential Commission on the Space Shuttle Challenger Accident (Roger's Commission), Washington: U.S. Government Printing Office, June 1986.

U.S. Congress, House of Representatives, Investigation of the Challenger Accident, House Committee Report 99-1016, Washington: U.S. Government Printing Office, October 29, 1986.

R. C. Cook, "The Challenger Report: A Critical Analysis of the Report to the President" (October 30, 1986).

H. Esch and T. Bell, "How NASA Prepared to Cope with Disaster," *IEEE Spectrum*, pp. 32-36 (March 1986).

Challenger—A Major Malfunction, New York: Doubleday & Co., 1987.

J. M. Logsdon, "The Space Shuttle Program: A Policy Failure?" *Science*, pp. 1099-1105 (May 30, 1986).

A. Bernstein and M. Kushment, *IEEE Spectrum*, p. 8 (April 1987).

J. B. Hammack, NASA-JSC Space Shuttle Safety Risk Management System (December 1986).

R. H. Kohrs, "Methodology for Conduct of NSTS Hazard Analysis," NSTS 22254 (May 1987).

A. D. Slay, "Shuttle Criticality Review and Hazard Analysis Audit," Interim Progress Report to National Research Council (July 1987).

C. S. Harlan, "Risk Management Approach for Manned Space Shuttle Operations," Society for Risk Analysis (October 1985).

Three Mile Island

Report of the President's Commission on the Accident at TMI (Kemeny Report), Washington: U.S. Government Printing Office, October 1979.

Three Mile Island Report to the Commissioners and to the Public (Rogovin Report) NUREG/CR 1250 (February 1980).

Electric Power Research Institute, Analysis of the TMI Accident NSAC-1 (July 1979, revised March 1980).

Investigation of TMI Accident, Office of Inspection and Enforcement, NUREG 0600 (August 1979).

E. L. Zebroski, "Nuclear Reactor Safety: Lessons of TMI," in *McGraw-Hill Yearbook of Science and Technology*, pp. 30-39, New York: McGraw-Hill, 1981.

L. M. Toth *et al.*, "The Three Mile Island Accident," American Chemical Society Symposium Series, Washington, (1986).

Chernobyl

U.S.S.R. State Committee on Atomic Energy, "The Accident at the Chernobyl Nuclear Power Plant and Its Consequences," Presented August 25-29, 1986, in Vienna (also English translations by IAEA and U.S. Department of Energy).

"Report on the Accident at the Chernobyl Nuclear Power Station," Joint report by U.S. DOE, EPRI, EPA, FEMA, INPO and NRC, NUREG-1250 (January 1987).

M. Goldman *et al.*, "Health and Environmental Consequences of the Chernobyl Nuclear Power Plant Accident," Report to U.S. DOE, DOE/ER-032 (June 1987).

J. Gittus *et al.*, "The Chernobyl Accident and Its Consequences," NOR-4200, London: UKAE (March 1987).

E. L. Zebroski, "The Nuclear Accident at Chernobyl U.S.S.R.," in *1988 McGraw-Hill Yearbook of Science and Technology*, New York: McGraw Hill (1987).

Game Strategic Behavior Under Conditions of Risk and Uncertainty

M. L. Livingston
University of Minnesota
St. Paul, MN

ABSTRACT

This paper analyzes game strategic behavior under conditions of risk and uncertainty when interest groups are participants in the decision making process (specifically in the context of impact assessment). The incentives that contribute to game strategic behavior are analyzed. It is argued that the inherent uncertainties in forecasting are critical in allowing strategic behavior to develop. Strategic manipulation of information is manifest in the form of (1) information overload and (2) support for extreme scenarios. These forms of "rent seeking" by interest groups produce serious transaction costs (costs attributable to decision making per se) and threaten to stymie decision making. Innovations in the treatment of risk and uncertainty are necessary to facilitate decision making.

KEYWORDS: Game strategic, participatory decision making, impact assessment, rent seeking

INTRODUCTION

Increasingly, in both private and public sectors, decision makers must consider input from various interest groups in decisions of regional and national import. Very often, interest group input comes in the form of contributions and responses to formal impact statements. As a rule, impact analysis is conducted in the context of risk and uncertainty. And uncertainty raises special issues regarding interest group behavior.

Economic models of interest group behavior constitute relatively new extensions in the domain of economic theory. Olson (1965) specifies the economic factors that influence interest group organization. Kreuger (1974) and Buchanan et al. (1980) have described the process whereby organized interest groups pursue favorable institutional arrangements at the expense of other interests and of economic efficiency in general. And Olson (1982) hypothesizes that when special interests proliferate, transaction costs result that stymie decision making.

The objective of this paper is to demonstrate the relevance of the models cited above to the problem of participatory decision making under uncertainty. Impact assessment is specifically addressed. The next section explains the incentive structure encountered in a participatory setting. Rent seeking theory is employed to address the typical range of interest groups involved in impact assessment and their distinct objectives. Subsequently, the critical role of uncertainty in rent seeking is discussed. Uncertain forecasts allow for manipulation of key assumptions, which results in support for extreme scenarios and information overload. This paper argues that strategic use of information combined with the veto power implicit in the participatory decision making process produces unprecedented transaction costs and threatens to stymie decision making.

THE INCENTIVE STRUCTURE ASSOCIATED WITH IMPACT ANALYSIS

Understanding the objectives sought by various interest groups and their attempts to influence decisions is critical in explaining the game strategic behavior under uncertainty. Rent seeking theory, as advanced by Kreuger (1980) and Buchanan *et al.* (1980), has high descriptive relevance in this regard. "Rent seeking theory" and the "logic of collective action" include systematic treatments of strategic behavior by interest groups in pursuit of favorable policies and decisions.

In the *Logic of Collective Action* (1965) Mancur Olson explains the impetus for individuals to organize in order to express economic or political power. Two key factors that figure in the success of individuals in organizing formal interest groups are (1) expected benefits, i.e., the magnitude of economic values at stake for each individual, and (2) the transaction costs associated with organization itself.

Benefit/cost analysis can be used to demonstrate that formal organizations are likely to exist for specialized groups with high expected benefits and few members. Similarly, the most zealous individuals in a group will tend to become leaders. The typical structure of benefits and costs can account for the existence of many "special" interest groups, and relatively few "public" interest groups.

Rent seeking theory is useful in explaining the behavior of interest groups once they are organized. This model relies on application of utility theory in a political rather than a market exchange context. Individuals are assumed to exhibit rational political behavior in pursuit of specified objectives. Given a solid notion of the objectives and decision rule faced by participants, rational behavior can be routinely deduced. The following paragraphs outline the typical range of interests involved in impact assessments distinguished both in terms of location and ideology.

Formally, a simple model of objectives pursued by the typical range of interest groups involved in impact assessment can be represented as follows:

$$U^E = f\left(\sum_{i=1}^{n} g_i, r, a\right) \tag{1}$$

where

U^E = utility of environmentalists,
g = growth in jurisdiction $i = 1...n$,
r = recreation opportunities,
a = aesthetic integrity.

Clearly,

$$\frac{dU^E}{g_i} < 0 , \qquad \frac{dU^E}{dr} > 0 , \qquad \frac{dU^E}{da} > 0 .$$

$$U^{D_i} = v(g_i, j_i, e_i) , \tag{2}$$

where

U^{Di} = utility of development officials in jurisdiction,
g_i = growth in jurisdiction i including secondary impacts,
j_i = jurisdictional power/independence,
e_i = excess of municipal revenues over expenditures in region i.

In this case,

$$\frac{dU^{Di}}{dg_i} > 0 \ , \qquad \frac{dU^{Di}}{dj_i} > 0 \ , \qquad \frac{dU^{Di}}{de_i} > 0 \ .$$

$$U^c = w(p) \ , \tag{3}$$

where

U^C = utility of resource competitor,
p = price of resource,

and

$\dfrac{dU^c}{dp} > 0$ under circumstances of industry contraction when the competitor is selling the resource,

$\dfrac{dU^c}{dp} < 0$ under circumstances of industry expansion when the competitor is purchasing the resource.

The foregoing model yields two insights that are key to understanding interest group strategy in a participatory setting. First, the range of goals sought is extremely diverse. Equation (2) in itself captures the diverse goals of city, suburban, and regional developers distinguished by location. Insofar as objectives reflect value systems, it is clear that interest groups are involved in a fundamentally moral disagreement.

Second, in accordance with the logic of collective action, interest groups will not mirror the general public. Organized interest will tend to represent "special" rather than "public" interests. Moreover, because of their perception of the stakes involved, the most zealous individuals tend to hold disproportionate power within interest groups. Moderates rarely become leaders.

The next section focuses on strategic behavior by organized interest groups and their interplay in the decision making process. The role of information is central in this regard. Through manipulation of uncertain forecasts, interest groups manage to transform a fundamentally moral disagreement into a factual dispute. A contest of wills becomes a contest of information.

STRATEGIC USE OF INFORMATION AND TRANSACTION COSTS

Typically, the purpose of impact assessment is to provide better information about the impact of potential decisions. The general problem of obtaining and evaluating information is enormously exacerbated by uncertainty. Uncertainty comes into play via an unavoidable element in impact assessment: forecasting. The future is inherently uncertain and successful planning depends directly on visions of the future.

The principle of diminishing marginal returns applies to forecasting. In the limit, additional time and effort devoted to forecasting does not guarantee any improvement in accuracy or certainty. In the final analysis, there is no method whereby one can prove many estimates key to a particular decision. If the problem were one of risk rather than uncertainty, expected values could be calculated and used to guide the decision according to whether they are risk averse, risk neutral, or risk seeking. Unfortunately in most cases probability distributions are not known.

The impossibility of perfect forecasts creates the opportunity for interest groups to strategically manipulate information. Uncertainty and the potential for manipulation constitute the crucial elements in producing the enormous transaction costs often associated with impact assessments. Each interest group fabricates scenarios of the future that support its particular position. Moral arguments are transformed into disagreements regarding the "facts."

Strategic use of information by interest groups often manifests itself in the form of: (1) information overload and (2) adversarial use of extreme alternatives. These two actions are bound to create serious transaction costs in terms of extreme expense and extraordinary delay. Of ultimate importance, the transaction costs that result are of a magnitude that threatens to stymie decision making.

Information Overload

Ironically, in pursuit of more accurate information, impact assessment often results in information overload which may heighten uncertainty. Additional information does not necessarily increase knowledge or understanding. In fact, quantity itself may become a problem insofar as it obscures the most relevant data and contributes to confusion and delay. The factors contributing to information overload are directly attributable to the incentive structure faced by interest groups and the rules governing their interaction.

In general, information overload reflects the diversity of interests involved and each group's desire to have all viewpoints favorable to their position presented and considered. In this sense information overload reflects the complexity of the conflict.

However, basic conflict alone is not enough to produce information overload. It must be fueled by uncertainty. Uncertainty serves to magnify the conflict enormously by allowing information to multiply through manipulation. Because forecasting is inherently uncertain, the implicit rewards in analysis are for "informed skepticism." An expert is bound to protect his/her reputation by questioning estimates rather than by positing answers. Because of inherent uncertainty, experts find it easier to discredit an estimate put forth by a professional colleague than it is to substantiate, much less prove, an original estimate.

Despite potential criticisms, definitive scenarios of the future and appropriate development are regularly provided by interest groups. Technical support for such scenarios can be garnered because experts themselves disagree due to objective considerations, personal convictions, and the "marketplace for ideas." In this case, the "marketplace for ideas" refers to professionals who supply the theoretical support for any idea that demands a high enough asking price.

Due to uncertainty, there is no limit to the rounds of criticism and refutation that interest groups and their hired technicians can impose upon each other. Impact assessment becomes a reiterative process of massaging forecasts and constructing assumptions that produce "facts" that support a group's preferred position while exposing the estimates of others to a barrage of criticism emphasizing their weaknesses.

Under these conditions, the consultants and technicians charged with obtaining forecasts (specifically, forecasts favorable to their employer) become a new interest group. Assuming the objective of consultants is to maximize income, and given that compensation is usually based on time spent, the added impetus to engage in protracted and complicated studies becomes clear.

Because, as with most monumental developments, scientists don't have the option to conduct and repeat objective experiments that will lay some assertions to rest, the disagreements that arise in an impact assessment are essentially unresolvable and throw doubt on the full range of information attendant to the process. Instead of reducing uncertainty, the proliferation of estimates exacerbates the information problem. Moreover, decision making becomes bogged down in information overload. Considerable time and expense must be devoted to developing, cataloging and criticizing information. In short, transaction costs explode.

Support of Extreme Scenarios

The second way in which game strategic behavior manifests itself is in the adversarial rather than cooperative stance taken by interest groups, especially in the form of supporting extreme rather than moderate scenarios depicting the future.

This strategy arises simply because it is rational for individual groups to support extreme, adversarial positions. Table 1 illustrates the general problem in terms of the standard payoff matrix associated with noncooperative game models (Hurwicz, 1987). For simplicity, only two typical interest groups are represented, but the model is applicable to the n group case. Using 0 as a base utility level for mutual compromise, the remaining possibilities can be evaluated on ordinal grounds.

Table 1. Payoff Matrix Relevant to Impact Assessment

		Position Taken by Environmentalists	
		Compromise	Extreme
Position Taken by Developers	Compromise	(0,0)	(− −, ++)
	Extreme	(++, − −)	(−, −)

If developers take an extreme position and environmentalists compromise, developers benefit and environmentalists suffer relative to mutual compromise. The reverse case also applies. If both groups take extreme positions, delay results and both groups lose relative to compromise insofar as all interests are left with considerable time and money invested with no indication, much less any guarantee, that any eventual decision will be to their liking.

For developers one would expect the resulting preference order to be R^D (++, − −) > R^D (0, 0) > R^D (−, −) > R^D (− −, ++). For environmentalists, the preference order is R^E (− −, ++) > R^E (0, 0) > R^E (−, −) > R^E (++, − −). The overriding point is that despite the rationality of compromise from a global viewpoint, extremity is rational. When either group analyzes rational action dependent on strategic behavior by the other, taking an extreme position is always preferred to compromise. This non-cooperative game leads, unfortunately, to a unique, Pareto inferior equilibrium of (−, −). This is the classic prisoners' dilemma.

Again, the existence of uncertainty allows interest groups to fabricate support for extreme scenarios. Typically, several forecasts must be combined in order to generate a scenario depicting future conditions and their implications for resource use. Interest groups can construct extreme scenarios by using the high or low estimate for *each* key forecast. Compilation of several extreme estimates compounds the extremity in the final scenario.

Extreme scenarios impose transaction costs on the process because they exacerbate the technical problem of analyzing alternatives. As demonstrated above, the extreme scenarios proposed by diverse interest groups have no common denominator for comparison. Extreme scenarios widen, rather than narrow, the gap between interest groups and impede informed decision making.

Transaction Costs

To summarize the preceding paragraphs, fundamental uncertainty allows interest groups to strategically manipulate information key to an impact assessment. Manipulation is most often manifest in the form of information overload and support of extreme and adversarial positions. The result is enormous transaction costs in terms of extraordinary delay and expense.

It must be recognized that game strategic behavior can produce the transaction costs described only under a sufficiently diffuse decision rule. A dictator (exercising the most concentrated decision rule) can largely ignore interest group input. In contrast, in participatory decision making requiring comprehensive analysis, unanimity is effectively established as a decision rule. Unfortunately, as noted by theorists (see Mueller, 1979; Runge and von Witzke, 1987), diverse interests coupled with a unanimity rule allow for "voting by veto" by which any single group can impose and enforce inaction. Participatory decision making allows strategic behavior to produce transaction costs of a magnitude that promise to stymie decision making. This phenomenon has been observed and analyzed in a national context by Olson (1982).

The problem can be illustrated in the context of Buchanan and Tullock's public choice model (1962). Referring to Fig. 1, with a decision rule in the range of D_1, one person or group exercises considerable discretion (imposing substantial external costs on other interests), and incurs low decision making costs. When an existing decision rule exposes one or more interests to extreme pressure and is perceived as unfair, the impetus for institutional change emerges (Runge and von Witzke, 1987; Ruttan, 1984).

Conceivably, participatory decision making is a reaction to a dictator-like decision rule that is perceived as unfair in terms of the external costs imposed on other interests. Unfortunately, the potential exists for the "solution" to become one of the opposite extreme. Participatory decision making may evolve into a decision rule effectively requiring consensus, like D_2. External costs are reduced but decision making costs reach unprecedented heights. Moderation is logical, because from a least-cost perspective a decision rule intermediate to dictatorship and unanimity (like D^*) is optimal. The challenge is to implement policy innovations that hold promise in balancing the tradeoff between external and transaction costs.

CONCLUSION

In conclusion, understanding (1) the diversity of objectives involved in participatory decision making, (2) the role of uncertainty in rent seeking, and (3) the consequences of

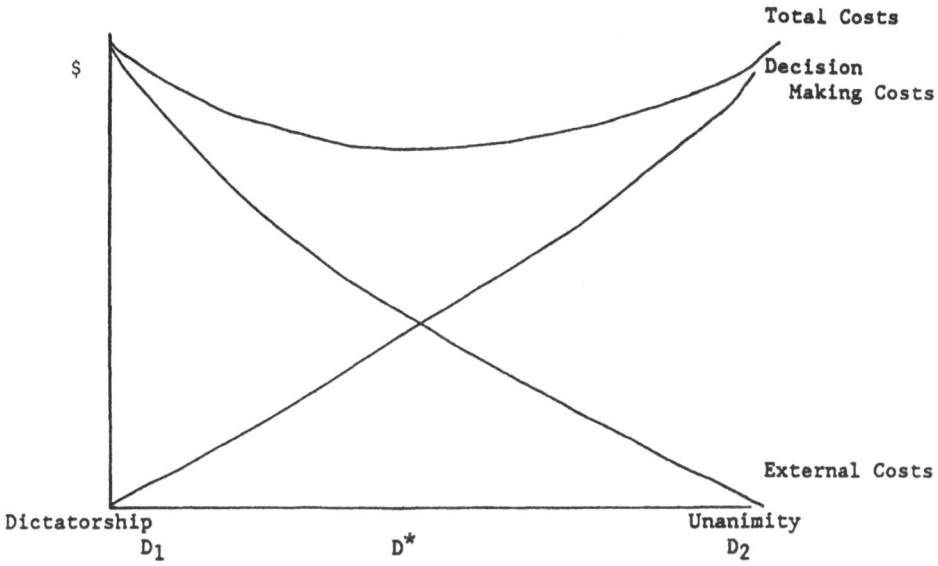

Fig. 1. Costs Associated with Alternative Decision Rules.

unanimity as a decision rule is critical in addressing decision making under uncertainty. Systematic analysis in the context of rent seeking and game theoretic models clarifies the incentives involved and provides the basis for intelligent discourse about potential solutions.

Institutional arrangements are not fixed and constant. The performance of a particular decision-making framework depends on the choice domain open (and closed) to individuals under certain circumstances, including their opportunities to manipulate the system. Uncertainty, especially as it relates to the forecasts necessary for planning, dramatically expands the opportunities for rent seeking by interest groups. The rules that apply to participatory decision making must explicitly address this reality in order to reduce transaction costs and facilitate decision making.

REFERENCES

Buchanan, J. M., and G. Tullock, *The Calculus of Consent*, Ann Arbor: University of Michigan Press, 1962.

Buchanan, J., R. Tollison and G. Tullock, Eds., *Toward a Theory of the Rent Seeking Society*, College Station: Texas A and M University, 1980.

Hurwicz, L., "Inventing New Institutions: The Design Perspective," *American Journal of Agricultural Economics* **69**, No. 2, May 1987.

Kreuger, A. O., "The Political Economy of the Rent Seeking Society," in *Toward a Theory of the Rent Seeking Society*, J. Buchanan, R. Tollison and G. Tullock, Eds., College Station: Texas A and M University, pp. 51-70, 1980.

Mueller, D. C., *Public Choice*, Cambridge: Cambridge University Press, 1979.

Olson, M., *The Logic of Collective Action*, Cambridge: Harvard University Press, 1965.

Olson, M., *The Rise and Decline of Nations*, London: Yale University Press, 1982.

Runge, C. F., and H. von Witzke, "Institutional Change in the Common Agricultural Policy of the European Community," *American Journal of Agricultural Economics* **69**(2):213-222, May 1987.

Ruttan, V. W., "Social Science Knowledge and Institutional Change," *American Journal of Agricultural Economics* **66**:549-559, 1984.

461

Conventional Wisdom on Risk Communication and Evidence from a Field Experiment[a]

F. Reed Johnson
U.S. Naval Academy
Annapolis, MD
and U.S. Environmental Protection Agency
Washington, DC

Ann Fisher
U.S. Environmental Protection Agency
Washington, DC

ABSTRACT

A recent comprehensive review of the literature identified a number of facts and principles governing risk communication. This paper evaluates several of these propositions using recent evidence from a field experiment in communicating the risks from radon in homes. At this point in the research, data relate primarily to the response of risk perceptions to different information treatments and different personal characteristics. The effect of various causal factors is sensitive to the particular test of risk perception applied. No information treatment was clearly superior for all tasks. An important implication of these findings is that risk communicators must determine what specific task or tasks the information program should enable people to do.

KEYWORDS: Risk communication, radon, risk perceptions,

INTRODUCTION

In a recent comprehensive review of the literature, Covello *et al.* (1987) identified a number of facts and principles governing risk communication. These facts and principles are based on evidence that varies from carefully controlled clinical experiments on risk perception to the intuition of practitioners with widely varying experiences of communicating risk to the public. Very little of what we "know" about risk communication is based on field data collected under controlled conditions.

This paper evaluates a number of propositions on risk communication in the literature using recent evidence from a field experiment in communicating the risks from radon in

---

a. This research was partially funded by U.S. Environmental Protection Agency Cooperative Agreements. The views expressed are the authors' and should not be attributed to the funding agency or their employers. This paper is to be published in the June 1989 issue of *Risk Analysis*. It is reprinted here with permission of the journal.

homes. Because this experiment is still in progress, insufficient time has elapsed to observe certain behavioral outcomes and some of our conclusions are still tentative. Because the research is not yet complete and because of the specific characteristics of the risk, we cannot evaluate all of the assertions made by various authors. Nevertheless, we can provide support for several important propositions and challenge others.

INDOOR RADON RISKS

The U.S. Environmental Protection Agency (EPA) estimates that radon causes more cancer deaths per year (5,000-20,000) than any other pollutant under its jurisdiction. Radon is a colorless, odorless gas that occurs naturally. It moves through the soil and becomes trapped in buildings. Since exposure occurs primarily in people's homes, conventional regulatory approaches are not appropriate. This situation has led EPA to turn to risk communication as a way of encouraging voluntary reductions in risk.

EPA has initiated several studies to investigate how people understand and react to new information on indoor radon risks. A general objective of these studies is to determine which approaches are most effective in communicating risk information. Radon provides a good opportunity for evaluating risk communication approaches for several reasons:

1. The risk is relatively unfamiliar. Unlike smoking or seat belts, radon has received significant media coverage for only a few years. In many areas where radon is suspected to be a problem, recent surveys and focus groups indicate that people know little or nothing about the origins of the problem, the health risk, or how to reduce their exposure to it. This means a risk communication program is likely to be the main source of information for the public.

2. Perceptions that individuals' rights to a safe, clean environment have been violated often complicate reactions to environmental hazards. Those affected tend to focus on determining who caused the problem and, therefore, who should be responsible for its solution. But because radon occurs naturally, there is usually no villain to blame for the problem.

3. While radon cannot be seen or smelled, testing for radon is simple and inexpensive. On the other hand, data on the cost and effectiveness of alternative mitigation techniques are limited. Sometimes mitigation is complicated and expensive. This means homeowners may be confronted with complex decisions that require considerable information.

4. Combined with data on actual exposures and survey data on risk perceptions, homeowners' mitigation choices can provide an objective measure of the impacts of a communication effort.

THE NEW YORK RADON STUDY

New York State's Energy and Research Development Authority (NYSERDA) sampled single-family homes to determine state-wide exposures to radon. NYSERDA placed three radon monitors in each of about 2300 homes. The first of these (placed in the living area) was to be sent back for analysis after 2-3 months, and the others were to be returned and analyzed after being in the homes for a year. This protocol would enable NYSERDA to judge whether the 2-3 month readings were acceptable approximations of annual averages. The homes were selected randomly within seven areas representing major geological formations across the state.

When homeowners agreed to participate in the study, they were promised the radon readings for their own homes, but it was not clear what information they would receive to interpret those readings. New York State officials were concerned about motivating households to take appropriate remedial actions without creating undue anxiety. This situation provided an opportunity to evaluate the effectiveness of alternative designs for the information materials used in communicating the risks from radon.

EPA provided the resources to inform these 2300 homeowners about potential health risks and ways of reducing these risks. Several alternative information "treatments" were designed to test the effectiveness of different formats. Telephone surveys provided data on changes in knowledge, perceptions, and intentions. The surveys were designed to answer three questions:

1. How much did people learn about radon and its associated risks?

2. To what extent are perceived risks consistent with objective measures of risk?

3. How much more mitigation is undertaken by those at higher risk, after controlling for other factors that might influence the perceived benefits and costs of mitigation?

The results reported here are based on homeowner reactions to the interim readings and information materials.[b] Homeowners were discouraged from taking action until the annual results were available, so we cannot yet address the third question of who mitigated and why.

The literature on risk communication clearly indicates that the way information is presented affects the way it is received. Despite agreement that how information is presented matters, there is no clear consensus in this literature about what features of information materials communicate risk concepts well.

The research design focused on testing two issues. First, do people respond better to risk information that is quantitative rather than qualitative? And second, do people respond better to a directive format that gives explicit instructions about what they should do under given circumstances (hereafter called "command" tone), or to a format encouraging judgment and evaluation in what might be considered a *Consumer Reports* framework (hereafter called "cajole" tone)?[c]

These considerations yielded four different radon risk information booklets incorporating the various combinations of quantitative/qualitative and command/cajole material. In August 1986 EPA published "A Citizen's Guide to Radon," which includes some quantitative and some qualitative information, and is partly directive and partly suggestive. This fifth booklet is a middle case across the dimensions of our design for the information treatments.

Because this social science experiment involves real people facing real risks, they all received the same basic information about radon risks. The core risk information is the same for the five booklets, and all of them are consistent with EPA's action guidelines. According to these guidelines, homeowners should take mitigative action within a few years if their readings are between four picocuries of radon per liter (pCi/l) of air and 20 pCi/l; within several months if readings are in the 20-200 pCi/l range; and within several weeks if they exceed this range.

---

b. The results summarized here are discussed in detail in Smith *et al.* (1987).

c. See Adler and Pittle (1984) for the source of this terminology.

Quantitative			Qualitative	

Radon Risk Chart			Radon Risk Chart*	
Lifetime exposure (pico-curies per liter)	Lifetime risk of dying from radon* (out of 1,000)	Comparable risks of fatal lung cancer (lifetime or entire working life)	Lifetime exposure (picocuries per liter)	Comparable risks of fatal lung cancer (lifetime or entire working life)
75	214 – 554		75	
40	120 – 380		40	
20	60 – 210	Working with asbestos	20	Working with asbestos
10	30 – 120	Smoking 1 pack cigarettes/day	10	Smoking 1 pack cigarettes/day
4	13 – 50		4	
2	7 – 30	Having 200 chest X-rays per year	2	Having 200 chest X-rays per year
1	3 – 13		1	
0.2	1 – 3		0.2	

*U.S. Environmental Protection Agency lifetime risk estimates. The National Council on Radiation Protection has estimated lower risk, but it still considers radon a serious health concern.

*Colors are based on U.S. Environmental Protection Agency lifetime risk estimates. The National Council on Radiation Protection has estimated lower risk, but it still considers radon a serious health concern.

Fig. 1. Radon Risk Charts.

The primary differences between the quantitative and qualitative versions are in the risk chart and its explanations. Figure 1 shows that both link radon concentrations to activities with comparable lifetime risks. Although not reproduced in Fig. 1, the actual booklets used a color scale to distinguish low increments to lifetime risks (green), moderate increments (yellow and orange), and high increments (red). The quantitative version also includes the range of incremental lifetime risk estimates at each radon level and has an example of how to interpret the ranges.

The differences between command and cajole versions are in tone, as illustrated in Fig. 2. The command version uses only the EPA action guidelines for radon concentrations, while the cajole version includes guidelines from the National Council on Radiation Protection and the Canadian government. Figure 2 also illustrates the more subtle differences in tone that appeared throughout the brochure, with the command version emphasizing what the reader should do and the cajole version emphasizing what the reader may want to consider in reaching a decision.

Command	Cajole

Action Guidelines (issued by the U.S. Environmental Protection Agency) .

Red: These levels are very high risks. You should act to reduce these levels, preferably within several months.

Orange: Living in these levels for many years presents a high risk. You should act within the next few years to reduce these levels.

Yellow: Living in these levels for many years still has some risk. You should see if it is feasible to reduce these levels.

Green: These are low levels and have lower risk. The average outdoor level is about 0.2 picocuries per liter. The average indoor level is about 0.8 picocuries per liter.

Because radon risk is cumulative, it usually is given as lifetime risk. This risk is based on two factors:

- **How long** you are exposed to your radon level: Lifetime risk calculations assume an average "lifetime" of 74 years in a house with a particular radon level.

- **Hours at home** each day: Lifetime risk calculations usually assume you spend about three-quarters of your time, or 18 hours, at home each day.

These assumptions will not fit you exactly, but you should use lifetime risk as a benchmark in making any decisions.

Should I have additional radon tests?

The monitors still in your home will measure the average amount of radon in your living area for an entire year. You will also get a reading for your basement, where radon levels are likely to be highest. Even if your risks are in the red or orange areas of the colored chart, you should have more than one test before spending any money to fix your home.

Are there any guidelines for radon levels?

Several government agencies and scientific groups have recommended that actions be taken at various levels.

Agency or Organization	Radon Level (picocuries per liter)	Action Guidelines
U.S. Environmental Protection Agency	20	Remedial action, preferably within several months
	4	Remedial action within next few years
National Council on Radiation Protection	8	Remedial action
Canadian Government	30	Prompt action
	4	Remedial action

What is a lifetime risk?

Because radon risk is cumulative, it usually is given as lifetime risk. This risk is based on two factors:

- **How long** you are exposed to your radon level: Lifetime risk calculations assume an average "lifetime" of 74 years in a house with a particular radon level.

- **Hours at home** each day: Lifetime risk calculations usually assume you spend about three-quarters of your time, or 18 hours, at home each day.

Because every household is different, you may want to adjust the typical risks to fit your circumstances. For example, if you had a reading of 10 picocuries per liter but spend only 9 hours inside your home on a typical day, you would multiply your risk from the colored risk chart on page 4 by one-half or .50. In this case, your risk would now range from as low as 15 out of 1,000 to as high as 60 out of 1,000. Your risk would now be in the beginning of the orange area of the risk chart. If you think lifetime risks are not appropriate for your situation, the next page shows a chart with risks for different exposure periods.

Should I have additional radon tests?

The monitors still in your home will measure the average amount of radon in your living area for an entire year. You will also get a reading for your basement, where radon levels are likely to be highest. In any case, it is a good idea to check the accuracy of a single test by having more tests before spending any money to fix your home.

Fig. 2. Major Differences in Brochure Tone.

New York State also developed a one-page fact sheet that included some background information and two paragraphs about radon risks. This fact sheet is typical of the information that households usually receive from state agencies and testing companies. It contains somewhat less information than the other information treatments. NYSERDA and the research team felt that this treatment should be limited to people at very low risk. Therefore, the fact sheet was sent only to homeowners with radon levels below one pCi/l. (One pCi/l is one-fourth of the minimum EPA action level; the national average exposure is 0.8 pCi/l.)

The final design included all six types of information treatments. Half of those with readings below one pCi/l received the fact sheet. One of the five booklets was randomly assigned to everyone else in the study. Those with readings above one pCi/l also received EPA's "Radon Reduction Methods: A Homeowner's Guide." In December 1986 we mailed the interim readings and information materials to the households who had returned the two-and-one-half month monitors. Shortly thereafter we interviewed the participating homeowners a second time to find out what they had learned and how they had reacted to the information they had received. (Baseline data on their knowledge and risk perceptions had been collected during the summer of 1986.)

SOME CONVENTIONAL WISDOM ON RISK COMMUNICATION

Covello *et al.* (1987) derive a number of facts or principles of risk communication from the literature. We take these propositions to represent the "conventional wisdom" on the subject. We have evidence at this point relating to several of these propositions, including:

- Subtle differences in the way risks are expressed can have a major impact on perceptions and decisions.

- People have difficulty interpreting probabilistic information. Individuals like to know how a risk can affect them personally.

- There are many publics, each with its own needs, interests, concerns, priorities, and preferences. Individual biases and limitations may lead to distorted and inaccurate perceptions of risk problems.

- Risks from dramatic causes of death (e.g., cancer) tend to be overstated.

- When informed about a particular hazard, people's concerns will generalize beyond the immediate problem to other related hazards.

It is evident from this list that much of the evidence from the New York study relates to risk perceptions. The remainder of the paper will discuss our evidence on each proposition.

EVIDENCE FROM THE EXPERIMENT

Subtle differences in the way risks are expressed can have a major impact on perceptions and decisions. With the exception of those with radon readings less than one pCi/l, homeowners received one of five brochures containing substantially the same information in different formats. We have analyzed responses to the baseline and followup surveys in three areas: performance on a radon quiz, congruence between objective and subjective risks, and willingness to pay for additional information. There is no single brochure that performs best in all three areas.

The cajole/qualitative brochure provided the best improvement in responses to quiz questions; the two quantitative brochures provided better congruence between objective and subjective risks, and the command and EPA brochures significantly reduced homeowners' demand for additional information. We conclude that the way information on health risks is presented has a measurable impact, but that performance varies depending on the particular measure of effectiveness employed.

People have difficulty interpreting probabilistic information. They like to know how a risk can affect them personally. The quantitative brochures provided information on the likely range of lung cancer cases per 1,000 people exposed to various levels of radon. The cajole/quantitative brochure also demonstrated how to adjust these population risks to conform to the household's actual exposure. The evidence on whether people had difficulty with the probabilistic information is of two kinds: whether receiving the numerical probabilities helped in identifying how serious their problem was on the colored risk chart and whether the numerical probabilities promoted better conformity between objective and subjective risks.

Although homeowners who received the NYSERDA brochures generally were more successful in identifying their correct placement on the risk chart than those receiving the EPA brochure, the quantitative brochures were not more effective than the qualitative brochures. However, those with higher readings were more likely to identify the "wrong" position on the risk chart and those receiving one of the command versions were more likely to identify the "right" position. Right was defined as the placement associated with the population risk related to their exposure. These results may imply that those at higher risk who were encouraged to adjust for individual circumstances actually identified a more appropriate place on the risk chart than the population risk implied.

Although the evidence on correct placement is ambiguous, the evidence on risk perception is quite clear: the quantitative information treatments were statistically significant in reducing discrepancies between objective and perceived risks.

There are many publics, each with its own needs, interests, concerns, priorities, and preferences. Individual biases and limitations may lead to distorted and inaccurate perceptions of risk problems. Ideally one would like to know in advance which personal characteristics are associated with what kinds of perceptual problems. If these characteristics are shared by an identifiable group, then information treatments could be designed and targeted for that group's specific needs. The New York study sheds some light on this issue.

The baseline survey obtained data on age, sex, income, education, number of children, years lived at address, ease with working with numbers, and responses to a set of attitude and personality questions. Table 1 indicates which of these variables was statistically significant for various tests. Education is significant in three of the four principal tests. In each case the sign is as expected. Holding other factors constant, including information treatment, better educated respondents were more likely to do better on quiz questions, use the risk chart correctly, and have a smaller discrepancy between objective and subjective risks.

Age and response to an attitude question about health concerns are significant in two tests. Older people are less willing and people with health concerns are more willing to pay for more information. Older people were also more likely to do poorly on the radon quiz, while people with health concerns were more likely to overstate their actual risk.

The remaining personal characteristics were significant in only one test. Women were more likely than men to use the risk chart correctly. Whites overstated their risks less than non-whites. As expected, higher income respondents were more willing to pay for additional information.

While these results confirm the conventional wisdom that personal characteristics matter, they do not offer encouragement that such characteristics might serve to identify target "publics" for particular information programs. Moreover, the effect of such characteristics varies among risk responses. It is not clear which of these tests is most relevant from a public health or risk management perspective.

Table 1. Effects of Personal Characteristics

	Age	Sex	Income	Education	Race	Ask Doctor
Performance on radon quiz	−			+		
Correct use of risk chart	−			+		
Diff. between obj. and subj. risk				−	−	−
Willingness to pay for more information	−		+			+

Notes on definitions of variables:

Sex:	Dummy variable = 1 if male.
Race:	Dummy variable = 1 if white.
Ask Doctor:	Dummy variable = 1 if respondent indicated that the statement "you always ask your physician a lot of questions or regularly read articles about health" described himself very or fairly well.

Risks from dramatic causes of death (e.g., cancer) tend to be overstated. This proposition is usually interpreted to mean that people's assessment of the probability of dying from cancer is greater than the technical risk assessment would predict. However, the technical risk estimate is typically a statistical construct that does not account for individual differences in exposure or vulnerability. Furthermore, the risk perception literature emphasizes that people find certain ways of dying more repugnant than others. Preferences about cause of death and other risk characteristics appear to be expressed in the form of perceived likelihood of occurrence.

In trying to devise a quantitative test of this hypothesis, it is important to discriminate between the perceived message and the perceived risk. Risk perceptions result from combining preexisting attitudes and knowledge with the information treatment. We have modeled this problem as a Bayesian process that specifies the perceived seriousness of risk in the followup survey to be the weighted average of the baseline seriousness and the perceived message in the information treatment. Letting the (unobserved) risk message be a function of information treatment and individual characteristics, we obtained maximum likelihood estimates of the relevant parameters. One of several models tested provided the following results:

$$\text{SRISK}_f = .277 + .294 \, \text{SRISK}_b + .016 \, \text{RADON} - .085 \, \text{COQUANT}$$

$$- .028 \, \text{COQUAL} - .087 \, \text{CAQUANT} - .060 \, \text{CAQUAL} - .037 \, \text{EPA}$$

$$- .004 \, \text{EDUC} - .006 \, \text{AGE} - .161 \, \text{RACE} + .004 \, \text{YEARS}$$

$$+ .001 \, \text{TIME} + .039 \, \text{DOCTOR} + .499 \, \text{MILLS}$$

$$\log(L) = -672.05$$

The subscripts f and b indicate the respondents' subjective risk in the followup and baseline surveys, respectively; RADON is the radon reading; the next five variables are dummies indicating the four NYSERDA brochures and the EPA brochure (the fact sheet is the omitted treatment); YEARS is years resided at address; TIME is length of time spent reading the materials; DOCTOR is an attitudinal dummy indicating concern about health; and MILLS is a correction for selection bias. Only COQUAL, EPA, and EDUC fail to have statistically significant parameters. Note that people with higher radon readings are likely to perceive risks as more serious and the quantitative and cajole treatments tend to reduce concern relative to those who received the fact sheet.

These coefficients suggest an average perceived risk message about ten times larger than the technical risk estimates, adjusted for life expectancy (based on age and sex) and length of exposure. This comparison relies on numerous assumptions and imperfect measures of risk perceptions. Nevertheless, the data seem to indicate a systematic upward bias in the way that respondents decoded the risk information they received.

When informed about a particular hazard, people's concerns will generalize beyond the immediate problem to other related hazards. Both the baseline and followup surveys contained a question about the seriousness of risks the household faces from auto accidents, hazardous wastes, and radon. Seriousness was measured on a one-to-ten scale. Table 2 compares responses to this question before and after the respondents received information on their radon test results and one of the information treatments.

Table 2. Mean Seriousness of Risks from Various Sources

	Baseline	Followup	Percent Difference
Auto Accidents	5.7	5.3	7%
Hazardous Waste	4.3	3.6	16%
Radon	4.2	3.4	20%

Respondents initially perceived their personal risk from radon and hazardous wastes to be about equally serious, with auto accident risk being considerably more serious. The radon readings for this sample of New York homes were generally quite low. Respondents' perceptions changed in the appropriate direction, with mean radon seriousness falling by 20%. The respondents received no new information on hazardous waste or auto accident risks. Nevertheless, the perceived seriousness of these risks also declined in the followup survey. The decline in perceived hazardous waste risk was more than twice that of auto accidents, perhaps because the characteristics of hazardous waste risk are more similar to those of radon. The evidence here appears to confirm the conventional wisdom, although the literature has focused on increases rather than decreases in perceived risks.

CONCLUSION

Evidence from the New York social experiment generally confirms the qualitative propositions of the conventional wisdom. An important exception is that numerical probabilities demonstrably improve performance on tasks involving comparing relative risks, evaluating the seriousness of risk exposures, and adjusting population risks for individual circumstances. Furthermore, we have obtained new insights by treating the conventional wisdom as hypotheses subject to specific quantitative tests.

At this point in the research, we have data primarily on the response of risk perceptions to different information treatments and different personal characteristics. The effect of various causal factors is sensitive to the particular test of risk perception applied. No information treatment was clearly superior for all tasks. Neither was there a single set of personal characteristics that identified a group with a clearly defined set of perceptual problems. Again, different personal characteristics were important for different perceptual tasks.

The most important implication of these findings is that risk communicators must determine what specific task or tasks the information program should enable people to do. The usual strategy of simply reducing anxiety may not be consistent with educating the public about risk facts, helping them to identify their personal risks, or to improve their perceptions about relative risk exposures. In short, the conventional wisdom that risk communication itself is a complicated, hazardous undertaking is quite correct.

REFERENCES

Adler, R., and D. Pittle, "Cajolery or Command: Are Education Campaigns an Adequate Substitute for Regulation?," *Yale Journal on Regulation* **1**:159 (1984).

Covello, Vincent T., Slovic, Paul, and von Winterfeldt, Detlof, "Risk Communication: A Review of the Literature," National Academy of Sciences (1987).

Smith, V. Kerry, William H. Desvousges, Ann Fisher, and F. Reed Johnson, "Communicating Radon Risk Effectively: A Mid-Course Evaluation," prepared for the Office of Policy Analysis, U.S. Environmental Protection Agency, under Cooperative Agreement No. CR-811075, by Vanderbilt University, Nashville, Tennessee, and Research Triangle Institute, Research Triangle Park, North Carolina (1987).

Managing Uncertain Risks Through "Intelligent" Classification: A Combined Artificial Intelligence/Decision-Analytic Approach

Louis Anthony Cox, Jr.

U S West Advanced Technologies
Englewood, CO

ABSTRACT

Many tasks in engineering risk analysis can be interpreted as generalized classification problems. Hazard identification, fault diagnosis and prediction, and feedback control rules can all involve aspects of classification. This paper reviews a class of algorithms, called "recursive partitioning algorithms," that has proven useful for solving such problems heuristically when costs of observations can be ignored. A general decision-analytic framework is also introduced that takes observation costs into account and that provides a basis for studying and comparing recursive partitioning algorithms and competing approaches.

Two recursive partitioning algorithms that are embodied in commercially available programs are reviewed: ID3, which forms the basis of the artificial intelligence/machine-learning package EXPERT-EASE; and CART, which has been developed and applied by computational and bio-statisticians concerned with health, safety, reliability, and environmental applications. The "classification trees" produced by CART and by ID3 are interpreted as approximations to "optimal" statistical decision strategies for sequential experimentation and decision making. The paper concludes with a summary of several methodological research directions that promise to make the generalized classification paradigm more useful to practicing risk managers.

KEYWORDS: Artificial intelligence, learning, recursive partitioning algorithms, decision rules, classification, classification trees

INTRODUCTION

Classification is a key component of many engineering risk management strategies. It can be applied to risk and hazard *identification* based on recognition of patterns (classes) in observed cases; to fault *diagnosis* (classification of observed symptoms in terms of their underlying causes); to fault *prediction* (classification of systems or components in terms of their probable future states); and to steady-state feedback *control* of systems (classification of observable states in terms of the control actions that they imply). Algorithms that learn effective classification rules from observed data have tremendous potential value throughout the applied fields of risk analysis and medical decision-making.

Parallel lines of research in computational statistics[1,2,7,10] and artificial intelligence[3-5] have recently suggested a class of algorithms, called *"recursive partitioning"* algorithms, that learn classification rules from a "training set" of cases whose true classes are known. The output of such an algorithm is a "classification tree" specifying a questioning strategy for gathering information about any new case to be classified and a decision rule for assigning a case to a class, based on the answers to the questions. This output can then be reduced to two sets of rules implemented in an expert system.[5] One set of rules determines which question to ask next, given the answers received so far. The other set recommends conclusions and confidence factors. In many applications, conclusions can be interpreted as risk identification, classification, or control recommendations, and a successful learning algorithm can be seen as learning rules for risk management from the examples in the training set.

This paper reviews these developments, critically contrasts the statistical and artificial intelligence literatures on recursive partitioning algorithms, and summarizes some extensions that could make this methodology more useful for practical risk management decisions.

CLASSIFICATION PROBLEMS

A typical classification problem contains the following elements:

- A set C of *cases* to be classified. These are units that are observed and about which decisions are made.

- A set Y of possible *observations* (also called "signals" or "measurements") that can be made on each case;

- A set X of *states* that a case may be in. Each case is assumed to have exactly one true state;

- A set A of *actions* or classification decisions that can be applied to a case;

- A set E of possible *experiments*. An experiment e in E is a conditional probability density function $e(y;x)$ giving the probability density of observing signal y if the true state of the case being observed is x.

- A *loss function* $L(a,x)$ giving the loss from applying decision a to a case c whose true state is x.

- A *cost function* $u(e)$ giving the cost of experiment e.

- A set H of *hypotheses* about the population of cases. Each hypothesis h in H specifies (at least) the prior probability $p(x)$ that the next case observed will have a true state of x.

Two important decision problems are (i) What experiment in E should be performed on each case? and (ii) What decision in A should be made about each case, given what has been observed?

Figure 1 displays decision problem (ii), assuming that the experiment e is fixed. A case c is submitted to the decision process by some external process. "Cases" could consist of patients entering a cancer clinic for treatment; hazardous waste sites awaiting investigation and remedial action; or chemicals being classified for carcinogenic potency on the basis of their chemical structures. The true but unknown state of case c is denoted by

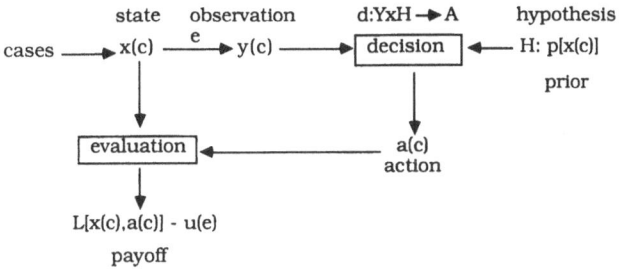

Fig. 1. A Generalized Classification Decision Problem.

$x(c)$. Although $x(c)$ is unknown, there may be (and, for Bayesian inference, must be) some prior hypothesis about its probable value, captured in a prior probability density function (p.d.f.) $p[x(c)]$. It is often reasonable to assume that $p[x(c)]$ is the same for all cases, i.e., that $p[x(c)] = p(x)$ for all c. We will make this assumption here.

Once $x(c)$ has been generated, the corresponding observation $y(c)$ is determined from $x(c)$ through the probabilistic mapping $e(y;x)$. $y(c)$ and $p(x)$ are then fed into a decision rule d that maps them into an act, $a(c)$, that is applied to case c. (In Bayesian inference, $y(c)$ and $p(x)$ are also mapped into an updated hypothesis in H, i.e., d is a mapping $d:Y \times H \rightarrow A \times H$ from observations and hypotheses to acts and updated hypotheses.) Finally, the true state $x(c)$ and the selected act $a(c)$ are fed into a loss function, say $L[x(c),a(c)]$, which evaluates the consequence of applying act $a(c)$ to a case with state $x(c)$. (If there is a choice of experiments, and if the cost of implementing experiment e is $u(e)$, then the evaluation function must be generalized, e.g., to $L[x(c),a(c)] - u(e)$.) In the special case of a "pure" classification problem, the action or decision set A consists of the state labels, $1,..,N$, and the *loss function* is $L(x,a) = 0$ for $a = x$; $L(x,a) = 1$ for a not equal to x. But if other acts (e.g., prescriptions of medical treatments) are possible, then $L(x,a)$ may be an arbitrary $M \times N$ loss table. If x and a are continuous, then $L(x,a)$ may be a function such as $L(x,a) = |x - a|$. The problem of "optimal" classification (in the sense of statistical decision theory) is to select a decision rule d minimizing the expected loss $E\{L[x(c),a(c)]\}$ given $p(x)$.

Let D be the set of all possible decision rules. For example, if classification is to be performed automatically by an expert system, then D might consist of all decision rules of the form "IF [y] is observed, THEN take action [a]" that can be expressed in the expert system language. The problem is to find that d in D that minimizes some error criterion—either the expected loss or a heuristic approximation of it. The training set on which learning is based is a subset C' of cases such that both $x(c')$ and $y(c')$ are known for each case c' in C'.

In passive learning, the cases in the training set are selected outside the control of the learner. In active learning, the observation $y(c)$ made on a case c is partially controlled by the experimenter/learner via choices of experiments. An important kind of experiment, which both the statistical and the artificial intelligence literatures on recursive partitioning learning algorithms have concentrated on, consists of lower-dimensional *projections* of the full data vector $y(c)$. Rather than observing the whole vector $y(c) = [y_1(c),...,y_n(c)]$, the learning algorithm sequentially chooses which *components* of $y(c)$ to observe, when to stop observing, and what terminal action $a(c)$ to take. Different components typically cost different amounts to observe, and the loss function must include observation costs.

Let $Y = \{Y_1,...,Y_n\}$ denote the set of n observable attributes of a case. In the sequential observations context, Y is no longer a static set of "signals," but a set of observation opportunities (attributes) from which the learner may select sequentially. The cost of observing the value of attribute Y_j for a case will be denoted by u_j.

A decision rule for classifying a case c begins by specifying which attribute (component of $y(c)$) to observe first. (Assume for the moment that it is optimal to make some observation before choosing $a(c)$.) Call this first attribute $Y(1)$, and suppose that its value for c is observed to be $y(1)$. The decision rule must next specify either a second attribute, say $Y(2)$, to observe next, given that $y(1)$ has been observed, or a terminal action $a(c)$ to implement in lieu of further observations. If $Y(2)$ is observed and has value $y(2)$, then the decision rule next maps the sequence of observations $y(1)y(2)$ into either a choice of a third attribute, $Y(3)$, to observe, or else a terminal action.

More generally, a sequential classification decision rule is a function $d:Y^* \rightarrow Y \cup A$, where Y^* denotes the set of all possible sequences (including partial sequences) of observed attribute values, d maps each sequence of observations y^* in Y^* into either a next attribute in Y to observe (which will extend the sequence of observations y^* by one new observation, again giving a member of Y^* for the rule to operate on), or a terminal act $a(c)$ in A. Given observation costs $u_1,...,u_n$ for attributes $Y_1,...,Y_n$ and a joint p.d.f. $p(y_1,...,y_n,x)$ on the components of $y(c)$ and the state $x(c)$, an "optimal" learning algorithm seeks a sequential classification decision rule that minimizes expected loss, defined as the sum of the observation costs and the loss $L[x(c),a(c)]$.

A SIMPLE EXAMPLE

To fix the above ideas, consider the example decision problem in Table 1. Here, each case has two attributes, $Y1$ and $Y2$. Each attribute has two possible values, coded as 0 and 1. There are four possible states, corresponding to the four possible combinations of attribute values; in effect, the "state" of a case consists of its attribute values. (More generally, each state in X would determine only a joint p.d.f. over attribute values, rather than the values themselves.) The state set is $X = \{1,2,3,4\}$, where $(y_1,y_2) = (0,0)$ for state 1; $(0,1)$ for state 2; $(1,0)$ for state 3; and $(1,1)$ for state 4. All four states are *a priori* equally likely. There are two possible acts, $A1$ and $A2$. Instead of a 4×2 *loss matrix*, Table 1 shows an equivalent 4×2 *payoff matrix*. [This can be converted to an equivalent loss table by subtracting each payoff from the best one that could have been achieved in that state, as follows: row 1 changes from $(3,5)$ to $(-2,0)$; row 3 changes from $(4,1)$ to $(0,-3)$; row 4 becomes $(-1,0)$; and row 2 becomes $(0,0)$.] When the data are presented as payoffs rather than losses, the goal is to *choose a strategy that maximizes expected net payoff* (payoff minus observation costs).[9]

Table 1. An Example Decision Problem

State	Attributes		Acts		Prior Prob.
	Y1	Y2	A1	A2	
1	0	0	3	5	0.25
2	0	1	1	1	0.25
3	1	0	4	1	0.25
4	1	1	1	2	0.25

$u_1=0.5$, $u_2=0.75$ Payoffs
Observation Costs

The experimenter/decision maker (d.m.) has several possible strategies to choose among in this example. Each act has a prior expected payoff, in the absence of any observations, of 2.25, which is thus the expected value of a completely random choice. Alternatively, the d.m. could observe attribute $Y1$, and then choose the act having the greater expected value (conditioned on the observed value of $Y1$). If the value of $Y1$ is $y_1 = 0$, then the d.m. should choose act $A2$ (whose payoffs of 5 and 1 weakly dominate the payoffs of 3 and 1 for act $A1$), for a conditional expected payoff of 3.0. But if $y_1 = 1$, then the d.m. should choose act $A1$ for a conditional expected payoff of 2.5, rather than act $A2$, which would have a conditional expected payoff of 1.5. This sequential decision strategy has an expected net payoff of $2.75 - u_1$, where u_1 is the cost of observing the value of $Y1$. The *expected value of the information* (EVI)[9] from observing the value of $Y1$ is thus $(2.75 - u_1) - 2.25 = 0.5 - u_1$.

Similar reasoning can be used to calculate EVI's for all the different strategies shown in Table 2. The two bottom strategies, involving sequential decisions, strictly dominate the "observe everything" strategy in which the d.m. pays up front to observe the values of both attributes, and then chooses the act giving the (at that point known) higher payoff. Any of the other strategies can be made optimal by appropriate choice of the observation costs u_1 and u_2. If $u_1 = 0.5$ and $u_2 = 0.75$, then the optimal strategy is the second one from the bottom in Table 2, in which the d.m. starts by observing the value of $Y1$, chooses $A2$ if $Y1 = 0$, and otherwise observes the value of $Y2$ and chooses the better act (which is $A1$ if $y_2 = 0$ and $A2$ if $y_2 = 1$).

Table 2. Expected Values of Alternative Strategies

Strategy	Expected Increase in Payoff
Observe nothing, choose either act	0
Observe $Y1$, choose $A1$ iff $y_1 = 0$	$0.5 - u_1$
Observe $Y2$, choose $A1$ iff $y_2 = 0$	$0.25 - u_2$
Observe $Y1$ and $Y2$, choose better act	$0.75 - u_1 - u_2$
Observe $Y1$, then choose $A2$ if $y_1 = 0$. If $y_1 = 1$, then observe $Y2$ and choose the better act.	$0.75 - u_1 - 0.5u_2$
Observe $Y2$, then choose $A1$ if $y_2 = 0$. If $y_2 = 0$, then observe $Y1$ and choose the better act.	$0.75 - u_2 - 0.5u_1$

Figure 2 displays this strategy graphically as a *classification tree*. Each non-terminal *node* of a classification tree corresponds to an *attribute* to be viewed if that point in the tree is reached. Each *branch* corresponds to a possible *value* of the attribute. And each *terminal node* corresponds to a terminal *act*. There is thus a one-to-one correspondence

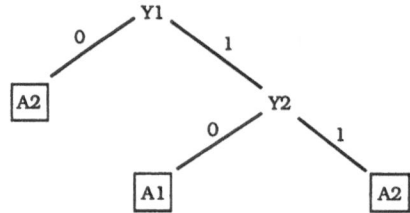

Fig. 2. An Optimal Strategy.

between classification trees and risk management strategies (i.e., sequential decision rules for observation and action).

Each classification tree determines an expected net payoff. The fundamental question is, *How can a classification tree with maximum expected net payoff (or, equivalently, with minimum expected net loss), or a reasonable approximation of such a tree, be constructed?* Possible answers range from exhaustive enumeration to stepwise growth (expanding one node at a time to get its successors) to stepwise pruning of the full set of nodes. We will now examine the stepwise growth, or "recursive partitioning," approach in detail.

RECURSIVE PARTITIONING ALGORITHMS: THE EXAMPLE OF ID3

Recursive partitioning methods provide a simple but empirically successful heuristic approach for constructing classification trees when observation costs can be ignored. In the artificial intelligence literature, the best-known such method is the ID3 algorithm of Quinlan.[3] ID3 is designed for application to cases where there are only a few attributes, each with only a few, discrete possible values or levels, and in which the state set X consists of a few discrete (unordered) states. The training set C' consists of pairs $[x(c'),y(c')]$, i.e., for each case c' in C', the full vector of attribute values and the corresponding correct classification (state) are known. The learning algorithm is as follows:

1. Given the training set of observations $[x(c'),y(c')]$ for all c' in C', compute the *sample entropy* of C'. This is defined as

$$M(C') = - \sum_x P(x)\log_2 P(x),$$

where $P(x)$ is the sample proportion of cases in C' that belong to class x.

2. For each attribute Y_j, $j = 1,...,n$, compute the *expected entropy reduction* from an observation made on attribute Y_j. If Y_j has N possible levels or values, say $y_{j1},...,y_{jN}$, then the expected entropy reduction from observing the value of Y_j is $M(C') - E[M(C';Y_j)]$, where

$$E[M(C';Y_j)] = \sum_k \Pr[Y_j = y_{jk}][M\{c' \text{ in } C': Y_j(c') = y_{jk}\}].$$

The set $\{c' \text{ in } C':Y_j(c') = y_{jk}\}$, abbreviated $\{C':y_{jk}\}$, is the subset of cases in the training set C' that have y_{jk} as the value of attribute Y_j. The subsets $\{C':y_{jk}\}$, $k = 1,2,...,N$ constitute a *partition* of C': every case in C' belongs to exactly one of these subsets. Each of the N subsets is called a *block* of the partition.

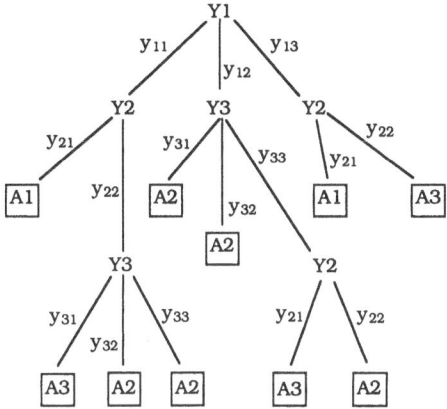

Fig. 3. Example of an ID3 Tree.

3. Let Y_j' be the attribute giving the greatest expected entropy reduction. (Break ties arbitrarily.) Make Y_j' the first node of the classification tree. If Y_j' has N possible values, then create N branches coming out the Y_j' node. Branch k corresponds to the block $\{C':y_{jk}'\}$ of cases in the training set with $Y_j' = y_{jk}'$.

4. Attach a new node to each of the N branches just created. To attach a node to branch k, treat the kth block of the partition, $\{C':y_{jk}'\}$, just as C' itself was treated, i.e., compute its sample entropy, find the remaining attribute (excluding Y_j', which has already been used) that reduces expected entropy the most, and attach that attribute to the kth branch as a new node. For each possible value of the selected attribute, create a new branch, and recursively repeat the entire procedure.

5. Stop expanding nodes when either all attributes have been observed, or the conditional probability measures for the state, conditioned on the different possible values of the next attribute that would be added, are not statistically significantly different. Assign to each terminal node the act that maximizes conditional expected payoff, given the sequence of observations that has led to it.

This is the basic ID3 algorithm. (It differs a little from the original ID3, e.g., by using conditional expected payoff as a basis for assigning terminal decisions to terminal nodes in Step 5. The original ID3 was designed for pure classification problems, and the terminal decision selected the conditionally most likely class (state) as the algorithm's prediction.) Figure 3 presents an example of an ID3 tree. Note that ID3 applies when there are few attributes, each with only a few possible levels. It does not apply to continuous numerical attributes. The CART algorithm described below covers this case.

ID3 has two important features that can be varied to obtain other recursive partitioning algorithms. One is its use of expected entropy reduction as its *node selection heuristic*. Other node selection heuristics, such as expected error rate reduction or myopic (one-step look-ahead) EVI could be just as useful. Second, the choice of a *stopping rule* for node expansion, based on statistically significant differences between conditional loss distributions, can be implemented in various ways, e.g., by using a chi-square test, as in (3), a Kolmogorov-Smirnoff test, or a variety of other statistical tests for significant differences between distributions. Or, in place of a stopping rule, it might be desirable to depart from strict recursive partitioning by growing a very large tree and "pruning" it back optimally. This strategy is followed in the CART methodology, to which we now turn.

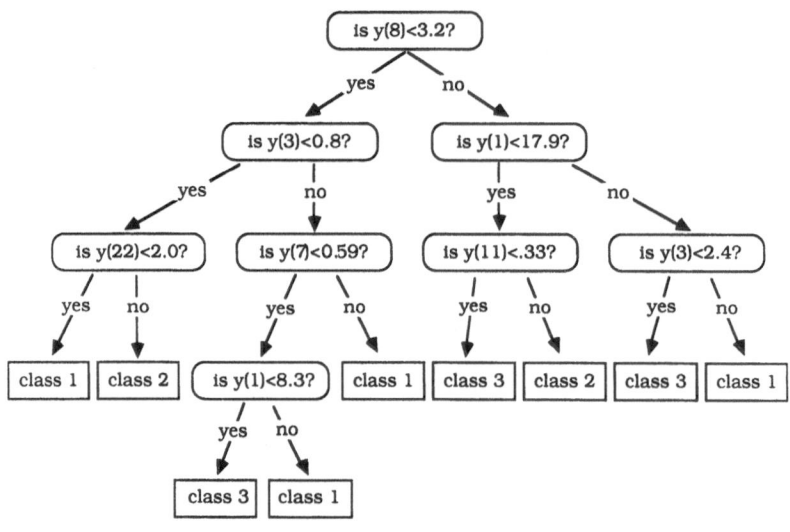

Fig. 4. Example of a Cart Tree.

CLASSIFICATION WITH CONTINUOUS ATTRIBUTES: CART

ID3 was developed for observation sets that can be factored as Cartesian products of n attributes, each with a few unordered levels. If Y is the Cartesian product of n continuous numerical attributes, a new approach is needed; it is impossible to create a branch in the classification tree for each possible value of a continuous attribute. The CART methodology[1,2,7,10] solves this problem by choosing a *critical threshold* for each attribute. Instead of asking "What is the value of attribute Y_j?" and creating a branch for each possible answer, CART asks "Is the value of Y_j less than K_j?," where K_j is the threshold selected for attribute Y_j. Each node in a CART tree corresponds to one such question. Each node partitions the cases reaching it into two sets—those with $Y_j < K_j$ and those with $Y_j \geq K_j$. A node is therefore also called a *split*.

To generate a new node, CART first calculates the "best" value of the threshold K_j for each attribute Y_j. (This can be the most informative, maximally separating, entropy minimizing, etc., value, depending on the node selection heuristic used. The commercial CART software allows two selection heuristics: the "Gini Coefficient" and the "twoing criterion.") Next, the "best" attribute, assuming that the best threshold value is used for each attribute, is selected for the next node. The pair (K_j^*, Y_j^*) giving the best value of the node selection heuristic is turned into the question "Is Y_j^* less than K_j^*?", and this question becomes the next node or "split" in the CART classification tree.

Figure 4 shows an example of a CART tree. Like ID3, CART was designed for pure classification tasks ($A = X$) rather than more general decision problems. Unlike ID3, CART grows only binary trees; moreover, it relies on growth followed by pruning to develop the final tree. A substantial contribution of the CART methodology is the use of statistical *cross-validation* (or, in some cases, hold-out samples) to give quantitative, validated estimates of the performance characteristics of the final tree. This statistical methodology allows the error rate expected from using the final tree to be estimated even when the prior distribution for states, $p(x)$, is unknown.

In practice, CART classification trees have given comparable or superior performance to more traditional parametric statistical methods, such as logistic regression or discriminant analysis, in applications ranging from medical diagnosis and prognosis[1,7] to failure analysis of turbine blades. The algorithm is nonparametric, since its node selection criteria depend only on ordinal properties of the training data, and it can be applied to ordered categorical attributes as well as to continuous ones. A version based on exhaustive enumeration has also been developed for dichotomous attributes. The CART methodology is well developed, and variations for missing data and exploratory data analyses, as well as for sensitivity analysis, are described in Ref. 1. Extensions to censored survival data found in cancer risk assessment are reported in Ref. 2.

SUMMARY: RESEARCH DIRECTIONS IN RECURSIVE PARTITIONING ALGORITHMS

This paper has formulated a class of decision problems, generalizing the pure classification problem, that has many potential applications in engineering risk analysis. The decision-analytic solutions to such problems take the form of classification trees. Recursive partitioning algorithms useful for growing such trees have been surveyed. Although these algorithms represent the state of the art in machine-learning and computational statistics, they leave some fundamental questions unanswered. Effective algorithms for growing classification trees have yet to be developed for situations with realistic complications such as costly observations, delayed (or noisy) feedback about the true states of cases, or non-stationary processes generating the observed cases.[8] Algorithms that will accommodate such conditions are one target of current research at U S WEST Advanced Technologies.[11]

A second line of research deals with understanding and improving the evaluation and optimization heuristics used to grow classification trees, and with developing more effective heuristics for realistically complex problems.[6] For example, why does CART's twoing criterion out-perform the myopic EVI criterion in problems with costless observations, as reported in (1)? Can recursive partitioning be used to generate initial trees that are then improved on by other methods?

Finally, neither the A.I. nor the statistics literatures on recursive partitioning algorithms has yet provided a sharp theory of *training set design*, i.e., how to choose "informative" cases to learn from. For example, if each case (or each attribute) can only be observed during an interval of time, then how should these constraints be incorporated in sequentially choosing which cases to look at next?

On-going research at U S WEST Advanced Technologies is starting to address some of these questions.[11] The goal is to develop improved algorithms for detecting and managing failures in complex engineering systems by developing new algorithms and heuristics for generating classification trees, and by using simulation to understand which heuristics work best on different kinds of classification tasks. Classification trees appear to offer a potentially useful paradigm for many risk management tasks and problems, and progress in tree construction methodologies is likely to have payoffs in a variety of engineering risk analysis applications.

REFERENCES

1. L. Breiman, J. H. Friedman, R. A. Olshen, and C. J. Stone, *Classification and Regression Trees*, Wadsworth International, Inc., Belmont, CA (1984).
2. L. Gordon and R. A. Olshen, "Tree-Structured Survival Analysis," *Cancer Treatment Reports* **69(10)**:1065-1069 (1985).

3. R. Quinlan, "The Effect of Noise on Concept Learning," in R. S. Michalski *et al.*, *Machine Learning: An Artificial Intelligence Approach*, Morgan Kaufman, Los Altos, CA (1986).

4. J. C. Schlimmer and D. Fischer, "A Case Study of Incremental Concept Induction," in *AAAI-86 Proceedings*, Vol. 1, Morgan Kaufman, Los Altos, CA, (1986).

5. R. Quinlan, "Generating Production Rules from Decision Trees," *IJCAI '87*, Vol. 1, Morgan Kaufman, Los Altos, CA (1987).

6. J. Pearl, *Heuristics: Intelligent Search Strategies for Computer Problem Solving*, Addison Wesley, Reading, MA (1984).

7. M. R. Segal, "Recursive Partitioning Using Ranks," Technical Report No. 15, Laboratory for Computational Statistics, Department of Statistics, Stanford University (1985).

8. J. C. Schlimmer and R.H. Granger, Jr., "Beyond Incremental Processing: Tracking Concept Drift," *AAAI-86 Proceedings*, Vol. 1, Morgan Kaufman, Los Altos, CA (1986).

9. H. Raiffa, *Decision Analysis*, Addison Wesley, Reading, MA (1971).

10. K. A. Grajski *et al.*, "Classification of EEG Spatial Patterns with a Tree-Structured Methodology: CART," *IEEE Transactions on Biomedical Engineering* **BME-33(12)**:1076-1086, 1986.

11. L. A. Cox, Jr., "Designing Interactive Expert Classification Systems That Acquire Expensive Information 'Optimally'," in J. Boose *et al.* (Eds.), *Proceedings of the European Knowledge Acquisition Workshop for Knowledge-Based Systems*, Gesellschaft fur Mathematik und Datenverarbeitung MBH, GMD Studien Nr. 143, Bonn, Germany, 1988.

Managing Fire Safety Problems Through Decision Analysis to Evaluate Fire Safety Hazards

Craig Van Anne
The Hartford Steam Boiler Inspection
  and Insurance Company
Walnut Creek, CA

ABSTRACT

Historically, the determination of optimum fire protection engineering solutions has been a predominantly subjective based process. An engineering method presently exists for numerically evaluating a relative level of risk in any building, and thus consistently comparing the fire safety qualities of various buildings in a particular jurisdiction. The flame movement part of the method involves the engineering determination of the probability of success in terminating a fire within each space of a building based on automatic suppression, flame propagation and manual response. The effectiveness of each barrier surrounding the space, whether it be a floor, ceiling, wall, or an empty void, is also evaluated.

An illustrative case study demonstrates a procedure for choosing the optimum fire protection solution for a particular decision maker by incorporating the engineering method into a decision analysis model of possible alternatives. This decision analysis model incorporates as key input the mathematically described risk aversion constant of the decision maker and the probability of flame termination for a given building in a given space in that building.

KEYWORDS: Fire protection, evaluation, risk aversion, expected value

THE ENGINEERING METHOD AND RISK MANAGEMENT

There exists today a detailed engineering method which, regardless of size or occupancy, can evaluate any building in a consistent manner. Through the application of this engineering method, an evaluation is performed which yields relative assessments of risk. Once the fire risk of a particular building is quantified, the appropriate parties can analyze possible risk management solutions.

This paper will concentrate its use of the engineering method on the Flame Movement Analysis. We wish to evaluate the likelihood that a fire will be limited to an area of the building. This likelihood is based on four engineering method parameters: an evaluation of the hazard present (the I-curve); automatic suppression (the A-curve);

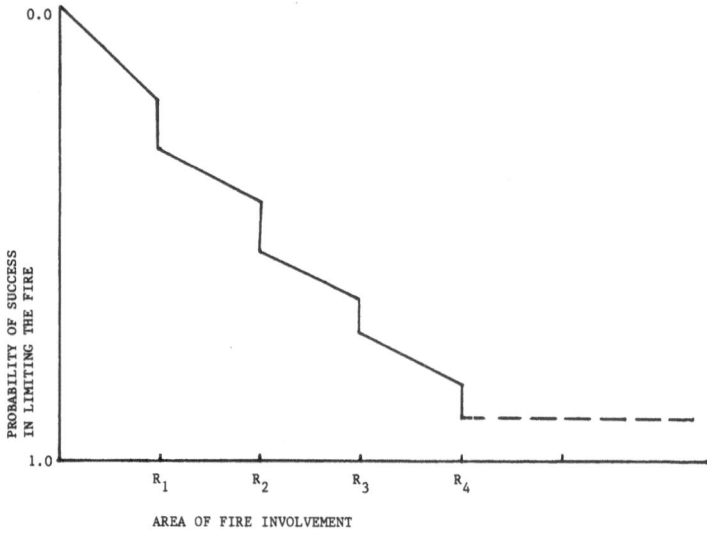

Fig. 1. Typical L-Curve.

notification and manual fire fighting (the M-curve); and barrier effectiveness (B), either physical or spatial.

An evaluation of these components yields a probability value in the form of an L-curve. The L-curve is a graphical representation of the cumulative probability that a fire will be limited to the area being considered.

Limited, in this sense, means that the fire will not propagate beyond the area which has been evaluated. A typical L-curve is shown in Fig. 1. The Y axis is in units of probability, ranging from 0.0 at the upper left to 1.0 at the lower left. The X axis represents the area being considered. The study area can be expressed in terms of square footage, defined spaces (rooms), or interruption time as a result of a fire.

The most common usage to date has been to describe the likelihood of termination in terms of spaces or rooms involved. In all applications, the vertical lines of an L-curve represent the strength of a barrier. The longer the line, the greater the probability that the barrier will prevent propagation into the next space. This is dependent on the specific type of fire that will attack the barrier, and, of course, barrier construction. A barrier can either be a physical or a spatial separation. A fire must penetrate the barrier to cause ignition in the next space.

RISK MANAGEMENT APPLICATIONS

The thrust of this paper is a field evaluation of a building. This evaluation is then used to analyze fire protection options offered to improve the fire safety qualities of the building, and thus reduce risk.

By simply inserting a zero value for any component (I-, A-, M-, & B) into a network calculation, the effect which that component has in the L-curve can be negated. The ease of this "what if" situation capability is ideal for an effective comparison of various loss prevention alternatives and the resources required to implement them.

As mentioned earlier, the engineering method offers as a final product a probabilistic description of the extent of fire loss to the area of study in the building. The extent of loss under different risk management situations is of primary importance if risk decisions are to be made. The description of a fire loss potential of any building is based on the field assessed fire safety probabilities. These field probabilities are then used to calculate the probabilistic evaluation of the building. Therefore, the manner in which field probabilities are assessed is critical. The value of any probabilistic risk assessment will be severely questioned if a logical and consistent framework does not exist from which the inspecting engineer can comfortably bring to bear his/her expertise with confidence.

DETERMINING THE PROBABILITIES: DECOMPOSITION

The decision analysis technique of decomposition is not a new one. The more a system can be explicated on paper, the more tests of coherence can be applied, the better to communicate an analysis, and the more comfortable the individual is likely to be with his understanding of the problem. Thus, the more a probability assessment can be split up into a sequence of interrelated factors and casual links, the more effective becomes the analysis in decision terms. This principle is used in the following example of a decomposition into hierarchical parts to calculate the event probability through the use of Bayes' theorem. This decomposition technique was put to substantial use by the U.S. Advanced Projects Research Agency in the 1970s to determine a country's likelihood to develop a nuclear weapons capability.

Deterministic equations to assess probabilities, ideally, are desirable to make a risk analysis procedure truly analytical. However, in many risk management situations, time and expense will not allow for a thorough deterministic engineering evaluation. In many situations, the additional confidence in the answer from a deterministically based risk analysis may not be appreciably greater than a subjectively assessed approach, particularly in view of the additional resources required. Engineering judgment takes into consideration the assumptions and limitations of the present state of scientific knowledge in fire protection engineering. Engineering judgment is the link relating available knowledge to field conditions; the link from science to engineering practice. The risk assessment technique in this paper has specifically been developed for field use. In addition to the approach taken here, objective and deterministic probabilities may be used without altering the risk analysis procedure contained within.

Research has shown that information available is very conservatively used because the way in which it is handled is improper. That is, the diagnostic implications of a single datum can be judged quite well, but the way interrelated data are put together is handled poorly. Typical laboratory experiments have shown that 50 to 80 percent of the information available to decision makers is wasted. Nearly every decision maker has the feeling that more information is needed to eliminate uncertainty. An optimal rule which wastes nothing for information processing is Bayes' theorem; this will be used in conjunction with the probability decomposition technique.

The decomposition technique breaks down collected information into an engineering based analysis framework in which each datum's influence in the overall picture is incorporated. Then, this information is used to generate input for Bayes' theorem to process, at the top of the framework hierarchy, resulting in the desired objective.

This probability decomposition has been developed for use in describing the risk of a telephone central office exchange building. The life safety and property protection philosophy historically in this occupancy has been superior construction, rapid detection, first aid fire response, and passive fire protection offered from compartmentalization through fire barriers and sealed penetrations. Figures 2 and 3 illustrate simple probability

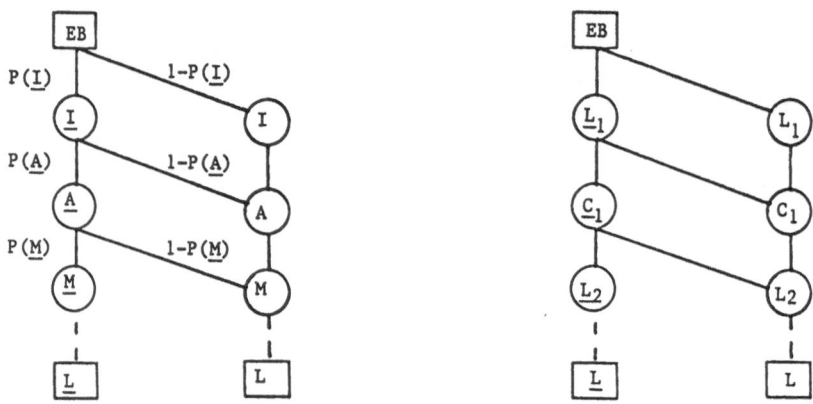

EB = Established Burning—The point at which an incipient fire is a threat.

Fig. 2. Room Probability Network.   **Fig. 3. Building Probability Network.**

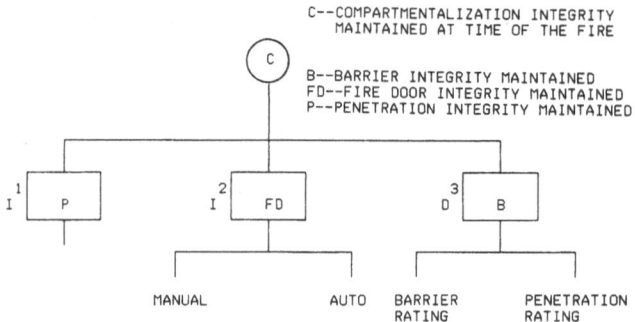

Fig. 4. Hierarchical Structure for Compartmentalization at Time of Fire.

networks which describe the cumulative probability of a fire room, and the cumulative probability of success in limiting flame movement within sequential rooms along a specific path of fire propagation, respectively. Consequently, a numerical level of risk for all parts of the building, such as in a specific area, can be generated.

The probability assessment of an entire telephone central office exchange will not be addressed in this paper in the interest of briefness. However, the passive fire protection offered by compartmentalization will be addressed to illustrate the process. This section of the evaluation has been chosen because of the important role it plays in the fire safety design of the exchange and the evaluation of both technical and human element areas. Figure 4 shows the breakdown of precursive indications which have been determined to impact on the effectiveness of compartmentalization. Note how the objective C has been clearly defined, as well as its complement $\bar{C}$. Immediately following are precursive data (D^i) and indicators (I^i). Events labeled D^i are those which have been observed, such as open cable penetrations or the lack of a penetration inspection program. Events labeled I^i are those which are not known with certainty at the time of the analysis. An example of this situation is the condition of barrier penetrations at the time of a fire. A further probability decomposition of penetrations, I^1, which contributes to the success of compartmentalization, is shown in Fig. 5.

This technique also differs from the more conventional Bayesian determination of the uncertain quantity by focusing on conditioning events such as D^i or I^i to assess separate

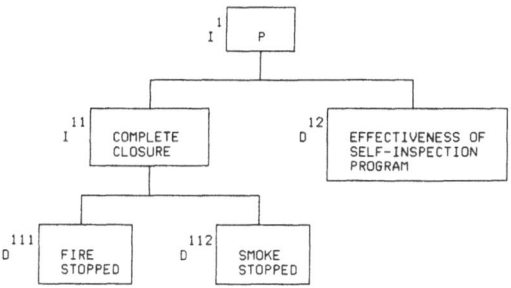

Fig. 5. Penetration Decomposition Network.

predictors $P(C/D^i)$, $P(C/I^i)$, $P(C/\underline{D}^i)$ and $P(C/\underline{I}^i)$. Given that all predictors and conditioning probabilities of C are known, then:

$$P(C) = P(C/D^i)P(D^i) + P(C/\underline{D}^i)P(\underline{D}^i) +$$

$$P(C/I^i)P(I^i + P(C/I^i)P(\underline{I}^i) \tag{1}$$

This results in obtaining, based upon each factor, a set of predictors for Pj(C) where j = 1,2,3,4.... A major problem in this approach, however, is assessing $P(C/D^i)$, the probability of successful compartmentalization (C) based upon considering the event D^i alone. It would be very difficult and an uncomfortable task for an engineer to quantify his/her opinion of success of an objective based solely on one contributing factor. Also, the engineer who doesn't understand why each predictor must be treated separately [refer to Eq. (1)], would soon become frustrated.

This problem can be sidestepped. Instead of focusing on the predictor $P(C/D^i)$, the likelihood of $P(D^i)$ can be deduced by an engineering analysis of the role D^i plays in the event of C's success (also in the event of C not occurring). This is done by assuming a situation in which successful compartmentalization is achieved, and conversely, not achieved.

Also, assuming a given datum $P(D^i/C)$ could be cumbersome in some instances because the probability would most likely be very small. In this case, it would be much simpler to assess the likelihood ratio:

$$\frac{P(D^i/C)}{P(D^i/C)}$$

This is very applicable when involving data which have been observed, such as the discovery of improperly closed cable penetrations. Assigning probabilities in light of the inspecting engineer's assessments yields Fig. 6.

It is judged that at a particular location a well implemented and effective inspection program to control closure of cable penetrations is in place. The engineer assigns the following probability, or likelihood in Fig. 6:

	I^1	$\underline{I}^1$
D^{12}	10	1

This matrix states that given I^1 is successful: no passage of flame, smoke, or products of combustion at the time of a fire, a successful cable penetration inspection program is 10

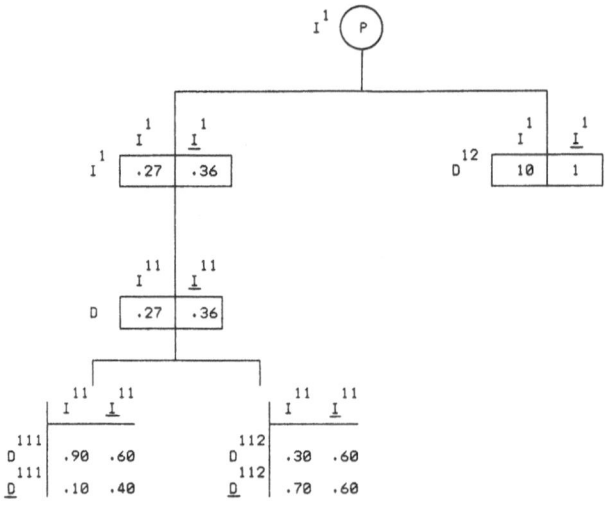

Fig. 6. Inductive Structure for Penetration at Time of Fire.

times more likely than an unsuccessful one. The actual ratio chosen could be made in light of corporate guidelines to ensure consistent selection.

I^i events are not known with certainty at the time of the analysis and thus must be assigned a probability of occurring or not occurring in the event of C or $\underline{C}$.

	I^1	I^1
I^{11}	1.0	0.0
I^{22}	0.0	1.0

The engineer's assessment of likelihoods in this deductive fashion results in Fig. 6. Note that Fig. 6 is related to Fig. 5 by way of superscripts assigned to the hierarchical components. Once these probabilities have been obtained from the engineer, the usual Bayesian calculations can be performed.

For each indicator, I^i, of Fig. 6, the likelihood of all data upon which it is contingent (all data below) must be obtained. That is, of I^{11} two pieces of data are contingent: D^{111} and D^{112}. Incomplete closure (open penetration) can occur by penetrations not being fire stopped or by not being smoked stopped. These two events are conditionally independent, so their joint likelihood is obtained from:

$$P(D^{111}, D^{112}/I^{11}) = P(D^{111}/I^{11})P(D^{112}/I^{11})$$
$$= .90(.30) = .27$$

This result conveys, if complete closure is achieved at the time of a fire, that there is a 27% probability that cable penetrations in the fire area are properly fire stopped for flame and smoke, and that spread of fire is successfully limited for "X" amount of time. The time frame used in any evaluation must be determined as part of the fire safety goal. A small matrix is now formed based on these results:

	I^{11}	I^{11}
D	.27	.36

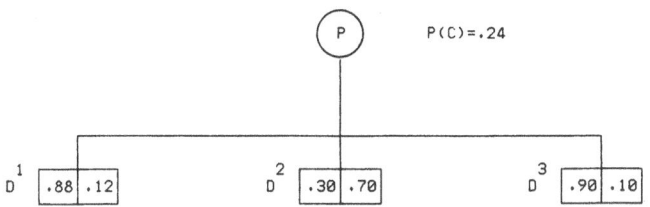

Fig. 7. Inductive Structure for Compartmentalization at Time of Fire.

which is placed at point I^{11} in Fig. 6. Note that D represents all the data below a certain point on a branch, and in this case it implies D^{111} and D^{112}.

Using Bayes' theorem and proceeding back up the tree in this way results in $P(I^1/D)$ = .88. Incorporating this probability into the network of Fig. 7, the likelihood of successful compartmentalization is .24.

RISK MANAGEMENT SELECTION

The results of individual networks such as Fig. 7 are used to calculate the probabilistic description, the L-curve, of the building being evaluated. At this point it should be emphasized that the engineering logic which goes into the decomposition networks of Figs. 4 and 5 is critical to the validity of the risk quantification generated in Fig. 3.

This value represents the status quo situation of our analyzed building. The L-curve of our building and its associated probability provide a quantification of risk at the rooms along the evaluated path of propagation. Figure 8 shows two L-curves for our building.

Curve 1 shows the status quo situation and Curve 2 reflecting the effect of some measure under consideration to improve the fire safety of the status quo. Curve 2 is generated in the same manner just discussed. The assessment of the mitigating measure is placed in the appropriate location of the network decomposition and a new L-curve reflecting the change is produced. A comparison of increased fire safety from various options can now be made.

Point "X" on the abscissa marks the extent of an "acceptable fire loss." This acceptable extent of loss could be expressed in building area (ft^2), rooms, or other variables. For each of these categories, the point "X" is determined by the risk attitude of the decision maker. The computer generated loss curve must originate in the worst room and "propagate" toward the appropriate area of concern. The worst room would be the area, given ignition, which would pose the greatest fire threat to the remaining building. The worst room can be designated, or be allowed to be chosen by the computer based on the field assessment. The computer program can generate a loss curve for a specific request, or a series of curves which reflect all possible paths of propagation.

Decision Analysis and the Attractive Alternative

With the fire safety assessment of our building now quantified, a decision analysis model can be used to support fire safety decision making. Engineering method generated probabilities will be used together with a decision analysis model to support fire safety decision making in the "two room" telephone central office building of Fig. 9. In conjunction with this model, a major parameter in identifying the best alternative is the decision maker's risk attitude.

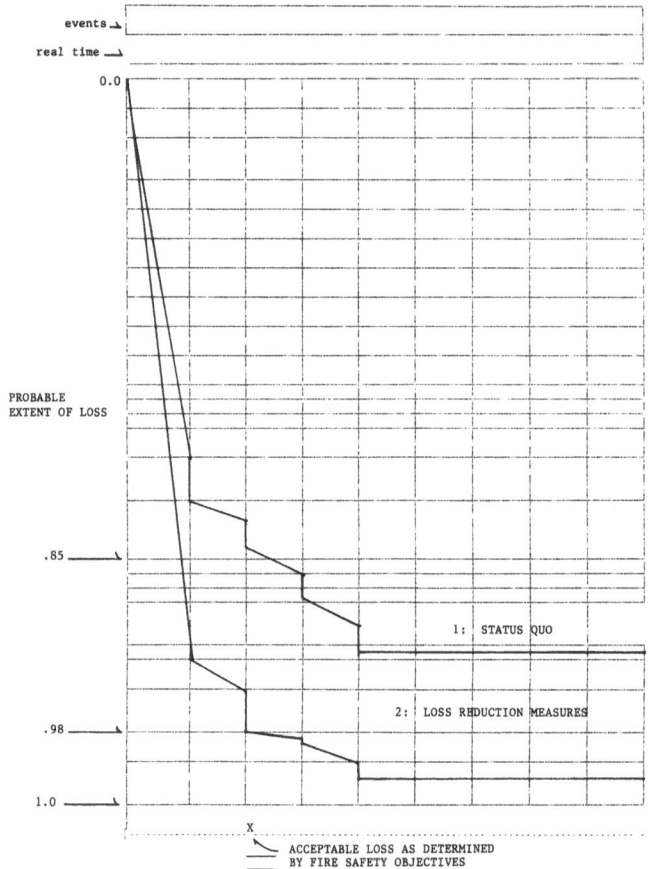

Fig. 8. Acceptable Fire Loss.

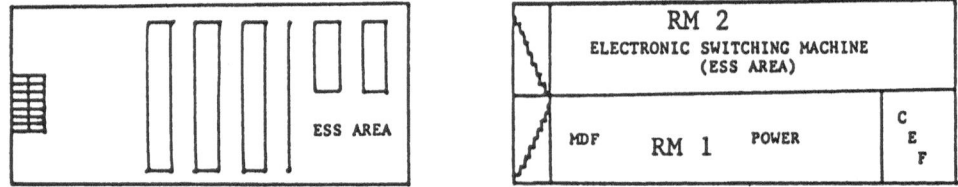

Fig. 9. Plan and Section of Typical Telephone Central Office Exchange.

The objective of this illustration is to show the overall process of applying decision analysis using probabilities generated by an engineering analysis. A fire is assumed to start in Room 1 and proceed from left to right. If the fire penetrates the barrier it must propagate to Room 2. This is a simple condition for the engineering method. In more complicated situations, the engineering method's computer program is capable of identifying the worst case scenario fire which would pose the greatest threat to any room or area of interest in the building.

The probability of limiting a fire at individual rooms is obtained from network diagrams. With this probability we can then determine the probability of limiting the fire to

490

"Z" sq. ft. in space X, given the fire was *not* limited to "Y" sq. ft. in space X, where Z is greater than Y:

$$P(L1 @ Z / \underline{L}1 @ Y) \qquad \text{Where } \underline{L} = \text{not limited}$$

For example,

$$P(L1 @ 3000/FIRE) = .919$$

$$P(L1 @ 6000/\underline{L}1 @ 3000) = 8.1 \times 10^{-6} = 0$$

are obtained from an evaluation of Room 1 using the engineering method for the partial floor areas. Likewise, this same evaluation would yield sector probabilities for Room 2.

The seemingly unwarranted number of significant figures are included above to illustrate a conservative assumption. If a fire is limited to Room 1, the vast majority of that probability of being limited would fall within the 3000 sq. ft. floor area, the area of initial fire growth. Therefore, the succeeding sector probabilities would be much smaller, approaching zero at the overall room area.

DECISION MAKING AND THE ATTRACTIVE ALTERNATIVE

With the knowledge of the probabilities describing the extent of fire propagation a decision can now be made concerning which fire protection engineering alternative to implement.

It can be shown that in every decision there are three types of attitudes toward risk:

1) risk aversion;
2) risk neutrality;
3) risk seeking.

When a decision involves uncertainty, a rational decision maker is not risk neutral. A decision maker in a business environment, again, if rational, is not risk seeking and therefore is risk averse. Depending upon uniquely impacting variables, different degrees of risk aversion will be evident. The risk attitude of the decision maker can be described by a risk constant, r, which enables a normalizing of all risk management alternatives.

Once all alternatives have been normalized, the most attractive alternative is identified based on maximizing gain or minimizing loss. The risk constant reflects the risk attitude of the decision maker. As the risk constant changes, the degree of the risk attitude exponentially changes. If $r>0$, the decision maker is risk averse; if $r=0$, risk neutral; and if $r<0$, risk seeking. Weighting is done using the following exponential utility relationship to determine the certainty equivalent (CE), or risk adjusted factor, which reflects the risk attitude of the decision maker for an alternative:

$$CE = (-1/r)\ln[EV(\exp -rz)] \qquad (2)$$

where EV is the individual expected values of each alternative.

Alternatives are normalized by determining a CE. Utility is the "worth" of a particular alternative when compared to the risk associated with that alternative. Expected value (EV) is simply an outcome of all possible outcomes, losses or gains, weighted by their probabilities. The risk attitude of the decision maker is not considered in the EV calculation. Therefore, the CE of an alternative is desired.

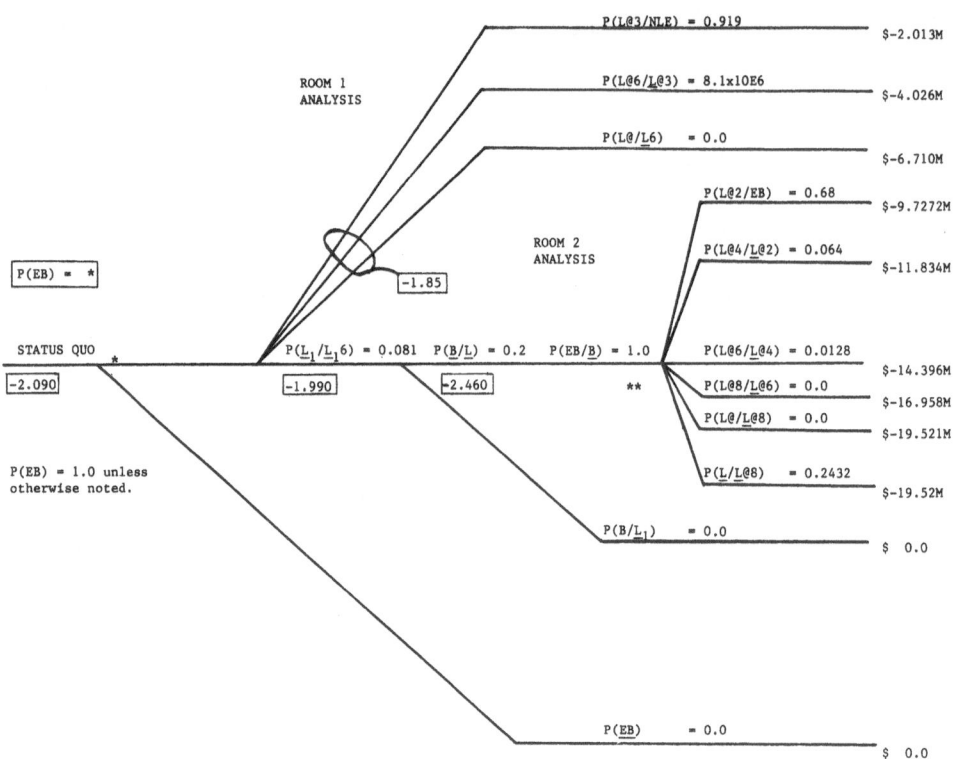

Fig. 10. Loss Analysis of Status Quo Situation.

Decision making based on expected value criteria shows the risk neutral attitude. But risk management decisions must consider associated uncertainty. Given the risk constant of the decision maker, the basis on which a decision is made can now be rationally changed. Identification of the most attractive risk management alternative can now be made based on the incorporated risk aversion of the decision maker. The CE calculation incorporates the decision maker's attitude toward relative amounts of money, or, in other words, risk. The new decision criterion is now that of maximizing gain/minimizing loss.

Evaluating the Alternatives

The decision tree reflecting value exposed in the STATUS QUO fire situation is provided by Fig. 10.

The respective values at the end of each branch are negative because they represent expenditures either as a fire loss or as the cost to carry out a fire protection engineering recommendation, as the case may be. Three engineering solutions have been offered to reduce the fire loss potential of this building. Each alternative can be implemented either alone or together, at some respective cost. How does the decision maker evaluate the cost effectiveness and relative worth of each alternative and thus quantify the resulting reduction in risk? What is the cost effectiveness of an engineering solution?

A cost effectiveness analysis is done in exactly the same way as that already done under the STATUS QUO situation. The post-survey engineering analysis of the building results in fire safety recommendations. These must be evaluated to determine which offers the most fire safety improvement for the money spent. The following recommendations modify probability assessments made in the STATUS QUO situation and thus change the building's L-curve (the probabilistic description).

ALTERNATIVE 1

The fuel loading (stored combustibles) of Room 1 is eliminated or rearranged such that potential flame spread is eliminated.

COST TO IMPLEMENT: $0

ALTERNATIVE 2

Improvement in detection is made by the installation of smoke and flame detectors.

COST TO IMPLEMENT: $50,000

ALTERNATIVE 3

Additional fire rated barriers are constructed to limit flame spread potential.

COST TO IMPLEMENT: $25,000

These modifications result in new network probability calculations of the engineering method. The resulting new probabilities are inserted into decision trees of the respective alternatives. As in Fig. 10, the trees are rolled back to obtain an expected value at the front of each tree for respective cases of the above three alternatives.

An expected value ranking of these three options and the STATUS QUO is as follows:

EV RANKING

1) ALTERNATIVE 1: $-2.0616M
2) ALTERNATIVE 3: $-2.0639M
3) STATUS QUO   : $-2.0904M
4) ALTERNATIVE 2: $-2.1046M

If the decision maker is neutral towards risk (r=0), the most attractive option of the four alternatives is Alternative 1 because it shows the least loss. However, as previously discussed, if uncertainty is involved a rational decision maker is not risk neutral. The present EV ranking does not consider the aversion to the uncertainty associated with loss potentials of the various fire protection recommendations.

Risk aversion of the decision maker must now be included in the evaluation of the four alternatives. Given that the range of loss potentials for the building of Fig. 9 are:

$-19.52M: total amount subject to loss

and

$-50,000: greatest expenditure of any one alternative,

two different risk constants, r, are determined:

$r(1) = 1.25 \times 10^{-7}$
$r(2) = 1.00 \times 10^{-5}$

Substituting into Eq. (2) respective EV values for the four alternatives, based on respective risk constants, the four options under consideration now fall into a new ranking of attractiveness:

$r = 1.00 \times 10^{-5}$ MORE CONSERVATIVE $r = 1.25 \times 10^{-7}$ LESS CONSERVATIVE

1) ALTERNATIVE 3:	CE = $-2.069M	1) ALTERNATIVE 1:	CE = $-2.0617M
2) ALTERNATIVE 1:	CE = $-2.075M	2) ALTERNATIVE 3:	CE = $-2.0697M
3) ALTERNATIVE 2:	CE = $-2.116M	3) STATUS QUO :	CE = $-2.0908M
4) STATUS QUO :	CE = $-2.136M	4) ALTERNATIVE 2:	CE = $-2.1057M

Additional r values could yield yet different rankings of these alternatives. If the risk constant accurately describes the risk attitude of the decision maker, then the most attractive alternative has been identified.

SUMMARY

An engineering method exists today for numerically evaluating a relative level of risk in any building. The flame movement part of the method involves the determination of the probability of success in terminating a fire within each space of a building. The effectiveness of each barrier surrounding the space, whether it be a floor, ceiling, wall, or an empty void, is also evaluated. A computer model exists which will describe any or all possible fire propagation paths from any specified room of origin and quantify the threat to any space along the path. The model then will calculate coordinates of the L-curve in time and space for each fire propagation path.

Once the probabilistic description of how the building in the example would react in a fire is determined, a decision analysis of possible fire protection alternatives is performed. This decision analysis model evaluates the effectiveness of each fire protection alternative against its cost to implement, and then compares it to the status quo by incorporating the decision maker's willingness to accept risk associated with minimum and maximum loss potentials.

REFERENCES

Van Anne, Craig, "Evaluation of the Risk Problem and the Selection of the Optimum Risk Management Solution," First Int'l Symposium on Fire Safety Science, National Bureau of Standards, 1985.

Van Anne, Craig, "Managing Risk Through Decision Analysis to Evaluate Fire Safety Hazards and Extent of Probable Loss by Use of an Engineering Method," Masters Thesis, Worcester Polytechnic Institute, 1985.

Fitzgerald, Robert, "Risk Analysis Using the Engineering Method for Building Analysis," Worcester Polytechnic Institute, 1984.

Can Risk Assessment and Risk Prioritization Be Extricated from Risk Management?

Mary R. English

Energy, Environment, and Resources Center
University of Tennessee
Knoxville, TN

ABSTRACT

This paper argues for an open, reflexive, iterative process of risk assessment, prioritization, and management. It maintains that, while quantitative risk assessments are helpful starting points, the risks of many situations cannot be understood without taking into account uncertainties in alternative risk management approaches. It also maintains that decisions on risk prioritization and management should take into account how risks will be distributed. If either uncertainty or inequity is ignored, intractable conflicts may arise which can derail the necessarily political process of assessing, prioritizing, and managing risk.

KEYWORDS: Risk management, risk assessments, risk prioritization, inequity, uncertainty

INTRODUCTION

The National Research Council has recommended that risk assessment and risk management activities be clearly distinguished and separated institutionally. According to an article in *Science* (1987) by Milton Russell, until recently an assistant administrator at the U.S. Environmental Protection Agency (EPA), and Michael Gruber of EPA's Office of Solid Waste, this separation has been accepted in principle by EPA.

Russell and Gruber comment that the separation between risk assessment and risk management has not been easy to maintain, because there is no hard-and-fast distinction between the policy considerations inherent in each. They also comment that risk-based approaches to priority-setting have not been easy to implement, because politics—especially budgetary politics—have often intervened. Despite these obstacles, EPA is trying to follow a rational sequence for selecting target risks and then for deciding how stringently to control them. This sequence includes (1) quantitative risk assessments using guidelines arrived at after considering alternative scientific approaches to risk estimation; (2) prioritization based upon the level of estimated risk relative to the costs of its control; and (3) selection of appropriate management techniques. EPA's risk-based approach does not automatically assume that either the best available technology (BAT) or pollutant levels approaching zero are the answer.

This type of approach has a number of merits. First, it recognizes the methodological and data uncertainties associated with estimating risk. As Russell and Gruber and others have noted, full objectivity is not possible, even with a scrupulously methodical approach: there are a lot of scientific controversies about what counts as sufficient evidence in making risk assessments, and positions on these controversies are necessarily infused with subjective values (see, e.g., Reaven, 1987; Shrader-Frechette, 1986). Second, the risk-based approach recognizes that funds available for risk reduction are finite: choices must be made in determining which risks to reduce and to what extent, and choices must also be made about whether to spend additional resources on environmental risk reduction or on other societal needs (some of which are intended to reduce risks in other ways). Third, this approach recognizes that there are trade-offs in risk allocation, and that in reducing one risk another more severe one may occur (e.g., asbestos may cause more harm when removed than it would have if left in place).

But, if taken as a model, this approach has two major shortcomings: it leaves out important elements, and it is linear rather than reflexive. I will concentrate my comments on the first but will return to the second in the conclusion.

I contend that, while quantitative risk assessments are helpful starting points, the risks of many situations cannot be understood without taking into account uncertainties in the risk management process. I also contend that decisions on risk prioritization and management should take into account how the risks will be distributed. If either uncertainty or inequity is ignored, intractable conflicts may arise which can derail the necessarily political process of assessing, prioritizing, and managing risk.

THE RISK MANAGEMENT PROCESS: SOURCES OF UNCERTAINTY

Uncertainty and hazard are inherent in risk: for example, the interplay of these two characteristics has led to one commonly used dichotomous classification of risk situations as "high probability, low consequence" or "low probability, high consequence." If risk is accepted, a degree of uncertainty is accepted, both about whether an event will occur and about the accuracy of occurrence predictions. But both hazard and uncertainty have thresholds of acceptability, and these thresholds vary from individual to individual and from group to group. Although societal debate often centers on the appropriate threshold of hazard, conflicts over the appropriate threshold of uncertainty can be at least as great.

Uncertainties about the probability of a hazardous event occurring and about the extent of damage resulting can arise from one or more of the following sources:

1. Lack of *simplicity* and *testability* of the risk management process.

Constructing a good bicycle helmet requires technical expertise, but the product is simple and fairly easily testable. Wearing a bicycle helmet is even simpler. Choosing between airbags and seat belts is more complicated. So is constructing, testing, and operating a nuclear power plant. Simplicity and testability in the risk management process minimize uncertainty over whether the best approach is being used, resulting in less controversy among experts and more confidence in the experts by laypeople.

2. Lack of *familiarity with* or *confidence in* the risk management technology and the risk managers.

A car is a complicated piece of machinery and riding in one is hazardous because of the potential for mechanical failure, driver error, or mistakes by others on the road. But, because automobile technology and driving are familiar and traffic patterns are usually predictable, the uncertainties of travel by car are reduced—especially if you are at the wheel and it's your own car. Furthermore, while the consequences of an automobile accident are dreaded, they are usually immediate and well-understood. In contrast, activities such as genetic engineering are at the frontier of human understanding, and their ramifications are debated by experts. The identities and competencies of the risk managers may be unknown, especially if the risk must be managed indefinitely into the future, and the consequences of a mishap may be mysterious and distant, sometimes with long latency periods.

INTRACTABLE CONFLICTS IN THE FACE OF UNCERTAINTY

If the level of uncertainty surpasses thresholds of acceptability, conflicts arise. These conflicts are intractable in that they involve principles, not interests, stemming from deeply held beliefs about knowledge and society. The following are two examples:

1. *One-shot solutions* versus *eternal vigilance.*

Through technology, human operator error can be minimized in one of two ways: by having alerts that something is about to go wrong, or by having fail-safe mechanisms that require no human attention. With both, the man/machine interface is crucial, but one-shot solutions require strong belief in the soundness of the initial design and execution, whereas eternal vigilance requires strong belief in the alert systems and the continued capacity of humans to respond to them.

For example, during the 1970s one debate over high-level radioactive waste (HLW) involved whether HLW should be permanently disposed of in a geologic repository or whether it should be kept above ground in a monitored retrievable storage facility. The geologic repository proponents argued that it is better to have a one-shot, permanent solution—one that would not put future generations at risk or rely on their institutional capabilities. The monitored retrievable storage proponents argued that because the geologic repository is a complex and unfamiliar risk management process, it would be better to proceed cautiously, even if this necessitated continued care by future generations. A similar debate is now going on with the disposal of low level radioactive waste (LLW).

2. *Autonomy* versus *epistemic dependence.*

With complex, unfamiliar technologies, even the ability of individual experts to understand the whole system on which they are working is limited, leading to questions about whether anyone is accountable and sufficiently competent—and possibly leading to a greater desire for autonomous control by those at risk. This may not be rational: rationality still may dictate leaving risk management decisions to risk management experts. (See, e.g., Hardwig, 1985.) But self-management satisfies our desire to view ourselves as adults, as independent and in control of our lives, especially if we lack trust in the managing institutions. And lack of trust is typical of risk situations that are characterized by uncertainty, particularly in a world that is increasingly interdependent and externally controlled and in an age that is skeptical of authority.

For example, some prospective host communities for hazardous waste facilities are beginning to voice demands for local control in the form of local monitoring and local power to close the facility in an emergency (Elliott, 1984; Bord, 1985). They doubt that the facility managers will be competent and concerned about community welfare, and, for similar reasons, they doubt that state regulations will be rigorously enforced. They seek technical assistance grants to do their own siting evaluations and to carry out ongoing oversight if they do host a facility.

THOSE AT RISK: EQUITY CONSIDERATIONS

In the discussion that follows, I will use equity to mean justice or fairness, not necessarily equality. An initial assumption is that for an equity issue to be involved in a risk management situation, the risk must be imposed by a human agent, not by an act of God or Fate. In addition, if the risk is clearly voluntary, as with activities such as smoking and drinking, equity issues don't enter in for the person at risk (although they may for society, if increased demands on medical services are created by these activities).

If an equity issue *is* involved in a risk situation, there appear to be three key questions:

1. Is the risk *necessary*—in other words, is it worth the benefits derived?

If not, then the next two questions are academic. Lack of need is sufficient to make the imposition of risk unfair.

2. Is the risk *compensible*?

Technological risks are concerned mainly with risks to life and health—only incidentally with risks to property. With life and health issues it is much more difficult to "raise someone to a level of indifference" through compensation. It is also much harder to generalize about what value to place on bodily losses.

3. Can the risk be *shifted* and/or *shared*?

If it can't, then there may be a kind of equity because of a unique capability to assume the risk—a counterpart to "deep pockets" in liability issues. (For example, each state being considered for the first-round HLW repository—Nevada, Texas, and Washington—would be more likely to accept the site if it were convinced that it had the one and only technically acceptable site in the country.) However, equity objections may be raised on other grounds, especially from a sense of unfairness at one having to bear the burden for all.

If the risk *can* be shifted and/or shared, three other scenarios are possible:

a. The risk cannot be shifted but can be shared (e.g., a two-repository approach to HLW disposal, with each repository having unique capabilities). This invokes a deep-pockets equity similar to the "can't shift, can't share" scenario, and, by sharing the risk, it lessens the sense of unfairness. (This would be true whether or not risk-*spreading* was involved.) From a shift/share standpoint, this may be the optimum scenario.

b. The risk can be shifted, but since only one "risky situation" is needed the responsibility cannot be shared (e.g., the one-repository-only approach to HLW disposal proposed in May of 1986 by the U.S. Department of Energy). This raises hard questions—in particular, how should the one be selected? Even if there is no "one and only" candidate for the risk, there still may be a "best" candidate—but what criteria should define "best," or even "technically acceptable," and how should they be applied? In the face of enormous epistemological and evidential difficulties, procedural fairness may become part of the answer. But "procedural fairness" also leads to questions on which there may be little agreement. In addition, as with "can't shift, can't share," an underlying sense may remain that it is unfair for one to be required to bear the burden for all. This scenario may raise the most equity objections.

c. The risk can be both shifted and shared (e.g., a two-repository approach, but with a number of qualified candidates). This complicates the evidential and distributional questions, but it lessens the sense of unfairness because the burden is shared.

INTRACTABLE CONFLICTS IN THE FACE OF INEQUITY

As with uncertainty in risk management, some inequity in risk distribution is expected. It is generally accepted that not everyone can bear an equal burden for each societal risk. Again, however, individuals and groups vary as to their levels of acceptable inequity, and if those thresholds are crossed, unresolvable conflicts can occur. At issue are deeply rooted principles—for example:

1. *Utilitarian* versus *deontological* distribution principles.

A straightforward utilitarian approach—"the greatest good for the greatest number"—can be used to justify individual dose rates that are higher than population dose rates. Utilitarian principles can also be used to justify regulatory guidelines that favor large populations over small populations. (For example, according to a recent article by Travis *et al.* (1987), EPA is considering a regulatory guideline where risks to small populations would be regulated only if they exceeded an individual lifetime risk level of approximately 1:10,000, whereas risks to large populations would be regulated if they exceeded an individual lifetime risk level of approximately 1:1,000,000.) Contrasting with a utilitarian approach are deontological principles that are more concerned with how rights and responsibilities are distributed and with who gets which risks and benefits. The deontological approach might accept higher costs, including the possibility of more total risk, in order to ensure a more equitable distribution of risks.

2. *Absolute* versus *relative* valuing of life.

Another source of conflict, both internal and external, lies in different attitudes toward the value of life. (See, e.g., MacLean, 1987.) Many people accord individual lives an absolute value—a value which is derived from religious or traditional sources and cannot be overridden. Heroic, expensive rescues exemplify this attitude. When lives are viewed in the abstract, however, there is less agreement. The value of a life may then be seen in a less than absolute sense: if necessary, the few may be sacrificed to save the many. (Although those sacrificed still may be seen as having intrinsic value—i.e., decisions about whom to save might still not be based on their extrinsic worth to society.) Conflicts arise both over the point at which the absolute value switches off and the relative value switches on and over the point at which extrinsic worth becomes a relevant criterion. Adding the dimension of future generations—is it better to save many lives in the future than a few lives today?—complicates the issue. One controversy concerns whether a discount rate should be applied, and what it should be. Uncertainty as to whether the hazardous event will occur and who will be affected adds to the dilemma.

CONCLUSIONS

If the above analysis is accepted, it follows (i) that the greater the uncertainty about the risk management process, the more intractable the conflict over risk assessments; and (ii) that the greater the potential inequity for those put at risk, the more intractable the conflict over risk prioritization and management choices. Given this, it appears that "uncertainty quotients" should be part of risk assessments and "inequity quotients" should be part of risk prioritization and management choices. This does not mean that "high uncertainty" and "high inequity" risk management approaches should be automatically excluded. Doing so might eliminate creative solutions in the first instance and might incur unacceptably high total risk and total cost in the second instance. But, as the conservative, pragmatic choice in order to avoid intractable conflicts, the bias should be against high uncertainty and high inequity.

Linear approaches to risk management such as that proposed by the National Research Council and adopted by EPA are intended as remedies to the "squeaky wheel" syndrome. In this regard they have merit: the wheel that squeaks the loudest is often the one that is the richest, best informed, and has the most political influence. What I'm suggesting here, however, is not a return to the squeaky wheel approach, but the inclusion of considerations about risk management in risk assessments and about those at risk in risk prioritization and management choices. This will necessitate a looped, iterative process rather than a linear one, and it will entail integrating public views into both risk assessments and management choices. This is bound to complicate an already complex process. It seems, nevertheless, that it is a necessary complication if the process is to be sound and politically legitimate, especially in involuntary risk situations.

REFERENCES

Bord, R., 1985, "Opinions of Pennsylvanians on Policy Issues Related to Low-Level Radioactive Waste Disposal," University Park, PA: The Pennsylvania State University Institute for Research on Land and Water Resources.

Elliott, M., 1984, "Coping with Conflicting Perceptions of Risk in Hazardous Waste Facility Siting Disputes," Ph.D. Dissertation, Massachusetts Institute of Technology, Cambridge, MA.

Hardwig, J., 1985, "Epistemic Dependence," *Journal of Philosophy* **82**, no. 7.

MacLean, D., 1987, "Comparing Values," Paper presented at June 1987 Conference on Valuing Health Risks, Costs, and Benefits for Environmental Policy Making held by the National Academy of Sciences.

Reaven, S., 1987, "The Methodology of Probabilistic Risk Assessment: Completeness, Subjective Probability, and the 'Lewis Report'," *Explorations in Knowledge*, Summer 1987.

Russell, M. and M. Gruber, 1987, "Risk Assessment in Environmental Policy-Making," *Science* **236**, no. 4799.

Schrader-Frechette, K., 1986, "The Conceptual Risks of Risk Assessment," *IEEE Technology and Society Magazine* **5**, no. 2.

Travis, C. C., S. Richter, E. Crouch, R. Wilson, and E. Klema, 1987, "Cancer Risk Management," *Environmental Science & Technology* **21**, no. 5.

Application of Risk Assessment Principles in Environmental Epidemiology

Stan C. Freni[a]
U.S. Department of Health and Human Services
Centers for Disease Control
Atlanta, GA

ABSTRACT

The application of methods of chronic disease epidemiology in environmental health studies needs a critical reevaluation. Quantitative risk assessment (QRA) of the health risk associated with environmental exposure is a rapidly developing methodologic approach to providing risk estimates used for regulatory purposes. Basic principles underlying QRA include the following: (1) disease predating exposure cannot be attributed to exposure; (2) exposure incurred after a health event is irrelevant; (3) current exposure levels are usually a poor indicator of past exposure; (4) the duration of exposure is as important as its level in assessing risk; (5) a dose is preferred over exposure level; (6) exposure to multiple chemicals via multiple routes is the rule rather than the exception; (7) quality assurance procedures need to be extended to the interview and to other data-gathering elements of the study; (8) statistically significant associations between risk factor and disease are likely to be spurious if they are unsupported by a dose-response effect and biological plausibility. Problems encountered when these QRA principles are applied in an epidemiologic study and the effect of these principles on the study outcome are discussed in the context of an epidemiologic study of the potential risk from exposure to volatile organic chemicals in drinking water.

KEYWORDS: Epidemiology, environmental risk, health assessment, epidemiologic modeling, exposure assessment

INTRODUCTION

Quantitative risk assessment (QRA) of health risks from exposure to chemicals or radiation is a new, rapidly developing scientific discipline. Just one decade ago, it was virtually unknown. Today, it is the basic tool used in the regulatory control of the safety of the occupational and ambient environment. As could be expected from any rapidly developing science with such a broad application potential, QRA is also the source of a large number of controversial issues, and, as a result, impinges upon many aspects of our society. For example, QRA is used to regulate the quality of river water and tapwater; it is used to condemn the building of incinerators to burn our household trash, but also to point

---

a. Current address: Center for Toxicological Research, Jefferson, AR.

501

out the danger of other kinds of waste disposal; and it has been used to ban substances, such as a particular kind of solvent for decaffeineting coffee and ethylene dibromide for cleaning milling machines, etc.

Its use and results are heavily challenged as being both too harsh on the economy and too soft on the chemical industry. These challenges do not come from considerations of self-interest and politics alone; they are perpetuated because of the scientific basis underlying QRA, a basis that often consists of policy decisions rather than hard facts.[6] One of the reasons policy decisions must be made is our still poor understanding of the processes leading to diseases. Along a broad front, scientists fight their way through the mysteries of life to unravel the cause of diseases. Epidemiology is a discipline that devotes its work entirely to this task. Risk assessors expect from epidemiologists both accurate and unambiguous conclusions that directly address specific problems in risk assessment, but hardly recognize that animal data are rarely accurate and unambiguous. It is expected from epidemiologic studies that (a) they address the health risk from a single chemical, (b) extreme care is taken to control confounding factors to compensate for the inability to experimentally subject people to exposure, (c) exposure from secondary routes is accounted for, (d) the risk is expressed as an estimate of absolute risk per dose unit, and (e) the dose-response relationship is evaluated. Some epidemiologic studies have yielded results of the greatest importance to risk assessors, including studies of asbestos, radiation, lead, vinyl chloride, and benzene. Contrary to what is often heard among risk assessors, epidemiologic studies are quite able to address QRA problems, but only on the condition that they are specifically designed for that purpose, which would require considerable, but rarely available, funding.

EPIDEMIOLOGIC STUDY DESIGNS

There are many approaches to designing and implementing epidemiologic studies. The design selected is not only determined by the subject of investigation, but also by the available resources, the size of the pool of potential study participants, and the time allowed. Basically, the studies can be categorized as follows:

1. Ecologic studies: the incidences or prevalences of diseases in exposed and unexposed populations are compared. Since exposure and disease are studied in populations rather than in individuals, the study results can by no means be interpreted as addressing causal relationships. It is often claimed that ecologic studies may help to generate hypotheses concerning the risk from chemicals. Although this may be true, the chance that such hypotheses may reflect a true association of exposure and health effects is so extremely remote that such studies can be dismissed for QRA.

2. Cross-sectional studies: a sample of a population is studied for the occurrence of disease and exposure in individuals. The analysis of the results may yield useful suggestions for further research. The weakness of this approach is that the study population is not sampled on the basis of individual exposure or health status. Differences in disease or exposure prevalence may, therefore, be due to chance or to unknown or not included risk factors, so-called confounding factors. No cross-sectional studies have yielded useful results for chemical-specific QRAs.

3. Case-control studies: people with the disease of interest (cases) are compared with others (controls) for individual exposure experience. The control group is usually selected to closely resemble the case group in risk factors other than exposure. Matching is often done on a one-to-one basis, which allows better control of confounding but is more time consuming and expensive because a larger number of people have to be screened before a match is found. With regard to risk assessment

objectives, several major problems still have to be solved: exposure levels that change over time; separating the key exposure from concurrent exposure to other related or unrelated chemicals; incorporating the time factor in dose-response evaluation; obtaining a sufficient number of exposed cases of the disease of interest; and expressing risk in absolute rather than in relative measures. A major disadvantage is also that only one disease can be studied. On the other hand, the strengths of this design are obvious: the study of incident cases guarantees good quality health data, better than that of animal data; very rare diseases can be studied, something that cannot be said of animal experiments; confounding factors, if known, can be controlled for in the design phase by matching or in the analysis; the time and cost involved, although substantial, are much less than that of a two-year chronic animal bioassay of the same size.

4. Cohort or followup studies: exposed and unexposed cohorts are observed for a certain period, and the incidence of diseases occurring in that period is compared. Prospective cohort studies are preeminently suited for risk assessment. The major problem is that the long observation period needed before a sufficient number of diseases has developed renders the feasibility of the study dependent on the availability of large funds and the willingness to wait that long for the results. The retrospective version circumvents this problem by looking at historical data, with the penalty of having to accept the quality of health and exposure data present in existing files, which is often poor. Both versions have the disadvantage that rare diseases require a large study size. The advantages of cohort studies are: absolute risk estimates can be derived directly from observed data; the design allows better quantification and classification of exposure and a more adequate evaluation of the dose-response relationship than in case-control studies; incorporation of multimedia exposure is easier; and multiple diseases can be studied simultaneously.

The characteristics of the study designs determine their usefulness for risk assessment: the case-control study is the first choice for qualitative risk assessment and very rare diseases, while QRA objectives are best served by cohort studies. This presentation highlights the application of risk assessment principles in a retrospective cohort study of the health effects of exposure to volatile organic compounds (VOCs) in drinking water.

EXPOSURE ASSESSMENT

As is clear from the study design, the first task is to estimate individual exposure. A correct exposure assessment includes estimating when exposure started, how its level changed over time, the total accumulated exposure, and the conversion of exposure to dose. Typically, environmental scenarios are characterized by the availability of a one-time measurement of individual exposure, in this case, one water sample. Very few examples have been found of modeling past exposure over time in the absence of monitoring data. Lawrence and Taylor,[5] in need of chloroform concentrations over time for use in a case-control study, modeled the chloroform concentrations in city water supplies from existing records from city water plants. They regressed the chloroform concentration on chlorine use, effluent chlorine residual, and the type of water source. Lagakos and coworkers,[4] with no direct evidence of current or past individual exposure, estimated past individual exposure from the results of a water distribution study by the city of Woburn, Massachusetts. In that study, a model of the city's network of water mains was used to assess the distribution of water from two contaminated wells, amidst a large number of clean wells constituting the well field, over the city's neighborhoods.

For this study a concentration-time (CT) curve was developed for each of the seven chemicals found in the contaminated well field of Battle Creek, Michigan.[2] This method essentially consists of statistically determining the slope of the CT-curve from monitoring

Table 1. Exposure Defined as Sum of Concentrations in Current Water Sample Versus Exposure Defined as the Integrated Area Under the Concentration-Time Curve, Individually Estimated

Current water sample		Concentration-Time INTEGRATION	
Proportion of People (%)	Sum of VOCs (ppb)	Proportion of People (%)	Accumulated VOCs (ppbmonths)
0	0	10	0
55	$1 - 99$	20	$1 - 99$
24	$10^2 - 10^3$	13	$10^2 - 10^3$
17	$10^3 - 10^4$	23	$10^3 - 10^4$
4	$10^4 +$	18	$10^4 - 10^5$
		12	$10^5 - 10^6$
		4	$10^6 - 10^7$

data on city wells and then applying this curve to the one water sample taken from adjacent private wells, in order to estimate when pollution of each of these residential wells started, and how the concentration C rose over time T. The exposure accumulated during the observation time was obtained by integrating the area under the CT-curve.

Figure 1 shows the important points in time for defining the integrated area: (a) when exposure started ($T1$), (b) when a person moved into the study area (TIN), (c) when exposure stopped ($TSTOP$), and (d) when a disease was first diagnosed ($TDIAG$). A disease diagnosed before TIN or $T1$ cannot be attributed to the exposure, and such a person should hence be considered unexposed, regardless of the exposure accumulated since $TDIAG$. This assessment of individual exposure was done for each of the VOCs found in the well water, and the results are shown in Table 1.

The exposure values in the table were derived by summing chemical-specific values to arrive at one composite value, based on equitoxicity of the chemicals for nonacute effects.[1] The values clearly reflect the big difference in the categorization of exposed people whether the current level of exposure or the exposure accumulated over time is used. From interview and clinical examination, the volume of unheated water drunk and the body weight were known, which allowed the conversion of accumulated exposure in units of ppbmonths into a dose unit of mg/kg. The next step was to estimate additional exposure from secondary sources, namely through inhalation and skin absorption during showering or bathing. As described earlier (1), this part of the study failed because of the lack of knowledge on the absorption rate, specific for each route and chemical, during the limited period of exposure at the prevailing low concentration levels. This is not a failure of the study design, but rather a failure of experimentalists to provide this basic information.

HEALTH ASSESSMENT

Health information was collected by means of interviewing, clinical examination, and retrieving and abstracting medical records. This is not the place to linger on the problems inherent in interviewing. Terms such as recall bias and interviewer bias are well known. Especially in the case of a sensitized public, interview responses are not always credible.

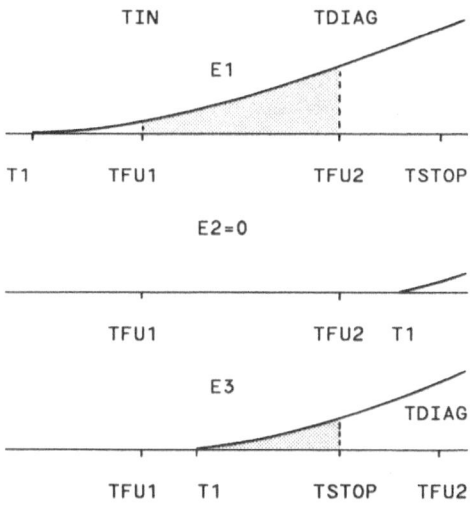

Fig. 1. Chemical-Specific Concentration-Time Curves with Different Points in Time That Determine the Total Accumulated Exposure (TAE) (Shaded Area Under the Curve). The Y-axis (omitted in the figure) represents the concentration of a chemical in water; the X-axis represents the time of followup (*TFU*). *T*IN is the date of moving to the study area, *T*STOP is when a person stopped drinking contaminated water, *T*DIAG is when a disease was first diagnosed, *T*1 is the date a well became contaminated, *TFU*1 and *TFU*2 are the starting and closing dates of the followup. *E*1, *E*2, and *E*3 represent TAE values for three individuals, each with an own set of time events. *E*1 is determined by *TFU*1 and *TFU*2, in which *TFU*2 was determined by *T*DIAG. *E*2=0 because *T*1 was after *TFU*2. The limits of *E*3 are *T*1 and *T*STOP. The concentration-time curve and *T*1 will differ for each chemical, resulting in a set of chemical-specific TAEs per person.

Suffice it to say that quality control procedures were not limited to laboratory tests but were extended to include the interview as well. The questionnaire contained questions on the same subject, repeated in different wording, as a tool to identify inconsistencies in responses. For instance, questions on use of prescribed drugs often yielded more information on disease occurrence than the response to a direct question on a specific disease. Direct questions about birth defects were less informative than when women were asked to list the outcomes and the dates of their pregnancies one by one.

From a risk assessment standpoint, validation of health data is of crucial importance. This was done by comparing the information from multiple sources within and among each other. The medical record data base was expected to provide the best information, but the multisource approach revealed a different picture.[3] Often, records were not found, were incomplete, or contained admission rather than discharge diagnoses. Patients who went to different hospitals for the same disorder might have been filed under different diagnoses. Diagnoses, even those made at discharge, were quite often descriptive rather than indicative of the primary disease. For instance, hematuria may point to bladder cancer, cystitis, or kidney stones. Comparing sources with each other showed that an intrinsic weakness of medical record data appeared to be the lack of certainty that the earliest recorded date of diagnosis was indeed the earliest one. In fact, interviewees reported dates that were many years earlier, sometimes supported by the dates of drug prescription. Records originating from private physicians' offices were of lesser quality than hospital records. Interview responses were more often wrong in dates of diagnosis and the kind of disease than the other sources were.

By limiting the abstracting of medical records to recording the diagnosis and date of diagnosis, much information was lost that could have been used to clarify uncertainties. On the other hand, keeping full copies of the records or making extensive summaries would have created difficulties in logistics (because of the volume of unnecessary data) and posed legal problems. It would also require extensive funding, because a physician would be required to make a meaningful summary. One important lesson from this study is that, despite the expense, quality control procedures have to be developed and implemented for retrieving and abstracting medical records.

STATISTICAL ANALYSIS

Statistical measures of effect include correlations, differences in means and proportions, odds ratios and relative risks, and multivariable analysis. Given the complexity of the relationship between chemical exposure and health effects, and the multicausality of virtually all chronic diseases, univariate analysis should be used for descriptive purposes only, and not as a tool to arrive at conclusions about the causality of an association observed between exposure and disease. A proper analysis of an epidemiologic study should be more than just running a computer program for multivariable analysis. It is important that an *a priori* hypothesis about the mechanism of pathogenesis should have shaped the study and its design, and be reflected in the statistical analysis. Knowing that related chemicals with a similar toxicologic profile act together in an additive fashion is not compatible with just stringing together chemical-specific values as covariates in the model. The result would be misleading because the combined effect as estimated with exponential models is a multiplicative effect. Available computer packages for cohort studies cannot handle multiple exposures, each of which changes in level during the observation period. Assuming a constant level, as is usually done, results in serious errors in the risk estimate.

For this study, a commercially available proportional hazard model was modified.[3] Simply stated, the modification consists of partitioning the observation period in one-year sections and the exposure and health status are assessed for each period and chemical. In other words, the experience of an individual in one period is considered an observation by its own, to be analyzed separately from the observations on the same person in other periods. Thus, one person with a followup of ten years, consisting of seven unexposed years followed by three years of increasing exposure levels, would constitute ten observations. The first seven would be seen as seven unexposed persons each with an age and other risk factors valid at the time of each specific period. The next three observations are treated like three persons with different levels of exposure. The statistical program is then applied to this aggregate of period-specific observations.

The great advantages of this approach are not only that it is more in harmony with reality than the conventional programs, it also facilitates (a) compliance with the risk assessment condition that exposure incurred after a disease is diagnosed is irrelevant; (b) computation of composite exposure values for chemicals prevalent at the time of a specific period for any number of chemicals with any kind of *CT*-curve; (c) introduction of the concept of lag time estimates for the period between initiation of disease and date of diagnosis; and (d) application of minimum thresholds for the duration of exposure. For instance, if a person developed a disease, all observations from periods following the diagnosis would simply be deleted.

CONCLUSION

In summary, there is an urgent need to reevaluate conventional methods in epidemiology, in the design as well as in the analysis. Environmental epidemiology needs a synthesis of components of biology and biochemistry, toxicology, medicine, environmental engineering, and risk assessment. Further progress in biochemistry and pharmacology will soon lead to the final stage of environmental epidemiology, the so-called molecular epidemiology. Until then, great improvements can be made now with today's tools, an open mind, and cooperation with all disciplines involved in environmental and human health.

REFERENCES

1. Freni, S. C., "An Epidemiologic Approach to Dealing with Multimedia Exposure to Multiple Related Chemicals," Proceedings Annual Meeting of the Society for Risk Analysis, Houston, Texas, November 1987.
2. Freni, S. C., and Phillips, D. L., "Estimation of the Time Component in the Movement of Chemicals in Contaminated Groundwater," *Environ. Health Perspect.* **74**:211-221 (1987).
3. Freni, S. C., and Bloomer, A. W., *Report on the Battle Creek Health Study*, Michigan Dept. of Public Health, 1989.
4. Lagakos, S. W., Wessen, B. J., and Zelen, M., "An Analysis of Contaminated Well Water and Health Effects in Woburn, Massachusetts," *J. Am. Stat. Assoc.* **81**:583-596 (1986).
5. Lawrence, C. E., and Taylor, P. R., "Empirical Estimation of Exposure in Retrospective Epidemiologic Studies," in *Environmental Epidemiology*, F. C. Kopfler and G. F. Craun, Eds., Lewis Publishers, Chelsea, Michigan, pp. 239-246 (1986).
6. Office of Science and Technology Policy, "Chemical Carcinogens: A Review of the Science and Its Associated Principles," *Federal Register* pp. 10371-10442 (1985).

A Comparison of the Remedial Priorities Determined Using More Likely Case and Worst Case Risk Estimates

Josephine A. Mauskopf, Terrence K. Pierson,
and Margaret E. Layne
Research Triangle Institute
Research Triangle Park, NC

ABSTRACT

This paper presents a simple method for incorporating uncertainty estimates into decision making based on the results of a cancer risk assessment. The method involves answering the general question: Would a decision based on worst case risk estimates be changed by choosing more likely parameter estimates and models for the risk assessments? An example is developed using data from seven industrial sites where the decision maker seeks the answers to two specific questions: Does the perception of a health risk at each site change according to the worst case or more likely case assumptions? Do priorities for cleanup vary under the alternative assumptions? The estimates presented indicate affirmative answers to both of these questions for some of the seven sites.

KEYWORDS: Uncertainty, risk assessment, quantitative ranking scheme, exposure analysis, exposure scenarios, more likely case, worst case risk estimates

INTRODUCTION

It is well known that lack of knowledge of the mode of action of toxic substances and limited availability of experimental data result in estimates of health risks attributable to exposure to toxic substances that have a high degree of uncertainty. In the past, regulators have adjusted for this uncertainty by choosing worst case (or reasonable worst case) values for the input parameters and conservative assumptions for estimation models, thus obtaining estimates of the health risks that are likely to overstate the true risks. Such an approach is reasonable if risk assessment is used only for setting safe standards and the costs associated with different standards are either negligible or thought to be of no importance. The levels set will almost certainly protect even the most sensitive maximally exposed human, and the difficulties involved in analyzing the magnitude of the uncertainty associated with the worst case assumptions would be avoided.

Such a worst case approach to regulation presents the decision maker with some rather complex issues when the costs of different standards must be considered, when the cost-effectiveness of different risk management alternatives is to be compared, or when priorities must be set for cleanup or regulation. Examples of problems that arise are the following:

- Use of worst case estimates can result in very stringent standards and, therefore, very high compliance costs.

- Use of worst case assumptions is likely to result in estimates of the benefits of the regulation (usually measured as avoided harmful health effects) that are overstated.

- Use of worst case estimates may result in misallocation of resources between different regulatory programs. For example, if the level of uncertainty of the adverse health effects varies between two toxic substances A and B, use of worst case estimates might show that chemical A is more dangerous than chemical B, whereas the use of more likely case estimates might show the reverse. A decision to allocate more resources to the control of chemical A rather than to the control of chemical B could be considered a misallocation of resources.

In this paper we propose a simple step that a decision maker or analyst might take to incorporate estimates of the uncertainty in risk assessment into the decision making process. This step involves asking the question: Would a decision based on worst case risk estimates be changed by choosing more likely parameter estimates and models for the risk assessment? If the answer is "no," then the decision based on the worst case estimates can be implemented with a greater degree of confidence. If the answer is "yes," the decision maker has three options: continue as planned; take steps to reduce the uncertainty in the risk estimates; or change the decision. Even if the decision maker chooses the first option, the uncertainty in the risk assessment process has been incorporated into the decision making process.

In order to illustrate how such a procedure might be applied, we will present an example of seven industrial sites that are contaminated with toxic substances. In this example, the decision maker has to determine how many of the seven sites are likely to significantly harm human health if no action is taken and to rank the hazardous sites in order of importance for cleanup. Thus the decision maker should seek the answers to the following two questions:

- Does the perception that there is a problem at each site vary according to whether worst case or more likely case assumptions are used in the risk assessment?

- Do priorities for cleanup vary under these alternative assumptions?

For the purposes of this paper we assume that the estimation of risk and the ranking of sites are based on the magnitude of the excess individual cancer risk expected for each site if no action is taken. In the discussion, we will briefly suggest how other harmful health effects might also be included in a quantitative ranking scheme. This approach is different from the Environmental Protection Agency's (EPA's) hazard ranking scheme in that the ranking of sites is based on a quantitative estimation of risk rather than a summation of ordinal rankings of various factors.

The sources of uncertainty modeled in our example are the following:

- The initial steady-state soil concentration.

- The inhalation rate.

- The duration of human exposure.

- The potency factors for carcinogens.

Other areas of uncertainty that could be included in the analysis if data were available include degradation rates for chemicals and human absorption rates for different routes of exposure.

SITE DESCRIPTION

The facility selected for this case study is an active chemical manufacturing plant located in a heavily industrialized area on the shore of a major river. The site covers 160 acres and is bordered by residences on the east and northwest sides. More than 250 organic and inorganic chemicals have been produced at this facility during its more than 50 years of operation. Currently about 40 chemical products are manufactured at the site.

A site investigation report was completed which focused on several active and inactive solid waste management units at the plant, including landfills, surface impoundments, tank and drum storage facilities, wastewater treatment plants, and an incinerator. The facility was divided into seven areas for the evaluation of remedial action priorities. Samples of soils, sediments, and wastes were collected in areas suspected of contamination. The sampling was nonrandom, with the sampling sites selected based on visual inspection and information about past activities in the area. Soil samples were collected both at the surface and at depths up to 18 to 24 inches. All samples were collected during a two-week sampling period.

The samples were analyzed for a wide range of organic and inorganic chemicals, and all areas sampled showed the presence of numerous contaminants. Up to 10 metals and more than two dozen organic compounds were identified, with each area typically containing more than 20 contaminants. Our exposure analysis was limited to two to four chemicals in each area. These indicator chemicals were selected based on their carcinogenicity or toxicity and the concentration measured at the site. The indicator chemicals selected by this method were generally the same as those which would be selected using the methodology described in EPA's Superfund Public Health Manual for selection of indicator chemicals.

EXPOSURE ROUTES

Our exposure analysis focused on vapor and dust inhalation by workers on-site and by nearby residents. Soil erosion and surface water runoff from the site are towards the river, where the dilution factor is quite large. It is unlikely that soil erosion and runoff would contaminate the soil of neighboring residences. Ground water flow is also towards the river, and residents are served by the local water utility rather than by private wells, so exposure via drinking water was not considered. However, a separate investigation was underway to determine the extent of ground water contamination. No agricultural activity occurs in the area, so human exposure through contaminated crops or livestock is unlikely. Access to the site is controlled, and workers in the active areas of the facility use protective clothing, so on-site dermal exposure and ingestion of soil by children were not evaluated.

EXPOSURE SCENARIOS

The first step in estimating human exposure to these contaminants is the determination of the average concentration of the chemical in the soil over a specified time period. The assumption that contaminant concentrations remain constant over the exposure period would give an overly conservative result, since the actual concentration of the contaminant will decrease over time through several mechanisms of degradation and transport. A slightly less conservative approach is to estimate the long-term average concentration of a contaminant in soil using a diffusion model which accounts for losses due to volatilization and ignores other losses such as chemical reaction, biological degradation, or leaching. This approach was used for our exposure analysis.

Model inputs include soil properties and parameters which describe the contaminants' tendency to sorb to the soil particles, dissolve in water, or vaporize. Average or typical values for these parameters are estimated based on characteristics of the site and the indicator chemicals. The initial concentration of each indicator chemical at time zero is assumed to be the maximum measured concentration in the soil for the worst case and the average measured concentration for the more likely case. The contaminants are assumed to be evenly distributed across a site. This same modeling approach also is used to estimate the emissions of the contaminant in the vapor phase.

The routes of human exposure addressed in our analysis include inhalation of toxic vapors and inhalation of dust particles contaminated with toxic chemicals. The dust emissions at the site were estimated based on soil characteristics, the fraction of bare soil exposed to the wind, and the average wind speed. The dust particles were assumed to have the same average contaminant concentration as that calculated in the soil. There were no differences in the values for these parameters for the worst case estimates and the more likely case estimates.

The concentrations of toxic vapors and dust in the ambient air were estimated using atmospheric dispersion models. A simple box mixing model was used to estimate on-site concentrations based on the average wind speed and site dimensions. A simplified Gaussian dispersion model was used to estimate downwind concentrations more than 100 meters from the source. Key input variables for this model are wind speed and atmospheric stability, a description of weather conditions important for atmospheric mixing.

Other considerations in developing the exposure scenarios for on-site and off-site inhalation of contaminated dust and vapors include body weight, average lifetime, duration of exposure, rate of inhalation, and absorption fraction, as well as the chemical specific potency factor for carcinogens. The average adult body weight was assumed to be 70 kg. The average lifetime was assumed to be 75 years. These factors were the same for both the more likely and worst case scenarios. The absorption fraction was assumed to be 0.5 for dust inhalation in the more likely scenarios and 1.0 for all other scenarios.

Inhalation rates have been shown to vary widely depending on activity levels. For this analysis, the inhalation rate for all more likely case scenarios except off-site outdoor dust inhalation was assumed to be 26.4 m^3/day, representing a relatively low level of physical activity such as office work. The reasonable worst case inhalation rate for all scenarios except off-site outdoor dust inhalation is 45.6 m^3/day, representing a moderate level of physical activity such as walking. The inhalation rates used for off-site outdoor dust inhalation represent a higher level of activity, such as performing yard work or heavy manual labor. These rates were 74.4 m^3/day for the more likely scenario and 103 m^3/day for the reasonable worst case.

We assumed a worker spends 40 hours per week at the facility for 50 weeks of the year for the duration of exposure for on-site scenarios. For the more likely case we assumed a worker will spend 10 years on the job at this facility, for a total exposure duration of approximately 800 days. For the reasonable worst case scenario we assumed that the worker will spend 25 years working at this plant for a total exposure duration of approximately 2000 days.

Off-site exposure durations vary for vapor and indoor and outdoor dust inhalation. The more likely case exposure duration for vapor inhalation and indoor dust inhalation was based on the assumption that the exposed individual lives in the neighborhood for 7 years and spends 73% of the time at home for a total of 1850 days of exposure. The reasonable worst case exposure duration for vapor and indoor dust inhalation was based on the assumption that the individual lives in the neighborhood for 25 years and spends 77% of the time at home for a total of 7042 days of exposure. Exposure to outdoor dust inhalation was assumed to occur for 2.6 hours per week for 7 years in the more likely case, giving a total of 40 days exposure, and for 3.3 hours per week for 25 years in the reasonable worst case, giving a total of 179 days of exposure.

POTENCY ESTIMATES

Potency factors were estimated using animal data from the Carcinogenic Potency Data Base published in *Environmental Health Perspectives* in 1984. We used the Crump Global86 computer model that assumes a multistage model for carcinogenesis to generate both maximum likelihood and upper 95 percent confidence limit estimates of potency. We intended to use the linearized 95 percent upper confidence limit estimates for the worst case scenario and the maximum likelihood estimates, allowing for nonlinear dose terms in the multistage model, for the more likely case scenario. However, most of the animal experiments for the chemicals present at the industrial sites had only one dose level apart from control and, therefore, only a one-hit model could be estimated. Thus, only relatively small differences in potency between the worst and more likely cases were estimated, with the ratio varying between 1.18 and 2.22 (see Table 1). For those chemicals for which nonlinear dose-response relationships have been estimated, use of these for the more likely case scenario will result in greater variability in the estimates of the potency factors. All of our potency estimates are based on body weight rather than surface area as a scaling factor.

Table 1. Estimated Potency Factors Used in the Analysis

	Worst Case	More Likely Case
Hexachlorobenzene	0.112	0.0923
Mirex	0.412	0.233
Aroclor 1260	0.647	0.549
DDD	0.00818	0.00358
DDT	0.0478	0.0323
β-Lindane	0.0679	0.0434
γ-Lindane	0.0279	0.0175

When the complete or linearized version of the multistage model is used, the differences in potency estimates are orders of magnitude smaller than those obtained when comparing different low dose extrapolation models such as the Weibull, multistage, one-hit, and Moolgavkar-Knudson-Venzon (M-K-V) models. However, with the possible exception of the M-K-V model, we as yet have little biological basis, or any other basis, to determine which of these models is most likely to approximate the truth.

RESULTS AND CONCULSIONS

Table 2 shows the estimates of lifetime excess cancer risks for the worst and more likely cases for exposed adults at each site aggregated across routes of exposure and chemicals as suggested by the Superfund Public Health Evaluation Manual. The excess lifetime risks vary between 1.7×10^{-3} (site M, on-site worst case) and 2.6×10^{-8} (site C, off-site more likely case). These estimates give an answer to the first question: Does the perception of a health hazard at the site change according to the risk scenario analyzed? A common measure of health hazard is an excess lifetime risk of cancer greater than 1.0×10^{-6}. Using this measure, six of the seven sites are health hazards for the on-site worst case scenario, but only four of the seven sites are health hazards for the more likely case on-site scenario. These same results are obtained if we look at each chemical and each exposure route separately as proposed by the U.S. Environmental Protection Agency's Office of Solid Waste. Thus the answer to the first question is: Yes, the perception of a health hazard changes for some but not all sites.

Table 2. Excess Individual Cancer Risk by Site

Site	On-Site Worst Case	On-Site More Likely Case	Off-Site Worst Case	Off-Site More Likely Case
C	3.0×10^{-6}	2.2×10^{-7}	5.9×10^{-7}	2.6×10^{-8}
D	2.1×10^{-4}	1.5×10^{-5}	3.8×10^{-5}	1.6×10^{-6}
F*	–	–	–	–
M	1.7×10^{-3}	2.3×10^{-4}	2.0×10^{-4}	1.7×10^{-5}
N	1.0×10^{-3}	6.4×10^{-5}	1.5×10^{-4}	5.6×10^{-6}
U	9.1×10^{-6}	8.1×10^{-7}	2.0×10^{-6}	1.1×10^{-7}
W	6.5×10^{-4}	1.8×10^{-4}	1.0×10^{-4}	1.7×10^{-5}

*No carcinogens were measured at site F.

Table 3 shows how the rankings of the seriousness of the sites differ for the worst and more likely case scenarios. For both the on-site and off-site scenarios, the ranking of sites W, N, U and C changed between the worst case and more likely case scenarios. Of particular note is the change in ranking of site W in the off-site scenarios. As this table illustrates, the ranking of the seriousness of the health hazards at each site is likely to be less sensitive to the magnitude of the uncertainty of the parameter estimates than is the perception of harm at each site. However, differences in ranking are more likely to occur where there are large differences among the chemicals in the degree of uncertainty of the input parameter estimates.

Table 3. Ranking of Sites on Basis of Excess Cancer Risk

On-site Worst Case	On-site More Likely Case	Off-site Worst Case	Off-site More Likely Case
M	M	M	M, *W
N	*W	N	
W	*N	W	*N
D	D	D	D
C	*U	U	*U
U	*C	C	*C
F	F	F	F

*Change in ranking of site between worst case and more likely case.

Finally, it is important to note than no estimates have been presented of the harmful health effects likely to be experienced as a result of exposure to noncarcinogenic toxic substances that have also been found at the contaminated sites. These substances include mercury, lead, and endosulfan. Comparisons can be made of exposure levels and the appropriate reference doses (RfDs) for each exposure route and chemical. (The reference dose is an estimate of a daily exposure to the human population, expressed in milligrams of chemical per kilogram of body weight per day, that is likely to be without an appreciable risk of deleterious effects during a lifetime.) However, the problem remains as to how to compare, quantitatively, the health hazards associated with exposures to a noncarcinogen above the RfD to exposures to a carcinogen that result in excess risks of cancer that are greater than 1.0×10^{-6}. The solution awaits the development of dose-response relationships and severity indexes for noncarcinogenic health effects.

Cancer Incidence Among Welders Due to Combined Exposures to ELF and Welding Fumes: A Study in Data Pooling[a]

R. M. Stern
World Health Organization Regional Office for Europe
Copenhagen Ø, Denmark

ABSTRACT

The demonstration of human risk due to occupational or general environmental exposures rests on the availability of epidemiological evidence. Frequently exposures are to such low levels, or exposed cohorts are so small that few or no studies can support statistical significance of evidence for the small resulting risk. For welders, exposure at the highest documented occupational levels for extremely low frequency electromagnetic radiation (ELF), a suspected leukemogen (magnetic fields of up to 0.1 Tesla), leads to repeated but not statistically significant observation of underincidence of leukemia, while exposure to high concentrations (up to 200 mg/m^3 total) of potentially carcinogenic metals leads to repeated observation of slight, but not statistically significant, excess risk of lung cancer. Pooling of the data via some form of meta-analysis indicates that these observed trends are apparently real, and suggests that statistical significance is not always a useful criterion for demonstrating the existence of small or highly localized risks among exposed cohorts. The use of meta-analysis cannot, however, identify the origin of the lung cancer excess, which might have a non-occupational (e.g., lifestyle) cause, or be due to a local, process dependent high risk hot-spot, but supports the need for a carefully controlled multicenter study to measure the risk among this large occupationally exposed population. Meta-analysis can also strongly support the lack of connection between VLF (magnetic field) exposure and leukemia.

KEYWORDS: Welders, ELF (extremely low frequency electromagnetic radiation), welding fumes, data pooling, leukemia, lung (respiratory-tract) cancer

Recent reports of excess leukemia incidence in several so-called electrical trades,[1] and similar results from studies of households situated near to the electrical power grid, have been interpreted as suggesting that some unknown component of extremely low frequency electromagnetic radiation (ELF) acts as a leukemogen. Since field measurements indicate that electric-arc and -resistance welders, who work in contact with cables and

---

a. Note added in proof: A complete analysis of this problem and of the data summarized in the Annex has been published elsewhere. See R. M. Stern, "Cancer Incidence Among Welders: Possible Effects of Exposure to Extremely Low Frequency Electromagnetic Radiation (ELF) and to Welding Fumes," *Environ. Health Perspect.* 76:221-229 (1987).

transformers carrying direct, alternating and pulsed electric currents of up to 100,000 amperes, have exposures to ELF magnetic flux densities among the highest recorded, ranging up to 100,000 micro Tesla, a study of leukemia incidence among welders might verify causality. Because of the difficulty in collecting individual cohorts large enough to sustain statistical significance, it is tempting to pool the results of numerous small epidemiological studies in an effort to test the model.[2] The 15 published studies on welders yield a total of 146 cases of leukemia observed (O) vs 159.46 expected (E), giving a risk ratio (RR) of 0.92, and where acute leukemia is studied separately, O=40 cases vs E=43.39; RR=0.92. Since only 10% of all welders use non-electrical techniques (e.g., gas welding), most historical studies are of cohorts exposed to ELF: this exposure is verified by the frequent observation of a statistically significant excess of respiratory tract cancer, presumably due to exposure to welding fumes generated by the arc. Pooling the results of 9 studies of proportional leukemia incidence or mortality in the "electrical trades" which together contain data on 13 different occupational categories, gives for all leukemia, O=519, E=443.77, RR=1.17, and for acute leukemia, where reported separately, O=227, E=158.5; RR=1.43. There are, however, little or no data to show that the "electrical trades" studied for leukemia have exposures to levels of ELF which are very much higher than those of the general working population: average exposures should be several orders of magnitude lower than those of welders. The absence of either elevation of the overall risk for leukemia or of an increase in the relative risk for acute leukemia among ELF exposed welders does not support the hypothesis that the observed excess risk of leukemia, or especially of acute leukemia, among workers in the "electrical trades" is due to their suspected ELF exposure.

Many recent studies have focused on a possible excess of respiratory tract cancer among welders.[3,4,5] Industry wide time weighted average (TWA) exposures are to 5 mg/cubic meter of fumes, mostly the oxides: of Fe for mild steel (MS), of Al for aluminum, and of Ni and Cr for stainless steel (SS) welding; concentrations above 200 mg/cubic meter frequently occur in confined spaces,[3] perhaps posing a high risk to a small fraction of the population. Some 36 studies of lung cancer among welders, with from 3 to 686 cases observed, give risk ratios from 0.7-7.0.[4]

Because lung cancer is common and the latency for the expression of a metal induced tumor is long, most epidemiological studies cannot support statistical significance for excess risks of less than 50-100%, although the pooled data from many small studies might. The 36 studies can be divided into three major classes: (a) cohort and case-control studies in non-shipyard working populations, (b) registry or death certificate based regional or local studies (these include some shipyard welders), (c) studies of shipyard welders (of these all but 4 regional studies involve 74 cases or less). Because of differences in exposures and design, the three classes should be examined separately, the pooled data yielding:

(a)(n=11 non-shipyard)   O=134 E= 98.8 RR=1.36 p<.0001,
(b)(n=7 regional)   O=217 E=151.1 RR=1.41 p<.000001,
(c)(n=7 shipyard)   O=122 E= 97.6 RR=1.25 p<.01.

Pooling may be justified, since bias against reporting of negative studies is avoided. Here, only 6 studies are designed on welding cohorts (only 24 cases); for the rest, welders are identifiable sub-cohorts in the general working populations examined. The 4 large regional studies give RR values from 1.35-1.46, p<.0001 for each, and the combined data for respiratory tract cancer yield:

B1)(n=4 regional)   O=1316 E=941.2 RR=1.40 p<.0000001,

almost identical to the excess risk found from the pooled data for the small studies, only several of which were significant at best at p<.05; pooling the small studies would have given the true value of excess risk even if the large studies were missing.[2]

For the cases of welders, it would appear that pooling the results of small epidemiological studies has demonstrated the absence of an excess risk for leukemia, while verifying the presence of small excess risk for respiratory tract cancer. The origin of the excess lung cancer risk remains obscure; it can be partly due to lifestyle (excess tobacco use), a small universal risk for inhalation of welding fumes, or restricted to one or more small sub-populations at high risk, e.g., those exposed to high TWA concentrations or to SS fumes, or at long latency or composed of highly susceptible individuals.[3] Pooling cannot be used to determine the origin of risk unless a common core protocol is used which will permit appropriate model testing; such a multicenter program is currently underway for welders under the auspices of the World Health Organization, Regional Office for Europe, coordinated by The Danish Welding Institute and The Clinic For Occupational Medicine, Copenhagen University Hospital, with mortality data pooling being undertaken by IARC, Lyon.[5]

REFERENCES

1. Milham, S. Jr., "Mortality in Workers Exposed to Electromagnetic Fields (1950-1982)," *Environ. Health Perspect.* **62**:297-300 (1985), and references therein.
2. Stern, R. M., "Cancer Incidence Among Welders: Possible Effects of Occupational Exposures to Extremely Low Frequency Electromagnetic Radiation (ELF) and Confounding Exposure to Welding Fumes," Report 86.xx, The Danish Welding Institute, DK 2605 BrØndby, 1986, p. 18.
3. Stern, R. M., "Process Dependent Risk of Delayed Health Risks for Welders," *Environ. Health Perspect.* **41**:235-253 (1981).
4. Stern, R. M., *Assessment, Management and Reduction of Risk for Welders in Health Hazards and Biological Effects of Welding Fumes and Gases*, R. M. Stern et al., Eds.; Excerpta Medica ICS 676, Amsterdam, 1986, pp. 535-552, and references therein.
5. Stern, R. M., "Health Effects of Occupational Exposure of Welders in Studies In Epidemiology Part 1, Health Aspects of Chemical Safety," Interim Document 15, WHO/Europe, Copenhagen, 1984.

ANNEX

SUMMARY OF EPIDEMIOLOGICAL STUDIES FOR LEUKEMIA AMONG ELF EXPOSED WORKERS, AND CANCER AMONG WELDERS

In addition to the references listed in the tables, occupational exposure of welders and the biological effects of welding fumes and their constituents have been discussed extensively in the literature (16,24,27,28,57,74,75,76,77,79,80). Cancer in welding populations at cites other than the respiratory tract and leukemia have been discussed in the literature (15,19,25,26,35,37,40,41,44,51,53,56,59,60,66,70).

Table AI. Studies of Leukemia Incidence in Welding Populations

ref.	All Leukemia			Acute Leukemia			Lung Cancer		
	Observed	RR	Expected	Observed	RR	Expected	O	RR	E
85)	6	0.96	6.24	m) 4	1.71	2.33	.	.	.
	-	-	-	l) 0	--	0.66	-	-	-
10)	20	0.83	25.3	a) 13	1.04	12.5	.	.	.
73)	7	2.25	3.1	.	.	.	.	.	.
18)	.	.	.	m) c	(3.8)	c	.	.	.
43,44)	19	0.89	21.3	a) 6	0.67	9.0	.	.	.
sub total	52	0.93	55.34	23	0.94	24.41			
5)	0	--	1.2	.	.	.	6	0.95	6.3
59)	0	--	1.56	.	.	.	17	1.5	11.3
67)d	4	4.2	0.94	.	.	.	10	2.2	4.5
3)	4	0.35	11.0	.	.	.	50	1.32	37.9
60)	1	2.5	0.4	.	.	.	7	1.38	5.1
56)	.	.	.	a) 6	1.81	3.3	27	0.99	27.3
70)	43	0.99	43.62	.	.	.	193	1.42	136.0
61) (all)	15	1.14	13.2	.	.	.	12	1.60	7.5
(high exp)	(4)	0.6	(6.7)	-	-	-	-	-	-
(low exp)	(11)	1.7	(6.5)	-	-	-	-	-	-
51)	27	0.85	31.6	m) 7	0.76	9.2	381	1.46	260.1e
	-	-	-	l) 4	0.63	6.4	305	1.27	240.6b
sub total	94	0.90	104.12	17	0.90	18.90			
Pooled Data	146	0.92	159.46	40	0.92	43.39	1008	1.34	736.6
		n.s.			n.s.			p<10^-10	

```
Sub total for studies with both all and acute leukemia data:
           72    0.86    84.4         34    0.85    40.09
```

d)gas welders, c)number of welders not given, e)<65a(SMR), b)65-75a(PMR), a)all acute leukemia, m)acute myeloid, l)acute lymphoid. All leukemia minus all acute leukemia= non-acute leukemia (where listed): O=38, E=44.37, RR=0.86. For studies without acute leukemia data, O=74, E=75.06, RR=0.99.

Table AII. Summary of Pooled Leukemia Data for 13 "Electrical Trades" (10,11,36,43,85) (PIR, PMR)

Job category (see text)	All leukemia observ	expect	Acute leukemia observ	expect	All - acute observ	expect	Acute Myeloid** observ	expect
1	16	19.3	9	6.79	7	12.5	5	3.48
2	23	11.94	9	3.82	14	8.12	3	1.52
3	159	142.87	68	58.54	91	94.33	25	26.19
4	71	61.1	34	23.92	37	37.2	17	12.52
5	32	22.9	15	7.43	17	15.5	7	3.41
6	9	3.46	4	1.03	5	2.16	1	0.14
7	20	10.6	11	4.3	9	6.3	-	-
8*	39	49.8	21	22.6	18	27.2	4	2.34
9	6	3.43	2	1.1	4	2.33	1	1.8
10	45	38.22	30	16.89	15	21.31	17	9.18
11	4	3	0	0	4	3	-	-
12	17	20.2	4	2.78	13	17.42	4	2.42
13	8	6.15	2	1.6	6	4.6	4	1.3
Total	449	392.97	213	150.8	236	252.0	88	64.3
Risk ratio	1.14 $p<0.002$		1.41 $p<10^{-4}$		0.94 $p<0,15$		1.37 $p<0.0003$	

* PIR, PMR low by about 5% due to high SMR for respiratory tract cancer
** Where reported

Relative increase in risk of All Acute Leukemia = RR(All Acute):RR(All) =1.41:1.14=1.24

Note that four additional studies (45,52,55,81) yield further data for several job classifications as follows: (55)(PIR):1,4,5,6,12/13; (45)(PMR):2, (52)(SMR):12; (81)(SMR):4,6. These studies contribute an additional 70 cases observed, 50.8 cases expected for all leukemia, and 14 cases observed, 7.7 cases expected for acute leukemia. The individual risk ratios follow the general trend among the occupational subclasses, and the additional data does not effect the overall pooled results appreciably. The lack of information concerning actual ELF exposures among these "electrical trades" has been criticized(64): exposure levels of welders have been measured however(34). In addition to occupational exposures, several reports have suggested that residential exposures to ELF from the distribution net might also give rise to an excess leukemia risk(eg 82, 83). Recent reviews (1,2,3,22,23,30,33, 48,65,71,84) do not provide evidence of the biological origin for such a response although some in vitro and invivo effects of ELF have been observed. The occupational subclases listed in Table AII above are as follows:
1) Electronic technicians (10,43,85), radio/radar mechanics (11,36)
2) Telegraph operators (43), radio/telegraph operators (10,11,36,85)
3) Electricians (10,11,36,43,85)
4) Power/telephone linemen (10,43,85), linemen/cable joiners (11,36), telephone installers/repairmen (11,36)
5) TV/Radio repairmen (10,43,85), electrical/electronic fitters/assemblers (11,36)
6) Power station operators (10,43,85)
7) Aluminium workers (43)
8) Welders, flamecutters (10,43,85)
9) Motion picture projectionists (10,43,85)
10) Electrical engineers (10,11,36,43,85)
11) Streetcar/subway motormen (10,43,85)
12) Electrical engineers (professional) (11,36)
13) Electronic engineers (professional) (11,36)

Table AIII. Cohort and Case Control Studies of Respiratory Tract Cancer Incidence in Non-Shipyard Welding Populations

Reference	Observed	Cohort studies Risk Ratio	Expected	Comments
13) Dunn and Weir (1968)	49	1.05	46.7	
17) Fletcher and Ades (1984)	8	1.46	5.5	
5) Becker et.al. (1985)	6	0.95	6.3	Stainless Steel(SS)
54) Ott et al.(1976)	2	1.0	2.0	
59) Polednak(1981)	17	1.5	11.3	SS
62) Redmond et al. (1979)	14	1.51	9.3	
68) Sjögren(1980)	3	4.41	0.68	SS

Reference	Cases(C)	Case control studies Risk Ratio(RR)	(Expected=C/RR)	Comments
9) Breslow et al. (1954)	14	7	2	
12) Decoufle et al. (1978)	9	0.9	10	
20) Gerin et al.(1984)	12	2.4	5	SS
29) Kjuus et al.(1986)	28	1.9	15.7	All welders
	(16)	(3.3)	(4.8)	(SS only)

Table AIV. Regional Respiratory Tract Cancer Incidence Studies Based on Death Certificates (Includes Some Shipyard Welders)

Reference	Observed(Cases)	Risk Ratio	Expected
14) Dunn et al.(1980)	19	1.12(SIR)	17.0
19) Gallager and Threefal(1983)	74	1.45(PMR)	51.0
21) Gottlieb(1980)	8	4.01(CC)	2(C/RR)
31) Lerchin and Samet (1984)	-	3.10(CC)	(unpub)
39) Menck and Henderson (1976)	48	1.37(SMR)	35.0
40-42,44,46) Milham(1976a 1976b,1981,1983,1985)	191	1.35(PMR)	141.5
47) Morgan and Treyve (1982)	31	1.78(SMR)	17.4
50) OPCS(1978)	246	1.51(SMR)	163.0
56) Peterson and Milham (1980)	27	0.99(PMR)	27.3
67) Silverstein et al. (1983)	10	2.20(SPMR)	4.5
69) Sjögren et al.(1982)	(96)	(1.44(SIR))	(66.7)
70) Sjögren and Carstensen(1986)	193	1.42(SIR)	136.0
51) OPCS(1986)	381	1.46(SMR)<65a	260.1
	305	1.27(PMR)65-75a	240.6

Table AV. Studies of Respiratory Tract Cancer Among Welders Occupied in Shipyards (Cohort and Case Control Studies)

Reference	Observed	Risk Ratio	Expected	Comments
3,4 Beaumont and Weiss(1980,1981)	50	1.32	37.9	shipyard and construction
38) McMillian and Pethybridge(1983)	5	0.96	5.2	
49) Newhouse et al.(1985)	26	1.13	23.0	Mesotheleoma SMR=1.84
60) Putoni et al.(1979)	7	1.38	5.1	
61) Putoni et al.(1985)	12	1.60	7.5	
63) Sheers and Cole(1980)	-	-	-	Mesotheleoma SMR=5.0
6) Blot et al.(1978)	11(cases)	1.05	10.5	RR=0.7 vs shipyard pop.
7) Blot et al.(1980)	11(cases)	1.25	8.8	

ANNEX BIBLIOGRAPHY

1. M. F. Barnothy, Ed., *Biological Effects of Magnetic Fields*, Vol. 2, p. 314, Plenum Press, NY-London (1969).
2. M. F. Barnothy, Ed., *Biological Effects of Magnetic Fields*, Vol. 1, p. 327, Plenum Press, NY-London (1984).
3. J. J. Beaumont and N. S. Weiss, "Mortality of Welders, Ship Fitters and Other Metal Trades Workers in Boilermakers Local No. 104 AFL-CIO," *Am. J. Epidemiol.* **112**:775-786 (1980).
4. J. J. Beaumont and N.S. Weiss, "Lung Cancer Among Welders," *J. Occup. Med.* **23**:839-844 (1981).
5. N. Becker, J. Claude, and R. Frentzel-Beyme, "Cancer Risk of Arc Welders Exposed to Fumes Containing Chromium and Nickel," *Scand. J. Work Env. Health* **4**:75-82 (1985).
6. W. J. Blot, J. M. Harrington, A. Toledo, R. Hover, C. W. Heath, and J. F. Fraumenti, Jr., "Lung Cancer After Employment in Shipyard During World War II," *N. Engl. J. Med.* **299**:620-624 (1978).
7. W. J. Blot, L. E. Morris, R. Stroube, I. Tagnon, and J. E. Fraumenti, Jr., "Lung and Laryngeal Cancers in Relation to Shipyard Employment in Coastal Virginia," *J. Natnl. Canc. Inst.* **65**:571-575 (1980).
8. D. Borgers and R. Menzel, "Physicians, Waitresses, Welders: An Analysis of Smoking Habits with Regard to Job and Occupation," *Munch. Med. Wochenschr.* (West Germany) **126**:1092-1096 (1984).
9. I. Breslow, L. Hoaglin, G. Rasmussen, and H. K. Abrams, "Occupations and Cigarette Smoking as Factors in Lung Cancer," *Am. J. Public. Health* **44**:177-181 (1954).
10. E. E. Calle and D. A. Savitz, "Leukemia in Occupational Groups with Presumed Exposure to Electrical and Magnetic Fields," *N. Engl. J. Med.* **313**:1436-1477 (1985).
11. M. Coleman, J. Bell, and R. Skeet, "Leukemia Incidence in Electrical Workers," *Lancet* **i**:982-983 (1983).
12. P. Decoufle, K. Stanislawczyk, L. Houten, I. D. J. Bross, and E. Viadana, "A Retrospective Survey of Cancer in Relation to Occupation," U.S. Dep. HEW. NIOSH Pub., pp. 77-178 (1978).

13. H. W. Dunn and V. M. Wier, "A Prospective Study of Mortality of Several Occupational Groups, Special Emphasis on Lung Cancer," *Arc. Env. Health* **17**:71-76 (1968).

14. J. E. Dunn, G. Linden, and L. Breslow, "Lung Cancer Mortality Experience of Men in Certain Occupations in California," *J. Nat. Canc. Inst.* USA **65**:571-575 (1980).

15. A. Englund, G. Ekman, and L. Zabielski, "Occupational Categories Among Brain Tumor Cases Recorded in the Cancer Registry in Sweden," *Ann. N.Y. Acad. Sci.* **381**:188-196 (1982).

16. Environmental Protectional Agency, Health Assessment Document for Chromium, EPA 600/8-83-019A, Washington, DC (1983).

17. A. C. Fletcher, and A. Ades, "Lung Cancer Mortality in a Cohort of English Foundry Workers," *Scand. J. Work Environ. Health* **10**:7-16 (1984).

18. U. Flodin, M. Fredricksson, O. Axelson, B. Pearsson, and L. Hardel, "Background Radiation, Electrical Work, and Some Other Exposures Associated with Acute Myeloid Leukemia in a Case Referent Study," *Arch. Env. Health* **41**:77-84 (1986).

19. R. P. Gallager and W. J. Threlfall, "Cancer Mortality in Metal Workers," *Can. Med. Assoc. J.* **129**:1191-1194 (1983).

20. M. Gerin, J. Siemiatycki, L. Richardson, and J. Pellerin, "Nickel and Cancer Associations from a Multicancer Occupation Exposure Case Referent Study: Preliminary Findings," in *Nickel in the Human Environment*, F. W. Sunderman Jr., Ed., IARC, Lyon, pp. 105-116 (1984).

21. M. S. Gottlieb, "Lung Cancer and the Petroleum Industry in Louisiana," *J. Occup. Med.* **22**:384-388 (1980).

22. H. B. Graves, T. D. Bracken, J. Griffin, J. de Lorge, and M. G. Morgan, and T. S. Teneorde, "Biological Effects of 60 Hz Power Transmission Lines," Florida Electric Power Coordinating Group, Tampa, FL (1985).

23. H. B. Graves *et al.*, Eds., *Biological and Human Health Effects of Extremely Low Frequency Electromagnetic Fields: Post 1977 Literature Review*, American Inst. Bio. Sci., Arlington, VA (1985).

24. K. Hansen, and R. M. Stern, "A Survey of Metal-Induced Mutagenicity In Vitro and In Vivo," *Journal of the American College of Toxicology* **3**:381-430 (1984).

25. S. Hernberg, P. Westerholm, K. Schultz-Larsen, R. Degerth, E. Kuosma, A. Englund, V. Engzell, H. S. Hansen, and P. Mutanen, "Nasal and Sinonasal Cancer," *Scand. J. Work Environ. Health* **9**:315-326 (1983).

26. G. R. Howe, J. D. Burch, A. B. Miller, G. M. Cook, J. Esteve, B. Morrison, P. Gordon, L. W. Chambers, G. Fodok, and G. M. Windsor, "Tobacco Use, Occupation, Coffee, Various Nutrients and Bladder Cancer," *J. Nat. Cancer Inst.* **64**:701-713 (1980).

27. IARC, "IARC Monographs on the Evaluation of Carcinogenic Risk of Chemicals to Humans: Some Metals and Metal Compounds," **23**:205-324 (1980).

28. IARC, "Chemicals, Industrial Process and Industries Associated with Cancer in Humans," IARC Monograph Supplement 4, IARC Lyon (1982).

29. H. Kjuus, R. Skjaerven, S. Langård, J. T. Lien, and T. Amondt, "A Case Referent Study of Lung Cancer, Occupational Exposures and Smoking," *Scand. J. Work. Environment Health* **12**:193-202 (1986).

30. B. G. Knave and S. G. Törnquist, "Epidemiological Studies on Effects of Exposure to ELF Electromagnetic Fields," A Review of Literature. Report 85E90.2 Int. U. Prod. Dist. Elect. Energy, Paris, p. 11 (1985).

31. M. Lerchin and J. Samet, "Occupation, Industry and Lung Cancer Risk in New Mexico Males," Proc. Soc. Epi. Res., Houston, Texas, June 1984, Unpub.

32. R. S. Lin, P. C. Dischinger, J. Conde, and K. P. Farell, "Occupational Exposure to Electromagnetic Fields and the Occurrence of Brain Tumours," *J. Occup. Med.* **24**:413-419 (1985).

33. J. G. Llaurado, A. Sances, Jr., and J. H. Battocletti, Eds., *Biological and Clinical Effects of Low Frequency Magnetic and Electric Fields*, C.C. Thomas, Springfield, IL, USA, 345pp (1984).

34. P. Lövsund, P. Oberg, and S. E. G. Nilsson, "ELF Magnetic Fields in Electro-Steel and Welding Industries," Report, Dept. Biomed. Eng. and Opthalmology, Univ. Linköping, Sweden, 14p (1983).

35. H. Malker and J. Weiner, "The Cancer-Environment Register: Example of the Use of Register Epidemiology within the Occupational Environment Sector," *Arbete och Hälsa*, (Solna) **9**:1-108 (1984).

36. M. E. McDowall, "Leukemia Mortality in Electrical Workers in England and Wales," *Lancet* i:246 (1983).

37. J. K. McLaughlin, "Epidemiology of Renal Cell and Renal Pelvis Cancer in Minneapolis - St. Paul, MN, Metropolitan Area, 1984-1979," Ph.D Thesis, Dept. of Epidemiology Univ. of Minn., Unpub. (1982).

38. G. H. C. McMillian and R. J. Pethybridge, "The Health of Welders in Naval Dockyards: Proportional Mortality Study of Welders and Two Control Groups," *J. Soc. Occup. Med.* **33**:75-84 (1983).

39. H. R. Menck and B. E. Henderson, "Occupational Differences in Rates of Lung Cancer," *J. Occup. Med* **18**:797-801 (1986).

40. S. Milham, Jr., "Cancer Mortality Patterns Associated with Exposure to Metals," *Ann. N.Y. Acad. Sci.* **172**:243-249 (1976).

41. S. Milham, Jr., "Occupational Mortality in Washington State 1950-1974," U.S. Dept. Health and Human Services, NIOSH, pp. 83-116 (1983).

42. S. Milham, Jr., "Proportion of Cancer — Washington State (1950-1979)," *Banbury Report* **9**:513-522 (1981).

43. S. Milham, Jr., "Mortality from Leukemia in Workers Exposed to Electric and Magnetic Fields," *N. Engl. J. Med.* **307**:249 (1982).

44. S. Milham, Jr., "Mortality in Workers Exposed to Electromagnetic Fields (1950-1982)," *Env. Health Prospect.* **62**:297-300 (1985).

45. S. Milham, Jr., "Silent Keys: Leukemia Mortality in Amateur Radio Operators," *Lancet* **1**:812 (1985).

46. S. Milham, Jr., "Occupational Mortality in Washington State, 1950-1971," U.S. Dep. Pub. NIOSH, 76-175-a (1986).

47. W. Morgan and E. Treyve, "Histological Differences in Occupational Risks of Lung Cancer Incidence," *Am. J. Ind. Med.* **3**:441-457 (1982).

48. National Academy of Science, "Biological Effects of Electric and Magnetic Fields Associated with Proposed Project Seafarer," NRC, Washington, DC, USA (1977).

49. M. C. Newhouse, D. Oakes, A. J. Wooley, "Mortality of Welders and Other Craftsmen at a Shipyard in Northeast England, UK," *Br. J. Ind. Med.* **42**:406-410 (1985).

50. Office of Populations Censuses and Surveys, "Occupational Mortality, 1970-1972," England and Wales, Decennial Supplement, HMSO, London (1978).

51. Office of Populations Censuses and Surveys (OPCS), "Occupational Mortality," Great Britain 1979-1980, HMSO, London (1986).

52. R. Olin, D. Vågero, and A. Ahlbom, "Mortality Experience of Electrical Engineers," *Br. J. Ind. Med.* **42**:211-216 (1985).

53. J. Olsen, S. Sabroe, M. Lajer, "Welding and Cancer of the Larynx: A Case Control Study," *Eur. J. Cancer Clin. Oncol.* **20**:639-643 (1984).

54. M. G. Ott, B. B. Holder, and R. R. Langer, "Determinants of Mortality in an Industrial Population," *J. Occup. Med.* **18**:171-177 (1986).

55. N. E. Pearce, R. A. Sheppard, J. K. Howard, J. Fraser, and B. M. Lilley, "Leukemia in Electrical Workers in New Zealand," *Lancet* i:811-812 (1985).

56. G. R. Peterson and S. Milham, Jr., "Occupational Mortality in the State of California, 1959-1961," U.S. Dep. HEW. Pub. NIOSH, pp. 80-104 (1980).

57. J. Peto, "Cancer Morbidity and Mortality Studies of Welders," in *Health Hazards and Biological Effects of Welding Fumes and Gases*, R. M. Stern *et al.*, Eds., Excerpta Medica, Amsterdam, N.Y., Oxford, pp. 423-434 (1986).

58. R. Peto and M. Schneiderman, Eds., Banbury Report 9, "Quantification of Occupational Cancer," Cold Spring Harbour Lab., pp. 3-111 (1981).

59. A. P. Polednak, "Mortality Among Welders, Including a Group Exposed to Nickel Oxides, *Arch. Environ. Health* **36**:235-242 (1981).

60. R. Putoni, M. Vercelli, F. Merlo, F. Valerio, and L. Santi, "Mortality Among Shipyard Welders in Genoa," *Ann. N.Y. Acad. Sci.* **330**:353-377 (1979).

61. R. Puntoni, M. Vercelli, M. Ceppi, F. Valerio, F. di Giorgio, L. Gogioso, S. Bonassi, G. Alloro, R. Filiberti, and L. Santi, "Epidemiological Investigation on Causes of Death Among Dockyard Workers by Type and Length of Exposure (1960-1980)," Proc. Int. Conf. Risk Assessment of Occupational Exposure in The Harbour Environment, Inrc. Genoa, pp. 43-54, October 3-5, 1985.

62. C. K. Redmond, H. S. Wieand, and H. E. Rockette, "Long Term Mortality Experience of Steel Workers," Update: NIOSH Contract No. HSM-99-71-32 (1979).

63. G. Sheers and R. M. Coles, "Mesothelioma Risk in a Naval Dockyard," *Arc. Environ. Health* **35**:276-282 (1980).

64. K. Sheikh, "Exposure to Electromagnetic Fields and Risk of Leukemia," *Arch. Env. Health* **41**:56-63 (1986).

65. A. R. Sheppard and M. Eisenbud, *Biological Effects of Electric and Magnetic Fields of Extremely Low Frequency*, NY Univ. Press, N.Y. (1977).

66. D. T. Silverman, R. N. Hoover, S. Albert, and K. M. Grate, "Occupation and Cancer of the Lower Urinary Tract in Detroit," *J. Nat. Can. Inst.* **70**:237-245 (1983).

67. M. Silverstein, N. Maizlish, R. Park, and F. Mirer, "Mortality Among Workers Exposed to Coal Tar Pitch Volatiles and Welding Emissions: An Exercise in Epidemiologic Triage," *Am. J. Public Health* **75**:1283-1287 (1985).

68. B. Sjögren, "A Retrospective Cohort Study of Mortality Among Stainless Steel Welders," *Scand. J. Work. Env. Health* **6**:197-200 (1980).

69. B. Sjögren, C. Hogstedt, and H. Macker, "Chromium and Asbestos as Two Probable Risk Factors in Lung Cancer Among Welders," *J. Occup. Med* **24**:1874-1875 (1982).

70. B. Sjögren and J. Carstensen, "Cancer Morbidity Among Swedish Welders and Gas Cutters," in *Health Hazards and Biological Effects of Welding Fumes and Gases*, R. M. Stern *et al.*, Eds., Elsevier Biomed. Press, Excerpta Medica, pp. 461-464, Amsterdam (1986).

71. D. H. Sliney, R. Bishop, *Magnetic Field Hazards Bibliography*, 2nd Ed., US Army Env. Hygiene Agency, Aberdeen Proving Ground, Md., USA (1985).

72. T. D. Sterling and J. J. Weinkam, "Smoking Characteristics by Type of Employment," *J. Occup. Med.* **18**:743-754 (1986).

73. F. B. Stern, R. A. Waxweiler, J. J. Beaumont, S. T. Lee, R. A. Rinsky, R. D. Zumwade, W. E. Halperin, P. J. Bierbaum, P. J. Landrigan, and W. E. Murray, "A Case Control Study of Leukemia at a Nuclear Shipyard," *Am. J. Epidemio.* **193**:980-989 (1986).

74. R. M. Stern, "Process Dependent Risk of Delayed Health Effects for Welders," *Env. Health Perspect.* **41**:235-253 (1981).

75. R. M. Stern, "Chromium Compounds, Production and Occupational Exposure," in *Biological and Environmental Aspects of Chromium*, S. Langård, Ed., Elsevier Biomedical Press, pp. 5-48, Amsterdam (1982).

76. R. M. Stern, "Assessment of Risk of Lung Cancer for Welders," *Arch. Environ. Health* **38**:148-155 (1983).

77. R.M. Stern, "Health Effects of Occupational Exposure of Welders to Chromium," in: *Studies in Epidemiology Part 1*, WHO/Europe Interim Document 15, Copenhagen 1984.

78. R. M. Stern, "The Management of Risk: Application to the Welding Industry," *Risk Analysis* **5**:63-72 (1985).

79. R. M. Stern, "Assessment, Management, and Reduction of Risk for Welders," in *Health Hazards and Biological Effects of Welding Fumes and Gases*, R. M. Stern *et al.*, Eds., Excerpta Medica, pp. 535-552, Amsterdam-N.Y.-Oxford (1986).

80. F. W. Sunderman, Jr., Ed., "Nickel in the Human Environment," IARC, Lyon (1984).

81. S. Törnquist, S. Norell, A. Ahlbom, and B. Knave, "Cancer in the Electrical Power Industry," *Brit. J. Ind. Med.* **43**:212-213 (1986).

82. N. Wertheimer and E. Leeper, "Electrical Wiring Configurations and Childhood Cancer," *Am. J. Epidemiol.* **109**:273-284 (1979).

83. N. Wertheimer and E. Leeper, "Adult Cancer Related to Electrical Wires Near the Home," *Int. J. Epid.* **4**:344-355 (1982).

84. World Health Organization, "Extremely Low Frequency (ELF) Fields," *Env. Health Criteria* **35**, WHO, Geneva (1984).

85. W. E. Wright, J. M. Peters, and T. P. Mack, "Leukemia in Workers Exposed to Electrical and Magnetic Fields," *Lancet* **2**:1160-1161 (1982).

Role of Risk Assessment in New Hampshire State Government

Amy Juchatz, Colleen Schwalbe, and Susan Lynch

New Hampshire Division of Public Health Services
Concord, NH

ABSTRACT

Within the New Hampshire Division of Public Health Services there is an Environmental Health Risk Assessment Unit which provides health risk assessments for environmental exposures. By state statute, we have been designated the focal point for risk assessment in the state. We evaluate a variety of environmental exposures ranging from drinking water contamination to indoor air pollution to exposure at hazardous waste sites. To other state agencies, we provide input on the potential public health impacts to assist risk managers in setting priorities or selecting remedial actions. To the general public, we offer a focal point for addressing concern about the risks of exposure to a variety of substances. Our role with the general public is to communicate what we consider the magnitude and extent of the risks to be and to put these apparent risks into perspective. A good example of our role in risk assessment is our program of evaluating risks from contamination of private drinking water supplies. During the past year, we have evaluated approximately 275 analyses of contaminated non-municipal drinking water. Since current state and federal standards are not applicable to private water supplies, we provide a necessary service to both the Department of Environmental Services, which conducts the analysis, and to the resident whose well is affected. Once a water supply is found to be contaminated, the Department of Environmental Sesrvices forwards the results to us, requesting that we evaluate the potential risks of exposure. We then respond to the affected resident, outlining the basis of any concerns we have and providing recommendations regarding water usage. This program removes the burden of responding to questions regarding health effects from the Department of Environmental Services and enables a more thorough explanation of risks to the general public.

KEYWORDS: Risk assessment, risk management, risk communication, drinking water

INTRODUCTION

In New Hampshire, the health risk assessment process is centralized within the Division of Public Health Services. In fact the Division has been given primary authority for risk assessments by the state legislature. Chapter 125-H of New Hampshire Law states that "the Division of Public Health Services....shall be the lead agency for health risk assessment." This statute further directs other state agencies to consult with the Division for an evaluation of potential health risks associated with environmental exposures. The

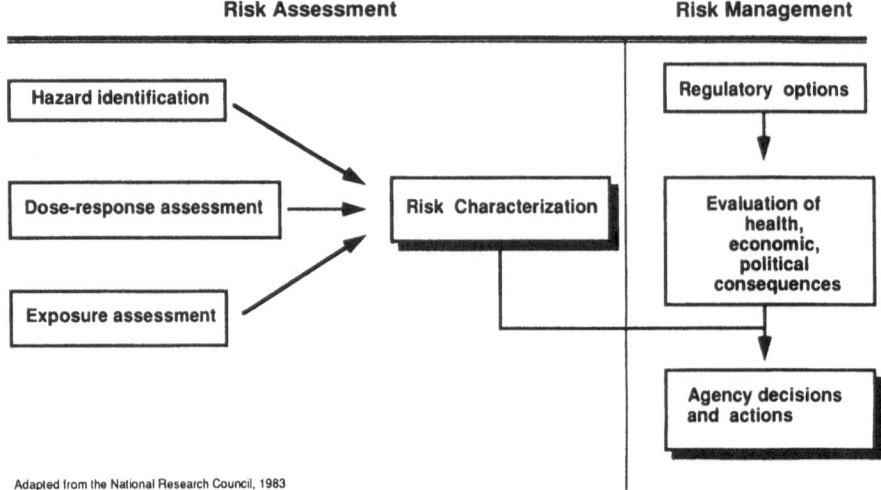

Fig. 1. Risk Assessment and Risk Management.

risk assessment function for the Division rests within the Environmental Health Risk Assessment Unit (EHRAU).

The EHRAU was established in 1984, initially to evaluate the public health impacts of hazardous waste sites. Because risk assessments are conducted in regard to environmental contamination, the state agency which the EHRAU interacts with the most is the Department of Environmental Services (DES). Housed within DES are the waste management programs, air and water pollution control agencies, and laboratory services. To a lesser extent, the EHRAU is also requested to perform health risk assessments by the Department of Agriculture, specifically the Pesticide Control Division, as well as by other programs within Public Health, such as radiological health, maternal and child health and environmental sanitation.

In general, we follow the basic guidelines established by the National Academy of Sciences in conducting risk assessments.[1] These guidelines separate the risk assessment process into four components: hazard identification, dose-response assessment, exposure assessment, and finally, the risk characterization (see Fig. 1). A brief summary of these components is provided in Table 1. These guidelines have been further developed by the U.S. Environmental Protection Agency and the Agency for Toxic Substances and Disease Registry.

The health risk assessments prepared by the EHRAU are intended to assist risk management agencies in making environmental regulatory decisions. Along with risk estimations, the EHRAU also provides some general recommendations. These recommendations usually address whether immediate action is necessary to protect the public from an imminent health threat, which exposure pathways are of greatest concern, and/or whether there is a need for more data to fully evaluate public health risks. With an objective health-based risk characterization and general recommendations available to them, risk managers (i.e., DES) can base their policy decisions on health risks as well as on technical, economic and political implications. By providing these health risk assessments, we assist the risk managers in setting remedial priorities that ensure public health consequences have been appropriately addressed.

Table 1. Components of Risk Assessment

Hazard identification:	Determination of whether exposure to a chemical of concern can lead to the development of an adverse health effect.
Dose-response assessment:	Characterization of the potential adverse effects associated with exposure to various levels (i.e., doses) of a toxicant.
Exposure assessment:	Estimation of the potential exposure to a population, including the dose, frequency, duration and route of exposure.
Risk characterization:	Compilation of the information obtained from the previous steps to determine the overall risk.

RISK ASSESSMENT OF CONTAMINATED DRINKING WATER SUPPLIES

The risk assessment activities within the EHRAU can be broadly categorized into five areas: contaminated drinking water supplies, hazardous waste site risk assessments, pesticide reviews, wildlife/food chain contamination and indoor air pollution. For the purposes of this paper, only the drinking water program will be discussed in detail, as an example of the role of risk assessment in New Hampshire.

When a private drinking water supply has been found to be contaminated, the DES requests that we evaluate the potential health risks associated with various uses of that water supply. (Since public water supplies are regulated by the DES, we do not evaluate public water supplies except under special circumstances.) The result of our evaluation of a private water supply is transmitted directly to the affected resident(s). Our risk characterization is summarized in a letter, along with various recommendations regarding the use of the water supply. Since we are often responding to individual residents, risk communication is a very important aspect of our drinking water program.

In assessing the public health risks of a contaminated water supply, we consider three routes of exposure: ingestion, inhalation and dermal absorption. Exposure from oral ingestion is handled in a fairly quantitative manner. We follow U.S. EPA guidelines in which 2 liters of water are assumed to be ingested per day with a relative source contribution from drinking water of 20 percent.[2]

Both the inhalation and dermal routes of exposure are evaluated in a much more qualitative manner. We do not consider dermal absorption a significant pathway of exposure unless contaminant levels are unusually high. In regard to inhalation, volatilization during showering or other household uses of water is considered a significant pathway for exposure once volatile organic chemicals exceed approximately 50 ppb. The selection of this level of exposure was based upon a pilot study in which we monitored the indoor air of homes with and without gasoline-contaminated water supplies. Air samples from the living room were taken before and after a 15-minute shower. Though this study was limited in the number of homes tested, it did indicate that in homes with water supplies containing greater than 80 ppb, increases in background indoor air benzene levels may be observed.[3]

Many of the drinking water contaminants that we evaluate have also been evaluated by the U.S. EPA. For such chemicals, a drinking water standard or criteria may have already been developed. These criteria may be either Lifetime Health Advisories,

Maximum Contaminant Level Goals (MCLG) or Maximum Contaminant Levels (MCL). We will typically review the U.S. EPA's evaluation and in most cases adopt its risk assessment and developed criteria. For most drinking water contaminants, if an MCL is available, that level will be used as our criterion for acceptability of the drinking water supply. A notable exception to this procedure is chloroform. The MCL for total trihalomethanes (THMs) is substantially based upon THMs being a byproduct of chlorination for disinfection purposes. We do not feel that the MCL for THMs is appropriate for private water supplies which are not chlorinated since the same risk/benefit analysis is not applicable.

When a MCL is not available, carcinogens and non-carcinogens are handled differently. For carcinogens (defined as a chemical classified in EPA's Group A or B), a 1 $\times$ 10^{-6} risk level (i.e., one excess cancer in a million people exposed) is used as a criterion when a MCL is not available. For non-carcinogens, a MCLG or lifetime health advisory is applied. If either of these criteria is not available, we conduct our own risk assessment after performing a literature search and following EPA risk assessment guidelines.

From October 1986 to October 1987, the DPHS has performed 165 drinking water evaluations. Of these, 55% have been associated with leaking underground storage tanks containing gasoline (see Fig. 2). The next highest category of associated sources was hazardous waste sites (27%). In 15% of the evaluations a source of contamination was not easily identifiable. Only a few percent of our evaluations were associated with natural contamination. Some caution should be exercised in interpreting these data as reflective of the type of groundwater contamination in New Hampshire. We know that natural contamination (arsenic, fluoride, radon, etc.) is found more frequently than is reported to us. In the case of radon, contamination is often referred to another agency within the Division of Public Health Services. In addition, the frequency distribution of contamination incidents is highly dependent upon the emphasis of the DES, which is actually collecting the samples. Figure 2 does, however, provide an indication of which sources of contamination state resources are primarily directed towards. A project we will be undertaking this coming year is to estimate which sources of contamination are responsible for the greatest risk to New Hampshire residents. We will be doing this by comparing the level of contamination found and the number of people affected to estimate health risks associated with various sources.

RISK COMMUNICATION

A very important aspect of the work of the Environmental Health Risk Assessment Unit is the communication of environmental risks to the general public. A great deal of emphasis is placed on risk communication because our efforts can only serve the residents of New Hampshire if our evaluation and recommendations are understood. In addition, since many of our risk assessments involve fairly emotional issues, such as hazardous waste sites or drinking water contamination, as environmental health officials we need to assist the general public in placing health risks into perspective. Since we are a non-regulatory program, residents must assimilate our information so that they can make their own, hopefully educated, decision on whether or not to follow our recommendations.

We communicate risk by way of public meetings, press releases, letters or telephone. Various forms of written material are prepared. In regard to our drinking water program, we prepare Health Information Summaries on each contaminant found in a drinking water supply. These sheets, though primarily used for drinking water correspondence, discuss in lay terms the potential toxicity of each compound by various routes of exposure. Background information on the uses of the chemical and on how it can get into the environment is also provided. In some cases we also prepare question and answer sheets which discuss various issues, such as radon or wood preservatives.

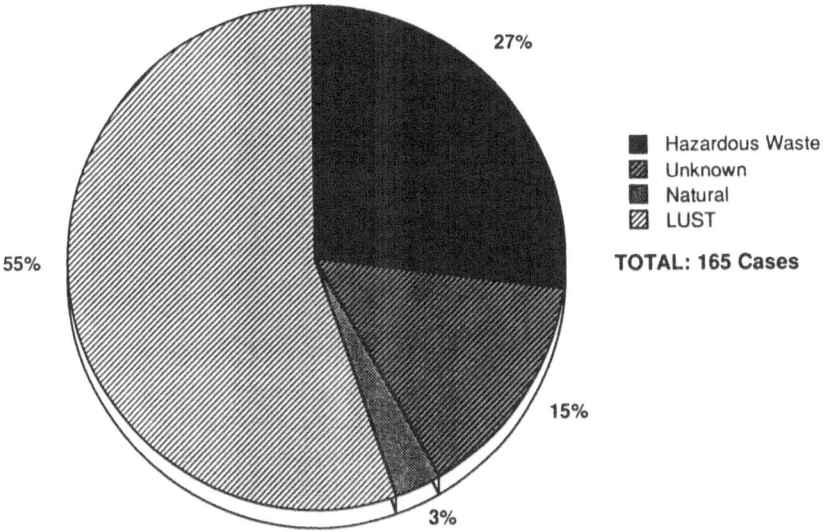

27%

Hazardous Waste
Unknown
Natural
LUST

TOTAL: 165 Cases

55%

15%

3%

Fig. 2. Sources of Contaminated Water Supplies from October 1986 to October 1987. Source: N.H. Division of Public Health Services.

We feel that an important aspect to our risk communication efforts is describing the uncertainty in the risk assessment process. We characterize the inherent uncertainty as consisting of the uncertainty in extrapolating from high-dose short-term exposure studies conducted on laboratory animals to low-dose chronic exposure in humans. We acknowledge that there is additional uncertainty if exposure is occurring to a mixture of chemicals and not just to a single compound and that individual sensitivity may vary from person to person. In regard to our drinking water program specifically, we point out that for carcinogens, there may be some degree of risk even below our drinking water criteria. The same acknowledgement is made for those chemicals whose MCL is above the MCLG. In our letters to residents with contaminated drinking water supplies, we also recommend that they consult their physician to share the information we have provided.

CONCLUSION

The creation of a centralized risk assessment unit within New Hampshire state government has allowed a clear separation between risk assessment and risk management. Perhaps more importantly it separates those who are assessing public health risks from those who are managing the risks and may be constrained by a regulatory framework. This separation is particularly important in the eyes of the public. A centralized risk assessment unit also provides a continuity in statewide risk assessment methodologies. We can ensure that the same underlying assumptions are made whether assessing risks that pertain to air, water, or soil.

The Environmental Health Risk Assessment Unit provides an expertise that is not available in risk management agencies. For this reason, we are particularly helpful to the management agencies when it comes to risk communication. Without a firm understanding of toxicological as well as risk assessment principles, it is difficult to address the emotional health concerns and questions that are raised by the general public. This program removes the burden of responding to questions regarding health effects from the Department of Environmental Services and enables a more thorough explanation of risks to the general public.

The overriding benefit of having a centralized risk assessment unit is that the Division of Public Health Services can ensure that public health considerations are adequately addressed in environmental management programs.

REFERENCES

1. National Academy of Sciences, "Risk Assessment in the Federal Government: Managing the Process," National Academy Press, Washington, DC (1983).
2. U.S. EPA, National Primary Drinking Water Regulations; Volatile Synthetic Organic Chemicals; Proposed Rulemaking, *Federal Register* **49(114)**:24330-24355 (1984).
3. New Hampshire Division of Public Health Services, Indoor Air Benzene Concentration Found in Six Homes with Known Gasoline Contamination of Drinking Water, Environmental Health Risk Assessment Unit (August, 1987).

A Method for Estimation of Fish Contamination from Dioxins and Furans Emitted by Resource Recovery Facilities

Helen M. Goeden and Allan H. Smith
Health Risk Associates
Berkeley, CA

ABSTRACT

A new method for calculating fish contamination resulting from dioxin and furan emissions has been developed. The method is based on the fact that the major determinant of fish dioxin and furan concentration is sediment. Estimates of the sediment concentration were made under the following assumptions: (1) bottom burial of sediment is the only significant removal pathway; (2) suspended sediment concentration is in equilibrium with the top layer of the bottom sediment; and (3) at steady state the newly deposited particulates will have the same concentration of dioxins and furans as the lake sediment. Under these assumptions it is only necessary to calculate the concentration of dioxins and furans in particulate matter entering the lake or pond through air deposition and runoff. Deposition modeling data were used to calculate the concentration of contaminants deposited directly on the lake surface and indirectly from runoff of the surrounding contaminated watershed area.

Once the concentration in sediment was calculated, fish-to-sediment ratios for the various dioxin and furan isomers from published field and laboratory studies were utilized to project fish concentrations.

The described method was incorporated into a risk assessment of a resource recovery facility. The analysis found that if an individual received 5 percent of an average daily freshwater fish intake from the potentially contaminated lake, this exposure pathway would comprise 40 percent of the total daily dioxin equivalent intake from all pathways of exposure. It is apparent that ingestion of contaminated fish may be an important exposure pathway and that potential contamination of fish should be closely evaluated in the risk assessment of dioxins and furans emitted from waste incineration plants.

KEYWORDS: Dioxins, fish contamination, resource recovery facilities

INTRODUCTION

Dioxins and furans in general exhibit low water solubility and are highly lipophilic. Because of their chemical characteristics they are persistent in the environment and bioconcentrate in organisms. Knowledge of human exposure to dioxins and furans is required to evaluate their risk. It is, therefore, necessary to assess the potential exposure to these substances through the ingestion of contaminated fish.

Dioxins and furans are formed by the combustion of chlorinated organic compounds (municipal and industrial wastes). The atmospheric transport of particulates appears to be the major source of dioxins and furans in lake sediment.[1] The objective of this presentation is to present a method for determining fish contamination and subsequent human exposure from combustion sources of dioxins and furans.

BASIS FOR MODEL

Currently, fish contamination is estimated by multiplying predicted water concentration and a bioconcentration factor or BCF. The BCF is traditionally defined as the concentration of a chemical at equilibrium in an organism divided by the mean concentration of the chemical in water. To date the BCF for 2,3,7,8-tetrachlorodibenzodioxin (TCDD) has been reported at values ranging from approximately 550 to 170,000. The wide degree of variation in BCF values reported indicates a high degree of uncertainty in predicting the BCF for highly insoluble compounds like TCDD.

The term BCF is usually used with the assumption that uptake from water is the chief source of the chemical concentrated in the organism. When contaminated sediment or food is the source of exposure, then a fish-to-sediment ratio or a fish-to-food ratio may be the appropriate term to describe the increases in the chemical concentration in the fish.[2] In light of the recent literature we believe that the use of a fish-to-sediment ratio may be the most appropriate for assessing dioxin and furan contamination of fish.

The exposure to humans from ingestion of fish from a lake potentially contaminated with dioxins and furans as the result of the operation of a resource recovery facility can be estimated by determining the concentration of these chemicals that could occur in fish.

The concentration of dioxins and furans attained in fish is dependent on three factors: (1) the amount of emitted dioxins and furans reaching the lake; (2) the concentration within the aquatic environment; and (3) the amount taken up by fish.

Dioxins and furans reach a lake either by direct deposition on the surface or in runoff from the surrounding watershed. Due to their chemical characteristics, particularly the strong affinity for the organic fraction of particulates, dioxins and furans remain adsorbed on the particulates in water. Therefore, any amount removed from the water by photolysis, oxidation, hydrolysis, and volatilization would be negligible. Removal by microbial degradation has also been shown to be negligible.

Sedimentation, however, would be an important removal pathway. The ideal method for estimating fish contamination based on data concerning incoming particulates would take into account bottom burial, or would not be influenced by that process. An important advantage of the proposed model is that at steady state it is not necessary to estimate the rate of removal by burial in sediment. The reason for this is that at steady state the newly arriving suspended solid matter and the active surface layer of sediment will have the same concentration as the sediment being buried and becoming inaccessible to resuspension and to fish consumption. The basis of the method is therefore to calculate the concentration of dioxins and furans in incoming particulate matter from direct and indirect sources, and to assume that this concentration would be the same as that in the sediment when steady state has been reached.

The amount of dioxins and furans taken up by fish depends on the sediment concentration. A bioavailability index (BI) is utilized to calculate fish concentration from sediment concentration. The bioavailability index is expressed as the ratio of fish concentration (pg/g lipid) to the sediment concentration (pg/g organic carbon).

To date only one published study reports adequate data to calculate BIs.[3] A bioavailability index for each 2,3,7,8-substituted dioxin and furan isomer was estimated from the field data. As additional data become available the BIs utilized should be modified, if necessary.

ILLUSTRATION OF MODEL

Table 1 presents an illustration of the application of BIs in calculating fish concentration from sediment concentration. The second column represents the distribution of the relative proportions of dioxin equivalents originating from the proposed facility that might reach the lake sediment. The values listed are purely for illustration purposes and are not representative of any given emission data. The calculation of equivalents would normally be based on the emission data used in the risk assessment and some scheme of 2,3,7,8-TCDD weighting factors (e.g., EPA dioxin equivalents). The various congener groups are divided into those with and those without chlorine atoms in the 2,3,7 and 8 positions, since only the 2,3,7,8-substituted isomers are taken up in fish.

Table 1. Illustration of the Calculations for Estimating Fish Uptake of Dioxin Equivalents

Compound	Dioxin Equivalents	Field Fish BI (pg/g lipid per pg/g organic carbon)	Fish Uptake of Equivalents
2,3,7,8-T4CDD	0.07	0.07	0.0049
Other T4CDD	0	0	0
2,3,7,8-P5CDD	0.45	0.013	0.0059
Other P5CDD	0	0	0
2,3,7,8-H6CDD	0.0096	0.004	0.00004
Other H6CDD	0	0	0
2,3,7,8-H7CDD	0	0.0004	0
Other H7CDD	0	0	0
2,3,7,8-T4CDF	0.396	0.026	0.010
Other T4CDF	0	0	0
2,3,7,8-P5CDF	0.528	0.025	0.013
Other P5CDF	0	0	0
2,3,7,8-H6CDF	0.012	0.015	0.0002
Other H6CDF	0	0	0
2,3,7,8-H7CDF	0	0.0006	0
Other H7CDF	0	0	0
TOTAL	1.4656		0.345

Since photolysis, hydrolysis and microbial degradation are assumed to result in negligible removal of dioxins and furans, column 2 also gives the relative proportions of the various congener groups which would be associated with the lake sediment carbon. The third column lists the fish bioavailability indices based on the field data gathered by Kuehl et al.[3] The fourth column is the product of the toxic equivalents, i.e., the second column and the BIs in the third column. The calculations indicate that, given this specific isomer distribution pattern, if there were 1.4656 units of dioxin equivalents associated with 1 gram of sediment organic carbon, then one would expect to find 0.0345 units of dioxin equivalents in fish lipid. The fish uptake factor (i.e., 0.0345/1.4656) represents the concentration of dioxin equivalents in fish lipid which would result from 1 unit of dioxin equivalents in lake sediment organic carbon. The uptake factor is directly affected by the isomer distribution pattern and, therefore, it is critical that the amount of equivalents taken up by fish be calculated for each isomer.

SUMMARY

Calculation of fish concentration requires only two steps. Step one involves the estimation of sediment concentration based on incoming particulate concentration. Step two concerns the calculation of fish levels based on the sediment concentration. The increase in simplicity has been achieved without loss in accuracy. In fact, the uncertainties should be reduced. It should also be noted that BIs are available for all major dioxin and furan isomers of concern, whereas this is not true for BCFs. Utilization of the bioavailability indices also allows one to estimate the concentration based on lipid content. The edible portion of fish is much lower in lipid content than the whole fish and therefore it is more appropriate to utilize values normalized for lipid content. A detailed description and illustration of the model is presented elsewhere.

REFERENCES

1. J. M. Czuczwa and R. A. Hites, "Sources and Fate of PCDD and PCDF," *Chemosphere* **15**:1417-20 (1986).
2. S. E. Bysshe, "Bioconcentration Factor in Aquatic Organisms," Chapter 5 in: *Handbook of Chemical Property Estimation Methods*, W. J. Lyman, W. F. Reehl, and D. H. Rosenblatt, Eds., McGraw-Hill, New York, pp. 5-1—5-30 (1982).
3. D. W. Kuehl, P. M. Cook, A. R. Batterman, D. Lothenbach, and B. C. Butterworth, "Bioavailability of Polychlorinated Dibenzo-P-Dioxins and Dibenzofurans from Contaminated Wisconsin River Sediment to Carp," *Chemosphere* **16**:667-79 (1987).

The Importance of the Hazard Identification Phase of Health Risk Assessments: Illustrated with Antimony Emissions from Waste Incineration Facilities

Allan H. Smith, Helen M. Goeden, and Jonathan Frisch
Health Risk Associates
Berkeley, CA

ABSTRACT

In the course of conducting a health risk assessment for emissions from municipal waste incineration facilities, the literature was reviewed for a variety of emissions which have previously been regarded as non-carcinogenic. The objective was to establish the lowest levels of exposure causing toxic effects in animals and humans with a view toward calculating safety margins with projected exposure levels. A journal paper was found reporting that antimony exposure caused lung cancer in rats. The same paper referenced a dissertation study with a similar finding. A report was subsequently found indicating a lung cancer excess among smelter workers exposed to antimony, although no journal publications were found for this study. This paper presents a dose-response analysis for the animal and human data. The human data analysis suggested that antimony has a cancer potency of the same order as that which the U.S. Environmental Protection Agency has published for nickel. However, the animal data suggested a much higher cancer potency. These cancer potency figures were included in cancer risk characterization for emissions from the facility. While antimony turned out not to be a major contributor to the overall cancer risk estimates for the municipal waste incineration facilities, the discovery that it is a probable human carcinogen illustrates the importance of the hazard identification phase of health risk assessments.

KEYWORDS: Hazard identification, antimony, carcinogen, risk assessment

INTRODUCTION

Health risk assessments involve several phases. These include hazard identification, exposure assessment, dose response analyses, and risk characterization. This paper illustrates the importance of the first phase—hazard identification.

HAZARD IDENTIFICATION

Hazard identification for the scenario being considered in a health risk assessment includes identification of potentially hazardous toxic agents. It is frequently confined to merely identifying the chemical agents involved. The health risk assessment may then only address health effects considered in past health risk assessments, or health effects identified by regulatory agencies for consideration.

Confining attention to health effects considered in the past fails to identify outcomes which more recent studies might indicate should be considered. In addition, regulatory agencies by necessity promulgate rules and guidelines with considerable lag periods. For these reasons, the hazard identification phase of a health risk assessment should include consideration of the scientific literature with regard to outcomes which may not so far have been linked with exposure to a particular chemical.

One important area for consideration is carcinogenesis. The reasons for this include both the severity of the disease and also the fact that it is theoretically possible that there may be no threshold exposure level below which cancer risks would not be increased. A health risk assessment using no threshold cancer models is therefore very sensitive to whether or not a particular agent is classified as a carcinogen.

The hazard identification phase should therefore include a review of possible health effects of each chemical, with particular attention to whether or not it might be classified as a carcinogen. If there is evidence to suggest that it is a carcinogen, then appropriate steps in dose-response analysis and risk characterization should follow.

HAZARD ASSESSMENT FOR MUNICIPAL WASTE INCINERATION

Since household garbage includes a large variety of materials, it is inevitable that combustion will produce a large variety of combustion products. While a large proportion of the products can be removed by modern pollution control equipment, some will escape into the atmosphere and human exposure will result. Known animal or human carcinogens which are emitted include the dioxins and furans, polychlorinated biphenyls (PCBs), polyaromatic hydrocarbons (PAHs), arsenic, beryllium, cadmium, chromium, nickel, formaldehyde, chlorinated benzenes and phenols, and vinyl chloride. Dose-response analyses have been undertaken for all of these chemicals or groups of chemicals with cancer potencies for use in health risk assessments published by regulatory agencies, particularly the U.S. Environmental Protection Agency.

In the course of the hazard identification phase for a health risk assessment, consideration must also be given to many other emissions. Recent risk assessment work on two resource recovery projects led us to conclude that antimony emissions should be regarded as potentially carcinogenic to humans, and should therefore be included as such in the risk characterization phase of health risk assessments. The next two sections of this paper will present the evidence that antimony should be treated as a human carcinogen.

HUMAN EVIDENCE FOR CARCINOGENICITY

Human evidence for carcinogenicity comes from observations of lung cancer among antimony smelter workers. No detailed epidemiological study results have been published. However, an increase in lung cancer deaths has been observed among antimony smelter workers. A retrospective epidemiology study of lung cancer deaths of smelter workers employed at the Newcastle, England, antimony and zinc works since 1925 revealed about a two-fold excess of deaths due to lung cancer.[1] The study revealed that 15 deaths from lung cancer of a total work force of 1081 had occurred up to 1971 (the time of the study) among smelter workers who had exposures of 7 years or longer. The length of occupational exposure ranged from 7 to 43 years with an average of 22 years. According to estimates made by the Employment Medical Advisory Service in 1973, there was about a two-fold excess over the number that would be expected according to local death rates for lung cancer.

A followup again showed a disparity between observed and expected deaths due to lung cancer in the antimony factory population (21 versus 14.03), most of which occurred in the factory subgroup engaged in smelting and related activities (18 versus 10.25).[2] The difference was even more pronounced in workers 45 to 64 years of age (13 versus 7.22).

One possible explanation of the excess lung cancer rates is the possibility of excess smoking among smelter workers. Among the 15 persons first ascertained to have lung cancer, six were heavy or very heavy smokers (43%), five were moderate smokers (36%), one was a very light smoker (7%), and two were non-smokers (14%). A detailed analysis of this situation indicated that these data are not consistent with smoking being the explanation of the increased lung cancer risks among these workers.[3]

ANIMAL EVIDENCE FOR CARCINOGENICITY

During a review of the literature concerning the toxic effects of antimony, two animal studies were found reporting cancer bioassays. The first study has been presented as a dissertation thesis, but has not yet been published in the scientific literature.[4] Three groups of 49 female Fisher rats were studied, an unexposed control group, a group exposed at 1.6 mg antimony per cubic meter as antimony trioxide, and a group exposed to 4.2 mg antimony per cubic meter.

The antimony trioxide had a chemical analysis of 99.4% antimony trioxide, 0.02% arsenic, and 0.2% lead, as well as trace amounts of a variety of elements. Animals were exposed for 6 hours per day, 5 days per week, for approximately 55 weeks. The lung neoplasms occurred at the end of exposure or later. The incidence of lung neoplasms in those animals alive and examined after the exposure period was 7 percent in controls, 3 percent in the 1.6-mg Sb/m^3 exposure group, and 61.8 percent in the 4.2-mg Sb/m^3 exposure group.

A recently published study by Groth et al. has also reported lung neoplasms in female rats.[5] Male and female rats were exposed to 45 to 46 mg antimony trioxide/m^3, 36 to 40 mg antimony ore concentrate/m^3, or filtered air. The antimony trioxide contained 80% antimony (i.e., 36 mg Sb/m^3) and 40 ppm arsenic. Antimony ore concentrate composition included 46% antimony (i.e., 16 to 18 mg Sb/m^3) and 792 ppm arsenic. Exposure was for 7 hours per day and 5 days per week for up to 52 weeks. Approximately 5 months after the termination of exposure all surviving animals were killed and autopsied.

The first lung tumor was observed in a female rat that died after 41 weeks of exposure to antimony ore. The first lung tumor in the antimony trioxide group was seen in a female rat killed as part of serial sacrifice at 53 weeks. In animals exposed for 41 weeks or longer the incidence of lung neoplasms was 27 percent in females exposed to antimony trioxide and 25 percent in females exposed to antimony ore concentrate. None of the male rats in any group or the female control rats developed lung neoplasms.

Both of these studies provide supportive evidence that antimony might be a human carcinogen. The studies have been assessed in greater detail elsewhere, including quantitative risk assessment analyses.[3]

QUANTITATIVE CANCER RISK ASSESSMENT

This section presents the steps in a quantitative risk assessment for antimony exposure.

Relative Risk Estimate

When attention was confined to workers in smelting and related activities, there were 18 lung cancer deaths with 10.25 expected. Details were not given concerning the calculation of the expected numbers but were presumably based on U.K. lung cancer mortality rates. In the absence of any evidence to the contrary, we will assume that the ratio of 18 over 10.25 is the relative risk estimate for this exposure. It should be noted that this excess is unlikely to be due to chance (90% confidence limits for the relative risk estimate of 1.76 can be calculated as 1.14 to 2.60.)

Exposure of Smelter Workers to Antimony

A small industrial hygiene survey of the Newcastle plant in 1963 found average levels of antimony (or trioxide) of 4.3 mg/m^3 with a range of 0.5 to 36.7. The second highest value was 5.3 mg Sb/m^3. However, in an analysis of samples in the period 1971 to 1975 the average level was found to be 9.5 mg/m^3 with a range of 0.3 to 56. Hence for the purposes of this risk assessment, it will be assumed that the average throughout was 9.5 mg/m^3, since it seems unlikely that earlier exposures were lower than those found in the more extensive industrial hygiene samples of the 1970s.

Actual exposures may have been higher or lower than this. The uncertainty of the actual exposure levels should not lead to rejection of the human data for use in risk assessment. In particular, it should be noted that extrapolation between species, such as from rodents to humans, could involve at least an order of magnitude error. The errors in exposure assessment are probably much less than an order of magnitude.

Exposure duration for the lung cancer cases was from 7 to 43 years with an average of 22 years. Of the 18 deaths, 13 were said to be age 45-64 at death. We have assumed the average age at death for the 18 cases was 60 years. Hence, the fraction of lifetime these cases were exposed was 22/60.

Adjustment to Equivalent Lifetime Exposure

The next step involves considering what the equivalent exposure level from birth would be for these workers in order to consider environmental exposure risks. A linear assumption is used throughout, as is usual in such risk assessments. The adjustment factors are 8/24 for hours per day; 5/7 for days per week; 46/52 for weeks per year, to account for sick leave and vacations; and 22/60 for fraction of lifetime exposure. The result of multiplying these by 9.5 mg/m^3 is 0.73 mg/m^3, which would be the equivalent lifetime exposure from birth for these workers under the simplifying linear assumption of risk with dose rate.

Lifetime Added Cancer Risk

In the U.S. one out of 19 deaths between 1960 and 1970 was from lung cancer. Under a steady-state assumption concerning age specific cancer risks and risks of competing causes of death, the lifetime risk of dying from lung cancer would be 1/19.

The relative risk for the above exposure is 1.76. Thus the lifetime added risk of dying from lung cancer is: $(1.76-1) \times 1/19 = 1/25$.

Cancer Potency

From the above, an equivalent lifetime exposure of 0.73 mg/m^3 would result in an added 1/25 lung cancer risk. Linear extrapolation indicates that exposure to 1 μg/m^3 would result in an added cancer risk of 55 per million people exposed. If the extrapolation used the upper 90% confidence limit (i.e., the level with 5% probability of being exceeded), the potency estimate would be that lifetime exposure to 1 μg/m^3 would result in 116 lung cancer deaths per million exposed.

CRITICAL ASSUMPTIONS

The most critical assumption is the linear extrapolation of cancer risk to low doses. This assumption is made in the absence of good empirical or theoretical evidence to the contrary. The true risks at low exposures may be zero. They could be in the range from zero to the value of 116 per million exposed at 1 μg/m^3, but they are unlikely to be higher in view of the conservative nature of the linear assumption. A second critical assumption is the linear adjustments to get lifetime equivalent exposures, and finally that relative risk is independent of age of exposure. Finally, the less critical assumption involves the exposure levels experienced by the workers. It should be noted that all but the last assumptions are also involved in extrapolating risks from animal bioassays to humans. Animals are not dosed to constant levels of exposure, and they are not normally exposed until they are fully grown.

CONCLUSION

While the antimony cancer risk estimates do not make a major impact on the overall cancer risks from resource recovery facility emissions, the importance of hazard identification in a health risk assessment has been shown by the possibility that significant cancer risks might have been found from an emission which is not widely known to be carcinogenic.

REFERENCES

1. H. E. Stokinger, "The Metals," in: *Patty's Industrial Hygiene and Toxicology*, 3rd Ed., G. D. Clayton and F. E. Clayton, Eds., pp. 1493-2060, Wiley-Interscience Publishers, New York (1981).
2. In: Public Health Service Center for Disease C, Ed., NIOSH Criteria for a Recommended Standard, Occupational Exposure to Antimony, DHEW(NIOSH) Publication Number 78-216 edn., Cincinnati, Ohio: U.S. Department of Health, Education and Welfare, September 1978.
3. A. H. Smith, H. M. Goeden, and J. Frisch, "Antimony Should Be Considered a Human Carcinogen in Health Risk Assessments," Submitted for publication 1988.
4. W. D. Watt, "Chronic Inhalation Toxicity of Antimony Trioxide: Validation of the Threshold Limit Value," Ph.D. Thesis, 1983.
5. D. H. Groth, L. E. Stettler, J. R. Burg, W. M. Busey, G. C. Grant, and L. Wong, "Carcinogenic Effects of Antimony Trioxide and Antimony Ore Concentrate in Rats," *J. Toxicol. Environ. Health* **18**:607-626 (1986).

The Utility of a National Food Survey in Assessing Dietary Risk and Exposure

C. A. Gregorio and B. J. Petersen
Technical Assessment Systems, Inc.
Washington, DC

ABSTRACT

Since the introduction of the EPA's current Tolerance Assessment System, field trial data have been used to estimate the expected level of pesticide residues in food and to estimate potential dietary exposure and risk. However, field trial studies are designed specifically to yield residue data for establishing tolerance values and, therefore, to reflect crop treatment that corresponds to the maximum proposed uses, such as 100% of the crop treated at the maximum labeled application rate and a minimum preharvest interval. It is extremely unlikely that 100% of the total acreage of crops registered for use are actually treated with the pesticide, and often much less than the maximum labeled rate is used. Therefore, these data do not always represent pesticide residues in foods as they are actually eaten. In order to provide such information, it is necessary to assay for actual pesticidal residue in foods in the market place. A National Food Survey is a nationally representative, statistically valid program designed to measure pesticide residues in foods as they are purchased. Store locations are selected on the basis of geography, population numbers, and store size to represent all major food distribution channels. Selected foods are taken directly from the grocery store shelves and represent the impact of all the variables which affect the level of real residues, such as the amount or rate of the pesticide application, the effect of processing, etc. Specifically, a National Food Survey program enhances the definition of anticipated dietary exposure for use in the EPA's Tolerance Assessment System and establishes a more realistic background for the interpretation of possible pesticide-related health risks for the population.

KEYWORDS: Exposure, risk, residues, nationally representative, statistically valid

INTRODUCTION

This paper presents an overview of an important new approach for evaluating the safety of today's food supply by enhancing the definition of anticipated dietary exposure and establishing a more realistic interpretation of possible pesticide-related dietary health risks. Dietary exposure to a chemical or chemical contaminant for the general population is generally determined as the product of how much of a given food is eaten and the amount or concentration of a chemical or contaminant in the food.

In the past, regulatory agencies such as the Environmental Protection Agency had conducted dietary exposure analyses on pesticides assuming that a pesticide-treated food contains the maximum residue concentration legally allowed, e.g., "tolerance" residue level. However, by definition, a tolerance residue level reflects the most strenuous conditions of pesticide use, i.e., maximum application rate, the maximum number of applications and as close to harvest as the label permits. In other words, tolerance represents a level which is used for regulatory enforcement purposes and is a level that is not expected to be exceeded, and, in fact, is a level which legally cannot be encountered in food. Recognizing the potential to vastly overestimate possible dietary exposure, more refined methods for determining or approximating "actual" or likely residue levels in foods as they are eaten have been investigated. For example, "processing" studies in which various methods of preparing foods for consumption (rinsing, cooking, freezing, etc.) have been conducted to provide more realistic estimates of residues in foods.

Some limited studies, such as the FDA's total diet study (Pennington, 1982) have reported major differences between theoretical estimates of residue concentrations and levels actually found. In some instances, theoretical "worst case" estimates may be higher than actual residues due to a number of circumstances, such as:

1. The chemical may have been applied at less than the maximum rates.

2. The interval between treatment and harvest may have been longer than the minimum allowed.

3. Not every field may have been treated.

4. Residues may have been removed or degraded during processing and cooking.

Conversely, actual residues may be higher than "worst case" predictions because they have entered the food from other sources (such as lead during processing) or because they concentrated in a particular fraction during processing.

Although these types of studies do demonstrate the potential differences between "theoretical" and "real" residue levels in foods, they were not designed to be statistically representative of the nation's food supply and therefore should not be used to determine anticipated dietary intake. A more refined program, National Food Survey, has been developed to incorporate statistically valid *and* nationally representative methodologies for collecting residue data which enhance the definition of dietary exposure and establish a more appropriate background for interpretation of possible chemically related health risks. This design incorporates maximum utility with minimum expenditure of resources and should not be confused with "Market Basket" or "Monitoring Surveys."

STUDY DESIGN

The purpose of a National Food Survey program is to establish a more accurate definition of dietary exposure. More specifically, the studies are designed to:

- Determine the residues of a particular chemical in selected foods in retail and/or whole sale outlets;

- Determine the mean, percentile, or other statistical measurements by using the residue data in combination with food intake or consumption data; and

- Determine the probability that individuals within the population could have chemical intakes at or exceeding acceptable levels.

546

A National Food Survey protocol must be individually designed based upon the specific use patterns of the chemical, the specific toxicological issues (is chronic or acute toxicology the primary concern?) and the chemical's properties (solubility, stability, etc.). Food samples to be collected are selected based on a preliminary "worst case" exposure assessment utilizing existing residue information (i.e., tolerance levels, processing data, etc.). Generally, there are five major elements in the protocol development:

1. Develop a statistically representative sampling.

    a. Select a statistically based sampling procedure.
    b. Identify the parameters to be assessed (foods to be analyzed, regions of the country, seasons, etc.).
    c. Outline standard operating and survey practices to ensure that the study will meet the criteria of current quality assurance principles.

2. Identify stores or warehouses to be sampled and determine their representativeness in the total potential universe.

3. Provide precise instructions to the product pick-up agents and outline a precise chain of custody, etc.

4. Pick up products and ship to designated laboratory for residue analysis.

5. Analyze results and prepare reports.

The results are analyzed and appropriate statistical weights are applied to determine each food sample's proportional share of national and/or regional anticipated residue levels. The resulting residue levels are then combined with food consumption data (U.S. Department of Agriculture, 1979–1986) for the entire U.S. population and subgroups (e.g., children 1–6, nursing infants, etc.), and dietary exposure profiles are generated (Fig. 1). Finally, the probability of any individual's actually being exposed to the estimated dietary levels is analyzed and determined.

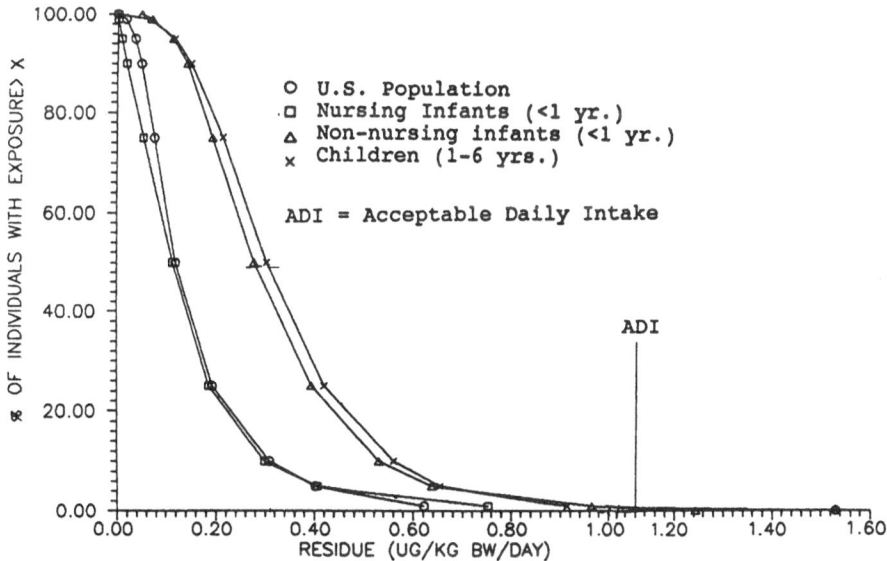

Figure 1. National Food Survey Dietary Exposure Analysis.

APPLICATIONS

The results from a National Food Survey have a number of potential applications, including the following:

1. Quantifying actual exposure.

2. Defining actual problem areas.

3. Highlighting differences between "theoretical" hypotheses and actual residue levels (Fig. 2).

4. Focusing research and development on important contributors to exposure and mechanisms for reducing exposure.

5. Serving as a basis for regulatory strategies and action (i.e., to reduce the need to regulate a "chemical" generically by allowing selective regulation of the most important contributors to exposure).

6. Checking "efficiency" of federal and state monitoring data.

    a. Are their findings representative?
    b. Are they finding the commodities in which residues actually are expected?
    c. Are "hits" representative, or do they reflect isolated cases of misuse?
    d. What is the range of exposure from products containing common food ingredients, e.g., from all potato products?

7. Verifying market projections of pesticide use on specified crops.

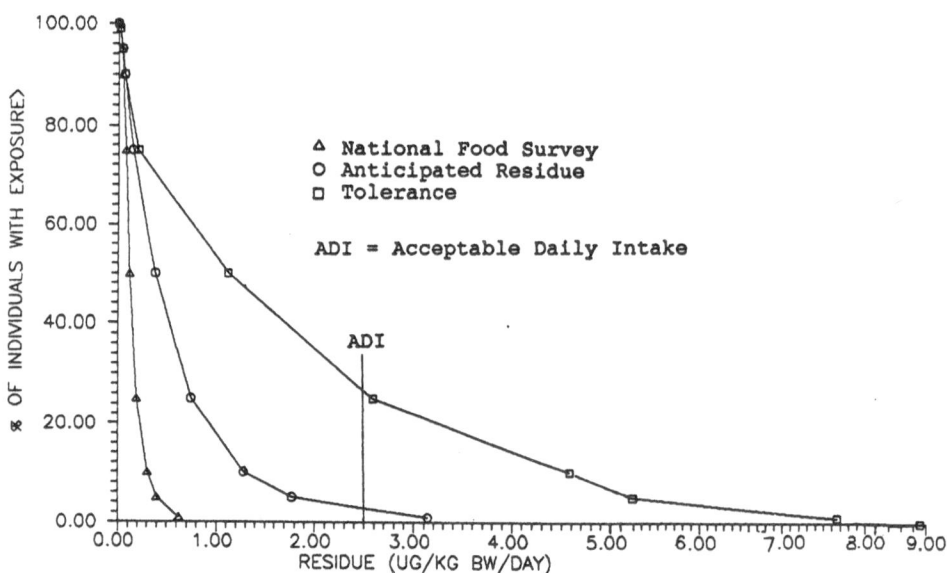

Figure 2. Dietary Exposure Analysis U.S. Population.

REFERENCES

Pennington, J.A.T., (1982), "Total Diet Study," *J. Am. Diet. Assoc.* **80**:166-173.

U.S. Dept. of Agriculture, Nationwide Food Consumption Survey (1979-1986), "Preliminary Report No. 1-10."

The Carcinogenic Risk of Some Organic Vapors Indoors: A Theoretical Survey

M. Tancrède, R. Wilson, L. Zeise,* and E. A. C. Crouch

Harvard University, E.E.P.C. and Department of Physics
Cambridge, MA

*California Department of Health Services
Berkeley, CA

ABSTRACT

This exploratory report examines the risk of selected organic air pollutants measured in homes in the United States and the Netherlands. After several theoretical assumptions, estimates are made for the carcinogenic potency of each chemical; combined with the exposure measurements, these give estimates of cancer risk. These estimates are compared with risks of these same pollutants outdoors and in drinking water and also with other well-known indoor air pollutants: cigarette smoke, radon gas, and formaldehyde. These comparisons indicate priorities for action. Some suggestions are made for future studies.

KEYWORDS: Risk estimates, volatile organic compounds (VOCs), chronic exposure, carcinogenic potency, uncertainties

INTRODUCTION

Several recent studies of organic air pollutants have found that common outdoor contaminants are usually present at much greater concentrations indoors. This paper is an initial attempt to determine how hazardous exposure to some of these indoor contaminants might be. It is far from complete: only relatively few of the many pollutants are considered, all are organic vapors, and the number of samples of each pollutant is small. The object is to gain some insight into the importance of various indoor air pollutants.

A variety of adverse health effects may be caused by indoor air pollution. Carbon monoxide, for example, can cause acute toxic effects at high enough concentration. Reducing exposure well below an effective threshold reduces the risks of such acute effects to negligibly small values or zero. However, the possible delayed effects of chronic exposure to low concentrations of air pollutants are much more difficult to evaluate.

In this exploratory paper, we make conservative estimates of the risks of one type of chronic effect—cancer. We deliberately make overestimates of cancer risks to individuals by assuming, first, that every chemical poses such a risk, and second, that there is a linear relationship between risk and exposure.

We emphasize that the word "risk" is being used here in a general sense. When no direct data on carcinogenesis in humans exist for a chemical, we use (indirect) data on carcinogenesis in animals. In the absence of any such data, it is improper to assume a risk of zero, or equivalently to ignore the chemical. There is a finite chance that the chemical is a carcinogen and so it poses a finite risk. We use a variety of even more indirect methods to estimate the risk from such a chemical, with appropriately large estimates of uncertainty. This approach thus differs from that of the U.S. Environmental Protection Agency (Anderson *et al.*, 1983) and other (U.S.) regulatory agencies, which make similar assumptions of linearity but ignore any chemicals not proved to be carcinogenic.

METHODOLOGY

The methodology used in this paper has been described previously (Crouch *et al.*, 1983; Zeise, 1984; Zeise and Crouch, 1984; Zeise *et al.*, 1984 and 1986; and Tancrède *et al.*, 1986). It is assumed that the probability of cancer is proportional to dose at sufficiently low doses. Then the estimate of risk is simply the product of two factors—the carcinogenic potency β and the dose d. Carcinogenic potency is the slope of the dose-response curve at low doses. Since the potency cannot be directly measured in humans, and actual doses are not in practice precisely measured and vary from individual to individual, we treat both potency and dose as random variables which can take on a range of values, some more probable than others. Then the risk estimate R must also be treated as a random variable. This emphasis on random variability is an important difference from the approach used by the Carcinogen Assessment Group (CAG) of the EPA (Anderson *et al.*, 1983).

Carcinogenic Potency

Of the chemicals included in this study, only benzene has been proved to be carcinogenic in humans, and only a few have been proved to be carcinogenic in test animals. For these, the risk assessment proceeds using reported values of carcinogenic potency and an interspecies factor, K_{ah} (Crouch and Wilson, 1979; Anderson *et al.*, 1983; Gaylor and Chen, 1986). Since for the chemicals of interest the animal studies are usually performed by the oral route and a potency for inhalation is required, an additional factor I is included to account for extrapolation between routes. The carcinogenic potency in humans (β_h) is estimated by

$$\beta_h = K_{ah} I \beta_a , \qquad (1)$$

where β_a is the estimate of potency in the test animals.

To estimate an upper bound on the possible carcinogenic potency of those compounds not currently proved carcinogenic, we have developed a number of indirect methods (Crouch *et al.*, 1983; Zeise *et al.*, 1984; Fiering and Wilson, 1983). These involve comparisons of chemicals not tested in carcinogenic bioassays to similar chemicals already so tested, on the basis of toxicity, mutagenicity, and other biological information. We call this evaluating risk by "analogy." Thus we assign an estimate of carcinogenic potency for some chemicals that others might ignore.

The chemicals studied, their carcinogenic potency estimates (β_h), and the method by which they are derived are listed in Table 1, together with our estimate of their uncertainty expressed as the standard deviation (σ_h) of a lognormal distribution.

We make the assumption that the uncertainties can be reasonably represented by using lognormal distributions. Furthermore our main interest is in making conservative estimates, for which the upper ends of the distributions are most important.

Table 1. Chemicals Included in Indoor Air Pollution Study: Estimates of Potency, Uncertainties, and Estimation Method

No.	Chemical	Potency $\beta^{(a)}$ $[\text{kg–day/mg}]^{-1}$	$\sigma_h^{(b)}$	$\sigma_x^{(b)}$	Derivation of $\beta^{(c)}$	
1	- n-hexane	2.6E-5	2.7	1.5	Analogy	
2	- n-heptane	3.1E-5	3.4	2.5	Analogy	
3	- n-octane	3.1E-5	3.4	2.5	Analogy	
4	- n-nonane	3.1E-5	3.4	2.5	Analogy	
5	- n-decane	3.1E-5	3.4	2.5	Analogy	
6	- n-undecane	3.1E-5	3.4	2.5	Analogy	
7	- n-dodecane	3.1E-5	3.4	2.5	Analogy	
8	- n-tridecane	3.1E-5	3.4	2.5	Analogy	
9	- n-tetradecane	3.1E-5	3.4	2.5	Analogy	
10	- n-pentadecane	3.1E-5	3.4	2.5	Analogy	
11	- n-hexadecane	1.1E-5	2.8	1.8	Analogy	
12	- 3-methylpentane	4.6E-4	2.7	1.5	Bioassay	(Inhalation)
13	- 2-methylhexane	4.6E-4	2.7	1.5	Bioassay	(Inhalation)
14	- 3-methylhexane	4.6E-4	2.7	1.5	Bioassay	(Inhalation)
15	- Cyclohexane	4.6E-4	2.7	1.6	Analogy	
16	- Methylcyclohexane	1.1E-5	2.9	1.9	Analogy	
	Dimethylcyclopentane:					
17	isomer 1,1	7.1E-5	2.9	1.9	Analogy	
18	isomer 1,2	7.1E-5	2.9	1.9	Analogy	
19	isomer 1,3	7.1E-5	2.9	1.9	Analogy	
20	- Limonene	2.3E-4	2.9	1.9	Analogy	
21	- Benzene	1.0E-3	2.3	N.A.*	Human data	(Inhalation)
22	- Toluene	9.0E-4	2.2	0.4	Bioassay	
23	- Xylenes	9.9E-4	2.2	0.4	Bioassay	
23a	- o-xylene	9.9E-4	2.2	0.4	Bioassay	
23b	- m- and p-xylenes	9.9E-4	2.2	0.4	Bioassay	
24	- Ethylbenzene	3.4E-4	2.3	0.6	Bioassay	
25	- n-propylbenzene	2.0E-4	2.6	1.5	Analogy	
26	- isopropylbenzene	5.5E-4	2.6	1.5	Analogy	
27	- o-methylethylbenzene	7.5E-4	2.7	1.6	Analogy	
28	- m-methylethylbenzene	7.5E-4	2.7	1.6	Analogy	
29	- p-methylethylbenzene	3.2E-4	2.7	1.6	Analogy	
	Trimethylbenzene:					
30	1,2,3- isomer	1.1E-3	3.2	2.3	Analogy	
31	1,2,4- isomer	4.6E-4	2.7	1.6	Analogy	
32	1,3,5- isomer	4.6E-4	2.7	1.6	Analogy	
33	- n-butylbenzene	3.8E-4	2.7	1.6	Analogy	
34	- p-methyl-isopropyl-benzene	7.5E-4	3.2	2.3	Analogy	
35	- naphthalene	1.2E-3	2.7	1.6	Analogy	
36	- 1-methylnaphthalene	2.7E-4	2.7	1.5	Analogy	
37	- Tetrachloromethane	6.1E-3	2.2	0.2	Bioassay	
38	- Trichloroethylene	1.7E-3	2.2	0.2	Bioassay	
39	- Tetrachloroethylene	9.2E-3	2.2	0.2	Bioassay	
40	- Chlorobenzene	1.4E-3	2.3	0.7	Bioassay	
	Dichlorobenzenes:					
41	- o-dichlorobenzene	7.6E-4	2.3	0.5	Bioassay	
42	- m-dichlorobenzene	2.6E-3	2.7	1.5	Analogy	
43	- p-dichlorobenzene	6.7E-4	2.3	0.6	Bioassay	(Inhalation)
	Trichlorobenzenes:					
44	1,2,3- isomer	3.5E-3	3.2	2.3	Analogy	
45	1,2,4- isomer	3.5E-3	2.6	1.5	Analogy	
46	1,3,5- isomer	3.5E-3	3.2	2.3	Analogy	
47	- Chloroform	1.4E-2	2.2	0.1	Bioassay	
48	- 1,2-dichloroethane	1.3E-1	2.2	0.3	Bioassay	
49	- 1,1,1-trichloroethane	1.7E-5	2.3	0.8	Bioassay	
50	- Styrene	1.9E-3	2.3	0.6	Bioassay	
51	- 1,4-dioxane	5.5E-3	2.2	0.2	Bioassay	(Drinking water)
52	- α-pinene	3.6E-4	2.6	1.5	Analogy	
53	- Radon	1.5E-5/WLM	0.7	N.A.*	Human Data	(Inhalation)
54	- Formaldehyde	1.1E-1	1.5	0.18*	Bioassay	(Inhalation)
55	- Passive tobacco smoke	0.3	1.0	N.A.*	Human Data	(Inhalation)

*N.A., Not Applicable.

(a) <u>Potency</u> (at low doses) is defined as the ratio of the excess lifetime cancer risk (R) to the average daily dose (d, in units of mg/kg body weight) received by the population. The table lists median estimates.

(b) σ_h and σ_x are estimates for humans and animals, respectively, of the standard deviations of the lognormal distributions for potency estimates (using natural logarithms).

(c) A description of the derivation of potency for each of the substances listed is given in Tancrede et al., 1986 and 1987. <u>Analogy</u>, potency estimated indirectly; <u>Bioassay</u>, potency estimated from cancer bioassays. All bioassays are by gavage unless otherwise specified.

Table 2. Characteristics of the Three Studies Selected

Authors	Location	Season	Characteristics[a]	Number of VOCs sampled[b]
Pellizzari et al., 1986a	USA New Jersey	Winter 83 Jan&Feb	Overnight personal[c]	
	Bayonne		n=27; EPS=48,503	9
	Elizabeth		n=22; EPS=45,541	9
Pellizzari et al., 1986b	USA California	Winter 84	Overnight personal[c]	
	Los Angeles		n=112; EPS=359,492	19
Lebret et al., 1984	The Netherlands Ede	July to Feb 1982 - 1984	Indoor air 4 houses n/house=15(houses A & D) n/house= 14(houses B & C) EPS=134 (nearby houses)[d]	45

(a) n, number of samples; EPS, Estimated Population Sampled.

(b) VOCs, volatile organic compounds.

(c) Overnight personal air samples are assumed to be reasonably representative of indoor air concentrations (Lance Wallace, personal communication).

(d) Representative of a larger group of 20,000 houses (Erik Lebret, personal communication).

Doses

Three studies of exposure to indoor air are used; their characteristics are listed in Table 2. To estimate dose it is assumed that people are continuously exposed to the measured air concentrations and that the organic vapors breathed are rapidly absorbed through the lungs, so that the dose may reasonably be approximated by the amount inhaled indoors, the environment in which most time is spent (Spengler and Sexton, 1983). We are interested in estimating individual risk, so we use the observed variability of doses when making risk estimates.

Table 3 lists the chemical measured, their concentration in indoor air, and the variability of their concentration, for the three studies selected.

Estimates of Risk

The details of the risk calculations and the uncertainties involved have been described elsewhere (Crouch et al., 1983; Wilson et al., 1985; Zeise and Crouch, 1984; Tancrède et al., 1986). To be conservative, we assume that 100% of the amount inhaled is absorbed.

Table 3. Geometric Mean Concentrations (μg/m³) and Logarithmic Standard Deviation (in Parentheses) for 52 Indoor Air Pollutants

No.	Chemical	NEW JERSEY		CALIFORNIA	FOUR DUTCH HOUSES			
		Bayonne	Elizabeth	Los Angeles	House A	House B	House C	House D
1	n-hexane				1.8(0.3)	5.5(1.5)	2.6(0.7)	51.0(0.7)
2	n-heptane				1.8(0.2)	1.9(0.4)	3.3(0.6)	4.4(1.5)
3	n-octane			3.4(1.2)	1.5(0.3)	0.7(0.5)	1.1(1.1)	1.7(1.0)
4	n-nonane				13.0(0.5)	3.6(0.8)	15.0(0.5)	6.2(0.9)
5	n-decane			1.3(1.8)	45.0(0.5)	11.0(0.7)	29.0(0.4)	9.8(0.9)
6	n-undecane			2.4(1.3)	110.0(0.4)	5.2(0.6)	7.5(0.6)	3.2(0.7)
7	n-dodecane			1.3(1.2)	100.0(0.5)	0.9(0.6)	1.2(0.6)	1.0(0.2)
8	n-tridecane				50.0(0.8)	0.8(0.4)	0.4(0.8)	0.8(0.2)
9	n-tetradecane				8.6(0.1)	1.9(0.3)	0.6(0.6)	1.4(0.2)
10	n-pentadecane				2.0(0.3)	1.3(0.2)	0.5(0.5)	0.5(1.3)
11	n-hexadecane				1.0(0.5)	0.5(0.8)	<0.3(0.3)*	0.3(1.0)
12	3-methylpentane				1.3(0.3)	2.6(1.2)	1.6(0.6)	17.0(0.6)
13	2-methylhexane				1.3(0.3)	2.1(0.5)	1.0(0.5)	4.0(1.5)
14	3-methylhexane				1.1(0.3)	1.6(0.5)	0.8(0.5)	3.1(1.5)
15	cyclohexane				0.8(0.4)	2.0(1.5)	0.5(0.6)	6.6(1.3)
16	methylcyclohexane				1.3(0.3)	1.3(0.4)	0.6(0.5)	2.0(1.4)
17	1,1-dimethyl--cyclopentane				0.3(0.3)	0.3(0.5)	<0.3(0.3)*	0.4(1.5)
18	1,2-dimethyl--cyclopentane				0.3(0.2)	0.3(0.4)	<0.3(0.3)*	0.6(1.6)
19	1,3-dimethyl--cyclopentane				0.6(0.3)	0.6(0.5)	0.3(0.6)	1.0(1.5)
20	limonene				2.6(0.7)	11.0(1.3)	27.0(0.6)	39.0(0.9)
21	benzene			13.6(1.0)	2.8(0.5)	4.6(0.4)	5.0(0.3)	5.8(0.3)
22	toluene				2.0(0.4)	29.0(0.5)	16.0(0.7)	45.0(1.2)
23	xylenes				7.1(0.5)	4.9(0.6)	7.5(0.5)	6.7(0.5)
23a	o- ---	8.7(0.8)	4.8(2.0)	8.9(1.0)				
23b	m&p- ---	25.5(1.0)	13.0(2.2)	20.3(0.8)				
24	ethylbenzene	8.8(1.1)	4.4(2.2)	7.0(1.1)	1.3(0.5)	1.3(0.9)	2.0(0.5)	1.7(0.8)
25	n-propylbenzene				0.3(0.9)	0.3(0.6)	1.3(0.7)	0.5(0.8)
26	isopropylbenzene				0.2(0.9)	0.3(0.5)	0.4(0.5)	0.4(0.6)
	Methylethylbenzenes							
27	o- ---				1.8(0.8)	0.4(1.0)	3.2(0.7)	1.8(0.9)
28	m- ---				3.3(0.5)	1.4(0.7)	4.8(0.6)	2.1(0.8)
29	p- ---				2.4(0.6)	0.6(0.8)	2.4(0.5)	1.1(1.0)
	Trimethylbenzenes							
30	1,2,3- ---				0.5(1.1)	0.3(1.1)	1.5(0.9)	0.6(1.0)
31	1,2,4- ---				8.1(0.5)	3.3(0.6)	8.9(0.5)	3.5(0.8)
32	1,3,5- ---				2.5(0.5)	1.1(0.6)	2.5(0.6)	0.9(0.8)
33	n-butylbenzene				2.8(0.5)	0.4(1.0)	3.1(0.4)	1.0(0.9)
34	p-methyl--isopropylbenzene				1.2(0.8)	1.5(1.6)	1.1(0.6)	0.7(0.7)
35	naphthalene				0.6(0.9)	0.3(0.9)	0.3(1.0)	0.2(0.9)
36	1-methylnaphthalene				0.6(0.4)	<0.3(0.3)*	<0.3(0.3)*	<0.3(0.3)
37	tetrachloromethane			0.7(0.8)	<3.9(0.3)*	11.0(1.6)	<3.9(0.3)*	0.8(1.6)
38	trichloroethylene	1.7(0.9)	1.0(3.1)	1.2(2.0)	2.9(0.4)	2.4(0.6)	1.1(0.8)	5.0(1.5)
39	tetrachloroethylene	11.4(2.0)	3.4(2.0)	8.9(0.9)	3.5(0.5)	2.8(0.6)	1.4(0.8)	6.3(1.5)
40	chlorobenzene				<0.4(0.3)*	<0.4(0.3)*	<0.4(0.3)*	<0.4(0.3)*
	Dichlorobenzenes							
41	o- ---			0.1(1.1)				
42	m- ---	3.5(3.4) m&p	8.9(3.1) m&p	3.5(1.9) m&p	<0.6(0.3)*	<0.6(0.3)*	<0.6(0.3)*	<0.6(0.3)
43	p- ---							
	Trichlorobenzenes							
44	1,2,3- ---				<0.8(0.3)*	<0.8(0.3)*	<0.8(0.3)*	<0.8(0.3)
45	1,2,4- ---				<0.8(0.3)*	<0.8(0.3)*	<0.8(0.3)*	<0.8(0.3)
46	1,2,5- ---				<0.8(0.3)*	<0.8(0.3)*	<0.8(0.3)*	<0.8(0.3)
47	Chloroform	3.7(1.1)	1.0(2.0)	1.5(1.0)				
48	1,2-dichloroethane			0.2(1.4)				
49	1,1,1trichloroethane	18.9(1.5)	11(3.4)	26.6(1.8)				
50	Styrene	1.6(0.9)	1.0(1.5)	2.4(1.3)				
51	1,4-dioxane			0.2(1.8)				
52	α-pinene			3.2(1.2)				

*Where no estimate of coefficient of variation was made in the study of Lebret et al. (1984), we have assumed a logarithmic standard deviation of 0.3.

Note: For further information on this table, see Tancrede et al., 1987.

Provided that the doses are small enough, the excess lifetime risk (R) due to a carcinogen is the product of the dose multiplied by the carcinogenic potency in humans. Thus when the amount of carcinogen in the air is given in units of μg/m³, an estimate of lifetime risk for a 70-kg man breathing 20 m³ per day is:

$$R = 3 \times 10^{-4} \beta_h d, \qquad (2)$$

where

β_h = carcinogenic potency in humans, in units of $[\text{mg/kg-day}]^{-1}$;
d = air contaminant concentration, in μg/m³.

Uncertainties

Usually human cancer potency is estimated from animal data. Replacing β_h by its new value [Eq. (1)] in Eq. (2), we approximated the uncertainties in each of the factors by assuming a lognormal probability distribution for each factor. We assume the factors β, K_{ah}, I, and d (carcinogenic potency, interspecies factor, route extrapolation factor and dose, respectively) are independently lognormally distributed, so that R is also lognormally distributed, with a standard deviation σ given by adding the individual standard deviations in quadrature:

$$\sigma^2 = \sigma_x^2 + \sigma_y^2 + \sigma_t^2 + \sigma_z^2 , \tag{3}$$

where

- σ_x = uncertainty in measuring potency in animal experiments,
- σ_y = uncertainty in extrapolating from animal to humans, assumed to be $\log_e(4.5) = 1.5$ (Crouch, 1983),
- σ_t = uncertainty due to the different routes of administration, assumed to be $\log_e(5) = 1.6$,
- σ_z = uncertainty (variability) in dose.

We emphasize that we *assume* lognormal distributions for the items in Eqs. (1) and (2). This assumption is not contradicted by data and has the advantage of leading to simple rules for combining errors. The outcome is a risk estimate represented by a distribution of possible values. It is usually useful to quote summary statistics.

As discussed in Tancrède *et al.* (1986), it is reasonable to assume that the median of both I and K_{ah} is unity. The median lifetime risk is simply obtained as the product of medians. Thus the following summary statistics can be calculated:

Summary Statistics (Lifetime Risk)

Median:	$R = (3 \times 10^{-4}) \beta_h d$
Mean (Average):	$R = \beta_h d = R \exp(\sigma^2/2)$
98th Percentile:	$R_{98} = R \exp(2.0537\sigma)$

Using an upper limit corresponds in intent to the EPA procedure (Anderson *et al.*, 1983). However, the EPA takes upper limits of the various terms before multiplying them and assumes that the interspecies factor K_{ah} is fixed. Our procedure accounts more correctly and completely for the uncertainties; but the EPA procedure often gives estimates close to our upper 98th percentile (partly because of the different animal to human extrapolation factor used).

When more than one chemical is present, some assumptions must be made about the carcinogenicity of the mixture. Some agents (such as asbestos and cigarette smoke) are known to act synergistically and their excess relative risks multiply when both are present in large amounts. When they are present in small amounts, the existence of a background makes it likely that the absolute risks add (Reif, 1984). This is the assumption made here. The overall estimate of risk is obtained by summing several variables, each with a lognormal distribution. The average (mean) value is simply the sum of the means, but in order to obtain the various percentiles of the distribution, the distribution of the sum must be determined. This is done by a Monte Carlo procedure.

We assume for simplicity that the various pollutants are independent of one another. This study does not take into account the potential diversity of the population at risk.

Table 4a. Median Estimate of Lifetime Risk from Indoor Air Pollution

No.	Chemical	Bayonne	Elizabeth	Los Angeles	House A	House B	House C	House D
		N E W J E R S E Y		**CALIFORNIA**	**F O U R**	**D U T C H**		**H O U S E S**
1	n-hexane				1.3E-08	4.1E-08	1.9E-08	3.8E-07
2	n-heptane				1.6E-08	1.7E-08	2.9E-08	3.9E-08
3	n-octane			3.0E-08	1.4E-08	6.4E-09	9.3E-09	1.5E-08
4	n-nonane				1.2E-07	3.1E-08	1.3E-07	5.5E-08
5	n-decane			1.2E-08	4.0E-07	9.9E-08	2.6E-07	8.7E-08
6	n-undecane			2.1E-08	9.7E-07	4.6E-08	6.6E-08	2.9E-08
7	n-dodecane			1.1E-08	9.1E-07	7.6E-09	1.0E-08	8.6E-09
8	n-tridecane				4.4E-07	7.5E-09	3.3E-09	7.0E-09
9	n-tetradecane				7.6E-08	1.7E-08	5.2E-09	1.2E-08
10	n-pentadecane				1.8E-08	1.1E-08	4.7E-09	4.6E-09
11	n-hexadecane				3.3E-09	<9.1E-10	<9.1E-10	7.9E-10
12	3-methylpentane*				1.7E-07	3.4E-07	2.1E-07	2.2E-06
13	2-methylhexane*				1.8E-07	2.7E-07	1.3E-07	5.7E-07
14	3-methylhexane*				1.4E-07	2.1E-07	1.1E-07	4.1E-07
15	cyclohexane				1.1E-08	2.6E-08	6.4E-09	8.7E-08
16	methylcyclohexane				4.0E-09	4.0E-09	1.9E-09	6.2E-09
17	1,1-dimethylcyclopentane				5.8E-09	5.5E-09	<5.9E-09	8.4E-09
18	1,2-dimethylcyclopentane				5.9E-09	5.6E-09	<5.9E-09	1.1E-08
19	1,3-dimethylcyclopentane				1.2E-08	1.1E-08	5.1E-09	2.0E-08
20	limonene				1.9E-06	8.3E-07	2.0E-06	2.9E-06
21	benzene*			3.9 E-06	8.0E-07	1.3E-06	1.4E-06	1.7E-06
22	toluene*				5.7E-06	7.5E-06	4.1E-06	1.1E-05
23	xylenes*				2.0E-06	1.4E-06	2.1E-06	1.9E-06
23a	o-xylene*	2.5E-06	1.4E-06	2.5E-06				
23b	m&p-xylenes*	7.2E-06	3.8E-06	5.7E-06				
24	ethylbenzene	8.5E-07	4.3E-07	6.8E-07	1.3E-07	1.3E-07	2.0E-07	1.6E-07
25	n-propylbenzene				1.9E-08	1.9E-08	7.2E-08	2.9E-08
26	isopropylbenzene				3.2E-08	4.2E-08	6.9E-08	6.6E-08
27	o-methylethylbenzene				3.8E-07	9.1E-08	6.8E-07	3.8E-07
28	m-methylethylbenzene				7.2E-07	3.0E-07	1.0E-06	4.5E-07
29	p-methylethylbenzene				2.2E-07	5.8E-08	2.2E-07	1.0E-07
30	1,2,3-trimethylbenzene				1.6E-07	8.2E-08	4.7E-07	1.8E-07
31	1,2,4-trimethylbenzene				1.1E-06	4.4E-07	1.2E-06	4.6E-07
32	1,3,5-trimethylbenzene				3.3E-07	1.4E-07	3.3E-07	1.2E-07
33	n-butylbenzene				3.1E-07	4.6E-08	3.4E-07	1.0E-07
34	p-methylisopropylbenzene				2.6E-07	3.2E-07	2.5E-07	1.5E-07
35	naphthalene				2.0E-07	9.3E-08	8.4E-08	7.1E-08
36	1-methylnaphtalene*				4.2E-08	<2.2E-08	<2.2E-08	<2.2E-08
37	tetrachloromethane*			1.1E-06	<6.8E-06	2.0E-05	<6.8E-06	1.3E-06
38	trichloroethylene*	8.3E-07	5.1E-07	5.7E-07	1.4E-06	1.2E-06	5.2E-07	2.4E-06
39	tetrachloroethylene*	3.0E-05	9.1E-06	2.4E-05	9.3E-06	7.4E-06	3.8E-06	1.6E-05
40	chlorobenzene				<1.6E-07	<1.6E-07	<1.6E-07	<1.6E-07
41	o-dichlorobenzene*			2.0E-08				
42	m-dichlorobenzene*	2.6E-06	6.6E-06	2.6E-06	<4.3E-07	<4.3E-07	<4.3E-07	<4.3E-07
43	p-dichlorobenzene*							
43	1,2,3-trichlorobenzene				<7.8E-07	<7.8E-07	<7.8E-07	<7.8E-07
45	1,2,4-trichlorobenzene				<7.8E-07	<7.8E-07	<7.8E-07	<7.8E-07
46	1,3,5-trichlorobenzene				<7.8E-07	<7.8E-07	<7.8E-07	<7.8E-07
47	chloroform*	1.5E-05	4.0E-06	6.0E-06				
48	1,2-dichloroethane*			8.2E-06				
49	1,1,1-trichloroethane*	9.2E-08	5.2E-08	1.3E-07				
50	styrene*	0.7E-07	5.3E 07	1.3E-06				
51	1,4- dioxane*			3.5E-07				
52	α-pinene			3.3E-07				

*--------------------
* Cancer potency estimated from animal carcinogenesis bioassay or from human epidemiology.

Results

The contributions (R_i) for *each of the selected pollutants* to an estimate of individual lifetime risk are listed in Table 4. The total risk for the mixture of contaminants is described in Table 5. To give perspective, Table 5 includes typical risks of other indoor air pollutants not explicitly studied here (radon gas, passive tobacco smoke, formaldehyde) (for further details see Tancrède *et al.*, 1987).

Estimated Contributions to Lifetime Risk

The risk estimates presented here do not represent population estimates but rather risk estimates for individuals.

COMMENTS

In this paper we have deliberately chosen winter air concentrations of the indoor air pollutants in order to obtain conservative risk estimates. In addition, we have assumed that

Table 4b. 98th Percentile Estimate of Lifetime Risk from Indoor Air Pollution

No.	Chemical	Bayonne	Elizabeth	Los Angeles	House A	House B	House C	House D
		NEW JERSEY		CALIFORNIA	FOUR	DUTCH	HOUSES	
1	n-hexane				3.3E-06	2.3E-05	5.6E-06	1.1E-04
2	n-heptane				1.6E-05	1.8E-05	3.2E-05	7.3E-05
3	n-octane			4.4E-05	1.4E-05	6.7E-06	1.3E-05	2.1E-05
4	n-nonane				1.2E-05	3.9E-05	1.4E-04	7.1E-05
5	n-decane			2.7E-05	4.2E-04	1.1E-04	2.7E-04	1.1E-04
6	n-undecane			3.2E-05	1.0E-03	5.0E-05	7.3E-05	3.3E-05
7	n-dodecane			1.6E-05	9.6E-05	8.2E-06	1.1E-05	8.6E-06
8	n-tridecane				5.3E-04	7.7E-06	3.9E-06	6.9E-06
9	n-tetradecane				7.6E-05	1.7E-05	5.7E-06	1.2E-05
10	n-pentadecane				1.8E-05	1.1E-05	5.0E-06	7.4E-06
11	n-hexadecane				1.2E-06	6.7E-07	<3.1E-07	3.6E-07
12	3-methylpentane*				4.3E-05	1.4E-04	5.7E-05	6.0E-04
13	2-methylhexane*				4.3E-05	7.0E-05	3.3E-05	3.0E-04
14	3-methylhexane*				3.4E-05	5.5E-05	2.7E-05	2.2E-04
15	cyclohexane				1.1E-06	6.2E-06	6.9E-07	1.7E-05
16	methylcyclohexane				1.6E-06	1.7E-06	8.2E-07	4.8E-06
17	1,1-dimethylcyclopentane				2.3E-06	2.3E-06	<2.3E-06	6.6E-06
18	1,2-dimethylcyclopentane				2.3E-06	2.3E-06	<2.3E-06	9.9E-06
19	1,3-dimethylcyclopentane				4.6E-06	4.6E-06	2.2E-06	1.6E-05
20	limonene				8.7E-06	5.5E-06	8.8E-06	1.4E-03
21	benzene*			6.7E-04	9.9E-05	1.6E-04	1.7E-04	1.9E-04
22	toluene*				6.1E-06	8.4E-06	5.0E-06	2.2E-03
23	xylenes*				2.2E-04	1.6E-04	2.3E-04	2.1E-04
23a	o-xylene*	3.3E-04	6.8E-04	3.8E-04				
23b	m&p-xylenes*	1.1E-03	2.5E-03	7.6E-04				
24	ethylbenzene	1.6E-04	2.8E-04	1.2E-04	1.6E-05	2.0E-05	2.5E-05	2.5E-05
25	n-propylbenzene				5.9E-06	5.0E-06	2.0E-05	8.5E-06
26	isopropylbenzene				9.7E-06	1.0E-05	1.7E-05	1.7E-05
27	o-methylethylbenzene				1.3E-04	3.4E-05	2.1E-04	1.3E-04
28	m-methylethylbenzene				2.1E-04	8.9E-05	3.0E-04	1.5E-04
29	p-methylethylbenzene				6.5E-05	1.9E-05	6.4E-05	3.9E-05
30	1,2,3-trimethylbenzene				1.7E-04	9.1E-05	4.5E-04	1.8E-04
31	1,2,4-trimethylbenzene				3.0E-04	1.3E-04	3.3E-04	1.5E-04
32	1,3,5-trimethylbenzene				9.2E-05	4.2E-05	9.7E-05	4.0E-05
33	n-butylbenzene				8.7E-05	1.7E-05	9.3E-05	3.6E-05
34	p-methylisopropylbenzene				2.3E-04	5.0E-05	2.1E-04	1.3E-04
35	naphthalene				7.1E-05	3.2E-05	1.3E-05	2.4E-05
36	1-methylnaphtalene*				1.0E-05	<5.3E-06	<5.3E-06	<5.3E-06
37	tetrachloromethane*			1.4E-04	<6.6E-04	5.2E-03	<6.6E-04	3.8E-04
38	trichloroethylene*	1.1E-04	1.2E-03	2.5E-04	1.4E-04	1.3E-04	6.6E-05	5.8E-04
39	tetrachloroethylene*	1.4E-02	4.0E-03	3.2E-03	9.0E-04	8.0E-04	4.7E-04	3.9E-03
40	chlorobenzene				<1.8E-05	<1.8E-05	<1.8E-05	<1.8E-05
41	o-dichlorobenzene*			3.5E-06				
41	m-dichlorobenzene*	1.4E-02	2.8E-02	2.2E-03	<1.1E-04	<1.1E-04	<1.1E-04	<1.1E-04
42	p-dichlorobenzene*							
43	1,2,3-trichlorobenzene				<5.9E-04	<5.9E-04	<5.9E-04	<5.9E-04
44	1,2,4-trichlorobenzene				<1.8E-04	<1.8E-04	<1.8E-04	<1.8E-04
45	1,3,5-trichlorobenzene				<5.9E-04	<5.9E-04	<5.9E-04	<5.9E-04
47	chloroform	2.4E-03	1.9E-03	8.8E-04				
48	1,2-dichloroethane*			1.7E-03				
49	1,1,1-trichloroethane*	2.7E-05	2.7E-04	5.2E-05				
50	styrene*	1.4E-04	1.4E-04	2.7E-04				
51	1,4- dioxane*			1.2E-04				
52	α-pinene			1.3E-04				

* Cancer potency estimated from animal carcinogenesis bioassay or from human epidemiology.

every chemical poses a cancer risk of a magnitude consistent with experimental evidence, even for those which have not been proven carcinogenic. To indicate the effects of this last assumption, Table 5 (column 3) also shows for comparison a comparable risk estimate obtained by ignoring all chemicals for which there are no data on carcinogenicity from animal bioassay or human epidemiology. Air concentrations of pollutants at other times of the year will presumably be lower than those in winter, leading to lower risk estimates. However, no single house had measurements of all 52 chemicals listed here, so that a more complete assessment would give higher risk estimates. Considering the differences in the number of chemicals sampled and in sampling conditions, the total risks emerging from the several studies considered in this paper are quite similar (see Table 5).

The upper 98th percentile risk estimate for each of the 52 chemicals (Table 4b) is greater than 10^{-6} per life in at least one location. This risk value is used by the Food and Drug Administration and EPA as an action level for accidental food additives or pollutants in water and air. The corresponding estimate of total risk is much greater than 10^{-6}/life (Table 5). Even ignoring those chemicals for which cancer potency was estimated by "analogy," mean lifetime risks are estimated to be of the order of 10^{-4} and higher. This suggests that indoor air deserves more public attention than outdoor air, water or food

Table 5. Mean, Median, and Upper 98th Percentile Estimates of Risk from Indoor Air Pollution

	ANNUAL MEAN[d]	- L I F E T I M E - MEAN	MEAN Direct Data only[e]	MEDIAN	98TH PERC.
NEW JERSEY					
Bayonne	2.7×10^{-4}	1.9%	0.3%	0.04%	4.6%
Elizabeth	4.3×10^{-4}	3.0%	0.3%	0.03%	4.8%
CALIFORNIA					
Los Angeles	2.6×10^{-5}	0.2%	0.1%	0.04%	1.1%
THE NETHERLANDS					
HOUSE A	2.9×10^{-5}	0.2%	0.03%	0.05%	1.1%
HOUSE B	2.7×10^{-5}	0.2%	0.1%	0.04%	1.2%
HOUSE C	1.9×10^{-5}	0.1%	0.02%	0.04%	0.8%
HOUSE D	3.2×10^{-5}	0.2%	0.1%	0.06%	1.3%
AVERAGE US HOUSE:					
RADON[a]					
smokers	7.3×10^{-4}	5.2%	5.2%	2.2%	32%
non-smokers	7.9×10^{-5}	0.6%	0.6%	0.2%	3.4%
PASSIVE SMOKING[b]					
At home	2.2×10^{-5}	0.2%	0.2%	0.1%	0.7%
At home & work	1.1×10^{-4}	0.8%	0.8%	0.5%	3.8%
FORMALDEHYDE[c]					
Average house	4.8×10^{-4}	3.4%	3.4%	1.1%	24%
All US houses	(4.8×10^{-4})	(3.4%)	(3.4%)	(0.2%)	(26%)

(a,b,c) For further details on risk estimation, see Tancrède et al., 1987.

(d) Lifetime risk/70 years.

(e) Average lifetime risk from only those chemicals for which there is either animal cancer bioassay data or human epidemiology data.

---

additives, as emphasized in Table 6 where estimated risks from drinking water contaminated with organics commonly found in drinking water are compared to those we have just calculated for the exposure to the same pollutants in indoor air. The estimates for indoor air risks are much greater. These risks are estimated using similar procedures.

None of the mean estimates of risk for particular chemicals, nor the total mean estimate of risk, reach 7×10^{-4}/year (5.2%/life), the risk from exposure to average levels of radon gas in homes (Table 5). Action to reduce radon gas exposures therefore appears to be more important than action to reduce concentration of organic vapors. Such action might reduce concentrations of organic pollutants simultaneously, clearly a desirable synergism.

The 52 chemicals listed and studied in the present paper are only a small fraction of the possible chemicals present in indoor air (Pellizzari et al., 1984). The total risk may exceed the estimates presented here. It therefore is important to study other air pollutants.

Pollutant concentrations in the U.S. and in the four Dutch houses examined are probably representative of indoor air contamination in the U.S. in general and should be cause for concern. The calculated increase in pollutant concentrations produced by energy conservation leads to an estimated increase of risk which, when applied to the whole U.S. population, exceeds the effect on public health of all actions taken by EPA since its inception. The assumption that contaminant concentrations measured in a few houses characterize those throughout the U.S. housing stock needs to be studied; i.e., an accurate characterization of indoor air in U.S. homes is needed. Until that becomes available, it seems prudent to couple fuel and energy conservation measures with other measures that *reduce* concentrations of indoor pollutants.

Table 6. Comparison of Risks from Polluted Drinking Water and Indoor Air

Chemical	Proposed MCL[a] (µg/l)	MEAN ESTIMATES OF LIFETIME RISK/10^{-6}		
		MCL Risk[b]	Highest risk found 25 cities[c]	INDOOR AIR RISK
Trichloroethylene	5-500 --> 0.3-31.0	0.14		8^d - 690^g
Tetrachloroethylene	5-500 --> 0.7-70.0	0.11		60^d - 2700^h
Tetrachloromethane	5-500 --> 1.7-170	14		$<59^d$ - 790^d
1,1,1-Trichloroethane	1000 --> 1.9	0.02		2^e - 300^g
1,2-Dichloroethane	1-100 --> 0.9-88.0	1.4		18^f - 240^i

(a) Proposed drinking water standard: Maximum Contaminant Level (MCL; EPA, 1982).

(b) Risks computed for exposure to 2 liters/day of drinking water contaminated at the maximum contaminant level. For details, see Crouch et al., 1983.

(c) Of the 25 U.S. cities considered in Crouch et al. (1983), the risk associated with the highest concentration listed.

(d) In four Dutch houses (from Table 4a).

(e-i) Details are given in Tancrède et al., 1988; e,f refer to this study; g,h,i refer to other studies.

It also seems important to try to extend this study to other chronic health effects and other pollutants. Large epidemiological studies of indoor air pollution in nonoccupational settings are difficult, since few building characteristics appear to correlate with the indoor air pollution level and since extraordinarily large numbers of people are needed to observe differences from background rates for the common cancers (Pochin, 1976). Thus for organic chemicals we must rely on indirect methods similar to those presented here, noting that the assumptions and procedures are only slight extensions of those currently used by the EPA and others for standard setting.

ACKNOWLEDGEMENTS

M. Tancrède thanks the Cabot Corporation for financial support during the period of this work. We are also grateful for grants from the General Electric Foundation and Dow Chemical Company.

REFERENCES

Anderson, E. L., and the Carcinogen Assessment Group of the U.S. Environmental Protection Agency, 1983, "Quantitative Approaches in Use to Assess Cancer Risk," *Risk Anal.* 3:277-295.

Carcinogen Assessment Group, CAG, 1979, *Carcinogen Assessment Group's Final Report on Population Risk to Ambient Benzene Exposures*, U.S. Environmental Protection Agency, Washington, DC.

Crouch, E. A. C., and Wilson, R., 1979, "Interspecies Comparison of Carcinogenic Potency," *J. Tox. Env. Health* 5:1095-1118.

Crouch, E. A. C., 1983, "Uncertainties in Interspecies Extrapolation of Carcinogenic Potency," *Environmental Health Perspectives* 50:321-327.

Crouch, E. A. C., Wilson, R., and Zeise, L., 1983, "The Risks of Drinking Water," *Water Resources Res.* **19**:1359-1375.

Environmental Protection Agency, EPA, 1982, "National Revised Primary Drinking Water Regulations, Volatile Synthetic Organic Chemicals in Drinking Water; Advanced Notice of Proposed Rulemaking," *Federal Register* **47**:9350-9358.

Fiering, M., and Wilson, R., 1983, "Attempts to Establish a Risk by Analogy," *Risk Anal.* **3**:207-216.

Gaylor, D. W., and Chen, J. J., 1986, "Relative Potency of Chemical Carcinogens in Rodents," *Risk Anal.* **6**:283-290.

Lebret, E., van de Wiel, H. J., Bos, H. P., Noij, D., and Boleij, J. S. M., 1984, "Volatile Hydrocarbons in Dutch Homes," in *Indoor Air, Volume 4, Chemical Characterization and Personal Exposure*, D19 1984, Swedish Council for Building Research, B. Berglund, T. Lindvall, and J. Sundell (Eds.), Stockholm, pp. 169-174.

Pellizzari, E. D., Sheldon, L. S., Sparacino, S. M., Bursey, J. T., Wallace, L., and Bromberg, S., 1984, "Volatile Organic Level in Indoor Air," in *Indoor Air, Volume 4, Chemical Characterization and Personal Exposure*, D19 1984, Swedish Council for Building Research, B. Berglund, T. Lindvall, and J. Sundell (Eds.), Stockholm, pp. 303-308.

Pellizzari, E. D., Perritt, K., Hartwell, T. D., Michael, L. C., Sparacino, C. M., Sheldon, L. S., Whitmore, R., Leninger, C., Zelon, H., Handy, R. W., and Smith, D., and Project Officer Wallace, L., 1986a, *Total Exposure Assessment Methodology (TEAM) Study: Elizabeth and Bayonne, New Jersey, Devils Lake, North Dakota and Greensboro, North Carolina*, Volume II (sections 1-7 and 8), Final Report, Part I, U.S. EPA Office of Research and Development, Washington, DC.

Pellizzari, E. D., Perritt, K., Hartwell, T. D., Michael, L. C., Whitmore, R., Handy, R. W., Smith, D., and Zelon, H., and Project Officer Wallace, L., 1986b, *Total Exposure Assessment Methodology (TEAM) Study: Selected Communities in Northern and Southern California*, RTI/2391/00-03F, Volume III, Final Report, Part I, U.S. EPA Office of Research and Development, Washington, DC.

Pochin, E. E., 1976, "Problems Involved in Detecting Increased Malignancy Rates in Areas of High Natural Radiation Background," *Health Physics* **31**:148-151.

Reif, A. E., 1984, "Synergism in Carcinogenesis," *J. Nat. Cancer Inst.* **73**:25-39.

Spengler, J., and Sexton, K., 1983, "Indoor Air Quality: A Public Health Perspective," *Science* **221**:9-17.

Tancrède, M., Wilson, R., Zeise, L., and Crouch, E. A. C., 1987, "The Carcinogenic Risk of Some Organic Vapors Indoors: A Theoretical Survey," *Atmospheric Environment* **21**(10):2187-2205.

Tancrède, M., Wilson, R., Zeise, L., and Crouch, E. A. C., June 1986, "The Carcinogenic Risk of Organic Vapors Indoors: A Survey," Discussion Paper Series #E-86-06 Energy and Environmental Policy Center, John F. Kennedy School of Government Harvard University, Cambridge, MA.

Wilson, R., Crouch, E. A. C., and Zeise, L., 1985, "Uncertainty in Risk Assessment," in Banbury Report No. 19: *Risk Quantitation and Regulatory Policy*, D. G. Hoel, R. A. Merrill, F. P. Perera (Eds.), Cold Spring Harbor Laboratory, Cold Spring Harbor, pp. 133-147.

Zeise, L., 1984, "Surrogate Measures of Human Cancer Risk," Ph.D. Thesis, Harvard University, Department of Engineering Sciences, Cambridge, MA.

Zeise, L., and Crouch, E. A. C., 1984, "Experimental Variation in the Carcinogenic Potency of Benzo(a)pyrene," Energy and Environmental Policy Center, Harvard University, Cambridge, MA.

Zeise, L., Wilson, R., and Crouch, E. A. C., 1984, "Use of Acute Toxicity to Estimate Carcinogenic Risk," *Risk. Anal.* **4**:187-199.

Zeise, L., Wilson, R., and Crouch, E. A. C., 1985, "Reply to Comments: On the Relationship of Toxicity and Carcinogenicity," *Risk Anal.* **5**:265-270.

Zeise, L., Crouch, E. A. C., and Wilson, R., 1986, "A Possible Relationship Between Toxicity and Carcinogenicity," *J. Amer. Coll. Toxicol.* **5**:137-151.

Risk Assessment and National Standards: Philosophical Problems Arising from the Normalization of Risk

David Geoffrey Holdsworth

York University
North York, Ontario, Canada

ABSTRACT

Risk assessment is nominally a scientific activity. As such, it is committed to the discovery and transmission of true propositions. Standards-setting is nominally a socio-political activity. As such, it is committed to the identification and promotion of norms which reflect community preferences. This simple distinction gives rise to seemingly simple philosophical questions: When, and to what extent, are the true propositions of scientific risk assessment relevant to the norms of social choice? Are the objective facts of risk analyses ever sufficient to determine the norms that a rational agent ought to accept? Is the implied distinction between the objectivity of analysis and the subjectivity of choice a useful one in the context of the facilitation of political decision? In this paper I argue that the now-orthodox distinction between objective analysis and subjective choice is misleading—that risk assessment is irreducibly normative in ways which significantly affect its relevance to national standards-setting. This is not to say that risk assessment has no relevance. Indeed, I argue the further point that, understood in thoroughly normative terms, risk assessment methodology has much to offer, not only to social and political choice, but also to a philosophically mature understanding of science itself.

KEYWORDS: Methodology, objectivism, philosophy of science, rationality, standards-setting

INTRODUCTION

I must begin with a disclaimer. I do not intend to address explicitly the theme of the primary title. Rather, implications for national standards-setting are implicit in my material. I shall be more interested in addressing the theme of the secondary title. The task that I have set for myself is to argue—stated very broadly—that there are important parallels between the recent history of the philosophy of science and the history of risk analysis in Anglo-American culture.

Moreover, these parallels give rise to a reciprocal and mutually beneficial relationship. Indeed, risk analysis and philosophy of science can learn a great deal from each other. For this reason I have set as the broad goal of my paper to take one step towards persuading risk analysts and regulators that there is a rich and important relationship between these two fields.

THE RATIONALITY DEBATES IN PHILOSOPHY

I want to start by discussing briefly some trends in contemporary philosophy (not only philosophy of science) which are relevant to risk analysis. I shall necessarily leave out a great deal and highlight the things that have struck me as important. Currently, the dominant trend in philosophy is its preoccupation with questions of rationality. The notion of rationality is extremely problematic both from a philosophical and from a practical—for the purpose of this context, from a regulatory—point of view. There are many traps that one can fall into and the history of risk analysis has demonstrated that some of these traps can be quite perilous in a regulatory setting.

Cognitive versus General Rationality

For example, there is a tendency among contemporary scientists to identify general rationality with cognitive rationality and to fall into the trap of believing that just because something is logically rigorous it is therefore irrational not to accept its conclusions. In one sense this may well be true. The philosophical notion that rationality has something to do with the idea that one should abide strictly by one's own principles of rational inference, rational belief and rational choice, has to be correct. To hold a clearly articulated principle of rationality but to self-consciously ignore it even when its conditions of application are believed to be filled is surely a *prima facie* case of irrationality. But the problem is more complicated when it is the rationality principles themselves which are at issue or when the principles are highly contextual.

Subjectivism versus Objectivism

A second pitfall that risk analysis has sometimes encountered is to accept uncritically the common distinction between the subjective and the objective. There is a widely held intuition that the distinction between the subjective and the objective corresponds exactly to the distinction between the normative and the empirical. There is thus established an alignment of principles. The empirical, the objective and the cognitive are aligned against the normative, the subjective and the general. In many cases the former group is taken to characterize the rational, whereas the latter group is taken to characterize the irrational.

There are compelling arguments for being skeptical about these alignments. In the philosophy of science literature today there are many sophisticated attempts to articulate these arguments. But what they all have in common is quite simple. They are usually based on attempts to show that the theoretical instruments of science are irreducibly present in all observation—in the current jargon, that all observation is theory-laden. If this can be established, then the subjectivity of empirical research is also established. Moreover, if it is accepted that theory choice is a subjective process requiring the same kind of political and institutional compromise usually present in normative choice, then it would also be established that the empirical and the normative are not totally distinct.

THE CARTESIAN HERITAGE

The Rise of Method

We have inherited from Western philosophy difficult—but important—conceptual distinctions and cognate principles of rationality. These are distinctions about which we must be very careful! One distinction that I have been emphasizing goes back (in conventional historical terms at least) to Descartes.[a] It is (to use a variety of contemporary idioms) the distinction between the empirical and the normative, between the given and modes of thought about the given, or equivalently, between the data and the process of reasoning.

Similar distinctions arose for Descartes out of an interest in developing a philosophical method based on systematic doubt. But Descartes' goal was to identify at least one thing about which we can be certain in human experience and to produce a method that would enable us, first, to recognize when we have found things about which we can be certain and, second, how to advance from these certain things to a broader framework of general knowledge.[b]

The Separation of the Normative from the Empirical

Another version of the normative/empirical distinction is present in the philosophy of Kant. This is made evident by the firm distinction that he wanted to maintain between the foundations of knowledge and the foundations of morals. Without going into how Kant intended to accomplish this goal, it is important to recognize that this period in the history of Western philosophy was a very important one because it explicitly introduced a new way of understanding a strong separation between the normative and the empirical—between what is the case and what ought to be done.

The Separation of the Normative from the Rational

That the normative/empirical distinction and the belief that it is important have survived is shown quite clearly by the work of the logical positivists at the beginning of this century. One sees it clearly in the earlier work of Bertrand Russell. It is the distinction between some form of the given—the manifest truth[c]—and a preoccupation in Russell's case with how to understand logic, the relation of logic to sets and in turn the relation of all these structures to reason. In other words, he was preoccupied with the way in which we are to construct the rest of truth from manifest truth using logic alone.

As is well known, for many of the later logical positivists this gave rise to a rejection of moral philosophy as meaningful.[d] This is more radical than Kant's position, which had held that moral philosophy could be grounded in experience. With logical positivism we have not only the radical separation of the normative from the empirical. The process is completed by removing the normative from the domain of the possibly rational.[e]

---

a. Many of my remarks here presuppose a conventional account of the history of philosophy. The attributions made here, while broadly accurate, should not be regarded as historically precise propositions.

b. It is ironic that although Descartes was a rationalist (i.e., he believed that knowledge could be based on reason alone) his philosophy has had such far-reaching influence on modern empirical science.

c. For Russell this was the immediate data of perceptual experience, colour patches in one's visual field, or what he simply called sense data.

d. This was based on the principle of verification, which asserted roughly that for a proposition to be meaningful it had to be possible in principle to verify it. It is an easy step to the conclusion that all of metaphysics and moral philosophy is meaningless. And an even shorter step to the conclusion that moral action is not rational—but not necessarily irrational.

e. This move ignores the perfectly reasonable possibility of considering logic to be a normative discipline. The recognition that logic is a normative discipline is not necessarily inconsistent with positivism if we give logic a privileged status. From a post-positivist perspective this tactic simply appears as an equivocation.

The Rise of Methodology

In recent philosophy of science there have been many new developments. I shall mention philosophers that risk analysts and regulators may well have heard of. Thomas Kuhn and his notion of a scientific paradigm has been cited more than once in the risk analysis literature. Other philosophers who are perhaps less well know are Karl Popper[f] and Paul Feyerabend,[g] but they too have been mentioned by others at this meeting. In the work of Feyerabend one sees a reaction to positivism which perhaps—I would argue does in fact—goes too far. There is a critical dialogue happening in Western culture today among philosophers, but not only among philosophers.[h] It is a dialogue about the nature of rationality itself.

More recent trends in philosophy of science have been self-consciously sociological. This sociological turn began with Kuhn, but since then it has become a robust research paradigm itself. In its extreme form it introduces the radical thesis that all of science is a social construction—equivalently, that scientific objectivity is a myth. As some of these researchers have seriously argued (Collingridge and Reeve, 1986), if this were true there would be little hope of science playing a constructive role in policy making. Indeed, Collingridge and Reeve argue that the institutional nature of modern science forces it to give rise to endless cycles of unproductive criticism that paralyzes decision making.

Below, I shall suggest a general strategy for avoiding these conclusions.

THE METAPHOR OF SEPARATION

The Metaphor of Alien Cultures

I want to turn to a passage that I have taken from a book by Richard Bernstein. It introduces a metaphor that I want to develop. The passage is this:

> To construe the main issue as the problem of defining standards of rationality, which is how most of the participants in the debate construe it, is misleading and mystifying. The vital issue here is really the question of what is involved in understanding, interpreting and explaining alien societies and not just their rationality or lack of rationality. How are we to do justice to the strangeness that we discover when we encounter alien types of activities, beliefs, rituals, institutions and practices without falsifying or distorting them? (Bernstein, 1985, p. 28)

The connection that I want to make between Bernstein's thesis about rationality and questions relevant to risk analysis is simply this: that it would be equally as misleading to try to arrive at an analytic definition[i] of risk as it would be to attempt an analytic definition of rationality. The difficulty with either attempt at definition is related to the preoccupation that our culture has had with the separation between the normative and the empirical. This

---

f. Popper is well know for political writings as well as for his work on objectivity and the logic of discovery (Popper, 1972).

g. Feyerabend has argued rather provocatively for the appropriateness of science operating in a totally anarchistic mode (Feyerabend, 1975).

h. The rationality debate is prominent in many of the social sciences—especially anthropology.

i. The expression "analytically true proposition" is used in philosophy to mean "an expression which is true by virtue of the meanings of its terms." Thus, if we stipulate that "risk" means "probability times consequence" then it is an analytically true proposition that risk increases linearly as the consequence increases linearly.

is because our understanding of the normative/empirical distinction has given rise to the notion that rational activity and rational belief are somehow grounded entirely in the empirical side of the distinction. The problem, as more and more philosophers today are beginning to see it, is simply that the distinction—the separation—between the empirical and the normative is a false distinction. It is, at the very least, a distinction that we have failed to maintain with the kind of clarity that it was once believed should be possible.

To scientists who accept a radical version of the normative/empirical distinction the world of ordinary citizens is an alien culture. The claim that one often hears that the man on the street is irrational because he does not accept the conclusions of risk analysis[j] is an instance of the kind of conflict of alien cultures that Bernstein is talking about.

This is exactly the kind of thing I mean when I say that there are parallels between what is happening in risk analysis and what is happening in the broader dialogue in philosophy about rationality. There are very real parallels in the respective treatments of these issues. The movement away from taking the risk management/risk assessment distinction as seriously as it used to be taken illustrates this quite clearly.

Objectivism and Relativism as Metaphor[k]

Another important theme of this paper flows from the issues I have been talking about. It is the idea that all of science is ultimately based on metaphor—that all of human knowledge and experience is ultimately suspended in metaphor. Although I cannot argue the thesis here, I shall at least try to illustrate it in terms of the metaphors which have already played a role in this paper. For example, I have used freely a metaphor of separation to express tensions between the normative and the empirical.

The normative/empirical tensions can be alternatively expressed as the separation between total objectivism and total relativism. By total relativism is meant the radical thesis that everything is a social construct—that nothing is a reflection of anything that we might sensibly call reality. By total objectivism is meant the radical thesis that there is some absolute frame of reference against which we can measure cognition—some way of coordinatizing knowledge against an absolute reference frame. This is the kind of separation or distinction that I have in mind.

Now, to say that our knowledge is ultimately suspended in metaphor is not necessarily to argue for relativism. It is to argue for a shift in perspective on the problem and for approaching it as the problem of identifying good metaphors while rejecting bad ones. The tendency in contemporary thought towards relativism[l] creates an important challenge for applied areas like risk analysis, as well as for theoretical areas like philosophy of science, to seek ways of avoiding the slide to total relativism. My own philosophical project is an attempt to contribute to that effort—to articulate a philosophical theory of objectivity that is nontrivial but which is not reducible to objectivism in the sense I have been defining it.

---

j. It is these sorts of claims that originally motivated my interest in rationality and risk.

k. I follow Bernstein here in talking about objectivism and relativism instead of objectivism/subjectivism or relativism/absolutism. His reasons for believing that total absolutism and total subjectivism are not taken seriously by anyone are to be found in his book (Bernstein, 1985, pp. 12-13).

l. It is the general tendency in our culture to move towards an "anything goes," "true for me is false for you" kind of relativism that has worried many scientists and which fans the flames of the "two cultures" approach to regulation and national policy-making.

THE PRESENTATION OF METAPHOR

A Formal Presentation of "Separation"

I have claimed that we are suspended in irreducible ways in metaphor. To support this claim, I shall introduce another level of metaphor into the discussion by invoking some simple models from quantum physics.

Systems in quantum mechanics are represented by vector spaces and states are represented by vectors. For the purpose of this paper, we can think of these as ordinary geometric vector spaces. Spin states are represented in a two-dimensional space. The dynamical state is represented by an infinite-dimensional space. The total system is represented by the tensor product of the spin space and the dynamical space.

Now spin states and dynamical states have a very important relationship to each other called independence. Independence is represented in the theory by keeping the representation of these states totally separate. When vectors are combined into linear combinations of vectors from the spin space and the dynamical space, the resultant vectors are simply inadmissible as states; i.e., they do not represent possible states.[m]

The implicit metaphor here is the metaphor of radical separation: spin states and dynamical states are somehow logically independent of each other in a very strong sense.

Separation and Combination

The quantum mechanical metaphor of separation is much more complete or radical than the classical metaphor of separation. Classically we would present the concept of separation by a spatial image in which two halves of a plane are separated by a line. In quantum mechanics the device of vector space models allows us to present an image of multidimensional spaces separated by a point, namely the empty space or geometric origin of the spaces. Yet at the same time the quantum mechanical image enables us to present combination in a more robust or holistic way. For example, when two spin 1/2 particles combine, the product is represented by a four-dimensional tensor product in which new states (e.g., singlet spin state) emerge that could not be predicted or even described classically.

The Presentation of Models

So it is clear that the metaphor of separation itself can be presented by alternative higher-order metaphors. Even fundamental concepts like separation and identity can be presented metaphorically in completely different idioms. These idioms can be obtained from many sources, including the most current and esoteric scientific theories. But these idioms are robust and can often be used in other contexts to enhance understanding.

I wish to use the quantum mechanical presentation of separation as a higher order metaphor in the context of risk analysis. As I have argued above, there is a tendency in western culture to think of normative issues and empirical issues as being totally separated and independent. I want to use the quantum mechanical metaphor to interpret the metaphor of separation as it arises in regulation, standards-setting and policy-making.

---

[m]. Spin-orbital coupling is an exception.

Separating and Combining the Normative and the Empirical

In my view it would be an error to think that the normative and the empirical are related like spin states and dynamical states in quantum mechanics. This would be the wrong version of the metaphor. The correct version of the metaphor would involve thinking of the normative and the empirical in terms of the metaphor of quantum mechanical combination. While it is true that in some contexts pairs of electrons exist in separate states, in other contexts they are best seen as dynamically interacting and losing their separate realities. The normative and the empirical interact, metaphorically speaking, in much the same way.

The second version of the metaphor is the kind of thing that Mr. Yosie was using at the plenum address to the Society for Risk Analysis here in Houston when he talked about the regulatory and the science sides of national policy-making being embedded in each other. This idea raises a host of new metaphors that I think are productive and useful. The whole notion of an embedding elicits images from the theory of homomorphic mappings in abstract algebra.[n] So there are many new metaphors. The metaphor of thermodynamics that Slovic and his colleagues were talking about at these meetings is another example. These are new productive metaphors which I think are important.

Perspective

An idea suggested by the quantum mechanical metaphor is the representation of cost benefit concepts, not as a linear one-dimensional continuum of properties, but, rather, as something that is visualized as a vector space. This representation suggests that benefits are really just one coordinatization of the space and what we call costs are really just another coordinatization of the space. This is the metaphor of perspective.[o] Different people have different perspectives on the same thing and what is one person's cost will be some other person's benefit. Perhaps quantum probabilities might even turn out to be applicable. That is something I have not worked on but the idea is an intriguing one.

Context

A metaphorical dimension already discussed above is the notion of context. Quantum mechanics is contextual rather than deterministic and absolute. The contextuality of quantum mechanics has metaphorical value in the context of regulatory science.

CONCLUSIONS

I have argued that several distinctions have converged in the history of philosophy. In particular, that the concepts of the subjective, the normative, the emotive and the irrational have come to be seen as virtually identical, as have the objective, the empirical, the cognitive and the rational. Risk analysis has sometimes been a victim of this tendency. Both philosophy and risk analysis have recently begun to question these convergences. The result is a new vision of science and its relationship to policy which acknowledges that science is both normative and empirical, both objective and subjective.

It was further argued that science is an irreducibly metaphorical activity. To illustrate this claim, it was pointed out that even the most fundamental categories of scientific

---

n. Transformation theory is the formal or theoretical language of invariance, an important aspect of the concept of objectivity. Leibniz was the first to make this observation.

o. Perspective is also important for a theory of invariance and its relation to objectivity. It was a central part of Kant's theory of the transcendental object.

thought can be presented in radically different idioms. It follows that if there is some absolute framework of validity in which to express fundamental scientific concepts, we have not yet found it. It is therefore reasonable to entertain the philosophical hypothesis that no such foundation exists.

The alternative presentation of the metaphor of separation is one which enables us to see in what sense the normative and the empirical are not the opposite of each other and to understand in what sense science can be both normative and empirical. This shifts the emphasis in science and policy-making from the traditional emphasis on objectivity to a new emphasis on the dynamics of theory choice and social choice. The problem becomes a problem for methodology.

Finally, it is important to remember that there are good metaphors and there are bad metaphors. I have argued that science is irreducibly metaphorical, but I have not argued that all metaphors are good. There is a bad metaphor that I found myself falling into the trap of using. It is tempting to say that the notion of metaphor is the new way of grounding our understanding of science. But that is a bad metaphor. Grounding something is the foundationalist metaphor of positivism, the one that is being rejected. What I really should say is: let's suspend science as a cultural activity and accept its metaphorical (and normative) dimensions.

QUESTIONS

1. Question:

The first question raised the point that there can be a kind of intrinsic rationality to the risk analysis side of risk assessment. It drew attention to the question of valuing human life in the risk management process, drew analogies between model building in toxicology and in physics, and finally drew attention to the work of Chomsky, in particular, Chomsky's move away from looking for rationality structures in the semantics of language towards an emphasis on syntactic structures.

Answer:

I am very sympathetic to the whole thrust of what you just said. It was not my intention to try to answer my own questions here. However, to tell you what my biases are, I think that an interesting way to avoid the slippery slope from objectivism to relativism is to take very seriously a notion that goes back to Leibniz—the notion of invariance under radical translations among perspectives. I have done some work trying to explicate the idea in terms of category theory—a general theory of transformations in abstract algebra—and I believe that philosophically that is a productive strategy towards doing exactly the kind of thing you are saying. I think what is important or interesting about Chomsky's work in this context is that he is saying similar things about the universality of structure.

2. Question:

You can apply any method to many systems. But I would like to know how you judge the validity of the analogy.

Well, I think I have to give a somewhat glib answer to that. Of course, it is true that we have to be able to identify the good metaphors. I am not claiming that definitive solutions are being offered to the philosophical problems. What is being argued is that there is a productive shift that can and should be made from talking about objectivity to talking about methodology. The whole notion, as I intend it, is that the answer lies in invariance. This is a methodological response to the question. It is not an ontological response to the question. My claim is that it is a productive shift in turning to methodology when trying to answer it.

REFERENCES

Bernstein, Richard, 1985, *Beyond Objectivism and Relativism: Science, Hermeneutics and Praxis*, University of Pennsylvania Press, Philadelphia.

Collingridge, David, and Reeve, Colin, 1986, *Science Speaks to Power*, Saint Martin's Press, New York.

Feyerabend, Paul, 1975, *Against Method: Outline of an Anarchistic Theory of Knowledge*, New Left Review, London.

Popper, Karl R., 1972, *Objective Knowledge: An Evolutionary Approach*, Oxford University Press, Oxford.

The Use of Pharmacokinetic Models in the Determination of Risks for Regulatory Purposes

Frederic Y. Bois
University of California
San Francisco, CA

Lauren Zeise
California Department of Health Services
Berkeley, CA

Thomas N. Tozer
University of California
San Francisco, CA

ABSTRACT

Pharmacokinetic analyses have recently been incorporated in risk assessments, with resultant risks sometimes lower and associated "allowable" exposures much higher than would have been otherwise calculated. Preliminary investigations of the predictions of such models as they have been used for regulatory purposes are presented here. Precision is studied by treating parameters as random variables and performing Monte Carlo simulations and determining the range of estimates once parameter uncertainties are considered. Physiologically based pharmacokinetic models cannot be validated at low doses because low dose data in humans and test species are generally not available. Accuracy is discussed in terms of the assumptions and biases pertaining to low dose and interspecies extrapolation.

KEYWORDS: Pharmacokinetic models, multistage models, cancer, tetrachloroethylene

INTRODUCTION

The coupling of physiologically based pharmacokinetic models (PBPK) to multistage models of carcinogenesis has been recently used as a means to take dose, route and species differences into account in risk assessment from exposure to xenobiotics. Physiologic models describe the distribution and the biotransformation of chemicals and potentially provide estimates of the target tissue exposure (e.g., rate of presentation of the active metabolites to the affected tissue). These models have been proposed to scale exposures among different species, different experimental conditions or different doses (Ramsey and Andersen, 1984).

The risk of cancer in humans is determined by fitting a dose-response model for carcinogenesis, usually the Crump multistage polynomial model (Anderson *et al.*, 1983; Crump, 1984), to the target tissue exposure and experimental cancer data. A relationship for excess cancer risk as a function of target tissue exposure is then obtained. The dose-response relationship is assumed to be the same for humans when exposure is expressed in terms of rate of presentation to the target tissue, also calculated by a physiological model. Typically, the uncertainties associated with model misspecification or parameter variability are not explicitly taken into account, and the model output is assumed to provide exact estimates of risk. Quantitative knowledge of these uncertainties should aid the user of this approach to risk assessment in determining the confidence to be placed on the results. In this paper, a statistical approach is used to evaluate the precision of a risk assessment based on the coupling of PBPK and multistage models. For illustration purposes, risk assessment is performed on tetrachloroethylene, a widely used volatile solvent which has been classified by the U.S. Environmental Protection Agency (1985) as a probable human carcinogen and by the International Agency for Research on Cancer (1979) as a known animal carcinogen. The cancer bioassay data used were from the National Toxicology Program (1986) experiments in male F344/N rats.

Model bias is discussed in terms of more general assumptions pertaining to low dose and interspecies extrapolations.

METHODS

The Models

Physiological Model. The model used (Fig. 1) is the same as that of Ramsey and Andersen (1984), Andersen *et al.* (1987), and Ward *et al.* (1987). The tissues are grouped as follows: slowly perfused fat, muscles and skin, vessel-rich tissues, such as brain and kidneys, and metabolizing tissues. The model is mathematically described by a set of mass-balance differential equations. For tetrachloroethylene, metabolism is assumed to occur only in the liver and to be saturable, according to the Michaelis-Menten relationship. The formation of an epoxide by the cytochrome P450 system is a common step to the known major metabolic pathways, and is assumed to be the saturable mechanism. The epoxide is considered to be the effective carcinogen (U.S. EPA, 1985). As such, its daily rate of formation at steady state or, equivalently, the rate of formation of metabolites, is taken as the measure of the exposure to tetrachloroethylene.

Values for most of the parameters are independent of the chemical in question and are found in the literature (e.g., organ volumes, blood flows, and pulmonary ventilation rates). Another set of parameters depends on the compound studied and is estimated from either in vitro (partition coefficients) or in vivo (V_{max}, K_m) experiments. The latter case generally involves estimation of the parameter values by fitting the full model (Fig. 1).

Multistage Model. The so-called linearized multistage, or Crump polynomial, model (Anderson, *et al.*, 1983; Crump, 1984) is used to relate the lifetime probability or "risk" of cancer, $R(E)$, to the effective exposure, E, to the carcinogen, namely the rate of formation of the epoxide or its surrogate measure, the rate of metabolism:

$$R(E) = 1 - \exp\left[-\left(a_0 + a_1 \cdot E + ... + a_{n-1} \cdot E^{n-1} + a_n \cdot E^n\right)\right], \qquad (1)$$

with $a_i > 0$ for all i.

The coefficients a_0 through a_n are estimated by the maximum-likelihood package "Mstage" (Crouch, 1985). Using the approach taken by the EPA (1986), the upper 95% confidence limit of the coefficient a_1 is defined as the "carcinogenic potency" of the

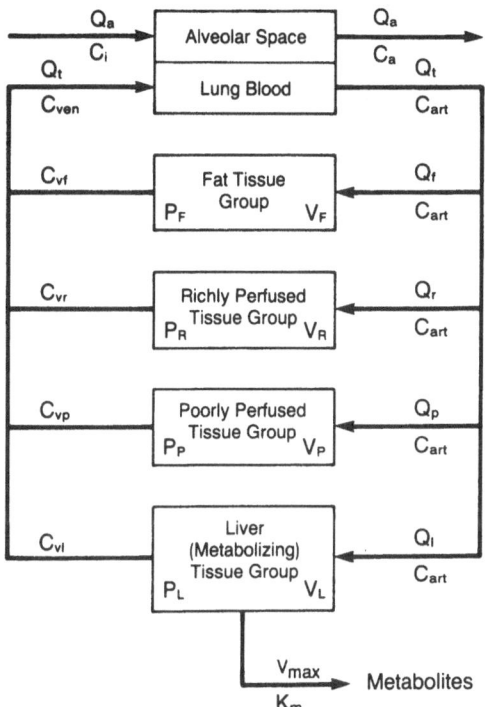

Fig. 1. Schematic Representation of the Physiological Model Used to Simulate the Distribution and Metabolism of Tetrachloroethylene. The symbols used are: Q_a, alveolar ventilation rate; Q, blood flow rates to the tissue; Q_t, total blood flow; V, volumes; P, tissue/air partition coefficient; V_{max}, maximum rate of metabolism; K_m, Michaelis constant for the kinetic model of metabolism.

chemical studied. "Extra risk," R_i, from exposure E is taken as the difference between the lifetime risk $R(E)$ and the background risk in the absence of exposure $R(0)$, relative to the fraction of the population expected to be cancer-free in the absence of exposure to the chemical (Abbott, 1925; CDHS, 1985), and is given by:

$$R_i = [R(E) - R(0)] / [1 - R(0)]. \tag{2}$$

At low doses,

$$R_i = a_1 \cdot E.$$

Probability Distributions of the Parameters of the PBPK Model

Parameters Except V_{max} and K_m. All the parameters are subject to uncertainty. The mean values for rat and human parameters are provided by the EPA (1986). These values are indicated in Table 1. Several of the parameters were scaled to body weight as indicated in the table (EPA, 1986). In the simulations the parameters were all assumed to vary according to given probability distribution functions. The cumulative distribution assumed for the value, P, of any parameter is given by:

$$F(x) = (1 - \cos(\pi \cdot x)) / 2 , \tag{3}$$

Table 1. Scaling Coefficients and Parameter Values Used for Tetrachloroethylene in Rats

	Parameter	Rat	Human
Body weight	Bw	(1)	70
Alveolar ventilation rate	$Q_a/\text{Bw}^{0.74}$	0.180	0.147
Total blood flow	$Q_t/\text{Bw}^{0.74}$	0.226	0.186
Blood flows			
metabolizing tissue group	Q_1/Q_t	0.37	0.37
fat tissue group	Q_f/Q_t	0.09	0.09
richly perfused tissue group	Q_r/Q_t	0.42	0.42
poorly perfused tissue group	Q_p/Q_t	0.12	0.12
Volumes			
metabolizing tissue group	V_1/Bw	0.04	0.02
fat tissue group	V_f/Bw	0.09	0.20
richly perfused tissue group	V_r/Bw	0.04	0.05
poorly perfused tissue group	V_p/Bw	0.73	0.62
Blood/air partition coefficient	P_a	18.9	10.3
Tissue/air partition coefficients			
metabolizing tissue group	P_1	70.3	38.3
fat tissue group	P_f	2060	1123
richly perfused tissue group	P_r	70.3	38.3
poorly perfused tissue group	P_p	20	10.9
Maximum rate of metabolism	$V_{max}/\text{Bw}^{0.7}$	0.01235	0.003
Affinity constant	K_m	2.937	0.3
Intestinal absorption constant	k_i	0.01	0.01

(1) Variable for different simulated experiments. Units: the flows (Q) are in L/min, the volumes (V) in L, V_{max} in mg/min, and K_m in mg/L. Source: U.S. EPA, 1986.

where

$$x = (P - \text{lower-bound}) / (\text{lower-bound} - \text{upper-bound}).$$

This function closely approximates the normal distribution function and has the advantage of being bounded.

The body weights of the male rats in the carcinogenicity study of the National Toxicology Program (NTP) varied from 300 to 500 grams during the second year of the experiment (NTP, 1986). In deriving the distribution of body weight (Eq. 3), a lower-bound of 300 grams and an upper-bound of 500 grams were assumed. The other parameters, before scaling, were assumed to vary within 10% of their mean values. The distribution function for V_{max} and K_m, metabolic parameters, is a special case as these parameters are jointly fitted to in vivo data.

V_{max} **and** K_m. V_{max} and K_m are derived by fitting the physiologic model (Fig. 1) to in vivo data on rate of metabolite formation in response to tetrachloroethylene exposure. For the rat, their values were estimated from the inhalation experiments of Pegg *et al.* (1979) and for humans from the occupational survey of Ikeda *et al.* (1972).

The fitting process involved maximum-likelihood techniques to derive the distribution functions of V_{max} and K_m (Edwards, 1972). The fitting procedure and the structure of the model are expected to result in a strong correlation between V_{max} and K_m. Because estimates of V_{max} and K_m covary, their joint distribution function must be determined. This distribution function is assumed to be proportional to the likelihood of particular values of V_{max} and K_m, given the data observed.

To derive the likelihood function, the difference between the true value and the experimental observation of the rate of metabolite formation is assumed to follow a t-distribution. To determine possible values for metabolite formation under the exposure conditions of the studies, the physiologic model was run with trial pairs of V_{max} and K_m, all other parameters being fixed to their mean values. The likelihood was obtained for a large set of regularly spaced pairs of values (V_{max}, K_m). The joint distribution function was constructed from the likelihood function. The probability of selecting a given couple (V_{max}, K_m) in the Monte Carlo simulation was set to be proportional to the likelihood associated with that couple.

Monte Carlo Simulations

For the Monte Carlo simulations, the parameter values of the physiologic model were picked at random, according to their distributions. The uniform random number generator, provided with the Silicon Valley Software® Fortran version 2.5, was used. A linear interpolation between the tabulated points of the distribution functions was made for both the one- and the two-dimensional cases. Monte Carlo runs were stopped when the running mean of the recorded results stabilized. This typically involved 500 simulations.

Determination of Uncertainty in the Risk Estimates

Monte Carlo simulations were used to compute the distribution of the average daily rate of formation of total metabolites in rats during the NTP inhalation bioassay (NTP, 1986). The formation of metabolites was simulated for the two exposures assayed (200 ppm and 400 ppm in air, 6 h/day, 5 days/week). Each simulation represented five days of exposure; following the EPA (1986), the rate of metabolism on the last day was taken as an estimate of the steady-state value. To correct for the body surface area, the rate of metabolite formation is divided by the body weight to the two-thirds power. These rates were used as the exposure in fitting the multistage model (Eq. 1) to the cancer response data (leukemia incidence). In this way the carcinogenic potency was derived in terms of metabolite exposure. For humans the distribution of the daily rate of formation of total metabolites after 7 days of exposure, corrected for surface area, for a continuous exposure of 1 $\mu g/m^3$, was obtained by Monte Carlo simulations (steady state assumed in all simulations). Following Eq. 2, the extra risk for humans is equal to the product of the rate of formation of total metabolites E by the carcinogenic potency a_1: The probability distribution of the risk is then obtained through the convolution of the distributions of the rate of metabolism and carcinogenic potency. This convolution was achieved by sampling for potency and rate of metabolism in humans in accordance with their parent distributions and forming their product.

RESULTS

Figures 2 and 3 present the results of the Monte Carlo simulations for the rate of metabolite formation in the rat for exposures to 200 and 400 ppm of tetrachloroethylene. At 200 ppm the median rate of metabolite formation is 10.6 mg/d per $kg^{2/3}$ (with 99% confidence limits of 8.2 and 12.0). On doubling the exposure to 400 ppm, the median rate of formation increases to a value of 13.0 mg/d per $kg^{2/3}$ (with 99% confidence limits of 8.6 and 18.5). The medians do not increase proportionately with dose due to the saturation of metabolism.

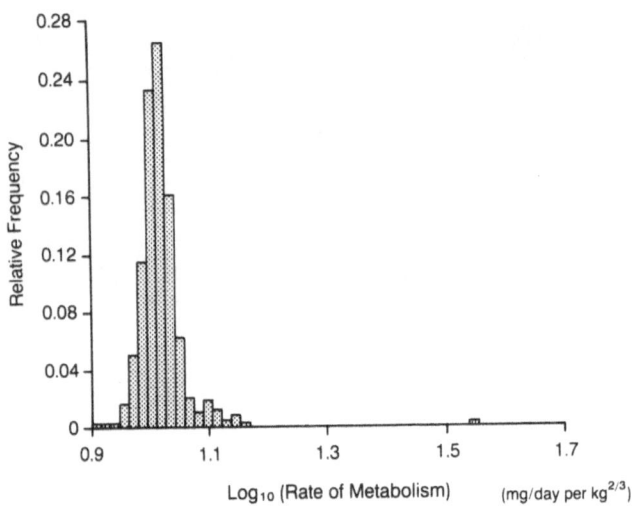

Fig. 2. Monte Carlo Generated Conditional Distribution (Over the Model Parameter Values) of the Logarithm (Base 10) of the Rate of Formation (mg/day per kg$^{2/3}$) of Metabolites in Rats. Exposure conditions of the NTP (1985) inhalation bioassay of tetrachloroethylene (200 ppm) were simulated. To obtain the rate in units of mg/d, multiply these results by the body weight to the 2/3 power.

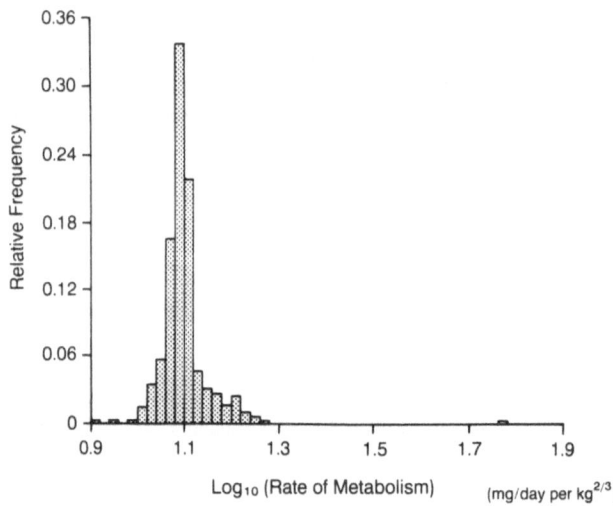

Fig. 3. Monte Carlo Generated Conditional Distribution (Over the Model Parameter Values) of the Logarithm (Base 10) of the Rate of Formation (mg/day per kg$^{2/3}$) of Metabolite in Rats. The conditions of exposure in the NTP (1985) inhalation bioassay of tetrachloroethylene (400 ppm) were simulated. To obtain the rate in units of mg/d, multiply these results by the body weight to the 2/3 power.

The frequency distribution of the carcinogenic potency, for rats, is presented in Fig. 4. The multistage model has been fitted to the leukemia data obtained in the rat, using the previously determined rate of formation of metabolites as a measure of the effective exposure. The median potency value is 0.076 (mg/d per $kg^{2/3})^{-1}$ with 99% confidence limits of 0.055 and 0.09. Figure 5 shows the distribution obtained for the rate of formation of total metabolites in humans. For a continuous exposure to 1 ng/L of tetrachloroethylene, the rate of metabolism at steady state (after surface area correction) has a median of 1.6×10^{-4} mg/d per $kg^{2/3}$ and 99% confidence limits of 6.2×10^{-5} and 2.2×10^{-4} mg/d per $kg^{2/3}$. Due to the nonlinear dependence of the computed rate with respect to V_{max} and K_m, the variability of the rate of formation of metabolites is much lower than might be expected from the variability of these parameters. Figure 6 presents the distribution of the theoretical cancer risk estimates for humans. The risk estimates have a median of 12 per million with 99% confidence limits of 3.0 and 19 per million. These results take into account variation in both the rate of formation of the total metabolites and the carcinogenic potency.

DISCUSSION

The analyses presented here address the degree to which model precision is influenced by uncertainty in parameter values. It is assumed that the structure underlying the coupled model is sound and unbiased. Whether particular assumptions, such as Michaelis-Menten kinetics, the epoxide as the sole active metabolite, and the multistage dose-response model are correct or not, was not investigated.

Precision

The 99% confidence limits for the theoretical cancer risk estimate were found to differ by a factor of approximately 6 for humans continuously exposed to 1 ng/L of tetrachloroethylene. This uncertainty may be understated to a considerable degree, due to the simplifications that were made in specifying the distribution functions of the physiologic model parameters. The relationships used to scale parameters on the basis of body weight were assumed to be known with certainty. This assumption leads to an underestimate of the variance associated with the parameters of the model. Yet, it appears that uncertainty in these parameters, with the notable exception of V_{max} and K_m, have a limited effect on the results. A joint distribution function was obtained for V_{max} and K_m. It was assumed that these parameters would not covary significantly with the other model parameters, except for V_{max} with body weight (via the scaling procedure). Again, this assumption may have led to an understatement of the uncertainty. Most importantly, only limited data sets with few dose groups or individuals were used for fitting V_{max}, K_m, and the parameters of the multistage model. This may be the largest source of underestimation of parameter uncertainty. For example, V_{max} and K_m values for humans were derived from data in a racially homogeneous population of healthy adult male workers (Ikeda, 1972). For the animal experiments, only two doses were assessed, with three animals per dose group (Pegg et al., 1979).

Bias

Because low-dose data on either metabolism in humans and animals or data on low-dose cancer risk are not available, the models cannot be fully validated and the bias directly assessed. Bias in the use of the multistage polynomial equation for carcinogenesis models has been discussed by a number of authors (for a review, see Zeise et al., 1987). It is unlikely that the mechanism of carcinogenesis for tetrachloroethylene will be elucidated soon. The issue here is whether the PBPK model produces better (i.e., more accurate) estimates of risks. EPA (1986) has pointed out that PBPK models fail to take into account pharmacodynamic differences across species. A similar problem faces the naive user of multistage models.

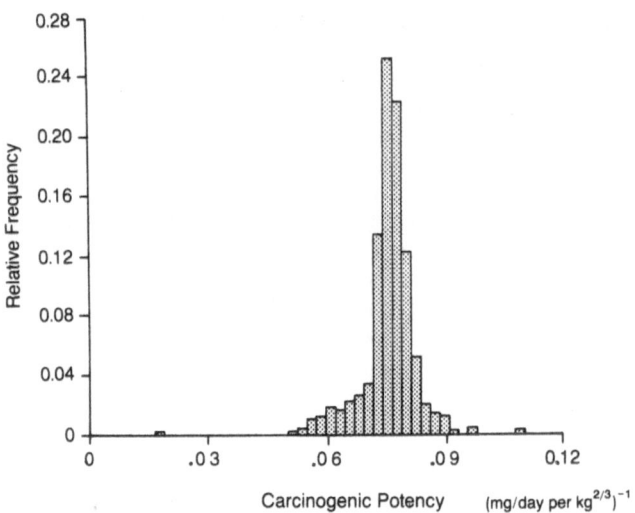

Fig. 4. Monte Carlo Generated Distribution of the Carcinogenic Potency of Tetrachloroethylene in Rats. NTP (1985) leukemia data were used to fit the multistage model. The variability in potency incorporates the expected variability in the rate of formation of metabolites (cf. Figs. 4 and 5).

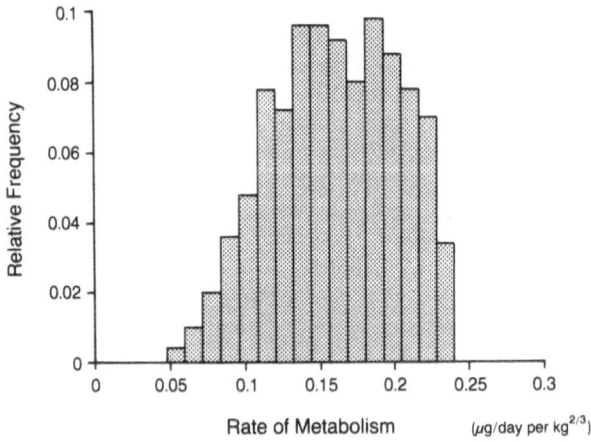

Fig. 5. Monte Carlo Generated Conditional Distribution (Over the Model Parameter Values) of the Rate of Formation (μg/d per $kg^{2/3}$) of Metabolites in Humans. A continuous inhalation exposure to 1 ng/L of tetrachloroethylene was simulated.

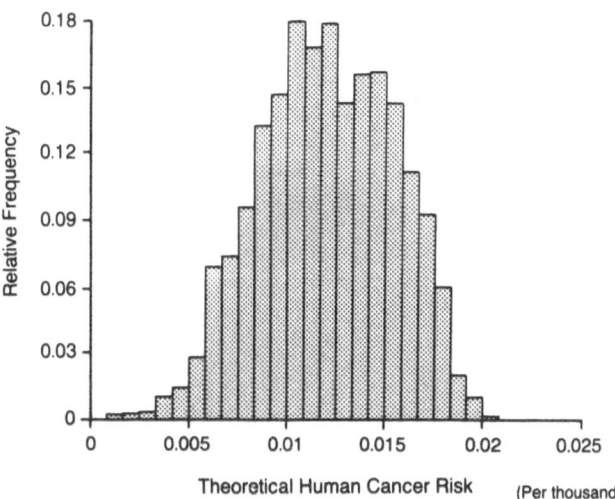

Fig. 6. Distribution of the Theoretical Human Cancer Risk Estimates Associated with a Continuous Inhalation Exposure to 1 ng/L of Tetrachloroethylene.

A host of simplifications are made and several potentially nonlinear phenomena were not considered. For example, in the physiologic model presented, the definition of the metabolizing tissue group is broad and results pertaining to this group are difficult to extrapolate across species. It is assumed that metabolism occurs in the liver, that the active metabolite is too unstable to be transported far, and yet leukemia is observed. Furthermore, detoxification reactions are not taken into account; the possibility of autoinduction of metabolism is not examined; important cellular (and pharmacokinetic) processes, such as DNA repair or DNA methylation, are ignored; in vitro data on partition coefficients for homogenized tissues are assumed to be representative of the in vivo case; and every relevant enzymatic reaction is assumed to follow Michaelis-Menten kinetics. Finally, general procedures for extrapolating metabolic parameters from one species to another have not been validated. Little is known about the importance of these simplifications or the bias that they introduce. The direction (under- or overestimation) of the bias is frequently not certain. All these simplifications are due to our ignorance of the processes involved, our inability to describe them adequately, and a lack of relevant data.

CONCLUSION

The output of a coupled pharmacokinetic and multistage polynomial model of carcinogenesis for tetrachloroethylene provides a theoretical cancer risk estimate of 12 per million, with 99% confidence limits of 3 and 19 per million. In deriving this range it is assumed that the basic structure of the coupled model is correct. The degree to which the models themselves introduce bias in estimates of risk at low dose has not been addressed here. Regarding the multistage model of carcinogenesis, alternative mechanistic models of carcinogenesis are currently receiving attention, yet considerable time may elapse before scientific understanding of the processes involved allows for the confident use of any particular one. The use of pharmacokinetics is another area in which considerable improvement over the current approach might be achieved. As long as one can be certain of the activity of particular metabolites, one can design experiments and models to determine levels with relative precision. Because the experimental data are still inadequate, these results can not be considered entirely trustworthy for tetrachloroethylene. Our study shows that the use of pharmacokinetic models in risk assessment may prove promising, once the models are adequately validated. The coupling of pharmacokinetic models to multistage models can provide relatively precise theoretical risk estimates.

ACKNOWLEDGEMENTS

This work was funded by the California Department of Health Services, IMA 85-87088 and, in part, by grant ES 04705 from the National Institute of Environmental Health Sciences.

BIBLIOGRAPHY

Abbott, W. S., 1925, "A Method for Computing the Effectiveness of an Insecticide," *J. Econ. Entomol.* **18**:265.

Andersen, M. E., Clewell, H. J., III, Gargas, M. L., Smith, F. A., and Reitz, R. H., 1987, "Physiologically Based Pharmacokinetics and the Risk Assessment for Methylene Chloride," *Toxicol. Appl. Pharmacol.* **87**:185.

Anderson, E. L., and the Carcinogen Assessment Group of the U.S. Environmental Protection Agency, 1983, "Quantitative Approaches in Use to Assess Cancer Risk," *Risk Analysis* **3**:277.

California Department of Health Services (CDHS), 1985, "Guidelines for Chemical Carcinogens Risk Assessments and Their Scientific Rationale," State of California Health and Welfare Agency.

Crouch, E. A. C., 1985, Manual Version 1.1 for MSTAGE Version 1.1, Personal Communication, Edmund Crouch, 44 Radcliffe Road, Somerville MA 02145.

Crump, K. S., 1984, "An Improved Procedure for Low Dose Carcinogenic Risk Assessment from Animal Data," *J. Environ. Pathol. Toxicol.* **5**:339.

Edwards, A. W. F., 1972, *Likelihood*, Cambridge University Press, Cambridge.

International Agency for Research on Cancer (IARC), 1979, *Some Halogenated Hydrocarbons*, IARC Monograph, Vol. 20, International Agency for Research on Cancer, Lyon.

Ikeda, M., Ohtsuji, H., Imamura, T., and Komoike, Y., 1972, "Urinary Excretion of Total Trichloro-Compounds, Trichloroethanol, and Trichloroacetic Acid as a Measure of Exposure to Trichloroethylene and Tetrachloroethylene," *Brit. J. Industr. Med.* **29**:328.

National Toxicology Program (NTP), 1986, NTP Technical Report on the Toxicology and Carcinogenesis Studies of Tetrachloroethylene (Perchloroethylene) (CAS No. 127-18-4) in F344/N Rats and B6C3F1 Mice (Inhalation Studies), Research Triangle Park, North Carolina, NTP TR 311, NIH Publication No. 86-2567.

Pegg, D. G., Zempel, J. A., Braun, W. H., and Watanabe, P.G., 1979, "Disposition of Tetrachloro(14C)ethylene Following Oral and Inhalation Exposure in Rats," *Toxicol. Appl. Pharmacol.* **51**:465.

Ramsey, J.C., and Andersen, M. E., 1984, "A Physiologically Based Description of the Inhalation Pharmacokinetics of Styrene in Rats and Humans," *Toxicol. Appl. Pharmacol.* **73**:159.

United States Environmental Protection Agency (EPA), 1985, Health Assessment Document for Tetrachloroethylene (Perchloroethylene): Final Report, U.S. Environmental Protection Agency, Office of Health and Environmental Agency, Washington, DC, EPA/600/8-82/005F, PB85-249704.

United States Environmental Protection Agency (EPA), 1986, Addendum to the Health Assessment Document for Tetrachloroethylene (Perchloroethylene), Updated Carcinogenicity Assessment for Tetrachloroethylene (Perchloroethylene, PERC, PCE): U.S. Environmental Protection Agency, Office of Health and Environmental Agency, Washington, DC, EPA/600/8-82/005FA.

Ward, R. C., Travis, C. C., Hetrick, D. M., Andersen, M. E., and Gargas, M. L., 1987, "Pharmacokinetics of Tetrachloroethylene," *Toxicol. Appl. Pharmacol.* (submitted for publication).

Zeise, L., Wilson, R., Crouch, E. A. C., 1987, "Dose-Response Relationships for Carcinogens: A Review," *Environ. Health Perspect.* **73**:259.

Ethylene Oxide Residues on Sterilized Medical Devices

Stephen L. Brown and Joseph V. Rodricks
ENVIRON Corporation
Washington, DC

ABSTRACT

Ethylene oxide is used to sterilize a variety of medical devices, and patients who use these devices may be exposed to residues of the ethylene oxide left on them. Ethylene oxide is associated with cancer in laboratory animals and can cause a variety of other toxic effects at sufficiently high doses, as can its reaction products, ethylene glycol and ethylene chlorohydrin. Patient exposures to these chemicals can be estimated by considering the numbers of devices sold in the U.S., the numbers of procedures performed that involve devices, the duration and nature of contact with the patient, and other factors. This paper describes a methodology for quantitative assessment of the risks from exposure to ethylene oxide residues on medical devices.

KEYWORDS: Ethylene oxide, ethylene glycol, ethylene chlorohydrin, medical devices, residues

INTRODUCTION

Ethylene oxide (EtO) is the sterilant of choice for many medical devices, some of which cannot be effectively sterilized with radiation, steam, or any other alternative. Many disposable and reusable medical devices, ranging from the common adhesive bandage to pacemaker systems and dialyzers (see Table 1), are sterilized with EtO by the manufacturer or custom sterilization companies prior to shipment for use.

After sterilization, some of the EtO may remain adsorbed on the surface of the device or dissolved into its materials of construction. These residues dissipate with time at a rate depending on the materials, the degree of aeration, and other factors. During sterilization, some of the EtO may react with water vapor to form ethylene glycol (EG) or with chlorinated compounds to form ethylene chlorohydrin (ECH), either of which may also remain on the device.

Each of these three compounds has been shown to cause chronic toxicity or reproductive effects in laboratory animals at relatively high doses. In addition, EtO has tested positive in animal cancer bioassays and has been associated with human cancers in epidemiological studies of uncertain significance. When the potencies of the three compounds are compared among the various endpoints, it appears that protection against EtO residues will generally protect against the other two compounds.

Table 1. Example Medical Devices That May Be Sterilized with Ethylene Oxide

Bags, blood administration	Pacemakers and accessories
Bags, IV administration	Prostheses, breast
Bandages, adhesive	Reservoirs, cardiotomy
Catheters, intravascular	Sponges, gauze and cotton
Catheters, urological	Syringes, hypodermic
Dialyzers and accessories	Syringes, insulin
Filters, blood	Tips, surgical suction
Gloves, surgical	Tubes, infant feeding
Lenses, intraocular	Tubing, blood or IV
Needles, hypodermic	Valves, heart, prosthetic

Given these facts, it is reasonable to ask whether the residues of EtO on sterilized medical devices pose unacceptable risks to patients using those devices. The Health Industry Manufacturers Association contracted with ENVIRON Corporation to investigate this question. The scope of our study included only risks to patients, not to medical personnel or employees of manufacturers or sterilizing companies, and only the risks associated with medical devices delivered sterile to hospitals, physicians, or the patient, not the risks of devices that are sterilized or resterilized in the hospital or clinic.

TOXICITY ASSESSMENT

The carcinogenicity of EtO is described in terms of its potency or Unit Cancer Risk (UCR), the slope of the dose-response relationship projected to hold at low doses. All of the noncancer effects are described by the No Observed Effect Level (NOEL), the chronic daily dose below which no effects of consequence have been seen in the most sensitive species studied, or the Lowest Observed Effect Level (LOEL), also from the most sensitive species, if no NOEL has been clearly identified.

Carcinogenicity

The cancer potency of EtO is based on studies in rats and mice that showed excesses of leukemias, brain tumors, and other cancers after exposure to air concentrations of 100 ppm or somewhat lower (Snellings, 1984; Lynch et al., 1984; NTP, 1986). These data were analyzed by the method of Gaylor and Kodell (1980), which produces an upper confidence limit on the low-dose response to a carcinogen, assuming equal potency in humans and animals on the basis of daily dose per unit body weight. The UCR of EtO predicted by this method is 4.8×10^{-2} (mg/kg/day)$^{-1}$.

Chronic and Reproductive Toxicity

EtO can cause neurologic and hematologic effects as well as damage to lungs, liver, kidney, and testes (USEPA, 1985; Sprinz et al., 1982; Snellings et al., 1984a, 1984b). Snellings et al. (1984b) showed that no effects were observed in rats exposed to 10 ppm for 6 hours per day, 5 days per week. The corresponding NOEL in humans is calculated to be 2 mg/kg/day. The reproductive NOEL for EtO is estimated to be 9 mg/kg/day, based principally on a study by Snellings et al. (1982), in which no adverse reproductive outcomes were seen in rats exposed to 33 ppm for 6 hours daily.

EXPOSURE ASSESSMENT

Methods for estimating exposure to EtO residues on sterilized medical devices have not been well developed. We know of only one other attempt to do so (RIVM, n.d.). Novel methods have been developed especially for this project.

Two distinct ways of estimating exposures have been employed. One, the "composite individual" method, is designed to characterize exposure to the entire U.S. population for the purpose of assessing the overall carcinogenic risks of EtO. The other, the "specific procedures" method, is designed to estimate the risk of either cancer or other effects from exposure to medical devices in the course of specific medical procedures.

Composite Individual

A "composite individual" is a hypothetical person who uses exactly his or her share of all the medical devices sold in the United States every year. For example, if 6 billion adhesive bandages are sold annually in the U.S. and the U.S. population is 240 million, then the composite individual uses 6,000/240 or 25 bandages per year. The corresponding numbers for infrequently used devices such as intraocular lenses will be considerably lower; even when we multiply the annual use rate by an expected lifetime of 73 years, the *lifetime* use rate for the composite individual may still be less than 1. The composite individual thus represents a compromise between people who use one or more devices of a specific type over a lifetime and those who use none at all.

Using the numbers of medical devices sterilized with EtO each year and an estimate of the mass of EtO received in the course of one use of each device, we can estimate the total dose received over a lifetime by our hypothetical composite individual. Then, using an average weight of 60 kg and a typical lifespan of 26,650 days, we can estimate his/her lifetime average daily dose (LADD, in mg/kg/day) as

$$\text{LADD} = \Sigma (D_d \cdot N_d)/(60 \cdot 365 \cdot P) \ , \tag{1}$$

where

D_d = dose, in mg, from one use of device d,
N_d = number of devices used annually,
60 = weight of the composite individual, in kg,
365 = number of days per year,
P = U.S. population, taken to be 240 million.

Device Use. No comprehensive source of data on the use of medical devices in the United States came to our attention. We therefore estimated annual usage through a combination of complementary approaches.

For a very limited number of devices, the Census of Manufactures (1985) presents data on number of units sold, total sales in dollars, or both. Unit or dollar sales are also available from the files of trade publications that serve the biomedical product industry. Prices of devices can be obtained from catalogs or industry sources and used with total dollar sales volume to estimate unit sales. Sales volume is a satisfactory surrogate for use volume except for items that enjoy widespread use for other than medical services.

Device use can also be estimated from statistics on disease rates and medical procedures. The National Center for Health Statistics (NCHS, 1986) compiles data on the number of surgical, diagnostic, and therapeutic procedures undertaken in short-term care hospitals (Table 2). These rates can be multiplied by the number of devices of different types used per procedure to project the annual use of devices in these procedures. Estimates of device use per procedure are sometimes obvious (one pacemaker system per

Table 2. Example Procedures Performed in Short-Term Care Hospital

	Annual Number of Procedures (thousands)
Appendectomy	294
Arthroscopy of knee	237
Bypass anastomosis for heart revascularization	202
Cesarean section and removal of fetus	814
Dilation and curettage of uterus	744
Excision or destruction of intervertebral disc	203
Insertion of cardiac pacemaker system	170
Left heart cardiac catheterization	220
Partial excision of large intestine	166
Repair and plastic operations on the nose	235
Total cholecystectomy	484
X-ray of urinary system	480
IV administration	20,000*

*ENVIRON estimate
Source: NCHS, 1986

pacemaker implant operation) but are best provided by medical professionals. We turned to active hospital nurses as our consultants.

The NCHS does not provide separate data on certain procedures, such as anesthesia, that are considered a routine part of another procedure, or on procedures that are rarely done inside the hospital, such as dialysis or insulin injections. The frequency of these kinds of procedures can sometimes be estimated from the prevalence in the population of persons with the corresponding disorders (kidney failure and diabetes, respectively).

Exposure per Use. For cancer risk assessment, the standard method for treating intermittent exposures is to estimate the total systemic dose (mg) per kg of body weight and spread it out evenly over the expected lifetime. Although this procedure often overestimates risk, it can also underestimate risk modestly (Kodell *et al.*, 1987).

We used the standard method and estimated the systemic dose, in mg, per use of each device. The quantity of EtO *available* for exposure (mg) is equal to the residue on the device (ppm by weight) times the weight of the device (g). The actual exposure is estimated by applying a "reduction factor" that accounts for the tendency of EtO to escape to other media (air or fluids leaving the body) or to remain on the device after use because the time of contact with the patient was insufficient for total transfer.

Lacking a readily available data base of residues on devices at the time of use, we must rely on an assumption about residues to make a plausible risk assessment. We relied on the levels proposed by the FDA (1978) as limits for residues. These proposals, which were never finalized by the FDA, are shown in Table 3. Our adoption of these levels for the purposes of illustration does not necessarily imply our endorsement of either the values or the way in which they are expressed. We were also unable to find any systematic compilation of device weights. For easily obtained devices, we weighed typical samples. We also obtained estimates from manufacturers or judged them from similarity to other devices for which weights were known.

Table 3. FDA-Proposed Residue Limits

	Concentration Limits (ppm)		
	EtO	ECH	EG
Implant			
Small (<10 grams)	250	250	5,000
Medium (10-100 grams)	100	100	2,000
Large (>100 grams)	25	25	500
Intrauterine device	5	5	10
Intraocular lenses	25	25	500
Devices contacting mucosa	250	250	5,000
Devices contacting blood ("ex vivo")	25	25	250
Devices contacting skin	250	250	5,000
Surgical scrub sponges	25	25	250

Empirical information on reduction factors is scanty. We were unable to find direct information on the transfer of EtO from sterilized medical devices to patients or even to experimental animals. Our estimates are thus based on measurements of the rate of transfer to a fluid (water or blood) (Kroes et al., 1985) or on the rate of loss of EtO during aeration of medical devices (Vink and Pleijsier, 1986; Anderson, 1971; Baan, 1976; Henna et al., 1984; McGonnigle et al., 1975; Simpson and White, 1979). The loss appears to occur with half-lives of a few hours or less to more than a week, depending on materials of construction.

The importance of this variation is reduced because those devices that transfer EtO rapidly to the body also rapidly lose EtO between release for sale and actual use. We used a linearized version of the loss function for a half-life of 24 hours, which results in a rate of transfer of about 3% per hour of exposure.

Times of contact of devices with patients range from a second or two for a gauze sponge to many years for implants. We assume that any device remaining in place longer than 36 hours transfers all of its EtO residues, either to the patient or to external fluids (air, urine). Transfer for shorter periods of contact is assumed to be proportionally lower. We obtained estimates of time of contact from our nurse consultants, from industry sources, or from our own observations. Devices applied to the skin or mucosa are assumed to transfer less ethylene oxide to the body than a device used invasively for the same period. Table 4 shows assumed transfer reduction factors by type of exposure for several periods of exposure.

Specific Procedures

The appropriate measure of exposure for chronic toxic effects other than cancer is not the LADD but the daily dose over a shorter period depending on the specific type of effect. In the case of teratogenesis, the critical period may be measured in days; for most other chronic effects, it is much longer.

To be conservative, we assumed that all the dose from all the devices used in a procedure was delivered in a relatively short time, generally one day unless we knew that

Table 4. Assumed Transfer Reduction Factors

Time of Contact	Devices Contacting		
	Blood or Internal Organs	Mucosa or External Conduits	Skin
1 second	7.0×10^{-6}	3.5×10^{-6}	2.0×10^{-6}
1 minute	4.0×10^{-4}	2.0×10^{-4}	1.3×10^{-4}
1 hour	2.5×10^{-2}	1.3×10^{-2}	8.0×10^{-3}
1 day	6.0×10^{-1}	3.0×10^{-1}	2.0×10^{-1}
prolonged	1.0	5.0×10^{-1}	3.3×10^{-1}

several devices were used over a period of several days. The maximum daily dose, MDD_p, from a given procedure, p, is thus given by

$$\text{MDD}_p = \Sigma(n_{dp} \cdot D_d)/60 \ , \qquad (2)$$

where n_{dp} is the number of devices of type d used per procedure p, D_d is the dose per use of device d, as defined earlier, and 60 is the weight of the average person, in kg. The units of MDD are thus also mg/kg/day. The identities and numbers of devices used per procedure were provided by our nurse consultants.

The information on exposures for specific procedures can also be use as a semi-independent check on the exposure estimated for the composite individual. The number of procedures accomplished per year in the U.S. (NCHS, 1986) is multiplied by the expected number of years in a lifetime (73) and divided by the population of the U.S. (240 million) to obtain an estimate of the expected number of procedures for the composite individual over a lifetime. That estimate can then be multiplied by the exposure from all the devices used in the procedure to obtain an estimate of the lifetime exposure from the expected number of procedures. For example, about 120,000 mastectomies are performed annually; that projects to a lifetime risk of mastectomy of $(0.121 \times 73)/240$ or 0.04 (one in 25). The contribution of mastectomies to lifetime exposure for the composite individual is obtained by multiplying the exposure for one mastectomy by the factor 0.04.

When these estimates are summed over all the types of procedures considered, a second estimate of lifetime dose for the composite individual is obtained. This estimate is not entirely independent of our earlier estimate because the same estimates of exposure per use of one device are employed and because our estimates of the annual number of devices used were sometimes obtained from the annual number of procedures performed. The procedure method will tend to underestimate total exposure because not all procedures can be included in a list of reasonable length and because uses of medical devices occur outside hospitals, especially in doctors' offices and homes. These omissions may result in underestimation by a factor of approximately 3.

RISK CHARACTERIZATION

The nature of any risks posed by exposure to residues of EtO on medical devices can be understood by combining the results of the toxicity assessment and the exposure assessment. This process is conducted differently for the risks of cancer and of health effects other than cancer.

Cancer

The analysis of the risks of cancer from exposure to EtO residues assumes conservatively that some risk of cancer exists at any non-zero level of exposure and that, for low exposures, risks increase linearly with increasing LADD. The constant of proportionality is the potency of EtO, 0.048 $(mg/kg/day)^{-1}$. Therefore, the lifetime risk, R, of developing cancer from EtO on medical devices is

$$R = 0.048 \times LADD . \quad (3)$$

This equation can be evaluated either for all devices combined [LADD from Eq. (1)] or for all the devices used in a specific procedure [dose from Eq. (2) averaged over a lifetime (LADD = MDD/26,650)]. The resulting lifetime risk can then be compared with criteria set by decision makers.

Chronic and Reproductive Toxicity

The risks of health effects other than cancer are evaluated by comparing the MDD [Eq. (2)] with some criterion of minimal toxic potential, in our case the NOEL (LOEL, if a NOEL is not available). The ratio of the NOEL or LOEL to the MDD is defined as the margin of safety, MOS:

$$MOS = NOEL/MDD . \quad (4)$$

If the margin of safety drops below unity, concerns about possible health effects in humans should be raised. If the margin of safety exceeds the uncertainty factor usually applied to a NOEL derived from animal experiments to estimate a safe exposure in humans, typically a factor of 100, then essentially no concern at all is justified. In between, the acceptability of the margin of safety must be judged by the quality and quantity of data used to fix the NOEL, the likelihood that sensitive groups of humans may be more susceptible to the effects of EtO than any of the animal species and strains tested, and the degree of conservatism invested in the estimate of the MDD.

ACKNOWLEDGMENTS

We gratefully acknowledge the Health Industry Manufacturers Association for supporting this work and our colleagues at ENVIRON, especially Susan Rieth, Andrew Fragen, and Resha Putzrath, for their contributions.

REFERENCES

Anderson, S. R., 1971, "Ethylene Oxide Toxicity," *J. Lab. Clin. Med.* 77:346-356.
Baan, E., 1976, "Ethylene Oxide Absorption and Desorption of Elastomers and Plastics," *Bull. Parenteral Drug Assoc.* 30:299-305.
Census of Manufactures, 1985; Census of manufactures, 1982; Industry Series, MC82-1-38B, *Medical Instruments; Ophthalmic Goods; Photographic Equipment;*

Clocks, Watches, and Watchcases, U.S. Bureau of the Census, U.S. Government Printing Office, Washington, DC.

FDA, Food and Drug Administration, 1978, "Ethylene Oxide, Ethlyene Chlorohydrin, and Ethylene Glycol: Proposed Maximum Residue Limits and Maximum Levels of Exposure," 43 FR 27474 (June 23).

Gaylor, D. W., and Kodell, R. L., 1980, "Linear interpolation Algorithm for Low Dose Risk Assessment of Toxic Substances," *J. Environ. Path. Toxicol.* 4:305-312.

Henne, W., Dietrich, W., Pelger, M., and Sengbusch, G. V., 1984, "Residual Ethylene Oxide in Hollow-Fiber Dialyzers," *Artificial Organs* 8:306-309.

Kodell, R. L., Gaylor, D. W., and Chen, J. J., 1987, "Using Average Lifetime Dose Rate for Intermittent Exposures to Carcinogens," *Risk Analysis* 7:339-346.

Kroes, R., Bock, F., and Martis, L., 1985, "Ethylene Oxide Extraction and Stability in Water and Blood," Travenol Laboratories, Morton Grove, IL.

Lynch, D. W., Lewis, T. R., Moorman, W. J., Burg, J. R., Groth, D. H., Khan, A., Ackerman, L. J., and Cockrell, B. Y., 1984, "Carcinogenic and Toxicological Effects of Inhaled Ethylene Oxide and Propylene Oxide," *Toxicol. Appl. Pharmacol.* 76:69-84.

McGonnigle, R. G., Renner, J. A., Romano, S. J., and Abodeely, R. A., Jr., 1975, "Residual Ethylene Oxide Levels in Medical Grade Tubing and Effects on an in vitro Biologic System," *J. Biomed. Mater. Res.* 5:273-283.

NCHS, National Center for Health Statistics, 1986, Detailed diagnoses and procedures for patients discharged from short-stay hospitals, United States, 1984, Series 13, No. 86, DHS Publication No. (DHS)86-1747, U.S. Government Printing Office, Washington, DC.

NTP, National Toxicology Program, 1986, NTP Technical Report on the Toxicology and Carcinogenesis Studies of Ethylene Oxide (CAS No. 75-21-8) in B6C3F1 Mice (Inhalation Studies), Board draft, NTP TR 326, NIH Publication No. 86-2582.

RIVM, Rijksinstituut voor Volksgezondheid en Milieuhygiëne. n.d., "Ethylene Oxide Residues in Medical Devices," National Institute for Public Health and Environmental Hygiene (Netherlands).

Rowe, V. K., and McCollister, S. B., 1982, "Alcohols," in *Patty's Industrial Hygiene and Toxicology*, G. D. Clayton and F. E. Clayton, Eds., 4675-4684, John Wiley and Sons, New York.

Simpson, R. C., and White, C. H., 1979, "A Dissipation Study of Residual Ethylene Oxide from Various Medical Device Thermoplastics and Rubbers, Technical report No. 24, Plastics Division, The West Company.

Snellings, W. M., Weil, C. S., and Maronpot, R. R., 1984a., "A Subchronic Inhalation Study of the Toxicologic Potential of Ethylene Oxide in B6C3F1 Mice," *Toxicol. Appl. Pharmacol.* 76:510-518.

Snellings, W. M., Weil, C. S., and Maronpot, R. R., 1984b, "A Two-Year Inhalation Study of the Carcinogenic Potential of Ethylene Oxide in Fischer 344 Rats," *Toxicol. Appl. Pharmacol.* 75:105-117.

Snellings, W. M., Maronpot, R. R., Zelenak, J. P., and Laffoon, C. P., 1982a, "Teratology Study in Fischer 344 Rats Exposed to Ethylene Oxide by Inhalation," *Toxicol. Appl. Pharmacol.* 64:476-481.

Snellings, W. M., Zelenak, J. P., and Weil, C. S., 1982b, "Effects on Reproduction in Fischer 344 Rats Exposed to Ethylene Oxide by Inhalation for One Generation," *Toxicol. Appl. Pharmacol.* 63:382-388.

Sprinz, H., Matzke, H., and Carter, J., 1982, *Neuropathological Evaluation of Monkeys Exposed to Ethylene and Propylene Oxide*, Midwest Research Institute, prepared for National Institute for Occupational Safety and Health, Cincinnati, OH, NIOSH Contract No. 210-81-6004, MRI Project No. 7222-B, PB83-134817.

USEPA, U.S. Environmental Protection Agency, 1985, Health Assessment Document for Ethylene Oxide, Office of Health and Environmental Assessment, Washington, DC, EPA/600/8-84/009F.

Vink, P. and Pleijsier, K., 1986, "Aeration of Ethylene Oxide-Sterilized Polymers," *Biomaterials* 7:225-230.

Tetrachloroethylene Metabolism Resulting from Domestic Respiratory Exposure: Pharmacokinetic Considerations Relevant to Risk Assessment

Kenneth T. Bogen and Thomas E. McKone
Lawrence Livermore National Laboratory
Livermore, CA

ABSTRACT

Physiologically based pharmacokinetic (PBPK) models describing the uptake, metabolism, and excretion of xenobiotic compounds are now proposed for use in regulatory health-risk assessments. In this study we compare how different scenarios for domestic respiratory exposure to tetrachloroethylene (PCE) from ground water influence the extent of PCE metabolism predicted using a PBPK model. Indoor exposure patterns we use as input to the PBPK model are realistic ones generated from a three-compartment model describing volatilization of PCE from domestic water into household air. Values we use for the metabolic parameters of the PBPK model are estimated from data on urinary metabolites in workers exposed to PCE. It is shown that for respiratory PCE exposure due to typical levels of PCE in ground water, use of time-weighted average air concentrations with a steady-state PBPK model yields estimates of total metabolized PCE similar to those obtained using completely dynamic modeling, despite considerable uncertainty in key exposure- and metabolic-model parameters. These findings suggest that, in this case, risk estimation taking pharmacokinetics into account may be accomplished using simple analytic methods.

KEYWORDS: Tetrachloroethylene, indoor respiratory exposure, pharmacokinetic models

INTRODUCTION

Multicompartment models known as physiologically based pharmacokinetic (PBPK) models are used increasingly in health-risk assessment for volatile organic compounds (VOC's). These models predict the uptake, metabolism and excretion of such chemicals in exposed animals or people. In current applications of such models, however, input has consisted of simplistic exposure scenarios that often do not represent realistic environmental exposure conditions of regulatory concern. Recent studies have shown that exposure to VOC's in drinking-water supplies can result not only from direct ingestion of contaminated water, but also from inhalation and dermal absorption.[1-3] McKone has shown that daily personal exposure to VOC's from tap water through inhalation can range from 1.5 to 6.0 times the exposure attributable to ingestion of 2 L/d of water.[4] This daily exposure is dominated by large, short-term exposures during specific activities such as showering or bathing and time spent in bathrooms. One question raised by this observation is whether the metabolized dose is sensitive to the temporal exposure pattern. To address this question, we explored how realistic scenarios for indoor respiratory exposure to tetrachloroethylene (PCE) from ground water affect the extent of PCE metabolism, as predicted by a PBPK model. We focused on the degree to which dynamic, as opposed to steady-state, modeling is actually necessary to predict PCE metabolism and on whether conclusions in this regard are significantly affected by uncertainty in estimates of parameters that govern PCE metabolism in humans.

To place our analysis in the context of risk assessment, we note that PCE has been shown to cause tumors in rodents and is considered to be a suspected human carcinogen for purposes of environmental regulation.[5] At very low doses of environmental concern, a conservative risk-assessment approach widely in use views the carcinogenic risk posed by PCE as an approximately linear function of the amount of this compound that is metabolized.[5-7]

METHODS

Predictions of indoor respiratory exposure to PCE and of PCE pharmacokinetics were made using the mass-balance models described below. Analytic treatment of the PBPK model at steady state is reported below in the section on "Results." The PBPK model was fit to two sets of available data on PCE metabolism in humans occupationally exposed to PCE, and the resulting estimates for model parameters governing metabolism are used with other information for a simple analysis of current uncertainty in the extent to which humans metabolize PCE.

A Three-Compartment, Mass-Balance Model for Indoor Respiratory Exposure

To model indoor respiratory exposure, we used a three-compartment model that simulates the transfer and distribution of VOC's inside homes.[4] The three compartments in this model are the shower/bath stall, the bathroom, and the remaining household volume. Figure 1 illustrates the major components of the model and shows the mass-flow pathways that are addressed in the model.

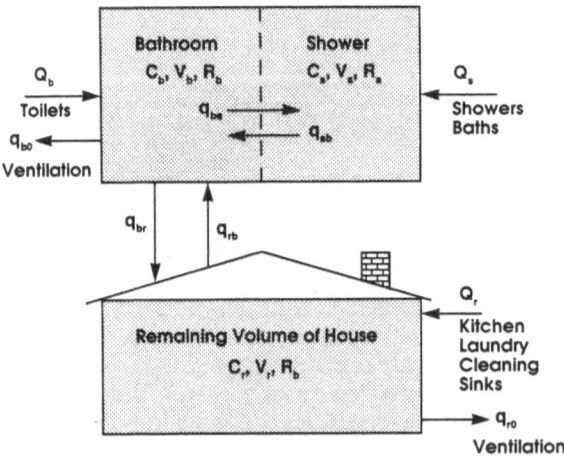

Fig. 1. A Three-Compartment Model for Simulating the Transfer of Pollutants from Tap Water to Indoor Air.

The mass-balance equations for the three compartments are:

$$V_s \dot{C}_s = Q_s + q_{bs}C_b - q_{sb}C_s \qquad \text{(shower stall).} \qquad (1)$$

$$V_b \dot{C}_b = Q_b + q_{sb}C_s + q_{rb}C_r - (q_{bs}+q_{bo}+q_{br})C_b \qquad \text{(bathroom).} \qquad (2)$$

$$V_r \dot{C}_r = Q_r + q_{br}C_b - (q_{rb}+q_{ro})C_r \qquad \text{(remainder of the house).} \qquad (3)$$

In these equations, the C's, V's, Q's and q's refer to concentrations in mg/L, volumes in L, sources in mg/min, and air-exchange rates in L/min, respectively. For notational convenience, the dependence of the state variables (C's and Q's) in this model on time, t, is suppressed, and dot notation is used to represent differentiation with respect to time (e.g., $\dot{C}_s = dC_s/dt$). The subscripts s, b, and r are used to indicate the shower, bathroom, and remaining household compartments, respectively, while the subscript o denotes outside air. In Fig. 1, the subscripted R's represent the residence time of air mass in each compartment. The air-exchange parameters, q, and the parameters V_s, V_b, V_r, R_s, R_b, and R_r are taken from and summarized in a report by McKone, where representative values and value ranges are provided.[4]

Daily concentration profiles obtained for C_s, C_b, and C_r demonstrate that the concentration in all three compartments is driven largely by the shower source, and that individual exposure (concentration × time) occurs predominantly in that compartment. We calculate two sets of 24-h concentration profiles: a "reference" profile (based on a set of typical values assigned to the model parameters) and an "upper bound" profile (obtained by assigning conservative values to the model parameters).[4] The time-dependent concentration profiles of daily human exposure were calculated using the following expression

$$C_{in} = F_s C_s + F_b C_b + F_r C_r \; , \qquad (4)$$

where C_{in} = personal air concentration at time t (mg/L); F_s = probability that an individual is in the shower at the time t; F_b = probability that an individual is in the bathroom at the time t; and F_r = probability that an individual is in the remaining household volume at the time t. Dynamic simulations using the exposure model were carried out by simultaneous numerical integration of the sets of differential equations involved, using the PREMOD/MODAID computer program for dynamic systems simulation.[8]

A daily concentration profile for PCE in personal air is illustrated in Fig. 2 for an individual who takes a "reference" 10-min shower using 75 L of tap water, spends an additional 15 min in the bathroom after the shower, spends 24 h/d in the home, and makes three additional 5-min visits to the bathroom. In the calculation, the PCE concentration in the water supply was taken to be 1 mg/L. The average personal-air concentration derived from Fig. 2 is 4.0×10^{-4} mg/L (0.058 ppm) and the corresponding peak concentration in the shower is about 0.1 mg/L (14 ppm). The "upper bound" conditions we considered for adult showers involved an individual who each day takes one 20-min shower using 75 L water. Both "reference" and "upper-bound" showering conditions were assumed to involve four consecutive showers per household-day, using a total of 300 L of water.[4] The corresponding predicted PCE concentrations in air may be linearly scaled to reflect an actual concentration of PCE in tap water. For example, a survey of 869 of 1426 wells in 79 large water-supply systems serving 7.2 million people in California revealed that water from 199 of the tested wells contained PCE at concentrations up to 166 µg/L, with a median concentration of 1.9 µg/L.[9,10] Using the highest measured water concentration of 166 µg/L as input, the indoor exposure model described predicts peak air concentrations in a shower stall, during the last of four consecutive showers, to be less than 1 ppm (see Fig. 3).

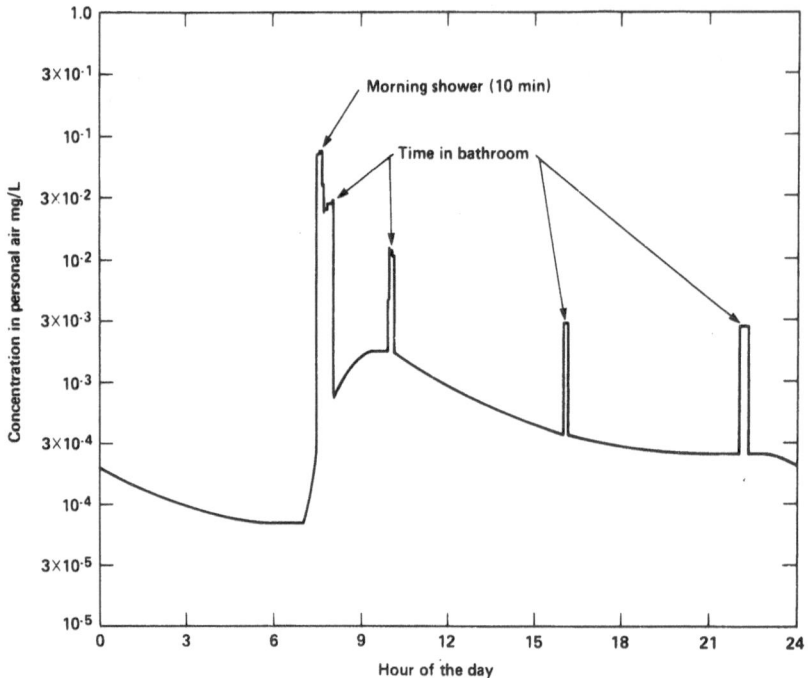

Fig. 2. Concentration Profile of PCE in Personal Air Derived from the Indoor Exposure Model.

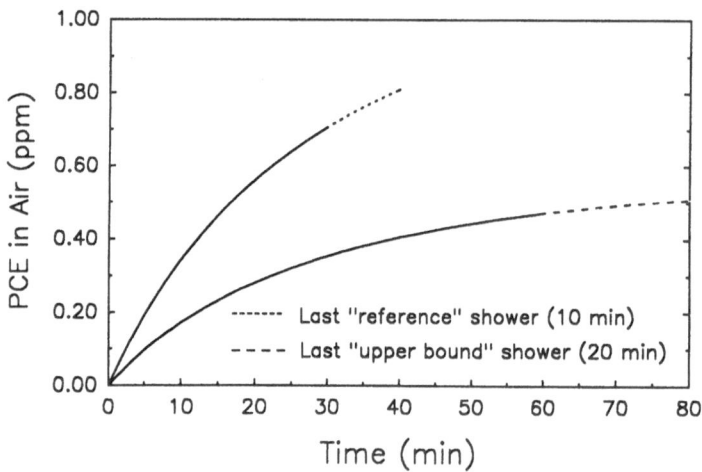

Fig. 3. Concentration of PCE in Shower-Stall Air Over Time Predicted by the Indoor Exposure Model. The model assumes each shower uses 75 L of tap water containing 166 μg PCE/L.

PBPK Model for PCE

The pharmacokinetic model used here was developed by Ramsey and Andersen to describe the uptake, metabolism and excretion of styrene in rats and humans.[11] The structure of the model is shown in Fig. 4, and its parameter definitions are given in Table 1. This type of model has been applied to the study and prediction of the pharmacokinetics of other volatile hydrocarbons in animals and humans[12-14] and to investigate PCE pharmacokinetics in particular.[6,15-17] According to this model, pulmonary uptake of a chemical occurs continuously such that alveolar concentration, C_a, is in instantaneous equilibrium with arterial blood governed by the blood/air partition coefficient, P_b, in accordance with the relation $C_a = B_a/P_b$. Similarly, the concentrations, C_i, of chemical in each of four tissue compartments are presumed to be in instantaneous equilibrium with the concentrations, B_i, in venous blood exiting the corresponding tissue, governed by the corresponding tissue/blood partition coefficients such that $B_i = C_i/P_i$. The amount of chemical in any given tissue compartment is given by $A_i = C_i V_i$. Again, the dependence of state variables (C's, B's and A's) on time, t, is suppressed, and dot notation is used to represent differentiation with respect to time.

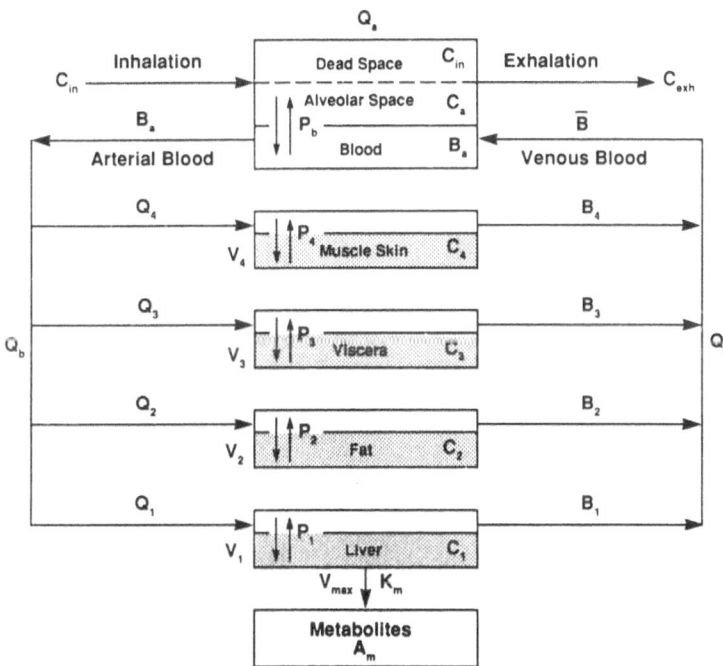

Fig. 4. Ramsey-Andersen Physiologically Based Pharmacokinetic (PBPK) Model for Inhalation of Volatile Organic Compounds. The model assumes that four "well stirred" tissue compartments collect or release a compound at rates governed by air concentration, air and blood flows, blood concentrations, compartment volumes, tissue/blood partition coefficients, and metabolism. (See Table 1 for definition of model parameters.)

Table 1. Compartment and Parameter Definitions for the Ramsey-Andersen PBPK Model

Abbrev.	Definition	Unit
C_{in}	Concentration in air inhaled	mg/L air
C_a	Concentration in alveolar air	mg/L air
C_{exh}	Measured concentration in expired breath	mg/L air
Q_a	Alveolar ventilation rate	L air/h
Q_b	Cardiac output	L blood/h
P_b	Blood/air partition coefficient	L air/L blood
B_a	Arterial blood concentration	mg/L blood
B	Venous blood concentration	mg/L blood
A_m	Amount metabolized in liver	mg
Q_i	Blood flow rate to compartment i	L blood/h
V_i	Volume of compartment i	L ($\equiv$ kg)
C_i	Concentration in compartment i	mg/L
B_i	Concentration in venous blood leaving compartment i	mg/L blood
A_i	Amount in compartment i	mg
P_i	Tissue/blood partition coefficient for compartment i	L blood/L tissue i
V_{max}	Maximum metabolite rate	mg/h
K_m	Michaelis constant = $\{B_1 \mid (dA_m/dt = V_{max}/2)\}$	mg/L blood

Compartmental subscripts:

i = 1 Liver (metabolizing tissue group)
    2 Fat tissue (very poorly perfused)
    3 Richly perfused tissues (brain, kidney, viscera)
    4 Poorly perfused tissues (muscle, skin)

In any given interval, dt, chemical delivered to the lung via respiratory retention and via returning venous blood is balanced in this model by the chemical mass exiting the lung via exhalation and via arterial blood, such that $Q_a(C_{in} - C_a) = Q_b(B_a - B)$, yielding

$$B_a = \frac{Q_a C_{in} + Q_b \bar{B}}{(Q_a/P_b) + Q_b} \ . \tag{5}$$

Equation (5) specifies B_a (and thus C_a and C_{exh}) to be at each instant a flow-weighted average of C_{in} and B. Similarly, the concentration B in venous blood returning from each compartment is presumed to be the instantaneous flow-weighted average:

$$\bar{B} = \frac{1}{Q_b} \sum_{i=1}^{4} Q_i B_i \ . \tag{6}$$

For the non-metabolizing tissues, the amount $Q_i B_a dt$ of chemical entering the ith compartment via arterial blood during any given interval dt is set equal to the amount dA_i gained by that compartment plus the amount $Q_i B_i dt$ leaving in venous blood, for $i = 2,3,4$. The chemical concentration for each of these compartments is therefore defined by the state equation

$$\dot{B}_i = \frac{Q_i}{V_i P_i}(B_a - B_i) \quad i = 2,3,4 \ . \tag{7}$$

The rate of chemical metabolism, assumed to take place only in liver, is given by the Michaelis-Menten relation

$$\dot{A}_m = \frac{V_{\text{max}} B_1}{K_m + B_1} , \tag{8}$$

in which K_m is defined as the concentration in venous blood from liver at which the liver's metabolic velocity $\dot{A}_m$ is half its maximum value, V_{max}. Thus, the state equation for venous liver-blood concentration is given by

$$\dot{B}_1 = \frac{Q_1}{V_1 P_1} (B_a - B_i) - \frac{\dot{A}_m}{V_1 P_1}. \tag{9}$$

The system of Eqs. (6–9) represents the PBPK model for inhalation of a volatile chemical, and for any given time its compartmental amounts A_i, or corresponding concentrations C_i or B_i, are found by simultaneous numerical integration of the system.

Of particular interest in the context of risk assessment is the metabolized fraction of the total quantity of chemical potentially available through alveolar ventilation for absorption and metabolism. Under steady-state conditions, the corresponding quantity of interest is the fraction, f_m, of the maximum plausible metabolic rate, given C_{in}:

$$f_m = \dot{A}_m / (Q_a C_{in}) , \tag{10}$$

which may be contrasted with the assumption, sometimes made in the context of carcinogen risk assessment,[7] that 100% of a chemical entering the lungs through total respiratory ventilation (or that contained in approximately 20 m^3/d for a reference 70-kg man) is absorbed and potentially available for metabolism.

PBPK Model for PCE Metabolism in Occupationally Exposed Humans

The Ramsey-Andersen PBPK model described above was adapted, as described below, in order to fit data of Ikeda et al.[18,19] and Ohtsuki et al.[20] on urinary metabolites produced by Japanese workers exposed to PCE. These two studies are the only ones available in which metabolites were measured in humans over sustained exposure periods. The Ikeda et al. data are based on surveys of workplace air and urine samples of 34 males in 7 workshops where PCE was present in air at 8-h time-weighted average (TWA) concentrations ranging from 0 to 400 ppm, and where workshop-specific concentrations were reported to be relatively constant over the common 8-h/d, 6-d/wk occupational schedule of the workers studied.[18] Urine samples were passed at about 1:00 pm in the "latter half" of the week.[18] Concentrations of the PCE metabolites trichloracetic acid (TCA), trichloroethanol, and total trichloro-compounds (TTC) in urine (adjusted to a specific gravity of 1.016) were measured for each worker along with a corresponding TWA concentration of PCE in air.

The Ohtsuki et al. data considered here are based on surveys of urine samples and PCE in personal workplace air of 20 males employed in several dry-cleaning shops and 16 males engaged in removing dye-repellent glue from silk in a *kimono* production shop; in all of the shops PCE was used as the solvent.[20] TTC concentrations were measured in samples of end-of-shift urine (adjusted to a specific gravity of 1.016), and measured TWA concentrations of PCE in personal air ranged up to 630 ppm. Otherwise, the exposure scenario for the workers in this study was the same as that in the Ikeda et al. study.

The values we used for the parameters listed in Table 1 pertaining to a reference, 70-kg male worker are given in Table 2. These parameter values were used by Ward et al.[17] in a similar analysis of the Ikeda et al. data on PCE metabolism in humans; the partition coefficients P_b and P_i were obtained experimentally using human tissues.[13]

Table 2. Parameter Values Used in PBPK Model for TCE

Parameter		Unit	Reference male human[a]	Japanese male worker[b]
W (body weight)		kg	70	55.2
Q_a		L/h	353.5	299.3
Q_b		L/h	371.6	314.7
P_b			10.3	10.3
Q_i/Q_b	i = 1		0.25	0.25
	2		0.05	0.04
	3		0.51	0.52
	4		0.19	0.19
V_i/W	i = 1		0.04	0.027
	2		0.20	0.15
	3		0.05	0.10
	4		0.62	0.61
P_i	i = 1		6.82	6.82
	2		159	159
	3		6.82	6.82
	4		7.77	7.77

[a] Physiological parameter values taken from Ward et al.[17]; values for partition coefficients are based on gas uptake experiments.[13]

[b] Parameter values are scaled from those corresponding to reference male human, based on the anatomical model of Kerr[23] for reference Japanese adults.

We assumed that a reference 70-kg male produces an average of 0.0583 L urine per hour during the day, for urine adjusted to a specific gravity of 1.016.[21] We also assumed, in the absence of more specific data, that urine samples were passed at 1:00 pm and 5:00 pm by all the workers studied by Ikeda et al. and Ohtsuki et al., respectively, on Thursdays, Fridays, and Saturdays of each week, and that a prior urination (emptying the bladder) occurred 4 h before samples were passed on each collection day for all workers. It was further assumed that for 70-kg males, all PCE metabolites, reflected by measured concentrations of TTC in urine, were removed exponentially from blood to collect in urine with a rate constant of 0.0137 h^{-1}, based on the assumption that TCA is the primary PCE metabolite in urine[5] and on data indicating that blood concentration of TCA in (TCA-dosed) humans decreases exponentially with a half-life of 50.6 h.[22]

For the purpose of modeling the occupational exposures considered here, we scaled the reference values given in Table 2 to approximate those applicable to a typical Japanese male worker. To this end, body weight was set at 55.2 kg and the tissue volumes, V_i, were adjusted to the new values shown in Table 2, based on an anatomical model developed specifically for Japanese adults.[23] The values for the blood-flow fractions Q_2/Q_b and Q_3/Q_b were changed slightly to reflect the altered tissue volumes. The reference flow rates Q_a and Q_b and that for urinary output were all decreased by the factor $(55.2/70)^{0.7}$, and the reference rate constant for metabolite clearance to urine from a given volume of blood was increased by the factor $(70/55.2)^{0.3}$, in accordance with the assumption that physiological parameters vary with basal metabolic rate in approximate proportion to body surface area.[6,11,12]

To approximate conditions of dynamic equilibrium, the occupational exposure scenario was run for a simulated 5-wk period. Workers were assumed to be exposed to TCE from 8:00 am to 12:00 pm and 1:00 pm to 5:00 pm on Monday through Saturday of each week. Calculated urinary concentrations of TTC on each of the 3 urine-collection days in the fifth week of the simulation were averaged to yield a predicted urinary metabolite concentration corresponding to any given values for the input parameters C_i, V_{max}, and K_m. Urinary metabolite concentrations were expressed in mg/L of TTC as TCA, as was done in the Ikeda et al. and Ohtsuki et al. studies,[18-20] which approximates the unit of mg/L PCE because the molecular weights of PCE and TCA are approximately equal. Numerical integration of the system of differential equations involved was performed on a VAX 11/750 computer using a variable-step Gear method.[24]

Metabolic Parameter Uncertainty

The existence of alternative sets of data on PCE metabolism in humans and animals allowed us to perform a crude analysis of uncertainty associated with our PBPK-based estimates of V_{max} and K_m. The human data were used directly to provide two study-specific alternative estimates of K_m. Corresponding estimates of V_{max} were first multiplied by the factor $(70/55.2)^{0.7}$ to scale the rate parameter to apply to a reference 70-kg human (see preceding section), and then multiplied by an "excretion factor" (f_e) to account for non-urinary PCE metabolites in humans. In the absence of quantitative information on the extent to which humans metabolize PCE to non-urinary metabolites, estimation of f_e was based on the data available for mice and rats[25-29] summarized in Table 3. On the basis of this animal data, it is reasonable to assume that it is unlikely that less than a third of metabolized PCE would be excreted in the form of urinary products in humans, but that this amount might be as high as 100%. We used the midpoint of this range as our best estimate, so that in this analysis $f_e = 1.5$ (range: 1.0 to 3.0).

RESULTS

Analysis of the Steady-State PBPK System

For this PBPK system using the reference parameter values listed in Table 2, steady-state conditions are approximately attained with constant PCE input after roughly 10 to 15 d. Since at steady state $B_a = B_i$ for $i = 2,3,4$, it follows that

$$\bar{B} = \frac{1}{Q_b} \left[Q_1 B_1 + B_a (Q_h - Q_1) \right] , \tag{11}$$

$$B_a = \frac{Q_a C_{in} + Q_1 B_1}{(Q_a/P_b) + Q_1} , \tag{12}$$

and

$$\frac{V_{max} B_1}{K_m + B_1} = Q_1 (B_a - B_1) , \tag{13}$$

so that the solution for venous liver blood concentration, given input C_{in}, is given by

$$B_1 = Y + \sqrt{Y^2 + Z} , \tag{14a}$$

Table 3. PCE Metabolism to Products Excreted in Urine by Mice and Rats

Species (Sex)	Strain	Total Applied Dose or Concentration	Metabolites (mg/kg)	Metabolites in Urine as a Percent of Total Metabolites (%)	Reference
Mice (M)	B6C3F1	500 mg/kg[a]	85.0	82.8	(25)
		10 ppm–6 h[b]	14.8	79.8	
Mice (M)	B6C3F1	900 mg/kg[c]	200	64.6	(26)
Mice (F)	NMRI	800 mg/kg[a]		66.1	(27)
Rats (M)	Spr.–Dawley	1 mg/kg[a]	0.28	57.1	(28)
		500 mg/kg[a]	51.0	45.7	
		10 ppm–6 h[b]	1.87	58.9	
		600 ppm–6 h[b]	36.4	49.9	
Rats (M)	Spr.–Dawley	8.1 mg/kg[d]	0.98	61.0	(29)
Rats (M)	Osb.–Mendel	1000 mg/kg[c]	23.9	46.9	(26)
Rats (F)	Wistar	800 mg/kg[a]	61.6	31.2	(27)

a Single gavage dose of radiolabeled PCE in corn oil vehicle.

b Single 6-h respiratory exposure to radiolabeled PCE.

c Single gavage dose of radiolabeled PCE in corn oil vehicle, after similar dosing with unlabeled PCE for 4 wk, 5 d/ wk.

d Administered in drinking water (approx. 150 ppm PCE) for 12 h.

in which

$$Y = \frac{1}{2}\left[C_{in}P_b - V_{\max}\left[\frac{P_b}{Q_a} + \frac{1}{Q_1}\right] - K_m\right], \tag{14b}$$

and

$$Z = K_m P_b C_{in} . \tag{14c}$$

The steady-state metabolic rate is thus given by using Eqs. (14a–c) to evaluate B_1 in Eq. (8), and likewise in the expression for the fraction f_m of maximal metabolic rate for PCE given by Eq. (10). Of greater interest in the context of environmental risk assessment is the limiting value of f_m that is approximated at very low exposure levels that might be typical of non-occupational respiratory exposure to PCE. It can be shown that this limiting value is given by

$$f_m^* = \lim_{C_{in}\to 0} f_m = \left[\frac{Q_a}{P_b}\left(\frac{K_m}{V_{\max}} + \frac{1}{Q_1}\right) + 1\right]^{-1} . \tag{15}$$

Also of interest is the maximally conservative assumption that infinite metabolism is approached as steady state $C_{in} \to 0$, yielding the new limit

$$f_m^{**} = \lim_{V_{max} \to \infty} f_m^* = \left[\frac{(Q_a/Q_1)}{P_b} + 1 \right]^{-1} , \qquad (16)$$

representing the physiologically determined upper bound on the fraction of intake capable of being metabolized. This limiting value is a function of just three physiological parameters, only one of which is influenced by the particular chemical under consideration. For PCE, using the reference parameter values given in Table 2, f_m^{**} equals 0.73 and is invariant with respect to body weight.

Metabolic Parameter Estimation

The relationships between the predicted concentration of PCE in urine and PCE concentration in workplace air based on our PBPK simulations for the workers from the Ikeda et al. and Ohtsuki et al. studies are shown in Figs. 5 and 6, respectively, along with the corresponding best-fit values for V_{max} and K_m. The values 0.19 and 6.1 mg/L were used directly to reflect a range of uncertainty in K_m, whereas the estimates for V_{max} were scaled as described above (see "Metabolic Parameter Uncertainty") to yield the corresponding best-estimate (range) values of 7.6 (5.1 to 15) mg/h and 12 (8.3 to 25) mg/h, respectively, for V_{max} based on the two studies considered.

Impact of Metabolic Parameter Uncertainty at Steady State

Figure 7 shows the relationship between C_{in} and f_m based on Eqs. (10), (14a-c), and (15) and using the estimates of V_{max} and K_m that we obtained for a reference 70-kg adult. This figure illustrates how the 60-fold range of uncertainty obtained for the ratio V_{max}/K_m (1.4 to 79 L/h) translates into a 15-fold range of uncertainty in the low-dose metabolic limit, f_m^* (0.038 to 0.55). Taking this uncertainty into account, f_m^* was estimated to be between 5% and 75% of its upper physiological bound, f_m^{**}, defined by Eq. (16).

Figure 7 also clearly illustrates that it may be very misleading to generalize about human capacity to metabolize very low doses of PCE on the basis of data on urinary metabolites produced by humans experimentally exposed to PCE for short periods at relatively high (>10 ppm) concentrations in air. It has been inferred from such experimental studies that f_m^* ranges from 2% to 4%,[5,20,30,31] whereas we, using an analytic PBPK approach, and other investigators,[6,16] using a numerical PBPK approach, have shown that f_m^* may actually be greater by a factor of 10 to 20.

PCE Metabolism Due to Indoor Exposure

Under time-varying exposure conditions, Eq. (8) implies that when varying air concentrations remain very small, such that $B_1 \ll K_m$ at all times, the average metabolic rate R_T over a given exposure period T is approximated by the value $R_T \approx (V_{max}/K_m)\bar{B}_1$, where $\bar{B}_1$ is the average value of B_1 over that period. Noting that

$$\frac{d\dot{A}_m}{dB_1} = \frac{V_{max}K_m}{(K_m + B_1)^2} , \qquad (17)$$

it follows that, so long as $B_1 < K_m$ over an exposure period T, then $R_T < R_T/4$, or metabolism is approximately linear (within a factor of 4, or, e.g., a factor of 1.1 if $B_1 < 0.1 K_m$). In the case that $B_1 \gg K_m$, a given exposure, $C_{in}Q_a T$, pulsed for a time T within a larger averaging period of length τ will result in a much lower metabolized dose than the "equivalent" time-weighted average exposure $\bar{C}_{in}Q_a \tau = C_{in}Q_a T$, where $\bar{C}_{in} < C_{in}$.

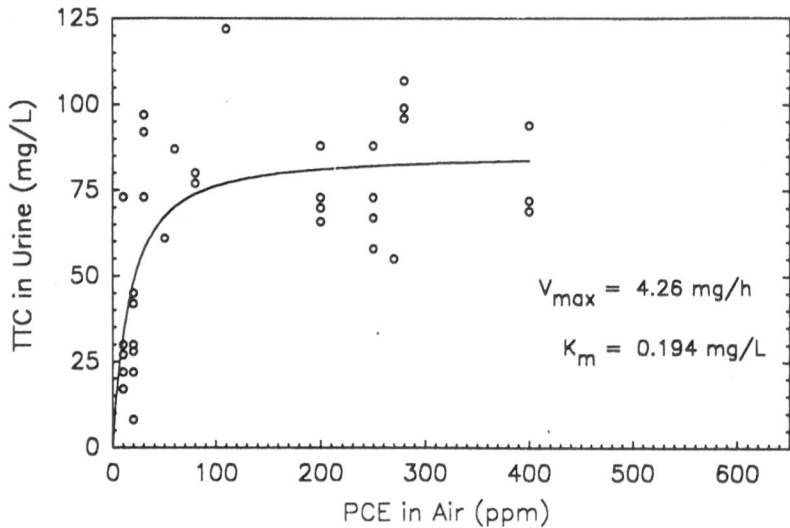

Fig. 5. Least-Squares Estimates of V_{max} and K_m Obtained by Fitting the Output of PBPK Model (Solid Curve) to Data (o) of Ikeda *et al.* Data on human metabolism of PCE to total trichloro-compounds (TTC) measured in urine (adjusted to a specific gravity of 1.016) at different occupational exposure levels.

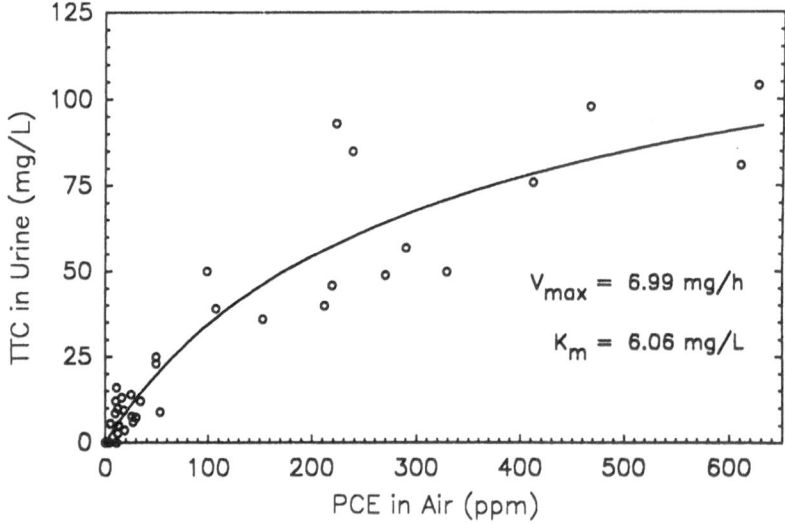

Fig. 6. Least-Squares Fit of V_{max} and K_m Obtained by Fitting the Output of PBPK Model (Solid Curve) to Data (o) of Ohtsuki *et al.* Data on human metabolism of PCE to total trichloro-compounds (TTC) measured in urine as a function of occupational exposure level.

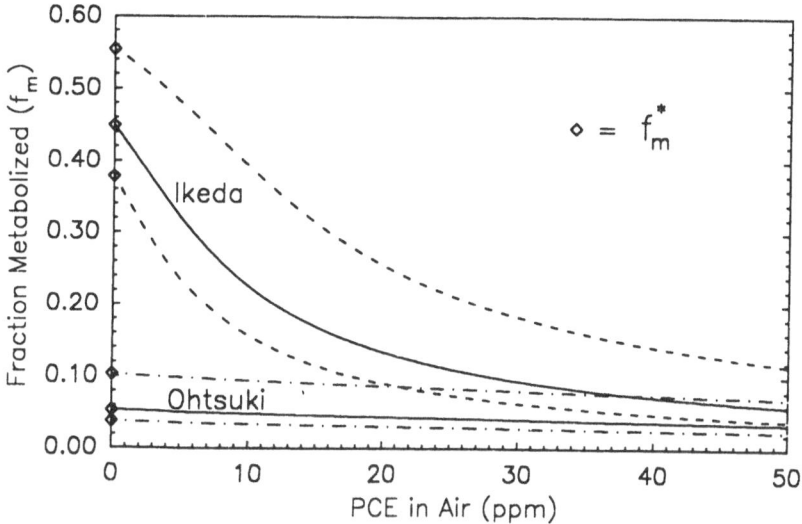

Fig. 7. Relation of the Metabolized Fraction (f_m) of Respired PCE to Alternative Constant PCE Concentrations in Air and Alternative Metabolic Parameter Values, As Predicted by the PBPK Model for a "Reference" 70-kg Male. Solid Curves(—) labeled "Ikeda" and "Ohtsuki" assume that the parameters (V_{max}, K_m) are best estimated by (7.6 mg/h, 0.19 mg/L) and (12 mg/h, 6.1 mg/L) based on the data of Ikeda *et al.* and Ohtsuki *et al.*, respectively. The corresponding lower and upper bounds on f_m shown as dashed curves (- - - for Ikeda, - · - for Ohtsuki) assume that V_{max} is 2/3 and 3/2 of its corresponding best estimate, respectively.

To determine whether nonlinear metabolism is relevant at the levels of inhalation exposure associated with PCE concentrations in public water supplies, we simulated PCE metabolism in a reference 70-kg adult taking the last of four consecutive showers, using the two air concentration patterns shown in Fig. 3 as inputs to the PBPK model. The dashed curves in Fig. 8 show the corresponding increases in B_1 with time under the assumption that $V_{max} = 5.1$ mg/h and $K_m = 0.19$ mg/L, which represents that set of values, among the six sets obtained earlier (see Fig. 7), generating the largest final value of the ratio B_1/K_m. Even under the conservative assumptions used here regarding PCE concentration in household water (0.166 mg/L) and metabolic parameter values, Fig. 8 reveals that B_1 attains less than 2% of the corresponding value of K_m, ensuring that virtually linear metabolism occurs in the context of domestic respiratory exposure to PCE.

CONCLUSIONS

Steady-state analysis of the PBPK model used here provides several simple relationships that are directly relevant to environmental risk assessment for VOC's subject to metabolism. In particular, it provides an upper bound on the amount of inhaled chemical that is capable of being metabolized. We have further demonstrated that, for respiratory PCE exposure at indoor air concentrations arising from typical levels of PCE contamination in well water, metabolized dose can conveniently be approximated using estimated TWA air concentrations of PCE with a PBPK model under steady-state assumptions, despite the large temporal fluctuations in air concentration associated with a realistic indoor exposure scenario and the large degree of uncertainty in parameter estimates for PCE metabolism in humans.

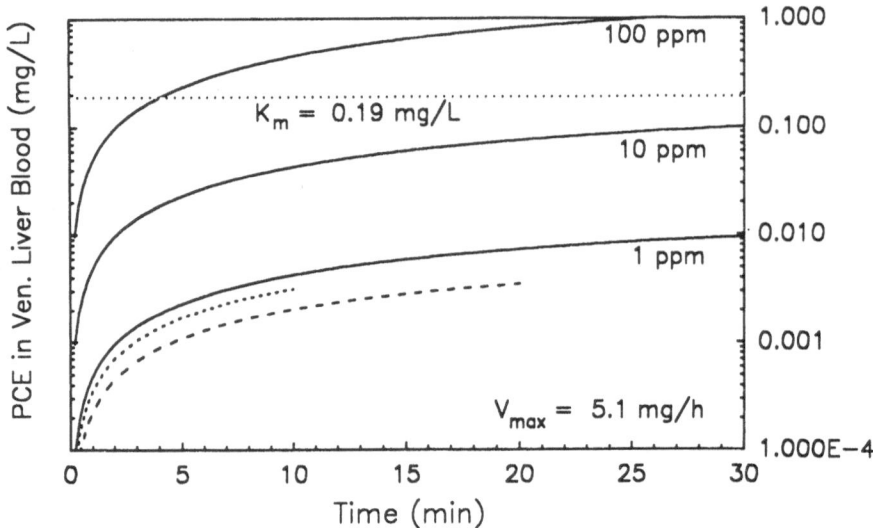

Fig. 8. PCE Concentration (B_1) in Venous Liver Blood Over Time in Response to Exposure to the Two Different PCE Concentrations in Air. Concentrations associated with the "reference" and "upper bound" estimates of shower exposures shown in Fig. 3, as predicted by a PBPK model using parameter values for a "reference" 70-kg male and the values of V_{max} and K_m shown. Also shown, for comparison, are the increases in B_1 predicted to result from continuous respiratory exposure to 1, 10, and 100 ppm PCE in air.

ACKNOWLEDGMENTS

This work was performed under the auspices of the U.S. Department of Energy by the Lawrence Livermore National Laboratory under Contract W-7405-Eng-48. Funding was provided under contract AF/DOE AML/86 with the U.S. Air Force Aerospace Harry G. Armstrong Medical Research Laboratory, Toxic Hazards Division.

REFERENCES

1. C. R. Cothern, W. A. Coniglio, and W. L. Maracus, *Techniques for the Assessment of the Carcinogenic Risk to the U.S. Population Due to Exposure from Selected Volatile Organic Compounds from Drinking Water via the Ingestion, Inhalation and Dermal Routes*, U.S. Environmental Protection Agency, Office of Drinking Water (WH-550), Washington, DC, PB84-213941 (1984).
2. J. B. Andelman, "Human Exposures to Volatile Halogenated Organic Chemicals in Indoor and Outdoor Air," *Environ. Health Perspect.* **62**:313-318 (1985).
3. H. S. Brown, D. R. Bishop, and C. R. Rowan, "The Role of Skin Absorption As a Route of Exposure for Volatile Organic Compounds (VOCs) in Drinking Water," *Am. J. Pub. Health* **74**:479-484 (1984).
4. T. E. McKone, "Human Exposure to VOC's in Household Tap Water: The Indoor Inhalation Pathway," *Envir. Sci. Technol.* **21**:1194-1201 (1987).
5. U.S. Environmental Protection Agency, Office of Health and Environmental Assessment, *Health Assessment Document for Tetrachloroethylene*

(Perchloroethylene): Final Report, Washington, DC, EPA/600/8-82/005F, NTIS Report No. PB85-249704 (1985).

6. U.S. Environmental Protection Agency, Office of Health and Environmental Assessment, *Addendum to the Health Assessment Document for Tetrachloroethylene (Perchloroethylene), Updated Carcinogenicity Assessment for Tetrachloroethylene (Perchloroethylene, PERC, PCE)*, Research Triangle Park, NC, EPA/600/8-82/005FA (1986).

7. E. L. Anderson, R. E. Albert, R. McGaughy, L. Anderson, S. Bayard, D. Bayliss, C. Chen, M. Chu, H. Gibb, B. Haberman, C. Hiremath, D. Singh, and T. Thorslund, "Quantitative Approaches in Use to Assess Cancer Risk," *Risk Anal.* 3:277-295 (1983).

8. T. B. Kirchner and J. M. Vevea, "PREMOD and MODAID: Software Tools for Writing Simulation Models," Third International Conference on State-of-the-Art in Ecological Modeling, Colorado State University, Fort Collins, CO (1982).

9. California Department of Health Services, Health and Welfare Agency, *Chemical Contamination of Large Public Water Systems in California*, Sacramento, CA (1986).

10. K. T. Bogen, L. C. Hall, T. E. McKone, D. W. Layton, and S. E. Patton, *Health Risk Assessment of Tetrachloroethylene (PCE) in California Drinking Water*, Lawrence Livermore National Laboratory, Livermore, CA, UCRL-15831 (1987).

11. J. C. Ramsey and M. E. Andersen, "A Physiologically Based Description of the Inhalation Pharmacokinetics of Styrene in Rats and Humans," *Toxicol. Appl. Pharmacol.* 73:159-175 (1984).

12. National Academy of Sciences, Safe Drinking Water Committee, *Drinking Water and Health* (National Academy Press, Washington, DC), Vol. 6, p. 168-225 (1986).

13. M. L. Gargas, H. J. Clewell, III, and M. E. Andersen, "Metabolism of Inhaled Dihalomethanes *in vivo*: Differentiation of Kinetic Constants for Two Independent Pathways," *Toxicol. Appl. Pharmacol.* 82:211-223 (1986).

14. M. E. Andersen, H. J. Clewell, III, and M. L. Gargas, F. A. Smith, and R. H. Reitz, "Physiologically Based Pharmacokinetics and the Risk Assessment Process for Methylene Chloride," *Toxicol. Appl. Pharmacol.* 87:185-205 (1987).

15. R. H. Reitz and R. J. Nolan, *Physiological Pharmacokinetic Modeling for Perchloroethylene Dose Adjustment: Draft comments on U.S. Environmental Protection Agency's Addendum to the Health Assessment Document for Tetrachloroethylene*, Dow Chemical Co., Midland, MI (1986).

16. D. Hattis, S. Tuler, L. Finkelstein, and Z. A. Luo, *A Pharmacokinetic Mechanism-Based Analysis of the Carcinogenic Risk of Perchloroethylene*, Report for Center of Technology, Policy and Industrial Development, Massachusetts Institute of Technology, Cambridge, MA (CTPID 86-7) (1987).

17. R. C. Ward, C. C. Travis, D. M. Hetrick, M. E. Andersen, and M. L. Gargas, "Pharmacokinetics of Tetrachloroethylene," *J. Toxicol. Appl. Pharmacol.*, submitted (1987).

18. M. Ikeda, H. Ohtsuki, T. Imamura, and Y. Komoike, "Urinary Excretion of Total Trichloro-Compounds, Trichloroethanol and Trichloroacetic Acid As a Measure of Exposure to Trichloroethylene and Tetrachloroethylene," *Br. J. Ind. Med.* 29:328-333 (1972).

19. M. Ikeda, "Metabolism of Trichloroethylene and Tetrachloroethylene in Human Subjects," *Environ. Health Perspect.* 21:239-245 (1977).

20. T. Ohtsuki, K. Sato, A. Koizumi, M. Kumai, and M. Ikeda, "Limited Capacity of Humans to Metabolize Tetrachloroethylene," *Int. Arch. Occup. Environ. Health* 51:381-390 (1983).

21. International Commission on Radiological Protection (ICRP), *Report of the Task Group on Reference Man*, ICRP No. 23, pp. 359-360, Pergamon Press, New York (1975).

22. G. Müller, M. Spassovski, and D. Henschler, "Metabolism of Trichloroethylene in Man, II. Pharmacokinetics of Metabolites," *Arch. Toxicol.* 32:283-295 (1974).

23. G. D. Kerr, "Organ Dose Estimates for the Japanese Atomic Bomb Survivors," *Health Physics* **37**:487-508 (1979).

24. A. C. Hindmarsh, "ODEPACK, A Systematized Collection of ODE Solvers," *IMACS Transactions on Scientific Computation* **1**:55-64 (1983).

25. A. M. Schumann, J. F. Quast, and P. G. Watanabe, "The Pharmacokinetics and Macromolecular Interactions of Perchloroethylene in Mice and Rats as Related to Oncogenicity," *Toxicol. Appl. Pharmacol.* **55**:207-219 (1980).

26. C. Mitoma, T. Steeger, S. Jackson, K. Wheeler, J. Rogers, and H. Milman, "Metabolic Disposition Study of Chlorinated Hydrocarbons in Rats and Mice," *Drug Chem. Toxicol.* **8**:183-194 (1985).

27. W. Dekant, M. Metzler, and D. Henschler, "Identification of S-1,2,2-Trichlorovinyl-N-Acetylcysteine As a Urinary Metabolite of Tetrachloroethylene: Bioactivation Through Glutathione Conjugation As a Possible Explanation of Its Nephrocarcinogenicity," *J. Biochem. Toxicol.* **1**:57-72 (1986).

28. D. G. Pegg, J. A. Zempel, W. H. Braun, and P. G. Watanabe, "Disposition of Tetrachloro-(^{14}C)-ethylene Following Oral and Inhalation Exposure in Rats," *Toxicol. Appl. Pharmacol.* **51**:465-474 (1979).

29. S. W. Frantz and P. G. Watanabe, "Tetrachloroethylene: Balance and Tissue Distribution in Male Sprague-Dawley Rats by Drinking-Water Administration," *Toxicol. Appl. Pharmacol.* **69**:66-72 (1983).

30. J. Fernandez, E. Guberan, and J. Caperos, "Experimental Human Exposures to Tetrachloroethylene Vapor and Elimination in Breath After Inhalation," *Amer. Industr. Hyg. Assoc.* **37**:143-150 (1976).

31. A. C. Monster, G. Boersma, and H. Steenweg, "Kinetics of Tetrachloroethylene in Volunteers; Influence of Exposure Concentration and Work Load," *Int. Arch. Occup. Environ. Health* **42**:303-309 (1979).

The Role of Speciation and Pharmacokinetics in Risk Assessment

R. M. Stern

World Health Organization Regional Office for Europe
Copenhagen Ø, Denmark

ABSTRACT

The role of speciation and pharmacokinetics in determining the potency of individual metal compounds is reviewed. An extremely wide range of responses per unit dose of substances which contain the same metallic ion in common is found for in vitro toxicity and genotoxicity studies of chromium and nickel. This suggests that regulation of metallic compounds must be based on risk assessment for individual exposures rather than on an assumption of similar risk for all exposures to the same metallic ion.

KEYWORDS: Speciation, pharmacokinetics, chromium, nickel, in vitro genotoxicity

INTRODUCTION

For the purposes of a discussion on the mediators of biological effects of trace metals, speciation is defined as the description of the chemical, physical, and morphological state of the elements of interest as they appear in various compartments in the environment, with the understanding that such characteristics affect the distribution and flow along various transport and uptake pathways that connect the sources of interest (environmental, extracellular, etc.) with the endpoints of interest (appearance of the element in body fluids, ultimate intracellular concentration, binding to DNA, etc.). Thus, speciation and pharmacokinetics are essential to any attempt at quantitative risk assessment of health effects of environmental exposures, especially for metals.

Interest in speciation of metals arises because, as opposed to well-defined organic entities, exposures to different substances which have a given metallic element in common frequently appear to have very different biological consequences. Thus, substances like chromium metal, potassium dichromate, calcium chromate, and chrome chloride must be treated separately, and generic discussions are for the most part inappropriate, especially for the question of regulation of exposure or a determination of toxic effects. This gives rise to questions concerning mechanisms of action, which ultimately require a description of the species involved, since different species may participate in different mechanisms. Similarly, the issue of dose requires an understanding of the mechanism of delivery, which is also target and species dependent. Although thermodynamics provides, in general, an adequate description of equilibrium between species in a closed compartment, few compartments are closed, passive cross-compartment flux terms are important, and reaction kinetics,

609

especially when permeable boundaries between compartments are present (i.e., cell membranes), involves active biological terms (i.e., ion channels and pumps).

METAL SPECIATION

The role of speciation in mechanism and dose can best be illustrated by a simple example of environmental exposures involving different uptake pathways, e.g., the controversy concerning the carcinogenicity of chromium compounds. Human respiratory tract cancer risk has been associated with occupational exposure to dust in the primary chromium industry (reduction of chromite ore), the pigment industry (exposure to lead and zinc chromates), ferrochrome industry, stainless steel welding, and perhaps plating, but not with tanning. Exposures are typically to a mixture of soluble and insoluble compounds of trivalent and hexavalent chromium. Frequently the substances are poorly defined in terms of purity, extent of hydration, morphology, particle size, and formulation and process technology. Animal studies on pure compounds and some industrial products suggested that only moderately water soluble compounds such as calcium, strontium, and zinc chromates were experimental carcinogens. Based on early studies, an exposure criteria document prepared by NIOSH (the U.S. National Institute of Occupational Safety and Health) declared that based on the human evidence, highly soluble chromate [Cr(VI)] compounds were noncarcinogenic, while certain sparingly soluble chromate compounds were human carcinogens. Since that time, two separate arguments have developed: one concerning mechanism (i.e., which chromium species are active at the level of DNA) and one concerning potency (how does speciation affect delivery to target molecules).

In vitro studies have demonstrated that Cr(VI) is stable for long periods (i.e., one week) under biological conditions, progressing through short-lived states of Cr(V) and Cr(IV) to stable Cr(III) complexes, although free Cr^{+++} is not soluble under physiological conditions. All oxidation states are active with free DNA, the ultimate lesion involving bound Cr(III). Additional studies have demonstrated that soluble Cr(VI) can produce tumors in animals, provided the dose is continually repeated during the experiment.

Re-examination of the animal data for chromium carcinogenicity demonstrates that there is an extremely wide range of potency for the various chromates and industrial material, suggesting that the highly insoluble chromates may be carcinogens but of such low potency as to be undetectable because of statistical limitations due to the presence of naturally occurring cancers in the control groups. The question of soluble chromates in animal studies is complicated by the fact that the experimental protocols involve single dose implantations which do not provide for long-term continuous exposures. A re-examination of the human exposure data suggests that risk is attributable to cumulative occupational exposure to "soluble" Cr(VI) regardless of the nature of the compound.

TOXIC AND GENOTOXIC POTENCY OF METALS

Thus, potency of metallic carcinogens becomes the issue, rather than carcinogenicity per se; for a given situation, potency is determined by delivered dose and mediated by the length of time of "delivery" and the ultimate concentration at the target. In vitro and in vivo evidence demonstrates that free and bound Cr(VI) can penetrate cellular membranes, but that, with few exceptions, Cr(III) does not.

For the case of inhalation toxicity of chromates, a number of problems arise: (a) stability of species in the air, especially in occupational settings where pyrolytic reactions may produce short-lived active species; (b) place of deposition in the respiratory tract where multidisperse aerosols can contain size fractions with different chemistry; and (c) clearance mechanism, whereby large particulates can be cleared by upward mobility in the mucocilliary sheath to the mouth and hence to the gut, while small particles may either dissolve or be phagocytosed by macrophages. Thus, initial speciation determines environmental fate, and speciation upon deposition in terms of solubilization in extra- and intra-cellular fluids determines compartmentalization and pharmacokinetics.

In addition to the ingestion and inhalation pathways described above, dermal uptake is important for chromium under certain circumstances. Cr(VI) in the form of CrO_4^- penetrates the skin easily, while Cr(III) in the form of $Cr(H_2O)_6^{+++}$ does not penetrate well; rather it hydrolyzes and binds to skin constituents. For commercial Cr(III) compounds, skin penetration is least for the nitrate, higher for the sulfate, and highest for the chloride. Anhydrous $CrCl_3$ is insoluble but the hydrated form $[CrCl_2(H_2O)_4]Cl(H_2O)_2$ is soluble. There is no oxidation of Cr(III) under physiological conditions, and the half-life of free Cr(VI) is of the order of one hour or more. Massive exposure to organic Cr(III) complex used in tanning has given rise to fatal Cr poisoning. The distribution of Cr(VI) species is extremely dependent on pH.

The effective dose of Cr under occupational exposures therefore is extremely sensitive to speciation. For the case of nonoccupational exposures, especially with respect to contact determatitis and Cr (and Ni) hypersensitization, solubility of the metal from the matrix (coins, jewelry, etc.) in sweat constituents (e.g., histidine) determines the potential dose. Understanding the potential for delivering an effective dose per unit external (dermal) exposure therefore requires a suitable description of the substance in question in terms of oxidation state, hydration, etc.

The significance of speciation in metal toxicity is determined by the time and place of interest, and it is different at the organ, tissue, cellular and molecular levels. Two separate classes of regions can be identified along the pathways from initial source to ultimate target. Within a given compartment (the surface of the lung, blood serum, kidney cell, etc.), one can consider what mechanisms give rise to speciation, e.g., the presence of metal binding proteins, pH, etc. At the interface between compartments (capillary wall, cell wall, nuclear membrane), one can consider which factors determine transport across the barrier and which mechanisms determine concentrations. In particular, it would appear that the cell plays an active role in creating the "ultimate" compartment for determination of toxicity and effective dose of the putative agent, since many metals have tissue-specific targets (Cd in the kidney; Ni, Cr in the lung). This must be a result of the details of relative ease and stability of ligand formation, the presence of specific binding proteins, and the degree to which the environmental agent can enter into natural reactions, replacing or complexing essential constituents, etc.

In the sense used above, the term speciation is extremely general, including all possible distinctions between chemical substances, some of which involve weak, transient and localized states, and some of which are robust and survive along the pathway from source to target. Because of the degree of complexity of the biological systems of interest, and the generality of the term "speciation," it is necessary to restrict interest to characteristics of importance only. Selection of significant species, pathways, etc., must always be based on a determination of the putative agent at the ultimate target, together with an inventory of the pathways and reactions which link this species to the external exposure, and one must distinguish between important and unimportant species, especially if the thrust of discussion is to determine the role of speciation in characterizing risk.

IN VITRO STUDIES

Because several types of mutational changes have been associated with tumor development, considerable interest has centered on the use of short-term in vitro genotoxicity tests as predictive assays for carcinogenicity. For the case of organic substances, although a wide range of responses is found among the 100 available assays, it is generally thought that use of a limited battery of test systems provides a good level of specificity and reliability for distinguishing carcinogenic from noncarcinogenic compounds. The case for metals is, however, unclear, since at least one compound of each of the almost 50 metallic elements tested is positive in at least one bioassay. Furthermore, there appears to be no simple relationship between the chemical and the genotoxic properties of the metal compounds. Animal laboratory studies and human occupational exposures suggest that certain compounds of only As, Be, Cd, Cr, Ni, and Pb are potential mammalian carcinogens. There is still considerable controversy over the data for Be, Cd, and Pb (the chromate is active); As is not an experimental animal carcinogen, although there is no doubt that exposure leads to cancer in humans; and the various Ni and Cr compounds and occupational exposures exhibit such a wide range of potencies that the classification of Cr and Ni compounds of various oxidation states and solubilities in terms of carcinogenic properties is still a subject of debate. Furthermore, only a limited number of the remaining metals have been tested experimentally or have been involved in significant human exposure, leaving the question of the carcinogenicity of the majority of the metals untested and unanswered.

In order to understand the possible relationships between the toxicity, genotoxicity and carcinogenicity of metals and metalloids, it is useful to examine three separate aspects of the problem: to review the results of genotoxicity assays as applied to metals, to study the details of the response of specific test systems and protocols to the characteristics of the material tested, and to investigate the specific mechanisms by which metals produce genetic damage.

The most widely used short-term assay, Salmonella typhimurium test (Ames), was designed to search for specific *his* point mutations (base pair substitutions and frameshifts), and is generally accepted to meet criteria established for predictive tests of carcinogenicity. Unfortunately, of the metals, only Cr(VI) appears directly active, suggesting that perhaps metal carcinogens may act by other mechanisms, or that there is only an uptake pathway for this metallic species in this particular assay. Many metals have, however, recently been shown to produce forward mutations in some bacteria and in mammalian somatic cells, e.g., salts of Pb, Mn, Ni, Cr, W, Mo, Be, Co, and As, although the results vary from assay to assay (e.g., HGPRT locus in Chinese hamster V79 cells, *lacI* gene mutation and lambda prophage induction in Escherichia coli, etc.), and from compound to compound of the individual metals, presumably due to variations in uptake and mechanism.

Structural chromosome anomalies have been closely related to different steps in carcinogenesis. In vitro studies of clastogenic properties of inorganic compounds have yielded positive results for salts of As, Cr(VI), and Ni; Cd, Pb and Hg salts were negative in these assays.

It has recently become possible to visualize sister chromatid exchanges (and chromosome breaks) in vitro. Although the mechanisms and biological significance are for the most part unknown, the technique appears to be the most sensitive available to demonstrate the cytological effects of chemical mutagens. Positive results have been obtained with salts of As, Be, Cr, Ni, and Cd; Pb and Hg salts tested negative.

Turbagenic compounds interfere with the spindle apparatus and induce disorders in chromosome distribution, resulting in aneuploîd cells and c-mitosis, mainly via inhibition of tubuline polymerization. Although mitotic poisons are not true mutagens, they can contribute to mutagenesis. The metal salts tested on plant material have yielded positive results (As, Cd, Cr, Pb, Hg, Ni, etc.).

The in vitro cell transformation assay has been used in some laboratories with success in assaying the carcinogenic potential of inorganic compounds. In a number of cases it has been demonstrated that the cells transformed in vitro by exposure to experimental carcinogens (i.e., nickel sub-sulfide) give rise to tumors when reimplanted in vivo. Positive results have been obtained for Ag(I), As(III), As(V), Be(II),Cd(II), Co(II), Cr(VI), Cu(II), Fe(II), Hg(II), Mn(II), Ni(II), Ni(III), Pb(II), Pt(II), Pt(IV), Sb(III), Tl(I), Zn(II); and negative results for Al((III), Ba(II), Cu(II), Li(I), Mg(II), Na(I), Sr(II), Ti(IV), and W(VI).

In general, the prevalence of positive results for the metals in many in vitro assays and the wide variation in response from compound to compound in a given assay have given rise to considerable debate as to the utility of in vitro testing of metallic and inorganic compounds.

QUANTITATIVE STUDIES WITH Ni AND Cr IN TRANSFORMATION ASSAY

Insight into the origin of the range of responses of in vitro assays when used with metals comes from investigating the details of the BHK-21 cell transformation assay, originally considered to have great promise for detecting carcinogens, but which, because of its instability, has not been widely used. The assay has the advantage that it is easy to score for both toxicity (survival) and transformation (growth of large colonies in soft agar). The assay also has separate uptake pathways for soluble (via membrane permeability) and insoluble (via phagocytosis) fractions.

Experiments made with a range of nickel compounds (several Ni oxide and metallic Ni powders from various industrial processes including MIG/Ni and MIG/SS welding fumes, nickel subsulfide, and nickel acetate) show a range of toxicity and transformation potency of a factor of 30 (in terms of test compound or nickel content). However, identical transformation potencies are observed at equitoxic (LD50%) doses. This implies that if a toxic dose is reached, it is indicative of the bioavailability of the test substance: if the toxicity is due to the test material [Ni(II)], then at equal bioavailable dose, Ni induces identical transformation rates, regardless of source.

When used with a soluble Ni compound (nickel acetate), the toxicity and transformation rates increase with time of exposure, indicating continued uptake via membrane permeability. When used with relatively insoluble particulates (<4 microns), no increase in response is observed after 6 to 8 hours of exposure, implying that uptake via phagocytosis is complete after this time. Most substances have limited solubilities, so that uptake is via both pathways. It is found, however, that separate testing of the soluble and insoluble fractions resulting from incubation of Ni metal or Ni oxide in the bioassay medium (which contains serum proteins) demonstrates that the soluble fraction is inactive; all activity is due to the insoluble particulates. This suggests that membrane permeability is highly selective and species dependent. Solubilized nickel bound to serum proteins is excluded from the assay.

Identical results are found for a range of Cr(VI) compounds of different solubility (MIG/SS and MMA/SS welding fumes, potassium dichromate, and the chromates of Ca, Zn, and Pb). Although the range of toxic dose is over a factor of 20, identical transformation potencies (different from that for Ni) are observed at equitoxic doses. Furthermore, for the case of lead chromate, its intracellular solubility is so low that exposure must continue for 6 to 8 days after phagocytosis is complete in order to achieve a LD50% dose. Negative results are found for Cr metal and particulates coated with Cr(III) and soluble Cr(III) salts, suggesting that there are no uptake pathways for these compounds through either the cell or the nuclear membrane.

The general conclusion from these studies is that interpretation of the results of in vitro bioassay for metals is difficult since potency is so strongly dependent on bioavailability, and that bioavailability is extremely species dependent, and dependent on bioassay protocol as well. Many assays do not contain a phagocytosis pathway, thus leaving insoluble fractions untested. Similarly, some solubilized but complexed species may also be excluded from the test system because of the lack of membrane permeability.

With respect to the specific problem of welding fumes, it would appear that when studied with a wide range of in vitro bioassay techniques, Ni and Cr as contained in such fumes behave as expected for Ni and Cr with the exception of questions of bioavailability as dictated by the welding fume matrix. Genotoxicity in welding fumes is apparently limited to these two metals, with one exception: in the micro-titre test for induction of lambda prophage in E. Coli, one MMA/MS fume is active, presumably due to the presence of Mn.

STUDIES ON Cr

A proper assay of genotoxicity for metalloids must therefore rely on questions of speciation and chemical properties, uptake mechanisms, metabolic changes, interaction with DNA and its environment, and DNA repair. The details of chromium genotoxicity are illustrative of the problem: chromate ions are found to be effectively transported through anion channels and Cr(VI) is an effective mutagen and carcinogen. Cation channels transport Cr(III) poorly due to their specificity, although the reduction of Cr(VI) within the cell traps Cr(III) and leads to considerable accumulation of Cr. Uptake of Cr(III) in human erythrocytes is considerably enhanced by complexation with hydrophobic ligands [i.e., $Cr(Bipyrridine)_2Cl_2^+$ or $Cr(Phenantroline)_2Cl_2^+$].

Cr(VI) produces DNA-protein-crosslinks in vivo and in culture; Cr(III) is found to bind to isolated nuclei and induce the same types of DNA damage when isolated nuclei are exposed to freshly solubilized $CrCl_3$-hydrate or $Cr(Glycene)_3$.

Recent results suggest that Cr(III) complexes with hydrophobic ligands are positive in Ames tests, presumably because of enhanced membrane permeability; certain Cr(III) compounds now show weak but significant activity in Chinese hamster V79 HGPRT assay as well.

Certain cations [Cu(II), Ni(II)] have been shown to dipurinize isolated DNA. Reduction of Cr(VI) to Cr(III) takes place over several steps, including the formation of a Cr(V) species with a free radical electron which is stable in vivo. It has now been shown that Cr(V)glutathione complex, created in vitro from chromate and glutathione, induces DNA breaks in supercoiled phage PM2 DNA, whereas potassium chromate and chromium trichloride are inactive.

Cr(III) and other metal ions decrease fidelity of bacterial and mammalian polyermerases. Both Cr(VI) and Ni(II) augment the mutagenic potency of UV light in CH V79 HGPRT mutation assay, as has been found for other metals (e.g., As). This implies that metals may act as indirect mutagens by inhibiting DNA synthesis or repair.

CONCLUSIONS

Thus, metal ions may interfere at several points with the structure and function of genes. The degree to which these interactions can be detected in vitro is strongly dependent on the protocol of the assay and the endpoint studied. Metal ions may also interfere with metabolic processes in vivo, and interfere, for example, with Ca(II) dependent regulation of cell proliferation, etc. Even though there is a considerable amount of in vitro information available for the action of Cr, the most well studied of the metals, little can be said about the relationship between the characteristics of Cr(VI) and Cr(III) compounds and their genotoxicity in vitro or in vivo; the activity of Cr(III) is a very recent observation and the predominant mechanism of Cr is still unknown. The situation is even less clear for the other metals.

The lesson to be learned from this discussion is that quantitative risk assessment for environmental exposures to metals must be based on studies of individual substances. Two compounds which have the same metallic ion in common cannot be assumed to represent the same potential risk per unit exposure, and generic regulation (e.g., for all Cr, all Ni compounds, etc.) is not an acceptable substitute for risk assessment of individual exposures.

ACKNOWLEDGMENT

This review is based on the author's summary reports from the Second Nordic Symposium on Trace Elements in Human Health and Disease, Odense, Denmark, 17-21 August 1987, and contains significant contributions from D. Beyersmann, A. Leonard, A. Astrup Jensen, and E. Nieboer, which are gratefully acknowledged.

BIBLIOGRAPHY (Selected Reading)

B. Buttner and D. Beyersmann, *Xenobiotica* **15**:735-741 (1985).

A. Kortenkamp, D. Beyersmann and P. O'Brien, *Toxicol. Environ. Chem.* **14**:23-32 (1987).

L. Tkeshelashvili, *Cancer Res.* **40**:2455-2460 (1980).

P. H. Connett and K. E. Wetterhan, *J. Am. Chem. Soc.* **107**:4282-4288 (1985).

IARC, Lyon, *Monograph* **23**:205-324 (1980).

K. Hansen and R. M. Stern, *In Nickel in the Human Environment*, F. W. Sunderman, Jr., Ed., IARC, Lyon 1984, pp. 193-200.

K. Hansen and R. M. Stern, *J. Appl. Tox.* **5**:306-316 (1985).

K. Hansen and R. M. Stern, *J. Am. Col. Tox.* **3**:381-430 (1984).

Assessing Risks of Cholinesterase-Inhibiting Pesticides

Lisa Y. Lefferts

University of North Carolina
School of Public Health
Chapel Hill, NC

ABSTRACT

The U.S. Environmental Protection Agency (EPA) generally uses cholinesterase inhibition (ChE-I) to characterize the risks of organophosphate and carbamate pesticides since ChE-I is a sensitive predictor of exposure. Assessing the risks of these widely used pesticides on the basis of ChE-I involves uncertainties which necessitate science policy decisions, including, "How large an uncertainty factor (UF) should be used when estimating a reference dose (RfD) or acceptable daily intake (ADI)?" Although EPA and other agencies recommend that a 100-fold factor be used when extrapolating from valid results from long-term studies on animals, EPA has frequently used an UF of 10 for ChE-I pesticides. The UF used should be large enough to account for the uncertainty in extrapolating from results in animals to results in humans (interspecies variation) and the uncertainty regarding different susceptibilities among humans (intraspecies variation). A review of the literature and a study of a small sample of regulated pesticides suggest that the use of an uncertainty factor as small as ten to account for inter- and intra-species variation of ChE-I may not be justified.

KEYWORDS: Risk assessment, cholinesterase inhibition (ChE-I), uncertainty factor

INTRODUCTION

Cholinesterase-inhibiting (ChE-I) pesticides, which include organophosphate and carbamate pesticides, are among the most widely used pesticides world-wide, both for agricultural and nonagricultural purposes.[1] For example, over half of the total volume of insecticides used in the U.S. are ChE-I pesticides,[2] and four of the top ten pesticides used by homeowners are ChE-I pesticides.[3] Furthermore, organophosphate pesticides are implicated in more human poisonings than any other class of pesticides.[1] Risk assessments conducted by the U.S. Environmental Protection Agency (EPA) generally characterize the risk of these compounds by their ability to inhibit cholinesterases (ChE's), since cholinesterase-inhibition is a sensitive indicator of exposure. This paper examines available data on intra- and inter-species variability in response to ChE-I pesticides in order to help resolve the debate on the size of the uncertainty factor used in deriving risk estimates for these compounds.

CHOLINESTERASES

In mammals there are two principal types of ChE's: acetylcholinesterase (AChE) (also called specific, true, or "e" type cholinesterase) and butyrylcholinesterase (BuChE) (also called nonspecific, pseudocholinesterase, or "s" type cholinesterase). ChE-I pesticides differ in their ability to selectively inhibit these enzymes. AChE is found in the central nervous system, motor end plates of skeletal muscle, and erythrocytes, and plays a key role in normal nervous system functioning by hydrolyzing the neurotransmitter acetylcholine. It is typically the AChE found in erythrocytes (and/or the BuChE found in plasma) that is measured to determine the toxicity of organophosphate and carbamate pesticides, since it is impractical to measure synaptic AChE activity. However, the function of erythrocyte AChE and plasma ChE is not well established.[4-8] Although BuChE and AChE share many similarities (mechanism of action, molecular shape), it is not known how, if at all, they are related physiologically.[4] Despite these uncertainties regarding the roles of AChE in erythrocytes and BuChE in plasma, substantial inhibition of these enzymes is associated with overt adverse effects in the whole organism following administration of organophosphates and carbamates and is currently used as a measure of toxicity and exposure to ChE-I pesticides.

UNCERTAINTY FACTORS

Although several approaches to using experimental dose-response data to obtain estimates of response at low, policy-relevant doses are available, as a matter of policy, EPA generally uses the NOEL-UF ("no observed effect level"-uncertainty factor) approach for systemic toxicants. In this approach, a reference dose (RfD), formerly known as an ADI or acceptable daily intake, is estimated by

$$RfD = NOEL/UF ,$$

where the NOEL is defined as the level (quantity) of a substance administered to a group of experimental animals which demonstrates the absence of adverse effects observed or measured at higher dose levels.[9] A UF is a number intended to account for the uncertainty in using a response at a single dose (the NOEL) from an experimental study to estimate a level which is likely to be without an appreciable risk of deleterious effects to the human population (including sensitive subgroups). Traditionally, UF's are multiples of ten.

Generally, EPA recommends that a 10-fold factor be used when extrapolating from valid experimental results from studies using prolonged exposure to average healthy humans and a 100-fold factor when extrapolating from valid results of long-term studies on experimental animals when results of human exposure are not available or inadequate. This 100-fold factor is really the product of two 10-fold factors: one 10-fold factor is intended to account for the variation in sensitivity among the members of the human population (intra-species variation), and the second 10-fold factor is intended to account for the uncertainty in extrapolating animal data to the case of humans (interspecies variation). These recommendations are similar to those made by the National Academy of Sciences, the Food and Agricultural Organization/World Health Organization, the World Health Organization Expert Committee for Pesticide Residues, and the U.S. Food and Drug Administration.[11]

However, EPA has frequently used an uncertainty factor of 10 when the NOEL is based on ChE-I by organophosphorus and carbamate pesticides. Does the evidence on intra- and inter-species variability in response to ChE-I pesticides support this selection of an uncertainty factor?

ANIMAL TO HUMAN EXTRAPOLATION (INTERSPECIES VARIATION)

In an attempt to quantitatively estimate the uncertainty in extrapolating experimental results from animals to humans, human NOELs for ChE-I were compared to other animal NOELs for nine pesticides for which studies in humans were available to EPA. Unfortunately, differences in study design and inadequate reporting of human data in several cases made valid comparisons difficult. Much of the human data were collected on prisoners or "volunteers" (usually male) and thus did not reflect the human population at large. The experiments on humans were typically of shorter duration than the animal experiments to which they were compared, making such comparisons questionable. In two instances human and animal studies of ChE-I NOELs (in mg/kg) were of comparable duration:

Malathion:
$$\frac{\text{Rat (erythrocyte) ChE-I NOEL (33 day)}}{\text{Human (erythrocyte) ChE-I NOEL (47 day)}} = \frac{5}{0.23} = 21.7 \text{[12]}$$

Dimethoate:
$$\frac{\text{Rat (erythrocyte) ChE-I NOEL (34 day)}}{\text{Human (erythrocyte) ChE-I NOEL (39 day)}} = \frac{0.7}{0.2} = 3.5 \text{[13]}$$

This implies that healthy adult men are 22 times more sensitive to ChE-I by malathion than rats, and adults (presumably healthy men) are nearly 4 times more sensitive to ChE-I by dimethoate than male rats. However, these results should be interpreted cautiously. For malathion, only 5 men (prisoners) were tested; body weights of men were not recorded (assumed 70 kg); the dosing procedure was not clear (e.g., not indicated whether dosed on a full or empty stomach, although both rats and humans were dosed orally); and the dosing schedule was somewhat unusual (i.e., administered 8 mg to each man every day for 32 days, gave no treatment for 3 weeks, then administered 16 mg/day for 47 days). For dimethoate, only 9 humans (gender not reported) were tested at the NOEL, and the route of administration differed (intraperitoneal injections for rat and oral aqueous solution for humans). Age of subjects (rats and humans) were not reported for either chemical. However, the human study of dimethoate was rated as "supplementary upgraded to minimum" by EPA's Office of Pesticide Programs and "High Confidence" by EPA's RfD workgroup. The human study of malathion is the basis for the ADI set by the World Health Organization.

Since opportunities to directly compare animal and human NOELs for ChE-I were limited, comparisons were also conducted between different animal species for studies of comparable length (see Table 1). If different species of test animals differ from each other in sensitivity, it is likely that humans will differ from test animals. Interspecies differences ranged from no difference (ratio = 1) to greater than a 25-fold difference, with a mean ratio of 5.

Due to the small number of comparisons which were made and differences between studies (e.g., in how, when, and for how many animals ChE levels were measured), no firm conclusions can be made from the data in Table 1. However, for the pesticides analyzed:

a. Humans were more sensitive to ChE-I than rats in the two instances where comparisons could be made;

b. Dogs are often (but not always) more sensitive than rats to ChE-I;

c. ChE-I measured at a particular site (i.e., plasma, erythrocytes, brain) did not vary more across species than at other sites; and

d. ChE-I measured after chronic exposure did not vary more across species than ChE-I measured after subchronic exposure.

Table 1. Comparisons of ChE-I NOELs Between Animal Species*

Pesticide	Species Compared	Subchronic Study[a]			Chronic Study[b]		
		Plasma	RBC	Brain	Plasma	RBC	Brain
Parathion	Rat/Dog	$\geq$0.4	$\geq$0.4	---	$\geq$2.5	---	$\geq$25.0
Malathion	Rat/Human	---	21.7[d]	---			
Chlorthiophos	Rat/Dog	$\geq$ 1	4	$\leq$1.9	0.4[c]	3.2[c]	---
Metasystox		(Studies Deficient)					
Ethion	Rat/Dog	6	---	---			
Chlorpyrifos	Rat/Monkey	1.9[e]	1.9[e]	---			
	Rat/Dog				10	1	1
Carbofuran	Mouse/Rat				3	3	3
	Mouse/Dog				6[f]	6[f]	---
Dimethoate	Rat/Dog	---	3.2	---			
	Rat/Human	---	3.5[g]	---			
Aldicarb Sulfoxide	Dog/Rat	> 2[h]	---	---			

---

* Derived from U.S. EPA Office of Pesticide Programs data.

[a] Subchronic studies are 90 day feeding studies, unless otherwise noted.

[b] Chronic studies are 2 years feeding studies, unless otherwise noted.

[c] Dog study was 1 year, rat study was 2 years.

[d] Human study was 47 days, rat study was 33 days.

[e] Monkey dosed by gavage.

[f] Both studies (mouse and dog) were 6 months.

[g] Rat dosed by intraperitoneal injection.

[h] "It does not appear that doses were established" in rat study, and plasma ChE-I was measured in dog study after 1 month, although it was a 3 month study.

VARIATION IN HUMAN SENSITIVITY (INTRA-SPECIES VARIATION)

Many factors are known or appear to influence sensitivity among individuals of the same species. For example, a genetically "atypical" cholinesterase has been identified, determined by a recessive gene (designated $E_1{}^a$). Persons who are homozygous for the recessive gene ($E_1{}^a E_1{}^a$), estimated to be about 4.5% of the population, have low BuChE activity (about 25% of normal, or $E_1{}^u E_1{}^u$). Heterozygotes have about 78% as much activity as normal.[14-16] Different races appear to have different frequencies of nontypical ChE's. According to Udsin,[15] a population of southern Eskimos studied had a high incidence of atypical homozygotes and heterozygotes, as well as "silent" gene individuals (with ineffective plasma ChE). The frequency of heterozygotes in Israel varies from 0.7% among North African Jews to 3.1% among European Ashkenazi Jews to 9.7% among Jews from Iraq and Iran.[15,17]

Besides genetic differences in ChE's, there are also genetic variants of arylesterases (also affected by ChE-I pesticides). Paraoxonase, an enzyme hydrolyzing paraoxon, the active metabolite of parathion, is an arylesterase which has been found to be polymorphically distributed in several populations.[18] About one half of the U.S. caucasian population is homozygous for a low activity allele (AA) which is speculated to place these individuals at higher risk of parathion poisoning than those with higher levels.[19]

In addition to genetic factors, there are sex-related differences in susceptibility to ChE-I pesticides. These are most likely to occur for those compounds (e.g., parathion) which require metabolic activation to produce ChE-I.[2] For example, Agarwal et al.[20] found female rats to be more susceptible than males to acute toxicity caused by parathion. Castration increased the susceptibility of male rats to a similar level as females. Sex differences also exist in absorption, distribution, and excretion. Among humans, males have higher ChE activities than females in their plasma and erythrocytes.[5] How ChE activity affects susceptibility in humans is not clear, although in the rat, the female is more susceptible and has higher ChE activity than the male.

Kacew and Reasor[21] report that "it is clear that neonates are more susceptible than adults to AChE inhibition." In the human, adult levels of AChE are not reached until about three to five months of age. Even though blood from neonates has a higher proportion (compared to adults) of young cells, which are associated with higher AChE activities than more mature cells, AChE activity in newborn circulating erythrocytes is less.[7] Thus, for a given concentration of ChE-I pesticide, more AChE is expected to be inhibited in newborns compared to adults.[21] This expectation seems to be borne out in animal studies, at least for some pesticides.[22-24] For example, chlorpyrifos is 30 times more toxic to calves than adult cattle.[21]

Nutritional deficiencies have been found in some cases to increase the susceptibility of test animals to ChE-I pesticides. Parathion, malathion, and banol each produced greater ChE-I in rats on low protein diets than on high protein diets.[25,26] Likewise, behavioral changes were noted more often in rats on low protein diets exposed to parathion or banol than in unexposed rats on low protein diets or exposed rats on high protein diets.[27] Dietary factors other than protein (casein), including calcium and magnesium intake[28] and food or water restricted diets,[29,30] also affect susceptibility to ChE-I by parathion.

Pregnancy also may affect susceptibility to ChE-I pesticides. Plasma ChE levels fall markedly during the first trimester of pregnancy and following delivery.[31,32] Do these low levels of plasma ChE affect susceptibility to ChE-I pesticides? Recall that it was unclear whether individuals possessing non-normal genotypes and low levels of plasma ChE would be more susceptible to ChE-I pesticides due to the decreased ability of plasma ChE to hydrolyze cholines, or less susceptible since their plasma ChE was found to be less sensitive to fluoride, dibucaine, and some other compounds that also inhibit cholinesterase. Weitman et al.[33] found pregnant mice to be *more* susceptible to single doses of parathion and paraoxon than virgin female controls. In pregnant mice, signs of cholinergic stimulation (tremor, weakness, lacrimation, salivation) were more intense, brain and plasma ChE activities were lower, blood and brain concentrations of parathion and paraoxon were higher, and serum paraoxonase activities were lower, compared to controls. Whether pregnancy-induced alterations of hepatic function, ChE activity, serum protein binding, serum esterases or a combination of these are responsible for the enhanced susceptibility is unclear.

The susceptibility to ChE-I pesticides can also be affected by exposures to physical factors (e.g., cold[2,34]), biological agents (e.g., viruses[35]), certain disease conditions,[17] and other toxic substances (e.g, pesticides,[36] drugs[1]).

RECOMMENDATIONS

The evidence adduced in the previous pages indicates that an uncertainty factor as small as ten is inadequate to account for the inter- and intraspecies variability in response to ChE-I pesticides. A UF of 10 is frequently applied by EPA to NOELs based on ChE-I, even when derived from an animal study.

In keeping with the recommendations of NAS, WHO, FDA, and EPA (for non-carcinogenic compounds other than ChE-I pesticides), it is recommended that an uncertainty factor of 100 be used when deriving a reference dose from a valid study in animals. In cases where available data demonstrate that the inter- and intraspecies variation is substantially different than 100, a modifying factor (MF) can be used. Documentation to justify the selection of a MF other than 1 should be included in the risk assessment.

REFERENCES

1. J. K. Marquis, "Contemporary Issues in Pesticide Toxicology and Pharmacology," *Concepts in Toxicology* **2**:1-108 (1986).
2. J. Doull *et al.*, *Casarett and Doull's Toxicology* (2nd edition), Macmillan Publishing Co., New York (1980).
3. U.S. General Accounting Office, *Nonagricultural Pesticides: Risks and Regulation*, GAO/RCED-86-97, Washington, DC (1986).
4. J. A. Edwards and S. Brimijoin, "Divergent Regulation of Acetylcholinesterase and Butyrlcholinesterase in Tissues of the Rat," *Journal of Neurochemistry* **38**:1393-1403 (1982).
5. J. H. Wills, "The Measurement and Significance of Changes in the Cholinesterase Activities of Erythrocytes and Plasma in Man and Animals," *CRC Critical Reviews in Toxicology* **1**:153-201 (1972).
6. T. Namba, "Cholinesterase Inhibition by Organophosphorus Compounds and Its Clinical Effects," *Bulletin of the World Health Organization* **44**:289-306 (1971).
7. F. Herz and E. Kaplan, "A Review: Human Erythrocyte Acetylcholinesterase," *Pediatric Research* **7**:204-214 (1973).
8. K. M. Kutty, "Review: Biological Function of ChE," *Clinical Biochemistry* **13(6)**:239-243 (1980).
9. U.S. Environmental Protection Agency, *Hazard Evaluation Division Standard Evaluation Procedure, Toxicity Potential: Guidance for Analysis and Evaluation of Subchronic and Chronic Exposure Studies*, EPA-540/9-85-020, Washington, DC (1985).
10. U.S. Environmental Protection Agency, *Integrated Risk Information System Supportive Documentation*, Appendix A, EPA/600/8-86/032a, Washington, DC (1987).
11. M. L. Dourson and J. F. Stara, "Regulatory History and Experimental Support of Uncertainty (Safety) Factors," *Regulatory Toxicology and Pharmacology* **3**:224-228 (1983).
12. H. Moeller and J. Rider, "Plasma and Red Blood Cell Cholinesterase Activity as Indications of the Threshold of Incipient Toxicity of EPN and Malathion in Human Beings," *Toxicology and Applied Pharmacology* **4**:123-130 (1962); and H. Golz and C. B. Schaffer, *Summary of Pharmacology and Toxicology*, American Cyanamid, New York (1956).
13. E. Edson *et al.*, "Safety of Dimethoate Insecticide," *British Medical Journal* **4**:554-558 (1967); and D. Sanderson and E. Edson, "Toxicological Properties of the Organophosphorus Insecticide Dimethoate," *British Journal of Industrial Medicine* **21**:52-64 (1964).
14. F. M. Williams, "Clinical Significance of Esterases in Man," *Clinical Pharmacokinetics* **10**:392-402 (1985).

15. E. Udsin, "Reactions of Cholinesterases with Substrates, Inhibitors, and Reactivators," *Anticholinesterase Agents, International Encyclopedia of Pharmacology and Therapeutics* 1:60-122 (1970).

16. T. M. Ashby, J. E. Suggs and D. L. Jue, "Detection of Atypical Cholinesterase by an Automated pH Stat Method," *Clinical Chemistry* **16(6)**:503-506 (1970).

17. A. Silver, "The Biology of Cholinesterases," *Frontiers of Biology* **36**:1-596 (1974).

18. B. La Du and H. W. Eckerson, "Could the Human Paraoxonase Polymorphism Account for Different Responses to Environmental Chemicals?," *Banbury Report* **16**:167-175 (1984).

19. J. Ortigoza-Ferado *et al.*, "Biochemical Genetics of Paraoxonase," *Banbury Report* **16**:177-182 (1984).

20. D. K. Agarwal *et al.*, "Influence of Sex Hormones on Parathion Toxicity in Rats: Antiacetylcholinesterase Activity of Parathion and Paraoxon in Plasma, Erythrocytes, and Brain," *Journal of Toxicology and Environmental Health* **9**:451-459 (1982).

21. S. Kacew and M. J. Reasor, *Toxicology and the Newborn*, Elsevier Press, New York (1984).

22. C. Cambon, C. Declume and R. Derache, "Effect of the Insecticidal Carbamate Derivatives (Carbofuran, Pirimicarb, Aldicarb) on the Activity of Acetylcholinesterase in Tissues from Pregnant Rats and Fetuses," *Toxicology and Applied Pharmacology* **49**:203-208 (1979).

23. C. E. Mendoza and J. B. Shields, "Effects on Esterases and Comparison of I50 and LD50 Values of Malathion in Suckling Rats," *Bulletin of Environmental Contamination and Toxicology* **17(1)**:9-15 (1977).

24. F. Lu, D. Jessup and A. Lavalee, "Toxicity of Pesticides in Young Versus Adult Rats," *Food and Cosmetic Toxicology* **3**:591-596 (1965).

25. J. L. Casterline, Jr. and C. H. Williams, "The Effect of 28-day Pesticide Feeding on Serum and Tissue Enzyme Activities of Rats Fed Diets of Varying Casein Content," *Toxicology and Applied Pharmacology* **18**:607-618 (1971).

26. I. Vaishwanar and S. Mallik, "The Effect of Malathion Dust on Certain Tissues of Male Rats Fed Varying Levels of Dietary Protein," *Indian Journal of Physiology and Pharmacology* **28(1)**:35-41 (1978).

27. J. L. Casterline, Jr., R. E. Brodie and T. J. Sobotka, "Effect of Banol and Parathion on Operant Learning Behavior of Rats Fed Adequate and Inadequate Casein Diets," *Bulletin of Environmental Contamination and Toxicology* **6(4)**:297-303 (1971).

28. J. L. Casterline, Jr. and C. H. Williams, "The Effect of Pesticide Administration on Serum and Tissue Esterases of Rats Fed Diets of Varying Casein, Calcium, and Magnesium Content," *Toxicology and Applied Pharmacology* **15**:532-539 (1969).

29. A. Baetjer, "Water Deprivation and Food Restriction on Toxicity of Parathion and Paraoxon," *Archives of Environmental Health* **38(3)**:168-171 (1983).

30. D. C. Villeneuve *et al.*, "The Combined Effect of Food Restriction and Parathion Exposure in Rats," *Archives of Environmental Contamination and Toxicology* **7**:37-45 (1978).

31. J. K. Howard, N. J. East, and J. L. Chaney, "Plasma Cholinesterase Activity in Early Pregnancy," *Archives of Environmental Health* **33**:2778-279 (1978).

32. R. T. Evans and J. M. Wroe, "Plasma Cholinesterase Changes During Pregnancy," *Anaesthesia* **35**:651-654 (1980).

33. S. D. Weitman *et al.*, "Influence of Pregnancy on Parathion Toxicity and Disposition," *Toxicology and Applied Pharmacology* **71**:215-224 (1983).

34. D. P. Chattopadhyay *et al.*, "Changes in Toxicity of DDVP, DFP, and Parathion in Rats Under Cold Environment," *Bulletin of Environmental Contamination and Toxicology* **29**:605-601 (1982).

35. M. K. Selgrade *et al.*, "Increased Susceptibility to Parathion Poisoning Following Murine Cytomegalovirus Infection," *Toxicology and Applied Pharmacology* **76**:356-364 (1984).

36. W. Aldridge *et al.*, "The Toxicological Properties of Impurities in Malathion," *Archives of Toxicology* **42**:95-106 (1979).

Nuclear Power Safety Goals in Light of the Chernobyl Accident

Chris Whipple and Chauncey Starr
Electric Power Research Institute
Palo Alto, CA

ABSTRACT

The recently adopted Nuclear Regulatory Commission safety goals include a proposed plant performance guideline limiting the frequency of large releases of radioactive materials. Analysis here indicates that the proposed plant guideline is potentially far more restrictive than the health objectives and goes well beyond previously established health objectives. The Chernobyl accident, which caused no off-site prompt fatalities, raised concerns that the health objectives do not limit the frequency of accidents sufficiently.

KEYWORDS: Safety goals, Chernobyl, health, Nuclear Regulatory Commission, nuclear power plants

INTRODUCTION

The U.S. Nuclear Regulatory Commission (NRC) has recently endorsed quantitative safety goals for nuclear power plants.[1,2] The safety goals communicate the safety objectives and philosophy of the NRC, as developed with public input, to the public, provide the NRC staff and industry with guidance to set specific requirements and to improve the predictability and consistency of requirements, and provide a framework for using the now familiar technique of probabilistic risk assessment (PRA) in reactor regulation.

Fortunately, severe nuclear power accidents are so rare that it is unlikely that in any reasonable time there will be sufficient data to establish statistically the probability of accidents in comparison to safety goals. An obvious implementation issue is the difficulty in determining whether a probabilistic criterion has been met. Because accident probabilities cannot be measured, application of quantitative safety goals will be inherently judgmental. The inability to measure nuclear power plant risk directly suggests that the useful purpose of the safety goals is as an expression of intent rather than of accomplishment.

The Chernobyl accident has raised questions about the safety goals. As analysis here indicates, the lack of public fatalities and low public exposures from the Chernobyl accident, relative to the magnitude of radioactive materials released, has cast doubt on the operational significance of the health objectives of the NRC safety goal limit accident. Concern that the health objectives of the safety goal do not necessarily lead to extremely low accident frequencies[3] has lead to NRC's effort to establish an additional safety

625

guideline of 1 in 1 million per year frequency for large releases of radioactive materials from nuclear power plants.

Although there is some perception that accident frequencies consistent with the current health objectives may be too high, this perception does not appear to be based on a judgment that the levels of individual risk from nuclear power as defined in the safety goals are too large. The health objectives recently approved by the Commission have remained basically unchanged throughout the development of the goals.[1,4]

THE CHANGING PERCEPTION OF NUCLEAR POWER PLANT RISK

There has been an evolution in the technical perception of nuclear power plant accident risk over the past several decades. Prior to the publication of WASH-1400,[5] the view within the nuclear safety community was that reactor risk was due predominantly to the potential for low-probability, catastrophic core melt accidents. The WASH-1400 analysis and subsequent PRAs have led to a different perception: that core melt accident probabilities are higher than was previously thought, but such accidents do not necessarily lead to large off-site releases.

The Three Mile Island accident confirmed that an accident that was severe in terms of the damage to the plant did not necessarily lead to significant off-site effects. The small release at TMI, given the nature of the accident, stimulated work to refine analysis of severe accidents and reduce uncertainties in accident source terms.

A second contributor to the lowered perception of prompt fatality risk was the Chernobyl accident. The initial skepticism in the U.S. towards reports of limited health consequences at Chernobyl illustrates the degree to which most parties in the U.S. have uncritically assumed that a large release would produce many prompt fatalities. Despite an exceptionally large release and delayed evacuation, no off-site prompt fatalities were reported. Because of design differences between the Chernobyl graphite-moderated reactor and U.S. plants, especially in the use of containment buildings and the abundance of water in U.S. plants, releases of the magnitude of the Chernobyl accident are thought to be extremely unlikely from accidents at U.S. plants. The implications of the Chernobyl accident for U.S. plants should be cautiously interpreted because, while it is probably true that the Chernobyl release was large in comparison to what might occur at a water reactor, doses to plant neighbors were moderated by favorable weather and a strong buoyant plume from the graphite fire. Had these conditions not occurred, the local exposures would probably have been greater, and distant exposures smaller. An important lesson of Chernobyl is that its post-accident risk management was capable of preventing any early public radiation fatalities despite the huge, uncontained release.

Given that radiation exposures and, presumably, health consequences, were less severe than would have been expected for the accidents at Three Mile Island and Chernobyl, the view has developed that the health objectives of the NRC safety goal are being met, often by a significant margin.

CURRENT QUANTITATIVE CRITERIA

The NRC safety goals address the risk of prompt (acute) fatality and induced cancer (delayed) to those living near a nuclear power plant. The prompt fatality safety objective is that the risk of prompt fatality to the average individual within a mile of a plant site boundary be increased by less than 0.1% due to the possibility of nuclear plant accidents; this is equivalent to a criterion for average individual risk of death of 5×10^{-7} per year. The NRC cancer objective is that the average risk of a cancer fatality in a population near a nuclear power plant (defined to be within 10 miles) be less than 0.1% of the cancer fatality risk from all other causes; this corresponds to a cancer risk of 2×10^{-6} per year.

The Policy Statement[1] endorsed for further staff examination a proposed general performance guideline that the projected mean frequency of a large release of radioactive materials to the environment from a reactor accident be less than 1 in a million per reactor-year. This supplants the earlier objective that core-melt frequency be less than 1 in 10,000 per reactor-year.

When the safety goals were first established, it was noted that the two health objectives and the plant performance guideline (at that time, the 1 in 10,000 per reactor-year limit for core melt) were not uniformly constraining on the operation of a nuclear power plant. Many observers noted that the cancer goal was unlikely to be a constraint; a plant which met the prompt fatality objective or a 10^{-4} per year objective for core melt accidents would meet the cancer risk objective.

CONSISTENCY OF HEALTH GOALS AND THE PLANT PERFORMANCE GUIDELINE

The following calculations indicate that the latent cancer objective is more easily met than the goals for prompt fatalities and plant performance. Depending on the definition for a "large release" in the plant performance guideline (limiting the frequency of large releases to 10^{-6} per year or less), this guideline ranges from being slightly more restrictive than the health objectives to more restrictive by many orders of magnitude.

For an individual, the risk of prompt fatality from a nuclear plant accident is the probability of a large release multiplied by the probability that a lethal exposure (one exceeding 4–5 Sv [400–500 rems]) is received, given such a release. Excluding sites where the local population is clustered in the path of a prevailing wind from the plant, wind direction is, on average, unfavorable for an average individual about one tenth of the time. It follows that an individual risk of 5×10^{-7} per year is achieved where there is a 1 in 10 chance of being downwind and a probability of a release sufficient to cause prompt fatalities of 5×10^{-6} per year.

Meeting the cancer objective also depends on accident probability and exposure given an accident. For example, the cancer objective would be satisfied but not exceeded for a situation in which there was a 10^{-3} per year probability of an accident that would cause average exposures resulting in a 2×10^{-3} lifetime chance of dying from cancer. Assuming that cancer risk is proportional to dose, with 1–2 latent cancer fatalities hypothetically occurring per 100 person-Sv [10,000 person-rems], the goal specifies an average individual risk level equivalent to a 10^{-3} year chance of being exposed to 0.1–0.2 Sv [10–20 rems].

Tables 1 and 2 summarize calculations dealing with cancer and prompt fatality risk in comparison to the proposed objective for large releases of 1 in 1,000,000 reactor-years. Two cases are considered, corresponding to two large release definitions under consideration by NRC. In these tables, the relative restrictiveness of the prompt fatality and cancer goals is compared to that of the proposed large release guideline. As these tables indicate, the proposed plant guideline is more restrictive than the health objectives in both cases considered.

Table 1 describes the case in which a large release is defined as that which would cause a prompt fatality at the site boundary. For a typical site, we estimate that the average downwind exposure within the first mile of a site boundary is about one-half that at the boundary, and that the 10-mile average downwind exposure is about one-tenth that within

Table 1. Accident Frequencies Consistent with NRC Safety Goals and Relative Stringency of the Goals: Case 1

Case 1: Prompt fatality at site boundary defines large release.

Exposure Assumptions:

Exposure downwind at site boundary:	>500 rem
Average exposure within 1 mile, downwind:	500 rem
Average exposure within 1 mile, all directions:	50 rem
Average exposure within 10 miles, all directions:	5 rem

Allowable accident frequencies:

Prompt fatality objective:

Risk = accident frequency × probability of exposure (wind direction)

Accident frequency = $[5 \times 10^{-7}$ risk/yr]/[0.1 risk/accident]

Allowable accident frequency = 5×10^{-6}/year

Latent cancer objective:

Risk = accident frequency × risk/accident

Assume cancer risk of 1/7500 per rem

Cancer risk/accident = 5/7500 = 6.7×10^{-4}/accident

Accident frequency = $[2 \times 10^{-6}$ risk/yr]/[6.7×10^{-4}/accident]

Allowable accident frequency = 3×10^{-3}/year

Proposed large release guideline:

Allowable accident frequency = 1×10^{-6}/year

Conclusions:

A large release guideline, defined as a release sufficient to cause a prompt fatality at the site boundary, is 5 times more stringent than the prompt fatality objective and 3000 times more stringent than the latent cancer guideline.

Table 2. Accident Frequencies Consistent with NRC Safety Goals and Relative Stringency of the Goals: Case 2

Case 2: 25 rem at site boundary defines large release.

Exposure Assumptions:

Exposure at site boundary downwind:	25 rem
Average exposure within 10 miles, downwind:	1–2 rem
Average exposure within 10 miles, all directions:	0.1–0.2 rem

Allowable accident frequencies:

Prompt fatality objective: Not applicable

Latent cancer objective:

Risk = accident frequency × risk/accident

Cancer risk/accident = risk from 0.1–0.2 rem $\approx 2 \times 10^{-5}$

Accident frequency = $[2 \times 10^{-6}$ risk/yr]/$[2 \times 10^{-5}$/accident]

Allowable accident frequency = 0.1/year

Proposed large release guideline:

Allowable accident frequency = 1×10^{-6}/year

Conclusions:

A large release guideline, defined as a release sufficient to cause a 25-rem exposure at the site boundary, is as much as 100,000 times more stringent than the latent cancer guideline.

---

one mile. For an accident producing an average exposure within 10 miles of 0.05 Sv [5 rem], the latent cancer objective is met for an accident frequency of 1 in 300 per year. Clearly this is not nearly as stringent in limiting accident frequency as is the prompt fatality objective which limits the frequency of such a release to 1 in 200,000 per year. More stringent still is the 1 in 1,000,000 per year objective for large releases proposed for study in the Safety Goal Policy Statement. For releases that would cause exposures above 5 Sv [500 rem] at the site boundary, the allowable accident frequency under prompt fatality objective and plant performance guideline would be unchanged, but that allowable for the cancer objective would be lower, and the inconsistency in restrictiveness less than described above.

Table 2 considers the case where a large release is defined as that which exposes any individual to 0.25 Sv [25 rem]. This case is based on the 0.25 Sv [25 rem] limit on whole-body exposure of 10 CFR Part 100. The prompt fatality objective does not apply to exposures too low to create a risk of acute effect. In this case, the cancer objective can be met if the accident frequency is roughly less than 1 in 10 per year.

THE CHERNOBYL ACCIDENT AND THE NRC SAFETY GOALS

Because the NRC safety goals apply to risks in prospect, not to exposures that have occurred, it is incorrect to ask whether the Chernobyl accident would have been permitted under the NRC safety goals. As an illustration of why this is so, consider a hypothetical goal by the Department of Transportation that the risk of death to passengers on trains be less than one per billion passenger-miles. It does not make sense to ask whether this goal would have been met in the January 4, 1987, fatal train accident near Baltimore; whether such a goal is being met is established by average risks, not by single accidents. Asking whether a particular crash exceeded a transportation safety goal is analogous to asking whether the Chernobyl accident was acceptable under NRC goals.

However, with certain assumptions and caveats, it is possible to calculate the permissible frequency of accidents similar to that at Chernobyl. As in Tables 1 and 2, it is necessary to assume that the Chernobyl accident is representative of the full risk spectrum of the plant. Additionally, one should note the distinction between the exposures that actually occurred versus those which would have been estimated to occur for such a release. This distinction is important for application of the prompt fatality goal to the Chernobyl accident. Because there were no prompt fatalities, it has been argued that the prompt fatality objective would not apply to an accident such as at Chernobyl.[3] However, under current NRC procedures, releases of the magnitude of that at Chernobyl generally are calculated to cause larger exposures than actually occurred.[6] How much of the discrepancy between what would be calculated and what actually occurred is due to favorable weather during the accident, and how much to other factors such as conservatism in the analytical procedure, is not clear. In any event, in attempting to examine the Chernobyl consequences in comparison to the safety goal, it is important to distinguish between calculated and actual consequences.

Were the general performance guideline for large releases in effect, the permissible frequency of such accidents is 1 per 1,000,000 reactor-years. The relation of the cancer fatality objective to the Chernobyl accident can be calculated by considering that the 24,000 most heavily exposed people at Chernobyl received an average exposure of 0.45 Sv [45 rems].[7,8] For an average exposure of this magnitude, the latent cancer objective is satisfied provided the accident frequency is less than 3×10^{-4} per year, that is, 1 per 3000 per reactor-year.

These calculations indicate that a plant with a frequency of 1 in 3000 per reactor-year of an accident with Chernobyl-like consequences would meet the NRC health objectives. However, this accident risk would exceed the frequency proposed for large releases by a factor of 300. Clearly, the various objectives of the NRC safety goal do not pose equivalent constraints on plant design and operation.

OBJECTIVES OF THE SAFETY GOALS AND CONCLUSIONS

The concern expressed by several members of the NRC is that this inconsistency means that the health objectives of the safety goals are insufficiently stringent. But the concern is not that individuals are at too great a risk; there has been little or no debate about whether the 0.1% of background risk criteria provide inadequate protection. The concern is that the health objectives can be satisfied by an accident frequency that appears too high.

What motivates the substantial safety effort in the industry and at NRC, if not concerns over health? This analysis indicates that the motivation to avoid accidents at NRC, e.g., that the frequency of large releases be less than 10^{-6} per reactor-year, cannot be explained solely by health objectives. Nor does concern about financial risk explain these performance objectives (see note below). A plausible answer is given by disbelief that the consequences at Chernobyl were as limited as they were. This suggestion is supported by the literature on perception of nuclear power risks.[9] That an accidental release of the magnitude of that at Chernobyl could have failed to have caused many immediate fatalities was a major surprise to reactor safety analysts.

It also appears that in working to achieve accident risk levels below those needed to meet health or economic criteria, the industry and its regulators appear to be seeking a level of performance which will restore public support to the technology. How closely the public perception of nuclear power risk relates to performance is not known. Attitudes towards nuclear power depend on many things in addition to perceived risks to health and the environment, but avoidance of accidents is obviously an important step toward establishing public support for nuclear power.

It is clear that accidents have an industry-wide effect which analysis of health risk and economic damage fails to capture. A reduction in public support for and acceptance of nuclear power and a corresponding set of new regulatory requirements followed the Three Mile Island accident. While the Chernobyl accident has not had such a dramatic effect on U.S. plants, the long-term prospects for the revival of nuclear power in the U.S. were damaged by the accident. It seems fair to say that survival of the nuclear power option, while in serious doubt for the short term but perhaps better in the long run, depends on the collective safety performance of nuclear power plants. The industry has now reached a world-wide size where a decade of operating experience provides several thousand reactor-years of information about the performance characteristics of nuclear power plants; it is therefore appropriate that public judgments are sensitive to the collective performance of the industry.

From this perspective, the appropriate level of safety is that which maximizes chances for preserving nuclear power as an option. If accidents are too frequent, public rejection of the technology may occur. Inefficient or overly stringent regulation can also foreclose the nuclear power option, if regulatory requirements are too expensive, as may currently be the case. The perception that nuclear power is held to a far more stringent performance standard than are other energy technologies may be true. But it may be in the long- term interests of the industry to accept such standards.

It is important that industry and the NRC both understand the factors which are at work in the present safety criteria. If the health objectives of the NRC goals appear to be easily met or exceeded by current plants, it does not mean that current regulations are too stringent and the plants "too" safe; objectives other than health are the controlling factor. But if the health goals are apparently exceeded by a wide margin, it also does not imply that these health objectives were set at too lax a level.

Note: A longer version of this paper appeared in *Nuclear Safety*, Jan-Mar, 1988.

REFERENCES

1. U.S. Nucl. Reg. Comm., 10CFR50, Safety Goals for the Operation of Nuclear Power Plants; Policy Statement, *Fed. Regist.* **51(149)**:28044-28049, August 4, 1986.
2. D. Okrent, "The Safety Goals of the U.S. Nuclear Regulatory Commission," *Science* **236**:296-300 (1987).

3. Commission meeting on Status of Safety Goal Implementation (Public Meeting), Washington, D.C., Thursday, January 8, 1987.

4. U.S. Nuclear Regulatory Commission, Office of Policy Evaluation, "Safety Goals for Nuclear Power Plants," NUREG-0880, Revision 1 for Comment, May 1983.

5. U.S. Nuclear Regulatory Commission, "Reactor Safety Study: An Assessment of Accident Risks in U.S. Commercial Nuclear Power Plants," WASH-1400, NUREG-70/014, October 1975.

6. U.S. Nuclear Regulatory Commission, Office of Nuclear Regulatory Research, Reactor Risk Reference Document, NUREG-1150, Draft for Comment, February 1987.

7. Committee on the Assessment of Health Consequences in Exposed Populations, "Health and Environmental Consequences of the Chernobyl Nuclear Power Plant Accident," DOE/ER-0332, 1987.

8. U.S. Nuclear Regulatory Commission, "Report on the Accident at the Chernobyl Nuclear Power Station," NUREG-1250, January 1987.

9. P. Slovic, S. Lichtenstein, and B. Fischhoff, "Images of Disaster: Perception and Acceptance of Risks from Nuclear Power," in *Energy Risk Management*, Goodman and Rowe, Eds., London, Academic Press, 1979.

PRA-Based Inspection for Oconee Unit 3

T. V. Vo

Pacific Northwest Laboratory
Richland, WA

ABSTRACT

Guidance for inspecting systems and components at the Oconee Unit 3 nuclear power plant has been developed using information from the plant's level 3 probabilistic risk assessment (PRA). This information will be used in conjunction with the U.S. Nuclear Regulatory Commission Inspection and Enforcement Manual and has been prepared in the form of a plant-specific appendix for incorporation into this manual.

An analytical method was developed for calculating public risk for affected plant systems. This was used to prioritize the Oconee-3 plant systems using the Fussel-Vesely importance measure. System fault trees were reanalyzed to identify and rank the risk significant components in each system.

The product of this effort is a prioritized listing of systems and components associates with 98% of the inspectable portion of the public risk from Oconee-3. This compilation will subsequently be used by NRC inspectors in the preparation of inspection plans addressing risk important systems and components at the Oconee-3 power plant.

KEYWORDS: Inspections, PRA, IE Manual, plant systems, prioritize

INTRODUCTION

This work was performed for the U. S. Nuclear Regulatory Commission (NRC) as part of an extensive program to develop information based on probabilistic risk assessments (PRAs) for use in the planning and performance of nuclear power plant inspections. Due to the broad scope of this program, the work has been divided among three national laboratories, each concentrating upon a particular reactor type: Brookhaven National Laboratory has analyzed BWR plants; Idaho National Engineering Laboratory has analyzed Westinghouse PWRs; and the Pacific Northwest Laboratory has analyzed PWRs from both Babcock and Wilcox and Combustion Engineering, because of the smaller number of plants from these vendors. This paper addresses PNL work for Oconee Unit 3.

In this project, information from the extensive PRA report for Oconee-3 (Sugnet *et al.*, 1984) was used to identify plant systems and components important to minimizing public risk and to identify failure modes for these components. This information was tabulated and correlated with inspection modules from the NRC Inspection and Enforcement (IE) Manual (USNRC 1984) used by inspectors in the planning and performance of inspections. The results of this analysis was published (Gore, Vo, and Harris, 1987) as a series of tables, organized by system and prioritized by importance to public risk, which identify components associated with 98% of the inspectable risk due to plant operation. External events, e.g., earthquakes, tornadoes, fires, and floods, are not included in the analysis.

Previous studies in this program (Hinton and Wright, 1986; Higgins, 1986) have addressed how PRA-based information may be best incorporated into inspection planning, performance and evaluation. The conclusion of this previous work was that the existing IE Manual provides a logical and effective framework for inspection planning. This manual contains an extensive sequence of inspection procedures, or modules, addressing functional areas such as calibration, surveillance, maintenance, emergency safeguards function (ESF) system walkdown, etc. It also contains a methodology for selecting the inspection modules to be performed, plus guidance on the frequency at which the modules should be performed. It was concluded that this manual should be retained as the general framework for inspection planning. PRA-based information, which is necessarily plant-specific, should be provided for each plant. This information should then be used in the inspection planning process to help focus on areas where public risk is most sensitive to performance degradation.

The NRC program is, therefore, directed toward the preparation of a series of plant-specific appendices to the IE Manual, each of which contains plant-specific information of similar safety significance. These appendices are structured according to a common format. Each appendix begins with a description of accident initiators and sequences important at the plant. This is followed by a listing of plant systems associated with 98% of the inspectable plant risk, which is ordered according to the importance of each system to public risk. For each system addressed, the components associated with 95% of the probability of system failure are identified and ranked according to importance. Three tables are presented for each system. The first identifies the failure modes by which each component contributes to risk. The second correlates each component with the IE inspection modules most related to ensuring component reliability. The third provides a modified system check-off list identifying the proper line-up of each component during normal operation. Examples from the plant-specific appendix developed for the Oconee-3 plant are presented in the body of this paper.

ANALYSIS OF THE OCONEE-3 PRA

The analysis of Oconee-3 consisted of three major steps. The first was the calculation of risk importance for each system from information in the PRA. This was used to select systems to be analyzed for component importances. The second step was the re-analysis of system fault trees from the PRA to identify component importances. The third step was the correlation of components with failure modes and with inspection modules relevant to maintaining component reliability. These steps are discussed below.

Calculation of System Importances

The selection of systems for detailed fault tree analysis required that they be ranked according to an appropriate measure of risk. The Oconee-3 PRA is a level 3 PRA; it addresses the probabilities of core melt, of subsequent containment failure, and of radionuclide releases and subsequent radiation doses to the public. Consequently, it was possible to base the determination of system importance on public risk, as measured by the expected annual radiation dose to the public. This is appropriate, since the mission of the NRC is the protection of public health and safety.

The Fussel-Vesely (F-V) importance measure (Henley and Kumamoto, 1981) applied to risk was selected to rank system and component importance in this study. It is the fraction of the total risk which results from failures involving the system or component of interest. Thus, high values of F-V importance identify systems which are the greatest contributors to risk. In addition, the increase in risk due to a given percentage increase in system failure probability is also highest for systems with highest F-V importance values. Thus, this measure identifies not only the systems which are the greatest contributors to risk, but also those for which risk is most sensitive to performance degradation.

The analyses in the PRA document address a wide variety of accident sequences which may lead to core melt, including both internal and external events. Internal events are those resulting from the failure of systems to function due to equipment or operational failures. External events (including earthquakes, tornadoes, fires and floods) are external to the system boundaries and can result in common-cause failures of redundant, safety-related equipment.

Our analysis did not address earthquakes, tornados, fires or external flooding for two reasons. First, inspection of system hardware cannot affect the initiation frequency of such events. Second, the probability of subsequent component failures, given occurrence of the initiating event, is altered by damage caused by the event, so that the beneficial results of inspections of system hardware are overcome by the event itself. Our analysis focused on event sequences involving failures associated with the operation and maintenance of system hardware and controls. We have referred to this as inspectable risk, in contrast to risk from external events which inspections cannot effectively protect against.

In addition to internal events involving failures associated with system hardware and operations, our analysis also addressed flooding due to internal events. Specifically, a leak or rupture in the condenser circulating water (CCW) system might release several hundred thousand gallons per minute of water into the basement of the turbine building which houses many systems of safety significance. Since the PRA-predicted contributions to core melt frequency for these internal flooding events were of the same magnitude as those for internal events, they were also included in the analysis. It is for this reason that the CCW system, which in most plants is of minimal safety significance, is found high on the list of prioritized systems in this analysis.

The PRA event sequences are grouped into six core-melt bins, which are related to general types of event progression and containment response. Examples include large LOCAs (loss-of-coolant accidents), small LOCAs, and various other transient types. Each of these core melt bins was associated with up to five plant damage bins, which were correlated with containment safeguard states. The probability of radionuclide release for each core melt event thus incorporated the probability of failure of the reactor building (RB) spray, RB cooling, low pressure injection (LPI) cooling, and timing and sequencing of such failures. Release magnitude and isotopic content were also correlated with event sequencing and system functioning (e.g., fission-product entrainment by RB spray, when functioning). Thus, each plant damage bin was correlated with up to six release categories, each of which was subsequently associated with a public risk per unit frequency of occurrence. Table 1 presents the results for the important systems identified from this study.

Calculation of Component Importances

In this study, a fault tree analysis technique was selected to identify the components most important to system failure. It was not possible to extract information with this degree of detail from the cut sets published in the PRA because, in general, the cut set elements were not basic events. Instead, many contained "module" elements, which combined the effects of several possible failures causing the final result, i.e., failure of a pump, or of its

Table 1. F-V System Importances

System	F-V Importance
Reactor Building Spray System	0.834
Reactor Building Cooling System	0.590
Condenser Circulating Water System	0.567
Safety Relief Valve System	0.423
Low-Pressure Injection System	0.284
Standby Shutdown Facility–High-Pressure Injection System	0.041
Low-Pressure Service Water System	0.031
Emergency Feedwater System	0.018

suction or discharge valves located in a single run of piping, any of which would prevent flow through the line.

For systems selected for analysis, the system fault tree published in the PRA was reanalyzed using the Integrated Reliability and Risk Analysis (IRRAS) computer code (Russell *et al.*, 1987) run on an IBM-PC. Other analysis methods were used for three systems: safety relief valves (SRVs), CCW, and standby shutdown facility–high pressure injection (SSF-HPI). IRRAS identified the dominant minimal cut sets and quantified the fault trees by ordering cut sets by probability. IRRAS also calculated the F-V importance of both cut sets and of system component failures. The calculated importance of the component failures was then used to select components for inclusion in the tables. For all systems analyzed, components comprising more than 95% of the total component importance were selected for tabulation, with an even higher proportion included for the more important systems.

Preparation of Tables

For each system presented in Table 1, the components selected for inclusion in the tables were grouped according to type for discussion of failure modes (e.g., pump suction and discharge MOVs in parallel trains). For many components, cut set elements indicated more than one failure mode (e.g., failure to operate, operator failure to initiate, inappropriate change of position). These failure modes were grouped and addressed for each component type in the system failure mode identification tables.

The characteristics of each component were assessed to determine what types of inspection would be most appropriate for ensuring component reliability. This information was then used to prepare a table for each system correlating each of the relevant IE inspection modules with components which should be addressed when the module is used in inspections of the system. Also included in this table was a cross correlation with the failure modes which would be minimized by the given type of inspection. For instance, pump failure to start and run is addressed in IE inspection modules for surveillance, operational safety verification, and ESF system walkdown. It is also addressed through the Maintenance module, in terms of minimizing unavailability due to maintenance scheduling and work.

For each system, tables show the important failure modes, the IE modules to address these failures, and a modified walkdown to inspect only the identified important components. Tables 2A, B, and C present the results for one of a total of eight systems

Table 2A. Reactor Building Spray System Failure Mode Identification

The reactor building spray system (RBSS) is a standby safety system and has no normal operating function. It is aligned in the standby mode during normal plant operation, with the pumps off and in automatic control, the discharge motor-operated valves closed, and all other motor-operated and manual valves open. The system is automatically activated on a signal of high reactor building pressure. At least one of the two trains of RBSS is required for system success. This system is important for public health, not for preventing plant damage.

Conditions that Lead to Failure

1. *Human Error–System Operation Inhibited, or Failure to Restore Valves 3BS-12, 13, 17, 18, and 21, or Failure to Restore P3A and P3B Switchgears After Test*

   These are failures of the operator, including misdiagnosis of spurious spray operation when it is actually required, or human error failure to realign a valve or restore a switchgear at the end of a test, plus failure to discover the error. The valves are pump discharge manual valves 3BS- 12, 3BS-17, and the manual test valves 3BS-13, 3BS-18 and 3BS-21 which may be left in the test position. These errors are addressed by emergency procedures, proper post-test surveillance, which should be reviewed and observed.

2. *Reactor Building Pumps 3A and 3B Fail to Start or to Run*

   Failure of pumps 3A and 3B will prevent waterflow from being provided to the spray headers. The important failure causes are random hardware failures, and human errors in following procedures. Operator awareness, surveillance and lineup for standby operation should be observed and reviewed.

3. *Failure of Motor-Operated Discharge Valve MOV 3BS-1 and MOV 3BS-2 to Open*

   These motor-operated valves must be opened to allow flow from the pumps to the spray headers. The dominant failure cause is random hardware failure. A contributing failure cause is human error failure to manually activate them. Operator awareness, surveillance and lineup for standby operation should be observed and reviewed.

4. *RBS Pump Trains Unavailable Due to Maintenance and Testing*

   This includes both scheduled and unscheduled maintenance and testing. The timely performance of maintenance and testing should be reviewed and observed to minimize this unavailability.

5. *RBS Pumps' Suction Valves MOV 3BS-3 and MOV 3BS-4, or Check Valves 3BS-11, 3BS-16, 3BS-14 and 3BS-19 Transfer Closed or Fail to Open on Demand*

   The valves are pumps' suction motor-operated valves 3BS-3, 3BS-4, and discharge check valves 3BS-11, 3BS-16, 3BS-14, and 3BS-19. Check valves should be locked open. The important failure causes are human error and random electrical or hardware failures. Operator awareness, surveillance and lineup for standby operation should be observed or reviewed.

Table 2B. IE Modules for Reactor Building Spray System Inspection

Module	Title	Components	Failure Mode[a]
61701	Surveillance (Complex)	Control Switches P3A,P3B	1
		Pumps 3A,3B	2
61726	Monthly Surveillance Observation	Control Switches P3A,P3B, 3BS-12, 3BS-13, 3BS-17, 3BS-18, 3BS-21	1
		Pumps 3A,3B	2
		MOV 3BS-1, MOV 3BS-2	3
		MOV 3BS-3, MOV 3BS-4, 3BS-11, 3BS-14, 3BS-16, 3BS-19	5
62700	Maintenance	Pumps 3A,3B	4
71707	Operational Safety Verification	3BS-12, 3BS-13, 3BS-17, 3BS-18, Switches P3A, P3B	1
		RBSS Pumps 3A, 3B	2
		MOV 3BS-1, MOV 3BS-2	3
		MOV 3BS-3, MOV 3BS-4, 3BS-11, 3BS-14, 3BS-16, 3BS-19	5
71710	ESF System Walkdown	3BS-12, 3BS-13, 3BS-17 3BS-18, Switches P3A, P3B	1
		RBSS Pumps 3A, 3B	2
		MOV 3BS-1, MOV 3BS-2	3
		MOV 3BS-3, MOV 3BS-4, 3BS-11, 3BS-14, 3BS-16, 3BS-19	5

[a]See Table 2A for failure identification.

Table 2C. Modified Reactor Building Spray System Walkdown

Component Number	Component Name	Loca-tion	Required Position	Actual Position
Electrical				
Pump 3A	RBSS Pump 3A Breaker	TC	Racked in	_____
Pump 3B	RBSS Pump 3B Breaker	TD	Racked in	_____
Pump 3A	RBSS Pump 3A Control Power Switch		On	_____
Pump 3B	RBSS Pump 3B Control Power Switch		On	_____
MOV 3BS-1	RBSS Pump 3A Discharge Valve Breaker	XS1	Closed	_____
MOV 3BS-2	RBSS Pump 3B Discharge Valve Breaker	XS2	Closed	_____
MOV 3BS-3	RBSS Pump 3A Suction Valve Breaker	XS1	Closed	_____
MOV 3BS-4	RBSS Pump 3B Suction Valve Breaker	XS2	Closed	_____
Valves				
MOV 3BS-1	RBSS Pump 3A Discharge Valve		Closed	_____
MOV 3BS-2	RBSS Pump 3B Discharge Valve		Closed	_____
MOV 3BS-3	RBSS Pump 3A Suction Valve		Closed	_____
MOV 3BS-4	RBSS Pump 3B Suction Valve		Closed	_____
3BS-11	RBSS Pump 3A Discharge Check Valve		Open	_____
3BS-12	RBSS Pump A Manual Discharge Valve		Open	_____
3BS-13	RBSS Pump A Manual Test Valve		Closed	_____
3BS-14	RBSS Pump 3A Discharge Check Valve		Open	_____
3BS-16	RBSS Pump 3B Discharge Check Valve		Open	_____
3BS-17	RBSS Pump B Manual Discharge Valve		Open	_____
3BS-18	RBSS Pump B Manual Test Valve		Closed	_____
3BS-19	RBSS Pump 3B Discharge Check Valve		Open	_____
3BS-21	RBSS BWST Manual Test Valve		Closed	_____

identified for this study. A system walkdown table (Table 2C) is the most important table. This table identifies the normal operating state or position of each component determined to be risk-significant from the PRA. It was compiled using information from the PRA, and also from plant systems descriptions, operator training information, and plant drawings. In most cases, it was possible to correlate and verify this information using system lineup tables from plant operating procedures. In general, these tables are considerably shorter than lineup tables in procedures. They therefore allow an inspector with limited time available for system walkdowns to concentrate on risk-significant components, while minimizing the possibility that he may overlook something important.

CONCLUSIONS AND RECOMMENDATIONS

In this project we have identified the systems and components most important to public risk during operation of the Oconee-3 power plant. Tables 2A, B, and C present the results for one of a total of eight systems identified for this study.

As indicated in Table 1, the RB spray and RB cooling systems are the most important systems for minimization of public risk. This is because most radionuclide releases from the plant involve failure of the containment, due to failure of one or both of these systems. The condenser circulating water system is also a very important system because of the possibility that breach of the large diameter piping and components in this system, which is fed by gravity flow from Lake Keowee, could result in uncontrolled flooding of the turbine building, causing common-cause failures of many important plant systems including feedwater, emergency core cooling systems, and the RB Spray and Cooling systems. Each of these three systems is involved in failures which contribute between 50 to 80% of the plant risk.

Two other systems are particularly important, and are involved in failures which contribute between 25 and 50% of the plant risk. These are the safety relief valves and the LPI system. They are both involved in high-consequence failures where core cooling fails and radionuclides escape via a steam generator tube rupture and stuck open steam safety valve, bypassing the containment.

Each of the other systems addressed is involved in less than 5% of the total plant risk. Cumulatively, more than 98% of the plant risk is addressed by the systems in the tables. Overall, the dominant contributor to public risk in all systems is human error in the operation, maintenance, or surveillance of these systems.

The information in these tables allows an inspector to identify quickly the components most important to public risk—a combination of failure probability and of the consequences of the failure. This information allows the inspector to direct attention to these components preferentially. In particular, by using the system walkdown tables the inspector can rapidly review the line-up of important system components on a routine basis. The inspector may also use these tables when selecting systems for the performance of more detailed inspection activities.

In using these tables, however, it is essential to remember that other systems are also important for other reasons. For example, the HPI and reactor protection systems are absent from the tables, primarily because of their intrinsic reliability and redundancy, despite the fact that they perform essential safety functions. If, through inattention, the failure probabilities of such systems were allowed to increase significantly, their risk significance might exceed that of systems in the tables. Consequently, a balanced inspection program is essential to minimizing plant risk. The tables allow an inspector to concentrate on systems of highest risk importance. In so doing, however, he must maintain cognizance of the status of systems performing other essential safety functions and ensure that their reliability is maintained.

REFERENCES

Gore, B. F., and Huenefeld, J. C., 1987, "Methodology and Application of Surrogate Plant PRA Analysis to this Rancho Seco Power Plant," NUREG/CR-4768, PNL-6032, USNRC Region 5, Walnut Creek, California.

Gore, B. F., Vo, T. V., and Harris, M. S., 1987, "PRA Applications Program for Inspection at Oconee Unit 3, NUREG/CR-5006, USNRC Region 1, King of Prussia, Pennsylvania.

Henley, E. J., and Kumamoto, H., 1981, *Realiability Engineering and Risk Assessment*, Prentice Hall Inc., Englewood, New Jersey.

Higgins, J. C., 1986, "Probabilistic Risk Assessment (PRA) Applications," NUREG/CR-4372, USNRC Region 1, King of Prussia, Pennsylvania.

Higgins, J. C., Taylor, J. H., Fresco, A. N., and Hillman, B. M., 1987, "Generic Safety Insights for Inspection Boiling Water Reactors," TANSAO 54, 235 American Nuclear Society, LaGrange Park, Illinois.

Hinton, M. F., and Wright, R. E., 1986, "Pilot PRA Applications Program for Inspection at Indian Point 2," EGG-EA-7136, Idaho, Inc., Idaho Falls, Idaho.

Russell, K. D., *et al.*, 1987, "Integrated Reliability and Risk Analysis," NUREG/CR-4844, Idaho National Engineering Laboratory, Idaho Falls, Idaho.

Sugnet, W. R., Boyd, G. J., Lewis, S. R., *et al.*, 1984, "Oconee PRA, A Probabilistic Risk Assessment of Oconee Unit 3," NSAC-60 Electric Power Research Institute, Palo Alto, California.

USNRC Inspection and Enforcement Manual, 1984, "Operations USNRC Office of Inspection and Enforcement," Chapter 2515, Washington, D.C.

Environmental Radiation Standards and Risk Limitation[a]

D. C. Kocher

Oak Ridge National Laboratory
Oak Ridge, TN

ABSTRACT

The Environmental Protection Agency and Nuclear Regulatory Commission have established environmental radiation standards for specific practices which correspond to limits on risk to the public that vary by several orders of magnitude and often are much less than radiation risks that are essentially unregulated, e.g., risks from radon in homes. This paper discusses a proposed framework for environmental radiation standards that would provide a more meaningful correspondence with limitation of risk. This framework includes (1) the use of limits on annual effective dose equivalent averaged over a lifetime, rather than limits on dose equivalent to whole body or any organ for each year of exposure, (2) consideration of dose and risk to younger age groups as well as adults, (3) limits on annual effective dose equivalent averaged over a lifetime for specific practices no lower than 0.25 mSv (25 mrem), (4) maintenance of all exposures as low as reasonably achievable (ALARA), and (5) establishment of a generally applicable *de minimis* dose for public exposures. Implications of the proposed regulatory framework for current standards for limiting public exposures are discussed.

KEYWORDS: Public exposures, risk limitation, radiation standards, radiation risks

INTRODUCTION

This paper presents a proposed regulatory framework for limiting radiation exposures of the public that would provide a reasonable correspondence with limitation of risk. The impetus for this proposal is the observation, supported herein, that current environmental radiation standards for specific practices, as promulgated by the Environmental Protection Agency (EPA) and the Nuclear Regulatory Commission (NRC), correspond to limits on risk that (1) may vary by several orders of magnitude, (2) often are much less than specific radiation risks that are essentially unregulated, e.g., risks from radon in homes, and (3) in some cases are less than radiation risks that are generally regarded as negligible. Thus, the effectiveness of the current system of radiation standards in relation to limitation of risk to the public is called into question.

a. Research sponsored by the U.S. Department of Energy under contract DE-AC05-84OR21400 with Martin Marietta Energy Systems, Inc.

CURRENT REGULATORY FRAMEWORK FOR
LIMITING PUBLIC EXPOSURES

In the U.S., the current regulatory framework for limiting radiation exposures of the public from routine releases includes (1) generally applicable radiation protection standards and (2) environmental radiation standards for specific practices.

Radiation Protection Standards

Radiation protection standards for the public are applicable to all sources of exposure, exclusive of natural background radiation and deliberate medical practices. These standards are based on an assumed limit on acceptable risk from radiation exposure and are regarded as necessary for protection of public health.

Current radiation protection standards for the public include (1) a limit on annual dose equivalent from uniform whole-body irradiation of 0.5 rem and (2) maximum permissible concentrations of radionuclides in air and water which correspond to limits on annual committed dose equivalents to whole body or any organ from inhalation or ingestion.[1] In addition, all exposures of the public should be maintained as low as reasonably achievable (ALARA).

Current recommendations on radiation protection standards by the International Commission on Radiological Protection (ICRP)[2] and the National Council on Radiation Protection and Measurements (NCRP)[3] replace limits on annual dose equivalent to whole body or any organ with limits on annual effective dose equivalent.[4] The recommendations for public exposures include (1) a principal limit on annual effective dose equivalent of 1 mSv (0.1 rem) for continuous exposure and (2) a subsidiary limit on annual effective dose equivalent of 5 mSv (0.5 rem) for occasional exposure, provided the annual effective dose equivalent averaged over a lifetime does not exceed 1 mSv (0.1 rem).

Environmental Radiation Standards

Environmental radiation standards apply to specific practices and are based primarily on (1) levels of public exposure that are judged to be reasonably achievable using best-available effluent control technologies or (2) reduction of environmental radioactivity to levels near ambient background. Thus, environmental radiation standards are not based on the need for limitation of risk *per se*; rather, they result essentially from application of the ALARA principle to standard setting itself. The importance of standards for specific practices is that they provide a practical set of requirements for limiting public exposures which ensure that generally applicable radiation protection standards will be met.[5]

Environmental radiation standards for several practices have been established by the EPA and the NRC.[6-12] Many of these standards include a limit on annual dose equivalent to whole body of 25 mrem.[6,8,10-12] Thus, a *de facto* environmental radiation standard for many practices is a limit on annual effective dose equivalent of 25 mrem (0.25 mSv). Federal agencies have also issued guidance that concentrations of radon in homes should not exceed 4 pCi/L in order to provide an acceptable limit on risk to the public from exposures to radon.[13]

LIMITS ON RISK ASSOCIATED WITH CURRENT RADIATION STANDARDS

This section discusses the correspondence between current radiation standards for the public and limitation of risk. We first consider a particular example.

Several environmental radiation standards contain limits on annual dose equivalent of 25 mrem to whole body, 75 mrem to thyroid, or 25 mrem to any other organ.[6,8,10,11] If we assume a risk factor for uniform whole-body irradiation of 2×10^{-4} rem^{-1} and weighting factors for specific organs recommended by the ICRP,[14] then the annual risk corresponding to these dose limits ranges from 5×10^{-6} for whole body to 2×10^{-7} for bone surfaces. Thus, the limits on risk corresponding to the standards for several practices may vary by as much as a factor of 30.

A general comparison of lifetime risks corresponding to radiation protection and environmental radiation standards, including the Federal guidance on radon in homes, is given in Table 1. The estimated risks assume exposure of adults over 70 years and are based on the risk factor and organ-specific weighting factors recommended by the ICRP.[14] The assumptions used in some cases are described below.

- The risk from radon in homes is the mean of the range given in the Federal guidance for a radon concentration of 4 pCi/L and an indoor residence time of 75%.[13]

- The risk from uranium and thorium mill tailings is based on the limits on (1) concentration of outdoor ^{222}Rn, (2) concentrations of ^{226}Ra in soil, (3) concentration of indoor ^{222}Rn daughter products, and (4) indoor gamma radiation level.[8,9] The risks from ^{222}Rn and daughter products are based on the mean annual effective dose equivalents per unit concentration or exposure for outdoor and indoor residence estimated by the ICRP.[15] The risk from ^{226}Ra in soil assumes external photon exposure above ground for daughter products in secular equilibrium with the parent, absorbed dose rates in air above ground per unit concentration of monoenergetic sources in soil[16] combined with the photon spectra of the parent and daughter radionuclides,[17] indoor and outdoor residence times of 85% and 15%,[15] respectively, a shielding factor during indoor residence of 0.7,[18] and a ratio of effective dose equivalent to absorbed dose in air of 0.8. The risk from indoor gamma radiation assumes an indoor residence time of 85%[15] and a ratio of effective dose equivalent to exposure in air of 0.7.

- The risks from drinking water are based on (1) the limits on concentration of ^{226}Ra plus ^{228}Ra and on annual dose equivalent to whole body or any organ from man-made, beta/gamma-emitting radionuclides,[7] (2) a daily water intake of 2 liters, and (3) effective dose equivalents and dose equivalents to specific organs per unit activity of radionuclides ingested.[19]

- The risk from high-level waste disposal is the average value for an individual in the current U.S. population, based on a limit of 1,000 health effects per repository over 10,000 years.[10]

In comparing the risk estimates in Table 1, it should be borne in mind that (1) radiation protection standards apply to all sources, excluding natural background, whereas the other standards apply to specific practices, (2) standards for drinking water and high-level waste disposal essentially limit risks to average individuals in the population, whereas the other standards apply to maximally exposed individuals, (3) the guidance on radon in homes and standards for drinking water and uranium and thorium mill tailings include contributions from natural background radiation, and (4) the guidance on radon in homes is not a standard for limiting public exposures. Nonetheless, the following conclusions may be drawn from Table 1.

- The risks corresponding to different environmental radiation standards for specific practices, excluding the standard for high-level waste disposal, vary by nearly four orders of magnitude.

**Table 1. Lifetime Risks Associated with Various Radiation Protection
and Environmental Radiation Standards for the Public**

Risk	Standard
3×10^{-2}	Radon in homes[a,b]
1×10^{-2}	Uranium and thorium mill tailings[a,c]
7×10^{-3}	Annual dose equivalent to whole body of 0.5 rem[d]
1×10^{-3}	Annual effective dose equivalent of 0.1 rem[e]
4×10^{-4}	Annual dose equivalent to whole body of 25 mrem[f]
6×10^{-5}	Ra-226 plus Ra-228 in drinking water[a,g]
6×10^{-5}	Annual dose equivalent to whole body of 4 mrem[g]
3×10^{-5}	Annual dose equivalent to thyroid of 75 mrem[f]
1×10^{-5}	Annual dose equivalent to bone of 25 mrem[h]
5×10^{-6}	Sr-90 in drinking water[a,g]
2×10^{-6}	I-129 in drinking water[a,g]
5×10^{-8}	Disposal of high-level wastes[a,i]

[a]See text for discussion of assumptions used in estimating risk.
[b]Federal guidance.[13]
[c]EPA and NRC standards.[8,9]
[d]NRC's radiation protection standard.[1]
[e]Radiation protection standard recommended by ICRP and NCRP.[2,3]
[f]Limit contained in several EPA and NRC standards.[6,8,10-12]
[g]EPA standards for drinking water.[7]
[h]Limit contained in several EPA and NRC standards.[6,8,10,11]
[i]EPA standard.[10]

- The risks corresponding to all environmental radiation standards for specific practices are less than the risk corresponding to the guidance on radon in homes, which is a radiation risk that is essentially unregulated.

- The risks corresponding to the guidance on radon in homes and the uranium and thorium mill tailings standards are higher than the risks corresponding to the NRC's radiation protection standards and the standards recommended by the ICRP and the NCRP.

- The risks corresponding to several environmental radiation standards for specific practices are comparable to or less than the negligible risk level of about 10^{-5} recommended by the NCRP,[3] which corresponds to an annual effective dose equivalent for continuous lifetime exposure of about 0.01 mSv (1 mrem).

Thus, current environmental radiation standards generally do not provide a meaningful correspondence with limitation of risk to the public.

PROPOSED REGULATORY FRAMEWORK FOR LIMITING PUBLIC EXPOSURES

This section presents a proposed regulatory framework for limiting radiation exposures of the public that would (1) provide a more meaningful correspondence with limitation of risk, (2) establish limits on risk for specific practices that are well above levels regarded as negligible, and (3) minimize adverse impacts on current standards.

Limitation of Lifetime Risk to the Public

An essential aspect of the proposed regulatory framework is the assumption that limitation of lifetime risk, rather than risk from each year of exposure, is the fundamental goal of standards for limiting exposures of individuals,[14] and that exposures of younger age groups as well as adults should be taken into account. Furthermore, for regulatory purposes, the concept of the effective dose equivalent[4] is assumed to provide a reasonable surrogate for risk from radiation exposure.

The importance of exposures of younger age groups in obtaining a reasonable correspondence between standards expressed as limits on effective dose equivalent and limitation of lifetime risk is discussed elsewhere.[20] Age-dependent calculations of dose from ingestion of radionuclides have shown that the dose from intakes by infants may be one-to-two orders of magnitude greater than the annual dose from intakes by adults. Thus, the usual practice of limiting dose to the public for each year of exposure (1) could lead to annual doses to infants that exceed limits in radiation protection standards if the dose is calculated only for adults, as is often the case, even when the limit on annual dose equivalent for a specific practice is as low as 25 mrem, and (2) could overestimate lifetime risk from chronic exposures, which are likely for many practices, by more than an order of magnitude if the dose limit is applied to infants. A much closer correspondence with lifetime risk is obtained if standards are expressed as limits on annual effective dose equivalent averaged over a lifetime.

Proposed Regulatory Framework

Based on the foregoing discussions, the following regulatory framework for limiting radiation exposures of the public is proposed.

- In generally applicable radiation protection standards, establish a principal limit on annual effective dose equivalent averaged over a lifetime of 0.1 rem (1 mSv) and a subsidiary limit on effective dose equivalent for any year of 0.5 rem (5 mSv), in accordance with current recommendations of the ICRP[2] and the NCRP.[3]

- Express environmental radiation standards for practices that do not primarily involve naturally occurring radionuclides in terms of limits on annual effective dose equivalent averaged over a lifetime.

- In environmental radiation standards that primarily involve naturally occurring radionuclides, specify limits on annual dose equivalent averaged over a lifetime to the extent practicable. Secondary limits on concentrations of radionuclides in the environment could also be specified, as in current standards,[7-9] but these limits should be related to dose.

- Take into account the age dependence of dose and organ-specific risks in evaluating annual effective dose equivalents and their average over a lifetime.

- Specify that all exposures should be maintained ALARA.[1-3]

- In environmental radiation standards, establish limits on annual effective dose equivalent averaged over a lifetime at levels no lower than 25 mrem (0.25 mSv) per practice, in order to avoid limits that correspond to negligible risks to the public. Maximum doses lower than the limit would be achieved by use of the ALARA principle on a site-specific basis.

- Establish a generally applicable *de minimis* dose for public exposures, i.e., a limit below which reductions in dose using the ALARA principle would be deliberately and specifically curtailed.[3] In accordance with current NCRP recommendations,[3] an annual effective dose equivalent averaged over a lifetime of about 1 mrem (0.01 mSv) is suggested as *de minimis*.[21] In conjunction with the proposed dose limit for specific practices described above, this *de minimis* dose would provide a substantial range over which the ALARA principle could be applied.

Implications of Proposed Framework for Current Standards

The proposed regulatory framework would require some changes in current radiation protection and environmental radiation standards. The most important changes are described below.

- All standards and methods for estimating dose would need to consider the age dependence of the effective dose equivalent.

- Environmental radiation standards which contain limits on annual dose equivalent of 25 mrem to whole body, 75 mrem to thyroid, and 25 or 75 mrem to any other organ[6,8,10-12] could be changed, with minimal adverse impact, to specify a limit on annual effective dose equivalent averaged over a lifetime of 25 mrem (0.25 mSv).

- In standards for uranium and thorium mill tailings,[8,9] limits on environmental concentrations of radionuclides could be retained as secondary standards, but they would need to be related to specified limits on annual effective dose equivalent averaged over a lifetime to the extent practicable.

- Current drinking water standards[7] would require significant modification. First, the standards would need to specify a single limit on annual effective dose equivalent averaged over a lifetime for all radionuclides. Second, if the dose limit were set at the proposed lower limit of 25 mrem (0.25 mSv) per practice, then the corresponding limit on risk would be considerably higher than the risks corresponding to current drinking water standards (see Table 1) which, in many cases, are generally regarded as negligible.[3] Doses from drinking water would be reduced below the proposed limit by use of the ALARA principle on a site-specific basis.

Recent Development in Radiation Protection Standards

A recent development in radiation protection standards for the public is to specify that the dose from individual sources should not exceed one-fourth of the limit from all sources.[3,5] This approach assumes that an individual is unlikely to receive significant exposures from more than four sources. If we assume a limit on annual effective dose equivalent averaged over a lifetime of 0.1 rem (1 mSv) from all sources, then this approach would result in a limit on annual effective dose equivalent averaged over a lifetime of 25 mrem (0.25 mSv) per source. Thus, the suggested source-related limit in radiation protection standards would correspond to the lower limit for practice-specific standards proposed in this paper.

REFERENCES

1. U.S. Nuclear Regulatory Commission, Part 20—Standards for Protection Against Radiation, p. 247 in "Code of Federal Regulations, Title 10, Parts 0 to 199," U.S. Government Printing Office, Washington, DC (1987).

2. International Commission on Radiological Protection, Statement from the 1985 Paris Meeting of the International Commission on Radiological Protection, ICRP Publication 45, *Ann. ICRP* **15**, No. 3, p. i (1985).

3. National Council on Radiation Protection and Measurements, "Recommendations on Limits for Exposure to Ionizing Radiation," NCRP Report No. 91, NCRP, Bethesda, MD (1987).

4. International Commission on Radiological Protection, A Compilation of the Major Concepts and Quantities in Use by ICRP, ICRP Publication 42, *Ann. ICRP* **14**, No. 4 (1984).

5. R. E. Alexander, A Logical Framework for Radiation Protection Standards for the Public, p. 295 in "Population Exposure from the Nuclear Fuel Cycle," American Nuclear Society and Oak Ridge National Laboratory, Oak Ridge, TN (1988).

6. U.S. Environmental Protection Agency, Part 190—Environmental Radiation Protection Standards for Nuclear Power Operations, p. 6 in "Code of Federal Regulations, Title 40, Parts 190 to 399," U.S. Government Printing Office, Washington, DC (1986).

7. U.S. Environmental Protection Agency, Part 141—National Interim Primary Drinking Water Regulations, p. 521 in "Code of Federal Regulations, Title 40, Parts 100 to 149," U.S. Government Printing Office, Washington, DC (1986).

8. U.S. Environmental Protection Agency, Part 192—Health and Environmental Protection Standards for Uranium and Thorium Mill Tailings, p. 16 in "Code of Federal Regulations, Title 40, Parts 190 to 399," U.S. Government Printing Office, Washington, DC (1986).

9. U.S. Nuclear Regulatory Commission, Part 40—Domestic Licensing of Source Material. Appendix A—Criteria Relating to the Operation of Uranium Mills and the Disposition of Tailings or Wastes Produced by the Extraction or Concentration of Source Material from Ores Processed Primarily for Their Source Material Content, p. 428 in "Code of Federal Regulations, Title 10, Parts 0 to 199," U.S. Government Printing Office, Washington, DC (1987).

10. U.S. Environmental Protection Agency, Part 191—Environmental Radiation Protection Standards for Management and Disposal of Spent Nuclear Fuel, High-Level and Transuranic Radioactive Wastes, p. 7 in "Code of Federal Regulations, Title 40, Parts 190 to 399," U.S. Government Printing Office, Washington, DC (1986); see also *Fed. Registr.* **50**:38066 (1985).

11. U.S. Nuclear Regulatory Commission, Part 61—Licensing Requirements for Land Disposal of Radioactive Waste, p. 658 in "Code of Federal Regulations, Title 10, Parts 0 to 199," U.S. Government Printing Office, Washington, DC (1987).

12. U.S. Environmental Protection Agency, Part 61—National Emission Standards for Hazardous Air Pollutants, p. 4 in "Code of Federal Regulations, Title 40, Parts 61 to 80," U.S. Government Printing Office, Washington, DC (1986).

13. U.S. Environmental Protection Agency and U.S. Department of Health and Human Services, "A Citizen's Guide to Radon," OPA-86-004, U.S. Government Printing Office, Washington, DC (1986).

14. International Commission on Radiological Protection, Recommendations of the International Commission on Radiological Protection, ICRP Publication 26, *Ann. ICRP* **1**, No. 3 (1977).

15. International Commission on Radiological Protection, Lung Cancer Risk from Indoor Exposures to Radon Daughters, ICRP Publication 50, *Ann. ICRP* **17**, No. 1 (1987).

16. D. C. Kocher and A. L. Sjoreen, "Dose-Rate Conversion Factors for External Exposure to Photon Emitters in Soil," *Health Phys.* **48**:193 (1985).

17. D. C. Kocher, "Radioactive Decay Data Tables," DOE/TIC-11026, U.S. Department of Energy, Oak Ridge, TN (1981).

18. U.S. Nuclear Regulatory Commission, "Regulatory Guide 1.109. Calculation of Annual Doses to Man from Routine Releases of Reactor Effluents for the Purpose of Evaluating Compliance with 10 CFR Part 50, Appendix I," Washington, DC (1977).

19. International Commission on Radiological Protection, Limits for Intakes of Radionuclides by Workers, ICRP Publication 30, Supplement to Part 1, *Ann. ICRP* **3**, No. 1-4 (1979).

20. D. C. Kocher, K. F. Eckerman, and R. W. Leggett, "On the Relationship Between Radiation Standards for the General Public and Limitation of Lifetime Risk," *Health Phys.* **55**:820 (1988).

21. D. C. Kocher, "A Proposal for a Generally Applicable *de Minimis* Dose," *Health Phys.* **53**:117 (1987).

The Anatomy of Safety Evaluations

C. Zervos[a]
Food and Drug Administration
Washington, DC

ABSTRACT

Health/environment statutes contain safety and effectiveness provisions which require use of the scientific method of inquiry. To be meaningful, terms in such provisions must be cast in a scientific frame of reference whose construction yields the component parts of safety evaluations, i.e., operational definitions; data collection and evaluation procedures; and compliance criteria. Generally, safety evaluation protocols evolve in systematic ways that yield cohesive and well-integrated approaches to meeting the requirements of health/environment statutes. The component parts of safety evaluations and the processes by which they evolve will be discussed.

KEYWORDS: Operational definitions, regulatory models

Governments use a mix of policy and science to decide the proper control of technology and its products and byproducts. The necessary risk/benefit decisions are controversial and generate policy issues which cluster loosely around decision credibility, proper allocation of research resources, and the nature and extent of testing.

Controversies often focus on specific data related to a technology or product. Just as often, however, they focus on the schemes of safety evaluation used by the government. Specific examples of the latter case are saccharin, benzene, tetrachlorodibenzodioxin, and polychlorinated biphenyl. For example, few if any scientists contested the bioassay data for saccharin during the height of the homonymic controversy, but neither was there a consensus about their meaning or about the way they were used in the attempt to ban the substance. This lack of agreement is often the case and it seems that the credibility of decisions concerning a specific product is frequently independent of the quantity and quality of specific data for the product itself. Instead, actual or contemplated government decisions are seen as defensible or not in a way that mirrors the strengths and weaknesses of their scientific frame or reference, i.e., their scientific context.

In what follows, I shall examine the structural and functional properties of the current frame of reference of safety evaluations. I hope to contribute to an understanding of why some decisions are supported by the public while others, similarly justified by experimental data, are not. I also hope to highlight the need for further study of these subjects, which I

---

a. The opinions expressed herein are those of the author alone and do not represent official opinions or policies of the FDA.

believe is likely to reveal new approaches to research planning and to economical uses of test resources.

I will begin by noting that the scientific frame of reference for safety evaluations is identified by a variety of names, the most common being safety evaluation protocols, risk assessment procedures, and acceptance procedures. Whatever their names, however, they owe their existence to two factors: a legislative intent to have science play a significant role in the societal control of technology and the operational nature of science.

By legislative design, for instance, key provisions in health and environmental protection statutes translate into questions that can be answered only by the scientific method of inquiry. Thus, administration of such statutes is a scientific endeavor, at least in part. Further, because of the nature of science, no technical or scientific provisions in a statute can be met consistently and coherently over time and across products and technologies without operational specifications.

Thus, safety evaluation protocols are aggregates of specifications which spell out the operational details of the when, why, and how of using experimental data to decide whether specific requirements of a statute have been met. They provide the scientific context of health and environmental protection decisions.

Specifications can be classified into three distinct groups: operational definitions of legal standards; procedures of collecting and reducing data; and criteria. Because of time limitations I shall focus exclusively on the first group, the operational definitions.

Over the years a process has evolved for the construction of operational definitions. It comprises a number of identifiable steps, including the identification of statutory provisions that require scientific specification; the interpretation of identified provisions in the light of current scientific potentialities and technical capabilities; and the development and articulation of proposed operational definitions. The process yields products with structural and functional properties that do not vary across substances, technologies, or statutes, and permits the systematic study of operational definitions.

It is perhaps unusual to start a discussion by saying what the subject matter is not. It needs to be said in advance, however, that operational definitions are not random accumulations of *ad hoc* interpretive statements of law or inference "do's" and "don'ts." They have a purpose, and their purpose dictates their structural and functional properties. Specifically, their purpose is to provide a unique and credible way of making certain types of predictions from a set of experimental observations. In formal parlance, operational definitions come under the definition of "calculus" and every operational definition of safety is, and could be called, a calculus of risks of a particular kind. Accordingly, every operational definition of safety comprises an axiomatic base upon which rests an edifice composed of the nature and the structure of that part of physical reality which is covered by the operational definition. The edifice is known as the regulatory model of the phenomenon of concern.

Axioms are extrascientific propositions. As a consequence, they cannot be improved or changed by research in the biological or physical sciences, and decisions which lack credibility because they are founded on incredible axioms are unlikely to improve as a result of such research.

What are regulatory models? They are broad-scope conjectures about the nature, the structure, and the properties of the natural phenomena addressed by the operational definition of safety. To paraphrase Ackoff *et al.*,[1] they accumulate and relate, in the way that best serves the purpose at hand, "the knowledge we have about a specific aspect of reality. They serve as instruments for explaining the past and the present and for predicting and controlling the future."

In a more practical sense, they are designed to accept some data and in return to yield estimates of decision variables, X_i, for decision problems of the general type

$$V = f(X_i, Y_j) \ ,$$

where V is the value of the decision to be made and Y_j represents the decision parameters.

It follows from their reason for being that within their confines, regulatory models must provide for identification, measurement, and comparison of metrics that characterize at least some of the important aspects of the thing being defined. They must also provide for the identification of possible correlations between the various metrics and for operations that permit predictions. As a minimum, a regulatory model must provide at least one identifiable relation between metrics, i.e., a function of metrics which can be used as the official risk prediction rule.

Regulatory models are not *de novo* creations of legislators, administrators, or risk assessors. If they sound like scientific constructs, it is because in many respects they are. They derive from *scientific conjectures*—some call them theories—which are native to the field of scientific inquiry. The main point I wish to make, however, is that the two (regulatory models and scientific conjectures) are not, and even under the best of circumstances cannot be, identical even though they are very closely related. This is an important point because much about the credibility of decisions, the optimality of research plans, and the economy of test resources depend on keeping the similarities and differences of the two in mind.

The main structural and functional features of regulatory models were discussed above. I will review next the salient features of scientific conjectures, and finally I will discuss the similarities and differences between the two.

Conjectures in general play an important role in our affairs, whether in reference to the scientific enterprise, to technical development, or to everyday life. Broad-scope conjectures about nature and its workings provide the indispensable frame of reference, the template or roadmap, so to speak, for the scientific enterprise. Thus, scientific conjectures are always much more than idle speculations about nature. Indeed, they have both characteristic structure and indispensable properties.

Without doubt, it is rarely if ever easy to discern the precise origins and the early evolution of scientific conjectures. Conjectures seem to originate in the unexplorable domains of intuition, creativity, and genius, although what can be observed and studied suggests that scientific conjectures constitute novel and harmonizing ways of looking at specific and incompatible aspects of accepted knowledge.

Compared to their origins and evolution, the apparent structure of scientific conjectures is easier to envision and understand. The centerpiece of each scientific conjecture seems to be a fundamental speculative premise about the specific natural process or phenomenon of concern. Complete scientific conjectures appear to be built deductively from the logical juxtaposition of a basic premise and accepted knowledge, in more or less the following way: the unconditional or conditional validity of the basic premise is assumed true in the context of accepted scientific facts, laws, and established theories. The process yields a primary set of consequences which can be expressed in the form of relations of any type between appropriate variables and constants. Some of these relations are stochastic, others are deterministic; some are strong, such as differential or integral equations, and others are simple classifications or set membership statements.

A secondary set of consequences is deduced next by repetition of the process, that is, by assuming, in the context of accepted knowledge, the conditional or unconditional validity of the basic premise and its primary consequences and then considering what follows logically. The process appears to continue until the predicted relations involve all the pertinent quantities, fundamental or derived, that can be observed by experimental techniques which are already available or can be developed for the purpose. Thus, it appears as if the complete structure of a scientific conjecture constitutes a hierarchy of sets of propositions or functions which trace their logical origins to the fundamental premise.

The important properties of scientific conjectures derive from their role in science. First, the fundamental premise of a scientific conjecture cannot be an outlandish speculation about nature. Rather, it is necessarily, the linchpin of a coherent resynthesis and a systematic extension of already accepted knowledge.

Second, the conjecture itself is *falsifiable in principle* in part or *in toto*. Let me explain this a bit because it is usually misinterpreted. What it means is that there must be clearly identifiable ways by which one can test the truth of the total conjecture or some of its specific predictions. This property derives from the scientific purpose of conjecturing, which is to compare predictions with actual experience. In practical terms, it means that a subset of the propositions, relations, or functions predicted by the scientific conjecture can be cast as statistical hypotheses which, in principle, can be tested experimentally. If there is no conceivable way to compare with experience at least some of the consequence of a conjecture, the conjecture is not scientific.

Third, a corollary to the property of falsifiability in principle is the expectation that scientific conjectures are temporary and transient constructs subject to constant modification and often total debunking.

Fourth, the universe of propositions, relations, and functions that inheres in the basic premise of any scientific conjecture is obviously open ended; the discovery and articulation of new propositions are matters of imagination, intuition, scientific insight, etc.; and in a scientific sense, a conjecture will be better as it leads to a greater number and variety of predictions that can be tested experimentally.

Finally, the propositions, relations, and functions of a scientific conjecture are not all of equal scientific worth. For instance, in accordance with their scope, functions have been classified as deterministic (causal), probabilistic (producer-product relationships), and correlative. The narrower the scope of a function, the more fact-like it is, and the broader its scope, the more it approaches a potential scientific law. Further, although not always true, the lower the order of a relation, the greater its scope and the more difficult it is to test experimentally. This property puts a high scientific premium on the identification of low-order propositions that can be tested by imaginative experimental means.

The preceding discussion focused on the structural and functional properties of what decision making (regulation) needs and what science offers. Given the often stated desire for science-based regulation, it would appear that scientifically and technically speaking, operational definitions of safety would be best if their regulatory models approximate, as closely as possible, the best scientific conjectures about the phenomenon of concern. An analysis and discussion of their properties will show the utopian nature of this approach, however.

First, to be scientific, a conjecture must, above all, be coherent. If it does not synthesize all pertinent knowledge into a coherent picture, a conjecture is not scientific. In contrast, in regulatory models coherence is not a sine-qua-non feature and although always desirable, it is often sacrificed to some degree in order to achieve other nonscientific but none-the-less important objectives, e.g., public health protection and administrative efficiency. In regulatory models, the place of coherence is to some extent taken by scientific respectability. A model must enjoy the respect of the scientific community; i.e., there must be a degree of scientific consensus that the loss of whatever coherence might have been sacrificed does not do unacceptable violence to the scientific enterprise and that the loss is worthwhile for the sake of attaining the other objectives of the regulatory model.

Second, regulatory models must be robust; i.e., they must remain valid over protracted periods of time. Operational definitions based on regulatory models which are subject to frequent modifications yield safety evaluations which create uncertainty in the marketplace, and the effects are considered detrimental even from the point of view of public health protection. Regulatory activities based on safety evaluations of transient validity are usually perceived by the public as unreasonable government hindrances to technical development. Thus, operational definitions should be constructed to last. This requirement is essentially antithetical to the requirement for falsifiability in principle, the fundamental property that makes scientific conjectures a superb tool of scientific research. Incidentally, the antithesis also diminishes the value of regulatory models as templates for scientific research.

Third, between otherwise similar regulatory models, the one that leaves the fewer unanswered questions is also the one that is better for the purposes of regulation. In contrast, the richer scientific conjectures are in terms of predictions, the better they are for the research enterprise, i.e., the greater the number of questions they open for exploration. This difference has practical importance for the matter of limits on testing requirements and is another powerful reason why, without dramatic modifications, scientific conjectures cannot be used to support operational definitions of safety. For instance, a substance whose safety must be ascertained for premarketing clearance can always be used to test the validity of a large number of the labyrinthine network of propositions, relations, and functions that make up even the simplest scientific conjectures used to support the various operational definitions of safety. Theoretically, once completed, testing governed by a scientific conjecture would increase to some extent the confidence of adverse health effect predictions which decision makers need to compare the outcomes of the available courses of action. The approach is impossible, however, for at least the following reasons: (1) The testing process would tend to alter significantly the operational definition of safety; (2) the testing process would be interminable, and few, if any, substances would ever be cleared for marketing; (3) some of the required tests are unlikely to be permissible on ethical grounds; and (4) there is no guarantee that the information or knowledge to be acquired from such a process would be worth the expense.

Fourth, a corollary to the preceding point is the sharp contrast in the focus of regulatory models and scientific conjectures. The focus of scientific conjectures is on the variety, the breadth, and the scope of hypotheses that can be tested experimentally. On the other hand, the focus of regulatory models is on one, or perhaps a small number, of the risk prediction relations that is acceptable to the scientific and regulatory communities at large.

Fifth, a consequence of the difference in the experimental focus of scientific conjectures and regulatory models is a contrast in the process used in the two cases to select hypotheses to test. In scientific research the selection is the choice of the researcher and is usually governed by the history of his research program, by his relative ranking of the scope of the hypotheses available for testing at a given time, and by the available resources. In the business of regulation, selection of hypotheses to test is not the choice of the experimenter and it is governed by the regulatory objective, i.e., to collect a modicum of data which can be subjected to specific manipulation to provide an estimate of risk. The research community places a premium on the "critical experiment." The regulatory community places a premium on limiting the amount and the type of data to be collected, i.e., on test standardization.

Sixth, whereas regulatory models must protect the public health to the degree mandated by the Congress that enacted the statute of concern, scientific conjectures are societally value-neutral.

Finally, whereas regulatory models must be administratively efficient, scientific conjectures are again neutral to principles of governance.

Because of these differences, society would be in a dilemma if it were to insist that scientific conjectures be used to construct the operational definitions that are necessary for the control of technology and its byproducts, and the way society avoids this dilemma is by using regulatory models. Regulatory models are converted scientific conjectures, the conversion consisting in the use of administrative fiat to remove uncertainties of the kinds and numbers sufficient and necessary to give the product the desired properties.

A detailed discussion of the evolution of regulatory models from scientific conjectures, a complex but orderly process, is beyond the scope of this presentation. Suffice it to say that it is driven by the need to identify the relation(s) that will be used as a risk prediction tool and to submit it to the concerned community (scientists, administrators, etc.) and to society at large for consideration and acceptance.

It was noted earlier that the identified relation plays a central role in risk assessment and that it must fulfill the specified requirements.

To summarize, it has been argued that implementation of the provisions of health and environmental protection statutes requires scientific and technical specifications known by such names as acceptance procedures, safety evaluation protocols, or risk assessments. The general structure and the salient properties of safety evaluation protocols have been described. It has been shown that in protecting the public health and the environment from the untoward effects of technology, what science can offer and what regulation requires are distinctly different. The differences are significant and bear on three important aspects of public health protection activities: decision credibility, optimal planning of research, and optimal utilization of test resources.

REFERENCE

1. R. L. Ackoff, S. K. Gupta, and S. J. Minas, *Scientific Method: Optimizing Applied Research Decisions*" Chapter I, John Wiley and Sons, Inc., New York, 1962

The Impact of Non-Drinking Water Exposures on Drinking Water Standards: A Problem of Acceptable Levels of Risk

Paul S. Price
American Petroleum Institute
Washington, DC

ABSTRACT

Under the Safe drinking Water Act, the Environmental Protection Agency (EPA) is required to establish highly protective drinking water standards for compounds which are likely to contaminant public water supplies. One of the steps in establishing standards for non-carcinogens is to adjust the dose permitted under the standard to allow for non-drinking water exposures. Because of the high levels of non-drinking water exposure, EPA's policy could, if rigorously applied, force the agency into establishing zero concentration drinking water standards for many non-carcinogens. The basic reason for the dilemma is the agency's failure to recognize that non-drinking water exposures need not meet the stringent level of acceptable risk under which drinking water standards are established. Similar problems occur in the standards setting procedures used in actions proposed under the Resource Conservation and Reclamation Act (RCRA) and the Clean Water Act (CWA).

KEYWORDS: Standards, drinking water, exposures, acceptable risk

INTRODUCTION

Risk and Standards

One of the maxims of regulatory toxicology is that it is the dose that an individual receives which determines whether a chemical is toxic. By this toxicologists have emphasized that risk is a function of both the toxicological properties of the substance and the level of exposure, or,

$$\text{RISK} = f\,(\text{TOXICITY, DOSE})\,.$$

By extension, a chemical standard should be understood as a function not only of the objectively determined toxicity, but also the level of risk under which the standard was established, or,

$$\text{STANDARD} = f\,(\text{TOXICITY, ACCEPTABLE LEVEL OF RISK})\,.$$

657

Thus, every chemical standard should be understood as the result of the application of a specific level of risk applied to a specific body of toxicological information. Because of the role of risk acceptance in standard setting, different organizations will establish very different standards based on the same toxicological information if they have different levels of acceptable risks.

The determination of what risks are acceptable has been the subject of considerable study.[1,2] Individuals and societies have and continue to make radically different decisions on risks in different areas of life.[3] Acceptance or rejection of a risk is usually thought to be based on specific factors unique to the nature of the risk and its context.[4] Because of this uniqueness, it is inappropriate to automatically extrapolate risk acceptance from one sphere of activity to another.

Drinking Water Standards

EPA is charged with the responsibility to establish drinking water standards under the Safe Drinking Water Act (SWDA), which calls for the establishment of standards called Maximum Contaminant Level Goals (MCLGs) and specifies stringent levels of acceptable risk. MCLGs are to "be set at the level at which no known or anticipated adverse affects on the health of persons occur and which allows an adequate margin of safety" Section 1412, SDWA.

Table 1 presents a general description of the steps which EPA goes through in setting drinking water standards for non-carcinogens.[5] In order to achieve this very low level of acceptable risk, EPA applies very conservative assumptions in each of these steps.

Table 1. Steps in Developing MCLGs Under the Safe Drinking Water Act

---

    I.   Toxicological review.

    II.   Determination of toxicological endpoint.

    III.   Determination of reference dose using uncertainty factors.

    IV.   Determination of the portion of the reference dose to be used in the MCLG.

    V.   Determination of the drinking water concentration which corresponds to the selected portion of the dose.

The first three steps in the development of an MCLG are the toxicological portion of the process and end with the development of the reference dose (RfD). In steps IV and V, EPA determines how the RfD should be related to the final drinking water concentration. In step IV, EPA apportions the RfD into the dose which may be permitted under the drinking water standard and the dose which is believed to be received from non-drinking water exposures. The goal of this action is to set the drinking water standard sufficiently low that persons who have non-drinking water exposures will not receive a total dose in excess of the RfD. Finally in V, EPA determines the concentration of a compound in tap water which will correspond to the portion of the RfD set aside for the drinking water standard. This has traditionally been done by assuming that individuals are exposed by consuming 2 liters per day of drinking water. The remainder of this paper will focus on step IV, the apportionment of the RfD.

NON-DRINKING EXPOSURE AND DRINKING WATER STANDARDS

EPA's policy for achieving step IV has been described in the proposal to establish standards for Synthetic Organic Chemicals,[5] and has been used to establish MCLGs for para-dichlorobenzene and 1,1,1-trichlorethane.[6] The policy calls for a determination of the contribution to total exposure from inhalation and diet. This dose is then subtracted from the RfD and the remainder is used to set the MCLG:[a]

$$\text{MCLG} = \frac{\text{RfD} - \text{(CONTRIBUTIONS FROM FOOD AND AIR)}}{\text{DRINKING WATER INTAKE}}$$

This equation was used to establish standards for several metals, including chromium, barium, and selenium. For these metals EPA determined that air intakes were negligible. The dietary intake was estimated based on FDA market basket surveys which are indicators of average intakes.[5] The policy also called for the use of a second equation for apportioning the RfD which was to be used only when there was insufficient data to accurately determine non-drinking water contributions to total exposure:

$$\text{MCLG} = \frac{\text{RfD} \times \textbf{RELATIVE SOURCE CONTRIBUTION}}{\text{DRINKING WATER INTAKE}}$$

The RELATIVE SOURCE CONTRIBUTION was assumed to be 20% for organics and 10% for inorganics. This equation was used for all of the standards proposed and established for organic compounds. In proposing the use of this second equation, EPA made the assumption that non-drinking water exposures were unlikely to result in a total intake which exceeds either 80% (for organics) or 90% (for inorganics) of the RfD.

This policy is consistent with other actions that the agency has taken in apportioning RfDs. The Water Quality Criteria established under the CWA proposed the following equation for taking into account exposure outside the control of the standard:[7]

$$\text{CRITERION} = \frac{\text{ADI} - \text{(DIETARY + INHALATION)}}{\text{DRINKING WATER + FISH}}$$

Where ADI is the same as the RfD, DIETARY is the non-fish dietary contribution to total exposure, INHALATION is the inhalation contribution to total exposure, DRINKING WATER is the assumed intake of 2 liters per day, and FISH is the intake of the compound as the result of consuming fish which have bioaccumulated the compound. This equation is equivalent to the first method of apportioning the RfD.

The second method of apportioning the RfD has also been practiced in other parts of the agency. In the 1987 RCRA proposed regulation for controlling the burning of hazardous waste in boilers and industrial furnaces, the dose permitted for the maximum exposed individual down wind from the facility is calculated using the following equation:[8]

$$\begin{array}{l}\text{REFERENCE} \\ \text{AIR} \\ \text{CONCENTRATION}\end{array} = \frac{\text{RfD} \times \begin{array}{c}\text{BODY} \\ \text{WEIGHT}\end{array} \times \begin{array}{c}\text{CORRECTION} \\ \text{FACTOR}\end{array} \times \begin{array}{c}\textbf{APPORTIONMENT} \\ \textbf{FACTOR}\end{array}}{\text{DAILY AIR INTAKE}}$$

In this equation, the Reference Air Concentration, which the maximum exposed individual receives, is established using an APPORTIONMENT FACTOR of 25%.

---

a. Before the June 19, 1986 amendments to the SDWA, MCLGs were referred to as Recommended Maximum Contaminant Levels or RMCLs.

PROBLEMS WITH EPA'S APPROACH

There are several problems with EPA's approach. First, EPA has not clearly identified whose non-drinking water exposures should be considered in determining the MCLGs. Available information on environmental exposures suggests that the wide range in ages, occupations, and lifestyles will result in a very wide range of non-drinking water exposures to the U. S. population. For example, the EPA TEAM study of New Jersey homes found that the differences between the median and 98th percentiles of indoor air levels of para-dichlorobenzene and 1,1,1-trichloroethane differ by two or more orders of magnitude.[9] A further example is the wide variation in the dietary intake of pesticides which occurs as the result of variations in individuals' diets.[10]

This wide range in intake poses a major problem for EPA. The agency may either set the MCLG on the basis of average intake and allow some individuals to receive total doses in excess of the RfD, or it may take into account the exposures which the highest exposed individuals receive and apportion only a very small portion of the RfD to the MCLG. In doing the latter, the agency may be forcing public water supplies[b] to spend additional resources to achieve standards which have been lowered because of other sources of exposure. This will lead to regulatory inefficiencies where equivalent levels of exposure reduction can be more readily achieved by controlling the other sources of exposure.

Second, and more importantly, EPA is faced with the problem that for most compounds some individuals receive non-drinking water exposures which are in excess of the RfDs. These exposures occur as the result of occupational exposure, indoor air pollution, and smoking.

Table 2 presents data on nine chemicals which have had MCLGs proposed or promulgated. The assumptions used in generating the table are given in Table 3. The table shows that current occupational standards permit exposures to occur which result in doses two to three orders of magnitude greater than the RfDs. Assuming that at least some individuals are exposed to a significant fraction of the occupational standards,[11] then occupation exposure will greatly exceed the RfDs. Based on the indoor air measurements reported in the New Jersey portion of the TEAM survey,[10] some individuals may receive exposures which also greatly exceed the RfD. Further, the indoor air measurements for the three aromatic compounds, styrene, xylenes, and toluene, are believed to reflect passive exposure to tobacco smoke.[12] This strongly suggests that exposure to the compounds from direct use of tobacco products will result in doses which exceed the RfDs for these compounds.

Based on this information, EPA's current policy of directly subtracting the known non-drinking water exposures from the RfD suggests that the agency will have to set the MCLGs for many non-carcinogenic compounds at zero. Clearly such stringent controls would have a dramatic impact on drinking water supplies. Further, since the MCLGs are also used in the RCRA and other programs, the economic impact of setting the MCLGs at zero would extend far beyond the drinking water utilities.

ACCEPTABLE LEVELS OF RISK AND EPA'S POLICY

The root of much of EPA's dilemma is its assumption that all sources of exposure to a given chemical should be judged under the stringent levels of acceptable risk which Congress instructed the agency to apply to drinking water exposures. EPA's policy directly

---

b. The MCLGs are not in themselves enforceable, however, the MCLGs must be set as close to the MCLG as is economically and technically feasible. Thus, a reduction in the MCLG will affect the MCL unless such a reduction goes below the "feasible" level.

Table 2. Non-Carcinogenic Drinking Water Standards

	MCLG (mg/1)	RfD	Occupational* (mg/day)	Indoor Air**
Para Dichlorobenzene	0.075	7	1,600	13
1,1,1-Trichloroethane	0.2	2	6,800	1,100
1,1-Dichloroethane	0.007	0.7	2,900	--
Ethylbenzene	0.68	6.8	1,600	7.5
Xylenes	0.44	4.4	1,600	50
Toluene	2.0	20	1,300	--
Styrene	0.14	1.4	800	33
Cis/Trans 1,2 Dichloroethylene	0.07	0.7	2,800	--

*1987 American Conference of Governmental and Industrial Hygienists Threshold Limit Values.
**1985 EPA TEAM Study, Maximum Concentration Reported.

Table 3. Assumptions Used in Generating Exposure Estimates

Occupational Exposures

- 50% of the compound inhaled is cleared from the lung.

- Workers inhale 10 cubic meters of air in an 8-hr work day.

- Workers are exposed to the TLV for 8 hr per day, 5 days per week.

Indoor Air Exposures

- 50% of the compound inhaled is cleared from the lung.

- Individuals inhale 10 cubic meters of air in a 12-hr day.

- Individuals are exposed to the maximum concentration surveyed for 12 hr.

assumes this by requiring that no individual should have a total exposure which exceeds the RfD. The introduction of the RfD should be understood to be a function of the level of acceptable risk permitted under the SDWA. Congress's directive to the agency on acceptable risk from chemical exposure does not automatically extend beyond drinking water. Individuals whose total exposures exceed the RfD must consider whether the risks associated with their total exposure are acceptable based on the circumstances of both the drinking water and the non-drinking water exposures.

Because the stringent requirements of acceptable risk may not apply, EPA need not apportion the RfD to accommodate all possible sources of exposures. EPA should apportion the RfD only where it is willing to make an explicit determination that the other sources of exposure should be controlled to the same degree as drinking water.

In the case of occupational exposures and smoking, both society and individuals are clearly willing to tolerate higher levels of acceptable risk than the Safe Drinking Water Act permits under drinking water standards. The level of risk under which occupational standards are established is described in the following quote:

> ...nearly all workers may be repeatedly exposed day after day without adverse effect. Because of wide variation in individual susceptibility, however, a small percentage of workers may experience discomfort from some substances at concentrations at or below the threshold limit; a smaller percentage may be affected more seriously by the aggravation of a pre-existing condition or by the development of an occupational illness.[11]

Clearly, the criterion for acceptable risk under occupational standards is far less stringent than for drinking water standards. While no formal statement on acceptable levels of risk has been developed for smoking, the wide-spread practice of smoking tobacco products in the presence of warnings on packages and advertisements clearly indicates that individuals tolerate levels of risk higher than those used to establish drinking water standards. Further, because our society is willing to allow its members to continue to expose themselves to such levels, higher risks are tolerated on a societal level. Because of the higher level of acceptable risk which individuals and society accept, it can be argued that EPA should not attempt to accommodate exposures from these sources when it apportions the RfD.

The agency may wish to consider apportioning the RfD to reflect exposures from diet and indoor air since reasonable arguments could be made that many individuals have low tolerances for risks from these sources of exposure. However, EPA should recognize that by accommodating these exposures, it is making a determination that these sources of exposure should be controlled to the same extent as if they occurred as the result of drinking water exposure. Such a determination goes beyond the direct authority of the SDWA and involves a separate finding of risk acceptability.

SUMMARY

EPA's policy on incorporating non-drinking water exposures in the setting of drinking water standards, if rigorously applied, could lead to the establishment of drinking water standards which are needlessly strict. This problem arises because the current policy does not take into consideration that the stringent levels of risk acceptability called for under the SDWA are not necessarily applicable to all sources of chemical exposure.

REFERENCES

1. W. W. Lowerance, *Of Acceptable Risk: Science and the Determination of Safety*, William Kaufman, Los Altos, CA, 1986.
2. P. Slovic, B. Fishhoff, and S. Lichtenstine, "Facts and Fears: Understanding Perceived Risk," in *Societal Risk Assessment: How Safe is Safe Enough?*, R. Schwing and W. Albers, eds., Plenum Press, New York, 1980.
3. J. D. Graham and J. W. Vaupel, "Value of a Life: What Difference Does It Make?," *Risk Analysis* **1**:89 (1981).
4. C. Hohenemser, R. W. Kates, and P. Slovic, "The Nature of Technological Hazard," *Science* **220**:378.
5. U.S.E.P.A., *National Primary Drinking Water Regulations: Synthetic Organic Chemicals, Inorganic Chemicals, and Microorganisms: Proposed Rule*, 50 FR 46936, 1984.
6. U.S.E.P.A., *National Primary Drinking Water Regulations-Synthetic Organic Chemicals, Monitoring for Unregulated Contaminants; Final Rule*, 52 FR 25690, 1987.
7. U.S.E.P.A., *Water Quality Criteria Documents; Availability*, 45 FR 79318, 1980.
8. U.S.E.P.A., *Burning of Hazardous Waste in Boilers and Industrial Furnaces*, 52 FR 16982, 1987.
9. U.S.E.P.A., *The Total Exposure Assessment Methodology (TEAM) Study: Summary and Analysis*, Vol. 1, Office of Research and Development, U.S.E.P.A., Washington D.C., 1986
10. U.S.E.P.A., *Tolerance Assessment System, Office of Pesticide Programs*, 1984.
11. American Conference of Governmental Industrial Hygienists, *Threshold Limit Values and Biological Exposure Indices for 1986-1987*, Cincinnati, OH, 1986.
12. L. A. Wallace and L. A. Pellizzari, "Personal Air Exposures and Breath Concentrations of Benzene and Other Volatile Hydrocarbons for Smokers and Nonsmokers," *Toxicology Letters* **35**, 1986.

Estimating Exposures and Health Risks for Alternative Ozone National Ambient Air Quality Standards[a]

Harvey M. Richmond and Thomas McCurdy
U.S. Environmental Protection Agency
Research Triangle Park, NC

ABSTRACT

The Environmental Protection Agency (EPA) is currently reviewing the ozone national ambient air quality standard (NAAQS) to determine if the existing standard "protects public health with an adequate margin of safety." In providing an adequate margin of safety, the Administrator is regulating not only to prevent pollution levels that have been demonstrated to be harmful, but also to prevent pollutant levels for which the risks of harm are considered unacceptable. In order to weigh these risks for margin of safety purposes, EPA assesses such factors as the nature and severity of the health effects associated with pollutant exposure, the degree of total human exposures (i.e., indoor and outdoor), and the risks (probabilities) of occurrences and expected number of specified adverse health effects being experienced when alternative ozone NAAQS are just met.

The general approach and structure of EPA's ozone NAAQS risk assessment is discussed. This effort involves both a probabilistic assessment of exposure-response relationships for several critical health effect endpoints and a probabilistic exposure analysis for alternative ozone NAAQS. A major feature of the ozone risk assessment is that major uncertainties will be treated explicitly and probabilistically. Key features of the ozone-specific NAAQS Exposure Model (NEM) are highlighted.

KEYWORDS: National ambient air quality standard (NAAQS), ozone exposure, ozone health risk assessment, NAAQS Exposure Model, Office of Air Quality Planning and Standards (OAQPS)

INTRODUCTION

The U.S. Environmental Protection Agency (EPA) is currently reviewing the ozone primary national ambient air quality standard (NAAQS) to determine if the existing standard protects public health with an adequate margin of safety. In order to evaluate whether alternative ozone NAAQS's provide an adequate margin of safety, EPA must assess such factors as the nature and severity of health effects associated with ozone

---

a. This paper was presented at the 1985 Annual Meeting of the Society for Risk Analysis held in Alexandria, Virginia, October 7–9.

exposure, the degree of total human exposure (i.e., indoor and outdoor) to ozone, and the risks (probabilities) of how many individuals would be adversely affected when alternative ozone NAAQS's are just attained. This paper briefly describes the general approach and structure of EPA's exposure and health risk assessment efforts for ozone which are designed to provide the above mentioned types of information.

The ozone exposure analysis and risk assessment projects are an outgrowth of a multiyear program by the Office of Air Quality Planning and Standards (OAQPS) to develop quantitative exposure and risk estimates to aid decisions on primary (health-based) NAAQS's (McCurdy and Richmond, 1983). EPA has already relied on exposure estimates generated as a result of this program in its review of the carbon monoxide and sulfur dioxide NAAQS's (Johnson and Paul, 1983; Paul and Johnson, 1985; Biller *et al.*, 1984). EPA's first on-line application of the risk assessment approach for NAAQS's is currently under way for the lead (Pb) NAAQS review through an interagency agreement with Argonne National Laboratory (Wallsten and Whitfield, 1985).

The Clean Air Scientific Advisory Committee (CASAC), part of EPA's independent Science Advisory Board, has been involved in the review of EPA's lead risk assessment efforts. The CASAC will also play an important role in reviewing and advising OAQPS on the ozone exposure and risk assessment projects now under way. Because the ozone work is in its initial stages, the ideas and approach described below are tentative and may be altered.

OVERVIEW OF THE OZONE EXPOSURE AND RISK PROJECTS

The objective of the ozone risk assessment project is to provide quantitative (probabilistic) estimates of the risk to public health associated with attainment of alternative ozone NAAQS's. The ozone risk assessment is not intended to replace scientific research nor to produce new scientific knowledge. Rather, it is intended to portray more explicitly the implications of and uncertainties in the existing scientific data base, making full use of existing scientific knowledge and expertise. The ozone risk assessment will deal with the major uncertainties that contribute to risk. These include uncertainties about: (1) pollutant levels that individuals will be exposed to upon attainment of alternative standards, (2) exposure-response relationships for the most susceptible populations, and (3) existence of effects in humans when they have been demonstrated only in animals. Probabilistic judgments will be elicited from health and air quality experts to represent the uncertainties in each of the above areas based on the current state of knowledge. These judgments will be used in models that consider both air quality and human activity patterns (exposure) and health information (exposure-response) to produce risk estimates for alternative NAAQS's. Risk estimates will be expressed as the probabilities that various fractions of the susceptible population will be adversely affected and the expected number of health effects and individuals affected in a given time period upon attainment of alternative standards.

Final judgments on the scope of the exposure analysis and risk assessment will be based on a careful review of the scientific and technical evidence presented and evaluated in the revised ozone "criteria document" and OAQPS "staff paper." The criteria document summarizes and evaluates the latest state of scientific knowledge concerning the effects of ozone on humans and the environment. The OAQPS staff paper reflects the staff's assessment of the key scientific evidence, uncertainties, and judgments that must be addressed in the standard-setting process based on the criteria document evaluation and other technical information and analyses (e.g., air quality analyses, population group estimates) (Padgett and Richmond, 1983). OAQPS's preliminary plans for the ozone exposure and risk assessment projects are described in the next section of this paper.

STRUCTURE OF THE OZONE RISK ASSESSMENT

The ozone risk assessment project has five major components, as depicted in Fig. 1. The first three components focus on estimating ozone exposures for the sensitive population. The fourth component concerns assessment of exposure-response relationships and the fifth component integrates the outputs from the exposure and exposure-response components to generate risk estimates. A more detailed description of each component follows.

Component Model 1 basically "allocates" existing air quality data to five micro-environments and then adjusts these data to fit a hypothetical future situation where an alternative ozone NAAQS is just attained at the critical monitor. Exposure and risk estimates will be provided for alternative 1-hour standards in the range 0.08 to 0.14 ppm, by increments of 0.02 ppm. [The current ozone primary standard is attained when the expected number of days per calendar year with maximum hourly average concentrations above 0.12 parts per million (ppm) is equal to or less than 1.] The primary output of Component Model 1 is a set of hourly micro-environment specific ozone values for each exposure district in the 10 urban areas that will be investigated. These 10 areas are:

Chicago	Los Angeles	Philadelphia
Denver	Miami	St. Louis
Houston	New York	Tacoma
		Washington, D.C.

The ten areas were chosen mostly on the basis of availability of monitoring data, but population and climatic factors were also important. The number of monitors used for analysis varies from three in Tacoma to 16 in Los Angeles; most of the areas have seven or more monitors with sufficient data to meet EPA's validity criteria for "continuously sampled" data (EPA, 1981).

Estimates of the 1980 populations for the ten urban areas vary between 0.5 million in Tacoma to 9.1 million in Los Angeles. Thus, a wide range exists in the sample with respect to population size. In addition, all but one EPA region (New England) is represented in the sample, so a broad range of geographical/climatic types are part of the ozone exposure analysis. EPA feels that the ozone study areas are a vast improvement over the sample of four used in the carbon monoxide (Johnson and Paul, 1983; Paul and Johnson, 1985) and other exposure analyses done in the past.

The micro-environments used for ozone are listed in Table 1. They are identical to those employed previously in ambient standard analyses. Multiplicative ratios relating the ozone level in each micro-environment to an ambient monitor were developed from a review of the ozone exposure literature (Ferdo, 1985). The ratios are the b_m term in a simple linear relationship that relates air quality in a micro-environment to air quality at a fixed-site monitor. The relationship is

$$x_{m,t} = a_{m,t} + (b_m * x_t) \ ,$$

where

$x_{m,t}$ = air quality in micro-environment m during hour t,

$a_{m,t}$ = hourly averaged pollutant concentration due to sources located in the micro-environment,

b_m = multiplicative ratio of the micro-environment concentration value to the monitored air quality value,

x_t = monitor-derived air quality value.

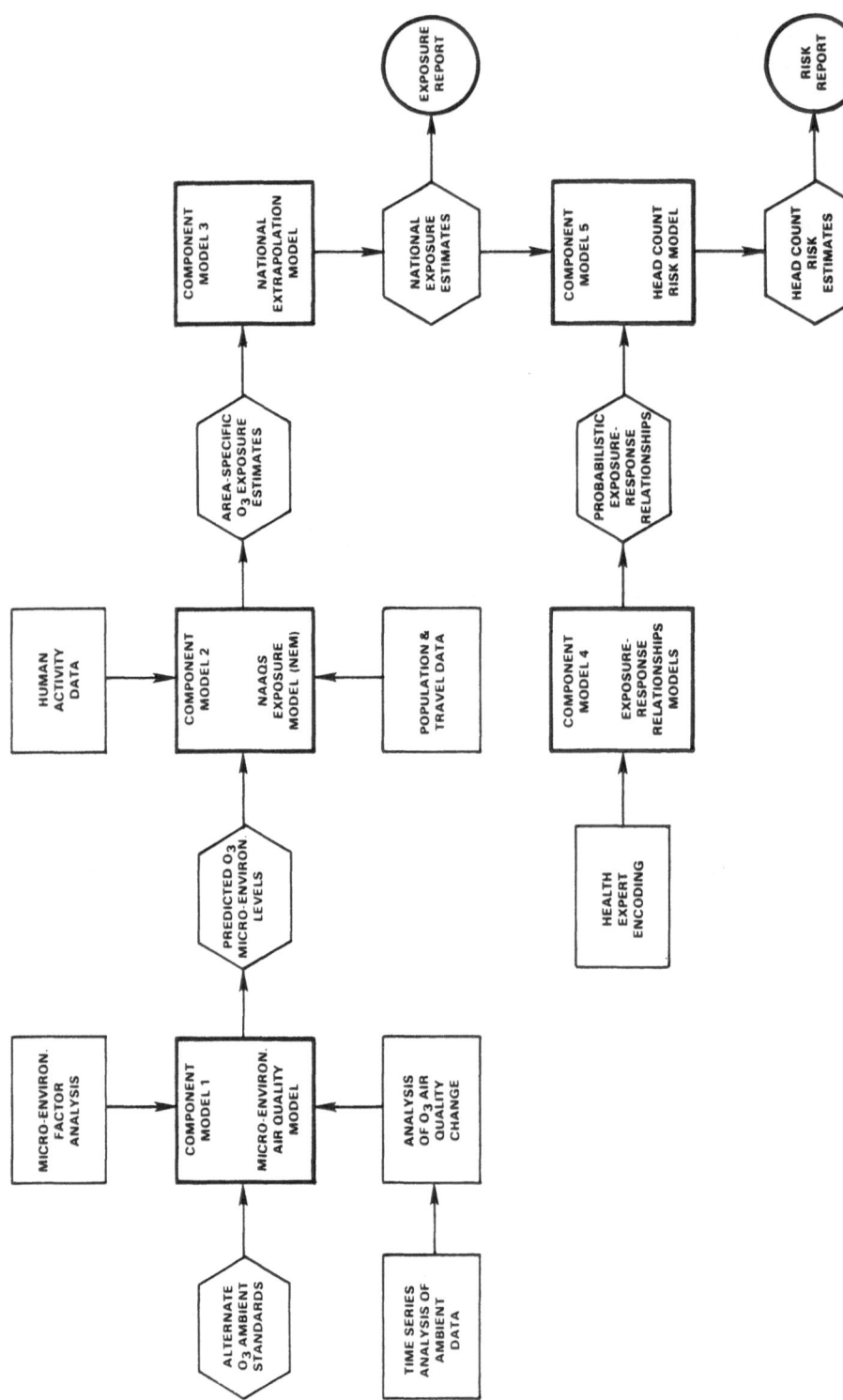

Fig. 1. Flow Chart of the Ozone NAAQS's Risk Assessment Project.

Table 1. Estimates of Multiplicative Micro-Environmental Factors (b_m)

Micro-Environment	Estimated Value (Unitless)		
	Low Estimate	Best Estimate	High Estimate
Indoors, residential	0.36	0.58	0.70
Indoors, other	0.40	0.52	0.62
Motor vehicle	0.00	0.11	0.30
Outdoors, near roadway	0.13	0.18	0.28
Outdoors, other	1.00	1.00	1.00

Because no significant sources of ozone were identified with any of the micro-environments, $a_{m,t} = 0$ for all environments. Estimated b_m values obtained from the literature are shown in Table 1. A cumulative frequency distribution of these values is depicted in Fig. 2.

Applying these ratios to the known x_t values produces the predicted ozone micro-environment output of Component Model 1. These outputs are utilized in Component Model 2, which is the NAAQS Exposure Model (NEM). NEM integrates the micro-environment specific air quality data with human activity, population, and transportation information. The human activity data come from a review of the current literature on the subject, which includes over 50 articles and books. A ten-minute time interval of activity is used in the ozone exposure analysis instead of the 1-hour interval used in previous NEM applications.

Different activity patterns were identified for 47 demographic subgroups, or cohorts (Pope, 1985). Within each of the cohorts, different activity patterns were identified for warm and cool weekdays, Saturdays, and Sundays; thus, there are six different patterns possible for each of the 47 cohorts. The patterns were further classified by location within the region (either a "home" or "work" district) and by exercise level (high, medium, or low). In addition, the cohorts are assigned to one of the five micro-environments noted above. An example of an activity pattern for the cohort "married female indoor worker with children" with a 20-minute commute time to work appears in Table 2 for part of a cool weekday.

NEM utilizes the human activity data by matching it with U.S. Census Bureau information relating to cohort-specific population estimates for "exposure districts." These districts are aggregations of census tracts surrounding each monitor used for the ambient air quality data that forms the pollutant concentration data basis of NEM. Obviously, exposure districts become smaller as more monitors are used. Having smaller districts allows NEM to better capture ambient pollutant concentration gradients, although a "boundary problem" (i.e., pollutant concentration discontinuity) will still exist across the district line. EPA will be investigating the use of Kreiging techniques to smooth out data between monitor locations in order to minimize the boundary problem. Because concentration gradients are

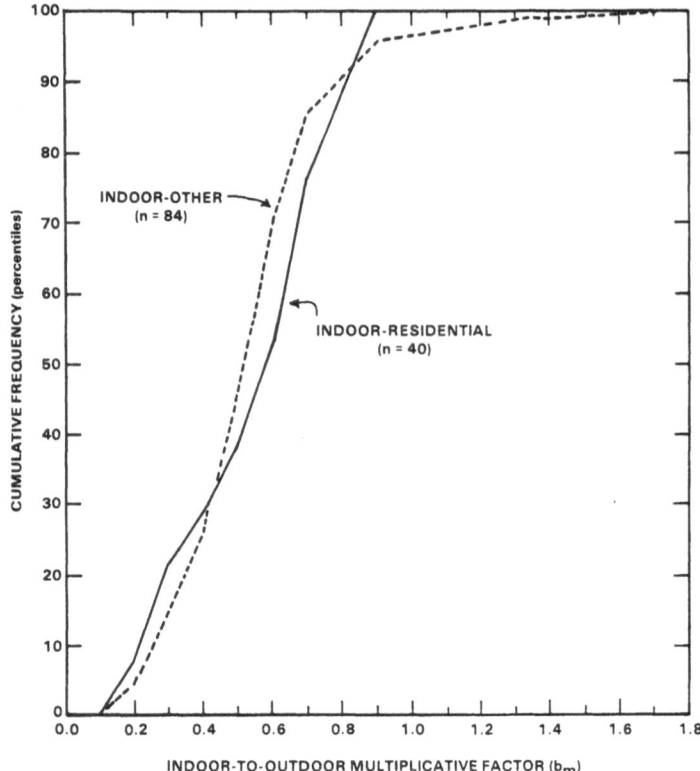

Fig. 2. Cumulative Frequency Distribution of the Indoor-to-Outdoor Multiplicative Factor. Source: Ferdo (1985).

relatively shallow, or flat, for ozone (generally less than 0.01 ppm per mile), the boundary problem is not as important for ozone as it is for other NAAQS pollutants that EPA deals with.

The transportation data used in NEM comes from the Census Bureau's "journey to work" surveys. These data are used in developing estimates of the origin-destination relationship among exposure districts in the 10 urban areas. There are n^2 possible one-way trips in an urban area of n exposure districts.

The output of NEM is a set of urban-area specific exposure estimates when alternative ozone standards are just met. While the aggregate of these estimates could be used to determine what happens to total exposure estimates in the sample when the ozone NAAQS is varied, many decision makers and public commenters want to see national exposure calculations. The task of Component Model 3 in Fig. 1 is to make these national estimates. This will be accomplished by "assigning" nonanalyzed urban areas to those that were analyzed and extrapolating NEM exposure results to the entire urban population of the U.S. (approximately 170 million persons, or 75% of the nation's total population). The assignment procedures to be used for ozone have not been developed to date.

The outputs of Component Model 3 are the national estimates, which will be incorporated into the ozone exposure report (see Fig. 1). This report will also explain all of the analytic work undertaken to develop the estimates, the assumptions made, and the

Table 2. Example of Completed Activity Pattern of a Cohort for Part of One Day-Type

<u>Cohort Definition</u>

Demographic group: *E*, indoor workers
Activity subgroup: Women, commute 20 minutes one way to work
Day and season: Cool weekday

Start Time	Number of 10-Minute Time Periods	Location[a]	Micro-Environment	Exercise Level[b]
0001	48	H	Indoors, residential	L
0800	2	H	Motor vehicle	L
0820	1	W	Near road	L
0830	20	W	Indoors, other	L
1150	1	W	Near roadway	L
1200	6	W	Indoors, other	L
1300	1	W	Near roadway	L
1310	23	W	Indoors, other	L
1700	1	W	Near roadway	L
1710	2	W	Motor vehicle	L
etc.				

[a]H = home district; W = work district.

[b]L = low exercise: activities involving sitting, standing, sleeping, light hand work, and desk work. (In work units, it is less than 300 kg · m/min of work for a 70-kg person.)

Source: Pope (1985).

extrapolation algorithms utilized. Results contained in the exposure report will become part of the ozone staff paper, mentioned earlier.

Component Model 4 involves the development of probabilistic exposure-response or dose-response relationships for the health effect indicators of interest. Exposure-response relationships will be sufficient for health endpoints based on evidence from human exposure studies. This is because response in the human studies are correlated with exposure at the subject's face and not with dose (i.e., level of ozone at various organ sites such as the lung alveoli). For health endpoints based on animal toxicology studies, it may be desirable to have a dosimetric model which transforms the ozone exposure estimates into distributions of ozone dose at various organ sites in the body. This allows one to better account for variation in uptake of ozone between species in estimating the degree of response that might occur in humans. For those endpoints where it is desirable to decompose the problem and utilize dosage estimates, Component Model 4 will focus on developing probabilistic dose-response relationships. For convenience sake, in the rest of this paper we refer only to exposure-response relationships, although most of the issues and principles also apply to development of dose-response relationships.

In order to fully represent uncertainties about the location and shape of various exposure-response relationships, we believe it is necessary to elicit or encode probabilistic judgments from recognized health experts (Feagans and Biller, 1981; Richmond, 1981). The use of judgmental probability encoding allows scientific experts to consider and weigh the full body of available evidence cited in the ozone criteria document. For example, the results from different studies of the same type on a particular endpoint can be considered (e.g., lung function changes in controlled human exposure studies), as well as results from different types of studies (e.g., controlled human exposure and field or epidemiological studies) relating to the endpoint of interest. This, of course, does not preclude health scientists from using, to the extent they feel is appropriate, available biological models or statistical analyses of data to aid them in their judgments. In addition to making judgments about the location and shape of exposure-response relationships, health experts will also provide judgments on the nature and severity of the effect in question and the reasoning or basis for their judgments.

From three to five health effect indicators will be chosen for which exposure-response relationships will be developed. Some of the health effect indicators under consideration are: pulmonary function effects, bronchial reactivity, symptomatic effects, aggravation of asthma, alterations in lung structure, increased susceptibility to infection, blood biochemical effects, and systemic effects. OAQPS will select the health indicators to be assessed based upon its review of the ozone criteria document and factors such as: (1) severity of the effect, (2) size of population groups potentially affected, and (3) degree of uncertainty associated with the scientific data base for the particular endpoint.

For each health indicator or endpoint which is to be assessed, a protocol for conducting the probability encoding will be developed. Protocols (1) provide a public record of how the encodings are accomplished, (2) provide a consistent method for encoding all judgments elicited for a particular endpoint, and (3) ensure that judgments of the experts are comparable. To avoid ambiguity over what the health experts' judgments mean and to ensure consistency, the protocols must define precisely the health effect indicator of interest and degree of response that is considered adverse (e.g., a certain percent decrement in FEV1), the population under consideration (e.g., asthmatics), and the exposure conditions that are assumed to prevail (e.g., exercise level, temperature, humidity, levels of other pollutants, etc.). The protocols are also designed to pose questions in a manner that minimizes motivational and judgmental biases based on research from the behavioral psychology field (Wallsten and Budescu, 1983).

Development of protocols for encoding is an interactive process involving review of the criteria document and consultation with EPA scientists, OAQPS staff, and health scientists who are familiar with the endpoints of interest. The protocol for each health endpoint will be pilot tested, probably with EPA scientists, prior to use with the experts whose judgments will be used in the assessment.

An important step associated with development of exposure-response relationships is selection of the health experts who will be encoded. EPA's goal is to select a set of experts for each endpoint to be assessed that spans the range of credible views on the health effects evidence. Criteria for selection of the experts will be developed in consultation with EPA's Office of Research and Development (ORD). OAQPS, however, is responsible for selecting the experts to be encoded. Our current plan is to encode approximately five experts for each endpoint examined.

Component Model 5 integrates exposure-response relationships from Model 4 with exposure distributions from Model 3 to produce the needed nationwide "head count" risk estimates. Head count risk is defined as the probability that a specified number of incidents of an adverse effect will occur, or a specified number of persons will suffer an adverse health effect, over a specified period of time when a given ozone NAAQS is just attained. Generally, a simulation approach using Monte Carlo techniques is required to estimate head count risk. An example of a head count risk estimate is that there is a 0.1 probability that 200,000 or fewer asthmatics per year will experience 15 percent or greater decrement in FEV1 due to ozone exposure upon attainment of a 0.12-ppm hourly average ozone standard. The head count risk measure accounts for uncertainties in exposure and exposure-response relationships. Other risk measures are also possible. For example, "benchmark risk" does not address how many sensitive people are affected but rather the risk that ozone levels will exceed the concentration that affects a specified fraction of the sensitive population if all individuals in the group were exposed. EPA will definitely generate head count risk estimates as part of its ozone risk assessment project and is considering other risk measures as well.

Results of the risk assessment project, including documentation of all major aspects of the project, will be contained in a technical report. A draft report will be made available to CASAC and the general public for review prior to any use of the results in developing alternative standards. Major results from the risk assessment will also be summarized in the second draft of the OAQPS staff paper, which is to be reviewed by CASAC.

REFERENCES

Biller, W. F., T. Johnson, R. Paul, T. Feagans, G. Duggan, T. McCurdy, and H. Thomas, 1984, "Estimation of Short-Term Sulfur Dioxide Population Exposures," *Proceedings of the 77th Annual Meeting of the Air Pollution Control Association*, Pittsburgh: Air Pollution Control Association.

Feagans, T. B., and W. F. Biller, 1981, "Risk Assessment: Describing the Protection Provided by Ambient Air Quality Standards," *The Environmental Professional* 3:235.

Ferdo, A., 1985, *Ozone Microenvironment Factors*, Durham, NC: PEI Associates, Inc.

Johnson, T., and R. A. Paul, 1983, *The NAAQS Exposure Model (NEM) Applied to Carbon Monoxide*, (EPA-450/5-83-003), Research Triangle Park, NC: U.S. Environmental Protection Agency.

McCurdy, T., and H. M. Richmond, 1983, "Description of the OAQPS Risk Program and the Ongoing Lead NAAQS Risk Assessment Project," *Proceedings of the 76th Annual Meeting of the Air Pollution Control Association*, Pittsburgh: Air Pollution Control Association.

Padgett, J., and H. M. Richmond, 1983, "Process of Establishing and Revising National Ambient Air Quality Standards," *JAPCA* 33(1):18.

Paul, R. A., and T. Johnson, 1985, *The NAAQS Exposure Model (NEM) Applied to Carbon Monoxide: Addendum*, (EPA-450/5-85-004), Research Triangle Park, NC: U.S. Environmental Protection Agency.

Pope, A., 1985, *Development of Activity Patterns for Population Exposure to Ozone*, Durham, NC: PEI Associates, Inc.

Richmond, H. M., 1981, "A Framework for Assessing Health Risks Associated with National Ambient Air Quality Standards," *The Environmental Professional* 3:225.

U.S. Environmental Protection Agency (EPA), 1981, *AEROS Manual Series; Volume III: Summary and Retrieval*, (EPA-450/2-76-009b), Research Triangle Park, NC: U.S. Environmental Protection Agency, National Air Data Branch.

Wallsten, T. S., and D. V. Budescu, 1983, "Encoding Subjective Probabilities: a Psychological and Psychometric Review," *Management Science* **29**(2):151.

Wallsten, T. A., and R. G. Whitfield, 1985, "Eliciting the Probabilistic Judgments of Experts about Lead-Induced Health Effects: A Methodology and Application," *Annual Meeting of the Society for Risk Analysis*, Alexandria, VA.

The Definition of Risk and Associated Terminology for Risk Analysis[a]

Prepared by the SRA Committee for Definitions
Lawrence B. Gratt, Chairman

IWG Corporation
San Diego, CA

INTRODUCTION

The broad field encompassing risk analysis through risk management involves a large number of different disciplines. The terminology used within the field has not been consistent. An attempt to create a set of definitions for use by the SRA was initiated by the formation of a committee on definitions. Initially, it was assumed that it would be easy to reach a consensus on the more commonly used terms. This assumption proved false. After about two years it was realized that a consensus was not being reached for the key definitions of risk, hazard, risk analysis and risk assessment. The Committee decided that rather than trying to establish "official" SRA definitions of these terms, it would recognize that different definitions are in use and prepare a discussion paper of different definitions. This process was started with emphasis on the single term "risk."

BACKGROUND

It is the hope of the Committee that this effort will produce a broader understanding and appreciation of the differences in the way risk terminology is used and encourage broader-based discussions among a more uniform usage. In the meanwhile, authors should clearly define how they use risk terms so as to avoid confusion.

The problem of consistent terminology for use by risk analysts has become a greater issue as more practitioners with different areas of expertise contribute their results to this expanding field. The National Research Council,[1] in its report on *Risk assessment in the Federal Government: Managing the Process*, addressed the problem of terminology. The report indicated that "despite the fact that risk assessment has become a subject that has been extensively discussed in recent years, no standard definitions have evolved, and the same concepts are encountered under different names." The authors of this NRC report spent a considerable initial effort on definitions and consequently adopted their own terminology for the reported study.

---

a. This paper was prepared as a discussion document for use in a special session at the 1987 Annual Meeting of The Society for Risk Analysis.

Gratt's 1985 review of terminology indicated that the terms "risk assessment" and "risk analysis" had been used in various ways and the need for a consistent set of definitions for use by practitioners in the field.[2] This resulted in a Committee for Definitions for the Society of Risk Analysis to address this problem. This paper will present some proposed definitions of risk for further consideration.

DEFINITIONS

The problem of defining the terminology used by risk analysts was indicated by the National Research Council which sought to clarify the differences between "risk assessment" and "risk management."[1] "Risk assessment" was defined as the qualitative or quantitative characterization of the potential health effects of particular substances on individuals or populations, and "risk management" as the process of evaluating alternative regulatory options and selecting among them. The NRC report made no attempt to define exactly what is "risk" or "risk analysis." As a result, the NRC was able to avoid the terminology controversy and provide widely accepted definitions.

Gratt,[2] attempted to synthesize definitions of all three terms. Gratt's proposed definitions were as follows:

- *Risk* is the potential for realization of unwanted, adverse consequences to human life, health, property, or the environment; estimation of risk is usually based on the expected value of the conditional probability of the event occurring times the expected consequence of the event given that has occurred.

- *Risk analysis* is defined as a detailed examination performed to understand the nature of unwanted, negative consequences to human life, health, property, or the environment; an analytical process to provide information regarding undesirable events; the process of quantification of the probabilities and expected consequences for identified risks.

- *Risk assessment* is the process, including both risk analysis and risk management alternatives, of establishing information regarding acceptable levels of that risk for an individual, group, society, or the environment.

These definitions view analysis as more restrictive than assessment: assessment includes analysis and judgment. Also, analysis is the scientific objective process of risk estimation; assessment is the combination of the objective analysis with subjective considerations. These definitions were based on common usage where analysis refers to the separation of a whole into its component parts (also an examination of a complex, its elements and their relationships) and where assessment is the act of determining a value. Opposing points of view state that analysis is the all-encompassing process that includes assessment.

SUMMARY OF POSITIONS

Gratt received numerous responses to his proposed definitions. Most of the replies stressed the fact that creating widely accepted definitions is a difficult problem. Many of these responses disagreed with Gratt's definitions of risk, risk analysis and risk assessment. The majority disagreed with his position that assessment is a much broader term than analysis.

There is a similar disagreement in the literature. Lawless, *et al.*, agrees with the choice of assessment as the broader term.[3] On the other hand, other authors have made the opposite choice. These include the authors of a DOD handbook called *Risk Assessment Techniques*.[4]

At this point, the Committee for Definitions met in conjunction with the 1985 Annual meeting and decided to focus efforts on a consensus definition of "risk." An attempt has been made to summarize some of the responses to the definition of risk. Fritz Seiler expressed his dissatisfaction with Gratt's proposed definition of "risk." He felt that this definition was neither concise nor disjunctive enough to be effective. He also disagrees with the "non-mathematical" approach Gratt had taken, saying that it possibly "obscured things more than it ought to." He continues to say that the main problem of the current definition is that it fails to admit that a risk is a probability. Seiler then states that since the "probability of occurrence is the product of the probability of the occurrence of the cause...multiplied by the conditional probability of the event...Both of these factors are really risks, so that the product is a composite quantity...[allowing] us to use a simpler definition of risk...The definition of the basic quantity 'risk,' however, has now been shifted to the lower level of a conditional probability for the occurrence of one single 'untoward' event, given that a series of other events has occurred. In my opinion, this is the simplest and most basic definition you can give."

Paul Slovic heartily agreed with the old definition of "risk," but not with the risk analysis/assessment definitions. He believes that risk analysis is the broader of the two ideas, and that assessment is just a part of analysis.

The suggestion of Edmund Crouch includes simple, curt definitions. He dislikes all of the (negative) qualifiers and adjectives, and he stresses the idea that "risk" must be defined quantitatively. For "risk," he suggests the following definition: An quantitative measure of the realization of a hazard. The related definition of "hazard" is: A condition with the potential of causing a consequence. For the risk analysis definition, Crouch again omits the negative qualifications as well as the "human life, health and property" phrase, saying "we might be interested in the effects of something on the duck population." He comments that the key work of the definition should be "hazards" instead of "risks." He then introduces a new idea---that "hazard analysis" should be introduced as the identification of hazards, and that "risk analysis" be replaced by "hazard assessment." Crouch states that risk assessment should not "contain judgments on what is or is not acceptable." He sums up "risk analysis" as "the process of breaking down and looking at all the pieces (the analysis), in order to find out how your risk measure depends on all the bits." He defines "risk assessment" as a "synthesis, taking account of how the bits are all related, and the uncertainty, etc., to predict the risk measure for the whole." He also states that risk management should not be considered a part of risk assessment, "except insofar as it is desirable to perform the synthesis in such a way as to be compatible with what it is going to be used for—possibly risk management."

Chris Whipple from EPRI commented that he prefers "risk analysis" as the broader field, "and 'risk assessment' as the more technical part." He then adds, however, that he has seen the terms used interchangeably, and that "based on the actual meaning of analysis and assessment, I agree that your definitions make a bit more sense."

Jeffery Roseman commented that he felt "the possibility of loss..." does not necessarily suggest something of "a probabilistic nature... Whether reality is basically probabilistic or deterministic is still a metaphysical question," he says. "The statement that an outcome is 'possible' may merely reflect human 'ignorance' of an essentially determined outcome rather than a probabilistic one."

Commenting on the definition, Lee Abramson stresses the importance of the specific severity of risk. He also states that the best way to represent a risk is in the form of a distribution which reflects the cause and accounts for the effect and the severity. He states his definition of risk as "a probability distribution over all possible consequences of specific cause which can have an adverse effect on human health, property, or the environment."

Later, the Committee arrived at a proposed definition for risk:

Risk is the probability of an adverse effect to human health, property and the environment and the severity thereof.

After the committee meeting, the definition was mailed to meeting participants and widespread rejection was received. Summarizing some of the comments on this definition it is useful to quote Ed Lawless, who found this definition inappropriate for several reasons. "First, the probability of adverse effects may not be the same for all items as implied." He states that "the probability of the severity is unclear, but is unlikely to be the same as the 'probability of an effect' in the first part of the definition." Lawless believes that the measure of interval must be included to make a quantitative statement of risk meaningful. His proposed definition is as follows: "Risk is a statement of the probability (or a qualitative expression of the uncertain potential) of incurring a specified adverse consequence during some stated measure of interval (e.g., time, distance, number of events, etc.)."

The Committee for Definitions met again in conjunction with the 1986 SRA Annual meeting. The problem of "seeming to get a consensus which falls apart upon reflection" was discussed. At that time, the Committee decided to assemble the favored definitions of risk and submit them to the SRA membership through the newsletter for comments, discussion and possible consensus by the whole Society.

The process of standardization of general use terminology may not be possible. Rayner and Cantor give the equation: Risk = (Probability × Magnitude)/Time. They state that "however and wherever it is discussed, it seems that there is a consensus that the essence of risk consists of the probability of an adverse event and the magnitude of its consequences."[5] The problem with this definition is that the formula defines risk as a probability rate, whereas the statement defines it as a probability.

Sam Morris says that "risk is the product of the probability of an adverse event times the consequence of that event were it to occur (dimensions units of consequence × time...)." These definitions go along with Seiler in that they are mathematical in nature. They are, however, more direct than Seiler's definition, and thus possibly simplified past the point of usefulness.

According to Cauncey Starr, "Risk is a measure of the exposure to a loss arising from an activity, with the usual descriptives of what, when, who and how much. The ambiguities associated with this simple definition arise from the variety of the losses that may be incurred,....the time period used for the probability statement,...the population exposed...and the uncertainty of the probability estimate." Starr states that risk calculation must consider the probability of an event, the magnitude of this event, and the resulting magnitude of the loss. Starr emphasizes the fact that a risk is neither only a probability nor only an end result, but a synthesis of the two in a certain time period and for a certain population. Starr also stated that assessment is a subset of analysis, and suggests a reversal in "the terminology...This would be consistent with the title of our Society of Risk Analysis and our journal."

Seiler states: "The problems that we encounter seem to root in the ambitious attempt to define the unqualified term 'risk.' In view of the wide diversity of phenomena considered, 'risk' can just mean to many things. However, these problems disappear as soon as we put this quantity in the proper context, something we almost always do or should do. As an example, the lifetime risk of lung cancer as a consequence of an exposure, or the risk of a loss within the next 5 years of 100 million dollars due to environmental damages are quantities which are clearly and unambiguously defined. The dependence on the time interval, the severity of the event or on any other parameters is thus relegated to the definition of the adverse event and the risk can quite generally be defined as a simple conditional probability."

Roseman has offered a categorization of risk definitions: non-probabilistic, probability of specified outcome with/without time specified, probability of specified outcome and a function of the consequences of the outcome with/without time specified, and "as a function of uncertainty." He believes this categorization indicates how many fundamentally different definitions of risk are of concern. Since the different fields which make up the Society use different definitions, Roseman feels "no single definition will satisfy all." He suggests the best approach is not to decide on one single definition for "risk," but to make it clear how many different definitions exist and, stressing the multidisciplinary nature of risk assessment, the use of "risk" along with its explicit definition.

Clearly there is a considerable disagreement on the most basic terms. Apparently, risk analysts know what they do, but can't agree on what to call it!

DEFINITIONS

Selected definitions of risk are listed below for consideration by the Society.

1. *Risk* is possibility of loss, injury, disadvantage or destruction; to expose to hazard or danger; to incur risk or danger.

2. *Risk* is an expression of possible loss over a specified period of time or number of operational cycles.

3. *Risk* (consequence/unit time) = Frequency (events/unit time) × Magnitude (consequence/event).

4. *Risk* is a measure of the probability and severity of adverse effects.

5. *Risk* is the conditional probability of an adverse event (given that the causative events necessary have occurred).

6. *Risk* is the potential for unwanted negative consequences of an event or activity.

7. *Risk* is the probability that a substance will produce harm under specified conditions.

8. *Risk* is the probability of loss or injury to people and property.

9. *Risk* is the potential for realization of unwanted, negative consequences to human life, health, or the environment.

10. *Risk* is the product for a probability of an adverse event time the consequences of that event were it to occur (dimensions of consequence × time).

11. *Risk* is a function of two major factors: (a) the probability that an event, or series of events of various magnitudes, will occur, and (b) the consequences of the event(s).

12. *Risk* is a probability distribution over all possible consequences of a specific cause which can have an adverse effect on human health, property or the environment.

13. *Risk* is a measure of the occurrence and severity of an adverse effect to health, property or the environment.

14. *Risk* .. [write in your own favorite]..

RECOMMENDATIONS

The next step in an attempt to arrive at a consensus is to have an open discussion of the above definitions and a vote. At this time, the consensus may be to have a set of fundamentally different definitions. Gratt will chair at a special session during the annual meeting to address this topic. Committee members will present short arguments in favor of their favorite definition or contrary opinions. Comments on the definitions and alternate definitions will be solicited from the floor. If appropriate, a vote can be taken so that opinions can be recorded. Recommendations for work on further terminology on behalf of the SRA membership will be solicited.

REFERENCES

1. National Research Council, *Risk Assessment in the Federal Government: Managing the Process*, National Academy Press, Washington, D.C., 1983.
2. L. B. Gratt, "Risk Analysis or Risk Assessment: A Proposal for Consistent Definitions," in *Proceedings of the Society for Risk Analysis International Workshop on Uncertainty in Risk Assessment, Risk Management, and Decision Making*, pp. 241-249 (1984).
3. E. W. Lawless, M. V. Jones, and R. M. Jones, *Comparative Risk Assessment: Toward an Analytical Framework*, National Research Institute, Kansas City, Missouri, January 1984.
4. Department of Defense, *Risk Assessment Techniques*, Defense Systems Management College, Fort Belvoir, Virginia, July 1983.
5. S. Rayner and R. Cantor, "How Fair Is Safe Enough? The Cultural Approach to Societal Technology Choice," *Risk Analysis* 7(1):3-9 (1987).

Author Index

Subject Index

A

acceptable exposure levels 86
acceptable risk 660
aflatoxin 235
age-dependence of cells 46,48,51
AIDS 432
air pollutants 160
air emission cancer risk 165
air toxic risk assessment 166
alcohol use 341,343,345
antimony 540
artificial intelligence 478

B

background mechanisms 59,62
background concentrations 92
benzene exposure 417
birth-death-mutation model 111
blood levels 282,283

C

cancer 573
cancer risk 224
carcinogen 540
carcinogen risk assessment 59
carcinogenic potency 223,552
cell proliferation 292,294
cellular proliferation 288,291,292,294
cellular proliferation (or adduct
    formation) 292
chemical emergency response
    planning 376
chemical industry 67
Chernobyl 392,394,625
chloroform 232
cholinesterase inhibition (ChE-I) 617
chromium 610,611,614
chronic exposure 551
chronic rodent bioassay 224
clarity 262
classification 473,474,475,476

classification trees 477,478,480
Clean Air Act 159
cleanup criteria 25,27
coal-fired power plant 165,166
common cause failure of perception 443
communication effectiveness 14
completeness 262
cost benefit 307
cost effectiveness 36,103,107
cost-effectiveness analysis 310
criteria for identifying hazardous
    chemicals 373
criticality analysis (CA) 379

D

dam failure 208,211
data pooling 519
decision rules 475,476
defense policy 135
definitions of risk 122
Delaney clause 263,264
de minimis 272
DES 393
dioxins 535
disaster magnification or impact
    magnification 205
dose-equivalent 76,80
dose response modeling 43,46
DNA adducts 288,289,290
drinking water 531,533,658
driver behavior 341

E

earthquakes 196
earth system 298,303
effective dose 280,281
ELF (extremely low frequency) 517
emotions 406
energy supply structure 307
engineering operations 444
environment 308

nuclear power 67
nuclear power plants 625,626,631
nuclear proliferation 152
Nuclear Regulatory Commission 625
nuclear wastes 233
nuclear weapons 151
numerical manipulation 359

O

objectivism 564,567,570
Office of Air Quality Planning and
    Standards (OAQPS) 666
occupant fatality 341,343,344
operational definitions 652
operations management 443
oral contraceptives 395
occupational health and medicine 242
organic compounds 83
ozone exposure 666
ozone health risk assessment 666
ozone layer 354

P

participatory decision making 455
pharmacodynamic models 292
pharmacokinetic models 292,573,593
pharmacokinetics 44,51,280,281,282,
    609,611
philosophy of science 563,564,566,567
plant systems 633
policy lesson 260
Portugal and the United States 251
prioritization 98,105,183
priority setting 159
PRA Oconee-3 633
prioritize 633
probabilistic risk assessment
    (PRA) 68,379
public exposures 643

Q

quantitative cancer risk assessment 43
quantitative ranking scheme 510

R

radiation 394
radiation risks 643
radiation standards 643
radioactive contamination 26
radon gas 234,464
ranking 160
rationality 564,566,567,570

recursive partitioning algorithms 474,478
regulatory compliance 241
regulatory models 652
relative risk 186
rent seeking 456
residues 545,585,587
resource recovery facility 536
risk 135,545
risk analysis 68
risk assessment 269,298,323,358,388,
    408,496,509,529,530,531,533,539,617
risk aversion 493
risk beliefs 432
risk communication 13,220,231,260,
    463,531,532,533
risk education 220,221,222
risk estimates 554
risk indexing 183
risk information 221,222
risk limitation 643
risk management 184,242,387,388,
    389,408,443,445,447,449,451,496,
    530,533
risk perception 251,387,388,389,
    432,468
risk prioritization 496
risk ratio 112
risk-risk 213,214,218
routes of exposure 79,88
rubber hydrochloride manufacturing 421

S

safety analysis 448,449,450,451,452
safety belt use 341,343,350
safety goals 625
SDI 151
seismic risk 196
short-term exposure 413
social impact 209,210
societal risk 119
soil cleanup guidelines 83
soil contamination 30
space industry 67
speciation 609,610,611,614
standards 657
standards-setting 563,568
statistical methods 59
statistically valid 545
strategic decision 138
stratospheric ozone depletion 354,356
susceptibility to chemicals 46,51
synthesis 333
systems analysis 311
Sweden 394,395